T0331134

Advancement of insects as food and feed in a circular economy

Series of Insects as Food and Feed

Series Editor

Arnold van Huis

The titles published in this series are listed at *brill.com/siff*

Advancement of insects as food and feed in a circular economy

Edited by

A. van Huis, B.A. Rumpold,
H.J. van der Fels-Klerx and J.K. Tomberlin

BRILL | WAGENINGEN ACADEMIC

Originally published, in open access, as Volume 7, Issue 5 (2021) in *Journal of Insects as Food and Feed*.

Cover illustration: Photo of *Hermetia illucens* adult made by Hans Smid | bugsinthepicture.com

Library of Congress Cataloging-in-Publication Data

Names: Huis, Arnold van, editor.
Title: Advancement of insects as food and feed in a circular economy :
  edited by A. van Huis, B.A. Rumpold, H.J. van der Fels-Klerx and J.K.
  Tomberlin.
Description: Leiden ; Boston : Brill : Wageningen Academic, 2025. | Series:
  Series of insects as food and feed, 2950-595X | "Originally
  published, in open access, as Volume 7, Issue 5 (2021) in Journal of
  Insects as Food and Feed"--Title page verso. | Includes bibliographical
  references.
Identifiers: LCCN 2024039141 (print) | LCCN 2024039142 (ebook) | ISBN
  9789004707672 (paperback) | ISBN 9789004707689 (ebook)
Subjects: LCSH: Edible insects. | Entomophagy. | Animal feeding. | Insect
  rearing. | Feed industry.
Classification: LCC TX388.I5 A38 2024 (print) | LCC TX388.I5 (ebook) |
  DDC 636.08/4--dc23/eng/20241024
LC record available at https://lccn.loc.gov/2024039141
LC ebook record available at https://lccn.loc.gov/2024039142

Typeface for the Latin, Greek, and Cyrillic scripts: "Brill". See and download: brill.com/brill-typeface.

ISSN: 2950-595X
ISBN 978-90-04-70767-2 (paperback)
ISBN 978-90-04-70768-9 (e-book)
DOI 10.1163/9789004707689

Contents

Introduction

A. van Huis[1]*, B.A. Rumpold[2], H.J. van der Fels-Klerx[3] and J.K. Tomberlin[4]

[1]Laboratory of Entomology, Wageningen University & Research, P.O. Box 16, 6700 AA Wageningen, The Netherlands; [2]Department of Education for Sustainable Nutrition and Food Science, Technische Universität Berlin, Marchstr. 23, 10587 Berlin, Germany; [3]Wageningen Food Safety Research, Akkermaalsbos 2, 6708 WB Wageningen, The Netherlands; [4]Department of Entomology, Texas A&M University, College Station, TX 77843-2475, USA; *editor-in-chief@insectsasfoodandfeed.com

In 2017, a book was published entitled 'Insects as food and feed: from production to consumption' (Van Huis and Tomberlin, 2017). However, the sector of insects as food and feed is developing so quickly that an update seems appropriate. There has been an exponential increase in publications dealing with the topic. For example, using the words 'edible insect' as key words in 'Web of Science' scored 421 hits during the last two years (2019 & 2020), an amount which equals the number recorded for the previous 20 years (1999-2018). We did consider publishing a new edition of the book, but concluded that the disadvantages outweighed the short-term results. We identified the following limitations: (1) it must be bought which limits its distribution; (2) the review process is less transparent and rigorous; and (3) the turnaround time for a book is much longer than for research articles. Thus, we felt that given the rate at which the industry is growing and diversifying, a second edition would possibly not be as up to date. In addition to the benefit of publishing up-to-date information quickly, we also concluded that the peer-review system would enhance its quality. Furthermore, publishing open access provides immediate engagement by parties globally in learning more about the industry or by enhancing their current facility. Fortunately, the authors and sponsors of such work were able to cover the open access costs.

As a means of transparency, it should be noted that the strategy adopted for assembling the topics and affiliated authors was done *a priori*. Authors with strong backgrounds on select topics were asked to contribute. Manuscripts were assigned to different Journal of Insects as Food and Feed editors, depending on their expertise. They were also charged with identifying reviewers as well as managing the review process. In some instances, editors were also authors or co-authors. However, they did not manage their own manuscripts in order to protect the integrity of the process.

All chapters dealt with relevant topics related to insects as food and feed, and most of the content of the articles is different from the 2017 book, reflecting developments in the field. In the description below, you will find the names of the authors who wrote the articles between parentheses.

So, the special issue starts with entotechnics, but instead of one chapter as in the 2017 book, four articles by the same author (Kok, 2021a,b,c,d), deal with the following topics: (1) overall mass and energy/heat balances; (2) organism kinetics, system dynamics and the role of modelling & simulation; (3) sub-process types and reactors; and (4) facility consideration. Facility designs and processes to rear insects are covered from artisanal operations to world-scale plants. We hope this will help potential 'entopreneurs' to plan their process and their enterprise before making a major commitment and investment. Then the environmental impact of insect as food and feed is dealt with, in which life-cycle analysis studies were compared (Smetana *et al.*, 2021). Increasingly the question is asked whether the industry can produce insects without compromising their welfare (Van Huis, 2021). When considering the environmental impact of insect mass production, the advantage of using insects as feed is the possibility to rear them on organic side streams. The question is what side streams can be used (Pinotti and Ottoboni, 2021). A topic that receives increasing attention is the role of microbes in transforming substrates into more acceptable materials for mass producing insects (Zhang *et al.*, 2021). Manure is a major waste stream of particular concern in many of the countries which advocate a circular economy. What are the prospects of upcycling the manure using insects (Cammack *et al.*, 2021)? A core issue with insect rearing and a major concern is colony health (Maciel-Vergara *et al.*, 2021). While insects can be mass produced, manipulating their nutritional value is a key question (Oonincx and Finke, 2021). Furthermore, what is the role of genetics in improving the performance of mass produced insects (Eriksson and Picard, 2021)? What about left-over substrates? Can they be used as fertiliser and are there plant protecting effects (Chavez and Uchanski, 2021)? One of the major advantages of using edible insects is the possible health effects for humans (Stull, 2021) and animals (Gasco *et al.*, 2021). Then, what is the effect of using insects as feed on the target animals: fish (Liland *et al.*, 2021), poultry (Dörper *et al.*, 2021), pigs (Veldkamp and Vernooij, 2021) and pets (Bosch and Swanson, 2021). Food safety deals with biological (Vandeweyer *et al.*, 2021) and chemical (Meyer *et al.*, 2021) contaminants. There is also the issue about allergies brought about by insect consumption (Ribeiro *et al.*, 2021). These food and feed safety issues will have a direct effect on legislative issues and regulatory frameworks (Lähteenmäki-Uutela *et al.*, 2021). The technique of processing is reviewed (Sinderman *et al.*, 2021) as are processing pathways and the extraction and utilisation of insect proteins, lipids and chitins (Ojha *et al.*, 2021). The final part of the special issue deals with consumer issues: how to design quality insect products (Reverberi, 2021) and how to convince consumers to buy insect products (Wassmann *et al.*, 2021). The final question is whether the insect industry is profitable. There are not many publications on this issue but in one article the question is addressed (Niyonsaba *et al.*, 2021). In the final chapter (Van Huis *et al.*, 2021) we try to give some perspective of the future for the industry. While this view is clearly limited due to the industry still being in its infancy when compared to other food and feed sectors, the chapter potentially provides some guidance as to hurdles to be addressed, opportunities to be seized, and new questions to be formulated.

References

Bosch, G. and Swanson, K.S., 2021. Effect of using insects as feed on animals: pet dogs and cats. Journal of Insects as Food and Feed 7: 795-805. https://doi.org/10.3920/JIFF2020.0084

Cammack, J.A., Miranda, C.D., Jordan, H.R. and Tomberlin, J.K., 2021. Upcycling of manure with insects: current and future prospects. Journal of Insects as Food and Feed 7: 605-619. https://doi.org/10.3920/JIFF2020.0093

Chavez, M. and Uchanski, M., 2021. Insect left-over substrate as plant fertiliser. Journal of Insects as Food and Feed 7: 683-694. https://doi.org/10.3920/JIFF2020.0063

Dörper, A., Veldkamp, T. and Dicke, M., 2021. Use of black soldier fly and house fly in feed to promote sustainable poultry production. Journal of Insects as Food and Feed 7: 761-780. https://doi.org/10.3920/JIFF2020.0064

Eriksson, T. and Picard, C.J., 2021. Genetic and genomic selection in insects as food and feed. Journal of Insects as Food and Feed 7: 661-682. https://doi.org/10.3920/JIFF2020.0097

Gasco, L., Józefiak, A. and Henry, M., 2021. Beyond the protein concept: health aspects of using edible insects on animals. Journal of Insects as Food and Feed 7: 715-741. https://doi.org/10.3920/JIFF2020.0077

Kok, R., 2021a. Preliminary project design for insect production: part 1 – overall mass and energy/heat balances. Journal of Insects as Food and Feed 7: 499-509. https://doi.org/10.3920/JIFF2020.0055

Kok, R., 2021b. Preliminary project design for insect production: part 2 – organism kinetics, system dynamics and the role of modelling & simulation. Journal of Insects as Food and Feed 7: 511-523. https://doi.org/10.3920/JIFF2020.0146

Kok, R., 2021c. Preliminary project design for insect production: part 3 – sub-process types and reactors. Journal of Insects as Food and Feed 7: 525-539. https://doi.org/10.3920/JIFF2020.0145

Kok, R., 2021d. Preliminary project design for insect production: part 4 – facility considerations. Journal of Insects as Food and Feed 7: 541-551. https://doi.org/10.3920/JIFF2020.0164

Lähteenmäki-Uutela, A., Marimuthu, S.B. and Meijer, N., 2021. Regulations on insects as food and feed: a global comparison. Journal of Insects as Food and Feed 7: 849-856. https://doi.org/10.3920/JIFF2020.0066

Liland, N.S., Araujo, P., Xu, X.X., Lock, E.-J., Radhakrishnan, G., Prabhu, A.J.P. and Belghit, I., 2021. A meta-analysis on the nutritional value of insects in aquafeeds. Journal of Insects as Food and Feed 7: 743-759. https://doi.org/10.3920/JIFF2020.0147

Maciel-Vergara, G., Jensen, A.B., Lecocq, A. and Eilenberg, J., 2021. Diseases in edible insect rearing systems. Journal of Insects as Food and Feed 7: 621-638. https://doi.org/10.3920/JIFF2021.0024

Meyer, A.M., Meijer, N., Hoek-Van den Hil, E.F. and Van der Fels-Klerx, H.J., 2021. Chemical food safety hazards of insects reared for food and feed. Journal of Insects as Food and Feed 7: 823-831. https://doi.org/10.3920/JIFF2020.0085

Niyonsaba, H.H., Höhler, J., Kooistra, J., Van der Fels-Klerx, H.J. and Meuwissen, M.P.M., 2021. Profitability of insect farms. Journal of Insects as Food and Feed 7: 923-934. https://doi.org/10.3920/JIFF2020.0087

Ojha, S., Bußler, S., Psarianos, M., Rossi, G. and Schlüter, O.K., 2021. Edible insect processing pathways and implementation of emerging technologies. Journal of Insects as Food and Feed 7: 877-900. https://doi.org/10.3920/JIFF2020.0121

Oonincx, D.G.A.B. and Finke, M.D., 2021. Nutritional value of insects and ways to manipulate their composition. Journal of Insects as Food and Feed 7: 639-659. https://doi.org/10.3920/JIFF2020.0050

Pinotti, L. and Ottoboni, M., 2021. Substrate as insect feed for bio-mass production. Journal of Insects as Food and Feed 7: 585-596. https://doi.org/10.3920/JIFF2020.0110

Reverberi, M., 2021. The new packaged food products containing insects as an ingredient. Journal of Insects as Food and Feed 7: 901-908. https://doi.org/10.3920/JIFF2020.0111

Ribeiro, J.C., Sousa-Pinto, B., Fonseca, J., Fonseca, S.C. and Cunha, L.M., 2021. Edible insects and food safety: allergy. Journal of Insects as Food and Feed 7: 833-847. https://doi.org/10.3920/JIFF2020.0065

Sindermann, D., Heidhues, J., Kirchner, S., Stadermann, N. and Kühl, A., 2021. Industrial processing technologies for insect larvae. Journal of Insects as Food and Feed 7: 857-875. https://doi.org/10.3920/JIFF2020.0103

Smetana, S., Spykman, R. and Heinz, V., 2021. Environmental aspects of insect mass production. Journal of Insects as Food and Feed 7: 553-571. https://doi.org/10.3920/JIFF2020.0116

Stull, V.J., 2021. Impacts of insect consumption on human health. Journal of Insects as Food and Feed 7: 695-713. https://doi.org/10.3920/JIFF2020.0115

Van Huis, A., 2021. Welfare of farmed insects. Journal of Insects as Food and Feed 7: 573-584. https://doi.org/10.3920/JIFF2020.0061

Van Huis, A., Rumpold, B.A., Van der Fels-Klerx, H. and Tomberlin, J.K., 2021. Advancing edible insects as food and feed in a circular economy. Journal of Insects as Food and Feed 7: 935-948. https://doi.org/10.3920/JIFF2021.x005

Van Huis, A. and Tomberlin, J., 2017. Insects as food and feed: from production to consumption. Wageningen Academic Publishers, Wageningen, The Netherlands, 448 pp. https://doi.org/10.3920/978-90-8686-849-0

Vandeweyer, D., De Smet, J., Van Looveren, N. and Van Campenhout, L., 2021. Biological contaminants in insects as food and feed. Journal of Insects as Food and Feed 7: 807-822. https://doi.org/10.3920/JIFF2020.0060

Veldkamp, T. and Vernooij, A.G., 2021. Use of insect products in pig diets. Journal of Insects as Food and Feed 7: 781-793. https://doi.org/10.3920/JIFF2020.0091

Wassmann, B., Siegrist, M. and Hartmann, C., 2021. Correlates of the willingness to consume insects: a meta-analysis. Journal of Insects as Food and Feed 7: 909-922. https://doi.org/10.3920/JIFF2020.0130

Zhang, J.B., Yu, Y.Q., Tomberlin, J.K., Cai, M.M., Zheng, L.Y. and Yu, Z.N., 2021. Organic side streams: using microbes to make substrates more fit for mass producing insects for use as feed. Journal of Insects as Food and Feed 7: 597-604. https://doi.org/10.3920/JIFF2020.0078

Preliminary project design for insect production: part 1 – overall mass and energy/heat balances

R. Kok

Bioresource Engineering, Macdonald Campus of McGill University, 21,111 Lakeshore Rd, Ste-Anne-de-Bellevue, QC H9X 3V9, Canada; robert.kok@mcgill.ca

Abstract

Preliminary project design (PPD) is an initial stage in project development that makes it possible for an entopreneur to gain insight into the feasibility and potential profitability of setting up an insect production facility. In this paper a simple, spreadsheet-based model is presented to facilitate the first step of PPD by estimating the overall mass and energy balances for a proposed project. The model calculates outputs on the basis of scientific data and estimated values for operating parameters for the system that is proposed. With the model it is easy to use a trial-and-error approach to investigate the effect of different parameter values on system operation. Thus, the entopreneur can enter values for parameters such as feed composition, temperature of the cooling air, etc. and see the effect on system productivity, conversion efficiency, energy requirements, etc. immediately. This facilitates the overall procedure of reaching final decisions about the organism, the feed, the processing approach, the scale of operation, etc. Normally, this is an iterative procedure that is based on 'trial-and-error', the two aspects being referred to here as the 'twin components of an iterative knowledge engine'. Thus, the outputs from the model will depend very much on the scientific data supplied and the values of the input parameters while, at the same time, use of the model will highlight what additional scientific data is needed and what alternate parameter values might prove profitable. Overall, the model allows the user to explore a large possibility space for both process constitution and operation much more quickly and easily than by experimental means alone. As such, it is a tool that can aid the entopreneur in thinking about a project and considering various alternatives, as well as in making decisions before a major commitment is made to any particular option. It is stressed here that PPD is only a preliminary stage in project development and that the investigation of overall process mass and energy balances is only the first step thereof. It is also stressed that results from modelling are invariably subject to empirical verification as well as 'common-sense filtering'. The model presented is general and thus not oriented to the production of any species in particular.

Keywords

mass balance – energy balance – entopreneur – modelling

1 Introduction

When one thinks about building a production facility in which insects will be reared or 'farmed' for commercial purposes two issues are of primary and immediate concern: (1) what is the organism/feed combination targeted; and (2) what will be the basic organisational scheme of the facility, i.e. how will it be structured and operated? These issues are central regardless of the scale of the planned project and final decisions about them must be made well before the actual design of a production facility is started. The organisational scheme is then implemented in the production facility or plant as the 'process', the latter term referring to the configuration of the equipment, the flows between various plant sections, as well as the actual operation of the plant itself. Regardless of which process type is chosen, for a facility to function properly, the process will need to be developed and designed for the specific organism/feed combination selected. This is a knowledge-based procedure and, unless expertise and technology are transferred from other parties, that knowledge must be acquired and organised through various scientific and engineering activities. Accordingly, entrepreneurial project development often passes through a sequence as presented in Figure 1.

In this sequence preliminary project design (PPD) is shown as a central activity linking early concept development, data acquisition and concept crystallisation to formal engineering design, construction, and ultimate plant operation. As a first step in PPD, in this paper (Part 1 of this set) the overall mass and energy/heat balances of a rudimentary larval rearing setup are explored by means of a simple model which is also made available to the reader. In Part 2 of this set of papers (Kok, 2021a) the effect of organism kinetics on process dynamics is discussed, together with how modelling and simulation can be used to elucidate this relationship. It is, therefore, largely an examination of what role the kinetics of organism growth plays in the PPD exercise. Kok (2017) mentioned a number of basic process types already in use for industrial insect production and these are dealt with in greater detail in Part 3 (Kok, 2021b) where aspects such as ease of process operation, labour requirements, equipment complexity, suitability for automation, etc. are also taken into consideration. In light of the former aspects, in Part 4 (Kok, 2021c) process dynamics as the convolution of organism kinetics and process choices are then discussed, together with some aspects of the physical facilities that are required to house the chosen process. The material presented is structured to be useful to entopreneurs whose ambitions range from very small scale, through medium-scale, to full industrial scale.

During the PPD phase of a project a large variety of possibilities should be investigated, the overall objective being to gain a 'fair understanding' of the advantages and disadvantages of a number of processing options for the targeted organism/ feed combination. This means that during PPD one reasons about a number of options on the basis of approximate, quantitative models and simulations, the term 'approximate' implying that the models will be based on simplifying assumptions in order to expedite the procedure. Obviously, such assumptions should always be clearly stated so they can be re-visited later for improvement or correction. 'Approximate' is also indicative of the fact that, although reasoning is a lot faster and cheaper than doing, there is no good substitute for empirical knowledge gained through experience. Thus, it must always be kept in mind that PPD is very useful as a guide for decision making but that reality is the ultimate model.

To illustrate and facilitate the exploration of the PPD possibility space, a simple model of the overall mass and energy/heat balances for a single tray of insect larvae growing on a bed of feed was developed (Excel spreadsheet based, available to the reader as electronic Supplementary Material S1; ESM). This model is a tool which allows the user to engage in 'trial-and-error' investigation with various process operating conditions and scientific values for the organism/feed combination. As such, it facilitates the user's exploration of a large possibility space for both process constitution and operation much more quickly and easily than by experimental means alone. As well, working with the model will often highlight the need for more data on the system under consideration and will also often lead the entopreneur to consider other operating options for the process. Use of such a model during a project's

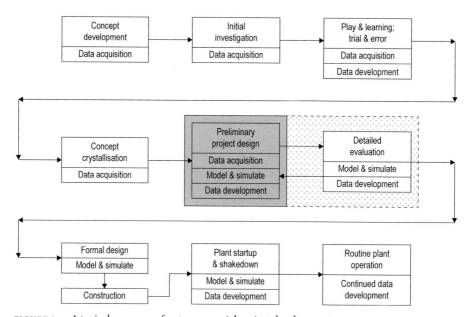

FIGURE 1   A typical sequence of entrepreneurial project development

PPD phase tends to lead to a 'step-by-step' iterative approach to process improvement. In the ESM file, by means of several worked examples, it is illustrated how model-based reasoning allows a user to quickly investigate a number of operating possibilities and see the consequences of the choices made. It is stressed here that the parameter and scientific values used in these worked examples are for illustration only and are not representative of any specific organism/feed combination. Thus, the model is a tool only and appropriate scientific data for the organism/feed combination under consideration, as well as values for the operating parameters, must be supplied by the user.

At the end of a project's PPD phase the aspiring entopreneur should have a good idea of the overall viability of the various process options, including the likelihood of financial success, and be able to make a decision on which approach to follow for the project. Hence, at this stage the entopreneur will have a fair understanding of the quantitative aspects of the chosen process option, including scale of operation, heat and mass flows, kinetics and dynamics, inputs and outputs, equipment requirements, plant operating scheme, labour requirements, etc. Basically, the PPD phase ends with an evaluation of the project's overall viability and a decision to proceed or not to proceed with the formal planning and design phase for the targeted organism/feed combination and the selected process type.

2   Basic considerations

Although the term 'design' can refer to a variety of activities, it is used here in the engineering sense. In this context design is a knowledge-based activity in which one starts off with certain, clearly stated objectives and that, ideally, results in the specification of a processing scheme, equipment choice, plant layout, operating methods and sequences, etc. To make such comprehensive design possible one needs both general knowledge as well as detailed data on the organism, the feed, the various operations, equipment capacities, and the supply and demand for feed components and products. In modern design it is desirable to understand both the static as well as the dynamic performance of a process, the latter being especially important for control purposes. This means that one needs to have available data not only on an organism's overall performance such as its conversion of feed into body mass, protein incorporation ratio, etc., but also on its kinetics, e.g. the larval growth curve, and how that is affected by temperature, body composition on a particular feed as the organism ages, etc. Overall, it is important to understand that true, formal design is possible only on the basis of adequate knowledge and data. Only once those are available is it possible to write models of the various parts of the process which can then be used in simulation to study both the static and dynamic behaviour of a proposed scheme. In essence, this approach is a formal extension of the natural thought process of humans based on the principle of 'think-before-acting'.

Formal, detailed design of a process and a facility, specification of equipment, etc. is the domain of professional engineers and in many jurisdictions may only be carried out by properly qualified professionals. Most process conceptualisation, development and preliminary design can, however, be done by entrepreneurs who themselves are not engineers. Accordingly, this chapter is oriented to guide those individuals interested in mass producing insects in their thinking about how this might be achieved and how the various possibilities and opportunities can be evaluated on a realistic basis, without transgressing the boundaries of professional responsibility.

Because insects pass through various life stages, a complete process for insect production will need to be multi-sectioned, i.e. it will consist of a set of sub-processes dealing with the various life stages such as egg production, egg incubation, larval growth, pupal incubation, adult maintenance, etc. (Kok, 1983; Kok *et al.*, 1988). Consequently, for the entrepreneur and process designer it is useful to think about the overall production process as an assembly of sub-processes, especially because it allows each section to be dealt with quite separately. As well, it allows for the concept of housing the various sub-processes in different physical locations. Thus, all the process sections or sub-processes may be housed in the same facility and share central ancillary services as illustrated in Figure 2 (ancillary services were described in some detail by Kok, 2017). Or they may be physically separated and occur in different locations. For example, pupal incubation, adult maintenance and egg production may occur in a specialised facility, the eggs then being delivered regularly to one or more larval rearing units. The division of a larger process into sub-processes and decision making about where each sub-process will be installed should occur reasonably early in overall project development.

As mentioned above, the term 'process' refers to the organisational scheme according to which an entire production plant is configured and operated, including the mass, energy and information flows between the various sections of the plant, between the different pieces of equipment, etc. Similarly, 'sub-process' refers to the same aspects, but on a more limited scale. Evidently, detailed knowledge and understanding of the various aspects will need to be developed for any sub-process that will be implemented in the entrepreneur's facility whereas less information will be required for sub-processes housed in external facilities. For instance, if eggs will be bought from a central supplier, less detailed knowledge is needed about their production than if they are generated in-plant. For the sake of simplicity and brevity only the sub-process of rearing larvae from egg to 'pre-pupa' is considered in this set of papers, the discussion being mainly applicable to the rearing of black-soldier fly (BSF) larvae, mealworms and other, similar kinds of organisms.

First of all, before any PPD activity can take place, the sub-process must be clearly identified and described. For instance, for rearing BSF larvae, one might start off with the idea to acquire eggs from a source external to the sub-process, put them on a well-defined early-growth feed, then manipulate their incubation temperature and feed regime so as to obtain the desired protein/fat ratio, and finally harvest them after a 10-day growth period. (It should be noted that the situation dealt with in the rest of the paper set is simpler than this.) During the PPD stage any number of alternatives to this scheme and any process options can then be considered and explored. In fact, one of the main functions of PPD is to allow for such exploration and investigation. But, regardless, of which specific process option is decided upon, the sub-process will be implemented in the plant on equipment that, generally, consists of a reactor through which pass mass and energy flows and to which is supplied information in the form of control signals and from which information is gathered as data (Kok and Lomaliza, 1986). As well, the mass flows will be modified while passing through unit operations such as grinding and sifting while various parts of the system will be heated, cooled, humidified, etc. A general organisational diagram of a larval rearing sub-process is shown in Figure 3. The model of the process mass and energy/heat transfer that is presented below is based on this general arrangement.

In this case there is a single reactor in which the larvae are reared (which might be composite, i.e. consist of a number of smaller units like bins) while there are three unit operations for grinding, mixing and separation of bed components.[1] Hence, the main mass inputs will be feed, water and air, cleaned air being used to supply oxygen to the bed, remove carbon dioxide and other emitted gases. The air flow will also carry away the heat of metabolism, both in sensible and latent form (as water vapor). Input energy is mostly used here for electric motors to power the

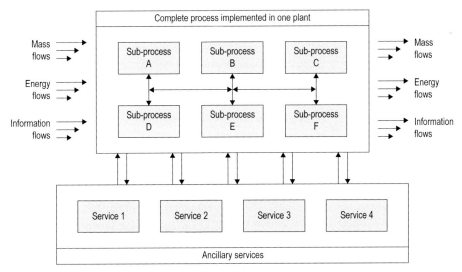

FIGURE 2   Conceptual diagram of a complete process implemented in a single plant

---

1   The three unit operations are not considered in the model.

grinding, mixing and separation, move bins around, etc. Normally this also results in a thermal load that is removed via an ambient air flow. Input information is principally for control purposes while data on system operation, mass flows, humidities, temperatures, etc. are the principal information outputs. The latter are mainly used for two purposes: (1) as inputs to an online control system which will then feed back control information to the sub-process; and (2) for archiving in a database on process operation and performance. The rationale and approach illustrated here for the larval rearing sub-process are equally applicable to other sub-processes but, obviously, the details for those would need to be worked out separately.

3   Mass and energy/heat balances

What is a mass balance? The mass balance of a process is a complete accounting of all the mass flowing into and out of that process. Basically: over time, mass in must equal mass out plus whatever accumulates in the system. Evidently, this principle holds as well for any sub-process. Additionally, if a process is dealt with in terms of a group or network of sub-processes, the mass balances of the various sections must match so that the outputs from one will equal the inputs to the connecting one(s). A mass balance can be very general, e.g. in terms of total mass in/out, or it can be as detailed as desired in terms of the various components of the mass. For instance, Kok *et al.* (1991) reported on both the total and the component mass balances of water, ash, carbohydrate, protein and fat for the rearing of *Tribolium confusum* larvae over a 19-day period, starting off with feed and eggs and ending up with larvae, waste material, and re-useable feed. In that mass balance only the solids were taken into account and air flow was not considered. The net mass loss in solids was assumed to be due to the metabolism of the larvae while growing, transforming part of the feed into carbon dioxide and water which then left the sub-process via the exhaust gas stream. Such composite mass balance results can be conveniently presented in easily-understood graphical form, as shown in Figure 4, in which all the quantities will normally be presented numerically rather than in symbolic form like W_1, C_1, TOT #1, etc.[2] The numbers might be derived experimentally or result from simulation, thus either representing reality as measured *in situ*, or a model thereof. In the latter case the usual purpose would be to plan or design a process, or to learn more about it in terms of its potential to attain specific objectives. This type of analysis and presentation can be very useful in evaluating and illustrating the efficacy of a process for destroying, preserving or synthesising a particular component. For example, BSF do not synthesise protein *de novo* (although they may metabolise it)

---

2   Thus, the symbolic names Wx, Cx, Px, Fx, Ax and TOT #x are used here to represent quantities; for instance, TOT #1 represents the total quantity of feed material and W_5 represents the total quantity of water present in the recycle material, etc.

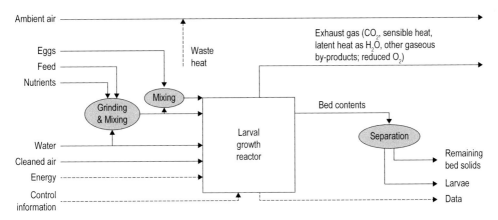

FIGURE 3   General organisational diagram of a simple larval rearing sub-process

and it is often helpful to track the protein mass flow separately throughout the rearing sub-process to ensure protein is maximally incorporated in the larvae.

Before PPD activity can be started on a project, substantial knowledge about both the organism/feed system and the total and component mass balances is needed. Sometimes such data is available in the literature (e.g. see Shumo *et al.* (2019a,b) for good quality data on BSF from which some approximate mass balance results can be calculated) or it has to be generated during earlier stages of project development (Figure 1). It should be noted that the actual mass balance of a process will be very dependent on the specifics of the organism/feed combination. For instance, Rehman *et al.* (2017) fed different combinations of dairy manure and soybean curd residue to BSF larvae and obtained substantially different compositional and quantitative results for the outputs. Diener *et al.* (2011) and Nyakeri *et al.* (2017) have also presented results that support this. Cammack and Tomberlin (2017) have reported on the impact of diet protein and carbohydrate on the life-history traits of BSF. As pointed out earlier, because the preliminary design phase is about evaluating a number of possible approaches for a project, it may happen that some basic factors such as diet composition are adjusted during the PPD, or even that the direction of the project is changed to a substantial degree. In either case it will probably be necessary to do more experimental work to obtain additional mass balance data for the new situation that is targeted. This aspect of PPD is addressed in some greater detail later in the chapter.

There are two major components of the energy/heat balance of a sub-process: (1) the chemical energy in the feed that is liberated through metabolism; and (2) mechanical and electrical energy that is used for motors, lighting, communications, etc. Only the former will be taken into account here. Some of the chemical energy will become incorporated within the organisms in components such as lipids and chitin, some will end up in constituent components in competing organisms such as bacteria, some will end up in process by-products, and some will become the thermal load that needs to be rejected from the process, normally as waste heat. As for the mass balance, over a given time period energy in must equal energy out plus whatever accumulates in the system. In dealing with the process heat/energy balance a major consideration is sufficient rejection of low-temperature heat, the main objective being the control of temperature of the larval environment. This is often achieved by blowing air over top of a bed in which larvae are growing, with the air increasing in temperature (gaining sensible heat) as well as moisture (gaining latent heat) while, at the same time, supplying oxygen to the metabolic process and carrying away carbon dioxide and other metabolic by-product gases (Figure 3).

4   Modelling of mass and energy/heat balances

PPD is usually based on a model of the process or sub-process under consideration. Such a model is a limited representation of the real situation and is based on a number of simplifying assumptions. It is used to compute process results (in simulation) according to the values that are supplied for a number of model parameters. Evidently, the assumptions and parameter values should correspond to a fair degree to physical reality and the model should frequently be 'tuned' by checking and

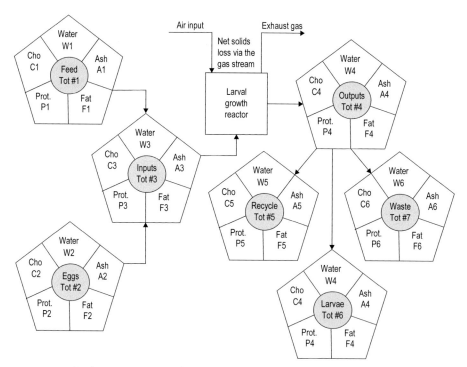

FIGURE 4   Graphic representation of components and total mass balances

correcting these, based on empirical knowledge as it becomes available. The major objective of this is to come to a reasonably complete understanding of the process that is adequate for PPD purposes.

To illustrate the concept and the procedure a simple model was developed for larval growth on a solid, granular substrate that, because of the addition of water, may be in the form of a mud or a slurry as is often fed to BSF larvae.[3] The model is for a single tray that contains a bed of substrate on which the larvae grow (as well as bacteria). This situation is seen as a 'base-case' for insect rearing and corresponds to the straight-batch sub-process described in greater detail in Part 2 (Kok, 2021a). It is illustrated in Figure 5.

A number of the model's parameter values are related to the organism's performance on the specific feed used (as are those of the bacterial population). For a real PPD exercise these would need to be obtained via experimental procedures and, as part of the iterative procedure, would have to be re-evaluated and re-entered every time the substrate composition is changed. Here, however, they were derived from a set of typal development curves for a single fly larva on a wetted-solids feed as presented in Figure 6. Although it is felt they are reasonably close to a physically real situation, these curves are hypothetical and presented for illustration only. In this case, as it grows, the larva first accumulates protein at a high rate and then, about halfway through its developmental period, while its other tissue growth slows, it begins to accumulate fat at an increasing rate. This type of delayed fat accumulation pattern was, for example, observed for the development of *T. confusum* larvae grown on wheat flour and yeast extract (Kok *et al.*, 1991). When ready to begin pupation, this hypothetical larva will be 48.5% protein, 33.7% fat, and 17.8% other material (all on a dry weight basis). At a final total weight of 181 mg it will contain 109 mg water so that, on a wet basis, it will be 19.2% protein, 13.3% fat, 7.0% other materials and 60.4% water.

In the model, calculations and discussion the variable names used are descriptive rather than purely symbolic and are therefore composed of meaningful terms. In some cases, in order to clarify the situation, these are connected by underscores as is often done in modern computer languages such as Python and JavaScript. For example, the name of a variable might be Other_Material_Type_A. One advantage of using variable names in that format is that it will facilitate further development and implementation of the model in other languages so that it can be made more complex and inclusive. This will be important if other process aspects need to be added to the model, such as reproduction, egg and pupal incubation, etc. In order to deal with inputs and outputs as well as intermediates in a uniform fashion, all low-level variables are defined in basic metric units, i.e. grams, seconds, metres and Joules (g, s, m, J). All calculations are, therefore, also carried out in these basic met-

---

3 The model is in the form of an Excel spreadsheet; approximately 575 lines, 39 adjustable parameter values. It is available to the reader as ESM.

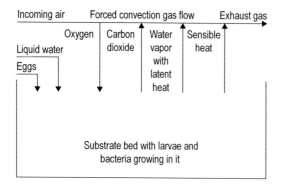

FIGURE 5   Rearing arrangement on which the model is based

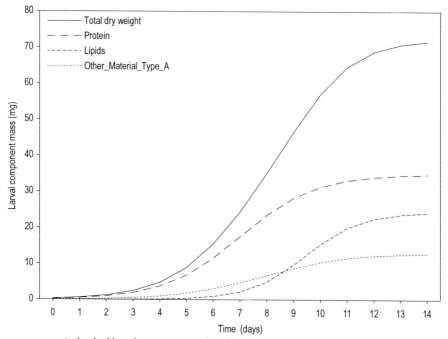

FIGURE 6   Individual larval component and total dry weight accumulation vs time

ric units. Derived units are, however, used at times for presentation to clarify the meaning of results. For instance, the Energy_content_of_Other_Material_Type_A in leftover substrate might be reported in basic units as 1.15600E+07 J, but, to make it easier to understand for readers, it will also be reported as 11,560 kJ.

The model is based on the process aspects and flows shown in Figure 5 and only those are taken into account. As is inevitable in any insect farming operation, there will be bacteria and some other organisms competing with the target organism. Hence, sub-process inputs are air, eggs, liquid water and fresh substrate while the outputs will be exhaust gas, larvae, bacterial mass, larval waste and leftover substrate material. The mass and energy content of the eggs are assumed to be negligible and therefore not considered in the calculations. The substrate dry matter is assumed to not contain any lipids and to be made up entirely of a generic, digestible carbohydrate (CHO), protein, and OM Type A (Other_Material_Type_A) which is not digestible by either the insect or any bacteria present. Normally, the larvae will be separated from the bed material when ready for harvest and the three other solid outputs will exit the sub-process as a fairly dry, granular mixture ('remaining bed solids' in Figure 3). Since OM Type A is indigestible, all of it will end up unchanged in the leftover substrate. The composition of the leftover substrate will therefore be quite different from the substrate itself.

In the model no *de novo* protein synthesis occurs, either by the bacteria or by the insect (or any of its internal symbionts), nor is there any protein metabolism or degradation as such. Thus, rather than deal with these complexities at this stage, protein from the substrate is simply allocated to the various fractions, albeit at an energy cost, that energy being supplied by CHO metabolism. This also means there are no nitrogenous compounds such as ammonia present in the exhaust gas. The energy content of all protein is taken to be 17 kJ/g.

Energy is derived solely from carbohydrate metabolism according to Equation 1 in which 180 g (one mole) of generic carbohydrate (CHO) combines with 192 g of oxygen to form 264 g of carbon dioxide and 108 g of water, liberating a total of 17 kJ/g of carbohydrate metabolised.

$$C_6H_{12}O_6 + 6\,O_2 \rightarrow 6\,CO_2 + 6\,H_2O \tag{1}$$

This energy then becomes available for chemical work like the transformation of protein from one state to another and the synthesis of cellular materials, as well as mechanical work for organism movement, etc. Accordingly, part of the energy liberated will be manifested as heat which must be removed from the sub-process in either sensible or latent form.

In the model a generic lipid (Reger *et al.*, 2010) is synthesised from carbohydrate as in Equation 2 in which 7,200 g of carbohydrate is converted into 2,628 g of lipid, accompanied by the release of 3,168 g of carbon dioxide and 1,404 g of water. The energy content assigned to lipids is 37 kJ/g so that out of 1.224E08 J of energy available in the carbohydrate 9.724E07 J ends up stored in the lipids (79%). The remaining 21% of the energy that is not stored is released in the form of heat.

$$40\,C_6H_{12}O_6 \rightarrow 3\,C_{56}H_{108}O_6 + 72\,CO_2 + 78\,H_2O \tag{2}$$

Altogether there are four types of Other_Material present in the system, OM Types A, B, C, and D, respectively constituents of the substrate, the larvae, the larval waste, and the bacteria. All four are assumed to have the same general chemical composition as the generic carbohydrate, as well as the same energy content, i.e. 17 kJ/g. There is, however, an energy cost charged for the transformation of generic carbohydrate into the various types of Other_Material. The specifics of these energy costs are all specified via the model's parameter values.

The mass and energy balances computed with the model for a base instance of the system (Case 0) are presented in Figure 7. For this case, ten thousand eggs (overall emergence and survival rate 50%) are added to a tray containing 8,500 g of wet substrate (60% water, 10% protein, bulk density 850 kg/m^3). The 340 g of protein initially present are then allocated to the larvae, larval waste, bacteria and leftover substrate in accordance with the parameter values supplied. Next, the carbohydrate is transformed into materials (lipids and OM B, C and D) and also consumed to generate energy for the various transformations, larval movement, etc. through metabolism. In Figure 7 symbols for inputs and outputs are attached to the boxes, inputs on the left and outputs on the right. The symbols O, W, C, E and H, respectively, refer to oxygen, water, carbon dioxide, energy and heat. For instance, to generate energy for larval metabolism and motion the inputs are CHO (indicated with an arrow) and oxygen (indicated with the O symbol). The outputs from that box are water, carbon dioxide and heat (indicated with the W, C and H symbols, the heat resulting from the metabolism and the motion). Similarly, for the allocation of protein to larvae, larval waste and bacteria, besides protein itself, chemical energy is required because an energy cost is assessed for these allocations. That energy is also derived from the metabolism of CHO. The energy is then used as inputs (E) to the three protein boxes for the larvae, larval waste and bacteria (but not for the leftover substrate) and ends up rejected as process heat (H) from all three.

As is usual for biological systems a substantial part of the total energy flux must be rejected, in this case to air. For Case 0 the incoming air is 21 C and 50% relative humidity while the exhaust gas leaves the system at 23 C and 60.9% relative humidity so that 76% of the heat is rejected as latent heat. The total air volume required is rather large in order to keep the carbon dioxide content of the exhaust gas within the specified limit (a parameter) and a value of 987 ppm was achieved (413 ppm incoming air). For a tray 50 cm wide, 40 cm long and 10 cm tall, containing a bed 5 cm deep with a 5 cm air space above it this means the average air velocity should be 2.6 cm/s. With these values the increase in gas temperature is quite small (2 C) and the decrease in oxygen content is negligible.

For Case 0, of the 340 g of protein initially present in the substrate 173.8 g ends up in the larvae (51%) while only 15% of the energy content is transferred from the substrate to the target organism, illustrating the potential of this type of model to estimate the capacity of an insect-based process to upgrade a feed in terms of protein content. One interesting statistic that is also generated by the model (not shown in Figure 7) is the productivity of the equipment. For Case 0 the production rate of larval material (dry) per real unit reactor volume is 7.77E-03 g/m^3/s (0.7 kg/m^3/day). To calculate this value, realistic allowances were made for tray floor thickness, walls, spaces between trays, space for tray supports, etc. (all as specified with adjustable model parameter values – see Case o in the ESM).

How one proceeds from this point depends very much on the objective(s) of the entopreneur in developing the process. As discussed above, the situation should first be clearly defined in terms of the targeted feed/organism combination and then the model should be tuned to that situation on the basis of empirical data. With other words, appropriate parameter values should be derived before the model can be used to investigate how the situation may be adjusted in order to improve process results. Through such a procedure incremental improvements can often be obtained quite easily and, as pointed out, it is not unusual for the data acquisition/simulation sequence to be iterative. Thus, through 'playing' with the model it may be possible to identify one or more a key variables whose manipulation will yield considerable improvement in a target variable. This will then require empirical verification, updating of the parameter values on the basis of experimental findings, and further work with the model. This 'hill-climbing' approach allows for the accumulation of small improvements and can, overall, result in substantial improvements in productivity. Evidently, it is best to engage in this activity during the early stages of PPD and project development, before any major commitments have been made and while concepts are still very flexible.

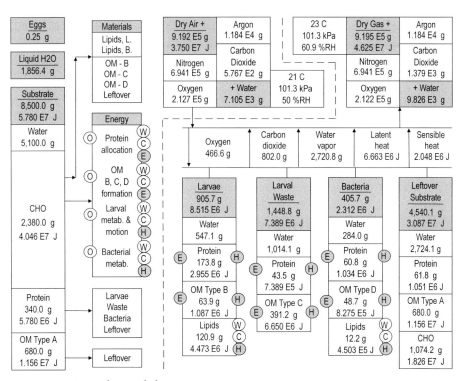

FIGURE 7   Mass and energy balances or Case o

Considering the situation of Case 0, for example, suppose the objective is to capture as much of the protein as possible in the target organism, the larvae, while maintaining the same protein:energy ratio as for Case 0. As it is, 51% of the protein ends up in the larvae, 13% in the larval waste, 18% in the bacteria, and 18% in the leftover substrate (Figure 7). Probably the easiest way to improve this is to capture as much as possible of the protein in the leftover substrate. Two straightforward methods to achieve this would be: (1) increase the number of eggs added to the tray; and (2) increase the survival rate of the larvae. Of these two, increasing the number of eggs seems the simplest. If the number of eggs added to the bed is increased from 10,000 to 12,000 (Case 1, spreadsheet available as ESM.) the model predicts that the protein captured in the larvae increases from 51 to 61% while the other three percentages shift from 13 to 15% (waste), from 18% to 22% (bacteria) and from 18% to 2% (leftover substrate). Thus, this approach looks promising!

To now make further gains in this regard one might investigate ways to inhibit bacterial growth so as to reduce the amount of protein incorporated in the bacteria. If a method can be found to reduce the bacterial growth by half, and the number of eggs added to the tray is now increased from 12,000 to 13,500 (Case 2, also available as ESM), the percentage of substrate protein captured in the larvae is predicted to be 69% while 17% is immobilised in the larval waste, 12% in the bacteria and 2% in the leftover substrate.

For all three cases discussed so far, the protein content of the substrate was 10% (dry weight basis) and the indigestible material (Other_Material_Type_A) was 20%. As is obvious from the results so far, this makes it a protein-limited system. If the substrate is upgraded by adding enough pure protein to raise the content to 13.5%, that would result in a feed containing 19.2% indigestible material and 67.3% CHO. If the number of eggs added to the tray is now raised to 18,000 (Case 3, available as ESM), 68% of the protein will be captured in the larvae, 17% in the larval waste, 12% in the bacteria and 3% in the leftover substrate (based on the assumption that there will be no major shift in metabolism due to these adjustments). In this instance, 26.5% of the substrate energy is captured in the larvae while the production rate of larval material (dry) per real unit reactor volume is 1.2 kg/m^3/day. In contrast, for Case 0 these numbers were 15% and 0.7 kg/m^3/day respectively.

The foregoing cases serve merely as an illustration of how a simple model of a process can be used to explore the operations possibility space very quickly and cheaply (once the model is written). In essence, it is an effective way to determine which process parameters are the most 'powerful' in determining yield and productivity improvements and are therefore likely to be the most interesting to investigate experimentally. In terms of optimisation and maximisation, it allows one to identify which variables have the largest partial derivatives for the 'slope-climbing' procedure. At the same time, the procedure will yield an indication of what range of parameter values might be most fruitfully targeted in further experimental work. If the model indicates that certain conditions will not yield good results one would probably be cautious about investing in experimental work over that range of conditions. For example, in Case 3 above the substrate protein content was increased to 13.5% but not higher because the model predicts that around 14% the system will run out of energy-yielding CHO.

Evidently, there are many other process parameters that can be easily investigated in a preliminary way with this type of model. One primary candidate would be the allowable carbon dioxide content of the exhaust gas since a higher value for the exhaust would allow the gas flow to be reduced. Another interesting variable is the substrate bed depth; deeper beds could lead to higher productivity per unit reactor volume although some heat transfer aspects inside the bed may then need to be included in the model.

5   Discussion and conclusions

During the PPD phase of project development system performance is predicted on the basis of biological knowledge of the proposed organism/feed combination, models thereof, process operation, equipment considerations and economic data, as well as experiential and intuitive insight. These are then all combined into a single reasoning stream which, once adequately developed, may take the form of a computer-based simulation with which the possibility space can be explored. The intent is that at the end of this phase the scale of the project, the details of the organism/feed combination, the processing approach and the financial aspects are reasonably fixed in the mind of the entopreneur. Evidently, it is the financial palatability of the project that will be the determining, final factor for decision making. It is only after this has been elucidated that the formal design phase of the project can be started.

So far, what has been discussed are the overall, 'beginning-to-end' mass and energy/heat balances of the larval rearing sub-process. In a more complete and detailed examination the other aspects of insect farming, such as reproduction, egg incubation and pupal incubation will be included in the discussion, if they are relevant. As well, the time variation of the variables will be considered. For example, for Case 0 above, the total air volume that needs to forced across the bed surface over the 13.5 day growth period is 772 m^3, resulting in an average air velocity in the cross-sectional area above the tray of 0.1 km/h – quite a modest value. However, during the 13.5 day growth period the growth rate of the larvae can vary greatly (as shown in Figure 6) and, together with that, so will the metabolic rate, the heat and carbon dioxide production rates and, therefore, the heat and by-product removal requirements. This implies that the rates of liquid water addition and incoming air flow (Figure 5) will need to be manipulated so as to control the situation. This is borne out in practice; preventing a dense bed of larvae in their final, maximum growth phase from overheating is often a major concern.

Thus, after dealing with the basic aspects of PPD such as the overall mass and energy balances one can proceed to more complex ones such as the kinetic aspects of organism development, the consideration of different process types and resulting process dynamics. In order to investigate the possibility space associated with many of the options and choices available here, one might construct a more sophisticated model in which organism kinetics as well as various process options and the dynamics associated with them can be taken into account. Such a model will allow for the time-based simulation of process operation and serve the same function as the model presented above, except in a larger space; it will be one of the twin components of an iterative knowledge engine, the second component being experimental exploration and verification. Evidently, models of different complexity can be constructed and it is up to the entopreneur to gauge what cost for dynamic model construction is justifiable in terms of the benefit likely to be derived. In Parts 2, 3 and 4 of this set of papers the interaction between organism kinetics and the dynamics of different process types are discussed in terms of exploration of the larger possibility space (Kok, 2021a–c).

Conflict of interest

The author declares no conflict of interest.

Supplementary material

Supplementary material S1. Model of the overall mass and energy/heat balances for a single tray of insect larvae growing on a bed of feed.

Supplementary material can be found online at https://doi.org/10.3920/JIFF2020.0055.

References

Cammack, J.A., Tomberlin, J.K., 2017. The impact of diet protein and carbohydrate on select life-history traits of the black soldier fly *Hermetia illucens* (L.). Insects 8: 56. https://doi.org/10.3390/insects8020056

Diener, S., Zurbrugg, C., Gutierrez, F.R., Nguyen, D.H., Morel, A., Koottatep, T., Tockner, K., 2011. Black soldier fly larvae for organic waste treatment – prospects and constraints. In: Alamgir, M., Bari, Q.H., Rafizul, I.M., Islam, S.M.T., Sarkar, G., Howlader, M.K. (eds.) Proceedings of the WasteSafe 2011 – 2nd International Conference on Solid Waste Management in the Developing Countries, February 13-15, 2011, Khulna, Bangladesh, pp. 1-8.

Kok, R., 1983. The production of insects for human food. Canadian Institute of Food Science and Technology Journal 16(1): 5-18.

Kok, R., 2017. Insect production and facility design. In: Van Huis, A. and Tomberlin, J.K. (eds.) Insects as food and feed: from production to consumption. Wageningen Academic Publishers, Wageningen, The Netherlands, pp. 142-172.

Kok, R., 2021a. Preliminary project design for insect production: part 2 – organism kinetics, system dynamics and the role of modelling & simulation. Journal of Insects as Food and Feed 7: 511-523. https://doi.org/10.3920/JIFF2020.0146

Kok, R., 2021b. Preliminary project design for insect production: part 3 – sub-process types and reactors. Journal of Insects as Food and Feed 7: 525-539. https://doi.org/10.3920/JIFF2020.0145

Kok, R., 2021c. Preliminary project design for insect production: part 4 – facility considerations. Journal of Insects as Food and Feed 7: 541-551. https://doi.org/10.3920/JIFF2020.0164

Kok, R. and Lomaliza, K., 1986. The insect colony as a food chemical reactor. In: LeMaguer, M. and Jelen, P. (eds.) Food engineering and process applications. Vol II. Unit operations. Elsevier, Amsterdam, The Netherlands, pp. 369-375.

Kok, R., Lomaliza, K. and Shivhare, U.S., 1988. The design and performance of an insect farm / chemical reactor for human food production. Canadian Agricultural Engineering 30: 307-317.

Kok, R., Shivhare, U.S. and Lomaliza, K., 1991. Mass and component balances for insect production. Canadian Agricultural Engineering 33: 185-192.

Nyakeri, E.M., Ogola, H.J.O., Ayieko, M.A. and Amimo, F.A., 2017. Valorisation of organic waste material: growth performance of wild black soldier fly larvae (*Hermetia illucens*) reared on different organic wastes. Journal of Insects as Food and Feed 3: 193-202. https://doi.org/10.3920/JIFF2017.0004

Reger, D.L., Goode, S.R. and Ball, D.W., 2010. Chemistry: principles and practice (3rd Ed.). Cengage Learning, Boston, MA, USA, 125 pp.

Rehman, K.U., Rehman, A., Cai, M., Zheng, L., Xiao, X., Somroo, A.A., Wang, H., Li, W., Yu, Z. and Zhang, J., 2017. Conversion of mixtures of dairy manure and soybean curd residue by black soldier fly larvae (*Hermetia illucens* L.). Journal of Cleaner Production 154: 366-373. https://doi.org/10.1016/j.jclepro.2017.04.019

Shumo, M., Osuga, I.M., Khamis, F.M., Tanga, C.M., Fiaboe, K.K.M., Subramanian, S., Ekesi, S., Van Huis, A. and Borgemeister, C., 2019a. The nutritive value of black soldier fly larvae reared on common organic waste streams in Kenya. Scientific Reports 9: 10110. https://doi.org/10.1038/s41598-019-46603-z

Shumo, M., Khamis, F.M., Tanga, C.M., Fiaboe, K.K.M., Subramanian, S., Ekesi, S., Van Huis, A. and Borgemeister, C., 2019b. Influence of temperature on selected life-history traits of black soldier fly (*Hermetia illucens*) reared on two common urban organic waste streams in Kenya. Animals 9: 79. https://doi.org/10.3390/ani9030079

Preliminary project design for insect production: part 2 – organism kinetics, system dynamics and the role of modelling & simulation

R. Kok

Bioresource Engineering, Macdonald Campus of McGill University, 21,111 Lakeshore Rd, Ste-Anne-de-Bellevue, QC H9X 3V9, Canada; robert.kok@mcgill.ca

Abstract

During preliminary project design (PPD) an entopreneur can investigate a variety of process types, organism/feed combinations, operating conditions and procedures, control approaches, etc. before deciding on a specific system arrangement and proceeding to a more formal design stage. With modelling and simulation the effort required to locate a high-value point within the overall possibility space for a system can be greatly reduced because much of the development work can be carried out in a virtual environment. Nevertheless, to formulate models of the organism/feed kinetics as well as other system aspects, this approach must be based on experimental data. And, of course, results of a simulation study must be verified empirically. A simulation-based approach to PPD allows an entopreneur to study the dynamics of a wide variety of system arrangements and to gain insight into how a given arrangement is likely to perform with different parameter values and disturbances.

Keywords

kinetics – dynamics – modelling – simulation

1 Introduction

The overall purpose of preliminary project design (PPD) is to facilitate an entrepreneur's exploration of the conceptual possibility space for a proposed project and to help them make some preliminary decisions about that project. In essence, PPD is the activity that links early concept development, initial data acquisition and project definition to formal engineering design, construction, and ultimate plant operation (Kok, 2021a; Figure 1). Although the principles of PPD are applicable to pretty well all types of projects, the discussion here is oriented to the mass production of insects on a commercial scale, and more specifically, the rearing of black-soldier fly larvae, meal-worms and other, similar kinds of organisms. Often, even before a project's PPD phase is started, some preliminary decisions will have been made about, e.g. the organism and feed, approximate scale of operation, process type, target market sector, etc. However, during PPD these aspects may be adjusted considerably, or even changed altogether. Thus, PPD is about evaluating the potential of a spectrum of options in order to obtain a fair understanding of the pros and cons of each and to select those options which best fit the entopreneur's intent and capabilities. In essence, PPD is a structured approach to decision making before any major, formal commitment is made to a project.

This paper is the second (Part 2) in a set of four. In the first paper (Part 1: Kok, 2021a) the overall mass and energy/heat balances of a rudimentary larval rearing setup were explored by means of a simple calculator model. The model was presented in spreadsheet form and is available to the reader as electronic supplementary material for use with their own data. Generally, in considering a production project, this will probably be the first item to address since it will yield some basic, if rudimentary, information about quantitative aspects. Then, based on this, an initial likelihood of a proposed project's financial success can be estimated. Evidently, if this likelihood is too far below an investor's hopes and expectations the project objectives and scope will need to be re-examined. In short, a calculator model of this type allows an investor and entopreneur to fairly quickly evaluate the potential of a proposed project while also providing them with the opportunity to investigate related alternatives, e.g. operation with a variety of different feeds, etc. In essence, the calculator model provides a framework for the entopreneur to work in and, at the same time, lets them know what data is required to make a rational decision. If adequate data cannot be obtained for an alternative, it is immediately clear that the reliability of the predictions is fairly low. One advantage of a simple calculator model of this type is that the results are available almost instantaneously so that the sensitivity of any of the outputs to variations in the inputs can be quickly established. Overall, the intent of providing the model is to facilitate the entopreneur and investor in making some preliminary decisions for a project about production volume, organism, feed, etc. on a rational basis.

After some of these preliminary decisions have been made, the type of process that will be employed and the type of reactor the process will be based on need to be considered. The choices related to these two items will be strongly influenced by a host of factors including the scope of the project, the target organism, technical capacity available, personal preferences, etc. A major item in these considerations is how the process will perform dynamically, i.e. how the system state and output will vary with time. Thus, whereas in Part 1 (Kok, 2021a) the overall process input/output balance was of primary concern, in this paper process dynamics are considered and, specifically, how they can be dealt with, thought about and anticipated.

Three main types of influences on the nature of the process dynamics are considered here: system factors, kinetic factors and control system actions. Of course, process inputs and disturbances will affect the specifics of the dynamics as well, but not the character of the system's response. The relationships between the factors are illustrated in Figure 1. They were also briefly discussed by Kok (2017). Two of the kinetic factors are directly related to the inherent properties of the organism while the other two are related to system operation. One of the latter two, environmental heterogeneity, is directly influenced by the process conditions which are the result of all the factors combined. Below, the interaction between the factors is examined with a stress on how a modelling and simulation approach can be used to understand and, at times, exploit the relationships between them. It should be noted that, as explained further below, 'modelling and simulation' here refers to a very general set of activities, including purely mental ones.

In Part 3 (Kok, 2021b) of this set a number of process types suitable for insect mass production at various scales are then explored in terms of their input/output dynamics, ease of process operation, equipment complexity, labour requirements, suitability for automation, etc. Finally, in light of these former aspects, in Part 4 (Kok, 2021c) overall process dynamics as the convoluted result of choices about the organism/feed combination and process type are discussed, together with some considerations of the physical facilities required to house the project. Altogether,

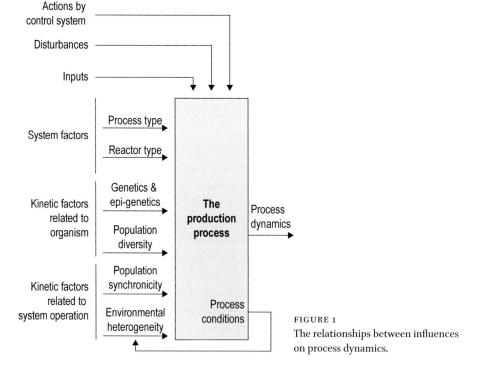

FIGURE 1
The relationships between influences on process dynamics.

the material is structured to help entopreneurs deal with the PPD of projects ranging from the very small, manually-operated enterprise to the full-scale, automated
industrial plant.

2   Modelling and simulation

During PPD one normally reasons about a project and the process on which it is
based according to some type of simulation. In its makeup such a simulation may
be situated anywhere between purely informal and formal. If the simulation of a
project is entirely informal it will probably consist of purely mental considerations
of a number of possible scenarios and the, again purely mental, evaluation of their
foreseen likelihood of success. Under such conditions the criteria for success may
never be explicitly stated and the final judgment may be made purely intuitively.
Whereas this approach has been quite successful in many instances (some people having 'good insight' and 'talent'), it can be difficult to justify judgments and
decisions, especially since the reasoning chain followed in intuitive decision-making is usually not accessible to even the individual themselves. Although this simulation method can be quite sophisticated (after all, the human brain is still the
most advanced computer we have available), without the presentation of a clear
and explicit reasoning chain at the conclusion it can be difficult to convince potential investors to underwrite a project. As an alternative to the purely informal
approach, a number of semi-formal simulation methods can be followed in which
'insight' and 'hunches', etc. can be supported with calculations and references to
other, similar projects that have already proven successful. Then, at the far end of
the spectrum lies the truly formal simulation. This is grounded in computer-based
reasoning combined with evaluation of all options on the basis of clearly stated
criteria so that the entire reasoning chain and all intermediate reasoning steps and
results are available for inspection. Thus, with a formal simulation approach it is
much easier to explain and justify the decisions that are made and to discuss the
advantages and disadvantages of various options with other parties. The disadvantage of the formal approach is that the reasoning followed is only as sophisticated
as the programming of the simulation (and the model on which it is based). Real-
istically, a modern, potential entopreneur will probably base their reasoning partly
on a formal simulation approach and combine the outcome of that with the results
from one or more semi-formal and informal methods so as to arrive at an ultimate
decision about whether to proceed and try to assemble the means to market the
project. After all, the concept will have to be promoted to potential investors who
will be convinced by the entopreneur's passion and enthusiasm for the project just
as much as by rational argument and cold facts about profitability. Only the formal
simulation approach is dealt with here.
     Any simulation of a process will be based on a model of that process and, in turn,

the model will be based on specific data about the system of interest, as well as more general knowledge. For a PPD exercise, really only a preliminary step in project development, the process model may be relatively coarse and incomplete and be based on simplifying assumptions in order to expedite the procedure. It should, however, be detailed enough to yield a 'fair impression' of process performance in the region of interest of the conceptual possibility space. Thus, part of the 'art' of PPD is choosing which aspects to model to what extent and how to interpret the results of the simulations and the ensuing reasoning based upon them. During a PPD exercise it will frequently occur that it becomes desirable to consider an option for which no data is available. For example, the system model may be based on larvae growing on a feed with soy meal as protein source. If it then seems interesting to partially replace that ingredient with cottonseed meal, more data will be needed to expand the range of that model. Thus, in this instance, PPD considerations lead to a data requirement. In another situation a simulation may suggest that system performance can be improved by varying the temperature during larval rearing. That would then need to be verified experimentally, leading to more data requirements, reasoning, etc. Therefore, in the approach presented here model-based reasoning and experimental exploration and verification are seen as the twin components of an iterative knowledge engine that one can use to explore the possibility space for a project during PPD. Generally, the overall objective is to identify a family of process operating conditions that will give rise to process stability as well as relatively high values for an objective function such as financial reward.

When the model of a process is made up only of simple algebraic relationships between independent variables (such as feed protein content) and dependent ones (such as productivity per unit reactor volume) there is essentially no difference between the simulation and the model on which it is based. This was the case in Part 1 of this set (Kok, 2021a) in which the computation of the overall mass and energy/heat balances of a single tray of larvae growing on a bed of feed was presented. But when the model is made up of more complex elements, some of which may be relationships between variables that are time based such as differential equations, the model and the simulation are quite different entities. Now the simulation is more like an exoskeleton within which the model is contained and integrated with respect to time. The structure of the model-simulation system is illustrated in Figure 2. As shown, some types of elements that may be included in a complex model are algebraic and differential equations as well as data, rules and cases respectively held in databases, rulebases and casebases, and even stories.

With simulation based on a dynamic system model time-varying inputs (and system disturbances) can be taken into account, as well as time-varying responses so that, in turn, dynamic, time-varying system states and outputs can be computed. This type of knowledge is very important because there can be very significant variations in process variables during, e.g. the rearing of a batch of larvae. Concomitantly, the wave form and magnitude of such variations are often strongly affected by the process type and the processing conditions. This is discussed in more detail in the sections below.

As mentioned above, a spectrum of simulation methods is available, ranging from the informal to the strictly formal, with any formal method being based on a computational approach. Evidently, a large number of languages, packages and simulation systems are available with which a model can be written and used in a simulation. It is the author's opinion, however, that for the modelling and simulation of a biological production system the use of an object-based approach provides for maximum inclusiveness and flexibility (Parrott and Kok, 2001a,b, 2002). Although this does require some investment it has the advantage that it readily allows for minor model adjustments as well as major modifications. Thus, a model (within a simulation) of this type can function as a pivotal instrument in the PPD exercise by facilitating a wide-ranging exploration of the possibility space, both for the process and for the control approach.

Although a formal simulation exercise as part of PPD comes at a cost, it also delivers major benefits. Clearly, those benefits come to light most eminently when the model reflects reality to a good degree, illustrating the fact that any computational effort is only one of the twin components of this type of knowledge engine, the second one being data and knowledge acquisition about the system under investigation.

3   Kinetic factors

In chemistry the term 'kinetics' refers to the rate of a reaction between molecules, that rate usually depending on conditions such as temperature, the concentrations of various reactants, the presence of a catalyst, etc. 'Kinetics' is also commonly used in reference to the growth of microorganisms like bacteria and yeasts. The mathematical treatment of such growth is well established, especially for production in submerged culture (Aiba *et al.*, 1973). Accordingly, it is used here in the same sense to refer to the reaction rate between a macroorganism such as an insect and its nutrients, subject to the environmental conditions of the situation. As is true for chemical and microbiological systems, the kinetics of an entotechnological system can vary substantially with the parameters of a process (e.g. Diener *et al.*, 2011; Nyakeri *et al.*, 2017) so that the details for each system and each set of circumstances must be determined experimentally.

The first kinetic factor considered here is related to genetics and epi-genetics which determine the organism's response to a given history of environmental conditions. Models of various complexity can be constructed of this response, with different variables being included or excluded. Thus, although it is mostly the organism growth rate that is dealt with in the discussion below, a comprehensive kinetic model will actually consist of a functional description of how the entire

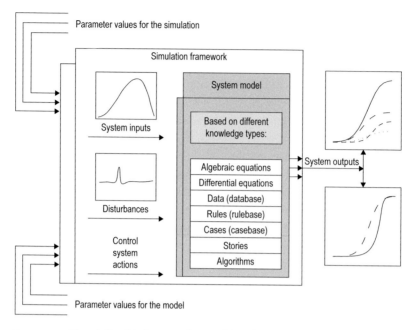

FIGURE 2   The relationship between the simulation framework and the dynamic system model

reaction will proceed. Hence, if such a model is reasonably complete, besides being a description of how a given quantity of eggs or young larvae will grow and how the composition of the population will change during its development, it will also be a functional description of the rates of substrate disappearance, oxygen consumption, carbon dioxide and heat production, the generation of various types of larval waste, etc. Also, there is a large spectrum of situations for which different kinetic models might be developed in accordance with how the environmental conditions will be varying during the rearing period. At the one end of the spectrum the control system is able to regulate the environmental conditions so they remain entirely stable throughout the rearing period. The general equations for this type of kinetic model are then as follows:

$$dX/dt = F1a \{ X, Z, A1 \} \tag{1a}$$
$$dY/dt = F1b \{ X, Z, A2 \} \tag{1b}$$

where:

$A1, A2$ = vectors of kinetic parameters

t = time

X = vector of larva-related variables including: larval weight and development; larval composition such as protein, fat, etc.

Y = vector of time-varying variables including: feed consumed; oxygen consumed; carbon dioxide given off; latent heat given off as evaporated

water; sensible heat given off; waste materials produced; and water
consumed.

Z = vector of non-time-varying variables including: feed type; feed
availability; temperature; water availability; oxygen availability; and carbon
dioxide presence.

In Figure 3 are depicted three component synthesis rate curves (protein, fat, and
other materials) corresponding to a hypothetical kinetic model as described by
Equation 1a. The curves are for an individual larva; the three rates were also integrated over a 14-day period and added, resulting in the larval development curve
(dry weight vs time) culminating in a final dry weight of 120 mg.

   At the other end of the spectrum, the environmental history in the reactor is
allowed to evolve freely and all variables allowed to find their own levels without interference of the control system. Larval growth may, for instance, generate heat and
humidity in the bed of substrate on which the larvae are being reared, causing the
environmental conditions during the rearing history to be time-varying to a greater
or lesser degree. This will lead to a substantially more complicated situation, the
general equations for this type of kinetic model being:

$$dX/dt = F2a \{ X, Y, B1 \} \hspace{4cm} (2a)$$
$$dY/dt = F2b \{ X, Y, B2 \} \hspace{4cm} (2b)$$

where:

B1, B2 = vectors of kinetic parameters

t = time

X = vector of larva-related variables including: larval weight and
development; and larval composition such as protein, fat, etc.

Y = vector of time-varying variables including: feed type; feed availability;
feed consumed; temperature; water availability; water consumed; oxygen
availability; oxygen consumed; carbon dioxide presence; carbon dioxide
given off; latent heat given off as evaporated water; sensible heat given off;
and waste materials produced.

Clearly, there are models that can be developed for situations intermediate to these
two extremes. For example, the environmental conditions might be adjusted once
or twice during the rearing history, but otherwise held steady by the control system.
Evidently, as the rearing situation becomes more complicated and variable more
experimental effort (or a very good literature source) will be required to obtain sufficient information to formulate an appropriate kinetic model. Regardless of the
type of model, the information about the kinetic relationships described by these
equations may be ultimately made available to a simulation framework in either
analytical or numerical format.

   In the models described above the first organism-related factor, genetics and

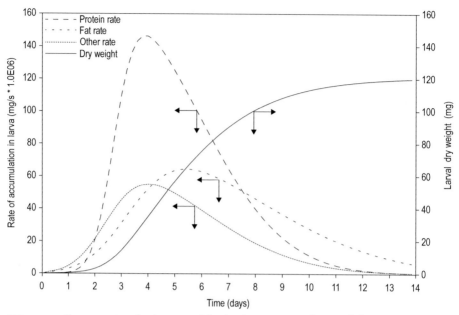

FIGURE 3   Component synthesis rates and the development curve for a single larva

epi-genetics, is dealt with, but not the second which is population diversity. Thus, referring to Equation set 1 and Figure 3, one would expect that if a thousand perfectly synchronised eggs (i.e. all exactly the same age) are subjected to the same environmental history (no heterogeneity), the same curves would be obtained, amplified a thousand times, the population attaining a total dry weight of 120 g after 14 days. As is true for pretty well all biological populations, however, the larval population is likely to be multifarious to a some degree in their genetic and epigenetic characteristics and this will result in diversity in their responses to the environmental history to which they are subjected. With other words, the rate curves for the individuals will not be exactly the same even if the larvae are subjected to the same histories so that the larvae will be distributed to some extent in their degree of development at the end of the 14-day incubation period. This is shown as Case 1 in Figure 4. Here the final development distribution is Gaussian in shape but that is for convenience of illustration only; the organisms' characteristics could have any distribution which would be reflected in the end result for the population. To take the population's genetic diversity into account in a simulation of the process it will need to be reflected in the organism kinetic model. This is best done by structuring some of the parameters in vectors A and B in equation sets 1 and 2 as probability functions. The shape of these functions would, of course, need to be determined experimentally.

Whereas the first two kinetic factors are directly related to the genetics of the population, the other two are of a more operational character. The third factor is related to it being quite unlikely to obtain a population of eggs that are exactly the same age. On the contrary, under real conditions it is much more likely that the eggs in a batch will be distributed in their ages. If this non-synchronicity of the eggs is taken into account at the same time as the organism's genetic diversity, the development distribution of the larvae at the end of the incubation period will be wider than when only the diversity is considered. As well, the shape of the distribution will probably be affected. This is shown as Case 2 in Figure 4, for which the development distribution is a slightly skewed normal curve.[4] As shown in Figure 4, it is likely that the widening of the development curve will lead to higher reject rates caused by both under-developed and over-developed larvae.

Whereas the third kinetic factor is related to the age distribution of the eggs being supplied to the process, the fourth is related to individuals being exposed to somewhat different environmental histories during their residence time in the reactor. Thus, even though a batch may be reared in the same tray or bin or on the same belt, at various times the organisms may well be subjected to quite different environmental conditions causing their environmental histories to not be the same. For example, a larva may spend its early life near the edge of a tray where it may be slightly cooler and drier than in the centre and that will affect its growth curve in accordance with its individual genetic and epigenetic characteristics. This applies to all larvae in the population so that the shape of the final development distribution will be affected in some way by heterogeneity in environmental conditions. In Figure 4 this is illustrated with Cases 3 and 4, for which the distributions are quite skewed, one somewhat more extreme than the other. Due to the greater spread of these development distributions as compared to Case 2, the reject rates due to larval under-development and over-development are again higher than before. As for the other kinetic factors, it is likely the environmental heterogeneity will be distributed in some way and the parameters for the distributions of the variables involved will need to be supplied to the simulation as part of the process model.

A key point to consider here is that the cumulative effect of the three causal distributions (diversity, non-synchronicity and environmental heterogeneity) is not merely additive. Instead, the effects will be interactive and will be convoluted.[5] A very significant feature of this phenomenon is that, although the three causal distributions may be fairly narrow in themselves, due to convolution the final distribution may become surprisingly wide and quite skewed, thus resulting in higher than anticipated reject rates from the final population, both as under-developed and over-developed individuals. Developing a good understanding of how these causal factors interact and how their distributions are shaped and convoluted while the larvae pass through the reactor is crucial for making decisions about a number of aspects of the process, ultimate process type choice and its design, as well

---

4 Note that this curve is hypothetical only and serves merely as an illustration for the discussion.
5 For a good discussion of convolution see https://en.wikipedia.org/wiki/convolution.

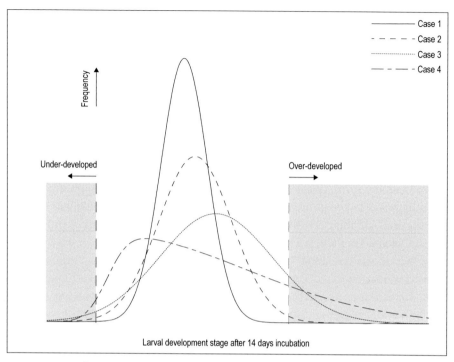

FIGURE 4   Development distributions for a larval population

as the structure of the control system. Manifestly, it is preferable to develop such insight during the earlier phases of a project's development since it can be useful when searching a possibility space for 'low-hanging fruit' type of opportunities to improve process performance and productivity.

4   The system model, simulation and process dynamics

'Process dynamics' refers here to the assembly of the histories of all process variables. Typically, for a bio-production process such as the rearing of an insect it is the result of the growth environment acting on the organism within the rearing reactor and associated equipment. The dynamics of the process come about as the interaction and convolution of the processing approach, the processing conditions and the kinetic factors of the organism. Studying these dynamics by means of simulation during the PPD phase of project development allows one to predict how a given configuration of a proposed system will perform and thus allows for discussion and planning well before any physical implementation takes place. A system model and simulation can also be very useful later on during project development to study how system performance might be modified and improved, etc.

As illustrated in Figures 1 and 2, in general, the rearing system relies for its functioning on its inputs but is subject to disturbances while being governed by a control system. This means that models of these various system components, including the inputs, must be made available to the simulation in order to predict how it will perform and respond dynamically to any disturbance. It does not mean, however, that these models need be very detailed or complicated, especially at the start of a project discussion. Thus, one might begin a simulation study for a proposed project with quite rudimentary models of all the system components and then gradually increase their sophistication as the need arises, in this way gradually developing the models as the study progresses. This procedure of gradual model development is discussed in more detail in the section below on the applications of simulation. Regardless of how coarse or how detailed the models are, in order to be reasonably representational, the components that will need to be supplied for the simulation are as illustrated in Figure 5.

As shown, besides the model of the rearing system, there will be ancillary models of the disturbances and the control system as well as of the inputs. The latter must encompass all materials and services being supplied to the system such as eggs, feed, water, air, energy being supplied, cooling, etc., including the states of intensive properties such as the temperatures and enthalpies of these. The same applies to the outputs. In this scheme the environmental histories to which the organisms are subjected and how they respond to these are determined by a combination of how the process is operated, how the reactor and its associated equipment is arranged, the organism kinetics, how the inputs and reactor conditions are manipulated by the control system, how some of the outputs can or will affect the environmental histories, and how disturbances will impact the production system. 'Other materials' refers to substances such as larval waste described in Part 1 (Kok, 2021a).

As mentioned, the system model components (and the ancillary models) can be as complicated or as simple or as desired. For instance, at the start of a simulation study one might simply not take into account any possibility of disturbances, set all controls to nil, assume stable inputs, define a simple processing scheme and reactor, ignore any environmental heterogeneity and any feedback from the outputs, and disregard any non-synchronicity, genetic diversity as well as the possible presence of competing organisms such as bacteria. In this case the simulation output will be a straightforward history of a uniform population growing under unrestrained conditions, the result mainly depending on the organism's genetics under the conditions dictated by the processing scheme and the equipment. The histories of organism development, feed depletion, other material accumulation, temperature, enthalpy of the outflowing air, etc. will then simply unfold in a natural, unbridled manner.

Then, as the simulation study progresses one can add elements to the models of the various system components in order to make it reflect reality better and, hopefully, elicit a dynamic response that matches. For example, one might wish to not ignore the presence of bacteria that will compete for feed with the larvae. To do so, a sub-model of bacterial kinetics will need to be added to the overall system model. This can be useful to examine how some of the inputs might be manipulated in order to reduce the 'bacterial composting effect'. This would also require the inclusion of a control module to carry out such manipulation.

In the relatively simple examples above the models are of the 'lumped-parameter' type meaning that the entire population of organisms is dealt with as a single entity, the same being true for the entire population of bacteria. If, however, genetic diversity, non-synchronicity and heterogeneity of the environmental conditions are to be taken into account, the system model components need to be more complex and of a distributed parameter type. As well, the processing scheme may be operated and the reactor set up to deal simultaneously with multiple, time-spaced batches of organisms (this is discussed in detail in Part 3 of this set; Kok, 2021b) so that this possibility will need to be accommodated in the model too. Although it is not always necessary to include all the above-mentioned types of variability and multiplicity in a system model, generally, one would include at least several of them and the modelling approach should reflect that. As pointed out, in the author's opinion, the object-based approach is very suitable for writing such models. It readily allows for a population or a phenomenon to be divided into groups at any

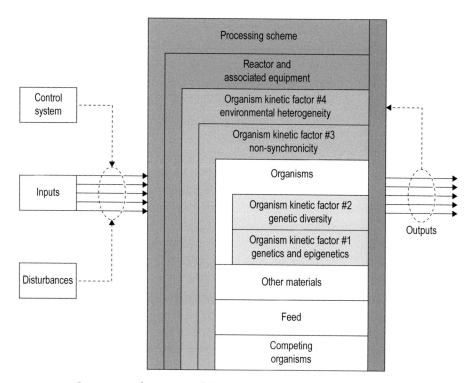

FIGURE 5   Components of a system model

resolution (even down to individuals), reflecting the distribution of the character-istic of interest. For example, a population of 100,000 eggs whose age distribution is known might be divided into a thousand objects whose ages have the same dis-tribution as the population. At the same time, although in reality the environmen-tal heterogeneity will be distributed continuously, it might be approximated with a hundred zones in which conditions are constant so that the thousand slightly different organism objects will variously experience somewhat different histories in a manner similar to that would happen in a physical reactor. This approach can be expanded to include any number of distributed phenomena in a model and is a simple and easy-to-understand way to approximate the convolution of a number of distributed phenomena. Thus, all in all it is a relatively easy way to model any number of trays or bins or groups of such to pass through the system while small, lumped-parameter batches of organisms are followed in detail as they are subject-ed to various inputs, environmental conditions, disturbances, etc. If necessary, this also allows for direct interaction between model elements, e.g. between organisms and the bacterial population. The overall progress of the process as well as inputs, outputs, etc. are then computed at the end of each simulation time step by add-ing the contributions of the small batches of larvae and competing organisms. All system mass and energy accounting also happens at that stage, with data being passed on to the simulation framework for bookkeeping and logging. At the end a complete and detailed picture of the system dynamics, including the histories of the mass and energy fluxes together with the system states will be available to the entopreneur. Generally, the more detailed the model the simulation is based on, the more representative the results will be of what to expect in reality. Also, evidently, as the model is made more complex the computational effort required to run the simulation will increase. And, most importantly, any result obtained by means of modelling and simulation should always be compared to reality so that their valid-ity can be verified.

5   Applications of system modelling and simulation

Modelling a proposed system and studying its anticipated dynamics by means of simulation can serve a number of purposes during project development. Having a reasonably good, dynamic model of the process available can also be very useful later, once the system has been built and is operational, when considering process modifications and improvements. Here only the issues of value in PPD are consid-ered.

The first impetus concerns knowledge and data about the kinetics of the or-ganism/feed combination that is being proposed for the project, together with a description of the environmental history to which it is planned to subject the lar-vae during their residence time in the rearing reactor. Thus, in order to construct a minimal kinetic model data is needed on the organism's growth curve and the various, related variables under the foreseen conditions. Depending on the type of process being planned, this corresponds to having available a limited version of Equations 1a and 1b or 2a and 2b, specific to a narrow set of operating conditions.[6] To create a more sophisticated kinetic model data will then be needed about the genetic and epigenetic diversity of the organism and what effect that has on the organism's growth curve as well as all the other, related variables under the proposed rearing conditions and any conditions that deviate from those. Essentially, every time another factor is taken into account in the model, a more complete version of the equations, or their equivalent, is required. To further expand the system model data on the age distribution of the eggs will also be needed. Hence, the decision to model the system and study its performance with simulation actually dictates the data requirement to make this possible and thus clarifies to the project planner what knowledge is essential for them to proceed in a deliberate and purposeful manner. Obviously, a fully developed kinetic model as expressed with Equation sets 1 or 2 that will yield an organism growth curve together with detailed data on all related variables as a function of organism age, genetic type, environmental history, time, etc., would be quite challenging to assemble, if only for the amount of experimental work required to obtain the data. As pointed out above it is, therefore, often more practical to initially start with a fairly simple model based on limited data and then expand the scope of the model gradually as data needs are identified. This principle applies to all the component models, i.e. of the competing organism, the disturbances, the control systems, etc., as well as that of the organism being reared.

The second purpose of modelling and simulating a system during PPD is probably the most important one: it is to facilitate and speed up the exploration of the possibility space for a project. In this sense, it is an extra tool for the project planner. Thus, the entopreneur might follow a traditional, evolutionary innovation path by starting off with a project concept and developing it by means of a combination of empirical and analytical means, learning from failure and success through a trial-and-error approach until it yields a satisfactory result, i.e. in this case, a combination of organism/feed, environmental conditions, etc. that works. This procedure can, however, be accelerated by having available a model of the system whose parameters can be changed at will so that one can, as it were, perform most of the trials in the virtual realm rather than the physical one. Generally, because it is much cheaper and faster to run simulations on a computer than to do experimental work with populations of organisms one can perform many more 'experiments' in this way. This means that the possibility space can be explored in a structured way in much greater detail than what is feasible with a strictly empirical approach so that it can be value mapped, giving the entopreneur more opportunity to locate any

---

6   Such information need not be available in analytical form; it can also be supplied in tabular format.

high-performance process operating areas and construct a path to an acceptable operating point, i.e. a combination of process type, organism characteristics, environmental conditions, etc. where the value of the process is acceptable.

Although there are many advantages to modelling and simulating a process, there are also some disadvantages. One disadvantage is that one needs to obtain sufficient data on the organism as well as the other system components to enable the construction of a reasonably representative model. This is partly offset by the gain of a good theoretical understanding of the whole system, and often a practical understanding as well. A second disadvantage is the effort and associated cost it takes to construct both the model and the simulation. Hopefully, that will be offset by a reduction in cost of the purely evolutionary, empirical approach because less experimental work will be needed to find a high-value process operating point. This also means the process development time will be shorter and the results will be better. The third main disadvantage of modelling and simulating is that one can easily become divorced from reality. This can and must be overcome by remaining critical of all simulation results and verifying all major conclusions empirically. In short, to obtain reliable results, during PPD a combination of the two approaches is necessary: experimental and simulation-based theoretical, the latter guiding the former and the former correcting the latter.

The dynamic modelling/simulation situation is actually not dissimilar to that dealt with in Part 1 of this set (Kok, 2021a) in which the overall heat and mass balances are estimated with a spreadsheet-based model whose parameters can be adjusted at will in order to investigate the process possibilities. The difference is that now the calculations are based on a dynamic model, but the intent is the same. In fact, considerable guidance on what parameter values to use for the simulation study will probably be derived from the overall mass and energy balance results. With simulation one can rapidly produce not only projected organism histories but also data on organism composition, overall process yield, by-product outputs, energy requirements, etc. for a large number of trial cases, i.e. for different control arrangements and for different environmental conditions that may be stable or that may vary with time, and that may be homogeneous or heterogeneous to some degree. In order to evaluate the success of a given set of trial parameters one usually combines the various process outputs into one or more objective functions which reflect the overall value of that combination. Basically, the objective functions will be a distillation of what the entopreneur wants to achieve with the project. For instance, the first objective function may result only in discrete values such as 0 and 1, respectively indicating that the process is simply not viable at all (0) or at least physically possible (1), while the second one might represent financial reward value to the entopreneur. The results for n independent process variables can then be mapped in a space of n dimensions, the first objective function defining the possibility space for the process, within which value contours of the second objective function can be shown. This is illustrated in Figure 6 in which, for practical reasons, there are only two independent variables present. In this instance the possibility space is shown to be contiguous, but that is not necessarily always so. Also, only the region comprising the inner contours is considered 'of interest' as judged by the entopreneur, i.e. here the process performance is considered worthwhile. Because the number of independent variables may be substantial and resources limited, this type of map will often be based on fairly sparse data and will therefore be most suitable to guide the entopreneur on their search path through the possibility space to an acceptable process operating point. Evidently, one might accomplish the same with a purely empirical approach and, accordingly, two search paths are presented in Figure 6, one simulation-guided and the other unguided. Here, the latter is shown as being longer and therefore requiring more effort while also ending up at a somewhat lower valued operating point for the process, although still well within the region of interest. Reality being capricious as it is, there is no guarantee that a search will always unfold in this manner but what is illustrated here corresponds to the major purpose of the entire modelling and simulation exercise: to shorten the search path and to arrive at a higher value operating point for the process than with an unguided approach.

The third purpose of modelling a process during PPD and simulating its performance is to allow the entopreneur to engage in reasoning that is somewhat more speculative. Thus, once a reasonably representative value contour map has been established and a feasible, high-value operating point for the process has been identified the entopreneur may start to wonder what other factors could possibly be altered so as to improve the situation to a yet-higher value point. For instance, could a different feed be found or could the feed composition be adjusted? Might varying the temperature during larval growth have a positive influence? Referring to Figure 4 in which the organism reject rates for Cases 3 and 4 are fairly high due to diversity, non-synchronicity and environmental heterogeneity questions would naturally arise about the feasibility of breeding a genetically less diverse strain, obtaining eggs with a narrower age distribution, arranging the reactor so that the environment will be less heterogeneous, etc. The potential impacts of any such possible adjustments can often be initially gauged with the existent models and data although, in order to address the queries in depth additional data may be needed. Thus, a speculative approach may lead one to identify and specify new data requirements which may then need to be fulfilled through more experimental work. This would be the case if, for example, different feeds were to be considered or if the option of varying the temperature during larval rearing were to be contemplated. Generally, in more speculative reasoning any of the system components, conditions or parameters can be addressed and adjusted including the models, the minimal requirements imposed by the entopreneur, the form of the objective functions and even the process type itself. In this flexibility the utility of simulation is manifested most clearly as a formal, structured approach to reasoning about a project and its associated possibility space. It should be noted here that, with fundamental aspects of the project being changed as part of the speculations, the possibility space and its value contours can undergo fairly dramatic changes in both shape and location, and even in dimension. Evidently, what might be possible under one set of restraints and goals might not be possible under another and *vice versa*.

The fourth purpose of utilising a simulation approach during PPD is to gain insight into the sensitivity of the process to two main sources of instability: *variation* in any of the main variables such as feed composition, genetic characteristics of the organism, a change in the competing organisms, etc. and *disturbances* such as short-term fluctuations in cooling air temperature or temperature distribution. Developing an understanding of the system's sensitivity to any of these factors and combinations of them facilitates the development of design specifications for the inputs, the system itself as well as the performance of the control system. For instance, a system's performance at a high-value operating point in the possibility space might turn out to be quite sensitive to the age distribution of the eggs, leading directly to the question whether it is feasible to obtain eggs with a very narrow age distribution – a specification for an input. Similarly, it may be sensitive at that point to genetic diversity, leading to the question whether it is possible to obtain eggs that are less dissimilar – another input specification. Larval development may be quite sensitive to environmental conditions during rearing so that the development distribution at harvest will depend heavily on constancy and homogeneity of those conditions – this would lead to specifications for the reactor and the control system, e.g. how far the bed temperature can be allowed to deviate from its set point.

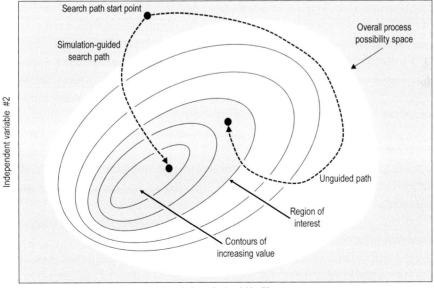

FIGURE 6   Value contours and development paths

As well, the posing of specific questions in response to gaining an understanding of sensitivities will lead to the formulation of more general queries about the system as a whole. These might be related to the feasibility and desirability of a breeding program, the design and development of sensors and other control components specifically for applications in entotechnology, the creation of artificially intelligent components to operate the process optimally so as enforce synchronisation, etc.

In summary, the overall motivation for using a simulation approach during PPD is to examine a larger possibility space for the project than is normally feasible with a purely experimental approach and to arrive at an acceptable operating point for the process. At the same time the entopreneur will gain considerable insight into anticipated design specifications for the process and its equipment as well as for the inputs and the control system. This will allow them to estimate what effort will be required and to assess the feasibility of the plan that is in place. As a first possibility it is not inconceivable that the initial search through the possibility space will lead to a process operating point well within the region of interest while the process at this point is not overly sensitive to variations and disturbances and, upon first examination, is also fairly easy to implement with off-the-shelf equipment. This is an unusually fortunate situation and one would probably proceed right away with experimental work to ascertain it is indeed true, attempt to improve the situation further through some more simulation work, and then proceed with the next project development step. As a second possibility, it is also quite feasible that within the possibility space which is initially established there lies a sub-space that appears promising but whose promise is contingent on a somewhat better definition of the situation, i.e. a more detailed description of the that region of the possibility space. This represents a data requirement and one would proceed in this case with obtaining those data on, for example, egg age distribution or an aspect of the kinetic model, and then compute and map a modified possibility space with new simulation runs. It is actually not unlikely that in reality this will happen a number of times so that during the simulation study the following sequence will be repeated: examination of the possibility space; path searching to a likely process operating point or region of interest (Figure 6); scrutiny and assessment of the system performance at this operating point; composition of questions and consideration of alternatives; setting further data requirements; fulfilling data requirements; compute and map new possibility space, etc. It should be taken into account, as a result of new data entering the system, the possibility space for the process can shift and change shape considerably.

What may also happen is that examination of the possibility space leads to doubts about the viability of the project as put forward. Thus, with the organism/feed combination and under the conditions that have been proposed, even after exploring a number of alternatives, there simply is no region of interest or, in extreme cases, not even a possibility space. Although this may be disappointing to the entopreneur planning the project, because much wasted effort and investment will be avoided, it is really a boon. At this point the entopreneur can either abandon the project altogether or engage in a much wider-ranging exploration of feeds, organisms, process types, by-products, etc., which will then lead to efforts to breed new varieties of organisms, to develop equipment, to innovate novel process types, and to engage in further simulation study based on those.

One issue that bears repeating at this point is that, whereas in the above almost all reference is to simulation performed in a formal manner, i.e. with mathematical and data-based models on computer platforms, it is also possible to do all this in an informal manner, i.e. with thought processes running on human brains. Some advantages of the latter approach are that brains are far superior in terms of applying intuitive reasoning to detect very complex patterns and to be creative so that it is more likely that far-flung prospects will be included in the considerations of alternatives. On the other hand, computer-based reasoning is often more consistent and repeatable so that computational results are usually easier to explain and justify. In many cases a combination of the two approaches will be fruitful.

6   Discussion and conclusions

The main point of this paper is: before an entopreneur commits to a specific plan for a project it's a good idea to think in detail about the dynamics of the system as well as the overall heat and mass balances. A modelling and simulation approach based on experimental data allows for the integration of knowledge about organism kinetics, reactor configuration, process performance and control system activity so that the system dynamics can predicted. This facilitates the rational analysis of possibilities for a project and the examination of alternatives.

At the conclusion of a project's modelling/simulation phase the aspiring entopreneur should have a fair understanding of all quantitative aspects of the project and the selected process type, including heat and mass flows, kinetics, inputs and outputs, dynamics, equipment requirements, etc. Some aspects of the various process types, equipment and operating schemes will be dealt with in Part 3 of this set (Kok, 2021b).

Conflict of interest

The author declares no conflict of interest.

References

Aiba, S., Humphrey, A.E. and Millis, N.F., 1973. Biochemical engineering (2nd Ed.). Academic Press, Cambridge, MA, USA, 434 pp.

Diener, S., Zurbrugg, C., Gutierrez, F.R., Nguyen, D.H., Morel, A., Koottatep, T., Tockner, K., 2011. Black soldier fly larvae for organic waste treatment – prospects and constraints. In: Alamgir, M., Bari, Q.H., Rafizul, I.M., Islam, S.M.T., Sarkar, G., Howlader, M.K. (eds.) Proceedings of the WasteSafe 2011 – 2nd International Conference on Solid Waste Management in the Developing Countries, February 13-15, 2011, Khulna, Bangladesh, pp. 1-8.

Kok, R., 2017. Insect production and facility design. In: Van Huis, A. and Tomberlin, J.K. (eds.) Insects as food and feed: from production to consumption. Wageningen Academic Publishers, Wageningen, The Netherlands, pp. 142-172.

Kok, R., 2021a. Preliminary project design for insect production: part 1 – overall mass and energy/heat balances. Journal of Insects as Food and Feed 7: 499-509. https://doi.org/10.3920/JIFF2020.0055

Kok, R., 2021b. Preliminary project design for insect production: part 3 – sub-process types and reactors. Journal of Insects as Food and Feed 7: 525-539. https://doi.org/10.3920/JIFF2020.0145

Kok, R., 2021c. Preliminary project design for insect production: part 4 – facility considerations. Journal of Insects as Food and Feed 7: 541-551. https://doi.org/10.3920/JIFF2020.0164

Nyakeri, E.M., Ogola, H.J.O., Ayieko, M.A. and Amimo, F.A., 2017. Valorisation of organic waste material: growth performance of wild black soldier fly larvae (*Hermetia illucens*) reared on different organic wastes. Journal of Insects as Food and Feed 3: 193-202. https://doi.org/10.3920/JIFF2017.0004

Parrott, L. and Kok, R., 2001a. Use of an object-based model to represent complex features of ecosystems. In: Minai, A.A. and Bar-Yam, Y. (eds.) Unifying themes in complex systems. Springer, Berlin, Heidelberg, Germany.

Parrott, L. and Kok, R., 2001b. A generic primary producer model for use in ecosystem simulation. Ecological Modelling 138: 75-99.

Parrott, L. and Kok, R., 2002. A generic, individual-based approach to modelling higher trophic levels in simulation of terrestrial ecosystems. Ecological Modelling 154: 151-178.

Preliminary project design for insect production: part 3 – sub-process types and reactors

R. Kok

Bioresource Engineering, Macdonald Campus of McGill University, 21,111 Lakeshore Rd, Ste-Anne-de-Bellevue, QC H9X 3V9, Canada; robert.kok@mcgill.ca

Abstract

This is a discussion about sub-processes suitable for the rearing of insect larvae on dry and semi-dry feeds. Three closely related, key aspects are dealt with as part of preliminary project design (PPD): the type of larval rearing sub-process to be employed, the reactor configuration and the operational approach to be used. A number of sub-process types and reactors are discussed. Because they are most commonly used in the industry today, all the sub-processes are 'plug-flow' (age stratified) and based on 'passive' reactors (contents moving through the reactor). Batch, semi-continuous as well as continuous sub-processes are dealt with and illustrated. Simulation to help the entopreneur chose the most appropriate type of sub-process is highly recommended.

Keywords

batch – semi-continuous – continuous – reactor – larval rearing

1 Introduction

Once someone has decided to become seriously involved in mass producing insects and thereby qualifying as an 'aspiring entopreneur', it is best for them to hold back and generate an overall plan of attack before investing too heavily in a specific concept or project. An outline of how one might proceed with this was presented by Kok (2021a, Figure 1). According to this scheme, one should start by negotiating several stages of early concept development supported by data acquisition while also gaining a minimum of practical experience through playing with some organisms and rearing methods. Although it may seem whimsical, 'playing' here is meant seriously, with the entopreneur engaging in investigative, creative activity without having a limited, fixed objective. 'Play is the highest form of research' (Calaprice, 2010). At the end of these preliminaries there should, hopefully, be a fledgling concept that is suitable for further development through preliminary project design (PPD). During the PPD stage the project concept will be investigated experimentally as well as by means of modelling and simulation resulting, first of all, in a decision to proceed or not to proceed with that project. If the decision is favourable PPD work will also result in decisions about the organism/feed combination on which the project will be based, the process type, the scale of operation, the system operating point and the control approach, etc. A method to estimate overall process mass and heat balances was presented in Part 1 of this series (Kok, 2021a) while process dynamics and how these depend on system kinetics and how they can be studied by means of modelling and simulation were discussed in Part 2 (Kok, 2021b). While, generally, the major features of the fledgling concept may survive, it is quite likely that during the activities described in Part 1 and 2 many of the other features will be substantially modified. This is normal and is to be expected. In fact, one of the main purposes of PPD is to locate a suitable operating point for the proposed system in a large possibility space and that may involve radical adjustments to the original concept. 'Suitability' of the operating point must always be judged according to a set of objective functions defined by the project developer, with one or more of those functions often being based on foreseen project profitability or return on investment (Kok, 2021b).

Although the author's main approach is to deal with design and development situations in a formal, computer-based manner it is also possible to do so in an entirely

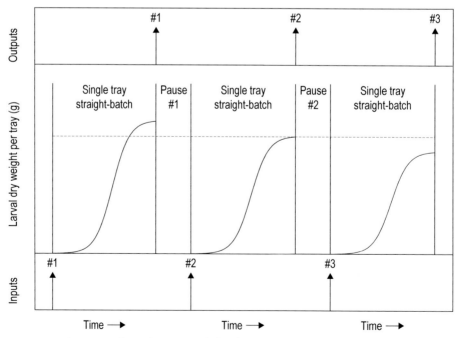

FIGURE 1   Dynamics of a single tray, straight-batch sub-process

'informal' manner. Thus, it is quite possible to analyse data and perform modelling and simulation *in situ* on human brains without having recourse to technological aids. In this way most of the PPD exercise might be carried out internally within the entopreneur and even intuitively so that the entopreneur them self may also not be actively aware of that occurring. A disadvantage of this latter approach is that it is difficult to work in a team in this way and that the results of the 'computations' are not easily justifiable. The formal approach, therefore, has major advantages for the development of larger projects.

In this paper a number of basic processes are discussed, mainly with application to the rearing of insect larvae on beds of dry or semi-dry material. Typically, any of the 'mealworms' (e.g. *Tenebrio molitor*) and the black soldier fly (BSF, *Hermetia illucens*) are reared under such conditions. Similar considerations apply to the rearing of other groups of insects such as crickets and cockroaches, but the methods and technologies discussed here are not directly applicable to them. In a complete insect-production process one would probably also deal with adult housing and egg production as well as egg and pupal incubation (for holometabolic insects) but only larval rearing is considered here. It seems most likely that, once the entotechnological sector develops sufficiently, the production of eggs will become a separate branch and will take place in a location different from where the larval rearing occurs. Accordingly, only larval rearing sub-processes are presented and discussed below.

Although they may be quite different in their operation and resulting dynamics, many of the rearing sub-processes are related in that the more complex ones are often multiplexes of simpler ones. A number of types of both simple and compound multiplexes are presented, collections and manifolds being the simplest types. Compound multiplexes may consist of collections of manifolds, etc. This terminology is explained in more detail below in the various sections on the sub-processes. More often than not, the simpler sub-processes will be the more suitable for smaller facilities and lower production volumes although a small-volume but fully continuous reactor and process are also described below.

2 The time-spaced, straight-batch, single-tray sub-process

In the most basic straight-batch arrangement eggs (or newly-hatched larvae) are added to a bed of substrate in a tray and simply left alone until they are ready to pupate. Depending on the species, it is often necessary to add some liquid water during the rearing period while, at the same time, water leaves the bed due to evaporation. The water vapor is then carried away as humidity in an air flow which can be due to either natural or forced convection. The air also supplies oxygen and removes metabolic carbon dioxide (as well as other gaseous emissions, e.g. from various *Tribolium* species (Duehl *et al.*, 2011; Roth, 1943)). Although this arrangement is labelled here as 'batch' it is, in reality only 'quasi-batch' because water, carbon dioxide and oxygen are exchanged continuously. It is correct, however, that the larvae are reared 'batchwise' in that they remain in the tray for the entire rearing period. The heat and mass transfer balances for this situation were dealt with in some detail in Part 1 of this set of papers (Kok, 2021a) and a fairly rudimentary spreadsheet-based calculator was presented (as electronic supplementary material) with which such balances can be calculated. That type of method can be useful to explore the possibility space in a rudimentary way to find a suitable process operating point. Minimally, it provides guidance for further study on the process dynamics through modelling and simulation because it yields some basic data on mass and energy flows as well as yields, thus also providing some insight into cost effectiveness. The dynamics of a very basic straight-batch process based on a single tray being filled and emptied repeatedly are illustrated in Figure 1. Note that the output from sequential trays may be slightly different due to variations in eggs, egg age, environmental circumstances, etc. This process will be used only by a very small operation or for experimental purposes.

Evidently, in the straight-batch arrangement sufficient substrate must be added at the beginning of the rearing period to last until the larvae have grown to their harvest stage (which is often, but not automatically coincident with pre-pupation). One advantages of this is the simplicity of the setup and ease of system operation. One major disadvantage, however, is that bacteria will be able to grow throughout all the feed for the entire rearing period and can therefore consume a substantial fraction of the substrate while also giving off heat and emitting carbon dioxide. This will lower the yield of larvae (g_dry_weight_ larvae_obtained / g_dry_weight_ substrate_disappeared) and increase the service requirements of the process. Also, productivity (g_dry_weight_larvae_obtained / unit_time × unit_reactor_volume) tends to be quite low in the straight-batch arrangement because, for much of the organism's residence time in the bed, the population density (g_dry_weight_larvae / unit_reactor_volume) will be fairly low. Of course, due to practical considerations, one can rear only a limited number of organisms in one tray. The tray itself can only be so large (actual limits are difficult to define, but there are limits) and the bed can only be so deep because oxygen needs to penetrate from the surface throughout its volume.

Consideration of the basic straight-batch system is useful mostly because many of the other larval rearing sub-processes are various kinds of multiplexes of this basic one. This means that most of the methods and reasoning developed for the basic straight-batch system apply equally to the more complex systems. Hence, the mass and energy/heat balances for these more sophisticated types of sub-processes, as well as the relationships between organism kinetics and process dynamics, are often closely related to those of the straight-batch type. Consequently, their balances and dynamics can usually be derived with similar methods and reasoning (and models and simulations) as long as the multifariousness introduced by collectivity

and manifoldness are taken into account. This is, however, not always true for some
of the continuous systems in which the populations are mixed-age.

3   The time-spaced, straight-batch, collection-based sub-process

Once a practical size limit is reached for a single tray, the most obvious way to increase production is to operate the sub-process with a simple multiplex that is a
collection of trays which are used in parallel and kept synchronised. They can be
arranged side-by-side on a bench (not unusual for very small operations) or as a
stack in a frame mounted on wheels (not unusual for slightly larger ones). The latter
facilitates loading and unloading of the trays and also makes it easy to move them
into and out of a controlled environment space (Kok, 2017). The facility is then
usually operated to deal with a time-spaced sequence of collections with pauses
in between for harvesting and processing the larvae, cleaning, feed preparation,
administrative tasks, etc. The dynamics of this sub-process type are illustrated in
Figure 2. In this case, there are 20 trays per collection, with some variation amongst
trays shown with dotted lines in the production graphs. This could be due to any of
the causes that can lead to widening of the distributions discussed by Kok (2021b).
Although productivity of this type of scheme is relatively low, the simplicity of the
physical arrangement as well as the ease with which it can be operated are very

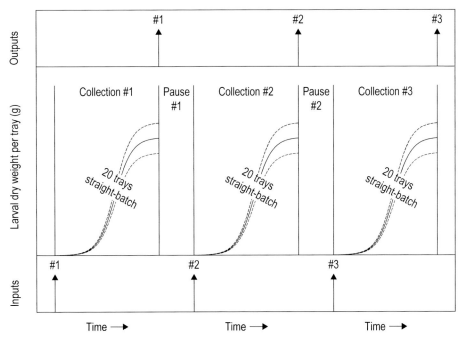

FIGURE 2   Dynamics of the time-spaced, straight-batch, collection-based sub-process

appealing for the small producer because all the activities are sequential. Thus, this type of sub-process can be operated by one person, even working on a part-time basis. All the product is available at the same time and can be separated, extracted, dried, etc. in a single work session. After that there is no immediate pressure to re-start the next collection; the pause between collections can be extended as needed so that tray preparation and inoculation can be timed for the convenience of the operator and the timing of the harvest can be planned for when labour will be available, e.g. during the weekend. Throughout the rearing period minimal effort is required; daily observation and watering are often sufficient.

4   The manifolded, straight-batch, collection-based sub-process

This kind of system is a compound multiplex in which collections of trays are manifolded, i.e. they are not sequentially spaced out in time as shown in Figure 2 but, rather, overlap in their residence time inside the reactor as illustrated in Figure 3. Generally, production will be higher and this arrangement is therefore more suitable for slightly larger-scale plants. If the collections are trays on rolling frames (Kok, 2017) one convenient arrangement is to move them sequentially through a tunnel reactor in which the environment is controlled. This type of reactor (which itself is static) is fairly easy and cheap to build and operate, with air flow coming in through a plenum from one side and out through another plenum on the other side. One advantage of this arrangement is that there can be different 'zones' along the length of the tunnel so that, if desired, the organisms can be subject to varying conditions during the rearing. The length of the tunnel will depend on the residence time of the frames and the desired output frequency. For example, if the trays need to stay in the reactor for 14 days and the desired output frequency is one collection per day, then the tunnel will have to accommodate 14 frames at a time. If each frame is 70 cm in depth, the tunnel will need to be about 10 m long, with a space allowance at both the entrance and the exit for loading and unloading. To attain a higher output frequency the tunnel can simply be made longer and the speed of the frames through the reactor increased. Evidently, this arrangement requires labour input more often than the simpler collection-based multiplex described above but, if the timing is set up properly, this system can still be operated on a part-time basis by one person who would need to process a collection and prepare a new frame of trays, e.g. every morning or evening.[7] One advantage of the manifold-based multiplex is that the equipment with which the larvae are processed (e.g. grinder, microwave dryer, separation unit, etc.) is used more often and is, therefore, easier to justify financially. This advantage becomes more significant as the production volume is increased.

---

7   Thus, not unlike having dairy cattle that require daily milking, feeding, etc.

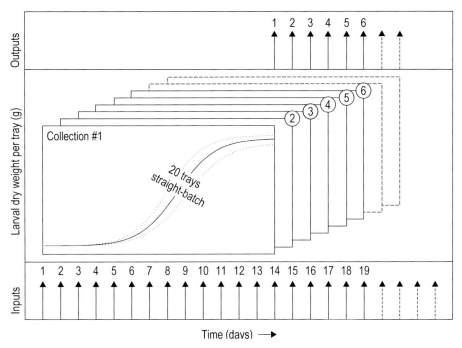

FIGURE 3   Dynamics of a manifolded, straight-batch, collection-based sub-process

Once a maximum practical limit for the size of a single reactor is reached, to augment production further one can increase the number of reactors. If the system is based on a tunnel-type reactor a convenient arrangement is to put the tunnels side-by-side with a service space in between. Such a multiplex of manifolded collection-based reactors employed in parallel would probably be operated as a manifold of manifolds, with the reactors as a whole being kept out-of-phase with one another so that the same machinery can be used at both the input and output ends to service all of them, the result again being better utilisation of that machinery. These considerations are dealt with in more detail in Part 4 of this set (Kok, 2021c), in the paper on facility considerations. Although it is evidently possible to expand such a system to a very large capacity while keeping both the physical arrangement as well as the operation of the process quite simple, for larger systems it will likely prove advantageous, albeit at a cost of increased operational complexity, to instead use one of the fed-batch (as opposed to 'straight-batch') methods described further below.

5   The manifolded, straight-batch, single-tray-based sub-process

In this type of multiplex it is not collections of trays that are manifolded but, rather, single trays or bins. Thus, in this case units are not prepared, inoculated and incubated in sets but, rather, as individual elements which will usually be spaced out evenly in phase (and time). For example, if one tray is prepared every 10 minutes and the required residence time in the reactor is 14 days, then there will continuously be 2,016 trays in the reactor, one emerging every 10 minutes to match the one being put in. Obviously, this arrangement calls for a specific kind of reactor. In this approach (used for the production of yoghurt, for example, where the fermentation takes place in the retail-level container) all the units are more-or-less kept moving continuously through the reactor while sitting on a belt or hanging from a chain. The situation is illustrated in Figure 4.

One advantage here is that the reactor's footprint can be kept fairly small because the reactor can be quite tall. Whereas a tall reactor is not impossible for the other multiplex types discussed above, in this case the situation is particularly appropriate. Another advantage is that a simple manifold-based process is very suitable for automation at both the input and output ends so that less manual labour is required per tray than for the previous two approaches. As well, because only small

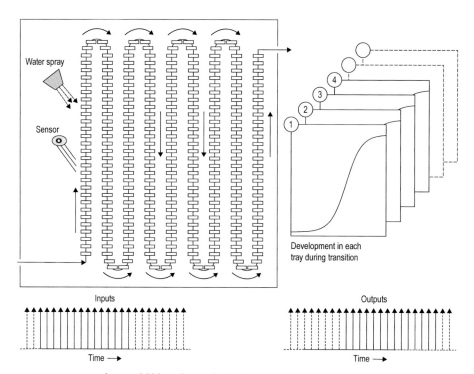

FIGURE 4   A simple manifold-based, straight-batch sub-process

quantities of materials need to be dealt with at any one time, relatively small-scale machinery can be employed for tray or bin preparation, product separation, grinding, drying, bagging, etc. That machinery will be used quite intensively and thus have a high utilisation factor. Monitoring of the process is relatively easy because the units will be moving past the sensors regularly and, similarly, taking corrective action such as spraying water onto a tray is also fairly straightforward. In Figure 4, the 2,016 trays (not all shown!) are spaced out along 9 vertical lines and 8 turns. Hence, there need to be about 210 trays per vertical line. If they are spaced at 10 cm vertically, it should be possible to fit the assembly inside a reactor of 25 m high, allowing for head room, etc.

Although automation has many advantages, there are also disadvantages. Thus, machinery of this type is continuously active and moving and cannot be left to operate by itself for extended periods as is possible with more passive equipment. This means that the process will need frequent attention and someone needs to be close by 24/7. Another disadvantage of this type of process and of reactors such as shown in Figure 4 is their vulnerability to disruptions. For instance, if there is an electrical power failure there are several thousand trays stuck in a 25 meter tall tower with no way to unload them in a hurry. For this type of system a sufficient air flow for cooling, oxygen supply and carbon dioxide removal also needs careful consideration. These aspects need to be taken into account during plant design and will be addressed in some detail in Part 4 of this set (Kok, 2021c), in the paper on facility considerations.

The process based on a simple manifold as illustrated in Figure 4 is most suitable for slightly larger operations; it requires more investment to build, it needs more labour to operate, and it normally has a larger capacity than the previously-mentioned processes. And once it is installed, the production capacity of the plant can be increased quite readily by adding more reactors in a side-by-side layout, with spacing in between to create a multiplex of reactors. Of course, this possibility should be taken into account when the basic plant layout is created for that to be readily achievable. In order to maximise equipment utilisation on the input and output ends one would probably operate such a multiplex of reactors as a manifold, i.e. out of phase with one another. For instance, for a set of two reactors of the kind described above being operated 180 degrees out of phase, one tray or bin would need to be prepared every 5 minutes and one tray would exit the assembly every 5 minutes. With several more reactors in parallel in the multiplex and operated equally out of phase production would really become 'semi-continuous' and approach the performance of a continuous process based on a plug-flow type of reactor as discussed further below.

6   The manifolded, immobile, fed-batch, collection-based sub-process

In the various straight-batch processes dealt with above the eggs (or very young larvae) were added to the tray or bin in which they were to be reared, together with the entire amount of feed they would need to reach harvest time (often at pupation, but not necessarily). Within this definition of 'straight-batch' is included the allowance that there will be an air stream through or over the surface of the bed and that water can be added as determined by the control system. The air stream will deliver oxygen to the biological reaction while also removing heat, carbon dioxide and, perhaps, some other gaseous by-products. The water that is added may contain some micro-nutrients. Although these processes are relatively straightforward in their operation, they are problematic in several ways. One problem is that all the feed is present during the entire rearing period while most of it is not consumed until fairly late during the rearing. Consequently, most of the feed is subject to degradation and bacterial attack for a considerable period before being consumed meaning that a certain fraction of it may go to waste through bacterial 'composting' which also results in extra heat and carbon dioxide production (Kok, 2021a). This is especially problematic if the larvae are raised in an environment whose water activity is relatively high as is, e.g. normally true for BSF. The second problem is that all the feed, while waiting to be consumed, takes up space in trays and bins so that this space is not productive much of the time. This results in low productivity for the reactor. The first problem can be addressed by using an 'immobile, fed-batch' sub-process while both problems are addressed simultaneously by using a 'mobile, fed-batch' approach. As is to be expected, the solution to one problem creates some new difficulties and that is also borne out in this instance. Both solutions are bought at the expense of increased process and equipment complexity.

In the immobile, fed-batch sub-process the larvae stay resident in the same tray or bin throughout the rearing period, but feed is added either regularly or as required. Hence, this may be implemented as a modification to any of the three straight-batch approaches discussed above: the time-spaced collection, the manifolded collection and the semi-continuous manifolded single tray systems. For the time-spaced collection system the modification is quite elementary. Thus, the operator will normally inspect and water the trays individually one or twice a day anyways so that for the fed-batch approach they will simply add some feed if this is judged necessary (quite similar to, e.g. traditional silkworm culture). For the manifolded-collection system the situation becomes more complicated since there are many more trays to look after and these may reside on frames or other structures inside long, tunnel-shaped or tower-shaped reactors with plenums on their sides for air supply and removal. A reasonably simple method to deal with this is to divide the reactor (or its equivalent) into a number of sequential segments and feed the trays during their transition between them. This is illustrated in Figure 5 in which a tunnel reactor is divided into three segments, respectively housing seven, three and

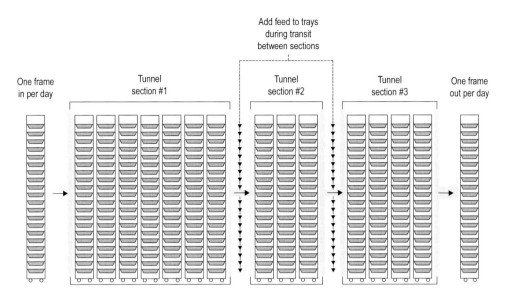

FIGURE 5   A manifolded, immobile, fed-batch, collection-based sub-process

four frames each of which hold 20 trays. Each day one frame exits and one frame is
added so that the larvae have a total residence time of 14 days in the reactor and receive extra feed on days 7 and 10. Therefore, the operator, on a daily basis, has to feed
two sets of 20 trays each, prepare 20 new trays and separate, dry, extract, etc. the
contents of 20 outcoming trays. The production capacity of such a system can easily be doubled or tripled as demand increases by adding tunnels in parallel. These
would then probably be operated as a manifold, i.e. somewhat out of phase so as to
spread the work load out over a longer period and increase machinery utilisation.

For the manifolded, single-tray system (Figure 4) the situation is, in principle, no
more complicated than for the time-spaced collection or for the manifolded-collection processes. Feed needs to be added to each tray at certain times during its
transition through the reactor. Since the manifolded, single-tray reactor is likely to
be automated already this may, in fact, be easier to accomplish than for the manifolded-collection arrangement. It will probably be necessary to add more sensors,
control equipment and error detection circuits together with mechanical or pneumatic components to achieve the desired feed addition.

7   The manifolded, mobile, fed-batch, collection-based sub-process

Whereas in the immobile fed-batch process the larvae will stay in the same bin
or tray during their entire development period, in the mobile fed-batch approach
this is not so. Thus, as the larvae grow from perhaps 20 μg to 70 mg (depending on

species, strain, feed, etc.) they will be transferred one or more times to different containers while also being fed on a regular basis. Although conceptually it is very appealing to have the young larvae initially resident in small containers and then transfer them to larger containers as they grow, there are substantial materials handling advantages to using a single size container throughout the process. This is especially true for larger operations in which container handling and cleaning will probably be entirely automated. Overall, in a 'mobile' system the larvae, as they grow, are given more room either by being moved to a larger container or by being distributed among more containers. As before, there are advantages as well as disadvantages to the mobile approach. The biggest advantage is that the very young larvae will now occupy much less space and therefore consume less resources so that the overall volumetric reactor productivity is increased and the amount of resources consumed per unit larval weight is decreased. With other words, in this way the larval weight per unit reactor volume can be maintained at a more uniform level throughout the rearing period than with the immobile approach. The main disadvantage is that, again, equipment and operational complexity are both increased meaning that, generally, the mobile approach is worthwhile only for larg-

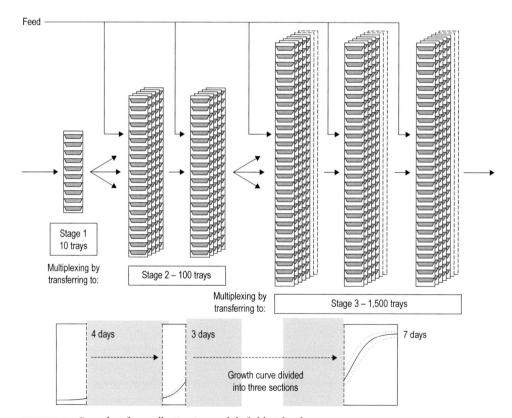

FIGURE 6   Procedure for a collection in a mobile fed-batch sub-process

er operations with manifolded systems that are collection-based. An example of an operating sequence for this type of sub-process is shown in Figure 6. It starts with a small collection of 10 trays containing very young larvae which, after four days, are then multiplexed to a collection of collections (the trays remain in phase!) by transferring the larvae to five sets of 20 trays each. Three days later these are then multiplexed further by transferring the growing larvae again, now to a collection of 50 sets (collections) of 30 trays each so that the total larval residence time is 14 days and the larvae are transferred twice while they are fed five times during their residence. For the sake of operational simplicity the same trays are used for all three stages of the sub-process, resulting in a total of 1,500 trays at the end. Obviously, for a real system the timing, etc. will depend on the species and the growth conditions.

In the above example only one initial collection (of 10 trays) of young larvae is considered. But normally a system of this type will be operated as a manifold with collections of young larvae and resultant sets of mature larvae respectively entering and leaving the system at regular intervals. For instance, one set of 10 trays could be added every day and a corresponding set of 1,500 trays removed from the system every day. Then, based on the numbers from Case 3 in the electronic supplementary material of Part 1 (Kok, 2021a), the daily output would be 2,445 kg (wet weight, 1.63 kg/tray) while every day 675 g of eggs will be needed to be added to the 10 starter trays (67.5 g/tray). Thus, in this system there will continuously be 14 manifolded populations of larvae present (mostly distributed in collections of collections!) on a total of 10,840 trays.

Because the organisms are at quite different points in their life cycles, it is likely that the three rearing stages of any population will occur in different reactors. It is not, however, always practicable to keep the members of a manifold who are at the same stage separated (N.B. these collections of different populations are at the same rearing stage, but out-of-phase). For instance, for the case mentioned above in which a collection of trays containing eggs or young larvae is added to the system every day there will be four collections of 10 trays each at Stage 1 of rearing, 24 hours out-of-phase, and these might very well be housed in the same reactor. Similarly, there will be three Stage 2 groups of 100 trays each (each group being a collection of five members, each of which is a collection of 20 trays) which might also be combined in one reactor while for the seven Stage 3 groups (each of which consists of 50 collections of 30-member collections of trays) it may actually be easier to keep them separated. A problem with combining collections in the same reactor space is that it increases the risk of contamination and will augment the magnitude of any disease outbreak that may occur.

Evidently, a mobile fed-batch system as described above can be expanded to almost any production scale by multiplexing in various ways. Two principal approaches are: (1) to increase the collection sizes; and (2) to decrease the manifold period. Thus, one might start off with an initial collection of 20 trays of eggs or very young larvae instead of 10 and in this way double the system's capacity. The same could be achieved by adding a 10-tray collection every 12 hours instead of every 24 hours, etc. Once the system is increased in capacity there are also productivity advantages to dividing the rearing period into more stages. It stands to reason that once demand goes beyond reactor capacity extra equipment must be installed. To double production of the above scheme, for example, a second set of reactors could be installed with the overall system manifold period being reduced to 12 hours so that 3,000 trays would need to be handled at the output end every day (4,890 kg/day wet larvae). If unloading is to be done during two four-hour work periods daily, that would result in a requirement to handle one tray every 9.6 seconds, resulting in a wet larvae stream of 170 g/s at the output. These quantities correspond to a small production facility (1,500-2,000 tonnes/year of wet material), with the rate and pattern of production approaching that of a semi-continuously operating plant.

A system of the type described here will require a substantial and robust materials-handling system to deal effectively with the various streams of solids like feed, bed contents and organisms. Dealing with live organisms and ensuring their survival during handling can represent some special challenges in this regard, e.g. during the transfers between stages. This is especially so since this type of system will need to be automated to a large degree, i.e. all the mechanical manipulations of materials will be carried out by machinery rather than humans. As well, almost all process regulation activities will be carried out by the control system, i.e. regulation of temperatures, humidities, carbon dioxide content, air flows, error detection, etc. Because the survival requirements of most living organisms are quite exacting, in any bio-production process the potential for disaster is much higher than for processes based on non-biological agents. Accordingly, the opportunities for catastrophic failure are many-pronged and ever-present. As well, it would seem, the will of fate is unbending and the involvement of a human operator 24/7 in some capacity in this type of process therefore unavoidable. It may be possible to reduce that requirement to some degree by the use of artificially intelligent (AI) systems for routine supervision of the process. A major benefit of involving AI in this way also is that it lends flexibility to the process operation. For instance, with machines that are not tied to a diurnal cycle in a supervisory role it is relatively easy to manipulate production by adjusting the manifold period. For instance, to decrease production slightly, the manifold period could be increased to 25 hours instead of 24, something that is much more difficult to achieve with a process that is under the direct supervision of humans who work in eight-hour shifts according to schedules. The same approach could be used to increase production by shortening the manifold period somewhat. Of course, such measures are subject to the capacity of the reactors and other equipment, but they do facilitate the direct matching of production to foreseen demand.

8   Semi-continuous sub-processes

There is not really a rigorous definition of 'semi-continuous' but the essential characteristics of a semi-continuous process are that its output occurs quite frequently and regularly, in discrete packets. The output patterns of the various sub-process types discussed above are shown together in Figure 7 together with that of
a truly continuous sub-process operating at steady-state, during which small, ir-
regular fluctuations do occur but which are corrected by the control system. From
this it is evident that the manifolded, single-tray (straight-batch or fed-batch) and
the mobile fed-batch sub-processes are likely the most suitable types to produce
a semi-continuous output. Of course, any of the other sub-process types can also
be made to produce semi-continuously by manifolding. But their advantages are
then often compromised, e.g. suitability for small-scale operation by one person on
a part-time basis. The main advantage of semi-continuous operation of a plant is
that it tends to lead to high utilisation factors for machinery and facilities, as well as
high productivity of labour. This also holds true to some extent for approximations
to semi-continuous operation, e.g. when the entire system output occurs in regular-
ly-spaced, discrete packets during one eight-hour daytime shift. Generally, though,
it is appropriate only for larger-scale systems. Thus, if a small-scale producer merely
wants to increase their capacity to a moderate degree it is usually easier to multiplex as a collection rather than as a manifold, i.e. install a parallel stream that is in
phase, or almost in phase, with the first one.

   At the end of the section on the manifolded, mobile sub-process an arrangement
was mentioned that was composed of a multiplex of two fed-batch streams with an
overall system manifold period of 12 hours, resulting in two four-hour unload periods per day during which an output rate of 170 g/s of wet larvae is achieved. That
would approximate semi-continuous operation with a total annual production
rate of 1,500-2,000 tonnes/year of wet larval material. In order to achieve full-time
semi-continuous operation four more streams could be added in parallel to that
multiplex and the system manifold period reduced to four hours so that every four
hours one of the six manifolded streams would be unloaded, resulting in an uninterrupted output stream. Total annual production will then be about 5,000 tonnes
per year of wet larval material, a volume comparable to that of a medium-sized
poultry farm (about 3 million birds per year).

   Worldwide, total (live) production volumes for poultry and seafood are about
100 million tonnes and 150 million tonnes per year respectively, and farm sizes
are increasing rapidly. To compete directly with these products or to position insect-based protein in the feed market it will likely be necessary to build production
facilities with a much larger capacity than 5,000 tonnes/year so as to achieve lower
unit costs. One might, for example, aim for a world-scale facility that outputs 200
times the volume of what has been mentioned so far, about a million tonnes/year.
To do that, several strategies can be followed. The first one is fairly straightforward:

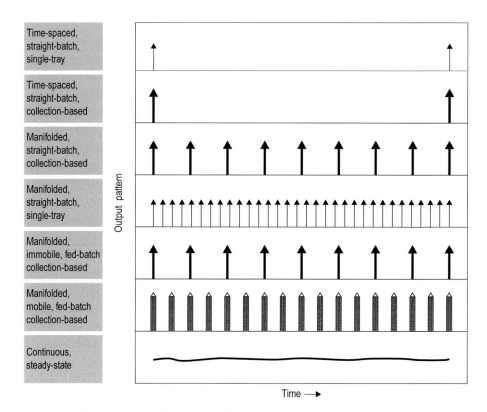

FIGURE 7   Output patterns of the various sub-process types

massively multiplex, both as collections and manifolds any of the sub-processes that have been mentioned. The second one is more challenging but also more
promising: change to a different processing approach altogether by moving to truly
continuous operation. The latter approach has been adopted by most large-scale
industries because, although it usually requires major investment to design and
develop a continuous process, the productivity per unit investment that can be attained will often yield a high rate of return. The type of output pattern that can be
expected from a continuous process operating at steady state is shown as the last
item in Figure 7.

9   Continuous sub-processes

In classical process engineering the two main types of reactors used for continuous operation are plug-flow and continuous stirred tank reactor (CSTR). For both
these types, when a system is fully functioning and operating at steady state, there
is an uninterrupted flow of material into the reactor and there is simultaneously an

equivalent, uninterrupted flow out. The main difference between the two is that in a plug-flow reactor materials of different ages do not mix whereas in a CSTR reactor all materials are constantly mixed. This means that if one takes a sample of the reactor contents, for a plug-flow reactor the sample material will have a very narrow age distribution whereas for a CSTR reactor it will have a wide age distribution. In reality, these two types are the extremes of a spectrum with most actual systems having some characteristics of both. Although there are some interesting aspects to continuous insect production with processes based on CSTR-type reactors, only the plug-flow case will be dealt with here because it is a direct continuance of the semi-continuous approach mentioned earlier. Moreover, only 'passive' sub-processes will be discussed, i.e. those in which the bed of feed containing larvae, etc. moves through the reactor on some kind of static support. In an 'active' sub-process it is the bed that stays in place and it is the machinery tending to it that will be mobile. Passive reactors can be either of the plug-flow or CSTR type. It is to be noted that plug flow reactors are most commonly employed for processing fluids in a confined space such as a pipe. However, as is done here, the concept can equally well be applied to a complex blend of finely divided solids, as well as mixtures of such blends with a liquid, reacting in a less confined space, such as on a belt. The texture of such a mixture may fall anywhere between that of a fine powder and a slurry, i.e. between that of dry corn starch or whole wheat flour and chunky peanut butter or concrete.

Thus, in true plug flow material travels through the reactor while its components interact. However, for the case of larval rearing, as for the other sub-processes discussed above, some components may leave or be added during the material's transit through the reactor. This is true, for example, for oxygen, carbon dioxide, water and even feed. Although this does require a slight modification of the classical definition of 'plug-flow', it interferes in no way with the essential feature thereof that, for any given element, there is little or no forward/backward mixing with elements that are younger or older. The result of this is that the age distribution of any element remains largely unchanged during its transit through the system except, of course, for the materials that leave or are added during that transit. The use of a plug flow reactor for larval rearing therefore tends to contribute very little to the widening of the organism's development distribution (Kok, 2021b). To assure or promote a longitudinal flow pattern of the bed through the reactor plug flow configurations customarily consist of a well defined, spatially confined path of much greater length than its diameter or width. A natural example of this (and probably the original prototype), is the intestine. In contrast, the CSTR is more like a stomach with its contents of all different ages being intimately mixed.

Whereas in all the previously-mentioned processes the bed of feed containing the growing organisms was confined on five sides by the bottom and walls of a tray or bin, in a continuous process based on a passive plug-flow reactor the bed is confined by walls on only three sides, namely, by the bottom and two sides of the bed support. The bed is then moved along the length of the support in one of the uncon-

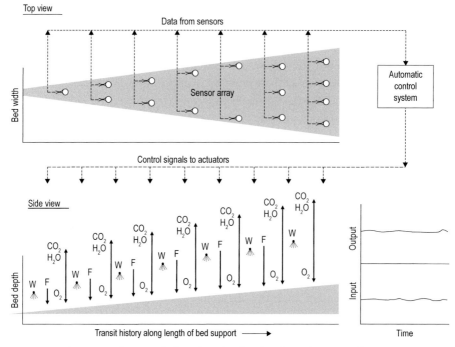

FIGURE 8   A conceptual portrayal of passive, plug-flow based continuous production

fined directions by some means, several of which are discussed below. A conceptual portrayal for passive, continuous production is illustrated in Figure 8, with both top and side views shown.

In this case the bed support consists of a single, flat, level surface that increases in width along the transition path and on which the bed slowly moves forward (at this point it is not significant how that would be accomplished). As it transits along the bed length, the bed gradually increases in both width and depth due to the addition of feed. Accordingly, the bed volume available per larva increases as the larvae develop and grow. One great advantage of continuous processes is that automatic control is a natural adjunct to such systems and, typically, a continuous system will be equipped with arrays of sensors to provide data to an automatic control system which then utilises actuators to implement various control actions such as feed addition, water spraying, ventilation, etc. In Figure 8 these are respectively denoted with F, W and the combination of O_2, CO_2, and H_2O. A carefully designed control system can handle most routine process operating tasks and in this way result in very low labour requirements for the productivity it can achieve. Nevertheless, as for larger fed-batch and semi-continuous systems, for a continuously operating reactor in which many tons of sensitive and fragile material are present at any one time, some kind of overall supervision will be required 24/7. At this point it may be possible to assign part of that responsibility to an AI but some human involvement will probably remain necessary until AI technology reaches a greater degree of maturity. In Figure 8 are also shown input and output traces for a continuous system operating at steady state, i.e. not subject to any major disturbance. Due to unavoidable fluctuations in many factors, for real processes such traces will almost always show some variation. Continuous production based on plug flow was also described in a general way by Kok (2017).

To meet the physical requirements of a passive plug flow process, a great many operating methods and reactor configurations are feasible (besides all the possibilities for active plug flow processes). Some of those were reviewed by Kok (1983) and only several other ones will be mentioned here. The bed support, rather than being constructed as a single unit taking up a lot of floor space as depicted in Figure 8, might consist of a set of stacked belts moving alternately in opposite directions from one another, with the bed material dropping from belt to belt. In such a setup the reactor would have a pyramid shape, with a narrow belt at the top receiving

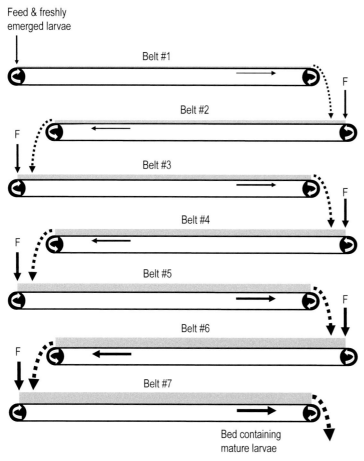

FIGURE 9   A plug-flow continuous reactor with seven stacked belts

the initial mixture of feed and freshly emerged larvae. Each belt below that would then be somewhat wider than the previous one with the bed resting on it being somewhat thicker also, due to feed addition at each transfer. In this way more bed volume will gradually become available to the larvae as they grow and move along the transit path.

A mechanically simpler arrangement than widening successive belts in a stack, but having the same effect, is to keep them the same width but run them at successively higher speeds. This is fairly easy to achieve with digitally-controlled motors whose speed can be varied over a wide range so that this approach provides for control flexibility. It would also not be difficult to integrate it with an AI-driven control system. At the same time as increasing the speed, the thickness of the bed can be increased by adding feed at each transfer, thus providing more volume per larva as they progress through the reactor. The use of these two methods combined is illustrated in Figure 9 for a stack of seven belts. For the sake of clarity, water spraying onto the bed and the removal and addition of gases by means of ventilation are not included in this diagram. Three major directions are available in this setup for the ventilation gas flow: length-wise, along the bed; sideways, across the bed; and vertically, upward or downward through the bed. As discussed below, the latter also offers some interesting possibilities for moving the bed along its transit path.

Rather than rely on the bed being moved along its support by mechanical means, e.g. on transport belts or with the use of scrapers, etc., it is also possible to have it slide down a gentle slope. On a smooth-surfaced bed this may be difficult to achieve in an organised fashion since the flow behaviour of powders and slurries is rather difficult to manage. What is somewhat easier to manage is to have the bed flow down a rough-surfaced slope under the encouragement of vibration and/or incipient fluidisation. To use the latter technique the top surface of the bed support needs to be porous so that short blasts of air can be blown through it in an upward direction, partially lifting the bed and allowing it to behave somewhat in a fluid-like manner. When this is done at regular intervals and in sequence along the length of the support, the bed can be made to flow down the slope smoothly while, as shown in Figure 8, feed and water are added to respectively increase the bed volume and regulate the water activity. This is a mechanically less complicated arrangement than a set of moving belts while it is also easier to widen the bed support along the transit path so as to increase the bed volume per larva as they develop. However, although the pneumatic approach is mechanically simpler than a set of belts, it is also more limited in its applicability because the bed consistency must correspond closely to the method. Thus, it is probably more appropriate for dealing with a bed of fairly dry solids, as used for mealworms, rather than a wetter slurry as often used for BSF.

A reactor whose operation is based on this pneumatic approach is illustrated in Figure 10. Its inner and outer containment walls are conical surfaces of different angles, connected by the bed support which spirals downward at a modest angle.

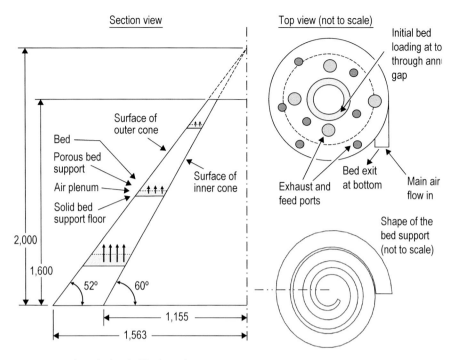

FIGURE 10   A conical, spiral-bed continuous reactor

The degree of incline of the support must, of course, be smaller than the angle at which the bed would slide down the slope unassisted. The support consists of two surfaces, the bottom one being solid and impermeable while the upper surface is porous, thus forming a plenum between them with the bed resting on top of the porous surface. The fluidisation air will pass through this plenum and whenever the pressure is increased it will flow at an increased rate through the porous surface, causing incipient fluidisation in the bed making it flow a certain distance down the slope.[8] This is done, section by section, in an upward sequence so that there is a gradual, slow descent of the bed together with its growing contents down the spiral slope and, ultimately, out of the reactor.

The support is therefore a sloping spiral that is level in the radial direction at any one point along its transit path, with the spiral gradually increasing in width from the top to the bottom of the reactor. The initial mixture of feed and freshly-emerged larvae (or eggs) is fed in at the top of the reactor and then, as the bed progresses down the slope, feed is added through ports in the top cone so that the bed gradually increases in depth as well as width along the path. If necessary, water can also be sprayed onto the mixture through nozzles mounted in the top cone. As for

---

8   A constant, small, upward flow will also need to be maintained to prevent particles from falling through the holes; this will also help to aerate the bed.

the other types of reactors, oxygen needs to be supplied to the organisms and heat and exhaust gases must be removed. This is mostly accomplished by continuously blowing air countercurrently over top of the bed, through the head space between the bed surface and the bottom surface of the spiral above. This main air flow (quite separate from the fluidisation air) comes in at the bottom of the reactor but, because the cross-section of the head-space becomes smaller along the path in the upward direction most of that air will be exhausted through ports in the top conical surface. This will ensure that the air velocity over the surface of the bed is kept well below the saltation velocity of the bed particles. The surface velocity should also be kept fairly low to prevent excessive bed drying and case hardening and this limits the amount of heat and exhaust gas that can be removed. In turn, that will limit the bed depth and the bed occupancy density. As discussed in Part 2 (Kok, 2021b) the possibility space for such an arrangement would have to be investigated by a combination of simulation and experimentation.

In order to convey an impression of the scaling of such equipment some dimensions are shown in Figure 10 (all in mm). In this case the two surfaces are circular cones of 52° and 60° respectively, cut off at a height of 1,600 mm (elliptical cones might have some advantages in terms of path length and equipment placement on a plant floor). As shown, the footprint at the bottom is 3,126 mm in diameter while the width of the bed at the bottom is 408 mm and the width of the annular space at the top is 82 mm. In this case the bed support is made to slope down at 5°, with the bed thickness increasing at 1°, giving a total bed length of 15 m and a bed surface of 4.2 m². If the initial bed thickness at the top is 4 mm it will be 30 mm at the bottom while the total bed volume will be 0.082 m³ or 82 l. This suggests that it may be possible to produce continuously on a relatively small scale with this method. As need increased, production could easily be ramped up by multiplexing such reactors.

Two fabrication methods of this type of reactor are foreseen. The first is a classical approach based on sheet metal work (galvanised, aluminium, stainless steel) in which the various pieces are cut from flat sheets and then rolled and bent, assembled, and fastened together. For the example above the angles, etc. were expressly chosen so that the spiral surfaces could be cut from single, flat sheets. This is illustrated in Figure 10 with the outline of the bed support shape. A more modern, and likely less costly, fabrication approach would be to 3D print most of the reactor as a single piece. Thus, the external and internal conical surfaces together with the air inlets and outlets, the support surfaces, the feed and exhaust ports, penetrations for water sprayers and monitoring equipment, etc. could all be printed at once. This would eliminate the necessity to cut the spiral support shape from a flat sheet and thus remove some of the constraints on this design. It would also greatly reduce the labour costs for fabrication and assembly while ensuring structural integrity of the unit. Assemblies of this type can be 3D printed in a wide variety of materials including various plastics and even metals.

10   **Discussion and conclusions**

From the above discussion it is evident that the arrangement of the larval rearing sub-process, its operational details and the configuration of the reactor that is central to its functioning are intimately related. Thus, although they represent different aspects, these three factors need to be considered simultaneously in a coordinated fashion during the PPD phase of a project. In order to arrive at a preliminary decision about what sub-process type to base the project on, the aspiring entopreneur will have to gauge what organism/feed combination is most interesting, what scale they want to start operating on, what potential for production expansion they want to take into consideration, what labour will be available, what operating schedule they want to follow, what other process features will be required, etc. All of the sub-processes presented above have advantages and disadvantages in terms of all of these as well as their implementation cost, simplicity of operation, volumetric productivity, energy efficiency, etc. In all cases it is a matter of finding a good match between the entopreneur's project concept, what a sub-process has to offer in terms of its pros and cons, and the operating point for any sub-process of interest. To evaluate the various alternatives that appear to be feasible for the conditions imposed, a modelling and simulation approach is very flexible and quite appropriate as long as it is supported by empirical verification. One very beneficial aspect of this is that it also allows the entopreneur to explore new methods and approaches that had not been thought about previously. Thus, as well as being a tool to evaluate existing options, simulation facilitates the exploration of a much larger possibility space and the discovery of new potentialities.

For all the sub-processes discussed above the reactors are more or less 'plug-flow', i.e. there is very little mixing of bed contents of different ages so that the beds containing the larvae are substantially stratified. As well, they are all 'passive', i.e. the bed in which the larvae are reared moves through the reactor. It is, however, also possible to rear insect larvae in reactors of the 'CSTR' type, i.e. in beds that are not age-stratified. And sub-processes can also be 'active' so that the bed remains in place and the reactor equipment is mobile. Consequently, larvae can also be reared in active plug-flow systems and in CSTR-type systems that can be either active or passive. These alternatives will be discussed elsewhere.

Conflict of interest

The author declares no conflict of interest.

References

Calaprice, A. (ed.), 2010. Probably not by Einstein. The ultimate quotable Einstein. Princeton University Press, Princeton, NJ, USA, 482 pp.

Duehl, A.J., Arbogast, R.T. and Teal, P.E.A., 2011. Density-related volatile emissions and responses in the red flour beetle, *Tribolium castaneum*. Journal of Chemical Ecology 37: 525-532.

Kok, R., 1983. The production of insects for human food. Canadian Institute of Food Science and Technology Journal 16: 5-18.

Kok, R., 2017. Insect production and facility design. In: Van Huis, A. and Tomberlin, J.K. (eds.) Insects as food and feed: from production to consumption. Wageningen Academic Publishers, Wageningen, The Netherlands, pp. 142-172.

Kok, R., 2021a. Preliminary project design for insect production: part 1 – overall mass and energy/heat balances. Journal of Insects as Food and Feed 7: 499-509. https://doi.org/10.3920/JIFF2020.0055

Kok, R., 2021b. Preliminary project design for insect production: part 2 – organism kinetics, system dynamics and the role of modelling & simulation. Journal of Insects as Food and Feed 7: 511-523. https://doi.org/10.3920/JIFF2020.0146

Kok, R., 2021c. Preliminary project design for insect production: part 4 – facility considerations. Journal of Insects as Food and Feed 7: 541-551. https://doi.org/10.3920/JIFF2020.0164

Roth, L.M., 1943. Studies on the gaseous secretion of *Tribolium Confusum* Duval II. the odoriferous glands of *Tribolium Confusum*. Annals of the Entomological Society of America 36: 397-424. https://doi.org/10.1093/aesa/36.3.397

Preliminary project design for insect production: part 4 – facility considerations

R. Kok

Bioresource Engineering, Macdonald Campus of McGill University, 21,111 Lakeshore Rd, Ste-Anne-de-Bellevue, QC H9X 3V9, Canada; robert.kok@mcgill.ca

Abstract

At the end of the preliminary project design (PPD) phase an aspiring entopreneur should have a clear idea of the scope and scale of the project being proposed. Hence, during PPD a number of aspects of the facilities must be considered. These are here dealt with in terms of basic questions that should first be answered about the project as a whole; project location issues; to what degree the facility is to be integrated or segregated into different units; what functionality is to be housed within the various building envelopes; and how safety and hygiene concerns can be addressed. As well, three ancillary issues are discussed that may affect facility design, construction and operation: the development of new production organisms that can grow on low-cost feeds; the possibility of disease evolution; and the matter of animal welfare.

Keywords

entotechnology – design – safety – hygiene

1   Introduction

In Parts 1, 2 and 3 of this set of papers a number of aspects of preliminary project design (PPD) for the mass production of insects were addressed including the calculation of some basic process heat and mass balances (Kok, 2021a), consideration of organism/feed kinetics and the use of modelling and simulation for process evaluation and design (Kok, 2021b), as well as the pros and cons of a number of process types and operating modes (Kok, 2021c). All this to help the aspiring entopreneur crystallise their project concept and arrive at a preliminary design (see Figure 1 of Kok, 2021a). For practical reasons the discussion has been chiefly oriented to the rearing of black soldier fly larvae (BSF, *Hermetia illucens*) and mealworms (*Tenebrio molitor* and relatives). The same holds true for this paper. Much of the reasoning and methods discussed are, however, also applicable to dealing with the eggs, pupae and adults of these two organisms as well as the production of other insects and related Arthropoda such as spiders and scorpions.

Before starting to consider actual facility requirements the entopreneur should have a clear idea of what exactly they want to do and accomplish in their project. Thus, as part of concept crystallisation and PPD they should have considered and formulated answers to a number of very basic questions about the project. Such questions would also be asked immediately by anyone interested in investing in the project but first wanting to scrutinise its viability in a preliminary manner. The answers to these questions will influence the facility arrangements very strongly and are therefore reviewed here first.

2   Basic questions

The first basic question concerns the nature of the project and the industrial sphere which it will touch most. Thus, what is the overall intent and purpose? At the moment there are three main spheres in entotechnology, as well as a host of minor ones: the production of food for human consumption; the production of feed for animal consumption, including pet food; and the conversion of a negative-value waste or a low-value by-product into an industrial material that will not enter the food or feed chain. A current example of insect-based human food is mealworm powder while many animal feeds and pet foods are now derived from BSF larvae, sometimes in fat-reduced form. BSF is also used to a limited extent for negative value reduction of substrates, e.g. for manure management. The production of insect-based fuels, chitin, enzymes and other bio-materials, although developing, is still in its infancy. Because they may be based on the same or similar organisms, the kinetics and the process types employed in these different applications can be quite similar. However, the hygienic and operating requirements for the facilities in which these processes would be housed may, at the same time, be very different. Thus, many of the organisational and physical aspects of a project will depend directly and strongly on its overall purpose. Some of these aspects will also depend to some degree on where the project will be located.

The second main question, obviously closely connected to the first, addresses the organism/feed combination that is chosen for the project, together with what the main project inputs and outputs will be. Is the entopreneur planning to rear mealworms on a mix of whole wheat flour, soybean and some additives to produce mealworm powder suitable for direct human consumption? And will they perhaps process (dry, grind, clean, etc.) it locally and package it in-plant in 100 g nitrogen-flushed, composite-walled, self-sealing baggies for distribution via the organic food division of large grocery retailers? Or perhaps the objective is to rear BSF larvae on-farm on a suitably fortified by-product that qualifies for inclusion in the food chain. In this case the eggs might be acquired on a daily basis from an external supplier and the larvae fed live to chickens well before they reach pupation. Etc. The possibility space for any of these endeavours is very extensive and can be investigated by means of simulation before any major commitment is made to a specific project (Kok, 2021b).

Once an overall direction has been chosen and an organism/feed combination selected, the third major question to address is which activities will be carried out locally. For instance, a complete process for BSF production will include feed preparation, adult maintenance, egg incubation, larval rearing and pupal incubation so that the entire reproductive cycle will be maintained locally, with the excess larvae being the main process output. As well, the larvae may even be processed into oil and high-protein feed at the same location so that they will be the main outputs. This is in sharp contrast to a plant receiving eggs and a pre-mixed feed on a daily basis from external suppliers and shipping out fully grown larvae, perhaps frozen, for processing elsewhere.

Another major question is at what scale to operate. At the very low end an entopreneur may wish to convert mixed by-products from a market-garden operation into insect mass to feed to fish, chickens or pigs. In many such cases, due to locality, restrictions on food chain entry don't apply, very simple processes and operating methods are employed, and a production capacity of 10 kg/day is entirely satisfactory. Once the production volume increases to 100 kg/day the entopreneur's classification shifts from 'very-low end' to 'small producer' and they will probably employ a more advanced process type and also start to rely on some automatic control (see Kok, 2021c). As production volume shifts further to 'intermediate', it becomes more or less imperative to utilise an industrial approach so that increasingly sophisticated process types and operating methods must be employed. All major plants currently being brought online fall into that category with production capacities in the range of 1000-10,000 kg/day, their processes based on multiplexed arrangements, as collections and as manifolds, usually both (Kok, 2021c). In addition, their control systems almost invariably rely on some artificially intelligent components. With this output capacity most are, essentially, industrial-scale demonstration plants that will serve as proof-of-concept for much larger scale developments.

A truly exciting prospect for entopreneurs (and especially for engineers working in this field!) is the creation of 'world-scale' plants with production capacities in the range of a million kg/day. If one only thinks about the protein requirements of the human population the following simple (approximate!) calculation illustrates the feasibility, and even the desirability, of building at this scale: for a human population of 8 billion individuals of average weight 40 kg, the total daily protein requirement will be 256 million kg/day (0.8 g/kg.day). If 10% of that demand is to be met with insect-derived material and the protein content of an insect is 20% (e.g. mealworms on a wet basis), then the total demand for insect mass (wet basis) to supply this much protein to the human population will be 128 million kg/day. Now, considering that the demand from the animal feed sector will likely be much larger than this it would seem that, on the world stage, there is definitely room for at least a hundred plants of this scale. After preliminary processing, a plant with a production capacity of one million kg/day live larva would have as major outputs about 200,000 kg/day of dry protein powder and 150,000 kg/day of fat (based on approximate mealworm composition). This is the equivalent of a loaded transport truck (20 tonnes net) leaving the plant once every hour during a two-shift work day. If a hundred plants of this capacity were built, together they would produce 36.5 million tonnes/year of wet material. Compared to other animal-derived food sources that would not be an overwhelming contribution but, because it might be additional to these other sources rather than a replacement, it could prove to be significant in terms of supplying high-quality protein to low-income groups and thus offset sub-optimal protein values in people's diets. In comparison to the number presented above for the 100 world-scale plants (36.5 million tonnes/year) some world production values are, in millions of tonnes/year: meat 350 (dressed carcass weight, including poultry); milk: 850; eggs: 80; both aquacultured and captured sea foods: each well over 100 (Ritchie and Roser, 2017, 2019). It is likely that industrial-type facilities will need to be built to realise production at the million kg/day scale if the following three objectives are to be met simultaneously: food safety, food quality, and low cost. In order to have a major impact on the global protein supply scene, the cost of insect-derived protein should ultimately be brought well below that of any of the other sources.

The fifth major question facing an aspiring entopreneur is what process type to employ in the project. Within this consideration are folded related questions about the heart of any process, the reactor, as well as the operating method and the control approach. Of course, the scale of the project will play a major role in any decision made about the process type and all related aspects. Will the process be based on plug-flow or a continuous stirred tank reactor approach? Will it be batch, semi-continuous or continuous? Will it be a simple process or multiplexed? If the latter, collection or manifold-based, or both? Will the reactor(s) be active or passive (see Kok, 2021b)? Again, answers to these questions will be closely intertwined with the aspects discussed above. And, evidently, they will also partly depend on the entopreneur's vision of the project and their preferences.

The final major question addressed here concerns the knowledge on which project decisions may be based. As discussed by Kok (2021b), such decisions may be arrived at through mentation pathways that are either accessible and inspectable, purely intuitive and as such not inspectable or, as is more usual, a combination of these two possibilities. The former type of mentation can be supported with a modelling and simulation approach while both types can be subject to a certain amount of emotional influence, e.g. due to an entopreneur fulfilling their 'dream'. Regardless of which decision-making pathway is mainly employed, specific and detailed knowledge of the kinetics of the proposed organism/feed combination is essential for the PPD exercise to yield useful results (Kok, 2021b). In addition, detailed knowledge of the functioning of the proposed process type, reactor configuration, operating methods and control approach are also necessary. Although in this paper the subject is not really dealt with beyond being mentioned, a good grasp of the micro-economics of the situation is obviously also required. Items that need to be dealt with in detail are: feed supply and cost and what impact the project might have on these; potential product demand; land and construction costs in the proposed locale; presence of transport infrastructure; labour availability and cost; environmental and safety legislation.

To sum up: facility considerations and decisions about facility arrangements will depend on a host of features of a proposed project. An entopreneur can decide to start a project and build a facility for many different reasons. Depending on the product and the situation, it is possible to build at a wide range of scales, use a variety of process types and reactor configurations, then operate and control the process in different ways. Concept development should be knowledge-based while it should be recognised that knowledge will need to be augmented as the concept shifts during development. Kok (2021a) has referred to these two aspects as the 'twin components of an iterative knowledge engine'. Project concept development can be supported, facilitated and enhanced with simulation based on a set of interactive models that are representative of the various concept components, e.g. the feed, the organism, the reactor, the process operation, the control system, and even some supply and demand aspects. In essence, the PPD exercise serves to explore the possibility space for a project while all decisions remain provisional and malleable until its conclusion. Thus, it is an iterative procedure during which any and all aspects of the project can initially be anchored but then re-considered as new opportunities come to light. Although most PPD work will concern addressing the basic questions outlined above, a number of ancillary but very practical issues will also arise and these are referred to here as 'facility considerations'. These issues will arise regardless of what industrial sector the project fits in, the organism/feed combination, the scale, the process type, etc. A number of these issues were also mentioned by Kok (2017).

3   Consideration 1: location issues

Once an overall project plan has been assembled one of the first question that arises is where it should be located. What factors should be taken into consideration in choosing the plant site? For small projects the answer is often obvious: it will be at or very near the residence of the entopreneur, especially if the process is to be manually operated and supervised. Although detailed data are not available on numbers, from the popular literature and media it is evident there are now thousands of such small production facilities around the world, on all continents. For slightly larger projects the answer requires some more consideration. For instance, how much land should be acquired initially for the project to be established and how much room should be allocated for subsequent expansion? This will often depend on the cost of land but also on the cost of moving a facility to a new location once it proves successful. In the long run it might be less costly to initially acquire more land so that later on it will be easier to expand production by multiplexing, i.e. adding a second and third identical reactor right beside a first one. Thus, the choice of location may be influenced strongly by the price and availability of land.

The second issue is reliable access to adequate feed at a low, stable cost. This is highly location sensitive and becomes more so as the scale of the project increases. For instance, for a world-scale plant (output 1 million kg/day live larvae, using the values of Liu *et al.* (2020) for mealworms which are: 64% water, 18% protein, 14% lipid, 3% carbohydrate and 1% ash) the input stream would need to be roughly(!) 1 million kg/day of dry material (starch or cellulose, taking into account undigestible materials, undigested feed, losses, etc.) or 365,000 tonnes/year. Taking a 'normal' yield of 0.7 kg/m^2 year dry mass for kernel corn as the basis for the calculation, supplying this amount of feed will take 52,000 ha of productive farm land or about 100,000 ha of land in total (accounting for roads, non-productive land, other uses, housing, etc.) to supply such a plant. This is equivalent to 1000 km^2. Thus, almost all available farm land in a circle of 18 km radius must now be dedicated to growing feed which must then be transported to the plant. If the supply density is lowered, i.e. if other crops are grown in the area as well as feed for the plant, the supply radius must be enlarged and transportation costs will increase. Although the calculations presented above are very approximate (!), they do provide some insight into how overall production of feed and the transportation infrastructure should be planned for a region if one is to install a major conversion facility. This holds for whatever the planned feed is. For instance, if the plan is to rear a cellulolytic organism on forestry waste on an industrial scale, the location should probably be chosen accordingly so that a sustainable supply (at a constant cost) and transport of that feed can be ensured in the longer term.

A third issue that is directly related to location is the air supply to the plant and, concomitantly, the exhaust. In many ways a rearing facility is no different from any other type of farm in which animals are confined while converting feed into more valuable materials. To do this, they require fresh air of good quality. Although, to facilitate the supply of feed, it may seem attractive to locate a plant in an agricultural area one danger of doing so is that there may accidental, airborne releases of insecticides that will enter the main air intake and then be distributed by the heating, ventilation and air conditioning system throughout the plant before ever being noticed. The way to offset this possibility, as well as to guard against air-borne organisms and other contaminants, is to thoroughly filter and wash all the intake air. Evidently, this comes at a cost! There is a similar problem with the exhaust from the plant because it will have an odour as well as a fairly high carbon dioxide content. And it may also contain small particles such as dust from the feed, insect setae, fungal spores, etc. The smell of the exhaust can be reduced by washing and the particles can be removed with filtering but there is no practical way to reduce the carbon dioxide content so that one must rely on adequate downwind dilution for this. These issues also limit the location for a plant, certainly making it very difficult to locate one in an urban setting. Concerns about exhaust from livestock facilities are not new and have been addressed in detail by many authors, e.g. Ubeda *et al.* (2013). Problems associated with the air supply and the exhaust will increase in direct proportion to the scale of the plant and, except for very small scale operations, must not be ignored, at the peril of the entopreneur. They should be dealt with during the PPD phase of the project rather than after the neighbours have started to complain to health authorities about odour overload or respiratory problems. At the same time, it should be investigated how compatible insect production is with the neighbourhood. For instance, what are the neighbours exhausting that could possibly end up in the air intake of the facility and what other sources of environmental contamination are nearby? Examples of such are large roads, railways, chemical plants, etc. The placement of the facility's main air intake and the relative placements of the air intake and the exhaust are discussed further in the section on facility segregation and integration below. Kok (2017) has also discussed some aspects of these issues.

The last issue mentioned here that is specifically location-related concerns the possibility of a sudden and dramatic failure of the plant envelope due to human error, natural disaster, terrorism, etc. Such a failure could lead to the instantaneous release of millions or even billions of insects which could have serious safety consequences. A large cloud of BSF adults, although not actually dangerous in any particular way, could blind drivers on a nearby highway and several million crickets moving together towards a neighbour's property might very well cause panic. There are anecdotes in circulation about a cockroach farm being accidentally demolished with some quite unpleasant consequences. In short, most neighbours would not appreciate a host of flying, hopping or crawling insects advancing upon them and this means that the location for the plant should be selected to minimise the impact of any major system failure. The entopreneur should therefore ensure that the property is large enough to allow for a safety and nuisance perimeter. This will also minimise the impact of incidentally released organisms which can be a routine bother to neighbours. For larger facilities an industrial zone as close as possible to the main source of feed is probably optimal.

A number of other project aspects are also somewhat, but not entirely, location-dependent. These are mostly related to the main liquid and solids inputs and outputs to and from the plant. The main liquid input to a plant will be water which, depending on the location, may be obtained from a central source or will have to be pumped, filtered, and purified locally. Kok (2017) discussed the water supply issue and how different water qualities are required for operation of the process, personnel, cleaning, etc. Standard technology is available to address any problems related to water supply but usually the lowest cost approach is to source from a central municipal supply, if available. A less trivial issue is the disposal of liquid wastes such as large volumes of heavily contaminated wash water with a high biological oxygen demand load. Again, for a larger facility, it is probably easiest and best to reject liquid wastes to a sewer system for central processing. If that is not available the plant will need to have its own treatment equipment – not an insignificant matter! Thus, whereas a small capacity plant could be located in a setting serviced by normal municipal infrastructure, for a plant of any significant capacity it will be easiest to locate it in a serviced industrial zone where large volume water demand and liquid waste disposal are normal requirements. If that is not possible, a water source will be needed nearby and adequate space on the property will need to be allocated for liquid waste treatment. Again, odour problems must to be taken into account.

Much of the same reasoning holds for the supply and management of solids. The main solids input will be feed and that will probably arrive at the plant by truck. Thus, adequate road infrastructure is needed to accommodate the traffic. And it is likely that high-value outputs like protein powder and lipids will leave the plant by truck also, using the same roads. For pretty well all organisms, however, a substantial amount of solids remains after the target material has been harvested (see Kok, 2021a). This material will be a mixture of non-digested feed, castings, frass, dead insects, etc. and can vary in consistency from a fairly dry granular mix to a thick sludge. It may be possible to separate a fraction of this and process it into other marketable products (e.g. insect frass) but a substantial amount of solids will remain to be disposed of. If the plant is located in an agricultural area and adequate space is available it may be possible to compost this material locally and then use it as a soil conditioner and fertiliser. If the plant is located in an urban setting or an industrial park it will, however, have to be removed by truck for offsite treatment, stabilisation and disposal. It should be noted that, whereas in many instances the leftover solids from insect culture are seen as a valuable resource with a lot of potential it is, at the same time, a heavily contaminated material whose bacterial load is uncertain and that may contain many other organisms of public health interest. It should, therefore, be treated with considerable caution while how it can be dealt with will depend partly on the location.

4   Consideration 2: degree of integration and segregation in the facility

Although the emphasis throughout the discussion of PPD has so far been on the rearing of the target organism, from the above it is evident that a considerable number of other aspects of a project must also be taken into consideration in making final or semi-final decisions about a project. Thus, factors such as the infrastructure that is available and the risk of establishing a plant at a proposed location will also play an important role. Once several primary candidate locations have been identified and it is clear what services will have to be provided locally it is possible to assess the required composition of the facility at each location and evaluate the risks and advantages. Obviously, all this will vary tremendously according to the scale, the process, the locale, etc. For example, a proposed project might be about a facility of intermediate scale (e.g. 10,000 kg/day), meant to digest mainly urban organic waste by means of BSF larvae, with access to a full complement of industrial infrastructure, receiving eggs daily from a specialised supplier, and able to send its output (larvae) to an external processor for chemical oil and protein extraction while its waste solids can be shipped to an external composting facility. In another scenario the plan may be to set up a relatively small plant in a non-industrialised area to convert a low-cost starch source into insect protein and fat so as to improve the quality of the local diet. In this case a number of infrastructure-related units may need to be included in the facility, e.g. a reliable water supply while, at the same time, some other factors will probably be less significant. In such situations issues like feed supply security, odour emission, mass releases of organisms, etc. are often not major items on the PPD agenda. Of course, quite on the contrary, if the project plan is to establish a world-scale plant they must be, simply because the implementation of the plan can have a major local effect on these factors and even a national or international impact. In all cases the questions to be asked during the PPD stage are: what units will need to be incorporated in the facility and what will be the relationship between them? And, to what degree should the facility be integrated and/or segregated?

Basically, 'integration and segregation' refers to which activities are to be carried out under the same or different roofs, within the same or in different envelopes. There are some major advantages to having all units comprising a facility within the same envelope, mainly related to convenience; ease of operation; cost of construction, maintaining and operating the plant; utilisation and management of personnel; communication, etc. But there are also some major disadvantages and it is a matter of deciding which factors outweigh the others. Thus, in any facility in which animals are raised in a high density the danger of disease striking and propagating very rapidly is ever present, some species being more resistant to this than others (see Eilenberg *et al.*, 2015; Joosten *et al.*, 2020). As well, the likelihood of the end product becoming contaminated with feed or waste is much larger if, e.g. larval processing takes place in the same building as feed preparation and larval rearing. Other activities, like waste treatment or standby-power generation, are entirely incompatible with processing. Thus, how the various units are to be segregated will depend, firstly, on what functionality is to be integrated in the facility and, secondly, on how serious the problems that will accompany integration are judged to be. Kok (2017) has addressed some of these issues, pointing out that for a 'complete' facility strain breeding and testing should probably be carried out in an entirely separate, auxiliary unit and that it would be wise to do the same for strain maintenance and inoculum preparation (i.e. eggs). This is in direct imitation of many other biotechnology industries such as brewing and enzyme production. That same approach is also used in the poultry industry for larger-scale facilities and even pork producers are moving in that direction. If it is necessary to treat them locally, both solid and liquid wastes should definitely be dealt with in auxiliary units that are far enough removed from the rearing and processing units (and downwind of them) so as to minimise the chance of contaminating the rest of the facility. Although it may be more difficult to arrange, ideally, exhaust gas treatment and cleaning should also be done a good distance removed from the main production area and the exhaust released so as to prevent 'short-circuiting' of the air flow, i.e. the exhaust gas should never be able to enter the intake. The air inlet should therefore always be put upwind from the exhaust, in accordance with the local, prevailing wind pattern. Another way to reduce short-circuiting is to put the air intake and exhaust at different heights. A location with fairly steady winds and a tall chimney for the plant exhaust are therefore a good idea. Even though standby power generation is generally not thought of as a major feature of a plant, for a rearing facility it is essential and sufficient generating capacity should be installed for the plant to keep operating throughout a fairly prolonged power interruption. When a full-sized, industrial genset (often diesel-powered) is running it generates a large amount of exhaust gas of variable quality and it is best to also vent that exhaust through a chimney so that it won't reach the inlet of the plant. It can, therefore, be situated near the exhaust treatment unit and vented through the chimney. Depending on how a plant is multiplexed and what its production scale is, it may also be desirable to clean the intake air in a separate, auxiliary unit. This allows for discrete, quality-controlled airflows into different sections of the plant. Obviously, by segregating the units the plant will have a larger footprint than when it is fully integrated. Also, more supply and transportation infrastructure is needed to provide services to the various units and to move materials between them. This will, inevitably, increase the cost of constructing the plant.

A sample layout for a production facility incorporating some of the aspects discussed is presented in Figure 1. It is not to scale (!) and only the main material flows are shown. In this instance the facility is quite segregated with primary inputs (feed and eggs) being received at one loading dock and the primary outputs (protein powder and oil) being shipped out via another loading dock. Water is obtained from a well and treated locally. The intake air is cleaned and conditioned in a separate unit before different streams are sent to rearing and processing. A third, separate air stream is allocated to the personnel unit which is kept far away from any of the production units. This is done to ensure worker safety as well as to reduce the danger of contaminating the process. Exhaust from all sources is combined for cleanup and is then disposed of via a chimney. Liquid wastes are treated locally while solid wastes are removed for off-site composting. Larval rearing and processing are done

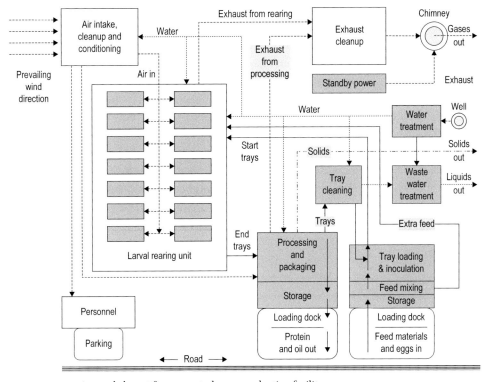

FIGURE 1   A sample layout for an egg-to-larvae production facility

in separate units with trays moving between them in a cyclical pattern. In this lay-out the main air intake and the personnel unit both face the prevailing winds and are upwind from all the other units. While working on the PPD stage of a project an entopreneur would probably generate many such diagrams while considering the possibilities and probabilities together with the pros and cons of each arrangement.

The question of segregation vs integration does not only apply to the facility as a whole; it is equally applicable to the organisation of the rearing and processing units where actual production activities take place. Thus, depending somewhat on the production scale and how the process is multiplexed it may be highly desirable to keep groups of organisms strictly isolated from one another. In Figure 1 the rearing sub-process illustrated is 'manifolded, immobile, fed-batch, and collection-based' (see Kok, 2021c), meaning that feed and eggs are initially loaded onto trays together and that this mixture stays on the same tray during the entire larval residence in the reactor, that an entire collection of trays is prepared almost at the same time so that their age distributions will be similar, that feed is added to each tray during its residence in the reactor, and that this procedure is repeated at regular intervals in an overlapping manner so that a manifolded multiplex is created. This will result in regular outputs so that the processing unit will have a high utilisation factor. In this case the main rearing unit contains fourteen compartments corresponding, for example, to fourteen days of egg/larval residence time. The question will be to what degree these rearing compartments should be kept isolated from one another. For smaller facilities this may very well be impractical but for larger ones it will probably be prudent to limit contact between them. Thus, each compartment should be supplied separately with air, water and feed. Low-level controls should be localised while reporting regularly to a centralised, high-level control and management system. This will augment system stability and resilience while reducing its vulnerability to infection, power fluctuations, physical breakdowns, etc. Movement of personnel through the rearing facility, and particularly through the different sections of the rearing manifold, should be severely restricted to reduce the incidence of disease transmission. As the magnitude of mass culture of organisms such as BSF, mealworms and crickets increases rather dramatically, the potential for serious economic damage due to infection also rises. This has certainly occurred in other, related industries (see Jones *et al.*, 2018; Pitts and Whitnall, 2019) and the entotechnological sector is no exception to this danger.

In some ways keeping the various units isolated from one another will increase the complexity of the facility although it can also make it simpler to expand its capacity. In the layout in Figure 1 there is no space allocated for expansion and once this organisational arrangement has been agreed upon in a general way the entopreneur may decide to modify it to allow for future increase in capacity. The easiest way to do that is to make sure the infrastructural units will be able to handle a larger load, to build the processing unit somewhat larger than needed at the start and to leave space on the property for extra rearing units. The great advantage of a 'plug-in' modular approach is that capacity can be increased relatively easily by multiplexing. For instance, the capacity of the facility illustrated in Figure 1 can be doubled simply by adding another rearing unit, as long as the air cleaning, feed mixing, tray preparation units, etc. can handle the increased demand.

5   Consideration 3: within the envelopes

In the above example all the major units comprising a facility are kept physically separate and at an adequate distance from one another. This approach is probably applicable to larger projects but not necessarily practical or affordable for smaller or intermediate-sized ones. Consequently, the question that often arises is which operations can be combined under the same roof, within one envelope, and which ones are entirely incompatible and must therefore be housed in separate building envelopes. The secondary question then becomes how to keep the various units that are housed within the same envelope adequately isolated from one another. The latter question is particularly important if the final product is intended for the human food chain since in most locales very complex rule sets apply to the manufacture of human food and all inputs to its manufacture. These rule sets may come in the form of laws or regulations, may originate at any number of different levels of government (e.g. federal, provincial, municipal) and may be simultaneously enforced in slightly different ways by a number of agencies. The purpose of all of these is to prevent contamination and to assure the quality and safety of the food supply. The aspiring entopreneur will do well to take all applicable rules into account very early during project planning while allocating activities to various units of the proposed facility. This should be part of the PPD exercise.

One key method to house different activities within one envelope is to subdivide it with walls that have very few, if any, penetrations except for those intended for essentials such as utilities, the transport of materials, communications, etc. That way air-borne dust and contaminants as well as most personnel are prevented from moving between areas, making it feasible to house, for example, personnel, air cleaning, feed mixing and tray loading under the same roof as rearing and processing. In this author's opinion it is best and safest to keep all waste treatment and major cleaning activities well away from any processing area. Also, because of the often amazing (!) penetration capacity of various types of rodents and stored product insects it is best to maintain physically separate units for shipping and receiving as well as any associated warehousing. Trucks load and unload through very large penetrations in the envelope that must seem like a challenge to any enterprising mouse or rat, the reward being an ample food supply. These units therefore need to be equipped with effective pest management systems. However, considering biological reality, no matter how good a pest control system is, it is never practical to bring the infestation rate down to zero and isolation of high-risk units is an effective way to reduce the incidence of unwanted biological penetration into the processing area even further. Obviously, to what degree the entopreneur needs to worry about this aspect will depend very much on the final product. For instance, if the objective is to produce non-food industrial oil, protein powder and chitin from a biological waste material the concern about contamination and infestation will be a lot less acute than if the product is chicken feed.

Although it will likely not be a primary consideration for most planners of entotechnological projects, the personnel unit is actually quite important not only because it is central to process management and control, but also for reasons of safety, hygiene and employee well-being. Depending on the scale of the production plant, office space for the following divisions will be needed: operations and control; management; sales, marketing and buying; human resources; general accounting, accounts payable, accounts receivable and payroll; janitorial and cleaning services; technical support including for IT; security. As well, space will be needed for bathrooms, showering facilities and laundry, a lunchroom or cafeteria, and even a day-care facility with a recreational area attached. So as to minimise health risks, in Figure 1 the personnel facility is located upwind and well away from the production area, it has its own air supply and a separate parking lot. Although not shown on the diagram, it will also need a separate, secure water supply that is thoroughly protected against contamination through backflow in the system. For intermediate-sized facilities some of the functions mentioned will probably be carried out in combination by a limited number of people. Thus, all financial transactions may be looked after by one person with assistance of an external accounting firm while operations, management, human resources, technical support, and janitorial services could be dealt with by a team of several partners supported by helpers. This would be typical for a family-owned business. For a very small facility most of the tasks will be carried out by one or two people, perhaps working part-time, fulfilling different roles while all activities will be carried out in the same space.

6   Consideration 4: safety, hygiene and cleaning

The raising of animals for economic purposes, husbandry, is generally not considered a dangerous undertaking for the humans involved. There are, nevertheless, a number of safety aspects to take into account and these are quite different for the three main approaches to farming: extensive, intensive and 'factory-farming'. Especially in the latter case are farm personnel often exposed to a number of industrial-type workplace hazards such as intense noise and odour, dust, ammonia and hydrogen sulphide, drugs and hormones, and even antibiotic-resistant pathogens. Air quality in intensive and factory-farm operations is often quite poor and personnel may run the risk of contracting respiratory diseases (Fox, 1980; Mitloehner and Calvo, 2008). Considering that entotechnology is mainly oriented to the 'factory-farming' method in which insects are reared in confinement at extremely high densities, similar hazard potentials apply to workers in entotechnological environments. A further hazard in this industry is the development of allergic reactions in workers. Bauer and Parnode (1981) have reported on this phenomenon and Kausar (2018) has stated that 'It is quite evident that insects contribute clinically important inhalant allergens to the air in respirable sized particles.' but also that '...the study on allergy caused by insects is limited.' Hence, apart from the mechanical and electrical hazards to which farm workers are generally subject, in this industry there are the inhalation of gases and insect particles as well as extended exposure to insects that can cause health problems. One approach to minimise worker contact and exposure is to automate most of the materials handling but this is not always possible in smaller operations. Sufficient ventilation with clean air is therefore very important for worker safety under all circumstances and in some cases it may be justified to have workers wear face masks and protective clothing to limit their inhalation and exposure. Concomitantly, as shown in Figure 1, in order to safeguard neighbours and passers-by, the exhaust from the rearing facility should be cleaned up to render it safe before it is discharged.

Whereas the consideration of safety is usually oriented to the health and well-being of humans, hygiene and cleaning are meant to secure the farmed species as well as their human farmers. Regardless of what animal is being reared, disease or the potential of disease is always there (Joosten *et al.*, 2020) and a robust system should be in place to minimise its impact when a disease strikes a production facility. Thus, the facility should be designed and constructed in the first place and then also operated so as to make it difficult for a disease organism to propagate. As mentioned above, this can be done by keeping the various production units physically separated as much as possible and to minimise both personnel and material movement between units. Also, in any material or equipment 'backflow' thorough cleaning is essential. For instance, in Figure 1 the trays on which the larvae are reared pass through the process in a cycle and thus flow back from the processing unit to the rearing unit, being cleaned in between. Under ideal circumstances those trays should be sterilised (in the 'commercial' sense) and a suitable material to withstand such repeated treatment would be stainless steel. This is, however, a rather expensive option and also not necessarily an optimal material choice for other reasons. Accordingly, during the PPD phase of the project considerable attention should be paid to the selection of materials that will simultaneously meet the various requirements of frequent handling, sterilisation, organism compatibility, etc. Persistent cleaning and 'commercial sterilisation' of other parts of the production facility are also an operating requirement if one is to avoid disease incidence. A well-trained and dedicated janitorial staff is therefore an essential part of the operations teams, no matter what the scale of the facility.

Prevention is the first defence against invasion – this holds for a unitary system such as a single larva but it also holds for large, composite systems such as a production facility which may hold a trillion larvae. Nevertheless, regardless of how well a system is designed and operated, sooner or later a predator or competitor will penetrate all passive defences. This is an unwritten law of nature. There should, therefore, be in place a robust monitoring system that is able to detect problems, coupled to a response system that is ready and primed to deal effectively with those problems. The detection system must rely on continuous and attentive observation of the process and, especially, the behaviour of the organisms. It is mostly deviations from the norm that will be the main response triggers. All operations personnel should be involved in process and organism monitoring and for larger facilities it may be appropriate to have specialised personnel for this, assisted by AI-driven systems. The use of neural networks to detect anomalies in race horse behaviour related to health status was discussed by Suchorski-Tremblay *et al.* (2001) and Flower *et al.* (2005) have used kinematic measures to detect hoof gait deviations related to dairy cow pathologies. Similar technologies could be used to generate early warning signals for a larval rearing facility, triggering a request for an entomological pathologist to come identify the problem and recommend action. Basically, the earlier an infection is noted and identified the more effective and the less costly the response will be. In the cattle industry foot-and-mouth disease can spread so rapidly that very fast detection, verification and response activation are essential to prevent major damage. Although it may seem extreme to some, this often involves culling all animals that have even a remote chance of having been infected. The same principles will hold for the factory farming of insects so that, once a disease problem has been identified and its severity established, it may be necessary to thoroughly sterilise an entire section of a rearing facility. To limit the overall damage it is therefore prudent to design, construct and operate a process in a defensive manner. Isolation of the various units from one another and strict hygienic operation lie at the basis of this approach.

7   Discussion and conclusions

In the end, the PPD exercise should result in a fairly complete description of the project in which all major design and operational features are dealt with in some detail. Thus, in the first section all the basic questions should be addressed: what is the overall intent and purpose of the project; what organism/feed combination is proposed and how is that justified; which activities will be carried out locally; what is the proposed scale of operation and what will be the impact on the local economic situation; what process type (and reactors) will be employed; what is the control approach; what knowledge is available about this system and what knowledge needs to be developed or expanded; and what will be the main reasoning/mentation approach followed for project development? Evidently, most of the answers will be dependent upon and interactive with similar questions about markets and demand for the product, but most of the details of these subjects are well outside the purview of this paper. If the reasoning and planning approach used is not entirely intuitive (so that it can be traced and inspected by external agents), ideally, the answers to the basic questions will be supported with calculations, simulations, results from experiments, etc. presented in appendices. Market study results and demand forecasts, etc. can also be presented there.

In the second section the facility issues can then be dealt with and the following presented: one or more specific project locations together with their characteristics, advantages and disadvantages; an infrastructure description containing a preliminary facility layout and a discussion of the integration/segregation approach followed; an outline of what functionality will be housed in which envelope and how the envelopes are to be connected; a detailed description of both passive and active safety features for the project that will have to be incorporated in the formal design of the facility and the specifications of the operating procedures (see Figure 1, Kok, 2021a). In the third section of the report a preliminary timeline for the project can be outlined together with items such as possibilities for future expansion, etc. Creating a fairly detailed PPD report as described above is invaluable for the aspiring entopreneur; it both forces them to collect and organise their thoughts into a coherent whole and, at the same time, provides them with 'ammunition' to sell the project to investors. The latter may very well find it an attractive proposition but then want to contribute to its final formulation, stressing the fact that PPD is an iterative procedure. Thus, the project and the report will remain subject to change until an equilibrium is reached between all stakeholders. PPD is, after all, first and foremost an investigation of the possibility space and all results thereof can be re-formulated in light of new opportunities and insights. Of course, how formal the interactions between parties and any investment negotiations will be will very much depend on the scale of the project, the relationship between the entopreneur and the investors as well as the overall mentation approach to the project.

So far, very little has been said about the financial aspects of the project and any financial reward that may accrue to investors. This is partially because during the PPD stage of a project the financial aspects are very difficult to gauge. Also, like market study and demand forecasting, etc., the subject is marginal to this paper. The author's approach is, however, that it is normal that investors should want a return on their investment and they will invest only if there is a reasonable chance of financial reward (this is not necessarily true for institutional or government investors who may be motivated more by political than financial reward). Doubtless, the project developer will think about it during PPD but a comprehensive assessment of prospective profitability has to wait until a fairly complete project plan is in place. Hence, this may be done when it is time to present to potential investors (or, for a small project, when the entopreneur needs to convince his banker). In dealing with potential investors the entopreneur should acknowledge that negotiations do not always gravitate towards a mutually satisfactory equilibrium point.

Besides the issues mentioned above, a number of other ones are appearing on the horizon of the entotechnological industry and one might keep these in mind when planning production facilities. The first one is the overall supply of feed. Thus far the main production organisms are BSF, mealworms and crickets. All of these can be used to upgrade and concentrate nutrients from lower grade feeds. But none of these insects are, however, able to synthesise protein, meaning that they are limited to extracting protein from their substrate. They are, therefore, very limited in what they can be grown on by the amount of protein present in their feed. If insects are to become a major item in the human food chain, either directly or indirectly, there will soon be a stress on the feed supply, causing prices to rise, etc. Consequently, the time is ripe to develop the mass culture of insects which can be grown on feeds which are abundantly available at low prices and high in energy, but low in protein content. Such materials are often composed of cellulose and lignin. In order to be useful as feed, they need to be: (1) mechanically degraded; and (2) digested by organisms that are also capable of synthesising protein rather than just extracting it. A number of beetles, some cockroaches and most termites are able to do much of this, aided by bacteria and protozoa living symbiotically in their guts (e.g. see

O'Brien and Slaytor, 1982). In some instances the feed can be fortified with low-cost, inorganic nitrates as a nitrogen source for protein synthesis while in some others the microorganisms, and by extension the insect, are able to synthesise protein *de novo* from atmospheric nitrogen (Täyasu *et al.*, 1994). The advantage of cheap, cellulosic feeds is that, although low in nutritional content, they are often not heavily contaminated and therefore not so difficult to have approved for entry into the human food chain.

'Nature abhors a vacuum' is a truism that certainly holds in biology! Thus, although an organism such as BSF is presently quite resistant to infection and disease (Joosten *et al.*, 2020), once it is reared in great quantities at high concentrations it is more than likely to prove an irresistible temptation to the forces of evolution, adaptation and procreation as well as fate. It is, therefore, to be expected that various parasitic and disease organisms will develop and, sooner or later, make a strong appearance in production facilities. As discussed, the impact of this can be partly offset by preventive design tactics combined with safe, hygienic operating procedures and continuous monitoring. The latter functionality should be able to call upon an incidence response team that includes an entomological pathologist. The point here is that the entotechnological industry should be stimulating and supporting the training of veterinary personnel specialised in production-related entomological issues. As well, it should be supporting research into pest and disease control methods. While it is a good strategy to both prevent and combat disease, it is also wise to develop approaches to avoid problems. A primary method to do that is 'to run faster than the opponent'. This relies on the ongoing development of new strains that are resistant to upcoming problems. Thus, by the time a pest or disease is on the verge of becoming a significant problem, production is switched over to a strain that is resistant to that specific organism. This obviously requires a serious commitment to strain development which should also be supported by the industry as a whole for all species of commercial interest. Kok (2017) has also discussed this issue, pointing out that the inclusion of new genetic material and the development of new strains should always happen in isolation, preferably in an entirely separate unit, well away from production.

A third issue that is now appearing on the entotechnological horizon concerns animal welfare. For a not-so-clear reason invertebrate creatures are generally regarded as not being able to 'suffer' so that animal welfare legislation is not applicable to them, the exception in this regard in many locales being the cephalopods. This situation may, however, change as attitudes about stress and the welfare of farmed insects mutate, as they have about traditionally farmed animals. The same arguments used earlier by mammalian and avian welfare proponents (e.g. Fox, 1980) to encourage people to change their attitudes may soon also be used by arthropodal welfare advocates. Indeed, this issue is already being actively discussed in the literature (e.g. Gjerris *et al.*, 2016). From a process and facility design perspective it will therefore be discreet to not draw attention to this aspect of entotechnology as well as to assure that insects are reared under conditions that will not be easy to interpret as causing them 'suffering'. Avoidance of problems of this type is probably the best approach. The industry should aim for 'happy larvae that are slaughtered as humanely as possible'.

Conflict of interest

The author declares no conflict of interest.

References

Bauer, M. and Parnode, R., 1981. Health hazard evaluation report HETA 81-0121-1421 – CDC – Insect rearing facilities. Agricultural Research Service, USDA, Washington, DC, USA. Available at: https://www.cdc.gov/niosh/hhe/reports/pdfs/81-121-1421.pdf

Eilenberg, J., Vlak, J.M., Nielsen-Leroux, C., Cappellozza, S. and Jensen, A.B., 2015. Disease in insects produced for food and feed. Journal of Insects as Food and Feed 1: 87-102. https://doi.org/10.3920/JIFF2014.0022

Flower, F.C., Sanderson, D.J. and Weary, D.M., 2005. Hoof Pathologies Influence Kinematic Measures of Dairy Cow Gait. Journal of Dairy Science 88: 3166-3173. https://doi.org/10.3168/jds.S0022-0302(05)73000-9

Fox, M.W., 1980. Factory farming. Miscellaneous studies and reports. Available at: https://www.wellbeingintlstudiesrepository.org/sturep/2

Gjerris, M., Gamborg, C. and Röcklinsberg, H., 2016. Ethical aspects of insect production for food and feed. Journal of Insects as Food and Feed 2: 101-110. https://doi.org/10.3920/JIFF2015.0097

Jones, P.J., Niemi, J., Christensen, J.-P., Tranter, R.B. and Bennett, R.M., 2018. A review of the financial impact of production diseases in poultry production systems. Animal Production Science 59: 1585-1597. https://doi.org/10.1071/AN18281

Joosten, L., Lecocq, A., Jensen, A.B., Haenen, O., Schmitt, E. and Eilenberg, J., 2020. Review of insect pathogen risks for the black soldier fly (*Hermetia illucens*) and guidelines for reliable production. Entomologia Experimentalis et Applicata 168: 432-447. https://doi.org/10.1111/eea.12916

Kausar, M.A., 2018. A review of respiratory allergy caused by insects. Bioinformation 14: 540-553. https://doi.org/10.6026/97320630014540

Kok, R., 2017. Insect production and facility design. In: Van Huis, A. and Tomberlin, J.K. (eds.) Insects as food and feed: from production to consumption. Wageningen Academic Publishers, Wageningen, The Netherlands, pp. 142-172.

Kok, R., 2021a. Preliminary project design for insect production: part 1 – overall mass and energy/heat balances. Journal of Insects as Food and Feed 7: 499-509. https://doi.org/10.3920/JIFF2020.0055

Kok, R., 2021b. Preliminary project design for insect production: part 2 – organism kinetics, system dynamics and the role of modelling & simulation. Journal of Insects as Food and Feed 7: 511-523. https://doi.org/10.3920/JIFF2020.0146

Kok, R., 2021c. Preliminary project design for insect production: part 3 – sub-process types and reactors. Journal of Insects as Food and Feed 7: 525-539. https://doi.org/10.3920/JIFF2020.0145

Liu, C., Masri, J., Perez, V., Maya, C. and Zhao, J., 2020. Growth performance and nutrient composition of mealworms (*Tenebrio molitor*) fed on fresh plant materials-supplemented diets. Foods 9: 151. https://doi.org/10.3390/foods9020151

Mitloehner, F.M. and Calvo, M.S., 2008. Worker health and safety in concentrated animal feeding operations. Journal of Agricultural Safety and Health 14: 163-187. https://doi.org/10.13031/2013.24349

O'Brien, R.W. and Slaytor, M., 1982. Role of microorganisms in the metabolism of termites. Australian Journal of Biological Sciences 35: 239-262. https://doi.org/10.1071/BI9820239

Pitts, N. and Whitnall, T., 2019. Impact of African swine fever on global markets. Agricultural Commodities 9: 52-54.

Ritchie, H. and Roser, M., 2017. Meat and dairy production. Available at: https://ourworldindata.org/meat-production

Ritchie, H. and Roser, M., 2019. Seafood production. Available at: https://ourworldindata.org/seafood-production

Suchorski-Tremblay, A.M., Kok, R. and Thomason, J.J., 2001. Modelling horse hoof cracking with artificial neural networks. Canadian Biosystems Engineering 43: 7.15-7.22.

Täyasu, I., Sugimoto, A., Wada, E. and Abe, T., 1994. Xylophagous termites depending on atmospheric nitrogen. Naturwissenschaften 81: 229-231. https://doi.org/10.1007/BF01138550

Ubeda, Y., Lopez-Jimenez, P.A., Nicolas, J. and Calvet, S., 2013. Strategies to control odours in livestock facilities: a critical review. Spanish Journal of Agricultural Research 11: 1004-1015.

Environmental aspects of insect mass production

S. Smetana*, R. Spykman and V. Heinz

German Institute of Food Technologies (DIL e.V.), Professor-von-Klitzing-Straße 7, 49610 Quakenbrück, Germany; *s.smetana@dil-ev.de

Abstract

Mass production of insects is calling for environmentally optimised and economically efficient insect value chains. It is a complex task considering a great variety in insect species, production scales, feed formulations, etc. Taking a challenge of environmental impact clarification, a few studies highlight on life cycle assessment (LCA) of insect production. The current study is aimed to systemise 24 selected previous studies to establish a modular framework for the determination of contribution of sustainability assessment factors of insect production chains. Reviewing published studies according to the elements of LCA, the study identified a feasible approach for the modelling of insect production chains, which can be used for the facilitation of comparability of further LCA studies. The approach is based on a modular analysis of insect production through a graphical mapping of value chains (allowed identification of precise system boundaries) supplemented with table analysis considering scale of production, reference (functional) unit, impact assessment methodology and type of LCA. Such an approach allows for consistency in LCA setting and further comparability of results.

Keywords

life cycle assessment – insect production chains – insect mass production – environmental optimisation – material flow analysis – sustainability

1 Introduction

Food production is facing the challenging task of assuring food security within the planet's carrying capacity. The environmental impact of the current food system should be substantially decreased (Willett *et al.*, 2019), which is extremely challenging considering the increasing demand for food in the next few decades (Gouel and Guimbard, 2019). The demand for protein sources and especially meat is projected to increase by 76% by 2050 in comparison to the basis year of 2005 (Alexandratos and Bruinsma, 2012), which will lead to critical environmental consequences. The search for alternative food and protein sources with lower environmental impact is becoming a vital task (Smetana *et al.*, 2015, 2020a; Van der Weele *et al.*, 2019) not only for the substitution of meat but also for animal protein feeds such as soybean meal and fishmeal (Van Huis *et al.*, 2013; Veldkamp *et al.*, 2012). Insects in this perspective are becoming an interesting potential solution not only for the challenges of protein supply for food and feed purposes (Van Huis *et al.*, 2013), but also for food waste treatment and nutrient recirculation in food systems (Gold *et al.*, 2020; Ites *et al.*, 2020; Mertenat *et al.*, 2019; Smetana, 2020).

While insects demonstrate the potential to deliver local and sustainable protein sources (Allegretti *et al.*, 2018; Smetana *et al.*, 2019b), the young industry of mass insect production in Western countries is still facing difficulties of setting up sustainable production starting from the design phase. There is a problem associated with data availability, which can be used for the comprehensive analysis. From one side small scale insect producers do not gather detailed data on the processes, so the data often do not exist. From another part, insect producers do not open data for the public use, so the data availability is lacking in comparison to well established feed and food industries (Bosch *et al.*, 2019; Ites *et al.*, 2020; Salomone *et al.*, 2017), which poses a challenge to the industry's efficient design of a sustainable production system (Ites *et al.*, 2020). At the same time assurance of sustainability (lower environmental impact) of insect products in comparison to conventional benchmarks on the market is crucial for the survival for insect mass production in Europe (Wade and Hoelle, 2020). Therefore, it is necessary to find ways for the reliable assessment of insect production at early industry development stages.

Life cycle thinking (LCT) (Fava, 1993) is a broad underlying concept, aimed at considering direct and indirect environmental impacts of complex systems, integrated into several related methods such as eco-efficiency, eco-design and life cycle assessment (LCA). Special attention to futuristic estimates is devoted in anticipatory (Guinée *et al.*, 2018), scenario-based (Fukushima and Hirao, 2002), consequential (Zamagni *et al.*, 2012), prospective or ex-ante LCA (Buyle *et al.*, 2019; Cucurachi *et al.*, 2018; Spielmann *et al.*, 2005; Walser *et al.*, 2011). They all are oriented to provide certain insights on the potential future outcomes. It should be considered that any type of prediction will include the accounting of the uncertainty, which would complicate the calculations. In order to deal with complexity issues prospective LCAs can be streamlined using a modular approach (Steubing *et al.*, 2016). Modularity of impact assessment allows for approximation of missing data values and automation of analysis (Steubing *et al.*, 2016). It is a basic concept integrated in modern LCA methods, which allows for the integrated and disintegrated analysis of production and supply chains, finding the impact hotspots and modelling of scenario effects. At the same time modularisation is not applied to insect production chains up to date.

Therefore, the main aim of this review was to establish a modular framework for the determination of environmental contribution of different parts of insect production chains (modules) based on recent literature devoted to environmental impact assessment of insect production.

In order to achieve this aim, the objectives of the paper were:

− to review current approaches to the environmental impact assessment of insect production systems, defining hotspots and the contribution of environmental impacts;
− to classify and systemise relevant environmental impact factors for modular sustainability assessment of insect production chains;
− to propose a modular framework for LCA of insect production and processing.

The article is structured to comprehensively review current literature sources dealing with production of feed for insects, insect farming, processing, and insect-based products use for food and feed purposes. It starts with explaining and illustrating the concepts of LCT and LCA. Further, the paper addresses the environmental aspects of insect production chains according to the aggregated stages of production with the aim to define the impact hotspots and assessment limitation factors. It is finalised with an overview on a potential modular framework for LCA of insect production and processing.

2   Study design

This study relies on available studies dealing with the environmental impact of insect production for food and feed, but also includes studies performing LCA on insect application for waste treatment.

The selection of studies was performed via 'Google Scholar', 'Mendeley' and 'WorldWideScience' in the beginning of 2020 using keywords: 'life cycle assessment' 'LCA' 'insect production'. The search yielded 125 studies, which were reviewed for the connection with analysis of environmental impact of insect production. Specific criteria set for the selection of studies were that is should contain an original LCA model of insect-based production of feed or food, waste treatment or other system aimed for insect production. One output of insect production system is the insect biomass, which is quantified. Design of insect production system should include the reproduction cycles and not rely on natural oviposition or natural insect harvesting. The review excluded the studies which do not include reflection on mandatory elements of LCA: goal and scope definition, life cycle inventory (LCI), functional unit (FU), life cycle impact assessment (LCIA) methods and interpretation. The references of the articles suitable under defined criteria were also explored for consistency. The review then concentrated on the analysis of selected articles (24). The selected articles reviewed in this manuscript follow the life cycle approach and include insect production stages, upstream and in some cases downstream processes. They all define goal and scope, system boundaries, provide inventory data and draw conclusions on the impact of a product as well as hotspots of production.

The review article will follow the conceptual approach applied to all the LCA studies. All the subchapters start with defining the goal and scope, specifying the reasons for separation and analysis of the production stage (module), boundaries for the module and limitations of such selection. Furthermore, data availability and requirements will be analysed as well as potential approaches for data collection, validation and aggregation for the input and output data to quantify material use, energy use, environmental discharges, and waste associated with each module. The final part of modules and sub-modules analysis will include the analysis on the use of LCIA methods, categories, indicators, equivalency and contribution factors. Critical analysis on the interpretation of results and drawn conclusions in current studies will be a closing part.

3   Life cycle assessment

LCT approach is aimed at a holistic conceptualisation of environmental issues (or other pillars of sustainability) at system level (Mont and Bleischwitz, 2007). LCT is a holistic approach, which examines the impacts of a product or a service through its entire life cycle from the extraction of raw materials (cradle), through production, use and final disposal or other end of life options (grave) (UNEP/SE-TAC, 2012). It means that the sustainability analysis considers required resources and expected impacts of all life cycle stages (design, production, use and end of life). LCT provides a comprehensive basis for the analysis of indirect and rebound effects, allowing to eliminate unintended negative consequences associated with higher rates of consumption of environmentally efficient and cheaper products (Hertwich, 2008; Mont and Bleischwitz, 2007). Classical applications of LCA relate to the determination of environmental impact of product through the entire life cycle of a product, identification of environmental hot spots (stages with the highest contribution), product comparison, eco-design and other aspects. The comparison of alternative products and minimisation of trade-offs between them, for the selection of less environmentally impacting options (Cucurachi et al., 2019), became a 'golden' standard in environmental assessment of products and services. Moreover, LCA is a basis for Environmental Product Declaration (Del Borghi, 2013; Schau and Fet, 2008) and Product Environmental Footprint (Bach et al., 2018), included in the guidelines of European Commission for the environmental impact assessment and declaration (Allio, 2007). The development of the European guidelines further triggers practical application of LCA for business and management strategies, marketing and product labelling (Mont and Bleischwitz, 2007; Rubik and Frankl, 2017).

LCA is a complex method, which requires a high-level knowledge, related to the method and dealing with environmental impact concepts and guiding factors. ISO standards define four main stages of LCA: (1) goal and scope definition; (2) LCI; (3) LCIA; and (4) interpretation (ISO 14040, ISO 14044; ISO, 2006a,b). Any LCA should include information on FU, system boundaries, impact assessment methods and timeframe (Thabrew *et al.*, 2009), but also details on assumptions, limitations, data quality and requirements, reference flows, etc. LCIA methods assign an impact factor to an elementary flow in the inventory, thus connecting the amount of resources used and emissions to the potential environmental impact caused (Zampori *et al.*, 2016). Moreover, two main approaches in completing LCA should be outlined: attributional (information on environmental burden associated with the specific product life cycle) and consequential (information on environmental burden appearing because of decision making with consequences of the market changes) (UNEP/SETAC, 2011). Goal and scope of the study define the type of approach to be followed. Some studies focus on an attributional LCA, others on a consequential LCA, and others on both, attributional and consequential. Such conceptualisation allows for certain consistency between different studies and standardisation of results. Further chapters include a few examples on the indicated components in relation to insect production chains.

Modularity of LCA is foreseen as a feasible way of dealing with many variants in a product's life cycle (Jungbluth *et al.*, 2000). The approach allows to assess multiple alternative value chains within a production system. In LCA a product's life cycle is modelled in various boundaries (depending on the goal and scope of the study: from cradle to grave, cradle to gate, gate to gate, etc.), to set some pre-conditions for the comparability of the studies in the same scope. To compare alternative value chains, each alternative life cycle needs to be modelled individually, even if changes were made only in one of the stages. By contrast, the fundamental idea of the modular LCA approach is to break down a production system or a product's life cycle into modules which can be recombined to form complete value chains (Steubing *et al.*, 2016). These modules are practitioner-defined and encompass life cycle stages or unit processes. Besides elementary flows, the modules only have input and output flows which link them to other modules of the studied production system. This is achieved by expanding a module's foreground process(es) to include all required background processes (e.g. utilities, waste treatment, infrastructure). This procedure is repeated until the entire production system is described in modular LCI. Based on these modular LCIs, LCIA is carried out, leading to individual LCIA results for each module. The LCIA result for a value chain is determined by aggregating the LCIA results of the involved modules (Rebitzer, 2005).

When several modules within a production system produce substitutable products, alternative value chains arise. A module-product matrix contains information on how the modules can be connected to form alternative value chains, taking scaling factors and interdependencies into account (Steubing *et al.*, 2016). An advantage compared to conventional LCA is that the modelling effort can be considerably lower, since it scales with the number of modules, not with the number of alternative value chains. However, the modular approach requires an up-front time investment for the modularisation, meaning the suitable definition of modules to represent key choices within the production system. Modular LCA can therefore streamline (aggregate) scenario analysis through the optimisation models, which allow for the identification of missing data points using optimisation algorithms (for example using Pareto optimisation). It can therefore not only enable optimisation of value chains by using the module data as inputs to optimisation models (Steubing *et al.*, 2016) but also can provide solutions for the data limitations in the assessment of emerging technologies (Thomas *et al.*, 2020).

4   Environmental hotspots of insect production

Overall, the available studies rely on a few approaches towards the LCA of insect production. Most of the studies use an attributional approach for the analysis, aimed at the identification of hotspots and comparison with similar products (Table 1). Only two studies take first steps towards consequential assessment of insect production, indicating the difficulties and high uncertainty rates associated with assumptions concerning product substitutions on the market (Smetana *et al.*, 2019b; Van Zanten *et al.*, 2018).

Most studies employ multiple impact categories and characterisation factors (coefficient units allowing equivalent aggregation of the environmental interventions to a particular impact category) to analyse the environmental impact of insect production. A wide variety of impact assessment methods is used. Separate indicators are mostly calculated in early studies (Joensuu and Silvenius, 2017; Komakech *et al.*, 2015; Oonincx and De Boer, 2012; Van Zanten *et al.*, 2015), while other studies rely on more aggregated methodologies, allowing for the inclusion of multiple indicators and end-point aggregation (Table 1).

Setting the goal and scope of the study is of outmost importance for any LCA study, as mistakes at the selection of FU or system boundaries could lead to wrong results and justifications (Rebitzer *et al.*, 2004). LCA studies of insect production chains are not exception. The general goal in the studies reflecting on the production of insects grown on conventional (commercial) feed is connected with identification of environmental impact of such production with certain comparison to similar protein production systems (Halloran *et al.*, 2017; Oonincx and de Boer, 2012; Smetana *et al.*, 2016, 2019b; Suckling *et al.*, 2020) for food and feed purposes. Strong comparative approach based on a few FUs to conventional 'traditional' protein and fat sources is taken in studies of Smetana *et al.* (Smetana *et al.*, 2015, 2016, 2019b, 2020a). Special attention in some studies is devoted to the identification of insect production impact if by-products are applied for the feeding (Bava *et al.*, 2019; Maiolo *et al.*, 2020) in these cases considering insect production for feed or petfood purposes mostly (Bava *et al.*, 2019; Maiolo *et al.*, 2020; Smetana *et al.*, 2016, 2019b; Thévenot *et al.*, 2018). On the other hand, insects as by-product of honey production are analysed in study of Ulmer *et al.* (Ulmer *et al.*, 2020). Separate goal is set in

TABLE 1   Life cycle impact assessment approaches in life cycle assessment studies of insect production

Study	Impact categories (characterisation factor)	Impact assessment method	Attributional/ consequential
Smetana et al., 2020a	Multiple mid and endpoint	IMPACT 2002+ Version 2.21	Atr
Suckling et al., 2020	Multiple midpoint	ILCD 2011 Midpoint+ method	Atr
Ites et al., 2020	Multiple mid and endpoint	IMPACT 2002+ Version 2.21	Atr
Maiolo et al., 2020	GWP, AP, EP; CED, WU	CML-IA baseline V3.05, CED (Frischknecht et al. 2007); AWARE	Atr
Roffeis et al., 2020	Single score	ReCiPe method (V 1.11)	Atr
Ulmer et al., 2020	Multiple mid and endpoint	IMPACT 2002+ (V 2.11)	Atr
Bava et al., 2019	Multiple midpoint	ILCD 2011 Midpoint V1.03	Atr
Smetana et al., 2019b	Multiple mid and endpoint	IMPACT 2002+ and IMPACT World for WF; ReCiPe for sensitivity	Atr, Cons
Van Zanten et al., 2018	GWP, EU, LU	Separate indicators	Atr, Cons
Mertenat et al., 2019	GWP	ReCiPe Midpoint (H)	Atr?
Bosch et al., 2019	GWP, LU, EU	Separate indicators	Atr
Thévenot et al., 2018	CED, CC, AP, EP, LU	CED was quantified using the Total Cumulative Energy Demand method v1.8 (VDI, 1997). CC, AP, EP, and LU were calculated according to the CML-IA baseline 2000 V2.03 method	Atr
Halloran et al., 2017	Multiple midpoint	ILCD method	Atr
Roffeis et al., 2017	Single score	ReCiPe method (V 1.11)	Atr
Salomone et al., 2017	GWP, LU, EU	CML 2 baseline 2000 method and GWP 100a v. 1.02 method (IPPC, 2007)	Atr
Joensuu and Silvenius, 2017	GWP	Separate indicators	Atr
Smetana et al., 2016	GWP, EU, LU; single score	ReCiPe V1.08 and IMPACT 2002+	Atr
Roffeis et al., 2015	ALO, WD, FD	ReCiPe 2008	Atr
Smetana et al., 2015	Multiple mid and endpoint	IMPACT 2002+	Atr
Van Zanten et al., 2015	GWP, EU, LU	Separate indicators	Atr
Komakech et al., 2015	GWP, EP, EU	Separate indicators	Atr
Oonincx and De Boer, 2012	GWP, LU, EU	Separate indicators	Atr

ALO = agricultural land occupation; AP = acidification potential; CC = climate change; CED = cumulative energy demand; EP = eutrophication potential; EU = energy use; FD = fossil depletion; GWP = global warming potential; LU = land use; WD = water depletion; WU = water use.

the studies dealing with insect production for waste treatment including manure treatment (Ites *et al.*, 2020; Komakech *et al.*, 2015; Mertenat *et al.*, 2019; Roffeis *et al.*, 2015; Salomone *et al.*, 2017) and further animals grown on insects (Van Zanten *et al.*, 2015). More prospective approach is presented in studies dealing with ex-ante and consequential assessment (Roffeis *et al.*, 2017; Smetana *et al.*, 2019b), and evaluation environmental performance of insect production in regional perspective (West Africa) (Roffeis *et al.*, 2020).

Analysed studies rely on various FU (Bava *et al.*, 2019; Salomone *et al.*, 2017; Smetana *et al.*, 2015, 2019b; Ulmer *et al.*, 2020). Weight-based units dominate in the studies; however, they reflect different aspects of insect production. Some studies account for the weight the input materials in case of waste or manure treatment to determine efficiency of biotransformation(Ites *et al.*, 2020; Komakech *et al.*, 2015; Mertenat *et al.*, 2019; Roffeis *et al.*, 2015; Salomone *et al.*, 2017). Studies dealing with production of insect-based feed (Maiolo *et al.*, 2020; Roffeis *et al.*, 2017, 2020; Thévenot *et al.*, 2018; Van Zanten *et al.*, 2015) or insect-based ingredients for food and feed purposes (Halloran *et al.*, 2017; Smetana *et al.*, 2015, 2016, 2019b; Suckling *et al.*, 2020; Ulmer *et al.*, 2020) rely on weight-based unit of output product (feed, meal, insects, dried insects). In order to consider nutritional properties of insects studies rely on comparison based on amount of proteins (Bosch *et al.*, 2019; Halloran *et al.*, 2017; Joensuu and Silvenius, 2017; Oonincx and De Boer, 2012; Salomone *et al.*, 2017; Smetana *et al.*, 2015, 2016, 2019b; Ulmer *et al.*, 2020), lipids (Salomone *et al.*, 2017; Smetana *et al.*, 2019b, 2020a) or energy (Smetana *et al.*, 2015).

The overall reliability of data in the analysed studies could be assessed as good, as a lot of studies relied on primary data for foreground processes of insect production (Bava *et al.*, 2019; Halloran *et al.*, 2017; Oonincx and De Boer, 2012; Roffeis *et al.*, 2017, 2020; Salomone *et al.*, 2017; Smetana *et al.*, 2019b; Suckling *et al.*, 2020; Thévenot *et al.*, 2018; Ulmer *et al.*, 2020), a few studies relied on mixed literature and primary measured data (Ites *et al.*, 2020; Maiolo *et al.*, 2020; Mertenat *et al.*, 2019; Smetana *et al.*, 2016). The studies, which have a hypothetical or review character relied on secondary modelled data or literature sources for the modelling of LCA (Bosch *et al.*, 2019; Komakech *et al.*, 2015; Roffeis *et al.*, 2015; Smetana *et al.*, 2015; Van Zanten *et al.*, 2015).

The differences in the goal and scope between different studies indicate that it is not viable to compare the results between all of them, as system set for the prime quality insect biomass production would be different from the system oriented solely on waste treatment. A bright example could be the studies of Roffeis *et al.* (2015, 2017, 2020) where the high impact of insect production could relate to a regional approach taken, not available in other studies. Selection of FUs applied in the studies demonstrate consistency with reliance on weight basis in most studies. Differences presented of weight units routed in the magnitude or concentration on various parts of insect production chains (input or output) can be levelled through the recalculation for the same unit. Reliability of data used for the LCA studies is assured using primary data from the production. If the direct measured data is not available, the studies rely on modelling or literature sources. In these cases, reliability and availability of data is of higher importance to be analysed.

4.1   *Feed for insects*

Type of feed selected (vegetable rests, compound feed, food waste, etc.) and properties of selected feed (nutrient content, moisture content) in a great degree define the performance and environmental impact of the insect production system (Bosch *et al.*, 2019; Ites *et al.*, 2020; Oonincx and De Boer, 2012; Smetana *et al.*, 2016, 2019b). And this relation is not straightforward. High quality of feed for insects in many cases results in higher environmental impact, but also comparatively short growing cycles. While lower nutritional quality of insect feed (which could have a lower impact of production) results in smaller size of insects, longer growing cycles and higher conversion ratio (Bosch *et al.*, 2019; Smetana *et al.*, 2016). This is the first trade-off which producers should consider. Moreover, the system is further complicated with the potential of insect application for waste treatment. Treatment of food waste may result in environmentally beneficial results especially if the feeding substrate is of good nutritional quality (Bosch *et al.*, 2019; Ites *et al.*, 2020; Salomone *et al.*, 2017; Smetana *et al.*, 2016). Environmental impact of animal manure treatment with insect technologies could also result in positive or negative environmental impact depending on the impact of avoided treatment processes (Roffeis *et al.*, 2017, 2020; Smetana, 2020; Smetana *et al.*, 2016).

4.1.1   Primary production of feed (feed ingredients)

LCA studies of insect mass rearing rarely pay attention to the variations of feed production or to the side-streams (by-product) allocation of impacts. A lot of insect producers rely on the commercial compound feeds due to the legislative limitations (Bosch *et al.*, 2019). Thus, it is necessary to pay careful attention to modelling of feed crops harvesting and feed production. In case of commercial feed production, the boundaries for insect feed are comparable to those outlined for animal feeds (Bava *et al.*, 2019; Bosch *et al.*, 2019; Halloran *et al.*, 2017; Smetana *et al.*, 2016). The boundaries should include the classic agricultural stages of sowing, growing and harvesting with further processing into animal feed. High availability of data and previously performed analyses (McAuliffe *et al.*, 2016; Papatryphon *et al.*, 2004; Poore and Nemecek, 2018) make the assessment of insects produced on conventional feeds somewhat easier and flexible in terms of selection of LCIA methods and indicators. However, the reliability of data and previous studies should be thoroughly analysed for the consistency and representability.

Additionally, the boundaries for insect feed production sometimes include side-streams and secondary products from food processing or agriculture (Bava *et al.*, 2019; Bosch *et al.*, 2019; Ites *et al.*, 2020; Maiolo *et al.*, 2020; Smetana *et al.*, 2016). It is envisioned that insect production in the future will even stronger rely on side-streams and secondary products from food processing or agriculture due to its potential for a constant supply (Smetana *et al.*, 2019b). In such cases the upstream production impacts should be allocated to the relevant by-product following physical or economic criteria (Ardente and Cellura, 2012). Formulations of feed for insects from food products at retail or consumer stages (wastes) should follow a dual approach – it should include all the impacts of upstream production and avoided waste treatment (Mondello *et al.*, 2017; Salomone *et al.*, 2017; Smetana *et al.*, 2016).

Depending on the type of feed the LCA of this stage should rely on careful allocation or system expansion which to a great degree would determine the overall impact of insect production and further use of insect for food or feed (Smetana *et al.*, 2019b; Van Zanten *et al.*, 2018). The importance of this stage is connected also with limited availability of data on the processing characteristics and allocated value and impacts to the by-products. Published studies rely mostly on attributional approaches and economic or nutritional allocation (Table 1) and rarely on system expansion and consequences on the market (Smetana *et al.*, 2019b; Van Zanten *et al.*, 2018).

4.1.2 Feed transportation and storage

Insect feed transportation is performed by usual means of dry and liquid feed transport relying on road infrastructure (trucks, lorries, and tractors). Most studies rely on rather short distances for the feed delivery in the scope of 100-300 km, which is justified due to the economic feasibility and availability of local suppliers (Bava *et al.*, 2019; Maiolo *et al.*, 2020; Smetana *et al.*, 2016), unless feed contains soy, grown overseas (Halloran *et al.*, 2017). It should be considered that in most cases transportation (along the whole production chain) contributes only a small share to the overall environmental impact, amounting to 2-6% of impact depending on the category (largest impacts in global warming potential, resource depletion and ozone depletion) and modelled scenario (distance, means of transportation). The share of feed transportation impact is increased in case of insect production on wastes – it can be responsible for up to 18% of GWP (Salomone *et al.*, 2017), however no increase is observed in absolute values. Most of the studies either model the delivery of feed or rely on existing databases for the analysis. Data availability is not a limiting factor for this assessment module.

4.1.3 Secondary feed production and processing

Feed and food processing stages are currently considered as very challenging to model due to the limitations in data availability, huge diversity of potential alternatives and diversity of application scales (Ites *et al.*, 2020; Smetana *et al.*, 2016, 2019b). Feed is mostly delivered to insects in dried or high-moisture forms. Dried feeds are supplied to mealworms, crickets, and grasshoppers, while moisturised feeds are prepared for larvae of flies. Dried feeds, sources from grains, require minimal processing which might consist of mixing, cutting and grinding. In some cases, additional sources of moisture like vegetables are supplied alongside the dried feeds (Halloran *et al.*, 2017; Oonincx and De Boer, 2012). Moisturised feeds, on the other hand, might be derived from wetting dried feeds or they might be delivered in moist form from the supplier (wet mills, breweries, farms in case of manure, etc.). Similar to transportation, the impact of the feed processing stage is minimal and that is why in most cases feed processing is considered in rather aggregated form. Recently investigated options for insect feed pre-treatment (Isibika *et al.*, 2019; Ravindran and Jaiswal, 2019) are not included in the examined LCA studies.

4.2   *Insect farming*

Insect farming together with insect biomass fractionation (separation of insect biomass into fractions including drying) are responsible for a considerable portion of environmental impact, which is reflected in the range of 15-70% depending on impact category and level of processing in LCA studies (Bava *et al.*, 2019; Halloran *et al.*, 2017; Maiolo *et al.*, 2020; Smetana *et al.*, 2016, 2019b; Thévenot *et al.*, 2018). Thus, high importance in the impact contribution is highlighted for energy consumption. Literature indicated that 18.4-37.6% of energy is used for insect farming, while the bigger part of cumulative energy demand (58.7-79.8%) is allocated to the production of main consumables (Maiolo *et al.*, 2020). Inclusion of processing in the farming stage rises the impact to 37-55% of overall energy consumed (Smetana *et al.*, 2019b). Separation of insect rearing and further biomass processing, depending on scenarios, results consumption of 50% of electricity used for insect rearing (Thévenot *et al.*, 2018). Insect biomass drying is responsible for a fraction of 7-45% of direct electricity used (Bava *et al.*, 2019) and fat separation for around 50% of electricity used or 48-55% of cumulative energy demand (Thévenot *et al.*, 2018; Van Zanten *et al.*, 2015). Heating with the gas could add another 22% to the energy use impact (Van Zanten *et al.*, 2015). Electricity and natural gas use at insect farming then are responsible for 51% of total GWP (Van Zanten *et al.*, 2015). Similarly, Suckling *et al.* (2020) indicates the impacts of rearing on global warming to be responsible for 59% of GWP with third part being allocated to heating. Therefore, the detailed analysis of insect farming and identification of improvement potential could play an important role. Direct metabolic emissions are indicated as neglectable (Ermolaev *et al.*, 2019; Mertenat *et al.*, 2019; Parodi *et al.*, 2020).

4.2.1   Insect cultivation

The properties of feed for insects to a great degree define the environmental performance of the whole production system due to three main characteristics: (1) type of insect feed and associated upstream impacts; (2) feed conversion ratio; and (3) residual biomass management (Table 2). LCA of insect production therefore should thoroughly, holistically and in a detail define these characteristics. Feed type and composition play an important role due to the upstream environmental impact associated with production (Halloran *et al.*, 2017; Oonincx and De Boer, 2012;

Thévenot *et al.*, 2018; Van Zanten *et al.*, 2015) or due to the amount of avoided impacts in case of waste treatment substitution (Bava *et al.*, 2019; Mertenat *et al.*, 2019; Salomone *et al.*, 2017; Smetana, 2020). Feed production for insects is responsible for a vast impact of insect production: 20-99% of contribution depending on the impact category (Bava *et al.*, 2019; Halloran *et al.*, 2017; Maiolo *et al.*, 2020; Oonincx and De Boer, 2012; Roffeis *et al.*, 2017, 2020; Smetana *et al.*, 2016, 2019b). Insect cultivation could be responsible for 95% of water resource depletion, 70% of land use, 45% of ozone depletion, 45% of freshwater eutrophication (Suckling *et al.*, 2020).

Furthermore, feed conversion (or bioconversion) defines the efficiency of insect feeding and growing, as the lower the feed conversion ratio (FCR) over specific timeframe the higher the performance of the production system and lower the environmental impact as long as the similar impacting feeds are considered. FCR, however, is not always transparently reflected in the studies. Different approaches include 'wet to wet' (Halloran *et al.*, 2017; Maiolo *et al.*, 2020; Oonincx and De Boer, 2012; Van Zanten *et al.*, 2015), 'wet to dry' (Roffeis *et al.*, 2017, 2020; Salomone *et al.*, 2017), 'dry to dry' (Bava *et al.*, 2019; Smetana *et al.*, 2019b) basis. Moreover, in some cases only 'ingested' feed was considered in the analysis of FCR (Bava *et al.*, 2019; Halloran *et al.*, 2017; Oonincx and De Boer, 2012; Thévenot *et al.*, 2018), while other studies include 'non-ingested' feed in the calculations, which jointly with the other factors could result in higher FCR. However, several of the reviewed studies indicate the generation of residual biomass, with further management through by-product allocation (Roffeis *et al.*, 2017, 2020; Smetana *et al.*, 2016, 2019b) or conventional waste treatment through anaerobic digestion (Smetana *et al.*, 2016). Modelling of residual biomass management would therefore include the allocation of part of the total environmental burden to a by-product (lowering the impact of the main insect biomass product) or it will add environmental burden to main product associated with waste treatment.

Energy use, GWP, and water use are the main contributors to the environmental impact of insect cultivation. While data is available from different studies, it is still fragmented, not always transparent and, in many cases, aggregated (thus difficult to reproduce). Data for the insect growing stage are also limited. Some studies indicate on the resources used for insect rearing (Thévenot *et al.*, 2018), climate system (Bava *et al.*, 2019; Smetana *et al.*, 2019b), and insect feeding (Smetana *et al.*, 2016). However, a complete detailed picture is not presented.

4.2.2   Reproduction

Insect reproduction is included in the production chain as a circulating component, separating a minor part of the adult population for mating and egg laying (Dossey *et al.*, 2016; Ites *et al.*, 2020; Salomone *et al.*, 2017; Thévenot *et al.*, 2018). Most of the studies analyse the larval stage of insect production, as this stage is the most nutritionally relevant for species from *Diptera* and *Coleoptera* orders (Table 3). For the *Orthoptera* order, on the other hand, the adult life stage is relevant, therefore, LCA of such insect production concentrated on adult stages.

TABLE 2   Insect feeding characteristics

Study	Feed type and components	Feed conversion	Residual biomass amount	Residual biomass management modelling
Oonincx and De Boer, 2012	Proprietary feed: fresh carrots, mixed grains supplemented with beer yeast	2.2 kg feed/ kg live weight	Not provided	Larvae manure as output, 100% allocation of impacts to insect output
Smetana et al., 2015	Used data from Oonincx and De Boer (2012)			
Roffeis et al., 2015	Pig manure (fresh/ dewatered)	2.8-7.4 kg substrate (DM) for 0.32-0.35 kg insects (DM)	1.8-6.4 kg (DM)	Packed residue substrate as one of outputs of system, all impacts allocated to FU manure reduction, none to insect output
Van Zanten et al., 2015	Mixed: food waste, laying hen manure, premix (vitamins and minerals)	4 kg substrate yield 1 kg fresh larvae	Not provided	Manure considered as fertiliser in consequential assessment, economic allocation in attributional part
Komakech et al., 2015	Organic waste and animal manure (theoretical)	Not provided	Not provided	Insect frass is a soil improver/ fertiliser, all impacts allocated to use of compost output; system expansion for (1) avoided fertiliser production and (2) avoided production of silver cyprinid for application in animal feed (fly larvae assumed to substitute silver cyprinid)
Smetana et al., 2016	Grains: rye meal, wheat bran Chicken manure Cattle manure Food processing by-product: beet pulp Food processing by-product: distiller's dried grains with solubles (DDGS) Municipal organic waste	22-109 kg / 1 kg of meal and 0.9 kg lipids	Not provided	Insect frass is a fertiliser or treated as waste, mass and economic allocation
Salomone et al., 2017	Food waste, average composition: 65% vegetal, 5% meat/ fish, 25% bread/pasta/rice, 5% other	10 t feed for 0.3 t dried larvae	10 t feed for 3.346 t manure	Larvae frass is a fertiliser, for FU1: system expansion - avoided compost production; for FU2, 3: impact of bioconversion fully allocated to insect output (economic allocation, lower compost price)

DM = dry matter basis; FU = functional unit; WM = wet matter basis.

TABLE 2   Insect feeding characteristics (*cont.*)

Study	Feed type and components	Feed conversion	Residual biomass amount	Residual biomass management modelling
Halloran *et al.*, 2017	Proprietary broiler feed: fish meal, soybean meal, grain maize, rice bran, palm oil, calcium carbonate, salt (optionally pumpkin)	1.47-2.5 kg feed ingested for 1 kg insects (WM)	Quantity of manure 72-85% of mass of harvested crickets (calculated)	Insect frass is a fertiliser, system expansion - avoided fertiliser production (amount based on full substitution of N, P, K in residual biomass)
Joensuu and Silvenius, 2017	Used data from Oonincx and De Boer (2012) and Oonincx *et al.* (2015)			
Thévenot *et al.*, 2018	Composite feed: cereal flours and meals, wheat bran, beat pulp	1.98 kg feed ingested for 1 kg larvae (WM)	3.85 kg per kg of meal	Insect frass is a fertiliser, all impact allocated to insect outputs; impacts from fertiliser out of scope
Bava *et al.*, 2019	Control hen diet	4.22 kg DM ingested feed for 1 kg DM larvae	3.056 kg DM residual feed and manure	Insect frass is a fertiliser, system expansion - avoided fertiliser production
	Food processing by-product: okara	2.80 kg DM ingested feed for 1 kg DM larvae	0.583 kg DM residual feed and manure	
	Food processing by-product: maize distiller's grains	2.81 kg DM ingested feed for 1 kg DM larvae	2.757 kg DM residual feed and manure	
	Food processing by-product: wet brewer's spent grains	3.30 kg DM ingested feed for 1 kg DM larvae	0.850 kg DM residual feed and manure	
Mertenat *et al.*, 2019	Organic waste: segregated household biowaste	Not provided	Not provided	Residues composted and sold on market; production of compost is part of functional unit; impacts allocated to waste treatment and compost production; system expansion to substitute larvae meal: avoided fishmeal production
Roffeis *et al.*, 2017, 2020	Chicken manure	40 kg for 1 kg dried larvae	28 kg residue	Insect frass is a fertiliser, FU1 system performance no allocation; FU2 economic allocation
	Sheep manure and fresh ruminant blood	37 kg for 1 kg dried larvae	16 kg residue	
	Chicken manure and fresh brewery waste	15.7 kg feed yielded 1 kg dried larvae	7.1 kg residue	

TABLE 2   Insect feeding characteristics (*cont.*)

Study	Feed type and components	Feed conversion	Residual biomass amount	Residual biomass management modelling
Smetana et al., 2019b	Side streams from milling, alcohol production, breweries	32.24 kg feed yielded 1.44 kg fresh puree, 1 kg meal, 0.34 kg lipids, 3.82 kg frass		Insect frass is a fertiliser; economic allocation for outputs (3.08:1 fresh insects:fertiliser)
Ites et al., 2020	Expired food products: organic waste	10 kg for 1 ton feed (WM)	60 kg for 1 ton feed	Insect frass is a fertiliser (hypothetical economic allocation)
	Food processing by-product: potato peels	0.32 t insect for 1 ton feed (WM)	0.11 ton for 1 ton feed	
	Food processing by-product: brewery grains	28.4 kg for 1 ton feed (WM)	0.12 ton for 1 ton feed	
Maiolo et al., 2020	Cereal by-products/grains	9.3 t of feeding substrate for 1.3 t larvae (live weight); 6 t of substrate for 1 t of meal	8,017 kg residue substrate and dead adult flies per 1000 kg insect meal	Insect frass is a fertiliser, economic allocation (low value to insect frass)
Suckling et al., 2020	Composite feed: wheat, meals, fats and oils, additives; plus peat	113.3 t feeding substrate (peat 17.46 t) for 12.5 t crickets (live weight)	107.5 t frass (WM); 98.5 t (DM)	Insect frass is a fertiliser, mass allocation; avoided production of fertiliser

Insect reproduction is usually separated from main feeding and growing into a separate facility. Sometimes reproducing population gets another type of treatment, which ensures better reproduction performance. Despite a special treatment, reproduction module is responsible for a minor impact in the scope of 2-8% (Salomone *et al.*, 2017; Smetana *et al.*, 2019b; Van Zanten *et al.*, 2015), which is often excluded from the boundaries of LCA studies (Ites *et al.*, 2020; Smetana *et al.*, 2016) or combined with main production (Bava *et al.*, 2019; Salomone *et al.*, 2017; Smetana *et al.*, 2019b). In some cases the impact of reproduction can reach up to 10% (Thévenot *et al.*, 2018). The highest impacts are associated with energy use and global warming potential. Similar to insect growing and feeding, only limited data is available for a transparent analysis.

4.3   *Insect products*
Insect production can result in a few potential products in the range from fresh live insects to fractionated and incorporated intermediates. All these products are a subject for a proper storage to assure their longer preservation and safety of further product development and distribution. These steps are rarely described in LCA studies, with a few exceptions (Smetana *et al.*, 2015, 2020a).

TABLE 3   Insects investigated in life cycle assessment studies

Study	Species	Family	Order	Life stage studied
Bava et al., 2019; Bosch et al., 2019; Ites et al., 2020; Komakech et al., 2015; Maiolo et al., 2020; Roffeis et al., 2017, 2020; Salomone et al., 2017; Smetana et al., 2016, 2020a	*Hermetia illucens*	Stratiomyidae	Diptera	larval/ pre-pupae
Mertenat et al., 2019; Smetana et al., 2019b	*Hermetia illucens*	Stratiomyidae	Diptera	larval/adult
Ites et al., 2020; Joensuu and Silvenius, 2017; Oonincx and De Boer, 2012; Smetana et al., 2015, 2020a	*Tenebrio molitor*	Tenebrionidae	Coleoptera	larval
Thévenot et al., 2018	*Tenebrio molitor*	Tenebrionidae	Coleoptera	larval/ pre-pupal
Oonincx and De Boer, 2012	*Zophobas morio*	Tenebrionidae	Coleoptera	larval
Van Zanten et al., 2015	*Musca domestica*	Muscidae	Diptera	larval
Roffeis et al., 2015, 2017, 2020	*Musca domestica*	Muscidae	Diptera	larval/adult
Ulmer et al., 2020	*Apis mellifera*	Apidae	Hymenoptera	pupa
Halloran et al., 2017; Suckling et al., 2020	*Gryllus bimaculatus*	Gryllidae	Orthoptera	adult
Halloran et al., 2017	*Acheta domesticus*	Gryllidae	Orthoptera	adult
Suckling et al., 2020	*Gryllus sigillatus*	Gryllidae	Orthoptera	adult

4.3.1   Primary processing

When insects reach the desired parameters in terms of size, age and composition they are processed in two stages. Primary processing aims to clean insect biomass and eliminate microbial load using operations like sieving, separation, blanching, decontamination or freezing while secondary processing targets improved properties of insect derived components through operations like milling, fractionation via centrifugation, drying and fat separation. The differences in processing result in a variation of end products sold to business or final consumers: whole live insects, whole fresh/frozen insects, fresh insect puree, dried whole larvae, defatted meal, insect oil, intermediate products containing insect components. Variations in processing depths allow to group the scope of most LCA studies into three main system boundaries: (1) cradle-to-farm gate; (2) cradle-to-processing gate; and (3) cradle-to-plate (Figure 1). For the first system boundaries (1) the resulting products are alive insects, suitable mostly for feed application (Bosch et al., 2019; Komakech

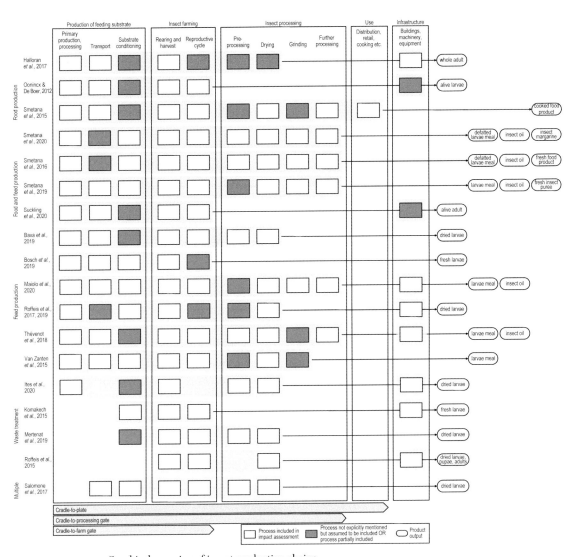

FIGURE 1   Graphical mapping of insect production chains

et al., 2015; Oonincx and De Boer, 2012; Suckling *et al.*, 2020). Application of alive larvae for direct food consumption is doubtful as currently there are no companies producing live larvae for direct human consumption. Whole larvae and adult insects, decontaminated and sometimes dried (2), on the other hand, are indicated to be applied directly for food and feed (Bava *et al.*, 2019; Halloran *et al.*, 2017; Ites *et al.*, 2020; Mertenat *et al.*, 2019; Roffeis *et al.*, 2015; Salomone *et al.*, 2017). Most studies, however, are concentrated on the assessment of fractionated or incorporated insect products in cradle-to-processing gate boundaries (Maiolo *et al.*, 2020; Smetana *et al.*, 2016, 2019b, 2020b; Thévenot *et al.*, 2018; Ulmer *et al.*, 2020; Van Zanten *et al.*, 2015).

4.3.2   Storage and packaging

Storage of insect products is of utmost importance for fresh (high-moisture) products like live, fresh-frozen insects and insect biomass incorporated in high-moisture products (Smetana *et al.*, 2015, 2016; Ulmer *et al.*, 2020). High water activity of such products usually requires low temperatures: cooled conditions for live insects and freezing for other products. Energy consumption in this case might be a considerable factor influencing the environmental impact of the insect product. Therefore, cooling and freezing storage periods should be limited to reduce the environmental impact. Insect based food and feed products of longer shelf-life nevertheless require specific storage conditions like dry rooms with climate-controlled conditions. Packaging is rarely mentioned in insect LCA studies and assumed to be comparable to the similar products in feed and food industry. Therefore, it can play a similar role of having a diverse ratio of impacts depending on the size of the product, weight of packaging, and material composition (Vignali, 2016).

4.3.3   Fractionation

Secondary processing of insect biomass relates to fractionation (separation) into a few fractions: water, lipid and protein. The purity of fractionation depends on the technology applied. Insect drying, relying on different technologies (heat drying, solar drying, freeze-drying) is indicated in a few studies as one of the most common processing techniques used (Bava *et al.*, 2019; Ites *et al.*, 2020; Mertenat *et al.*, 2019; Roffeis *et al.*, 2017, 2020; Salomone *et al.*, 2017; Smetana *et al.*, 2019b). It can have a relatively high energy demand and could results in high associated environmental impacts. In some cases, the heat used for insect production could be sourced as a by-product from other industries and have low environmental impacts (Joensuu and Silvenius, 2017). In other cases, when a high quality of insect biomass should be assured for food applications, e.g. drying is relying on lyophilisation technologies (freeze-drying of whole larvae (Bava *et al.*, 2019; Lenaerts *et al.*, 2018), which is associated with high energy expenses and relatively high impacts.

Further fractionation techniques include separation of water, lipid and protein fractions from fresh and dry biomass by centrifugation, cold or hot pressing (Alles *et al.*, 2020; Smetana *et al.*, 2019b, 2020a) and in some cases by supercritical liquid extraction (Purschke *et al.*, 2017). Emerging food processing technologies (such as pulsed electric fields) are also finding its niche in improvement of insect biomass fractionation (Alles *et al.*, 2020; Shorstkii *et al.*, 2020; Smetana *et al.*, 2020b). With such developments the dependence on energy is reduced as well as the relative environmental impact.

4.3.4   Integration (product development)

Further product development is associated with use of the concentrated protein fraction (insect flour, insect meal or defatted protein concentrate) and of insect lipids (fats and oils) as a part of a more complex matrix. The applications of both protein and lipid fractions are associated with both feed and pet-food (Gasco *et al.*, 2020; Surendra *et al.*, 2016; Zorrilla and Robin, 2019) as well as food applications (Smetana *et al.*, 2015, 2016, 2018b, 2020a; Tzompa-Sosa *et al.*, 2019). Protein fractions are incorporated into new products through mixing and baking (González *et al.*, 2019; Roncolini *et al.*, 2020); extrusion cooking and pelletising or high-moisture extrusion for meat substitutes production (Smetana *et al.*, 2018b, 2019a; Ulmer *et al.*, 2020). Diverse application possibilities result in several tested and marketed products: pelleted feeds, bars, pasta, spreads, etc. Mixing of fresh or dried insect biomass with plant material results in hybrid products, which potentially can have a lower environmental impact. However, high levels of processing (in case of isolates application) could increase the impacts of the final product.

Lipid fractions are either used as an additive for animal feed (Gasco *et al.*, 2019), for baking purposes (Delicato *et al.*, 2020; Tzompa-Sosa *et al.*, 2019) or as a part of complex fat products such as spreads and margarines (Smetana *et al.*, 2020a). There are only a couple studies, which performed LCA research in the scope of cradle-to-product application boundaries (Figure 1), even though cooking of the product at consumer stage may pose considerable environmental impacts associated with long preservation, excessive wasting and high energy use at inefficient cooking practices (Smetana *et al.*, 2015). Despite multiple challenges associated with suboptimal production chains, insect products are often assessed as having a similar or lower environmental impact compared to conventional food and feed products (Bava *et al.*, 2019; Halloran *et al.*, 2017; Oonincx and De Boer, 2012; Salomone *et al.*, 2017; Smetana *et al.*, 2016, 2020a).

5   Conceptualisation of a modular assessment framework for insect production

Despite quite a few studies presenting LCA results, there is an evident lack of systemised information, inhibiting comparisons with other insect products and production chains or with conventional products. Insect production LCA studies are often performed on lab-scale and small pilot-scale level, which does not allow for a direct result transfer to industrial scales (Table 4). Upscaling of insect production will likely decrease the environmental impact of insect product (Heckmann *et al.*, 2019; Smetana *et al.*, 2018a, 2019b).

Climate change is one of the most popular categories of assessment in the studies. Use of a control diet (standard, based on commercial or proprietary feed) is associated in with 2.3-3.1 kg CO_2 eq. per kg of fresh insects produced (Halloran *et al.*, 2017; Joensuu and Silvenius, 2017; Oonincx and De Boer, 2012). It corresponds well to the results presented for 1 kg of dried larvae: 5.76 kg CO_2 eq. (Bava *et al.*, 2019) and for 1 kg of protein: 3.9-7 kg CO_2 eq. (Bosch *et al.*, 2019; Halloran *et al.*, 2017). At the same time, some authors highlight carbon footprint as high as 21.1 kg CO_2 eq. per

TABLE 4   Characterisation of insect life cycle assessment studies according the production scale and environmental impacts

Studies	Unit	Scale of production/ assessment	Impacts		
			Control diet	Food processing by-product/ food waste	Manure diet
Bava et al., 2019	1 kg DM whole larvae	Lab: 1000 larvae per batch	CC: 5.76 LU1: 94.7 WRD: 1.26	CC: 0.7-2.0 LU1: 1.3-4.9 WRD: 0.8-1.1	n/a
Bosch et al., 2019	1 kg of protein	Lab: 100-1000 larvae per batch	GWP: 4-7 EU: 159-202 LU: 11-93	GWP: 1-5 EU: 18-77 LU: 0-1	GWP: 1-7 EU: 0-22 LU: 0
Halloran et al., 2017	FU1: 1 kg edible WW; FU2: 1 kg of protein in edible	Pilot: 36.7 tons of insects annually	FU 1 CC: 2.3-2.6 WRD: 0.42 FU 2 CC: 3.9-4.4 WRD: 0.71	n/a	n/a
Joensuu and Silvenius, 2017	FU1: 1 kg WM whole larvae; FU2: 1 kg of protein	Based on: Oonincx et al., 2015; Oonincx and de Boer, 2012	FU1: GWP2: 3.1 FU2: GWP2: 23-27	FU1: GWP2: 3.1	n/a
Ites et al., 2020	1 kg DM whole larvae*	Pilot mobile: 12.7-64 tons of insects annually	n/a	GWP1: -6.42 to 2.0 NRE: -108 to 8.9 LO: -16.8 to -0.006	n/a
Komakech et al., 2015	1 kg DM whole larvae*	Industrial hypothetical: 426 tons of insects annually	n/a	GWP2: 0.29 EU1: 0.36	n/a
Oonincx and De Boer, 2012	1 kg WM whole larvae	Pilot: 83 tons of fresh insects annually	GWP2: 2.7 EU1: 33.7 LU: 3.6	n/a	n/a
Roffeis et al., 2015	1 kg DM whole larvae*	Pilot: 1 tonne manure per week	n/a	n/a	FD: 5.9-9.7; ALO: 4.4-7.7; WD: 113.9-187.6
Roffeis et al., 2017, 2020	1 kg DM whole larvae	Pilot: 3.5-4.4 tons DM larvae annually	n/a	GWP3: 4.5-12; FD: 0.96-1.5 ALO: 5.5-61 WD: 8.5-11	
Salomone et al., 2017	1 kg DM whole larvae*	Pilot industrial: 110-329 tons DM larvae annually	n/a	GWP4: 1.0; EU2: 7.2 LU3: 0.022	n/a

TABLE 4 Characterisation of insect life cycle assessment studies (*cont.*)

Studies	Unit	Scale of production/ assessment	Impacts		
			Control diet	Food processing by-product/ food waste	Manure diet
Smetana *et al.*, 2016	1 kg DM dried defatted powder	Pilot industrial: 50 tons insect flour	GWP1,3: 1.36-15.1 NRE: 21.2-99.6 (A)LO: 0.0032-7.03		
Smetana *et al.*, 2019b	1 kg DM dried meal	Industrial: more than 1000 tons DM larvae annually	n/a	GWP1: 5.3 NRE: 84.2 LO: 1.9 WU: 2.8	n/a
Smetana *et al.*, 2020a	1 kg DM margarine (insect lipids)	Based on: Smetana *et al.*, 2019b; Thévenot *et al.*, 2018	n/a	GWP1:2.4-4.1 NRE: 16.4-54 LO: 2.4-3.7	n/a
Thévenot *et al.*, 2018	1 kg WM whole larvae	Pilot: 17 t WM larvae annually	n/a	EU2: 24.3 CC1: 0.99 LU3: 1.6	n/a
Ulmer *et al.*, 2020	1 kg of edible protein	Lab: 40 kg WM from 12 colonies annually	GWP1: 15-29 NRE: 248-425 LO: 1.1-17	n/a	n/a
Van Zanten *et al.*, 2015	1 kg DM dried meal	Pilot industrial?	n/a	GWP2: 0.77 EU1: 9.3 LU2: 0.032	
Suckling *et al.*, 2020	1 kg WM whole larvae	Pilot: 12.5 t WM insects annually	CC: 21.1 LU1: 157 WRD: 0.82	n/a	

ALO = agricultural land occupation in m^2yr (ReCiPe); CC1 = climate change in kg CO_2 eq. (CML2); CC = climate change in kg CO_2 eq. (ILCD 2011); DM = dry matter basis; EU = energy use in MJ (not specified method); EU1 = energy use in MJ (separate indicator); EU2 = energy use in MJ (CML 2); FD = fossil depletion kg oil eq. (ReCiPe); GWP = global warming potential in kg CO_2 eq. (not specified method); GWP1 = global warming potential in kg CO_2 eq. (IMPACT2002+); GWP2 = global warming potential 100 years in kg CO_2 eq. (separate indicator); GWP3 = global warming potential in kg CO_2 eq. (ReCiPe); GWP4 = global warming potential in kg CO_2 eq. (IPCC 2007); LO = land occupation m^2 org arable (IMPACT2002+); LU = land use m^2 (not specified method); LU1 = land use in kg C deficit (ILCD 2011); LU2 = land use in m^2, separate indicator; LU3 = land use in m^2a (CML 2); NRE = non-renewable energy consumption in MJ primary (IMPACT2002+); WD = water depletion in m^3 (ReCiPe); WM = wet matter basis; WRD = water resource depletion in m^3 water eq. (ILCD 2011); WU = water use in L deprived (IMPACT World+); * = recalculated for 1 kg DM of whole larvae.

kg of fresh larvae (Suckling *et al.*, 2020) or in the range of 15-29 kg CO_2 eq. per kg of protein (Ulmer *et al.*, 2020). The high impact in last studies is explained by inclusion of frass application to the field as emission factor (Suckling *et al.*, 2020) or by the analysis of a very different production system with low technology readiness level (Ulmer *et al.*, 2020).

Global warming potential impacts of insect production based on food processing by-products (food waste) can vary in a wide range from positive for the environment -6.42 to 5.3 kg CO_2 eq. for all the FUs (Bava *et al.*, 2019; Bosch *et al.*, 2019; Ites *et al.*, 2020; Joensuu and Silvenius, 2017; Komakech *et al.*, 2015; Salomone *et al.*, 2017; Smetana *et al.*, 2019b; Thévenot *et al.*, 2018; Van Zanten *et al.*, 2015). The only study differentiated form the majority indicates somewhat higher impacts of 4.5-12 kg CO_2 eq. per 1 kg DM (Roffeis *et al.*, 2017, 2020). Application of manure as feed for insects could have a great potential for the environmental improvement (Smetana *et al.*, 2016). However, reviewed studies indicated considerable impacts on the environment from 0.77-12 kg CO_2 eq. per 1 kg of dried insects (Roffeis *et al.*, 2017, 2020) to 1-7 kg CO_2 eq. per 1 kg of proteins (Bosch *et al.*, 2019).

Water footprint is assessed only in a few studies, indicating that with control diet 1 kg of fresh insects result in 0.42-0.82 m³ of water depleted (Halloran *et al.*, 2017; Suckling *et al.*, 2020). Similar impact is indicated for the protein-based unit: 0.71 m³ (Halloran *et al.*, 2017). Calculation of the results per dry matter content results in higher impacts 1.26 m³ (Bava *et al.*, 2019).

Production of insects on by-products (food waste) results in the contradictory amounts of water depleted from low 0.8-1.1 m³ per kg of dry matter content (Bava *et al.*, 2019) to high 8.5-11 m³ per kg of fresh insects produced (Roffeis *et al.*, 2017, 2020). Water footprint of insects produced on manure is also not indicative with ranges from low: 8.5-11 m³ per 1 kg of insect on dry matter basis (Roffeis *et al.*, 2017, 2020) to very high: 113.9-187.6 m³ (Roffeis *et al.*, 2015). There is a lack of studies indicating the water footprint of insects grown on food waste and manure. Higher impacts might be explained by more regionalised approach taken in the studies (Roffeis *et al.*, 2015, 2017, 2020).

Application of conventional diet results in quite high energy use impacts: 33.7 MJ per kg of fresh insects produced (Oonincx and De Boer, 2012); 159-425 MJ for 1 kg of proteins (Bosch *et al.*, 2019; Ulmer *et al.*, 2020). Energy use for insect production in case they are grown on by-products and food waste according to the reviewed studies is very diverse and ranges from rather positive -108 to 8.9 MJ per 1 kg of insect biomass on dry matter basis (Ites *et al.*, 2020; Komakech *et al.*, 2015; Salomone *et al.*, 2017) to high impacts of 24.3 MJ per kg of fresh insects produced (Thévenot *et al.*, 2018) or 18-77 MJ per 1 kg protein (Bosch *et al.*, 2019). Variations in the impacts can be explained by the differences in the modelling approaches (consideration of raw materials as by-products or wastes).

Use of energy for insect production on manure highlighted in a few studies, has moderate ranges: from 9.3-62.8 MJ per 1 kg of dry insects (Roffeis *et al.*, 2017, 2020;

Van Zanten *et al.*, 2015) and 0-22 MJ per 1 kg of proteins (Bosch *et al.*, 2019). However, a single study indicated a huge potential impact in energy use: 247-406 MJ per 1 kg of dry larvae (Roffeis *et al.*, 2015). The explanation might lay in the biological or geographical variations applied in the study.

Assessment of land use impact in the studies dealing with conventional diet is not straightforward. While production of fresh insects could result in 3.6 m² per kg of fresh insects produced (Oonincx and de Boer, 2012), land use impact calculated in other studies is much higher: 94.7 m² per 1 kg of insects dry matter basis (Bava *et al.*, 2019), 1.1-93 m² per 1 kg of proteins (Bosch *et al.*, 2019; Ulmer *et al.*, 2020). Land use impacts of insect production of by-products are indicated as low: 1.6 m² per kg of fresh insects produced (Thévenot *et al.*, 2018); -16.8 to 4.9 m² per 1 kg of insect on a dry matter basis (Bava *et al.*, 2019; Ites *et al.*, 2020; Salomone *et al.*, 2017; Smetana *et al.*, 2019b; Van Zanten *et al.*, 2015) and 0-1 m² per 1 kg proteins (Bosch *et al.*, 2019). Studies of Roffeis *et al.* (2017, 2020) highlight the possibility of higher land use impact: 5.5-61 m² per 1 kg on dry matter basis. Regionality might play here a high role as well. Land use impacts were low in studies dealing with manure treatment: 0.032 -7.7 m² per 1 kg of insect on dry matter basis (Roffeis *et al.*, 2015; Van Zanten *et al.*, 2015) and even neutral in some cases: 0 m² per 1 kg of proteins (Bosch *et al.*, 2019). And again, there was one outline study highlighting rather high impact associated with land use: 5.5-61 m² per 1 kg of insects produced (Roffeis *et al.*, 2015) potentially due to its concentration on a specific regional basis.

The variations in insect species, units of measurement, assessed scale of production and feeding diets do not allow for a straightforward evaluation of different production pathways (Table 4). However, the overall tendency indicates that the use of food processing by-products, wastes or manure for insect feeding can reduce the environmental impact of insect products (Bosch *et al.*, 2019; Ites *et al.*, 2020; Komakech *et al.*, 2015; Roffeis *et al.*, 2017, 2020; Salomone *et al.*, 2017; Smetana *et al.*, 2016; Van Zanten *et al.*, 2015). The impact of insect production furthermore can be reduced through the application of alternative energy sources (Smetana *et al.*, 2016, 2019b), use of insect for additional ecosystem services tasks (pollination, biotransformation) (Ulmer *et al.*, 2020), application of more efficient processing chains and use of passive heating and cooling methods or application of live insect with minimal processing.

The complexity of insect production chains does not allow for simple and straightforward answers about the environmental impact of insect-based products. The impact depends on the type of insect, compositions of the diet, optimisation of growing conditions, level of processing, type of distribution, etc. In order to analyse the relative efficiency, economic feasibility and environmental impact of insect production chains it is necessary to rely on a systematic holistic approach, which should include modularisation of insect production stages and their analysis on a standardised scale.

Based on the reviewed publications and available results of LCA studies we propose a modular framework, which should improve the performance of future LCA studies. It consists of 3 main components. The first component includes determination of system boundaries of insect production chains and relevant comparable studies via graphical mapping (Figure 1). For a comparative analysis insect production chains should be divided conceptually into five main groups representing: (1) the production of feeding substrate (as indicated to be of highest importance for the environmental impact); (2) insect farming; (3) processing; (4) overall infrastructure; and (5) application of insect product. These five groups combine represent the variability in the scope and boundaries of LCA – cradle-to-gate, cradle-to-plate, etc. (Figure 1), which leads to the identification of the type of insect production chain. Further LCA analysis should include more detailed modules of insect production. For example, insect farming could be further differentiated into modules of reproduction, fattening, supporting services (washing), and pre-processing (if located in the same facility). Thus, the characterisation of substrate production should include primary production or processing, transportation and on-site substrate conditioning. Insect farming consists of two main modules: rearing-harvesting and reproduction. Insect processing includes pre-processing, fractionation, grinding, and secondary processing for product development. The use phase should include distribution, retail, cooking and utilisation. Infrastructure consists of buildings, machinery and equipment relevant to capital investments. Such modular conceptualisation of LCA approaches to insect production chains allows to systematically consider the most important components and parameters relevant for a reliable analysis, determine the proper FU, scale of production and impact assessment methodology.

The second component of the framework should include the modularisation of insect production chain according to the modularisation scheme (Figure 1) and LCIA approaches (Table 1). Determination of modules should allow for the identification of important production chain elements and relevant data required. Moreover, setting up the analysis based on proposed modules should support the balance model thinking crucial for the correct results in LCA. Identification of relevant LCIA methodology should eliminate further hurdles of results interpretation and comparability with other studies.

The third component of modular framework includes the consideration of a FU and production scale (Table 4). Changes in both these parameters can affect the results and alter the final outcomes and conclusions. Such an approach provides a justified and solid basis for conducting state-of-the-art LCA and enables a reliable comparison of LCA studies of different insect production chains.

The proposed approach has certain limitations. First, it is based on the standardised approach to the LCA, which currently does not include the impact on biodiversity. There are a few methods being developed, which can find the application and can be also accounted in insect production LCA studies. Another indirect aspect which can be considered in the future is the direct potential impact of insect production especially dealing with waste treatment on the health of workers. Potential negative consequences might include allergies or intoxications. Currently there is not enough information on the potential effects, especially if the safety measures are considered. In future studies such information should be included. Animal welfare of insect production could also be a potential assessment category included in future LCA studies.

6 Conclusions

The study was oriented to review current scientific literature to establish a modular framework for the determination of environmental contribution of different parts of insect production chains.

Most LCA studies concentrated on attributional approach with results presented for several impact categories. Analysed studies relied on diverse impact assessment methods (LCIA) which can be grouped into ReCiPe, IMPACT 2002+, CML, ILCD and separate indicators. The goal of reviewed LCA articles deals with estimation of environmental impact of insect production for food and feed purposes to waste and manure treatment scenarios. Most studies rely on primary data from pilot insect production or on mix of primary data and information from the literature. There is a lack of studies, which would include the transportation and distribution of insect biomass/products, as most studies concentrate on cradle-to-gate approach.

The studies also reflected on environmental hot spots, which included production of feed (in case of commercial feed), insect cultivation and processing. Most impacts are associated with use of energy (electricity, fuel, natural gas). These factors are associated with high impacts in categories of global warming potential, non-renewable energy use, water and land use. Type of feed and modelling of its assessment was in many cases decisive for the determination of environmental impact of insects. Selection of by-product allocation rules, substitution criteria and waste scenarios determined the wide ranges of environmental impacts presented for food processing by-products, food waste and manure. Most LCA studies concentrated production of three insect species: *Hermetia illucens*, *Tenebrio molitor* and *Musca domestica*. Other five species are covered by single studies.

The analysis indicated that research literature is very diverse in the scope and boundaries of the LCA, selection of FU, LCIA methodologies, assessed insect species, scale of production and other aspects. For performing further LCA studies a systemised modular approach was suggested. It consists of three stages: (1) determination of system boundaries of insect production chains and relevant comparable studies via graphical mapping; (2) the modularisation of insect production chain according to the modularisation scheme and LCIA approaches; (3) the consideration of a FU and production scale which may affect the results and alter the final outcomes and conclusions.

Acknowledgements

Research is partially supported by the German Federal Ministry of Education and Research (BMBF), grant number 01DN17017 through Era-Net LAC project EntoWaste (ELAC2015/T03-0580), grant number 031B0934A through FACCE-SURPLUS project UpWaste, and partially by the European Union's Horizon 2020 research and innovation programme under grant agreement no. 861976 project SUSINCHAIN.

Conflict of interest

The authors declare no conflict of interest.

References

Alexandratos, N. and Bruinsma, J., 2012. World agriculture towards 2030/2050: the 2012 revision. FAO, Rome, Italy. Available at: http://www.fao.org/3/a-ap106e.pdf.

Allegretti, G., Talamini, E., Schmidt, V., Bogorni, P.C. and Ortega, E., 2018. Insect as feed: an emergy assessment of insect meal as a sustainable protein source for the Brazilian poultry industry. Journal of Cleaner Production 171: 403-412.

Alles, M.C., Smetana, S., Parniakov, O., Shorstkii, I., Toepfl, S., Aganovic, K. and Heinz, V., 2020. Bio-refinery of insects with pulsed electric field pre-treatment. Innovative Food Science & Emerging Technologies 64: 102403.

Allio, L., 2007. Better regulation and impact assessment in the European Commission. Regulatory impact assessment: towards better regulation. EC, Brussels, Belgium, pp. 72-105.

Ardente, F. and Cellura, M., 2012. Economic allocation in life cycle assessment: the state of the art and discussion of examples. Journal of Industrial Ecology 16: 387-398.

Bach, V., Lehmann, A., Görmer, M. and Finkbeiner, M., 2018. Product environmental footprint (PEF) pilot phase – comparability over flexibility? Sustainability 10: 2898.

Bava, L., Jucker, C., Gislon, G., Lupi, D., Savoldelli, S., Zucali, M. and Colombini, S., 2019. Rearing of *Hermetia Illucens* on different organic by-products: influence on growth, waste reduction, and environmental impact. Animals 9: 289.

Bosch, G., Van Zanten, H.H.E., Zamprogna, A., Veenenbos, M., Meijer, N.P., Van der Fels-Klerx, H.J. and Van Loon, J.J.A., 2019. Conversion of organic resources by black soldier fly larvae: legislation, efficiency and environmental impact. Journal of Cleaner Production 222: 355-363.

Buyle, M., Audenaert, A., Billen, P., Boonen, K. and Van Passel, S., 2019. The future of ex-ante LCA? Lessons learned and practical recommendations. Sustainability 11: 5456.

Cucurachi, S., Scherer, L., Guinée, J. and Tukker, A., 2019. Life cycle assessment of food systems. One Earth 1: 292-297.

Cucurachi, S., Van der Giesen, C. and Guinée, J., 2018. Ex-ante LCA of emerging technologies. Procedia CIRP 69: 463-468.

Del Borghi, A., 2013. LCA and communication: environmental product declaration. The International Journal of Life Cycle Assessment 18: 293-295.

Delicato, C., Schouteten, J.J., Dewettinck, K., Gellynck, X. and Tzompa-Sosa, D.A., 2020. Consumers' perception of bakery products with insect fat as partial butter replacement. Food Quality and Preference 79: 103755.

Dossey, A.T., Tatum, J.T. and McGill, W.L., 2016. Modern insect-based food industry: current status, insect processing technology, and recommendations moving forward. Insects as Sustainable Food Ingredients. Elsevier, Amsterdam, The Netherlands, pp. 113-152.

Ermolaev, E., Lalander, C. and Vinnerås, B., 2019. Greenhouse gas emissions from small-scale fly larvae composting with *Hermetia illucens*. Waste Management 96: 65-74.

Fava, J.A., 1993. Life cycle thinking: application to product design. Proceedings of the 1993 IEEE International Symposium on Electronics and the Environment. May 10-12, 1993. Arlington, VI, USA, pp. 69-73.

Fukushima, Y. and Hirao, M., 2002. A structured framework and language for scenario-based life cycle assessment. The International Journal of Life Cycle Assessment 7: 317.

Gasco, L., Biancarosa, I. and Liland, N.S., 2020. From waste to feed: a review of recent knowledge on insects as producers of protein and fat for animal feeds. Current Opinion in Green and Sustainable Chemistry 23: 67-79. https://doi.org/10.1016/j.cogsc.2020.03.003

Gasco, L., Dabbou, S., Trocino, A., Xiccato, G., Capucchio, M.T., Biasato, I., Dezzutto, D., Birolo, M., Meneguz, M., Schiavone, A. and Gai, F., 2019. Effect of dietary supplementation with insect fats on growth performance, digestive efficiency and health of rabbits. Journal of Animal Science and Biotechnology 10: 4.

Gold, M., Cassar, C.M., Zurbrügg, C., Kreuzer, M., Boulos, S., Diener, S. and Mathys, A., 2020. Biowaste treatment with black soldier fly larvae: Increasing performance through the formulation of biowastes based on protein and carbohydrates. Waste Management 102: 319-329.

González, C.M., Garzón, R. and Rosell, C.M., 2019. Insects as ingredients for bakery goods. A comparison study of *H. illucens, A. domestica* and *T. molitor* flours. Innovative Food Science & Emerging Technologies 51: 205-210.

Gouel, C. and Guimbard, H., 2019. Nutrition transition and the structure of global food demand. American Journal of Agricultural Economics 101: 383-403.

Guinée, J.B., Cucurachi, S., Henriksson, P.J.G. and Heijungs, R., 2018. Digesting the alphabet soup of LCA. The International Journal of Life Cycle Assessment 23: 1507-1511.

Halloran, A., Hanboonsong, Y., Roos, N. and Bruun, S., 2017. Life cycle assessment of cricket farming in north-eastern Thailand. Journal of Cleaner Production 156: 83-94.

Heckmann, L.-H., Andersen, J.L., Eilenberg, J., Fynbo, J., Miklos, R., Jensen, A.N., Nørgaard, J.V. and Roos, N., 2019. A case report on inVALUABLE: insect value chain in a circular bioeconomy. Journal of Insects as Food and Feed 5: 9-13. https://doi.org/10.3920/JIFF2018.0009

Hertwich, E.G., 2008. Consumption and the rebound effect: an industrial ecology perspective. Journal of Industrial Ecology 9: 85-98.

Isibika, A., Vinnerås, B., Kibazohi, O., Zurbrügg, C. and Lalander, C., 2019. Pre-treatment of banana peel to improve composting by black soldier fly (*Hermetia illucens* (L.), Diptera: Stratiomyidae) larvae. Waste Management 100: 151-160.

International Organization for Standardization (ISO), 2006a. ISO 14040 Environmental management – life cycle assessment – principles and framework. ISO, Geneva, Switzerland.

International Organization for Standardization (ISO), 2006b. ISO 14044 Environmental management – life cycle assessment – requirements and guidelines. ISO, Geneva, Switzerland.

Ites, S., Smetana, S., Toepfl, S. and Heinz, V., 2020. Modularity of insect production and processing as a path to efficient and sustainable food waste treatment. Journal of Cleaner Production 248: 119248.

Joensuu, K. and Silvenius, F., 2017. Production of mealworms for human consumption in Finland: a preliminary life cycle assessment. Journal of Insects as Food and Feed 3: 211-216. https://doi.org/10.3920/JIFF2016.0029

Jungbluth, N., Tietje, O. and Scholz, R.W., 2000. Food purchases: impacts from the consumers' point of view investigated with a modular LCA. The International Journal of Life Cycle Assessment 5: 134-142.

Komakech, A.J., Sundberg, C., Jönsson, H. and Vinnerås, B., 2015. Life cycle assessment of biodegradable waste treatment systems for sub-Saharan African cities. Resources, Conservation and Recycling 99: 100-110.

Lenaerts, S., Van Der Borght, M., Callens, A. and Van Campenhout, L., 2018. Suitability of microwave drying for mealworms (*Tenebrio molitor*) as alternative to freeze drying: impact on nutritional quality and colour. Food Chemistry 254: 129-136.

Maiolo, S., Parisi, G., Biondi, N., Lunelli, F., Tibaldi, E. and Pastres, R., 2020. Fishmeal partial substitution within aquafeed formulations: life cycle assessment of four alternative protein sources. The International Journal of Life Cycle Assessment 25: 1455-1471.

McAuliffe, G.A., Chapman, D.V. and Sage, C.L., 2016. A thematic review of life cycle assessment (LCA) applied to pig production. Environmental Impact Assessment Review 56: 12-22.

Mertenat, A., Diener, S. and Zurbrügg, C., 2019. Black soldier fly biowaste treatment – assessment of global warming potential. Waste Management 84: 173-181.

Mondello, G., Salomone, R., Ioppolo, G., Saija, G., Sparacia, S. and Lucchetti, M., 2017. Comparative LCA of alternative scenarios for waste treatment: the case of food waste production by the mass-retail sector. Sustainability 9: 827.

Mont, O. and Bleischwitz, R., 2007. Sustainable consumption and resource management in the light of life cycle thinking. European Environment 17: 59-76.

Oonincx, D.G.A.B. and De Boer, I.J.M., 2012. Environmental impact of the production of mealworms as a protein source for humans – a life cycle assessment. PLoS ONE 7: e51145.

Oonincx, D.G.A.B., Van Broekhoven, S., Van Huis, A. and Van Loon, J.J.A., 2015. Feed conversion, survival and development, and composition of four insect species on diets composed of food by-products. PLoS One 10: e0144601.

Papatryphon, E., Petit, J., Kaushik, S.J. and Van der Werf, H.M.G., 2004. Environmental impact assessment of salmonid feeds using life cycle assessment (LCA). AMBIO: A Journal of the Human Environment 33: 316-323.

Parodi, A., De Boer, I.J.M., Gerrits, W.J.J., Van Loon, J.J.A., Heetkamp, M.J.W., Van Schelt, J., Bolhuis, J.E. and Van Zanten, H.H.E., 2020. Bioconversion efficiencies, greenhouse gas and ammonia emissions during black soldier fly rearing – a mass balance approach. Journal of Cleaner Production 271: 122488.

Poore, J. and Nemecek, T., 2018. Reducing food's environmental impacts through producers and consumers. Science 360: 987-992.

Purschke, B., Stegmann, T., Schreiner, M. and Jäger, H., 2017. Pilot-scale supercritical CO_2 extraction of edible insect oil from *Tenebrio molitor* L. larvae – influence of extraction conditions on kinetics, defatting performance and compositional properties. European Journal of Lipid Science and Technology 119: 1600134.

Ravindran, R. and Jaiswal, A.K., 2019. Wholesomeness and safety aspects of irradiated foods. Food Chemistry 285: 363-368.

Rebitzer, G., 2005. Enhancing the application efficiency of life cycle assessment for industrial uses. EPFL PP, Lausanne, Switzerland. https://doi.org/10.5075/epfl-thesis-3307

Rebitzer, G., Ekvall, T., Frischknecht, R., Hunkeler, D., Norris, G., Rydberg, T., Schmidt, W.-P., Suh, S., Weidema, B.P. and Pennington, D.W., 2004. Life cycle assessment part 1. Environment International 30: 701-720.

Roffeis, M., Almeida, J., Wakefield, M., Valada, T., Devic, E., Koné, N., Kenis, M., Nacambo, S., Fitches, E., Koko, G., Mathijs, E., Achten, W. and Muys, B., 2017. Life cycle inventory analysis of prospective insect based feed production in West Africa. Sustainability 9: 1697.

Roffeis, M., Fitches, E.C., Wakefield, M.E., Almeida, J., Alves Valada, T.R., Devic, E., Koné, N., Kenis, M., Nacambo, S., Koko, G.K.D., Mathijs, E., Achten, W.M.J. and Muys, B., 2020. Ex-ante life cycle impact assessment of insect based feed production in West Africa. Agricultural Systems 178: 102710.

Roffeis, M., Muys, B., Almeida, J., Mathijs, E., Achten, W.M.J., Pastor, B., Velásquez, Y., Martinez-Sanchez, A.I. and Rojo, S., 2015. Pig manure treatment with housefly (*Musca domestica*) rearing – an environmental life cycle assessment. Journal of Insects as Food and Feed 1: 195-214. https://doi.org/10.3920/JIFF2014.0021

Roncolini, A., Milanović, V., Aquilanti, L., Cardinali, F., Garofalo, C., Sabbatini, R., Clementi, F., Belleggia, L., Pasquini, M., Mozzon, M., Foligni, R., Federica Trombetta, M., Haouet, M.N., Serena Altissimi, M., Di Bella, S., Piersanti, A., Griffoni, F., Reale, A., Niro, S. and Osimani, A., 2020. Lesser mealworm (*Alphitobius diaperinus*) powder as a novel baking ingredient for manufacturing high-protein, mineral-dense snacks. Food Research International 131: 109031.

Rubik, F. and Frankl, P., 2017. The future of eco-labelling: making environmental product information systems effective. Routledge, London, UK.

Salomone, R., Saija, G., Mondello, G., Giannetto, A., Fasulo, S. and Savastano, D., 2017. Environmental impact of food waste bioconversion by insects: application of life cycle assessment to process using *Hermetia illucens*. Journal of Cleaner Production 140: 890-905.

Schau, E.M. and Fet, A.M., 2008. LCA studies of food products as background for environmental product declarations. The International Journal of Life Cycle Assessment 13: 255-264.

Shorstkii, I., Alles, M.C., Parniakov, O., Smetana, S., Aganovic, K., Sosnin, M., Toepfl, S. and Heinz, V., 2020. Optimization of pulsed electric field assisted drying process of Black soldier fly (*Hermetia illucens*) larvae. Drying Technology. https://doi.org/10.1080/07373937.2020.1819825

Smetana, S., 2020. Life cycle assessment of specific organic waste-based bioeconomy approaches. Current Opinion in Green and Sustainable Chemistry 23: 50-54. https://doi.org/10.1016/j.cogsc.2020.02.009

Smetana, S., Aganovic, K., Irmscher, S. and Heinz, V., 2018a. Agri-food waste streams utilization for development of more sustainable food substitutes. Designing sustainable technologies, products and policies. Springer, Cham, Switzerland, pp. 145-155.

Smetana, S., Ashtari Larki, N., Pernutz, C., Franke, K., Bindrich, U., Toepfl, S. and Heinz, V., 2018b. Structure design of insect-based meat analogs with high-moisture extrusion. Journal of Food Engineering 229: 83-85.

Smetana, S., Leonhardt, L., Kauppi, S.-M., Pajic, A. and Heinz, V., 2020a. Insect margarine: processing, sustainability and design. Journal of Cleaner Production 264: 121670.

Smetana, S., Mathys, A., Knoch, A. and Heinz, V., 2015. Meat alternatives: life cycle assessment of most known meat substitutes. The International Journal of Life Cycle Assessment 20: 1254-1267.

Smetana, S., Mhemdi, H., Mezdour, S. and Heinz, V., 2020b. Pulsed electric field-treated insects and algae as future food ingredients. Pulsed electric fields to obtain healthier and sustainable food for tomorrow. Elsevier, Amsterdam, The Netherlands, pp. 247-266.

Smetana, S., Palanisamy, M., Mathys, A. and Heinz, V., 2016. Sustainability of insect use for feed and food: life cycle assessment perspective. Journal of Cleaner Production 137: 741-751.

Smetana, S., Pernutz, C., Toepfl, S., Heinz, V. and Van Campenhout, L., 2019a. High-moisture extrusion with insect and soy protein concentrates: cutting properties of meat analogues under insect content and barrel temperature variations. Journal of Insects as Food and Feed 5: 29-34. https://doi.org/10.3920/JIFF2017.0066

Smetana, S., Schmitt, E. and Mathys, A., 2019b. Sustainable use of *Hermetia illucens* insect biomass for feed and food: attributional and consequential life cycle assessment. Resources, Conservation and Recycling 144: 285-296.

Spielmann, M., Scholz, R., Tietje, O. and de Haan, P., 2005. Scenario modelling in prospective LCA of transport systems. Application of formative scenario analysis. The International Journal of Life Cycle Assessment 10: 325-335.

Steubing, B., Mutel, C., Suter, F. and Hellweg, S., 2016. Streamlining scenario analysis and optimization of key choices in value chains using a modular LCA approach. The International Journal of Life Cycle Assessment 21: 510-522.

Suckling, J., Druckman, A., Moore, C.D. and Driscoll, D., 2020. The environmental impact of rearing crickets for live pet food in the UK, and implications of a transition to a hybrid business model combining production for live pet food with production for human consumption. The International Journal of Life Cycle Assessment 25: 1693-1709.

Surendra, K.C., Olivier, R., Tomberlin, J.K., Jha, R. and Khanal, S.K., 2016. Bioconversion of organic wastes into biodiesel and animal feed via insect farming. Renewable Energy 98: 197-202.

Thabrew, L., Wiek, A. and Ries, R., 2009. Environmental decision making in multi-stakeholder contexts: applicability of life cycle thinking in development planning and implementation. Journal of Cleaner Production 17: 67-76.

Thévenot, A., Rivera, J.L., Wilfart, A., Maillard, F., Hassouna, M., Senga-Kiesse, T., Le Féon, S. and Aubin, J., 2018. Mealworm meal for animal feed: environmental assessment and sensitivity analysis to guide future prospects. Journal of Cleaner Production 170: 1260-1267.

Thomas, C., Grémy-Gros, C., Perrin, A., Symoneaux, R. and Maître, I., 2020. Implementing LCA early in food innovation processes: study on spirulina-based food products. Journal of Cleaner Production 268: 121793.

Tzompa-Sosa, D.A., Yi, L., Van Valenberg, H.J.F. and Lakemond, C.M.M., 2019. Four insect oils as food ingredient: physical and chemical characterisation of insect oils obtained by an aqueous oil extraction. Journal of Insects as Food and Feed 5: 279-292. https://doi.org/10.3920/JIFF2018.0020

Ulmer, M., Smetana, S. and Heinz, V., 2020. Utilizing honeybee drone brood as a protein source for food products: life cycle assessment of apiculture in Germany. Resources, Conservation and Recycling 154: 104576.

United Nations Environment Programme and the Society of Environmental Toxicology and Chemistry (UNEP/SETAC), 2011. Global guidance principles for life cycle assessment databases. A basis for greener processes and products. UNEP, Geneva, Switzerland.

United Nations Environment Programme and the Society of Environmental Toxicology and Chemistry (UNEP/SETAC), 2012. Greening the economy through life cycle thinking ten years of the UNEP/SETAC life cycle initiative. UNEP, Geneva, Switzerland, 64 pp.

Van der Weele, C., Feindt, P., Van der Goot, A., Van Mierlo, B. and Van Boekel, M., 2019. Meat alternatives: an integrative comparison. Trends in Food Science & Technology 88: 505-512.

Van Huis, A., Van Itterbeeck, J., Klunder, H., Mertens, E., Halloran, A., Muir, G. and Vantomme, P., 2013. Edible insects. FAO forestry paper. Food and Agriculture Organization of the United Nations, Rome, Italy, 201 pp.

Van Zanten, H.H.E., Bikker, P., Meerburg, B.G. and De Boer, I.J.M., 2018. Attributional versus consequential life cycle assessment and feed optimization: alternative protein sources in pig diets. The International Journal of Life Cycle Assessment 23: 1-11.

Van Zanten, H.H.E., Mollenhorst, H., Oonincx, D.G.A.B., Bikker, P., Meerburg, B.G. and De Boer, I.J.M., 2015. From environmental nuisance to environmental opportunity: housefly larvae convert waste to livestock feed. Journal of Cleaner Production 102: 362-369.

Veldkamp, T., Van Duinkerken, G., Van Huis, A., Lakemond, C.M.M., Ottevanger, E., Bosch, G. and Van Boekel, T., 2012. Insects as a sustainable feed ingredient in pig and poultry diets: a feasibility study. Wageningen UR Livestock Research, Wageningen, The Netherlands. Available at: https://www.wur.nl/upload_mm/2/8/0/f26765b9-98b2-49a7-ae43-5251c5b-694f6_234247%5B1%5D.

Vignali, G., 2016. Life-cycle assessment of food-packaging systems. In: Muthu, S.S. (ed.) Environmental footprints of packaging. Springer, Berlin, Germany, pp. 1-22.

Wade, M. and Hoelle, J., 2020. A review of edible insect industrialization: scales of production and implications for sustainability. Environmental Research Letters 15: 123013. https://doi.org/10.1088/1748-9326/aba1c1

Walser, T., Demou, E., Lang, D.J. and Hellweg, S., 2011. Prospective environmental life cycle assessment of nanosilver T-shirts. Environmental Science & Technology 45: 4570-4578.

Willett, W., Rockström, J., Loken, B., Springmann, M., Lang, T., Vermeulen, S., Garnett, T., Tilman, D., DeClerck, F. and Wood, A., 2019. Food in the anthropocene: the EAT-Lancet commission on healthy diets from sustainable food systems. The Lancet 393: 447-492.

Zamagni, A., Guinée, J., Heijungs, R., Masoni, P. and Raggi, A., 2012. Lights and shadows in consequential LCA. The International Journal of Life Cycle Assessment 17: 904-918.

Zampori, L., Saouter, E., Schau, E., Cristobal Garcia, J., Castellani, V. and Sala, S., 2016. Guide for interpreting life cycle assessment result. Publications Office of the European Union, Luxembourg.

Zorrilla, M. and Robin, N., 2019. Nutrition technologies: offering price competitive black soldier fly protein and oil to the animal feed and pet food sectors. Industrial Biotechnology 15: 328-329.

Welfare of farmed insects

A. van Huis

Laboratory of Entomology, Wageningen University & Research, Droevendaalsesteeg 1, 6708 PB Wageningen, The Netherlands; arnold.vanhuis@wur.nl

Abstract

The recent interest in using insects as food and feed is based on their capacity to be a sustainable alternative to other protein sources. When farmed as mini livestock, the question is raised as to whether they are 'sentient beings' (self-conscious)? In researching this topic, the problem is that humans often expect animals to have the same subjective experience as we do (anthropomorphic) and consider themselves as the centrc of the universe (anthropocentric). We discuss insects' sentience by looking at their brain, behaviour, and communicative abilities. The miniature brains of insects seem to be arranged in a very efficient functional way due to their very long evolutionary history. As for their behaviour, insects are capable of social and associative learning. Even dopamine, a neurotransmitter involved in reward and pleasure, plays a role. Human communication is mainly verbal, while for insects other means of information exchange are more important, such as tactile, chemical, visual, and vibrational. The distinction needs to be made between nociception and pain, the latter being an emotional experience. It is difficult to prove that insects can experience pain, although they have a large repertoire of withdrawal and defensive behavioural responses. The philosophical attitudes deal with how we view insects and their relations to humans. This also determines the ethical attitude and how we should treat them. Are they just there for our benefit or do we consider them as co-animals? Insects as food requires that many insects must be killed. However, the number killed may not be different when one chooses a plant-based diet. It is concluded that insects should be farmed and killed using the precautionary principle, which assumes that they can experience pain. To discuss the consequences for the industry sector that produces insects for food and feed, we used Brambell's five freedoms as a framework.

Keywords

edible insects – ethics – invertebrate welfare – nociception – pain

1   Introduction

Although most animal species on earth are insects (Mora *et al.*, 2011; Stork, 2018), welfare issues related to invertebrates have not received a lot of attention. This oversight has to do with assumptions that invertebrates do not experience pain and stress, reinforced by the public's negative view of invertebrates (Horvath *et al.*, 2013).

For more than four millennials insects have been domesticated, notably the silkworm and the bee. Insects are now also reared for a number of other purposes (Boppré and Vane-Wright, 2019): research, pollination, insect pest and weed management such as biological control and sterile insect technique, recreation and entertainment, medicine (e.g. maggot therapy) and as food for humans and feed for animals. In many tropical countries, insects have been harvested for food for centuries. However, it is only in the last ten years that the interest in using insects as food for humans and feed for animals has increased exponentially (Van Huis, 2020). This is because insects are increasingly being considered as a high-quality, efficient, and sustainable alternative protein source for both the food and feed sector. This means that billions of insects are and will be farmed, often in industrial settings.

Public concerns about the current livestock industry focusses on the conditions in which livestock are reared, transported, and slaughtered. For insects farmed as mini-livestock similar concerns are being expressed. This has a lot to do with the question of whether insects are considered 'sentient beings' (beings with consciousness). Therefore, welfare issues need to be addressed. For example: what are the requirements of rearing specific insects, what about their health, and how can they be killed 'humanely'? (Pali-Schöll *et al.*, 2018).

There is also a discussion on how to manage insect welfare in experimental research. However, we will not deal with this here but instead refer the reader to publications on this topic (Drinkwater *et al.*, 2019; Moltschaniwskyj *et al.*, 2007; Pollo and Vitale, 2019; Wilson-Sanders, 2011).

The problem when discussing the question of whether insects have feelings, is that insects are viewed in an anthropomorphic manner. We expect animals to have the same subjective experiences as a human. Words often used in this context are consciousness and sentience. For that reason, we wish to define those words. Consciousness has been defined as 'the quality or state of being aware of an external object or something within oneself'[9] while sentient has been defined as 'responsive to or conscious of sense impressions'. Consciousness, or simply the presence of subjective experience, can be studied in humans in three different ways: neurological, behavioural and verbal (Barron and Klein, 2016). Each of these topics will be discussed for invertebrates below. We will then discuss whether insects would be able to experience pain, before embarking on some philosophical theories. Based

---

9   https://www.merriam-webster.com/dictionary/consciousness?src=search-dict-box.

on the earlier deliberations, we are faced with the ethical issue of how to deal with insects. Lastly, we will discuss how the insect food and feed industry should deal with insect welfare.

2   Neurological arguments

Invertebrates have an evolutionary history of 500 million years, while humans (*Homo* spp.) have a history of 2.5 million years. Conscious processes have probably arisen independently on many occasions during the evolution of vertebrates and invertebrates (Le Neindre *et al.*, 2017: 119). According to Budelmann (1995) the nervous system of cephalopods and insects has the highest degree of centralisation ('cerebralisation') for any invertebrate. This considerably reduces the time for information processing between stimulus reception and behavioural reactions. Barron and Klein (2016) and Klein and Barron (2016) contend that the integrative functional arrangements in the insect brain have many parallels with that of the vertebrate midbrain and the insect brain is thus capable of the same neural modelling, and hence of conscious experience. The question is whether neuron numbers by themselves are critical for the neural implementation of sentience. There is the *a priori* and maybe erroneous assumption that small brains are unlikely to support cognition or sentience (Mikhalevich and Powell, 2020). The number of neurons in the brain of the bee, considered to have an impressive behavioural repertoire, is 960,000 (Menzel and Giurfa, 2001), compared to 75 million for a mouse (Herculano-Houzel *et al.*, 2006) and 85 billion (Herculano-Houzel, 2009) for a human. The estimated number of neurons in the brains of a mealworm was estimated to be only 25,000 (Scherer *et al.*, 2017). However, if sentience is a function of structural arrangements of neural operations, neuron numbers may not be the most important criterion. Sentience may then be present in brains of different sizes. The mechanism of sentience even in large brains may require more than a subset, even a small subset, of available neurons (Merker, 2016). The miniature brains of insects exhibit a sophisticated behavioural repertoire and cognitive capabilities often go beyond simple associative learning (Giurfa, 2013). Andrews (2011) mentions how difficult it is to identify functionally analogous pathways in invertebrates with fundamentally differently organised central nervous systems such as in vertebrates. Tomasik (2019) even wonders whether smaller animals have faster subjective experiences than larger ones, considering their greater temporal resolution of vision. Would a higher brain metabolic rate (neuronal firing being one component) relate to moral weight of the mind?

3   Behavioural arguments

In a 1986 review by Carew and Sahley (1986) on invertebrate learning and memory, the belief was stated that the distinction between vertebrate and invertebrate learning and memory will diminish as our understanding of underlying mechanisms increases. Social learning (acquiring new behaviours by observing and imitating others) may not be restricted to large-brained animals, which are often assumed to possess superior cognitive abilities. Insects are capable of social learning (Coolen *et al.*, 2005): take the example of wood crickets (*Nemobius sylvestris*) using congeners' behaviour by hiding under leaves to avoid wolf spiders (*Pardosa* spp.). Associate learning has also been demonstrated in different insect groups such as honeybees (Leadbeater and Chittka, 2007), ants and moths (Giurfa, 2013). Many experiments have been done with associative olfactory and visual learning in the fruit fly *Drosophila melanogaster* (Guo and Guo, 2005), and how dopamine is involved (Karam *et al.*, 2019). In humans the neurotransmitter dopamine is involved in motivational processes as well as reward, pleasure and addiction. Perry *et al.* (2016) demonstrated that bumblebees' behaviour is influenced by dopamine. Bees learned to fly quickly to one colour/cylinder location to obtain sucrose and slowly to a different colour/cylinder location that contained only water. When bees were induced with a positive affective state (giving them an unexpected 60% sucrose reward) they flew faster to a cylinder with ambiguous cues than non-rewarded bees. These behavioural changes disappeared with topical application of the dopamine antagonist fluphenazine. It has also been shown that bees are able to distinguish between human painting styles or to recognise faces (Mikhalevich and Powell, 2020).

However, whether 'emotion-like' states in insects are accompanied by emotional feelings is a question that remains unanswered (Mendl *et al.*, 2011; Mendl and Paul, 2016). Also, according to Mason (2011), the problem-solving and stimulus-recognition skills of invertebrates, and their cognitive abilities like responses to tissue-damaging stimuli and escape mechanisms, are no evidence for conscious affective states and the ability to suffer. Otherwise, he states, we should be worrying about the well-being of computers.

In mobile, spatially-orienting animals there should be a central convergent interface for behavioural decision-making for target and action selection, and motivational ranking, and there is often a delay between target selection and goal acquisition (Merker, 2016; Paul and Mendl, 2016). Barron and Klein (2016) mention examples of how insects process topographically organised visual information. For example, Tarsitano (2006) showed that the jumping spider, *Portia labiate*, can make complete detours in which it moves away from a goal (i.e. prey) before approaching it. Decision-making occurred gradually, during both the scanning and the locomotory phases in the form of vicarious trial and error attempts.

Giurfa (2013) issues a warning about experimental designs to study insect behaviour in which the restricted animal can only do what the experimenter allows it to do.

4   Verbal arguments

The effectiveness of human communication is often mentioned as evidence of humans being the most unique and highly-evolved animals. It is commonly assumed that there is no equivalent to human language in other animal species. Identifying meaning in non-human animal communication is probably the most difficult task in linguistics because of our limited ability to infer the real goals and intentions of non-human animals (Prat, 2019). One problem with effective communication is that tests are conducted in the expectation that animals will perform in a similar way to humans (Lockwood, 1987). In animal tests dealing with auditory communication, Prat (2019) considered it bothersome that humans in tests are assumed to be unique and superior, while studies of non-human animals' vocal communication provided results like those expected if similar methods were to be applied to human vocal behaviour. Human communication centres on vocal and visual means. However, insects not only have these communication means, but also use other means of exchanging information more than humans do, e.g. tactile, chemical (smell and taste), and vibrational (Hedwig, 2014; Roitberg, 2018; Yack, 2016).

5   Can insects experience pain?

In invertebrates the discussion centres on the difference between pain and nociception. The definition of pain according to the International Association for the Study of Pain is 'an unpleasant sensory and emotional experience associated with actual or potential tissue damage', while nociception has been defined as 'the ability to detect stimuli that elicit damage to the body or the potential for such damage', the difference being 'pain has both an emotional and a sensory component, and this latter component, nociception, refers specifically to the detection of damaging or potentially damaging stimuli' (Burrell, 2017). These nociceptive signals in vertebrates are processed by the central nervous system and perceived as pain. Some authors believe that insects cannot experience pain. Eisemann *et al.* (1984) was one of the first who embarked on the question and concluded that 'the evidence from consideration of the adaptive role of pain, the neural organisation of insects and observations of their behaviour does not appear to support the occurrence in insects of a pain state, such as occurs in humans.' They conducted an oft-cited experiment in which a locust continued to feed while itself being eaten by a mantis. However, Sherwin (2001) was not impressed because vertebrate prey species may also have a selective advantage in not showing pain or injury to a predator. Nociception often causes a withdrawal response, and a number of examples can be given such as the defensive responses of *Manduca sexta* larvae as a response to noxious stimuli (Walters *et al.*, 2001) or the directional rolling of *Drosophila* larvae to escape the attack of the parasitoid wasp *Leptopilina boulardi* (Hwang *et al.*, 2007), noxious heat or harsh mechanical stimulation (Im and Galko, 2012; Tracey *et al.*, 2003).

Bateson (1991) points to the fact that the less animals are like humans, the more difficult it is to assess pain in them. According to Elwood (2011) the rapid learning to avoid pain, coupled with a prolonged memory, seem to indicate that central processing is involved rather than simple reflexes. According to this author, invertebrates are capable of using complex information, suggesting a cognitive ability to have a fitness benefit from a pain experience. Although evidence is not conclusive, this seems to indicate that there is more than just nociception. Sneddon *et al.* (2014) have put forward an interesting idea: the principle of triangulation. They argue that when the neurological, behavioural, and verbal arguments are taken in isolation, it cannot be considered as definitive evidence of 'pain' in animals, but when using a multimodal approach it may well be.

6   Philosophical theories

Human empathy for insects in general is low, maybe because 'invertebrates look far less like us' (Smyth, 1978). Another reason could be that in Western cultures insects are considered dirty, disgusting, and dangerous (Looy *et al.*, 2014) and this negative attitude is reflected in idioms, proverbs, and slogans about insects (Meyer-Rochow and Kejonen, 2020). Mikhalevich and Powell (2020) also believe that the moral exclusion of invertebrates is caused by the distorting influence of the empathy gap and disgust response. In general, in the animal kingdom invertebrates are considered at the 'lower end' and humans at the extreme 'upper end' (Andrews, 2011). This is reinforced by the reputation of some insect species, like malaria mosquitoes or the desert locust, which can have a dramatic negative influence on human well-being. Others, such as houseflies, are considered a nuisance. The threat of insects to human welfare has already been mentioned in very early articles (Gossard, 1909).

The importance of insects has only been highlighted very recently because of the decline of insect biomass in protected areas of Europe (Hallmann *et al.*, 2017) and other parts of the world (Van Klink *et al.*, 2020). This has drawn attention to the importance of the ecological services that insects provide, such as pollination. Crop production to the value of 235-577 billion US$ is at risk because of pollinator loss (IPBES, 2019). Natural biological control by predators, parasitoids and entomopathogens of crop pests provides a value of 400 billion US$ per year (Costanza *et al.*, 1997). In addition, there are other services relating to provisioning (e.g. insects as nutrients), supporting (e.g. recycling) and cultural (e.g. recreation and bio-indicators). We humans may have to be a bit more modest given that *Homo sapiens* have only been around for half a million years, while insects have a history of 400 million years. There are about 5.5 million species of insects, of which one million have been described (Stork, 2018). Insects are very successful because of their design and processes (Van Huis, 2014). The American entomologist Edward O. Wilson indicated '... we need invertebrates, but they don't need us' and '... if invertebrates were to disappear, I doubt that the human species could last more than a few months' (Wilson, 1987). Considering the importance of insects in our ecosystem, they deserve our respect.

Homo sapiens and the fruit fly *D. melanogaster* had a common ancestor approximately 783 million years ago. A total of 15% of human genes and 46% of fly genes have orthologs (genes in different species that descended from the same ancestral sequence) to one or more fly and human genes, respectively (Shih *et al.*, 2015). Insights gained from model genetic systems such as *D. melanogaster* can be applied immediately to vertebrate systems, by for instance analysing the function of human disease genes (Bier, 2005). Many of the genetic pathways that guide basic developmental processes in vertebrates and invertebrates have remained largely intact during evolution.

According to Lestel and Taylor (2013), Westerners have lived in a culture that has constantly insisted on the man/animal opposition. According to these authors, 'It is wrong to say that we live with animals; it is more correct to say that we are animals and animals are us.' They quote Dominique Lestel: 'each species that disappears is a part of our imagination that we amputate perhaps irreversibly'. In their view, the major challenge is 'how we might consider the specificity of the human in proximity to other living beings (including plants and fungi) rather than setting strict boundaries.'

This brings us to the philosophical attitudes towards invertebrates. Mather (2011) mentions three: contractarian (animals are very different from humans; they are just machines and can be used for human benefit); utilitarian (looking at gains versus losses, e.g. bees should be protected as they assure pollination); and the rights-based approach (experiences and awareness of animals should be taken into account). Here comes the difference between animal welfare and animal rights. In animal welfare, animals have interests, but these can be traded away when human benefits justify that sacrifice if 'humane' guidelines are followed. In animal rights these interests cannot be sacrificed or traded away just because it might benefit others. Jena (2017) writes that in the animal welfare concept animals are provided with some comfort and freedom of movement in the period prior to being killed. However, there is no concern for the rights of animals; nor is there an interest in keeping the animals alive. For her, mere 'freedom', as formulated by Brambell (1965), is not sufficient and she opts for combining animal welfare and rights if the aim is to ensure their actual welfare.

Do insects have intrinsic value? This refers to an integrity that is independent of the animal's utility or use to humans (RDA, 2018). The consequence is that all instances of keeping and using animals must be justified in a manner that takes the animals' interests into account. In the Netherlands invertebrates fall under the Animals Act[10] as soon as they are farmed for production purposes. This means that the

---

10 https://wetten.overheid.nl/jci1.3:c:BWBR0030250&z=2013-01-01&g=2013-01-01.

government formally acknowledges their intrinsic value, thereby imposing clear welfare requirements (Section 1.3.2 and 1.3.3), such as the five freedoms of Brambell (1965): freedom from hunger and thirst, from discomfort, from pain, injury or disease, from fear and distress, and freedom to express normal behaviour.

7   Ethics

Ethics, also called moral philosophy, deals with concepts such as good and evil, and right and wrong. Therefore, the term 'food ethics' have been coined, which deals with animal welfare, the environment, health and fair trade (Barnhill and Doggett, 2018; McEachern, 2018; Nordgren, 2012). The consumption of animal products is rapidly growing in many developing countries because of increasing populations, urbanisation, and wealth. However, the livestock industry is considered an inefficient way of producing calories and nutrients (Alexander *et al.*, 2017) and is associated with environmental, welfare, and health problems (McClements, 2019). So, what about alternative more sustainable protein sources, such as insects?

Some people have moral objections to insect consumption. This may have to do with the uncertainty as to whether insects are 'sentient beings' or with the very large number of insects that have to be killed compared to, for example a pig, making it unethical (Knutsson, 2016; Scherer *et al.*, 2017). Therefore, plant-based diets may be preferred, although billions of insects need to be killed in order to allow for plant-based diets (Fischer, 2016, 2019).

There is also a question as to whether invertebrates or insects are taken as a group or whether we should have a species-specific approach considering the emotional and cognitive abilities and specific behavioural needs of certain insects (Gjerris *et al.*, 2015; Mather and Carere, 2019). This may influence informed decisions regarding insect welfare and the ability of legislators to formulate accurate regulations. Also, Schukraft (2020) discusses whether there are degrees in the moral status of animals: e.g. are there differences in self-awareness, intelligence, autonomy, communicative ability, creativity, sociability, etc? This would particularly apply to insects with an enormous diversity (even between larvae and adults) and probably differ in moral status and capacity for welfare. Just consider the number of neurons in mealworms (30,000) compared to a cockroach or bee (about one million).

When it comes to eating insects, Waltner-Toews and Houle (2017) indicate the complexity of the problem. On the one hand, they indicate that we do not want to cause pain or suffering to an animal in our care. On the other hand we are often unable to avoid causing their death: stepping on them when walking outside, eating fruits and vegetables that could have kept them alive, or eating them directly or inadvertently.

For economic reasons insect companies will try to put the insects in optimal conditions to obtain the highest production possible. So, normally, insect welfare is guaranteed. Increased insect welfare and economics can go together as shown by Adámková *et al.* (2017) when rearing mealworms: (1) death by freezing, compared to boiling, was found to be better from both a welfare and a nutritional perspective; and (2) nutritional deprivation affected 'welfare', but also negatively influenced the nutritional value and economics.

The ethical proposal put forward by Lockwood (1987) is that 'we ought to refrain from actions which may be reasonably expected to kill or cause nontrivial pain in insects when avoiding these actions has no, or only trivial, costs to our own welfare'.

8 Farmed invertebrates

8.1 *Justification*

There are insects that are harvested from nature, which is a traditional way of collecting food. What about when insects are reared for use as food or feed? This was practised in the Western world mainly to provide feed for reptiles, zoos, etc. However, over the last ten years there has been a steep rise in the number of companies that rear insects to produce food for humans and feed for animals. As discussed in food ethics, this is partly to do with all the problems associated with the current livestock production, which is not considered sustainable. Insect production is more sustainable than the production of beef, pork and poultry (Smetana, 2020; Smetana *et al.*, 2015; Tapanen, 2018; Van Huis, 2019; Van Huis and Oonincx, 2017). Furthermore, insects can contribute to a circular economy because certain insect species can be reared on organic side streams (Bortolini *et al.*, 2020; Cappellozza *et al.*, 2019; Heckmann *et al.*, 2019; Nava *et al.*, 2020); this is important, considering that one third of our food and agricultural produce is wasted (FAO, 2014).

8.2 *Precautionary principle*

Then there is the question of how to deal with insects from an animal welfare' perspective. Some vegetarians may eat insects if insect welfare is guaranteed. However, when an animal rights approach is used, people may refrain from consuming them. But how are we to deal with the uncertainty that insects may experience pain? There are indications that insects experience pain but extensive scientific knowledge on the issue is lacking. In these cases, politicians use the precautionary principle, giving them the benefit of the doubt, using the aphorism: 'Absence of evidence is not evidence of absence'. This has been mentioned by a number of authors discussing the welfare of farmed invertebrates (Adamo, 2016; Birch, 2000; Elwood, 2011; Fischer, 2019; Knutsson and Munthe, 2017), although Birch (2017) introduces the concept of 'appropriate burden of proof for sentience', which may have not been attained for many orders of arthropods. Monsó (2018) argues that insect sentience may not matter that much. The first argument is that sentience of insects does secure them a right not to be raised and slaughtered for food. The second argument put forward is that pain in insects may be less severe or less harmful than that in vertebrates, which would favour consuming the first rather than the latter (for her numbers do not matter). The third argument is that a commitment not to harm other types of animals, such as mammals, may imply an obligation to consume insects. Meyers (2013) also mentioned that it is not only morally acceptable to eat insects but that it can be morally good to do so. This is because industrial farming of livestock is detrimental to livestock species while that of mini livestock is not, or much less so.

8.3 *Farmed insects and Brambell's five freedoms*

It we consider the five freedoms of Brambell (1965), we can make the following remarks about insect farming for food and feed.

1. Freedom from hunger and thirst, by ready access to water and a diet to maintain health and vigour

Ready access to fresh water and sufficient diet is normally provided, to get them into optimal shape and maintain their full health and vigour. The house fly will die of dehydration within days if water is not provided, and will cease reproducing if food of insufficient quality is provided (Erens *et al.*, 2012). However, optimal production does not automatically mean species-appropriate conditions. For example, abiotic conditions (humidity, temperature, light) and biotic conditions (diet, crowding) may be modified to tailor the insect nutritional values to specific requirements of quantity and quality of the end product; these conditions may not be appropriate for the animal's well-being.

Starvation, i.e. one- or two-day fasting, is recommended to empty the insect gut before they are used as food or feed. Farmers want to sell insects, not the manure. This starvation should not be too long, as the animals may turn cannibalistic (for food safety reasons, starvation does not seem to be necessary (Wynants *et al.*, 2017)). The question should also be asked, when rearing insects on cheap organic side streams, whether the development will not be too long such as with black soldier fly on manure (Oonincx *et al.*, 2015) and/or survival too low such as with crickets on municipal-scale food waste and diets composed largely of straw (Lundy *et al.*, 2015).

2. Freedom from discomfort, by providing an appropriate environment

Insect housing must be such that it allows for species-specific movement needs, such as means of locomotion (crawling, walking, jumping, flying); social interaction (including cannibalism); reproduction; and concealment opportunities. Insects sometimes have specific pupation and egg-laying requirements, and food must not only have the right form and nutritional/other composition, but hygiene is also important (RDA, 2018). Live insects could be stored under cool conditions, and should be transported under appropriate abiotic conditions.

Normally, insect farming companies make sure that conditions for growth are optimal as this assures highest production, but as in livestock farming, crowding is an important issue. In insect farming, larvae (such as those of mealworms and the black soldier fly) are reared under crowded conditions, but often in higher densities than in a natural setting (Boppré and Vane-Wright, 2019). Mealworms produce metabolic heat and therefore the density (and with it the temperature) for optimal growth should not be too high or too low (Erens *et al.*, 2012). A dilemma raised by Boppré and Vane-Wright (2019) is that the plasticity of living conditions in commercial mass rearing is lower for insects than for common production animals, meaning that appropriate (a)biotic conditions are more crucial for the survival of the animals.

3. Freedom from pain, injury, or disease, by prevention or rapid diagnosis and treatment

What companies fear most is disease, as this may compromise their whole enterprise (Eilenberg *et al.*, 2017). To maintain a healthy stock, they will do their utmost by rearing insects under strict hygienic and optimal rearing conditions. The problem is that not much is known about pests and diseases when rearing insects. Veterinary science for insects is virtually absent, although there are now projects that are trying to mediate this (WUR, 2020).

As regards killing the insect, the main reason for choosing a particular killing method is often the quality of the product, e.g. for crickets the sensory qualities and physiochemical properties (Farina, 2017) and for the black soldier fly the lipid oxidation, colour and microbial load (Larouche *et al.*, 2019). However, these reasons do not necessarily coincide with the well-being of the insect. Hakman *et al.* (2013) recommended in a report to the Dutch government that in insect-rearing either for food or feed, killing methods should be quick and effective. The methods proposed were freezing (insects are cold blooded), heating (cooking or blanching), and shredding. Most companies do not reveal how they kill their insects, but some take into account their cold bloodedness and cool them before shredding. Putting them in the fridge is often considered more 'humane' (putting them to sleep) than freezing them (they die) and therefore cooling may precede freezing (Bear, 2019). In addition to these killing methods, Zhen *et al.*, (2020) used carbon dioxide and vacuum, and a method they called humane, i.e. carbon dioxide treatment followed by blanching.

It has been suggested that because of the large numbers of small insects on insect farms, farmers would be emotionally and ethically detached when killing (Bear, 2019). It is true that for small animals in large numbers it is more likely that the group rather than the individual is taken into consideration.

4. Freedom to express normal behaviour, by providing sufficient space, proper facilities, and appropriate company of the animal's own kind

When comparing how insects live in nature and how they are reared, Boppré and Vane-Wright (2019) argue that on insect farms 'natural stresses' are eliminated as optimal (a)biotic conditions are offered. But the question is whether those benign conditions can be considered natural. Normally, an insect production facility has a reproduction unit, delivering the eggs or larvae for the production unit. In the production unit the animals are not allowed to reproduce as they themselves are the commercial product.

There may also be a question about the use of strains or mutants. There are several mutants of the housefly (Hoyer, 1966), one mutant being the curly winged fly (Nickel and Wagoner, 1975) used as feed for captive pets such as frogs and lizards. Curly refers to the deformed wings preventing the animal from being able to fly. The mutant is also blind so that enemies can no longer be seen. Curly winged flies are very active and only make small jumps. They do not hide, and they bump into everything, making them easy prey. Companies that sell them say that they are 'farm bred in ethical conditions'. The question is whether the rearing of these mutants can be considered ethical.

The industry also produces mealworms that are artificially treated with juvenile hormones. This influences the moulting of the larvae, which induces them to grow beyond their normal length. They are called 'Mighty mealies' and are used as pet food. Does this constitute normal behaviour for the animal, and should it be classified as unethical?

The rearing conditions of locusts in cages severely limits their ability to fly when they would normally migrate over hundreds of kilometres. The aggregated way of rearing will cause them to occur only in the gregarious form, as the solitary form requires the rearing of the insects individually. The rearing of locusts severely limits their normal behaviour.

5. Freedom from fear and distress, by ensuring conditions and treatment that prevent mental suffering

There has been a discussion about how difficult it is to measure fear, mental suffering, pain, and stress in invertebrates. If the precautionary principle is used, then we act as if they would be able to experience pain. Insects show stress responses that include the release of stress hormones/neurohormones (such as adipokinetic hormone) to maintain optimal immunity (Adamo 2012, 2017). Stressed animals, for example those that have received too little food, may become weak; this will lower their immune status and make them susceptible to opportunistic pathogens causing disease (Joosten *et al.*, 2020). For example, stressed larvae of the super worm *Zophobas morio* are cannibalistic and rapidly spread the pathogen *Pseudo-*

monas aeruginosa (Maciel-Vergara *et al.*, 2018). A higher incidence of the *Acheta domesticus* densovirus in the house cricket was observed under stress vectors such as crowding, waste accumulation, low protein diets, temperatures above 35 °C, and high relative humidity (Szelei *et al.*, 2011).

8.4   *Guidelines for ethical behaviour*

In 2018 the Council of Animal Affairs of the Netherlands wrote a report entitled 'The emerging insect industry: invertebrates as production animals' (RDA, 2018). They recommended treating invertebrates as sentient beings, as future research may show that some species are indeed sentient. They considered investments in the welfare of invertebrates to be in the interests of the producer: 'To adapt farms to suit the needs and developmental stages of certain species as much as possible not only increases production, but is also important for the social acceptance of the insect industry.' The International Platform of Insects as Food and Feed, representing the interests of the insect production sector to EU policy-makers, states that all insect producers should adhere to high standards of animal welfare and ensure insect well-being (IPIFF, 2019).

Bear (2019) carried out a survey among insect farmers in the United Kingdom, mainly about killing insects. Farmers were concerned, not so much because they believe in the intrinsic value of insects, or that insects are 'sentient beings' or may experience pain, but more because of the way in which consumers may perceive this. However, all farmers were aware that they need to act in the face of scientific and ethical uncertainties. According to Bear (2019) further exploration of the understanding of 'good insect welfare' is necessary among farmers and their practices and actual or potential consumers.

Boppré and Vane-Wright (2019) propose creating an international Insect Welfare Charter – a framework that could be used to evaluate our current and future 'handling' of insects, based on species-appropriateness and respect towards all organisms while also considering environmental issues.

9   Conclusions

The farming of insects as feed and food has been gaining momentum for the last ten years, exemplified by the increase in start-up companies and the number of scientific publications. This is because insects are increasingly being considered as a high-quality, efficient, and sustainable alternative protein source. Insect production has a lower environmental impact than the production of livestock and insects can bio-transform poor quality organic waste streams, contributing to a circular economy. However, the rapid increase in this newly emerging agricultural sector prompts questions about insect welfare and the ethics involved.

However, invertebrates do not enjoy a lot of empathy and this has to do with their negative reputation as being a nuisance (e.g. houseflies) and/or being dangerous (e.g. malaria mosquitoes). Then there is the anthropocentric attitude of human beings who consider themselves the most significant entity in the universe, while invertebrates are at the lower end of evolution. Grimaldi and Engel (2005) have put this in perspective in their book about the evolution of insects: 'People gladly imagine a life without insects. But if ants and termites alone were removed from the earth, terrestrial life would probably collapse'. They go on to mention the hazardous consequences for planet earth of removing these insect groups. Moreover, invertebrate brains seem to comprise more than 99% of the brains that exist on earth, which is one reason why invertebrate cognition and sentience deserve attention (Mikhalevich and Powell, 2020).

Baracchi et al. (2017) feel that caution is needed when dealing with interpretations of emotions in invertebrates as they are often anthropocentrically biased. Appropriate emotion-related terminology may be required through careful experiments objectively analysing invertebrate behaviour. The word 'anthropodenial' has been coined, which is a blindness to the human-like characteristics of animals or the animal-like characteristics of ourselves (Jones, 2020). Sneddon et al. (2018) suggest that there may be motivational issues involved: 'sentience is at the heart of the decision about whether to provide animals with legislative protection'. For fish, sentience was denied at first, but there now seems to be enough evidence to suggest that they can experience positive and negative emotions. If we have underestimated the moral value of vertebrates, Schukraft (2020) believes that we may have to redirect resources to invertebrates.

In general, the precautionary principle should be used, whereby we assume that invertebrates can experience emotions. This has consequences for the insect industry, which should take the necessary precautions so that insects are well treated. According to Browning and Veit (2020), granting invertebrates moral status does not imply that we must make large sacrifices on their behalf, although it is open to question where a threshold of 'too demanding' should be set. This applies to how they are raised as well as the way in which they are killed.

The industry for insects as food and feed is in its infancy and the related welfare issues are increasingly being highlighted. This may require an international effort to propose guidelines or standards of care to the industry for the well-being and slaughter of farmed invertebrates involving both the biological and social sciences. As almost all research published to date on the welfare perception of invertebrates concludes that too little is currently known, an increased research effort on the cognitive and emotional capacity of invertebrate species is required.

Conflict of interest

The author declares no conflict of interest.

References

Adámková, A., Adámek, M., Mlček, J., Borkovcová, M., Bednářová, M., Kouřimská, L. and Josef Skácel, E.V., 2017. Welfare of the mealworm (*Tenebrio molitor*) breeding with regard to nutrition value and food safety. Potravinarstvo Slovak Journal of Food Sciences 11: 460-465. https://doi.org/10.5219/779

Adamo, S., 2012. The effects of the stress response on immune function in invertebrates: an evolutionary perspective on an ancient connection. Hormones and Behavior 62: 324-330. https://doi.org/10.1016/j.yhbeh.2012.02.012

Adamo, S.A., 2016. Do insects feel pain? A question at the intersection of animal behaviour, philosophy and robotics. Animal Behaviour 118: 75-79. https://doi.org/10.1016/j.anbehav.2016.05.005

Adamo, S.A., 2017. Stress responses sculpt the insect immune system, optimizing defense in an ever-changing world. Developmental & Comparative Immunology 66: 24-32. https://doi.org/10.1016/j.dci.2016.06.005

Alexander, P., Brown, C., Arneth, A., Dias, C., Finnigan, J., Moran, D. and Rounsevell, M.D.A., 2017. Could consumption of insects, cultured meat or imitation meat reduce global agricultural land use? Global Food Security 15: 22-32. https://doi.org/10.1016/j.gfs.2017.04.001

Andrews, P.L.R., 2011. Laboratory invertebrates: only spineless, or spineless and painless? Institute for Laboratory Animal Research (ILAR) Journal 52: 121-125. https://doi.org/10.1093/ilar.52.2.121

Baracchi, D., Lihoreau, M. and Giurfa, M., 2017. Do insects have emotions? Some insights from bumble bees. Frontiers in Behavioral Neuroscience 11: 157-157. https://doi.org/10.3389/fnbeh.2017.00157

Barnhill, A. and Doggett, T., 2018. Food ethics I: food production and food justice. Philosophy Compass 13: e12479. https://doi.org/10.1111/phc3.12479

Barron, A.B. and Klein, C., 2016. What insects can tell us about the origins of consciousness. Proceedings of the National Academy of Sciences 113: 4900-4908. https://doi.org/10.1073/pnas.1520084113

Bateson, P., 1991. Assessment of pain in animals. Animal Behaviour 42: 827-839. https://doi.org/10.1016/S0003-3472(05)80127-7

Bear, C., 2019. Approaching insect death: understandings and practices of the UK's edible insect farmers. Society & Animals 27: 751-768. https://doi.org/10.1163/15685306-00001871

Bier, E., 2005. *Drosophila*, the golden bug, emerges as a tool for human genetics. Nature Reviews Genetics 6: 9-23. https://doi.org/10.1038/nrg1503

Birch, J., 2017. Animal sentience and the precautionary principle. Animal Sentience 2: 17. Available at: https://tinyurl.com/y5dzyhd5.

Birch, J., 2020. The search for invertebrate consciousness. PhilSci-Archive 16931. http://phils-ci-archive.pitt.edu/16931/

Boppré, M. and Vane-Wright, R.I., 2019. Welfare dilemmas created by keeping insects in captivity. In: Carere, C. and Mather, J. (eds) The welfare of invertebrate animals. Springer International Publishing, Cham, Switzerland, pp. 23-67. https://doi.org/10.1007/978-3-030-13947-6_3

Bortolini, S., Macavei, L.I., Saadoun, J.H., Foca, G., Ulrici, A., Bernini, F., Malferrari, D., Setti, L., Ronga, D. and Maistrello, L., 2020. *Hermetia illucens* (L.) larvae as chicken manure management tool for circular economy. Journal of Cleaner Production 26: 121289. https://doi.org/10.1016/j.jclepro.2020.121289

Brambell, F.W.R., 1965. Report of the Technical Committee to enquire into the welfare of animals kept under intensive livestock husbandry systems. Command Papers 2836. Her Majesty's Stationery Office, London, UK. Available at: https://tinyurl.com/y64m6ldx.

Browning, H. and Veit, W., 2020. Improving invertebrate welfare. Commentary on Mikhalevich & Powell on invertebrate minds. Animal Sentience 5: 333. Available at: https://tinyurl.com/yxlls6ry.

Budelmann, B.U., 1995. The cephalopod nervous system: what evolution has made of the molluscan design. In: Breidbach, O. and Kutsch, W. (eds) The nervous systems of invertebrates: an evolutionary and comparative approach: with a Coda written by T.H. Bullock. Birkhäuser, Basel, Switzerland, pp. 115-138. https://doi.org/10.1007/978-3-0348-9219-3_7

Burrell, B.D., 2017. Comparative biology of pain: what invertebrates can tell us about how nociception works. Journal of Neurophysiology 117: 1461-1473. https://doi.org/10.1152/jn.00600.2016

Cappellozza, S., Leonardi, G.M., Savoldelli, S., Carminati, D., Rizzolo, A., Cortellino, G., Terova, G., Moretto, E., Badaile, A., Concheri, G., Saviane, A., Bruno, D., Bonelli, M., Caccia, S., Casartelli, M. and Tettamanti, G., 2019. A first attempt to produce proteins from insects by means of a circular economy. Animals 9: 5. https://doi.org/10.3390/ani9050278

Carew, T.J. and Sahley, C.L., 1986. Invertebrate learning and memory: from behavior to molecules. Annual Review of Neuroscience 9: 435-487. https://doi.org/10.1146/annurev.ne.09.030186.002251

Coolen, I., Dangles, O. and Casas, J., 2005. Social learning in noncolonial insects? Current Biology 15: 1931-1935. https://doi.org/10.1016/j.cub.2005.09.015

Costanza, R., d'Arge, R., De Groot, R., Farber, S., Grasso, M., Hannon, B., Limburg, K., Naeem, S., O'Neill, R.V., Paruelo, J., Raskin, R.G., Sutton, P. and Van den Belt, M., 1997. The value of the world's ecosystem services and natural capital. Nature 387: 253-260. https://doi.org/10.1038/387253a0

Drinkwater, E., Robinson, E.J.H. and Hart, A.G., 2019. Keeping invertebrate research ethical in a landscape of shifting public opinion. Methods in Ecology and Evolution 10: 1265-1273. https://doi.org/10.1111/2041-210X.13208

Eilenberg, J., Jensen, A.B. and Hajek, A.E., 2017. Prevention and management of diseases in terrestrial invertebrates In: Hajek, A.E. and Shapiro-Ilan, D.I. (eds) Ecology of invertebrate diseases. John Wiley & Sons, Ltd., Hoboken, NJ, USA, pp. 495-526. https://doi.org/10.1002/9781119256106.ch14

Eisemann, C.H., Jorgensen, W.K., Merritt, D.J., Rice, M.J., Cribb, B.W., Webb, P.D. and Zalucki, M.P., 1984. Do insects feel pain? – A biological view. Experientia 40: 164-167. https://doi. org/10.1007/bf01963580

Elwood, R.W., 2011. Pain and suffering in invertebrates? ILAR Journal 52: 175-184.

Erens, J., Van Es, S., Haverkort, F., Kapsomenou, E. and Luijben, A., 2012. A bug's life: large-scale insect rearing in relation to animal welfare. Project 1052 'Large-scale insect rearing in relation to animal welfare', Wageningen University & Research, Wageningen, The Netherlands.

Farina, M.F., 2017. How method of killing crickets impact the sensory qualities and physiochemical properties when prepared in a broth. International Journal of Gastronomy and Food Science 8: 19-23. http://dx.doi.org/10.1016/j.ijgfs.2017.02.002

Fischer, B., 2016. Bugging the strict vegan. Journal of Agricultural and Environmental Ethics 29: 255-263. https://doi.org/10.1007/s10806-015-9599-y

Fischer, B., 2019. How to reply to some ethical objections to entomophagy. Annals of the Entomological Society of America 112: 511-517. https://doi.org/10.1093/aesa/saz011

Food and Agriculture Organization of the United Nations (FAO), 2014. Mitigation of food wastage. Societal costs and benefits. FAO, Rome, Italy. Available at: http://www.fao. org/3/a-i3989e.pdf.

Giurfa, M., 2013. Cognition with few neurons: higher-order learning in insects. Trends in Neurosciences 36: 285-294. https://doi.org/10.1016/j.tins.2012.12.011

Gjerris, M., Gamborg, C. and Rocklinsberg, H., 2015. Entomophagy – why should it bug you? The ethics of insect production for food and feed In: Dumitras, D.E., Jitea, I.M. and Aerts, S. (eds) Know your food: food ethics and innovation. Wageningen Academic Publishers, Wageningen, The Netherlands, pp. 345-352. https://doi.org/10.3920/978-90-8686-813-1_52

Gossard, H.A., 1909. Relation of insects to human welfare. Journal of Economic Entomology 2: 313-332.

Grimaldi, D. and Engel, M.S., 2005. Evolution of the insects. Cambridge University Press, New York, NY, USA.

Guo, J. and Guo, A., 2005. Crossmodal interactions between olfactory and visual learning in *Drosophila*. Science 309: 307-310. https://doi.org/10.1126/science.1111280

Hakman, A., Peters, M. and Van Huis, 2013. Toelatingsprocedure voor insecten als mini-vee voor het plaatsen van nieuwe insectensoorten op de lijst voor productie te houden dieren. Wageningen University, Wageningen, The Netherlands.

Hallmann, C.A., Sorg, M., Jongejans, E., Siepel, H., Hofland, N., Schwan, H., Stenmans, W., Müller, A., Sumser, H., Hörren, T., Goulson, D. and De Kroon, H., 2017. More than 75 percent decline over 27 years in total flying insect biomass in protected areas. PLoS ONE 12: e0185809. https://doi.org/10.1371/journal.pone.0185809

Heckmann, L.H., Andersen, J.L., Eilenberg, J., Fynbo, J., Miklos, R., Jensen, A.N., Nørgaard, J.V. and Roos, N., 2019. A case report on inVALUABLE: insect value chain in a circular bioeconomy. Journal of Insects as Food and Feed 5: 9-13. https://doi.org/10.3920/JIFF2018.0009

Hedwig, B., 2014. Insect hearing and acoustic communication. Springer, Heidelberg, Germany.

Herculano-Houzel, S., 2009. The human brain in numbers: a linearly scaled-up primate brain. Frontiers in Human Neuroscience 3: 31. https://doi.org/10.3389/neuro.09.031.2009

Herculano-Houzel, S., Mota, B. and Lent, R., 2006. Cellular scaling rules for rodent brains. Proceedings of the National Academy of Sciences 103: 12138. https://doi.org/10.1073/pnas.0604911103

Horvath, K., Angeletti, D., Nascetti, G. and Carere, C., 2013. Invertebrate welfare: an overlooked issue. Annali dell'Istituto Superiore di Sanità 49: 9-17. https://doi.org/10.4415/ANN_13_01_04

Hoyer, R.F., 1966. Some new mutants of the house fly, *Musca domestica*, with notations of related phenomena. Journal of Economic Entomology 59: 133-137. https://doi.org/10.1093/jee/59.1.133

Hwang, R.Y., Zhong, L., Xu, Y., Johnson, T., Zhang, F., Deisseroth, K. and Tracey, W.D., 2007. Nociceptive neurons protect *Drosophila* larvae from parasitoid wasps. Current Biology 17: 2105-2116. https://doi.org/10.1016/j.cub.2007.11.029

Im, S.H. and Galko, M.J., 2012. Pokes, sunburn, and hot sauce: *Drosophila* as an emerging model for the biology of nociception. Developmental Dynamics 241: 16-26. https://doi.org/10.1002/dvdy.22737

Intergovernmental Science-Policy Platform on Biodiversity and Ecosystem Services (IPBES), 2019. Report of the Plenary of the Intergovernmental Science-Policy Platform on Biodiversity and Ecosystem Services on the work of its seventh session. IPBES, Bonn, Germany. Available at: https://ipbes.net/sites/default/files/ipbes_7_10_add.1_en_1.pdf.

International Platform of Insects for Food and Feed (IPIFF), 2019. Ensuring high standards of animal welfare in insect production. IPIFF, Brussels, Belgium. Available at: https://tinyurl.com/yd88udze

Jena, N.P., 2017. Animal welfare and animal rights: an examination of some ethical problems. Journal of Academic Ethics 15: 377-395. https://doi.org/10.1007/s10805-017-9282-1

Jones, R.C., 2020. Speciesism and human supremacy in animal neuroscience. In: Johnson, L., Fenton A. and Shriver A. (eds) Neuroethics and nonhuman animals. Advances in Neuroethics. Springer, Cham, Switzerland, pp. 99-115. https://doi.org/10.1007/978-3-030-31011-0_6

Joosten, L., Lecocq, A., Jensen, A.B., Haenen, O., Schmitt, E. and Eilenberg, J., 2020. Review of insect pathogen risks for the black soldier fly (*Hermetia illucens*) and guidelines for reliable production. Entomologia Experimentalis et Applicata 168: 432-447. https://doi.org/10.1111/eea.12916

Karam, C.S., Jones, S.K. and Javitch, J.A., 2019. Come fly with me: an overview of dopamine receptors in *Drosophila melanogaster*. Basic & Clinical Pharmacology & Toxicology 126: 56-65. https://doi.org/10.1111/bcpt.13277

Klein, C. and Barron, A.B., 2016. Insects have the capacity for subjective experience. Animal Sentience 1: 100.

Knutsson, S., 2016. Reducing suffering among invertebrates such as insects. Sentience Politics 1: 1-18. Available at: https://ea-foundation.org/files/reducing-suffering-invertebrates.pdf.

Knutsson, S. and Munthe, C., 2017. A virtue of precaution regarding the moral status of animals with uncertain sentience. Journal of Agricultural and Environmental Ethics 30: 213-224. https://doi.org/10.1007/s10806-017-9662-y

Larouche, J., Deschamps, M.-H., Saucier, L., Lebeuf, Y., Doyen, A. and Vandenberg, G.W., 2019. Effects of killing methods on lipid oxidation, colour and microbial load of black soldier fly (Hermetia illucens) larvae. Animals 9: 182. https://doi.org/10.3390/ani9040182

Le Neindre, P., Bernard, E., Boissy, A., Boivin, X., Calandreau, L., Delon, N., Deputte, B., Desmoulin-Canselier, S., Dunier, M., Faivre, N., Giurfa, M., Guichet, J.-L., Lansade, L., Larrère, R., Mormède, P., Prunet, P., Schaal, B., Servière, J. and Terlouw, C., 2017. Animal consciousness. EFSA Supporting Publications 14: 1196E. https://doi.org/10.2903/sp.efsa.2017.EN-1196

Leadbeater, E. and Chittka, L., 2007. Social learning in insects – from miniature brains to consensus building. Current Biology 17: 703-713. https://doi.org/10.1016/j.cub.2007.06.012

Lestel, D. and Taylor, H., 2013. Shared life: an introduction. Social Science Information 52: 183-186. https://doi.org/10.1177/0539018413477335

Lockwood, J.A., 1987. The moral standing of insects and the ethics of extinction. Florida Entomologist 70: 70-89. https://www.jstor.org/stable/3495093

Looy, H., Dunkel, F.V. and Wood, J.R., 2014. How then shall we eat? Insect-eating attitudes and sustainable foodways. Agriculture and Human Values 31: 131-141. https://doi.org/10.1007/s10460-013-9450-x

Lundy, M.E. and Parrella, M.P., 2015. Crickets are not a free lunch: protein capture from scalable organic side-streams via high-density populations of Acheta domesticus. PLoS ONE 10: e0118785. https://doi.org/10.1371/journal.pone.0118785

Maciel-Vergara, G., Jensen, A. and Eilenberg, J., 2018. Cannibalism as a possible entry route for opportunistic pathogenic bacteria to insect hosts, exemplified by Pseudomonas aeruginosa, a pathogen of the giant mealworm Zophobas morio. Insects 9: 88. https://doi.org/10.3390/insects9030088

Mason, G.J., 2011. Invertebrate welfare: where is the real evidence for conscious affective states? Trends in Ecology & Evolution 26: 212-213. https://doi.org/10.1016/j.tree.2011.02.009

Mather, J.A., 2011. Philosophical background of attitudes toward and treatment of invertebrates. Institute for Laboratory Animal Research (ILAR) Journal 52: 205-212.

Mather, J.A. and Carere, C., 2019. Consider the individual: personality and welfare in invertebrates. In: Carere, C. and Mather, J. (eds) The welfare of invertebrate animals. Springer International Publishing, Cham, Switzerland, pp. 229-245. https://doi.org/10.1007/978-3-030-13947-6_10

McClements, D.J., 2019. Towards a more ethical and sustainable edible future: one burger at a time. In: McClements, D.J. (ed.) Future foods: how modern science is transforming the way we eat. Springer International Publishing, Cham, Switzerland, pp. 323-361. https://doi.org/10.1007/978-3-030-12995-8_11

McEachern, M.G., 2018. Ethical food: transitioning towards sustainable meat consumption? Journal of Consumer Ethics 2: 26-32. Available at: https://tinyurl.com/yy7ad7jf.

Mendl, M., Paul, Elizabeth S. and Chittka, L., 2011. Animal behaviour: emotion in invertebrates? Current Biology 21: 463-465. https://doi.org/10.1016/j.cub.2011.05.028

Mendl, M.T. and Paul, E.S., 2016. Bee happy. Science 353: 1499-1500. https://doi.org/10.1126/science.aai9375

Menzel, R. and Giurfa, M., 2001. Cognitive architecture of a mini-brain: the honeybee. Trends in Cognitive Sciences 5: 62-71. https://doi.org/10.1016/S1364-6613(00)01601-6

Merker, B., 2016. Insects join the consciousness fray. Animal Sentience 1: 109. Available at: https://tinyurl.com/y5hm6grb.

Meyer-Rochow, V. and Kejonen, 2020. Could Western attitudes towards edible insects possibly be influenced by idioms containing unfavourable references to insects, spiders and other invertebrates? Foods 9: 172. https://doi.org/10.3390/foods9020172

Meyers, C.D., 2013. Why it is morally good to eat (certain kinds of) meat: the case for entomophagy. Southwest Philosophy Review 29: 119-126.

Mikhalevich, I. and Powell, R., 2020. Minds without spines: evolutionarily inclusive animal ethics. Animal Sentience 5: 329. Available at: https://tinyurl.com/y3ysge64.

Moltschaniwskyj, N.A., Hall, K., Lipinski, M.R., Marian, J.E.A.R., Nishiguchi, M., Sakai, M., Shulman, D.J., Sinclair, B., Sinn, D.L., Staudinger, M., Van Gelderen, R., Villanueva, R. and Warnke, K., 2007. Ethical and welfare considerations when using cephalopods as experimental animals. Reviews in Fish Biology and Fisheries 17: 455-476. https://doi.org/10.1007/s11160-007-9056-8

Monsó, S., 2018. Why insect sentience might not matter very much. In: Springer, S. and Grimm, H. (eds) Professionals in food chains. Wageningen Academic Publishers, Wageningen, The Netherlands, pp. 375-380. https://doi.org/10.3920/978-90-8686-869-8_59

Mora, C., D.P. Tittensor, S. Adl, A.G.B. Simpson, and B. Worm, 2011. How many species are there on earth and in the ocean? PLOS Biology 9: e1001127. https://doi.org/10.1371/journal.pbio.1001127

Nava, A.L., Higareda, T.E., Barreto, C., Rodríguez, R., Márquez, I. and Palacios, M.L., 2020. Circular economy approach for mealworm industrial production for human consumption. IOP Conference Series: Earth and Environmental Science 463: 012087. https://doi.org/10.1088/1755-1315/463/1/012087

Nickel, C.A. and Wagoner, D.E., 1975. Mutants on linkage groups 3 and 4 of the house fly. Annals of the Entomological Society of America 67: 775-776.

Nordgren, A., 2012. Ethical issues in mitigation of climate change: the option of reduced meat production and consumption. Journal of Agricultural and Environmental Ethics 25: 563-584. https://doi.org/10.1007/s10806-011-9335-1

Oonincx, D.G.A.B., Van Huis, A. and Van Loon, J.J.A., 2015. Nutrient utilisation by black soldier flies fed with chicken, pig, or cow manure. Journal of Insects as Food and Feed 1: 131-139. https://doi.org/10.3920/JIFF2014.0023

Pali-Schöll, I., Binder, R., Moens, Y., Polesny, F. and Monsó, S., 2018. Edible insects – defining knowledge gaps in biological and ethical considerations of entomophagy. Critical Reviews in Food Science and Nutrition 59: 2760-2761. https://doi.org/10.1080/10408398.2018.1468731

Paul, E.S. and Mendl, M.T., 2016. If insects have phenomenal consciousness, could they suffer? Animal Sentience 1: 128. Available at: https://tinyurl.com/y2aknsd8.

Perry, C.J., Baciadonna, L. and Chittka, L., 2016. Unexpected rewards induce dopamine-de-pendent positive emotion-like state changes in bumblebees. Science 353: 1529. https://doi.org/10.1126/science.aaf4454

Pollo, S. and Vitale, A., 2019. Invertebrates and humans: science, ethics, and policy. In: Car-ere, C. and Mather J. (eds.) The Welfare of Invertebrate Animals. Springer International Publishing, Cham, pp. 7-22. https://doi.org/10.1007/978-3-030-13947-6_2

Prat, Y., 2019. Animals have no language, and humans are animals too. Perspectives on Psy-chological Science 14: 885-893. https://doi.org/10.1177/1745691619858402

Raad voor Dieraangelegenheden (RDA), 2018. The emerging insect industry: invertebrates as production animals. Council on Animal Affairs in the Netherlands, The Hague, The Netherlands. Available at: https://english.rda.nl/publications/publications/2018/09/03/the-emerging-insect-industry

Roitberg, B., 2018. Chemical communication. In: Córdoba-Aguilar, A., González-Tokman, D. and González-Santoyo, I. (eds.) Insect behavior: from mechanisms to ecological and evo-lutionary consequences, pp. 145-157. Oxford University Press, Oxford.

Scherer, L., Tomasik, B., Rueda, O. and Pfister, S., 2017. Framework for integrating animal welfare into life cycle sustainability assessment. The International Journal of Life Cycle Assessment 23: 1476-1490. https://doi.org/10.1007/s11367-017-1420-x

Schukraft, J., 2020. Comparisons of capacity for welfare and moral status across species. Ef-fective Altruism Forum. Available at: https://forum.effectivealtruism.org/posts/EDCwb-DEhwRGZjqY6S/invertebrate-welfare-cause-profile.

Sherwin, C.M., 2001. Can invertebrates suffer? Or, how robust is argument-by-analogy? An-imal Welfare 10: 103-118.

Shih, J., Hodge, R. and Andrade-Navarro, M.A., 2015. Comparison of inter- and intraspecies variation in humans and fruit flies. Genomics Data 3: 49-54. https://doi.org/10.1016/j.gda-ta.2014.11.010

Smetana, S., 2020. Life cycle assessment of specific organic waste-based bioeconomy ap-proaches. Current Opinion in Green and Sustainable Chemistry 23: 50-54. https://doi.org/10.1016/j.cogsc.2020.02.009

Smetana, S., Mathys, A., Knoch, A. and Heinz, V., 2015. Meat alternatives: life cycle assess-ment of most known meat substitutes. The International Journal of Life Cycle Assess-ment 20: 1254-1267. https://doi.org/10.1007/s11367-015-0931-6

Smyth, D., 1978. Alternatives to Animal Experiments.Solar Press, London, UK.

Sneddon, L.U., Elwood, R.W., Adamo, S.A. and Leach, M.C., 2014. Defining and assessing an-imal pain. Animal Behaviour 97: 201-212. https://doi.org/10.1016/j.anbehav.2014.09.007

Sneddon, L.U., Lopez-Luna, J., Wolfenden, D.C.C., Leach, M.C., Valentim, A.M., Steenbergen, P.J., Bardine, N., Currie, A.D., D.M., B. and Brown, C., 2018. Fish sentience-denial: muddy-ing the waters. Animal Sentience 3: 21. Available at: https://tinyurl.com/y28rugto.

Stork, N.E., 2018. How many species of insects and other terrestrial arthropods are there on earth? Annual Review of Entomology 63: 31-45. https://doi.org/10.1146/annurev-en-to-020117-043348

Szelei, J., Woodring, J., Goettel, M.S., Duke, G., Jousset, F.X., Liu, K.Y., Zadori, Z., Li, Y.,

Styer, E., Boucias, D.G., Kleespies, R.G., Bergoin, M. and Tijssen, P., 2011. Susceptibility of North-American and European crickets to *Acheta domesticus* densovirus (AdDNV) and associated epizootics. Journal of Invertebrate Pathology 106: 394-399. https://doi. org/10.1016/j.jip.2010.12.009

Tapanen, T., 2018. Environmental potential of insects as food protein source. Master's Thesis, LUT University, School of Energy Systems, Sustainability Science and Solutions, Lappeenranta, Finland.

Tarsitano, M., 2006. Route selection by a jumping spider (*Portia labiata*) during the locomotory phase of a detour. Animal Behaviour 72: 1437-1442. https://doi.org/10.1016/j.anbehav.2006.05.007

Tomasik, B., 2019. Do smaller animals have faster subjective experiences? Available at: https://reducing-suffering.org/small-animals-clock-speed/.

Tracey, W.D., Jr., Wilson, R.I., Laurent, G. and Benzer, S., 2003. Painless, a *Drosophila* gene essential for nociception. Cell 113: 261-273. https://doi.org/10.1016/s0092-8674(03)00272-1

Van Huis, A., 2014. The global impact of insects. Farewell address upon retiring as Professor of Tropical Entomology at Wageningen University on 20 November 2014. Available at: https://edepot.wur.nl/410394.

Van Huis, A., 2019. Environmental sustainability of insects as human food. Elsevier Reference Collection in Food Science, pp. 1-5. https://doi.org/10.1016/B978-0-08-100596-5.22589-4

Van Huis, A., 2020. Insects as food and feed, a new emerging agricultural sector: a review. Journal of Insects as Food and Feed 6: 27-44. https://doi.org/10.3920/JIFF2019.0017

Van Huis, A. and Oonincx, D.G.A.B., 2017. The environmental sustainability of insects as food and feed. A review. Agronomy for Sustainable Development 37: 43. https://doi. org/10.1007/s13593-017-0452-8

Van Klink, R., Bowler, D.E., Gongalsky, K.B., Swengel, A.B., Gentile, A. and Chase, J.M., 2020. Meta-analysis reveals declines in terrestrial but increases in freshwater insect abundances. Science 368: 417. https://doi.org/10.1126/science.aax9931

Walters, E., Illich, P., Weeks, J. and Lewin, M., 2001. Defensive responses of larval *Manduca sexta* and their sensitization by noxious stimuli in the laboratory and field. Journal of Experimental Biology 204: 457-469.

Waltner-Toews, D. and Houle, K., 2017. Biophilia on the dinner plate: a conversation about ethics and entomophagy. Food Ethics 1: 157-171. https://doi.org/10.1007/s41055-017-0015-3

Wilson, E.O., 1987. The little things that run the world (the importance and conservation of invertebrates). Conservation Biology 1: 344-346.

Wilson-Sanders, S.E., 2011. Invertebrate models for biomedical research, testing, and education. ILAR Journal 52: 126-152. https://doi.org/10.1093/ilar.52.2.126

Wageningen University & Research (WUR), 2020. Insect doctors. WUR, Wageningen, The Netherlands. Available at: https://www.wur.nl/en/newsarticle/INSECT-DOCTORS-1. htm.

Wynants, E., Crauwels, S., Lievens, B., Luca, S., Claes, J., Borremans, A., Bruyninckx, L. and Van Campenhout, L., 2017. Effect of post-harvest starvation and rinsing on the microbial numbers and the bacterial community composition of mealworm larvae (*Tenebrio moli-*

tor). Innovative Food Science & Emerging Technologies 42: 8-15. https://doi.org/10.1016/j.ifset.2017.06.004

Yack, J.S., 2016. Vibrational signaling. In: Pollack, G., Mason, A., Popper, A. and Fay, R. (eds) Insect hearing. Springer handbook of auditory research, vol 55. Springer, Cham, Switzerland, pp. 99-123. https://doi.org/10.1007/978-3-319-28890-1_5

Zhen, Y., Chundang, P., Zhang, Y., Wang, M., Vongsangnak, W., Pruksakorn, C. and Kovitvadhi, A., 2020. Impacts of killing process on the nutrient content, product stability and *in vitro* digestibility of black soldier fly (*Hermetia illucens*) larvae meals. Applied Siences 10(17): 6099. https://doi.org/10.3390/app10176099

Substrate as insect feed for bio-mass production

L. Pinotti* and M. Ottoboni

Department of Health, Animal Science and Food Safety 'Carlo Cantoni',
University of Milan, via dell'Università 6, 26900 Lodi, Italy;
*luciano.pinotti@unimi.it

Abstract

Insects are able to convert organic material (i.e. waste and by products) into high-quality biomass, which can be processed into animal feed. Several studies have investigated the influence of growing substrates on the nutritional value of different insect species, particularly black soldier fly larvae and prepupae. This article reviews studies on how insects bioconvert different substrates, the effect of the substrate on the composition of insect meals, and on the development time (time needed to reach the harvesting state). All these studies indicate that insects convert low and high quality organic material (i.e. waste, by products, compound feeds) into high-quality insect biomass. The role and effects of selected nutrients, such as ether extract/fats, carbohydrates and fibre in the substrate, seem to be key factors in defining the features of the biomass as well as the time needed to reach the harvesting state.

Keywords

substrates – black soldier fly – circular economy – innovative feed ingredients

1   Introduction

One of the most important requirements for any animal-producing farm is the biomass that the farmer uses for feed. As insect farming is a relatively new and currently small but fast growing enterprise, there is little knowledge regarding the optimal nutritional content each farmed species needs for maximal productivity.

The concept of 'feed' in terms of insects is not very well defined. There are more than 900 different agricultural products that are used to produce/prepare animal feed (FAO, 2013), and usually the traditional feed classification includes the following categories: (1) roughage; (2) concentrate; (3) feed supplements; and (4) feed additives.

There are no such feed categories for farmed insects. The reason for this is related to the insect's ability to upgrade low nutrient substrates such as industrial co-by-products and vegetable waste (Dicke, 2018; Pinotti *et al.*, 2019). In fact, a wide range of organic materials can be used as a source of nutrients or as substrates for insect rearing. These depend on the availability, the legislative framework, applicability in the specific farming system, and the cost.

According to the definition of 'farmed animals' (EC, 2009b), insects bred for the production of processed animal proteins (PAPs) are considered as farmed animals, and are therefore subject to 'feed ban' rules (EC, 2001) as well as the rules of animal feeding (EC, 2009b). Thus, PAPs and insects are closely connected (EC, 2017; Ottoboni *et al.*, 2017), although in opposite ways.

This review covers the main different rearing substrates reported in the literature for farmed insects and provides a broad analysis of the different substrates used for all insects, with a specific focus on those used for black soldier fly (BSF) larvae.

2   Type of substrates

'Substrate' is the overall term applied for the materials used as insect feed. The use of substrates for the growth of insects is still an open issue. Currently there are two categories, those that are legally permitted and those that are not. The line between these two categories differs depending on the different authorities and countries.

In the EU, for instance, the use of several substrates such as ruminant proteins, catering waste, meat-and-bone meal and manure is prohibited (EC, 2009a), in line with regulations on transmissible spongiform encephalopathies and bovine spongiform encephalopathy. Furthermore, certain substrates, such as manure and intestinal contents, catering waste or former foodstuffs containing meat and fish, as well as human manure and sewage sludge, are also not allowed. Specifically, in its risk assessment (EFSA, 2015), the European Food Safety Authority (EFSA) considers the following seven categories (from A to G) of substrates:

A. Animal feed materials according to the EU catalogue of feed materials (Commission Regulation (EU) No 68/2013) and authorised as feed for food producing animals (EC, 2013).

B. Food produced for human consumption, but which is no longer intended for human consumption for reasons such as expired use-by date or due to problems of manufacturing or packaging defects. Meat and fish may be included in this category.

C. By-products from slaughterhouses (hides, hair, feathers, bones, etc.) that do not enter the food chain but originate from animals fit for human consumption.

D. Food waste from food for human consumption of both animal and non-animal origin from restaurants, catering, and households.

E. Animal manure and intestinal content.

F.   Other types of organic waste of vegetable nature such as gardening and forest material.

G.   Human manure and sewage sludge.

In Western countries the main substrates currently used for insect production thus include commercial animal feed, co-products from the primary production of non-animal origin food, and former foodstuffs (not containing meat and fish) for example production surplus, misshapen products, food with expired best-before-dates that have been produced in compliance with food legislation which have been already proposed as valid carbohydrates/energy sources in animal nutrition (Giromini *et al.*, 2017; Luciano *et al.*, 2020; Ottoboni *et al.*, 2019; Pinotti *et al.*, 2019, 2020, 2021; Tretola *et al.*, 2019). By contrast, in other countries (e.g. African countries), a wide range of bio-waste products are readily available and used as substrates for insect rearing, and these organic materials, for example animal manure, are typically composted or heat-treated before use (Münke-Svendsen *et al.*, 2017).

Other feed substrates commonly used by farmers include conventionally compounded meals made from seasonally available cereals and legumes such as maize, sorghum and soybeans, and indigenous vegetables such as collard greens, jute mallow, amaranth leaves, black nightshade, cowpea leaves, and spider plants.

Insect producers also use local substrates, such as forage leaves, as a substitute for vegetables during the dry season (Münke-Svendsen *et al.*, 2017). These include cassava, banana, sweet potato, tomato, pawpaw, and moringa leaves. These feed items are highly sought after by other farming activities as well, and are also used as food for humans (Münke-Svendsen *et al.*, 2017).

Insect producers select substrates based on a number of criteria, including the nutritional composition, expected effects on the insect species such as weight gain (also known as biomass production), feed efficiency (feed conversion ratio), finally the time needed to reach the harvesting stage, and steady year-round supply. In terms of the quality of the derived insect biomass, the key issues are protein concentration and amino acid profile, the fatty acid profile, micronutrients (such as minerals), the absence of hazards, or ease of removal during harvesting.

A further key issue in terms of efficiency and quality is the water content. A recent study (Dzepe *et al.*, 2020) in which five substrates moisture content levels (40, 50, 60, 70 and 80%) were tested, indicated that increasing the substrate moisture content reduces the larval feed reduction (mass reduction, %; Salomone *et al.*, 2017), wet weight, development time, and body size and body thickness of the larvae. This is in line with another study (Lalander *et al.*, 2020), which reported that adding too much water content reduced the biomass conversion ratio and survival rate of the larvae. In fact, substrates with a water content of between 80-90%, can only be used in combination with ventilation, while very wet substrates with a water content of over 90% were not considered suitable for BSF larvae compost even with active ventilation. These findings could have implications for the waste management sector interested in the fly larvae treatment of fruit and vegetable wastes.

This is because the BSF larvae composting of wet substrates is simpler if a dry separation of the larvae from residue is carried out after the completed treatment, thus rendering this option viable for a wider range of substrates (Lalander *et al.*, 2020). Efficient drying is one of the key elements in producing a profitable feed. Typically, dryers will use between 3,000 kJ/kg and 4,500 kJ/kg of evaporated water (Hamm, 2017). However, complete water removal is not required in insect rearing for substrate preparation, indeed certain species of insect, such as BSF, can be reared on substrates containing up to 80% moisture (Lalander *et al.*, 2020). This opens up new scenarios in insect substrate evaluation, since wastewater residues from the starch industry and cheese industry, for instance, can be considered as wet nutrient sources to rear insects, with benefits in terms of cost and performance. Accordingly, by adopting these strategies (incomplete or no drying for insect substrates), drying costs can be reduced or avoided.

3   Potential of insects in upgrading waste material and organic streams

Irrespectively of the legal basis, several substrates can be used in insect farming, namely: chicken feed, manure/faeces/sludge/abattoir waste, industrial by-products, fruit and vegetables, vegetable and cereal-based standard insect diets, restaurant and food waste, animal based products and by-products, straw, and algae. This is also in line with the 14 studies listed in Table S1, in which more than 50 different types of substrates have been tested on the same insect species (i.e. BSF larvae). Eight classes/categories were identified, namely chicken feed, manure digestate sludge, industrial by-products, fruit and vegetables, standard diet BSF, animal by-products, straw, and algae.

Figure 1 reports the mean, quartiles, minimum and maximum observations and outliers for normalised BSF biomass yield (g dry matter (DM) biomass yield from 100 young larvae) from the different substrate categories. From these records, even if case sensitive, it is clear that the best performing substrates are those characterised by 'balanced diets' in terms of nutrients composition and profile. This is the case of chicken feed and BSF standard diets proposed in different studies. Poultry diets, for instance, are composed primarily of a mixture of several feedstuffs such as cereal grains, soybean meal, fats, and vitamin and mineral premixes. These diets provide the essential energy and nutrients that are essential for fowl growth, reproduction, and health, namely proteins and amino acids, carbohydrates, fats, minerals, and vitamins. The energy necessary for maintaining the chicken's general metabolism and for producing meat and eggs is provided by the energy-yielding dietary components, primarily carbohydrates and fats, but also protein. Similarly, in insects, the best growing performance are obtained when a complete/balanced feeding media (i.e. chicken feed) is used. Indeed, such substrate in general are rich in energy and protein, while balanced for all nutrients (Barragán-Fonseca *et al.*, 2018). By contrast, straw and algae substrates did not guaranteeing adequate insect biomass yields, probably due the presence of different anti-nutritional factors (ANFs). Straw is a crop residue consisting of the dry stems and leaves left over after the harvest of cereals, and legumes (Heuzé, and Tran, 2015). In term of nutrient content, straw is considered a lignocellulosic biomass, indeed it contain cellulose (38%), hemicelluloses (25%), lignin (12%) (Gummert *et al.*, 2020). As feed for farm animals, all types of straw consist of coarse, high-fibre, low-protein and low-digestibility roughage (Gummert *et al.*, 2020). For insect however the situation is more controversial: the high quantity of rice straw in the substrate (daily feeding of 200 mg of ground rice straw per larvae) resulted in the highest prepupal dry weight, lowest (i.e. quick) developmental time, but lowest waste reduction efficiency (10.85%). The highest waste reduction efficiency (31.53%) was recorded by larvae fed at the lowest straw rate (12.5 mg of straw/larvae/day) (Manurung *et al.*, 2016). Another recent study (Liu *et al.*, 2018) reported that among the different types of fibre (acid detergent fibre; ADF, neutral detergent fibre; NDF, hemicellulose, cellulose, lignin) all of them, with the exception of hemicellulose, had negative effect on larval growth. Among ADF, NDF, cellulose, and lignin however, lignin has shown the strongest negative impact. This is in also explained by the fact that very often, in vegetable matrix, hemicellulose and cellulose are linked to lignin, which in turn protects them from digestion (Liu *et al.*, 2018). However, the ability of BSF larvae to bio convert fibre rich substrate may be partially explained by the presence, in the digestive tract of the insect, of intestinal bacteria able to degrade cellulose (Kim *et al.*, 2014b). Thus, based on these findings it seems that insects have some potential in digesting some fibre fraction (i.e. cellulose and hemicellulose), even though other factors might be important in defining larvae growth performance (e.g. larvae density; Barragán-Fonseca *et al.*, 2018).

The same low growth rate has been reported when algae are used as substrate: by increasing the inclusion of brown algae in the feeding medium, the BSF larvae had a reduced growth and feed intake. The reason for this is unknown although recent publications have reported the presence of several ANFs in algae in general, and *Ascophyllum nodosum* in particular. These marine materials contain alginate and fucoidan, which have several side effects. Alginate extract is used in reducing the appetite and energy intake in humans (Hall *et al.*, 2012), whereas Fucoidan is an efficient inhibitor of α-amylase and α-glucosidase (Kim *et al.*, 2014a). Both these ANFs, probably affect the feed efficiency of seaweed as a feeding media for insects, although their percentage in the diet is important in defining such negative effects. The reduced insect growth rate however is in contrast with the nutrient profile of seaweed which can also contain several micronutrients. Algae contain iodine, sterols, essential amino acids/fatty acids (e.g. eicosapentaenoic acid) and vitamin E (Liland *et al.*, 2017). Thus, insect larvae can act as carriers of such essential nutrients from sources not directly suitable for human or animal nutrition, in human or animal diets. This ability can be used to tailor the composition of the insect larvae to-

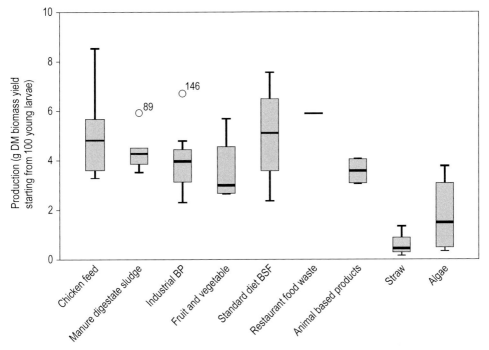

FIGURE 1   Mean, median, quartiles, minimum and maximum observations and outliers for larval biomass
             yield production (grams of larval biomass yield on the basis of 100 young larvae). DM = dry
             matter

wards the desired nutrient profiles to be used as livestock-feed ingredients (Liland
et al., 2017; Pinotti *et al.*, 2019).

Combining the findings of different studies (Supplementary Table S1) however,
the potential of insects to upgrade waste material and organic streams is clear. This
is confirmed when considering the three major nutrient groups including proteins,
lipids, and carbohydrates (P, L, and C) (Ortiz *et al.*, 2016). To calculate PLC ratios,
each major nutrient (P, L, and C) percentage is divided by the sum of the percentages of proteins, lipids, and digestible carbohydrates (total carbohydrate; fibre). The
sum of the three ratios (P, L, and C) should always be 1.

The contents of the major nutrient groups and PLC ratios of the food ingredients commonly used in diets for farmed insects are presented in Figure 2 and
3. The two figures show how insects are able to upgrade and convert unbalanced
carbohydrates based (70-90%) substrate into a more balanced biomass combining
carbohydrates, proteins and lipids. This is in line with data reported by Spranghers
et al. (2017), who reported that BSF larvae synthesise fatty acids, above all lauric
acid ($C_{12}:0$), using non-fibre carbohydrates contained in the substrate. This feature
seems to be specific for this insect species, in fact when comparing the fatty acid
profile of the prepupae with that of their respective substrates, it appears that the

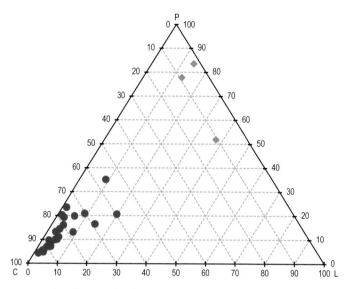

FIGURE 2   Ternary plot of protein, lipid, and carbohydrate (PLC) ratios of substrate used for rearing black soldier flies. Carbohydrates are determined by the following calculation on dry matter basis: carbohydrates = 100 − (% crude Ash) − (% crude fat) − (% crude protein). Substrate are classified as: vegetable substrate, circles (●); and animal based products, diamonds (♦). Data elaborated from Biancarosa *et al.* (2017); Bruno *et al.* (2019); Liland *et al.* (2017); Ma *et al.* (2018); Manurung *et al.* (2016); Meneguz *et al.* (2018); Nguyen *et al.* (2013, 2015); and Spranghers *et al.* (2017).

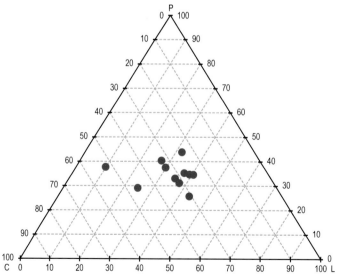

FIGURE 3   Ternary plot of protein, lipid, and carbohydrate (PLC) ratios of BSF prepupae biomass. Carbohydrates are determined by the following calculation on dry matter basis: carbohydrates = 100 − (% crude ash) − (% crude fat) − (% crude protein). Data elaborated from Biancarosa *et al.* (2017); Liland *et al.* (2017); Meneguz *et al.* (2018); and Spranghers *et al.* (2017).

substrate has a limited effect on the fatty acid profile of the prepupae. Interestingly, in the same study, lipids of the harvested prepupae were mainly composed of $C_{12:0}$, even when the substrate only contained this fatty acid in trace amounts.

4   Insect vs substrates

In order to facilitate a comparison, two representative studies were selected, namely Spranghers *et al.* (2017) and Meneguz *et al.* (2018), in which eight different substrates were tested (Table 1; Figure 4).

Combining substrate features with insect (BSFL) biomass led to various conclusions. The crude protein content, and the chitin corrected crude protein content in larval biomass both varied (mean 307-530 and 234-403 g/kg DM, respectively) among different types of substrate used for growing larvae. These figures are based on two assumptions: (1) all the nitrogen in the food is present as protein; (2) all food protein contains 160 g N/kg (McDonald *et al.*, 2012). Accordingly, the nitrogen-to-protein conversion factor, is set at 6.25; and second chitin is excluded when calculating the protein content, which implies the use of the nitrogen-to-protein conversion factor, of 4.76. In fact, the presence of non-protein nitrogen (NPN) in insects, i.e. chitin, may lead to an overestimation of the protein content. A nitrogen-to-protein conversion factor (Kp) specific for insect larvae (Kp = 4.76) has been proposed (Janssen *et al.*, 2017). This Kp value is calculated from the ratio of the sum of the amino acid residue weights to nitrogen content. This enables the 'real protein content' to be estimated, which is 24% lower than the protein content based on a Kp of 6.25. These values however, were obtained using a low protein substrate (mean 14% crude protein (CP) on DM basis).

However, considering a protein upgrade of the substrate for both values (crude protein and the chitin corrected crude protein contents), an increase of between 2.1 and 2.8 was recorded. In fact, in both studies, the larval biomass protein content was at least twice the protein content recorded in the substrate. These upgrade figures were 5.2 in the case of fats. Rearing substrates were generally characterised by a lower lipid input compared to larval biomass (86 and 287 g/kg DM, respectively). There was a small variability in the lipid content in the rearing substrates, whereas by contrast a large variability was observed in larval material (287 ± 151 g/kg DM). These figures, however, have to be considered with caution since are based on a limited dataset and can be different in other insect species (Dreassi *et al.,* 2017). Very recently Oonincx and Finke (2021) provided a very exhaustive overview of the nutrients content of insect and how selected nutrients can be manipulated starting from the rearing substrate.

The large variability in fat content in larvae can be explained by the ability of BSF larvae to synthesise fatty acids on the basis on different nutrients. In fact, insects can convert carbohydrates into lipids. In the presence of abundant non-fibre

TABLE 1   Selected studies on black soldier fly (BSF) considered in this review and the main inputs considered

Study	Substrates	W	CP	EE	ash	IDF/NDF*	NFC*	Energy*	Insect larvae (BSF)						
									Development time	Production DM	CP* (kp 4.76)	CP* (kp 6.25)	EE	Ash	Energy*
		g/kg	g/kg DM					kcal/kg DM	(day)		g/kg DM				kcal/kg DM
Meneguz et al., 2018	Brewery by-products	917	200	86	40	447	225	2,910	8	2.29	403	530	299	73	5,107
	Winery by-products	868	117	79	103	566	134	1,724	22.2	3.09	262	344	322	146	3,950
	Fruit waste	641	46	28	30	139	756	3,590	22	2.63	234	307	407	72	4,825
	Fruit and vegetable waste	768	120	26	91	178	585	3,028	20.2	3.29	319	419	263	130	3,994
Spranghers et al., 2017	Restaurant waste	738	157	139	45	41	618	4,451	19	5.87	328	431	386	27	5,548
	Fruit and vegetable waste	873	86	21	108	331	449	2,239	15.5	5.75	304	399	371	96	4,805
	Digestate	757	246	62	299	381	7	785	15	3.5	321	422	218	197	3,335
	Chicken feed	742	175	53	115	175	425	3,030	12.3	8.51	314	412	336	100	4,653
Mean		788	143	62	104	282	400	2,720	17	4	311	408	325	105	4,527

CP = crude protein; DM = dry matter; EE = ether extract; IDF/NDF = food fibre/neutral detergent fibre; NFC = non-fibre carbohydrates; W = water content.
* = estimated or calculated from Meneguz et al., (2018) and Spranghers et al. (2017).

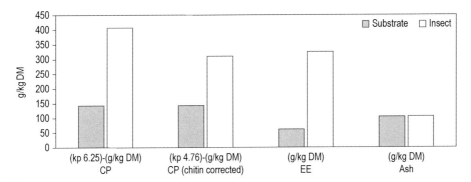

FIGURE 4   Nutrient content in four different rearing substrates and resulting black soldier fly larval
biomass (expressed on DM basis; data from Meneguz *et al.*, 2018; Spranghers *et al.*, 2017).
CP = crude protein; DM = dry matter; EE = ether extract

carbohydrates in the rearing substrate, BSF larvae mainly synthesise lauric acid
(over 50% of total fatty acids) (Arrese and Soulages 2010; Meneguz *et al.*, 2018; Oonincx *et al.*, 2015; Spranghers *et al.*, 2017), however in some cases this capacity was
also observed with high fibre substrates (Liland *et al.*, 2017; Meneguz *et al.*, 2018).
Incidentally, when fat quality (fatty acid profile) is considered, the situation is extremely variable (Pinotti *et al.*, 2019; Schiavone *et al.*, 2017). Moreover, insect fat content is also important in defining technological quality of the resulting insect meals
as reported elsewhere (Gasco *et al.*, 2020; Ottoboni *et al.*, 2018; Pinotti *et al.*, 2019).

The average ash content reported in these studies was 51±50 g/kg DM in rearing
substrates and 105±50 g/kg DM in larval biomass. It thus seems that BSF larvae tend
to accumulate minerals contained in the substrate. In fact, the average ash content
in larval biomass was twice that in the rearing substrate. Other studies, (Arango
Gutiérrez *et al.*, 2004; Newton *et al.*, 1977, 2005), have demonstrated that the ash
content of BSF larvae was extremely variable depending on the type of substrate
they were grown on. However, considering the mineral composition, Spranghers *et
al.* (2017) found no correlation between any of the elements in the larval biomass
and relative substrate. Although ash has no nutritional value, its content needs to
be monitored, since ash generally reduces the digestibility of feed and nutrients,
therefore limiting the biomass potential.

The interplay between mineral/ash/metal content in substrates and insect bioaccumulation is important for two reasons: (1) insects can accumulate minerals; (2)
bioaccumulation can occur for both desirable and undesirable minerals/metals. In
this respect some studies (Marone 2016; Purschke *et al.*, 2017) have already indicated that insects – both wild harvested and farm raised – are potentially vulnerable to
the accumulation of chemical contaminants, such as metals, ingested via contaminated feed or water.

Considering two (out of seven) of the main farmed insects authorised within the EU as fish feed ingredients, it seems that not only the mineral/metal substrate concentration is important in terms of their accumulation in the insect biomass but also the exposure time. In the case of *Tenebrio molitor*, Vijver *et al.* (2003) observed that Pb and Cd concentrations of mealworm larvae, reared on spiked soils linearly increase with exposure time. With regard to BSF, in larvae reared on experimentally spiked substrate (Cd, Pb and Zn), Diener *et al.* (2015) observed an initial accumulation of Cd in larvae followed by a decrease in adults ascribed to defecation prior to pupation. However no accumulation was observed in the case of Pb and Zn. In the same species, Purschke *et al.*, (2017), reported a significant bioaccumulation of Cd and Pb in BSF larvae, exposed to experimentally contaminated substrates. Of note the bio-accumulation factors were however different: higher than 9 for cadmium, and over 2 for lead. These results indicate that as with all livestock feed insect substrates need to be checked for contaminants in order to ensure feed and food safety standards throughout the value chain.

The fibre (expressed as a NDF, hemicellulose, cellulose, and lignin) substrate content can vary considerably. In fact, the findings of two recent studies, namely Spranghers *et al.* (2017) and Meneguz *et al.* (2018), are extremely revealing. They reveal that an insect biomass containing two to four times the substrate's energy content can be obtained from poor energy substrates such as wine/beer fibrous by-products. Specifically, when BSFs are used to bio-enhance a low energy substrate (i.e. digestate), the energy content of the rearing substrate is increased by over four times in the deriving larvae. By contrast, in the case of energy dense substrates (i.e. restaurant waste), the energy content is only increased by 25% in the larval biomass, which increases in any case (Figure 1).

These figures highlight that BSF larvae are able to efficiently bio-convert waste and by-products that are high in fibre content (38-55% NDF), accumulating an appreciable amount of lipids and proteins, without any detrimental effect on their growth performance. This is in confirmed in Meneguz *et al.* (2018), who reported a very good performance of BSF larvae reared on a high fibre substrate (NDF: 447 g/kg DM; ADF: 225 g/kg DM).

Figure 5 reports the energy content in rearing substrates and BSF larval biomass. Although the formulas used to calculate the energy content were designed for pigs, these figures give a further indication of the potential of insects to upgrade waste material and organic streams. Considering the energy content in substrates, values ranged between 700 kcal/kg DM in digestate and 4,500 kcal/kg DM in restaurant waste. On the other hand, the energy content varied from 3,000 kcal/kg DM and 5,500 kcal/kg DM in the larval biomass.

These figures highlight that larvae efficiently bio-convert waste and by-products with a high fibre content (38-55% NDF) without any detrimental effect on their growth performance. In addition, their ability to convert 'fibre in fat' enhances their potential and efficiency: it would seem that insects are able to return over twice the

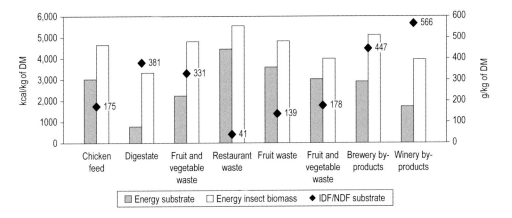

FIGURE 5   Energy content in substrates and insect biomass, insoluble dietary fibre (IDF) or neutral detergent fibre (NDF) content (♦) in the substrate (g/kg DM) in Spranghers *et al.* (2017) and Meneguz *et al.* (2018)

amount of energy than that introduced into the rearing system. The ability of BSF larvae to bio convert fibre rich substrate may be partially explained by the presence, in the digestive tract of the insect, of intestinal bacteria able to degrade cellulose (Kim *et al.*, 2014b).

5   Time needed to reach the harvesting stage

Although several environmental factors such as temperature, light, and humidity are known to affect growth and development of insect (Oonincx and Finke, 2021), a further factor that needs to be considered in defining the substrate is the 'time' needed to reach the harvesting stage. As reported in Figure 6, different substrates can accelerate or delay the development time needed to rear insects.

Balanced diets, as in the case of chicken feed, would seem able to accelerate the growing phase of BSF larvae, thus reducing the time needed to reach the harvesting state by up to 50%. Most other substrates are able to guarantee the harvesting within four weeks. The only exceptions are industrial by-products as in the case of the brewery by-product (BRE) substrate obtained during beer production (Meneguz *et al.*, 2018). The growth dynamics and waste reduction efficiency parameters of BSF larvae grown on BRE were found to be excellent. The BRE larvae in fact, showed a very good performance despite the high structural carbohydrate content of the relative rearing substrate (NDF: approximately 45% on a DM basis; ADF: 23%), reaching the harvesting phase in less than 2 weeks (Meneguz *et al.*, 2018). These results however clearly demonstrate that BSF larvae are able to efficiently bio convert waste and by-products characterised by a high fibre content, thanks

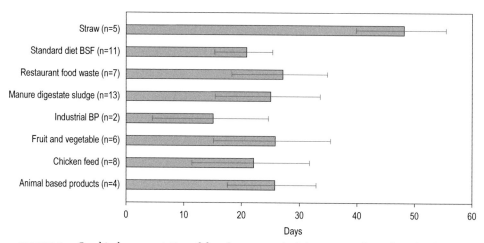

FIGURE 6   Graphical representation of days (mean, standard deviation, and number of replicates
'n') needed by black soldier fly (BSF) larvae for reaching the prepupal stage mean and
standard deviation according to relative rearing substrate class, namely, straw, standard
diet BSF, restaurant food waste, manure digestate sludge, industrial byproducts (BP),
fruit and vegetable, chicken feed, animal based products. Data elaborated from studies
reported in Table S1

to the presence, in their digestive tract of fibrolytic bacteria (Kim *et al.*, 2014b). By
contrast, straw slowed the growth of the larvae, which need almost 50 days before
they can be harvested. The role of lignin in these results is still unclear. The same
scenario was also confirmed when the insect biomass produced was plotted against
development time (Figure 7).

Another key nutrient in defining the time needed to reach the harvesting stage
seems to be the protein content in growing substrate. Oonincx *et al.* (2015) observed
that in the BSF the growing performances in term of feed conversion rate and efficiency of conversion of ingested foods were similar over dietary treatments, although they tended to use the low protein diets (13-14% CP) less efficiently than the
other diets. By contrast the same trial evidenced that BSF larvae developed faster
when a high protein and energy dense diet was administered (Oonincx *et al.*, 2015).
Similarly, Barragán-Fonseca *et al.* (2018) have observed that development time were
largely affected by selected nutrient concentration. Specifically, development time
increased with lower protein content. These two studies used both vegetable based
diet. However, the effect of substrate protein concentration on development time
was similar when animal sources were included (Nguyen *et al.*, 2013). Some of these
issue have been also addressed by Oonincx and Finke in a recent review (2020),
in which the authors suggested that variation in amino acid patterns between life
stages (i.e. age) of a species partially depends on whether that species undergoes
complete metamorphosis – as in the case of BSF-or incomplete metamorphosis.

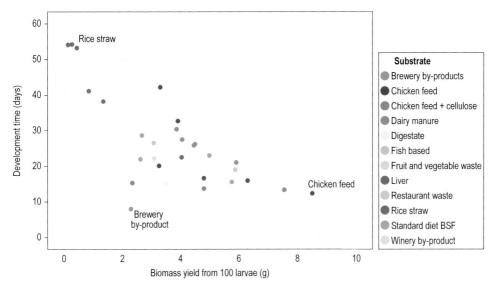

FIGURE 7   Biomass yield from 100 larvae (g) plotted against development time (days). Data elaborated from studies reported in Table S1

6   Conclusions

The aim of the present work was to review the main rearing substrates reported in
the literature for farmed insects. The focus was on offering a broad analysis of different substrates for all insects, and especially those used for BSF larvae.

   Although most of the data reviewed highlight the huge potential of insects to
upgrade 'low input' substrates with a high value biomass, several aspects still need
clarification. One of the main issues in insect livestock is the lack of standardisation, not only in practice but also during the research and development process.
For instance, in one day, BSF larvae can reduce 30 metric tons of food waste to ca.
10 metric tons (waste reduction 66%), while producing 930 kg of dry biomass (Chia
et al., 2019; Salomone *et al.*, 2017). Based on the figure efficiency can be very low,
even though in the balance it must be considered that insect are reared on organic
waste, which would otherwise end up in dumpsites, causing environmental pollution (Chia *et al.*, 2019). These figures clearly indicate that standard performance
indicators are needed to prevent any misunderstanding in the evaluation of the
micro-livestock system.

   An insect Substrate Conversion Rate needs to be devised, expressed in dry matter
in order to avoid confusion, along with specific normalisation units. For instance,
when the insect production/yield is reported, its normalisation is essential, for example, the grams of larval biomass yield (DM) on the basis of 100 young larvae.
These can be obtained not only through an extensive assessment of insect livestock
systems and conditions, but also by defining the common standard reference indicators that are generally recognised.

Considering the feeding rate, the optimal performance in terms of grams of larval biomass yield (DM) on the basis of 100 young larvae and time needed to reach the harvesting stage, is obtained when feeding 200 mg/day/larvae (Diener *et al.*, 2009). This quantity however has to be consider as indicative since as reported elsewhere (Oonincx and Finke, 2021) above the feeding rate other many possible factors can affect insect growth. Among those larvae density, for instance, has been reported as one of the main important (Barragán-Fonseca *et al.*, 2018).

In terms of nutritional value, this review shows how insects are able to upgrade and convert an unbalanced carbohydrates-based (70-90%) substrate, into a more balanced biomass that combines proteins, lipids, and carbohydrates (ratio protein, lipids, carbohydrates 34:33:33 ±5.2:9.6:8.9) depending on the diet.

However care needs to be taken regarding undesirable compounds in rearing substrate such as ANFs and contaminants. In the case of ANFs, marine materials may contain alginate and fucoidan, for which several side effects have been reported such as a reduced insect growth rate. In the case of contaminant like heavy metals, it seems that accumulation and excretion patterns show similarities at different levels such as order (i.e. also species) and life stages. Variation in dietary intake levels will to some extent affect insect concentrations (Oonincx and Finke, 2021). Furthermore, evidence indicates that that not only the mineral/metal substrate concentration is important for their accumulation in the insect biomass, but also the exposure time. For example for heavy metals, the bio-accumulation factor is higher than 9 for cadmium, and over 2 for lead. This thus indicates that insect substrates as with all livestock feed, require monitoring for contaminants in order to ensure feed and food safety standards throughout the value chain. A further point that has to be consider is 'what insect prefer'. In this direction a recent study (Morales-Ramos *et al.*, 2020) on self-selection of food ingredients and agricultural by-products by cricket, evidenced something new. Specifically, among several ingredients tested, some materials have shown a greatest selective/preference consumption by these species; rice bran (whole and defatted), corn dry distillers grain, buckwheat, and dry cabbage, were the most preferred by the crickets under experimental conditions. Accordingly, these five ingredients have been proposed as key ingredients for insect diet development in the future.

An additional aspect that has been not fully addressed in the literature whether the substrate should be changed during the various rearing phases. Even investigate in different scenario (the aim was to study how the substrate exposure can shape insect mycobiota; Varotto Bocazzi *et al.*, 2017), the findings showed that it is possible to 'design' the insect composition by combining the substrates and time of exposure. In summary, the substrate on which insects are maintained significantly contributes to shaping the nutrient composition of the resulting insect material, although the effect of time needs further investigation.

Acknowledgements

The present work has been done in the frame of the following project: Sustainable feed design applying circular economy principles: the case former food in pig nutrition (SusFEED-Rif. Pratica: 2018-0887); funded by the Fondazione Cariplo within the framework of the following call: Economia circolare: ricerca per un futuro sostenibile.

Conflict of interest

The authors declare no conflict of interest.

Supplementary material

Table S1. Studies on the suitability of different substrates for black soldier fly larvae.

Supplementary material can be found online at https://doi.org/10.3920/JIFF2020. 0110.

References

Arango Gutiérrez, G.P., Vergara Ruiz, R.A. and Mejía Vélez, H., 2004. Compositional, microbiological and protein digestibility analysis of the larva meal of *Hermetia illuscens* L.(Diptera: Stratiomyiidae) at Angelópolis-Antioquia, Colombia. Revista Facultad Nacional de Agronomía Medellín 57: 2491-2500.

Arrese, E.L. and Soulages, J.L., 2010. Insect fat body: energy, metabolism, and regulation. Annual Review of Entomology 55: 207-225. https://doi.org/10.1146/annurev-ento-112408-085356

Barragan-Fonseca, K.B., Dicke, M. and Van Loon, J.J., 2018. Influence of larval density and dietary nutrient concentration on performance, body protein, and fat contents of black soldier fly larvae (*Hermetia illucens*). Entomologia Experimentalis et Applicata 166: 761-770. https://doi.org/10.1111/eea.12716

Biancarosa, I., Liland, N.S., Biemans, D., Araujo, P., Bruckner, C.G., Waagbø, R., Torstensen, B.E., Locka, E.J. and Amlund, H., 2017. Uptake of heavy metals and arsenic in black soldier fly (*Hermetia illucens*) larvae grown on seaweed-enriched media. Journal of the Science of Food and Agriculture 98: 2176-2183. https://doi.org/10.1002/jsfa.8702

Bruno, D., Bonelli, M., De Filippis, F., Di Lelio, I., Tettamanti, G., Casartelli, M., Ercolini, D. and Caccia, S., 2019. The intestinal microbiota of Hermetia illucens larvae is affected by diet and shows a diverse composition in the different midgut regions. Applied and Environmental Microbiology 85: e01864-18. https://doi.org/10.1128/AEM.01864-18

Chia, S.Y., Tanga, C.M., Van Loon, J.J. and Dicke, M., 2019. Insects for sustainable animal feed: inclusive business models involving smallholder farmers. Current Opinion in Environmental Sustainability 41: 23-30. https://doi.org/10.1016/j.cosust.2019.09.003

Dicke, M. 2018. Insects as feed and the Sustainable Development Goals. Journal of Insects as Food and Feed 4: 147-156. https://doi.org/10.3920/JIFF2018.0003

Diener, S., Zurbrügg, C. and Tockner, K., 2015. Bioaccumulation of heavy metals in the black soldier fly, *Hermetia illucens* and effects on its life cycle. Journal of Insects as Food and Feed 1: 261-270. https://doi.org/10.3920/JIFF2015.0030

Diener, S., Zurbrügg, C. and Tockner, K., 2009. Conversion of organic material by black soldier fly larvae: establishing optimal feeding rates. Waste Management and Research 27: 603-610. https://doi.org/10.1177/0734242X09103838

Dreassi, E., Cito, A., Zanfini, A., Materozzi, L., Botta, M. and Francardi, V., 2017. Dietary fatty acids influence the growth and fatty acid composition of the yellow mealworm *Tenebrio molitor* (Coleoptera: Tenebrionidae). Lipids 52: 285-294. https://doi.org/10.1007/s11745-016-4220-3

Dzepe, D., Nana, P., Fotso, A., Tchuinkam, T. and Djouaka, R., 2020. Influence of larval density, substrate moisture content and feedstock ratio on life history traits of black soldier fly larvae. Journal of Insects as Food and Feed 6: 133-140. https://doi.org/10.3920/JIFF2019.0034

European Commission (EC), 2017. Commission Regulation (EU) No. 2017/893 of 24 May 2017 amending Annexes I and IV to Regulation (EC) No. 999/2001 of the European Parliament and of the Council and Annexes X, XIV and XV to Commission Regulation (EU) No. 142/2011 as regards the provisions on processed animal protein. Official Journal L 138/92: 1-25.

European Commission (EC), 2001. Commission Regulation (EU) No. 999/2001 of the European Parliament and of the Council of 22 May 2001 laying down rules for the prevention, control and eradication of certain transmissible spongiform encephalopathies. Official Journal of the European Union L 147: 1-40.

European Commission (EC), 2009a. Commission Regulation (EU) No. 767/2009 of the European Parliament and of the council of 13 July 2009 on the placing on the market and use of feed, amending European Parliament and Council Regulation (EC) No 1831/2003 and repealing Council Directive 79/373/EEC, Commission Directive 80/511/EEC. Official Journal of the European Union L 229(1): 1-28.

European Commission (EC), 2009b. Commission Regulation (EU) No. 1069/2009 of the European Parliament and of the Council of 21 October 2009 laying down health rules as regards animal by-products and derived products not intended for human consumption and repealing Regulation (EC) No. 1774/2002 (Animal by-products Regulation). Official Journal of the European Union L 300(1): 1-33.

European Commission (EC), 2013.Commission Regulation (EU) No. 68/2013 of the European Parliament and of the Council of 16 January 2013 on the Catalogue of feed materials. Official Journal of the European Union L29: 1-64.

European Food Safety Authority (EFSA) Scientific Committee, 2015. Risk profile related to production and consumption of insects as food and feed. EFSA Journal 13: 4257. https://doi.org/10.2903/j.efsa.2015.4257

Food and Agriculture Organisation (FAO), 2013. Edible insects: future prospects for food and feed security. FAO, Rome, Italy. Available at: http://www.fao.org/3/i3253e/i3253e.pdf

Gasco, L. Acuti, G., Bani, P., Dalle Zotte, A., Danieli, P.P., De Angelis, A., Fortina, R., Marino, R., Parisi, G., Piccolo, G., Pinotti, L., Prandini, A., Schiavone, A., Terova, G., Tulli, F. and Roncarati, A., 2020 Insect and fish by-products as sustainable alternatives to conventional animal proteins in animal nutrition. Italian Journal of Animal Science 19(1): 360-372. https://doi.org/10.1080/1828051X.2020.1743209

Giromini, C., Ottoboni, M., Tretola, M., Marchis, D., Gottardo, D., Caprarulo, V., Baldi, A. and Pinotti, L., 2017. Nutritional evaluation of former food products (ex-food) intended for pig nutrition. Food Additives & Contaminants: Part A 34: 1436-1445. https://doi.org/10.1080/19440049.2017.1306884

Gummert, M., Van Hung, N., Chivenge, P. and Douthwaite, B., 2020. Sustainable rice straw management. Springer International Publishing, Cham, Switzerland, 192 pp. https://doi.org/10.1007/978-3-030-32373-8

Hall, A.C., Fairclough, A.C., Mahadevan, K. and Paxman, J.R., 2012. *Ascophyllum nodosum* enriched bread reduces subsequent energy intake with no effect on post-prandial glucose and cholesterol in healthy, overweight males. A pilot study. Appetite 58: 379-386. https://doi.org/10.1016/j.appet.2011.11.002

Hamm, J., 2017. The economics of feed drying: processing. AFMA Matrix 26: 12-13.

Heuzé, V. and Tran, G., 2015. Rice straw. Feedipedia, a programme by INRA, CIRAD, AFZ and FAO. Available at: https://www.feedipedia.org/node/557.

Janssen, R.H., Vincken, J.P., Van den Broek, L.A., Fogliano, V. and Lakemond, C.M., 2017. Nitrogen-to-protein conversion factors for three edible insects: *Tenebrio molitor*, *Alphitobius diaperinus*, and *Hermetia illucens*. Journal of Agricultural and Food Chemistry 65: 2275-2278. https://doi.org/10.1021/acs.jafc.7b00471

Kim, E., Park, J., Lee, S. and Kim, Y., 2014b. Identification and physiological characters of intestinal bacteria of the black soldier fly, *Hermetia illucens*. Korean Journal of Applied Entomology 53: 15-26.

Kim, K.T., Rioux, L.E. and Turgeon, S.L., 2014a. Alpha-amylase and alpha-glucosidase inhibition is differentially modulated by fucoidan obtained from *Fucus vesiculosus* and *Ascophyllum nodosum*. Phytochemistry 98: 27-33. https://doi.org/10.1016/j.foodchem.2013.04.123

Lalander, C., Ermolaev, E., Wiklicky, V. and Vinnerås, B., 2020. Process efficiency and ventilation requirement in black soldier fly larvae composting of substrates with high water content. Science of The Total Environment 729: 138968. https://doi.org/10.1016/j.scitotenv.2020.138968

Liland, N.S., Biancarosa, I., Araujo, P., Biemans, D., Bruckner, C.G., Waagbø, R., Torstensen, B.E. and Lock, E.J., 2017. Modulation of nutrient composition of black soldier fly (*Hermetia illucens*) larvae by feeding seaweed-enriched media. PLoS ONE 12: e0183188. https://doi.org/10.1371/journal.pone.0183188

Liu, Z., Minor, M., Morel, P.C. and Najar-Rodriguez, A.J., 2018. Bioconversion of three organic wastes by black soldier fly (Diptera: Stratiomyidae) larvae. Environmental Entomology 47: 1609-1617. https://doi.org/10.1093/ee/nvy141

Luciano, A., Tretola, M., Ottoboni, M., Baldi, A., Cattaneo, D. and Pinotti, L., 2020. Potentials and challenges of former food products (food leftover) as alternative feed ingredients. Animals 10: 125. https://doi.org/10.3390/ani10010125

Manurung, R., Supriatna, A., Esyanthi, R.R. and Putra, R.E., 2016. Bioconversion of rice straw waste by black soldier fly larvae (*Hermetia illucens* L.): optimal feed rate for biomass production. Journal of Entomology and Zoology Studies 4: 1036-1041.

Marone, P.A., 2016. Food safety and regulatory concerns. In: Dossey, A.T., Morales-Ramos, J.A. and Guadalupe Rojas, M. (eds.) Insects as sustainable food ingredients. Academic Press, Cambridge, MA, USA, pp. 203-221.

McDonald, P., Edwards, R.A., Greenhalgh, J.F.D., Morgan, C.A., Sinclair, L.A. and Wilkinson, R.G., 2012. Animal nutrition, 7th edition. Pearson Education Ltd., Essex, UK, 762 pp.

Meneguz, M., Schiavone, A., Gai, F., Dama, A., Lussiana, C., Renna, M. and Gasco, L., 2018. Effect of rearing substrate on growth performance, waste reduction efficiency and chemical composition of black soldier fly (*Hermetia illucens*) larvae. Journal of the Science of Food and Agriculture 98: 5776-5784. https://doi.org/10.1002/jsfa.9127

Morales-Ramos, J.A., Rojas, M.G., Dossey, A.T., Berhow, M., 2020. Self-selection of food ingredients and agricultural by-products by the house cricket, *Acheta domesticus* (Orthoptera: Gryllidae): a holistic approach to develop optimized diets. PLoS ONE 15(1): e0227400. https://doi.org/10.1371/journal.pone.0227400

Münke-Svendsen, C., Halloran, A., Oloo, J., Henlay, J.O., Nyakeri, M., Manyara, E. and Roos, N., 2017. Technical Brief #2: insect production systems for food and feed in Kenya. GREEiNSECT. https://doi.org/10.13140/RG.2.2.24053.99040

Newton, G.L., Booram, C.V., Barker, R.W. and Hale, O.M., 1977. Dried *Hermetia illucens* larvae meal as a supplement for swine. Journal of Animal Science 44: 395-400. https://doi.org/10.2527/jas1977.443395x

Newton, L., Sheppard, C., Watson, D.W., Burtle, G. and Dove, R. 2005. Using the black soldier fly, *Hermetia illucens*, as a value-added tool for the management of swine manure. Animal and Poultry Waste Management Center, North Carolina State University, Raleigh, NC, USA.

Nguyen, T.T., Tomberlin, J.K. and Vanlaerhoven, S., 2013. Influence of resources on *Hermetia illucens* (Diptera: Stratiomyidae) larval development. Journal of Medical Entomology 50: 898-906. https://doi.org/10.1603/me12260

Oonincx, D.G.A.B. and Finke, M.D., 2021. Nutritional value of insects and ways to manipulate their composition. Journal of Insects as Food and Feed 7: 639-659. https://doi.org/10.3920/JIFF2020.0050

Oonincx, D.G.A.B., Van Broekhoven, S., Van Huis, A. and Van Loon, J.J., 2015. Feed conversion, survival and development, and composition of four insect species on diets composed of food by-products. PLoS ONE 10: e0144601. https://doi.org/10.1371/journal.pone.0144601

Ortiz, J.C., Ruiz, A.T., Morales-Ramos, J.A., Thomas, M., Rojas, M.G., Tomberlin, J.K., Yi, L.,

Han, R., Giroud, L. and Jullien, R.L., 2016. Insect mass production technologies. In: Dossey, A.T., Morales-Ramos, J.A. and Guadalupe Rojas, M. (eds.). Insects as sustainable food ingredients: production, processing and food applications. Academic Press, London, UK, pp. 153-201.

Ottoboni, M., Spranghers, T., Pinotti, L., Baldi, A., De Jaeghere, W. and Eeckhout, M., 2018. Inclusion of *Hermetia illucens* larvae or prepupae in an experimental extruded feed: process optimisation and impact on *in vitro* digestibility. Italian Journal of Animal Science 17: 418-427. https://doi.org/10.1080/1828051X.2017.1372698

Ottoboni, M., Tretola, M., Cheli, F., Marchis, D., Veys, P., Baeten, V. and Pinotti, L., 2017. Light microscopy with differential staining techniques for the characterisation and discrimination of insects versus marine arthropods processed animal proteins. Food Additives and Contaminants – Part A Chemistry, Analysis, Control, Exposure and Risk Assessment 34(8): 1377-1383. https://doi.org/10.1080/19440049.2016.1278464

Ottoboni, M., Tretola, M., Luciano, A., Giuberti, G., Gallo, A. and Pinotti, L., 2019. Carbohydrate digestion and predicted glycemic index of bakery/confectionary ex-food intended for pig nutrition. Italian Journal of Animal Science 18: 838-849. https://doi.org/10.1080/1828051X.2019.1596758

Pinotti, L., Giromini, C., Ottoboni, M., Tretola, M. and Marchis, D., 2019. Insects and former foodstuffs for upgrading food waste biomasses/streams to feed ingredients for farm animals. Animal 13: 1365-1375. https://doi.org/10.1017/S1751731118003622

Pinotti, L., Luciano, A., Ottoboni, M., Manoni, M., Ferrari, L., Marchis, D. and Tretola, M., 2021. Recycling food leftovers in feed as opportunity to increase the sustainability of livestock production. Journal of Cleaner Production 294: 126290. https://doi.org/10.1016/j.jclepro.2021.126290

Pinotti, L., Manoni, M., Fumagalli, F., Rovere, N., Luciano, A., Ottoboni, M., Ferrari, L., Cheli, F. and Djuragic, O., 2020. Reduce, reuse, recycle for food waste: a second life for fresh-cut leafy salad crops in animal diets. Animals 10(6): 1082. https://doi.org/10.3390/ani10061082

Purschke, B., Scheibelberger, R., Axmann, S., Adler, A. and Jäger, H., 2017. Impact of substrate contamination with mycotoxins, heavy metals and pesticides on the growth performance and composition of black soldier fly larvae (*Hermetia illucens*) for use in the feed and food value chain. Food Additives & Contaminants: Part A 34: 1410-1420. https://doi.org/10.1080/19440049.2017.1299946.

Salomone, R., Saija, G., Mondello, G., Giannetto, A., Fasulo, S., Savastano, D., 2017. Environmental impact of food waste bioconversion by insects: application of life cycle assessment to process using *Hermetia illucens*. Journal of Cleaner Production 140: 890-905. https://doi.org/10.1016/j.jclepro.2016.06.154

Schiavone, A., Cullere, M., De Marco, M., Meneguz, M., Biasato, I., Bergagna, S., Dezzutto, D., Gai, F., Dabbou, S., Gasco, L. and Dalle Zotte, A., 2017. Partial or total replacement of soybean oil by black soldier fly larvae (*Hermetia illucens* L.) fat in broiler diets: effect on growth performances, feed-choice, blood traits, carcass characteristics and meat quality. Italian Journal of Animal Science 16(1): 93-100. https://doi.org/10.1080/182805 1X.2016.1249968

Spranghers, T., Ottoboni, M., Klootwijk, C., Ovyn, A., Deboosere, S., De Meulenaer, B., Michiels, J., Eeckhout, M., De Clercq P. and De Smet, S., 2017. Nutritional composition of black soldier fly (*Hermetia illucens*) prepupae reared on different organic waste substrates. Journal of the Science of Food and Agriculture 97: 2594-2600. https://doi.org/10.1002/jsfa.8081

Tretola, M., Ottoboni, M., Luciano, A., Rossi, L., Baldi, A. and Pinotti, L., 2019. Former food products have no detrimental effects on diet digestibility, growth performance and selected plasma variables in post-weaning piglets. Italian Journal of Animal Science 18: 987-996. https://doi.org/10.1080/1828051X.2019.1607784

Varotto Boccazzi, I., Ottoboni, M., Martin, E., Comandatore, F., Vallone, L., Spranghers, T., Eeckhout, M., Mereghetti, V., Pinotti, L. and Epis, S., 2017. A survey of the mycobiota associated with larvae of the black soldier fly (*Hermetia illucens*) reared for feed production. PLoS ONE 12(8): e0182533. https://doi.org/10.1371/journal.pone.0182533

Veldkamp, T. and Eilenberg, J., 2018. Insects in European feed and food chains. Journal of Insects as Food and Feed 4: 143-145. https://doi.org/10.3920/JIFF2018.x006

Vijver, M., Jager, T., Posthuma, L. and Peijnenburg, W., 2003. Metal uptake from soils and soil-sediment mixtures by larvae of *Tenebrio molitor* (L.)(Coleoptera). Ecotoxicology and Environmental Safety 54: 277-289. https://doi.org/10.1016/S0147-6513(02)00027-1

Organic side streams: using microbes to make substrates more fit for mass producing insects for use as feed

J.B. Zhang¹*, Y.Q. Yu¹, J.K. Tomberlin², M.M. Cai¹, L.Y. Zheng¹ and Z.N. Yu¹

¹State Key Laboratory of Agricultural Microbiology, College of Life Science and Technology, Huazhong Agricultural University, Wuhan, 430070, China P.R.; ²Department of Entomology, Texas A&M University, 2475 TAMU, College Station, TX 77843, USA; *zhangjb@mail.hzau.edu.cn

Abstract

Microbes, combined with insects, convert organic waste into products of value. Resulting insects can be harvested and used as a high-quality protein resource, while the residues can be used as fertiliser. Microbes play an important role in the conversion process. This review's aim was focused on how microbes promote insects such as black soldier fly (*Hermetia illucens* L.), house fly (*Musca domestica* L.), waxworm (*Plodia interpunctella*) and yellow meal worm (*Tenebrio molitor* L.), to convert organic waste, while also harmlessly reducing organic waste pollution. The novelty is reflected in some core gut microbiota and their secreted enzymes degrade macromolecules such as protein, fat, polysaccharide, cellulose, polystyrene and polyethylene. Gut microbiota also could help insects degrade hazardous substances such as antibiotics, mycotoxin, odorous substances, and inhibit pathogens in organic wastes to make substrates more fit for insects.

Keywords

insects – gut microbiota – organic waste – conversion – degradation

1   Introduction

Globally, approximately 1.6 billion tons of food waste are generated annually (Ma *et al.*, 2020). However, due to challenges with collecting these materials, only a portion of the food waste is being treated. Such practices (i.e. not recycling such materials) pose environmental and economic concerns (Salihoglu *et al.*, 2018), as well as human health hazards (Sindhu *et al.*, 2019). Other common organic wastes are livestock and poultry manure. According to the forecast of the demographic website (https://populationstat.com/), the global human population will reach 9.7 billion by 2050. Such growth will inevitably increase the demand for food, especially live-stock and poultry products, which will increase the amount of livestock and poultry manure. Livestock and poultry manures are rich in nitrogen, phosphorus, and po-tassium (Dróżdż *et al.*, 2020). Unfortunately, large amounts of livestock and poultry manures are not treated or utilised properly potentially leading to eutrophication and soil degradation (Qian *et al.*, 2018).

Plastics in association with food production are also being mass produced and resulting in additional environmental concerns. Data from the Ministry of Agricul-ture and Rural Affairs of the People's Republic of China show that the use of agri-cultural plastic films nationwide reached 2.593 million tons in 2016. During the past 30 years, residual pollution of plastic film has become a serious issue worldwide. Plastic film remaining in the soil is difficult to degrade, and it will also lead to the reduction of crop production (Yan *et al.*, 2016).

Currently, pollution issues resulting from the previously described organic waste concerns demonstrate the need for sustainable methods of remediation. Using in-sects to convert organic wastes into products of value can be achieved. Further-more, integrating microbes as part of the process could enhance efficiency of such systems (Xiao *et al.*, 2018). The gut microbiome of insects and microbiota in organic waste are now being harnessed as a means to increase the ability of mass produced insects to recycle these materials.

2 Microbe enhanced conversion efficiency and insect growth

A number of insect species are used to digest the wastes previously described. The primary species are black soldier fly (BSF; *Hermetia illucens* L.), house fly (HF; *Musca domestica* L.) and yellow mealworm (YMW; *Tenebrio molitor* L.) (Čičková *et al.*, 2015).

Black soldier fly larvae (BSFL) convert different organic wastes, and while the structure of its gut microbes shift depending on diet, Firmicutes and Proteobacte-ria have been identified as the core phyla (Wynants *et al.*, 2019; Zhan *et al.*, 2020). These phyla may play an important role in promoting the ability of the BSFL to convert organic waste into biomass. For example, BSFL gut microbes produce carbohydrate-active enzymes that significantly enhance this process (Jiang *et al.*, 2019). Studies have confirmed that Firmicutes can secrete a variety of proteases and pectinases, allow the BSFL to digest animal faeces as well as indigestible carbohy-drates in straw-related compost (Sun *et al.*, 2015; Zhang *et al.*, 2018).

Gastrointestinal morphology of the BSFL plays an important role in shaping microbial load and diversity (Bruno *et al.*, 2019). Microbial diversity is greatest in the midgut and then gradually decreases towards the posterior regions. However, bacterial load was greatest in the posterior region. Xylanases produced by Bacteroidetes, one of the dominant phyla in BSFL gut directly involved in hemicellulose digestion (Bruno *et al.*, 2019).

Bacillus and associated yeasts are in high abundance in the BSFL intestine (De Smet *et al.*, 2018; Zhan *et al.*, 2020). *Bacillus* can degrade lignocellulose, while yeasts utilise metabolites and accumulate protein and polysaccharide (Li *et al.*, 2016). Rehman developed an efficient co-digestion strategy for using BSFL to treat dairy manure (DM) when mixed with an appropriate proportion of chicken manure (CHM) or soybean curd to reaching nutrition (C/N ratio) balance and buffering capacity (pH) of the animal wastes (ur Rehman *et al.*, 2017a,b). Furthermore, adding exogenous probiotics resulted in higher bioconversion efficacy (10.8%), survival rate (99.1%), food conversion ratio (4.5), manure reduction rate (48.7%), and fibre content reduction (cellulose; 72.9%, hemicellulose; 68.5%, lignin; 32.8%), than relying on associated gut microbiota alone. Analysis of the fibre structural and chemical changes with scanning electron microscopy and Fourier-transformed infrared spectroscopy indicates the structure and chemical composition of the fibre had been modified. The surface morphology of fibres of DM40:CHM60 when pre-treated with exogenous probiotics was relatively compact and rigid before BSFL digestion, but after digestion the material was porous and more corrugated. The degraded effects in the chemical composition of the fibre were more obvious in the treatment when BSFL were assisted with microbes (ur Rehman *et al.*, 2019). Clearly, microbes play an important role in the conversion of organic waste, by adding microbes to organic waste, the ability of insect hosts to degrade organic waste can be improved (Table 1). The HF also could convert animal manure (Van Huis *et al.*, 2013). Conversion of HF also defined as a microbial-driven process, which was dominated by *Entomoplasma somnilux*, *Proteobacterium*, and *Clostridiaceae* bacterium (Zhang *et al.*, 2012).

Waxworm (WW) (Yang *et al.*, 2014) and YMW (Zhang *et al.*, 2017) can degrade plastic wastes. Two bacterial strains capable of degrading polyethylene (PE) were isolated from the gut of WM, *Enterobacter asburiae* YT1 and *Bacillus* sp. YP1. Over a 60-day incubation period of the two strains with PE film degraded PE with approximately 6.1±0.3 and 10.7±0.2% (Yang *et al.*, 2014). When degrading polystyrene (PS) and PE, it was found that *Citrobacter* sp. and *Kosakonia* sp. in the gut of YMW are closely related to the degradation of PE and PS (Brandon *et al.*, 2018). The dominant strains in the YMW gut are mainly *Enterobacter*, *Lactococcus* and *Enterococcus*. The strains with biodegradable activity *Klebsiella*, *Pseudomonas* and *Serratia* were isolated from YMW digesting plastics (Urbanek *et al.*, 2020). The potential to enrich these strains *in vitro*, and then combined with the insect could enhance plastic degradation. Furthermore, by changing the diet of YMW, the microbial community structure in its gut could be shifted, thereby improving the efficiency of PS degradation (Peng *et al.*, 2019). Super worm (*Zophobas atratus*) can also degrade PS. Inoculating diet with antibiotics was found to inhibit associated gut microbes and the ability of the insect to degrade PS demonstrating the gut microbes play a role

TABLE 1   Black soldier fly larvae (BSFL) combined with microbes to convert organic waste

Substrate	Microbe source	Organic waste reduction rate (%) compared with the control group	Larval weight gain rate (%) compared with the control group	References
Chicken manure	*Bacillus subtilis* BSF-CL	13.4	15.9	Yu *et al.* (2011)
Chicken manure	*B. subtilis* BSF-CL (from gut bacteria)	12.7	15.9	Xiao *et al.* (2018)
Dairy and chicken manure mixed	Exogenous bacteria (from soil and pig manure fermentation products)	0.6-16.2	7.99-16.7	ur Rehman *et al.* (2019)
Soybean curd residue	*Lactobacillus buchneri* L3-9 (from Microbial Pesticide Key State Laboratory of HZAU)	13.6-26.9	15.8-23.5	Somroo *et al.* (2019)
Chicken manure	Companion bacteria (from the egg surface of BSF within 6 hours after spawning, and from the gut of 5th instar larvae)	0.9-7.2	5.79-28.57	Mazza *et al.* (2020)
Banana peel	*Trichoderma reesei* and *Rhizopus oligosporus* (from the Department of Molecular Sciences, SLU)	25.7-33.9	26.1-58.9	Isibika *et al.* (2019)
	A mixture of bacteria (from the BSFL gut)	4.3	6.7	

in such processes (Yang *et al.*, 2020). We speculated some gut microbes in WW and YMW could produce special emzymes to degrade the plastics such as PE and PS. The mechanism should be further explored.

Bacteria from other insects are known to play a role in host metabolism. *Acetobacter pomorum*, which was isolated from *Drosophila* (Diptera: Drosophilidae), has obvious growth promoting effect (Kim *et al.*, 2020). *A. pomorum* WJL (ApWJL) and *Lactobacillus plantarum* NC8 (LpNC8) can promote the growth of *Drosophila* larvae by shifting nutritional requirements of the host (Consuegra *et al.*, 2020). *A. pomorum* regulates insulin/ insulin-like growth factor signal through its pyroquinoline quinone ethanol dehydrogenase to regulate host intestinal homeostasis, which controls host development rate and energy metabolism (Shin *et al.*, 2015). *L.*

plantarum promotes systemic development of *Drosophila* by regulating hormone growth signals through the host nutrition sensing system dependent on target of rapamycin (Storelli *et al.*, 2011). Acetic acid, a metabolite of *L. plantarum*, binds to the diaminopropionic acid-type peptidoglycan-sensitive receptor PGRP-LC in the innate immune pathway of *Drosophila*-immune deficiency pathway, which stimulates the endocrine peptide tachykinin. Transcription, thereby optimising *Drosophila* lipid metabolism and activating insulin signals, promoting the growth and development of *Drosophila* larvae (Kamareddine *et al.*, 2018). Honey bee (*Apis mellifera*) gut microbes promote the growth of the host through the regulation of metabolites (Zheng *et al.*, 2017). *Bifidobacterium* and *Gilliamella* are the principal degraders of hemicellulose and pectin in the honey bee larval gut, providing nutrients for the host to promote growth (Zheng *et al.*, 2019). Sterile BSFL without gut microbe are impaired; however, after adding its gut microbe, such as *Bacillus subtilis* BSF-CL, BSFL growth and associated chicken manure conversion can be improved. The gut microbes of BSF can not only promote the growth of the host, but also improve the efficiency of the host to convert organic waste (Table 1).

3   Microbes help insect in antibiotic-resistant attenuation and antibiotics degradation

Using antibiotics as prophylaxis for poultry poses serious threat to human health due to resulting manure used as fertility spreading antibiotic-resistant bacteria (Tyrrell *et al.*, 2019). Antibiotic resistance genes (ARGs) can be transmitted from animal products or the environment to people (Landers *et al.*, 2012; Muloi *et al.*, 2018). In agriculture, manure and sludge are often used to increase crop yields, which is a key enabler of resistance gene flow into an ecosystem (Kivits *et al.*, 2018; Udikovic-Kolic *et al.*, 2014). Approximately 58% of veterinary antibiotics are transferred to the environment with most being introduced to soil through resulting manure used as a fertiliser (Xie *et al.*, 2018). Among the compost products of livestock and poultry manure, the potential bacterial hosts of ARGs are actinomycetes, such as *Leucobacter*, *Mycobacterium* and *Thermomonosporaceae* (Zhang *et al.*, 2020).

Developing methods to remediate ARGs in manure is crucial. Microbes in combination with insects, such as the BSFL, could be used to mitigate ARGs risk. For example, BSFL treatments could effectively reduce antibiotic resistance genes in chicken manure by 95.0%. Previous research has determined BSFL intestinal microbiota serve as a major mechanism regulating these processes (Cai *et al.*, 2018a). How the BSFL intestinal microbiota interact with their host to mitigate ARGs should be further explored. Furthermore, BSFL are capable of directly degrading antibiotics, such as tetracycline (TC) in chicken manure. Nearly 97% of TC was degraded within 12 days by BSFL. The gut microbiota of the BSFL plays a significant role in degrading TC, effectively and rapidly (Cai *et al.*, 2018b). How the BSFL intestinal microbiota interact with their host to degrading antibiotics also should be further explored. Similarly, swine manure digestion by HF larvae is a promising method for manure reduction and associated ARGS as well. Previous research determined 94 out of 158 ARGs were significantly mitigated (by 85%) following the conversion. ARG attenuation was significantly correlated with changes in the relative abundances of the family *Ruminococcaceae*, class bacilli, or phylum *Proteobacteria* and microbial community, especially reduction in Clostridiales and Bacteroidales (Wang *et al.*, 2015, 2017).

HF larvae also can degrade antibiotics. Previous research determined HF larvae could degrade monensin in chicken manure by increasing the number of *Stenotrophomonas* sp. and *Alcaligenes* sp. in the larval gut (Li *et al.*, 2019a). If insect larvae are combined with microbes, such as Classes *Gammaproteobacteria*, *Bacteroidia*, and *Betaproteobacteria*, and *Klebsiella* sp. SQY5, etc., to reduce antibiotics in organic waste, degradation efficiency may be improved. House fly larvae convert a mixture of swine manure and gibberellin fermentation residues (GFRs). The concentration of gibberellin in the converted residue decreased significantly, and with the increase of GFRs in the mixture, *Bacillus* became the dominant bacteria, indicating that *Bacillus* can promote gibberellin degradation (Yao *et al.*, 2020). There is already evidence that *Corynebacterium variabile* Q0029 combined with HF larvae can promote the degradation of gibberellin (Yang *et al.*, 2015).

4   Microbes help insect in pathogens inhibition

Livestock manure contain various pathogens, including bacteria, viruses, parasites, fungi, such as *Salmonella* spp., *Escherichia coli* O157:H7, Avian influenza virus, etc. (Alegbeleye and Sant'Ana, 2020). *E. coli* (STEC) O157:H7 in manure produces Shiga toxins that can contaminate surface water and cause haemolytic uremic syndrome in humans and animals (Tanaro *et al.*, 2018). These manure-borne pathogens pose a huge threat to public health and the environment (Spencer and Guan, 2004). Due to the expansion of livestock and poultry farming, associated manure in some instances is directly discharged into the environment without treatment, causing pathogens in the manure to threaten the safety of drinking water and cause other animals to be infected (Alegbeleye and Sant'Ana, 2020; Bicudo and Goyal, 2003; Ström *et al.*, 2018). In the aquaculture industry, the use of untreated livestock and poultry manure not only pollutes the water, but also results in fish disease and economic loss (Wanja *et al.*, 2020). Therefore, reducing pathogens in livestock and poultry manure is very important to public health and environment.

Insect larvae such as BSFL can significantly reduce the concentration of pathogens in swine manure and dairy manure (Erickson *et al.*, 2004; Liu *et al.*, 2008). Gut commensal microbes are in a competitive relationship with pathogens that enter the gut. Elimination of the commensal microbes results in pathogens proliferation and disease manifestation (Fast *et al.*, 2018). The gut microbiota can regulate the host immune cells and enhance the gut mucosal barrier function, so that the host can establish a strong immune response against invading pathogens, and the importance of gut microbiota in regulating the immune system is mainly to maintain the homeostasis of the gut (Alarcón *et al.*, 2016; Kogut *et al.*, 2020; Shi *et al.*, 2017). Insect gut microbes play an important role in maintaining gut homeostasis, and their ability to inhibit pathogenic bacteria has a positive effect on host health.

Interestingly, researchers found a novel, temperate *Escherichia* bacteriophage designated vB_EcoS_PHB10 (PHB10), which could lyse two out of 13 *Escherichia* strains tested (Chen *et al.* 2019). Accordingly, the BSF immune system might have adapted to mitigate pathogenic microbes. Furthermore, the genomes of BSF encode more secreted peptidoglycan recognition proteins (PGRPs) and Gram-negative binding proteins compared to those in other dipteran species. Further research found two BSF immune genes *BsfDuox* and *Bsf*TLR3 could regulate the gut key bacteria *Providencia* and *Dysgonomonas* homeostasis to depress zoonotic pathogens (Huang *et al.* 2020). The mechanism of bacteriophage and gut key bacteria in depressing zoonotic pathogens is still unknown and need further be explored.

5   Microbes help insect in odour removal of organic waste

Manure emits greenhouse gases and other volatile organic compounds during the decomposition process (FAO, 2009). Volatile organic compounds, such as 4-methylphenol was responsible for 67.3% of odour activity in dairy manure (Hales *et al.*, 2012). Processing organic waste with BSFL results in volatile fatty acids (VFAs) being reduced (Sundberg *et al.*, 2013). VFAs are short-chain fatty acids containing six carbon atoms or less, such as acetic, propionic acid, iso-butyric acid, n-butyric acid, iso-valeric acid and n-pentanoic acid (Agler *et al.*, 2011; Fang *et al.*, 2016). Lactic acid and acetic acid have a higher odour threshold than butyric acid and valeric acid (Rosenfeld *et al.*, 2007). VFAs is easily produced under anaerobic and low pH conditions (Liu *et al.*, 2020; Sundberg *et al.*, 2013).

BSFL can significantly reduce poultry, swine, and dairy manure release volatile organic compounds to the environment (Beskin *et al.*, 2018). BSFL can reduce the accumulation of VFAs in organic wastes by 10.12-28.50%. Interestingly, VFAs can be used as a carbon source for the survival of BSFL and natural microbiota (Liu *et al.*, 2020). Use an engineering microbial ecosystem to remove odorous gas from industrial waste, before treatment, the dominant strains are genera *Thiobacillus* and *Oceanicaulis*, after treatment, the dominant strains are genera *Acidithiobacillus* and *Ferroplasma* (Li *et al.*, 2019b). We speculated both exo- and endo-symbionts inhibited the microbes that produced odorous gas.

6 Prospects of research on the degradation of organic waste by insects and microbes

There are many ways to deal with organic wastes. BSFL is a representative insect that converts organic wastes into products of value. After the conversion, the content of many organic wastes decreased by 20.31-22.18%. Furthermore, VFAs are also decreased by 25.58-80.08%. Nitrogen content in the waste also is reduced by 6.08-14.37%. Total phosphorous, total Kjeldahl nitrogen, and total nutrients increased by 42.30-64.16, 45.41-88.17, and 26.51-33.34%, respectively (Liu *et al.*, 2019).

Black soldier fly larvae have broad application prospects for the treatment of livestock and poultry manure pollution. Current research is devoted to further improving the efficiency of BSFL to convert livestock and poultry manure. At present, combining BSFL and microorganisms can further improve the efficiency of BSFL converting livestock and poultry manure. Some of the microorganisms used include BSFL gut microbes, BSFL associated microorganisms (BSF egg surface microbes), and microorganisms separated from livestock and poultry manure composting (Mazza *et al.*, 2020; Xiao *et al.*, 2018). However, it is not clear that the mechanism of how microorganisms cooperates with BSFL to convert manure and further research is needed.

Microbes can not only improve the efficiency of insect hosts to degrade organic waste, but also promote the growth of insect hosts (Xiao *et al.*, 2018). In particular, the mutually beneficial symbiosis between intestinal microbes and the host has always been a research hotspot. Microbes play an important role in helping the host to resist adverse environments. The gut commensal bacteria *Klebsiella michiganensis* BD177 stimulates the metabolic activity of arginine and proline in the host to help the host resist the low temperature environment (Raza *et al.*, 2020). *Bifidobacterium animalis* enhances the host's ability to resist influenza by mediating the metabolism of valine and coenzyme A (Zhang *et al.*, 2020).

Conclusions

Food and Agriculture Organization of the United Nations (FAO) recommended several main insects such as BSFL, HF, and YMW to treat organic wastes (Van Huis *et al.*, 2013). They not only convert the animal manure to insect biomass, but also food waste, PS and PE. Scientists found their gut microbiota play important role in these organic side streams conversion. The novelty is reflected in that some gut microbes and their secreted enzymes degrade macromolecules such as protein, fat, polysaccharide, cellulose, PS and PE. For example, the phyla Firmicutes and Proteobacteria are main microbiota response to macromolecules degradation in organic waste. Other microbes, such as *Bacillus, Enterobacter*, and yeast can secrete some protease, cellulase, lipase and xylanase, and degrade the macromolecules in organic wastes into small molecules, which can be used as the nutrients of insects. Gut microbiota also could help insects degrade hazardous substances such as antibiotics, mycotoxin, odorous substances and inhibited pathogens in organic wastes. So insect gut microbiota could help their host to degrade macromolecules and hazardous substances in organic side streams and make substrates more fit for insects. But their mechanisms are not very clear and need further be explored.

Acknowledgements

This work was supported by the National Key Technology R&D Program of China (2018YFD0500203) and National Natural Science Foundation of China (31770136).

Conflict of interest

The authors declare no conflict of interest.

References

Agler, M.T., Wrenn, B.A., Zinder, S.H. and Angenent, L.T., 2011. Waste to bioproduct conversion with undefined mixed cultures: the carboxylate platform. Trends Biotechnology 29: 70-78.

Alarcón, P., González, M. and Castro, É., 2016. Rol de la microbiota gastrointestinal en la regulación de la respuesta inmune. Revista Médica de Chile 144: 910-916.

Alegbeleye, O.O. and Sant'Ana, A.S., 2020. Manure-borne pathogens as an important source of water contamination: an update on the dynamics of pathogen survival/transport as well as practical risk mitigation strategies. International Journal of Hygiene and Environmental Health 227: 113524.

Beskin, K.V., Holcomb, C.D., Cammack, J.A., Crippen, T.L., Knap, A.H., Sweet, S.T. and Tomberlin, J.K., 2018. Larval digestion of different manure types by the black soldier fly (Diptera: Stratiomyidae) impacts associated volatile emissions. Waste Management 74: 213-220.

Bicudo, J.R. and Goyal, S.M., 2003. Pathogens and manure management systems: a review. Environmental Technology 24: 115-130.

Brandon, A.M., Gao, S., Tian, R., Ning, D., Yang, S., Zhou, J., Wu, W. and Criddle, C.S., 2018. Biodegradation of polyethylene and plastic mixtures in mealworms (larvae of *Tenebrio molitor*) and effects on the gut microbiome. Environmental Science & Technology 52: 6526-6533.

Bruno, D., Bonelli, M., De Filippis, F., Di Lelio. I., Tettamanti, G., Casartelli, M., Ercolini, D. and Caccia, S., 2019. The intestinal microbiota of *Hermetia illucens* larvae is affected by diet and shows a diverse composition in the different midgut regions. Applied and Environmental Microbiology 85: 1-14.

Cai, M., Ma, S., Hu, R., Tomberlin, J.K., Thomashow, L.S., Zheng, L., Li, W., Yu, Z. and Zhang, J., 2018a. Rapidly mitigating antibiotic resistant risks in chicken manure by *Hermetia illucens* bioconversion with intestinal microflora. Environmental Microbiology 20: 4051-4062.

Cai, M., Ma, S., Hu, R., Tomberlin, J.K., Yu, C., Huang, Y., Zhan, S., Li, W., Zheng, L., Yu, Z. and Zhang, J., 2018b. Systematic characterization and proposed pathway of tetracycline degradation in solid waste treatment by *Hermetia illucens* with intestinal microbiota. Environmental Pollution 242: 634-642.

Chen, Y., Li, X., Song, J., Yang, D., Liu, W., Chen, H., Wu, B. and Qian, P., 2019. Isolation and characterization of a novel temperate bacteriophage from gut-associated *Escherichia* within black soldier fly larvae (*Hermetia illucens* L. [Diptera: Stratiomyidae]). Archives of Virology 164: 2277-2284.

Čičková, H., Newton, G.L., Lacy, R.C. and Kozánek, M., 2015. The use of fly larvae for organic waste treatment. Waste Management 35: 68-80.

Consuegra, J., Grenier, T., Baa-Puyoulet, P., Rahioui, I., Akherraz, H., Gervais, H., Parisot, N., Da Silva, P., Charles, H., Calevro, F. and Leulier, F., 2020. *Drosophila*-associated bacteria differentially shape the nutritional requirements of their host during juvenile growth. PLoS Biology 18: e3000681.

Dróżdż, D., Wystalska, K., Malińska, K., Grosser, A., Grobelak, A. and Kacprzak, M., 2020. Management of poultry manure in Poland – current state and future perspectives. Journal of Environmental Management 264: 110327.

De Smet, J., Wynants, E., Cos, P. and Van Campenhout, L., 2018. Microbial community dynamics during rearing of black soldier fly larvae (*Hermetia illucens*) and its impact on exploitation potential. Applied and Environmental Microbiology 84: e02722-17.

Erickson, M.C., Islam, M., Sheppard, C., Liao, J. and Doyle, M.P., 2004. Reduction of *Escherichia coli* O157:H7 and *Salmonella enterica* Serovar Enteritidis in chicken manure by larvae of the black soldier fly. Journal of Food Protection 67: 685-690.

Fang, W., Zhang, P., Gou, X., Zhang, H., Wu, Y., Ye, J. and Zeng, G., 2016. Volatile fatty acid production from spent mushroom compost: effect of total solid content. International Biodeterioration & Biodegradation 113: 217-221.

Food and Agriculture Organisation (FAO), 2009. The state of food and agriculture 2009. Livestock in the balance. FAO Agriculture Series. FAO, Rome, Italy.

Fast, D., Kostiuk, B., Foley, E. and Pukatzki, S., 2018. Commensal pathogen competition impacts host viability. Proceedings of the National Academy of Sciences 115: 7099.

Hales, K.E., Parker, D.B. and Cole, N.A., 2012. Potential odorous volatile organic compound emissions from feces and urine from cattle fed corn-based diets with wet distillers grains and solubles. Atmospheric Environment 60: 292-297.

Huang, Y., Yu, Y., Zhan, S., Tomberlin, J.K., Huang, D., Cai, M., Zheng, L., Yu, Z. and Zhang, J., 2020. Dual oxidase duox and toll-like receptor 3 TLR3 in the toll pathway suppress zoonotic pathogens through regulating the intestinal bacterial community homeostasis in *Hermetia illucens* L. PLoS ONE 15: e0225873.

Isibika, A., Vinnerås, B., Kibazohi, O., Zurbrügg, C. and Lalander, C., 2019. Pre-treatment of banana peel to improve composting by black soldier fly (*Hermetia illucens* (L.), Diptera: Stratiomyidae) larvae. Waste Management 100: 151-160.

Jiang, C., Jin, W., Tao, X., Zhang, Q., Zhu, J., Feng, S., Xu, X., Li, H., Wang, Z. and Zhang, Z., 2019. Black soldier fly larvae (*Hermetia illucens*) strengthen the metabolic function of food waste biodegradation by gut microbiome. Microbial Biotechnology 12: 528-543.

Kamareddine, L., Robins, W.P., Berkey, C.D., Mekalanos, J.J. and Watnick, P.I., 2018. The *Drosophila* immune deficiency pathway modulates Enteroendocrine function and host metabolism. Cell Metabolism 28: 449-462.

Kim, E.K., Lee, K.A., Hyeon, D.Y., Kyung, M., Jun, K.Y., Seo, S.H., Hwang, D., Kwon, Y. and Lee, W.J., 2020. Bacterial nucleoside catabolism controls quorum sensing and commensal-to-pathogen transition in the *Drosophila* gut. Cell Host Microbe 27: 345-357.

Kivits, T., Broers, H.P., Beeltje, H., Van Vliet, M. and Griffioen, J., 2018. Presence and fate of veterinary antibiotics in age-dated groundwater in areas with intensive livestock farming. Environmental Pollution 241: 988-998.

Kogut, M.H., Lee, A. and Santin, E., 2020. Microbiome and pathogen interaction with the immune system. Poultry Science 99: 1906-1913.

Landers, T.F., Cohen, B., Wittum, T.E. and Larson, E.L., 2012. A review of antibiotic use in food animals: perspective, policy, and potential. Public Health Reports 127: 4-22.

Li, L., Stasiak, M., Li, L., Xie, B., Fu, Y., Gidzinski, D., Dixon, M. and Liu, H., 2016. Rearing *Tenebrio molitor* in BLSS: dietary fiber affects larval growth, development, and respiration characteristics. Acta Astronautica 118: 130-136.

Li, H., Wan, Q., Zhang, S., Wang, C., Su, S. and Pan, B., 2019a. Housefly larvae (*Musca domestica*) significantly accelerates degradation of monensin by altering the structure and abundance of the associated bacterial community. Ecotoxicology and Environmental Safety 170: 418-426.

Li, W., Ni, J., Cai, S., Liu, Y., Shen, C., Yang, H., Chen, Y., Tao, J., Yu, Y. and Liu, Q., 2019b. Variations in microbial community structure and functional gene expression in bio-treatment processes with odorous pollutants. Scientific Reports 9: 17870.

Liu, Q., Tomberlin, J.K., Brady, J.A., Sanford, M.R. and Yu, Z., 2008. Black soldier fly (Diptera: Stratiomyidae) larvae reduce *Escherichia coli* in dairy manure. Environmental Entomology 37: 1525-1530.

Liu, T., Awasthi, M.K., Awasthi, S.K., Duan, Y. and Zhang, Z., 2020. Effects of black soldier fly larvae (Diptera: Stratiomyidae) on food waste and sewage sludge composting. Journal of Environmental Management 256: 109967.

Liu, T., Kumar, A.M., Chen, H., Duan, Y., Awasthi, S.K. and Zhang, Z., 2019. Performance of black soldier fly larvae (*Diptera: Stratiomyidae*) for manure composting and production of cleaner compost. Journal of Environmental Management 251: 109593.

Ma, Y., Shen, Y. and Liu, Y., 2020. Food waste to biofertilizer: a potential game changer of global circular agricultural economy. Journal of Agricultural and Food Chemistry 68: 5021-5023.

Mazza, L., Xiao, X., ur Rehman, K., Cai, M., Zhang, D., Fasulo, S., Tomberlin, J.K., Zheng, L.,

Soomro, A.A., Yu, Z. and Zhang, J., 2020. Management of chicken manure using black soldier fly (Diptera: Stratiomyidae) larvae assisted by companion bacteria. Waste Management 102: 312-318.

Muloi, D., Ward, M.J., Pedersen, A.B., Fèvre, E.M., Woolhouse, M.E.J. and Van Bunnik, B.A.D., 2018. Are food animals responsible for transfer of antimicrobial-resistant *Escherichia coli* or their resistance determinants to human populations? A systematic review. Foodborne Pathogens & Disease 15: 467-474.

Peng, B., Su, Y., Chen, Z., Chen, J., Zhou, X., Benbow, M.E., Criddle, C.S., Wu, W. and Zhang, Y., 2019. Biodegradation of polystyrene by dark (*Tenebrio obscurus*) and yellow (*Tenebrio molitor*) mealworms (Coleoptera: Tenebrionidae). Environmental Science & Technology 53: 5256-5265.

Qian, Y., Song, K., Hu, T. and Ying, T., 2018. Environmental status of livestock and poultry sectors in China under current transformation stage. Science of the Total Environment 622-623: 702-709.

Raza, M.F., Wang, Y., Cai, Z., Bai, S., Yao, Z., Awan, U.A., Zhang, Z., Zheng, W. and Zhang, H., 2020. Gut microbiota promotes host resistance to low-temperature stress by stimulating its arginine and proline metabolism pathway in adult *Bactrocera dorsalis*. PLoS Pathogens 16: e1008441.

Rosenfeld, P.E., Clark, J.J.J., Hensley, A.R. and Suffet, I.H., 2007. The use of an odour wheel classification for the evaluation of human health risk criteria for compost facilities. Water Science and Technology 55: 345-357.

Salihoglu, G., Salihoglu, N.K., Ucaroglu, S. and Banar, M., 2018. Food loss and waste management in Turkey. Bioresource Technology 248: 88-99.

Shi, N., Li, N., Duan, X. and Niu, H., 2017. Interaction between the gut microbiome and mucosal immune system. Military Medical Research 4: 14.

Shin, N., Whon, T.W. and Bae, J.W., 2015. Proteobacteria: microbial signature of dysbiosis in gut microbiota. Trends in Biotechnology 33: 496-503.

Sindhu, R., Gnansounou, E., Rebello, S., Binod, P., Varjani, S., Thakur, I.S., Nair, R.B. and Pandey, A., 2019. Conversion of food and kitchen waste to value-added products. Journal of Environmental Management 241: 619-630.

Somroo, A.A., ur Rehman, K., Zheng, L., Cai, M., Xiao, X., Hu, S., Mathys, A., Gold, M., Yu, Z. and Zhang, J., 2019. Influence of *Lactobacillus buchneri* on soybean curd residue co-conversion by black soldier fly larvae (*Hermetia illucens*) for food and feedstock production. Waste Management 86: 114-122.

Spencer, J.L. and Guan, J., 2004. Public health implications related to spread of pathogens in manure from livestock and poultry operations. In: Spencer, J.F.T. and Ragout de Spencer, A.L. (eds.) Public health microbiology: methods and protocols. Humana Press, Totowa, NJ, USA, pp. 503-515.

Storelli, G., Defaye, A., Erkosar, B., Hols, P., Royet, J. and Leulier, F., 2011. *Lactobacillus plantarum* promotes *Drosophila* systemic growth by modulating hormonal signals through TOR-dependent nutrient sensing. Cell Metabolism 14: 403-414.

Ström, G., Albihn, A., Jinnerot, T., Boqvist, S., Andersson-Djurfeldt, A., Sokerya, S., Osbjer, K.,

San, S., Davun, H. and Magnusson, U., 2018. Manure management and public health: Sanitary and socio-economic aspects among urban livestock-keepers in Cambodia. Science of the Total Environment 621: 193-200.

Sun, L., Pope, P.B., Eijsink, V.G.H. and Schnürer, A., 2015. Characterization of microbial community structure during continuous anaerobic digestion of straw and cow manure. Microbial Biotechnology 8: 815-827.

Sundberg, C., Yu, D., Franke-Whittle, I., Kauppi, S., Smårs, S., Insam, H., Romantschuk, M. and Jönsson, H., 2013. Effects of pH and microbial composition on odour in food waste composting. Waste Management 33: 204-211.

Tanaro, J.D., Pianciola, L.A., D'Astek, B.A., Piaggio, M.C., Mazzeo, M.L., Zolezzi, G. and Rivas, M., 2018. Virulence profile of *Escherichia coli* O157 strains isolated from surface water in cattle breeding areas. Letters in Applied Microbiology 66: 484-490.

Tyrrell, C., Burgess, C.M., Brennan, F.P. and Walsh, F., 2019. Antibiotic resistance in grass and soil. Biochemical Society Transactions 47: 477-486.

Udikovic-Kolic, N., Wichmann, F., Broderick, N.A. and Handelsman, J., 2014. Bloom of resident antibiotic-resistant bacteria in soil following manure fertilization. Proceedings of the National Academy of Sciences 111: 15202.

ur Rehman, K., Cai, M., Xiao, X., Zheng, L., Wang, H., Soomro, A.A., Zhou, Y., Li, W., Yu, Z. and Zhang, J., 2017a. Cellulose decomposition and larval biomass production from the co-digestion of dairy manure and chicken manure by mini-livestock (*Hermetia illucens* L.). Journal of Environmental Management 196: 458-465.

ur Rehman, K., Rehman, A., Cai, M., Zheng, L., Xiao, X., Somroo, A. Wang, H., Li, W., Yu, Z. and Zhang, J., 2017b. Conversion of mixtures of dairy manure and soybean curd residue by black soldier fly larvae (*Hermetia illucens* L.). Journal of Cleaner Product 154: 366-373.

ur Rehman, K., ur Rehman, R., Somroo, A.A., Cai, M., Zheng, L., Xiao, X., ur Rehman, A., Rehman, A., Tomberlin, J.K., Yu, Z. and Zhang, J., 2019. Enhanced bioconversion of dairy and chicken manure by the interaction of exogenous bacteria and black soldier fly larvae. Journal of Environmental Management 237: 75-83.

Urbanek, A.K., Rybak, J., Wróbel, M., Leluk, K. and Mirończuk, A.M., 2020. A comprehensive assessment of microbiome diversity in *Tenebrio molitor* fed with polystyrene waste. Environmental Pollution 262: 114281.

Van Huis, A., Van Itterbeeck, J., Klunder, H., Mertens, E., Halloran, A., Muir, G. and Vantomme, P., 2013. Edible insects: future prospects for food and feed security. Food and Agriculture Organization of the United Nations, Rome, Italy, pp. 59-61. Available at: http://www.fao.org/docrep/018/i3253e/i3253e.pdf.

Wang, H., Sangwan, N., Li, H., Su, J., Oyang, W., Zhang, Z., Gilbert, J.A., Zhu, Y., Ping, F. and Zhang, H., 2017. The antibiotic resistome of swine manure is significantly altered by association with the *Musca domestica* larvae gut microbiome. The ISME Journal 11: 100-111.

Wanja, D.W., Mbuthia, P.G., Waruiru, R.M., Mwadime, J.M., Bebora, L.C., Nyaga, P.N. and Ngowi, H.A., 2020. Fish husbandry practices and water quality in central Kenya: potential risk factors for fish mortality and infectious diseases. Veterinary Medicine International 2020: 6839354.

Wang, H., Li, H, Gilbert J.A., Li, H., Wu, L., Liu, M., Wang, L., Zhou,Q., Yuan, J. and Zhang Z., 2015. Housefly larva vermicomposting efficiently attenuates antibiotic resistance genes in swine manure, with concomitant bacterial population changes. Applied and Environmental Microbiology 81: 7668-7679.

Wynants, E., Frooninckx, L., Crauwels, S., Verreth, C., De Smet, J., Sandrock, C., Wohlfahrt, J., Van Schelt, J., Depraetere, S., Lievens, B., Van Miert, S., Claes, J. and Van Campenhout, L., 2019. Assessing the microbiota of black soldier fly larvae (*Hermetia illucens*) reared on organic waste streams on four different locations at laboratory and large scale. Microbial Ecology 77: 913-930.

Xiao, X., Mazza, L., Yu, Y., Cai, M., Zheng, L., Tomberlin, J.K., Yu, J., Van Huis, A., Yu, Z., Fasulo, S. and Zhang, J., 2018. Efficient co-conversion process of chicken manure into protein feed and organic fertilizer by *Hermetia illucens* L. (Diptera: Stratiomyidae) larvae and functional bacteria. Journal Environmental Management 217: 668-676.

Xie, W., Shen, Q. and Zhao, F., 2018. Antibiotics and antibiotic resistance from animal manures to soil: a review. European Journal of Soil Science 69: 181-195.

Yan, C., He, W., Xue, Y., Liu, E. and Liu, Q., 2016. Application of biodegradable plastic film to reduce plastic film residual pollution in Chinese agriculture [in Chinese]. Sheng Wu Gong Cheng Xue Bao 32: 748-760.

Yang, J., Yang, Y., Wu, W., Zhao, J. and Jiang, L. 2014. Evidence of polyethylene biodegradation by bacterial strains from the guts of plastic-eating waxworms. Environmental Science Technology 48: 13776-13784.

Yang, S., Xie, J., Hu, N., Liu, Y., Zhang, J., Ye, X. and Liu, Z., 2015. Bioconversion of gibberellin fermentation residue into feed supplement and organic fertilizer employing housefly (*Musca domestica* L.) assisted by *Corynebacterium* variabile. PLoS ONE 10: e0110809.

Yang, Y., Wang, J. and Xia, M., 2020. Biodegradation and mineralization of polystyrene by plastic-eating superworms *Zophobas atratus*. Science of the Total Environment 708: 135233.

Yao, Y., Zhu, F., Hong, C., Chen, H., Wang, W., Xue, Z., Zhu, W., Wang, G. and Tong, W., 2020. Utilization of gibberellin fermentation residues with swine manure by two-step composting mediated by housefly maggot bioconversion. Waste Management 105: 339-346.

Yu, G., Cheng, P., Chen, Y., Li, Y., Yang, Z., Chen, Y. and Tomberlin, J.K., 2011. Inoculating poultry manure with companion bacteria influences growth and development of black soldier fly (Diptera: Stratiomyidae) larvae. Environmental Entomology 40: 30-35.

Zhan, S., Fang, G., Cai, M., Kou, Z., Xu, J., Cao, Y., Bai, L., Zhang, Y., Jiang, Y., Luo, X., Xu, J., Xu, X., Zheng, L., Yu, Z., Yang, H., Zhang, Z., Wang, S., Tomberlin, J.K., Zhang, J. and Huang, Y., 2020. Genomic landscape and genetic manipulation of the black soldier fly *Hermetia illucens*, a natural waste recycler. Cell Research 30: 50-60.

Zhang, K., Hu, R., Cai, M., Zheng, L., Yu, Z. and Zhang, J., 2017. Degradation of plastic film containing polyethylene (PE) by yellow meal worms. Chemistry and Bioengineering 34: 47-49.

Zhang, L., Li, L., Pan, X., Shi, Z., Feng, X., Gong, B., Li, J. and Wang, L., 2018. Enhanced growth and activities of the dominant functional microbiota of chicken manure composts in the presence of maize straw. Frontier Microbiology 9: 1131.

Zhang, M., He, L., Liu, Y., Zhao, J., Zhang, J., Chen, J., Zhang, Q. and Ying, G., 2020. Variation of antibiotic resistome during commercial livestock manure composting. Environment International 136: 105458.

Zhang, Z., Wang, H., Zhu, J., Suneethi, S. and Zheng, J., 2012. Swine manure vermicomposting via housefly larvae (*Musca domestica*): the dynamics of biochemical and microbial features. Bioresource Technology 118: 563-571.

Zheng, H., Perreau, J., Powell, J.E., Han, B., Zhang, Z., Kwong, W.K., Tringe, S.G. and Moran, N.A., 2019. Division of labor in honey bee gut microbiota for plant polysaccharide digestion. Proceedings of the National Academy of Sciences 116: 25909.

Zheng, H., Powell, J.E., Steele, M.I., Dietrich, C. and Moran, N.A., 2017. Honeybee gut microbiota promotes host weight gain via bacterial metabolism and hormonal signaling. Proceedings of the National Academy of Sciences 114: 4775.

Upcycling of manure with insects: current and future prospects

J.A. Cammack[1]*, C.D. Miranda[1], H.R. Jordan[2] and J.K. Tomberlin[3]

[1]*EVO Conversion Systems, LLC, 5552 Raymond Stotzer Pkwy, College Station, TX 77845, USA;* [2]*Department of Biological Sciences, Mississippi State University, P.O. Box GY, Starkville, MS 39762, USA;* [3]*Department of Entomology, Texas A&M University, 410 Minnie Belle Heep, College Station, TX 77843, USA; *jacammack@evoconsys.com*

Abstract

An unavoidable by-product of any animal production system, be it vertebrate- or invertebrate-based, is the manure generated by the animals themselves. In this review, we focus on the role that insects, particularly the black soldier fly *Hermetia illucens* (L.) (Diptera: Stratiomyidae), could play in managing the mass amount of manure produced through animal agriculture, and the subsequent commodities that could be generated by such a system. Although the focus of this review is on the black soldier fly, we postulate that other species, including the lesser mealworm *Alphitobius diaperinus* (Panzer) (Coleoptera: Tenebrionidae) and the house fly *Musca domestica* L. (Diptera: Muscidae) are also well poised to help with the challenge of managing animal manure, while generating products of value.

Keywords

manure management – sustainable agriculture – lesser mealworm – house fly – black soldier fly – circular economy

1 Introduction

The insects as feed industry is positioned to alleviate many of the environmental burdens associated with livestock production, such as diverting plant protein to human consumption, reducing pressure on international fisheries for fish oil and fish meal, land use, and water required to generate protein. Although using insect products as components of livestock feed could allow for more sustainable (i.e. efficient and environmentally-conscious) meat production, one environmental issue that remains is manure management and its associated negative environmental impacts (e.g. runoff, eutrophication, groundwater contamination, greenhouse gas and volatile organic compound emissions, and proliferation of pest species and pathogens) (Adhikari *et al.*, 2005; Beskin *et al.*, 2018; Chang and Janzen, 1996; Cossé and Baker, 1996; FAO, 2006; Kyakuwaire *et al.*, 2019).

Over the past 25+ years, structural changes in animal production across the globe (i.e. reduction in total farm numbers and an increase in the number of animals per farm) have resulted in large quantities of manure in smaller areas, that must be properly managed. For example, from 1970 to 2006, the number of USA dairy farms decreased by 88%, while average herd size increased six-fold, and milk production twelve-fold (MacDonald *et al.*, 2007). Traditionally, a majority (70% in 1994) of swine produced in the USA was from small operations (< 5,000 head/operation). By 2014, large operations (5,000+ head/operation) accounted for 90% of annual production (United States Department of Agriculture, 2015). A similar trend has occurred in the broiler industry. From 1995 to 2010, the number of broiler firms decreased from 55 to 41 (Davis *et al.*, 2013), while chicken meat production increased by 47% (National Chicken Council, 2019b). By 2019, the number of USA broiler firms dropped to 30 (National Chicken Council, 2019a), but chicken meat production increased by almost 19% (National Chicken Council, 2019b). These trends are similar across the globe: in the European Union (EU), an estimated 90% of broilers were produced by large firms in 2018, with production increasing approximately 25% during the previous decade (Augére-Granier, 2019). From 1996-2006 in China, the number of poultry farms decreased by 67%, while production increased by over 75% (FAO, 2008).

The evolution of these industries to larger facilities has amplified concerns regarding waste management, as the volume of manure produced outweighs the amount that may be utilised as fertiliser on nearby land. While manure is typically considered a valuable resource, like anything else, too much can be a major issue, and current methods of manure management (e.g. lagoons, composting, spreading, etc.) are inefficient, environmentally challenging, and not fully sustainable (Adhikari *et al.*, 2005; Chang and Janzen, 1996; Edmonds *et al.*, 2003; Grossman, 2014; Kyakuwaire *et al.*, 2019; Welch and O'Hagan, 2010). Novel methods to manage manure that decrease the environmental impact and increase the sustainability of animal production are desperately needed; insects are likely ideal candidates for tackling such a global challenge.

2   Insects that consume manure

Although thousands of arthropods feed on animal manures, these relationships have been investigated mainly in natural habitats (Miller, 1954; Mohr, 1943; Nichols *et al.*, 2017) with few species studied for manure management. Dung beetles (Coleoptera: Scarabaeidae) are widely studied for their impacts on manure recycling as related to pasture health (Bertone *et al.*, 2006; Nichols *et al.*, 2008), whereas most fly (Diptera) species associated with manure are typically regarded as pests, and research efforts are typically focused on their control (Axtell, 1986; Barth, 1986; Hall and Foehse, 1980; Sheppard, 1983). While the following three species discussed for manure management are by no means exhaustive, the knowledge-base of these species has them positioned as the most promising fits for this monumental task.

2.1   *Alphitobius diaperinus*

The lesser mealworm, *Alphitobius diaperinus* (Panzer) (Coleoptera: Tenebrionidae) is a small, dark beetle, with global distribution. Eggs are deposited in larval resources such as grains, poultry litter, and manure (Rumbos *et al.*, 2019); resulting larvae complete development in 40 to 60 d, while pupae require 7-12 d to develop to the adult stage (Axtell and Arends, 1990). When one considers the insect species frequently recommended for recycling animal manure, or used as protein, the lesser mealworm is not typically highlighted. Worldwide, this species is considered a primary pest of poultry production. Its presence signifies the potential for vectored pathogens (Crippen *et al.*, 2009, 2012), reduced feed conversion by the birds, and damage to poultry houses (Lyons *et al.*, 2017; Tomberlin *et al.*, 2008). Thus, being labelled a pest is not a surprise, especially from the perspective of the poultry producer (Axtell and Arends, 1990).

However, the lesser mealworm might have more to offer than is appreciated (Rumbos *et al.*, 2019). This species is commonly associated with manure, can develop massive populations in poultry facilities, and has a similar nutritional value to yellow mealworm (*Tenebrio molitor* (L.)), which is currently mass produced worldwide for food and feed (Rumbos *et al.*, 2019). If facilities were designed to manage the population, lesser mealworms could possess the same potential as the black soldier fly *Hermetia illucens* (L.) (Diptera: Stratiomyidae) for recycling manure, while producing protein and associated compost.

Recall, the black soldier fly was considered a pest (Axtell and Edwards, 1970) long before it was recognised as the 'crown jewel' of the insects as feed industry (Tomberlin and Van Huis, 2020). Issues with this species ranged from destruction of poultry houses and inhibition of manure management, to causing myiasis (Tomberlin and Van Huis, 2020). While these are all accurate in terms of problems the black soldier fly can cause when present, harnessing its abilities eliminated these issues. Could the lesser mealworm be on the same path created by the industrialisation of the black soldier fly? Answering that question is challenging given its current pest status and lack of effort investigating its use to recycle waste and produce protein (Rumbos *et al.*, 2019). However, the stage has been set for the industrial development of this species, as the European Union has taken the first step by approving its use as a feed (Leni *et al.*, 2020). And, several studies have examined the nutritional value of this insect as a potential food and feed (Azzollini *et al.*, 2019; Roncolini *et al.*, 2020).

2.2   **Musca domestica**

Another insect known to feed on manure is the house fly, *Musca domestica* L. (Diptera: Muscidae). Owning up to its namesake, the house fly is commonly found in or around homes and is the most widely distributed insect, colonising every location where humans and domestic animals occur (Keiding, 1986). As such, its biology has been well described in an effort toward control. House flies fed milking calf, swine, or poultry manure can develop to pupae in as little as 6 d (Larrain and Salas, 2008), but manure type influences their performance and fitness (Khan *et al.*, 2012; Larrain and Salas, 2008). Still, the house fly can consume manure from a variety of animals, including those produced from concentrated animal facilities (e.g. swine, dairy, and poultry) (Miranda *et al.*, 2020a), which makes them a suitable candidate for manure management.

House flies can degrade large volumes of manure in a short amount of time. Compared to the lengthy time necessary for traditional composting, house fly larvae, depending on the density, have been demonstrated to digest over 100 kg of swine manure in one week (Zhang *et al.*, 2012) and over 400 kg of swine manure in two weeks (Čičková *et al.*, 2012). As larvae feed on manure, reductions in odours (Wang *et al.*, 2013) and moisture (Miller *et al.*, 1974; Wang *et al.*, 2013) occur. These reductions can be attributed to changes in the microbial community, but the presence of house flies is also beneficial as they mechanically aerate the substrate, allowing the environment to remain aerobic for increased loss of gases and water (Beard and Sands, 1973). House flies also reduce antibiotic resistant genes in swine manure and in doing so, provide a biofertiliser that is less of a threat to the environment than untreated manure (Wang *et al.*, 2015, 2017). Additionally, the presence of house fly larvae can reduce swine manure mass (wet weight) by up to 70% (Wang *et al.*, 2013), dry matter by 31-35%. (Miranda *et al.*, 2020a; Wang *et al.*, 2013) and nitrogen and phosphorous by up to 78% and up to 30%, respectively (Wang *et al.*, 2013). The extent of total mass reductions (wet weight) for dairy and poultry manure by house flies on a full scale is not known as it is for swine manure, but house flies can reduce dry matter in dairy and poultry manure on a small scale by up to approximately 20 and 40%, respectively (Miranda *et al.*, 2020a).

House flies reduce nitrogen and phosphorous in dairy and poultry manure as well. For example, Hussein *et al.* (2017) determined house flies reduce nitrogen in dairy manure by 25% and phosphorous by 6% on a dry matter basis. In poultry manure, house flies reduced nitrogen from 7.5 to 2.6% and phosphorous from 3.4 to 1.9% on a dry matter basis (Teotia and Miller, 1974). At the same rate, the biomass generated from house flies fed manure is valuable as lipids extracted from larvae can be processed for biofuel (Zi-zhe *et al.*, 2017).

Part of the challenge in using the house fly for waste management is that it is considered a pest, and like the lesser mealworm, is a vector of numerous pathogens (Keiding, 1986). There are multiple ways in which house flies harbour and disseminate pathogens. Since larvae partially rely on bacteria for development (Zurek *et al.*,

2000), all life stages are associated with microbe-rich habitats. As larvae develop, some harmful bacteria may remain within the gut and can be carried through subsequent life-stages (Zurek and Nayduch, 2016). As an adult, they must feed to mature reproductively (Morrison and Davies, 1964) and so pathogens may be spread via regurgitation and excretion. Sasaki *et al.* (2000) showed *Escherichia coli* O157: H7 remained in the adult house fly crop for 4 d and was excreted for 3 d after feeding on contaminated food (Sasaki *et al.*, 2000). Pathogens can also be disseminated by adults through mechanical contact via their tarsi (Wasala *et al.*, 2013). Adult house flies are opportunistic and will move indiscriminately among locations in search of food or a place to oviposit. Yet, despite their behaviour and pest status, house flies offer a promising solution to waste management because of their ability to digest manure in a short amount of time.

2.3   Hermetia illucens

Compared to the previous insects discussed, the black soldier fly is currently the best insect candidate for manure management. Found in temperate and tropical regions throughout the world, the black soldier fly is now viewed as a non-pest species and is highly regarded due to the wide range of wastes it can digest (Alyokhin *et al.*, 2019; Lalander *et al.*, 2019; Nguyen *et al.*, 2013). The life cycle of the black soldier fly takes approximately 40-43 d when fed an artificial diet, with the larval stage lasting 22-24 d (Cammack and Tomberlin, 2017; Tomberlin *et al.*, 2002), which is nearly twice as long as the entire house fly life cycle. However, black soldier fly larvae are larger, capable of consuming more resource (Čičková *et al.*, 2015), and possess a more diverse array of digestive enzymes when compared to house fly larvae (Kim *et al.*, 2011b).

Research regarding black soldier fly larvae fed manure has largely focused on dairy, swine, or poultry manures (Diener *et al.*, 2009; Miranda *et al.*, 2019; Oonincx *et al.*, 2015a; Zhou *et al.*, 2013). All three manure types are suitable for their development; however, poultry manure appears to be optimal (Miranda *et al.*, 2019; Zhou *et al.*, 2013), likely because it has higher nutrient concentrations than the other manure types (Chen *et al.*, 2003). Blending waste streams can increase the digestion and subsequent larval yield. This is particularly effective when feeding lower quality manure types, such as dairy manure. For example, feeding black soldier fly larvae a mixture of dairy and poultry manure (40 and 60%, respectively) decreased development time by 2 d, increased survivorship by 6%, and increased larval weight by ~75% when compared to larvae fed dairy manure alone (ur Rehman *et al.*, 2017a). Similarly, other wastes (e.g. kitchen wastes, poultry slaughterhouse waste, fish offal, and human excrement) can be mixed with cow manure to improve black soldier fly survivorship, weight gain, and dry matter reduction (Gold *et al.*, 2020; St-Hilaire *et al.*, 2007). Black soldier fly digestion of manure also results in reduced production of odorous volatile organic compounds (Beskin *et al.*, 2018), further contributing to their positive impact on the manure management system.

Currently, there is little information available on large-scale digestion of manure by black soldier fly larvae. Sheppard *et al.* (1994) developed a system where the innate behaviour of the black soldier fly larvae and prepupae could be harnessed to collect larvae produced from poultry layer manure that accumulated beneath the birds, and Newton *et al.* (2005) modified this system to manage manure generated from swine production (Newton *et al.*, 2005; Sheppard *et al.*, 1994). Industrialisation of the black soldier fly on different wastes certainly exists across the globe, but most research available on this species is studied on a small/ laboratory scale (hundreds of larvae fed grams of substrate). To our knowledge, there is only one study that has examined the development of black soldier fly larvae fed manure on a large scale (10,000 larvae fed 7 kg of swine, dairy, or poultry manure) (Miranda *et al.*, 2020b). Despite the lack of information available on large-scale production of black soldier fly larvae fed manure, there is great potential for this species to serve as a waste management agent. Given their voracious appetite for all things decomposing, biomass accumulation, literature availability, and the current global focus of mass production, the remainder of this manuscript will focus on using the black soldier fly for manure management.

3 Microbiology, black soldier fly, and manure

The increasing human population has led to more intensive livestock farming, and with this comes concern about pathogens, antibiotics, and other pharmaceutical residues in animal manure shed into the environment with potential for increased accumulation of antibiotic resistant bacteria or human allergies from environmental exposure. Several pathogens comprising bacteria, viruses, and protozoan species, naturally occur in animal manure and under certain circumstances may pose a risk to human health. However, owing to differences in microbial characteristics between and within these groups of organisms, pathogen survival in the environment or manure depends on the characteristics of the particular organism, the source, and chemical composition of manure (e.g. ammonium content), pH, dry matter, temperature, oxygen, microbial competition, and moisture of these materials (Haruta and Kanno, 2015; Manyi-Loh *et al.*, 2016; Sheng *et al.*, 2019). Bacterial pathogens that are most often associated with manure with regard to human health include *Salmonella* spp., *E. coli* O157:H7, *Campylobacter jejuni*, *Listeria monocytogenes*, *Yersinia enterocolitica* and *Clostridium perfringens*, some of which may be widely distributed in the environment and include over 2000 species and strains (Manyi-Loh *et al.*, 2016; Sobsey *et al.*, 2006). These bacterial pathogens may persist for long periods in animal manures under typical farm conditions. This may be extended when the temperatures are low, moisture remains optimal, and aeration is not used. For instance, *E. coli* O157:H7 survived for 4-6 months in animal manures and slurries kept at 1-9 °C, 49 times longer than if kept at 40-60 °C (Kudva *et al.*,

1998). *Salmonella enterica* serovar Typhimurium was also found to survive best at 5 °C for up to 40 d, compared to 15 and 25 °C (Garcia *et al.*, 2010). Other studies have shown that aeration of solid manure decreased survival times for *E. coli* O157:H7 and *Salmonella* by as much as 88%. The persistence of *L. monocytogenes* for several weeks in manure-amended soil, suggests it could be transmitted through soil, especially during the cold months (Jiang *et al.*, 2004). Survival of protozoa in animal manures may also be related to temperature, but the trends are not as prominent as those reported for bacterial pathogens (Olson, 2020; Olsen *et al.*, 1999).

Composting manure significantly reduces the risk of exposure to these pathogens, as the high temperatures (≥54 °C) achieved during the process kills many pathogens and parasites. However, because of the pathogen diversity in manure, composting may not kill all pathogens, particularly those capable of forming spores or surviving in biofilms (Manyi-Loh *et al.*, 2016; Stanford *et al.*, 2015; Usui *et al.*, 2017). Hence, multiple manure management practices should be used.

Insects, such as the lesser mealworm, house fly, or black soldier fly, may suppress many of these pathogens given their coprophagous nature. The larvae aerate the manure and also promote the growth of beneficial saprophytic microbes (Sarpong *et al.*, 2019). And, as previously mentioned, their larvae may reduce or eliminate bacteria through direct ingestion and digestion or through the generation of antimicrobial compounds.

3.1 *Lab studies show black soldier fly larvae reduce pathogens in manure*

One of the first studies determining the ability of black soldier larvae to reduce manure-associated pathogens was in 2004, when Erickson *et al.* inoculated gfp-labelled *E. coli* O157:H7 and gfp-*Salmonella enterica* serovar Enteritidis into dairy, swine, and poultry manure (Erickson *et al.*, 2004). They determined black soldier fly larvae significantly accelerated *E. coli* inactivation in poultry manure, while showing no effect in dairy manure. On the other hand, *E. coli* growth accelerated in swine manure. Storage conditions and pH likely impacted pathogen populations and black soldier fly antibacterial activity within the differing manure substrates. Larval feeding also reduced *Salmonella* in poultry manure, but the pathogen accumulated in the larvae. From these data, they speculated larval waste accumulation and increased moisture in older manure likely led to reduced larval feeding, and a decrease in *Salmonella* Enteritidis inactivation (Erickson *et al.*, 2004).

A study investigating the reduction of *E. coli* in manure was conducted in 2008 (Liu *et al.*, 2008), when gfp-*E. coli* was inoculated into 50-125 g sterile dairy manure with or without black soldier fly larvae for 72 h at 23, 27, 31 or 35 °C. Black soldier fly larvae significantly reduced *E. coli* counts in all manure weight treatments, but the capacity to reduce *E. coli* populations was temperature dependent, with greatest suppression occurring at 27 °C. Additionally, pathogen accumulation within the larvae was not determined.

Lalander *et al.* (2015) measured the effects of black soldier fly larvae feeding on the concentration of *Salmonella* spp., *Enterococcus* spp., the ΦX174 bacteriophage, and the parasite *Ascaris suum* in human faecal waste, and whether these accumulated in, or on, the prepupae (Lalander *et al.*, 2015). *Salmonella* spp. count in the waste was reduced by $6\log_{10}$ in eight days, compared with a $< 2\log_{10}$ reduction in the control. However, they found no significant reduction of *Enterococcus* spp., bacteriophage ΦX174 or *A. suum* ova associated with larval feeding. In contrast to Erickson *et al.* (2004), the *Salmonella* spp. concentration was 10^5 cfu/ml at the start of the experiment but decreased to below the detection limit in both the waste and prepupae as the experiment progressed, even though freshly inoculated material was continuously added. Concentrations of *Enterococcus* and bacteriophage ΦX174 remained between 10^5 and 10^6 cfu/ml in larval guts. Also, there was an overall lower concentration of organisms found inside the gut of the prepupae compared with the gut of the larvae, suggesting that the prepupae empty their gut prior to dispersal (Lalander *et al.*, 2015). These studies of pathogen inoculation have led to data supporting the utilisation of black soldier fly larvae for waste management. However, more studies are needed to determine industrial scale applications, as well as the accumulation of these and other manure-associated pathogens throughout black soldier fly life stages, and whether transmission occurs between generations. Research into the means to achieve absolute pathogen removal from the system when using insects to digest waste should also be considered.

For instance, it is well known that insect endosymbionts can confer ecologically relevant traits and contribute to the fitness of their host through outcompeting pathogens (Eleftherianos *et al.*, 2013; Kaltenpoth, 2009; Kaltenpoth *et al.*, 2005; Paniagua Voirol *et al.*, 2018). Symbiotic bacteria often facilitate degradation of complex macromolecules for utilisation and increasing nutrient quality for the insect. Yu *et al.* (2011) measured black soldier fly larval growth and development when fed poultry manure inoculated with four *Bacillus subtilis* strains isolated from black soldier fly larval guts or from feed. They found that bacterial amendment increased larval growth while shortening the development time. Mazza *et al.* (2020) explored the effect of companion bacteria when poultry manure was converted to insect biomass. They isolated nine bacterial species from black soldier fly eggs and larval gut and inoculated them into poultry manure along with black soldier fly larvae. Results indicated that black soldier fly larvae reared in manure with *Kocuria marina, Lysinibacillus boronitolerans, Proteus mirabilis* and *B. subtilis,* alone or in combination, had higher weight gain and protein content and increased manure reduction rates compared to the control (Mazza *et al.*, 2020). Both of these studies showed that bacterial amendment into the manure was beneficial with respect to black soldier fly larval development and manure reduction. This nutrient provisioning with probiotics, along with natural mechanisms of pathogen reduction within the black soldier fly can be exploited and represent a promising strategy for manure reduction, with potential for pathogen control.

3.2   *Feeding black soldier flies to livestock reduces gut pathogens*

Another avenue of research has been the investigation of insects used as feed to reduce pathogens in the digestive tract of traditional livestock through modulation of gut microbiota and their metabolites. For instance, Park *et al.* (2017) showed that laying hens fed black soldier fly larval meal had higher *Lactobacillus* counts, but coliforms, total aerobic bacteria, and *E. coli* were lower than in those not provided the insect-based diet. Another study by Kim *et al.* (2020) found that laying hens fed diets containing black soldier fly oil lowered *C. perfringens* counts in the ileum, but not cecum samples, by 13.7% at 30 d, indicating moderate, although nonsignificant, antimicrobial activity of black soldier fly larvae oil in broilers.

Borrelli *et al.* (2017) used 16s rRNA gene sequencing to investigate the effect of black soldier fly larval meal on gastrointestinal microbiomes and bacterial metabolite production in laying hens. Results showed strong differences between caecal microbiota of soybean and black soldier fly meal fed groups, both in type and relative abundance of microbial species. They found microbial DNA signatures with potential to degrade the insect meal that correlated with high levels of gut-associated short chain fatty acid (SCFA) production. The authors suggested the increase in SCFA concentrations may result from modulation in the microbial population induced by the insect-based diet and may be associated with SCFAs derived by microbial degradation of the insect meal. Because SCFAs regulate bacterial pathogenesis (Sun and O'Riordan, 2013, and references therein), they speculated the possible beneficial effects of a black soldier fly larval meal diet (Borrelli *et al.*, 2017). Though further studies are necessary to determine the impact of this feeding strategy on human and animal physiology and long-term health, these studies suggest that feeding animals with black soldier fly larvae meal or oil modulate gut microbes, and merits further investigation into the potential for pathogen reduction in resulting manure.

3.3   *Larval black soldier fly production of antimicrobial peptides*

Antimicrobial peptides (AMPs) are short and generally positively charged peptides with the ability to kill microbial pathogens directly, or indirectly by modulating host defences. These peptides are found in a wide variety of life forms from microorganisms to humans and are also part of the innate immunity of insects (Sierra *et al.*, 2017; Wu *et al.*, 2018). These insect peptides exhibit antimicrobial activity through disruption of pathogen membranes and have been associated with lower levels of antimicrobial resistance (Wu *et al.*, 2018). As a result, insects have been surveyed for novel AMPs, and several studies have shown that black soldier fly larvae could be a rich source. Elhag *et al.* (2017) identified seven gene fragments of three types of AMPs in black soldier fly larvae. One gene, named *stomoxynZH1*, was shown to have diverse inhibitory activity on *Staphylococcus aureus*, *E. coli*, fungus *Rhizoctonia solani* Khün (rice)-10, and fungus *Sclerotinia sclerotiorum* (Lib.) (Elhag *et al.*, 2017).

In another study, Vogel *et al.* (2018) fed black soldier fly larvae diets supplemented with four inoculated bacterial mixtures and variable organic plant waste and measured the immune related transcriptome. The authors identified 53 genes encoding putative AMPs, including 26 putative defensins, from the black soldier fly transcriptome. Larval protein and lipid extracts also showed a broad range of antibacterial activity. And, Alvarez *et al.* (2019) isolated peptides from *E. coli* challenged and control black soldier fly larvae, and tested their anti-*Helicobacter pylori* activity. The authors found over 90 peptides in their analysis, four of which they determined to have pronounced anti-*H. pylori* activity. But only larvae injected with *E. coli* were found to have anti-*H. pylori* antimicrobial peptides, suggesting that these peptides were produced in response to *E. coli* challenge (Alvarez *et al.*, 2019).

These studies showed that AMP activity was dependent on diet and associated with bacterial loads within the diet. Additionally, immune challenge was induced with a needle prick, with AMP activity measured between 24-72 h post infection (Alvarez *et al.*, 2019; Elhag *et al.*, 2017; Vogel *et al.*, 2018). It would thus be interesting to measure AMP activity of black soldier fly actively feeding on animal manures versus a control. This would allow measurement of black soldier fly AMPs under constant exposure to a wide array of microbial pathogens.

3.4 Black soldier fly larvae degrade pharmaceuticals and reduce antibiotic resistance genes in manure

Antibiotic residues are frequently detected in livestock wastes, where mechanisms of dissemination into the environment may be through the direct spreading of manure onto fields, processed in lagoon slurry, or spreading solids for livestock bedding (Wallace *et al.*, 2018). Often, antibiotic residues have been detected at levels much higher (as much as 28,000 times) than in background soil or water (Chee-Sanford *et al.*, 2009; Sim *et al.*, 2011). And, though the abundance of antibiotic residues in livestock waste varies among livestock farm types, and regional regulations, the most frequently detected antibiotic residues include those from the fluoroquinolone, sulphonamide, and tetracycline drug classes, though other classes such as trimethroprim and macrolides have also been detected (Chee-Sanford *et al.*, 2009; Prosser and Sibley, 2015; Wallace *et al.*, 2018; Yuan *et al.*, 2019). While further research needs to be conducted to determine the full impact on human, animal, and ecosystem health, there is general agreement within public and private entities that mitigation strategies should be implemented to prolong the effectiveness of currently used antibiotics while lessening the dissemination of antibiotic resistant bacteria and genes (Carbajal *et al.*, 2017; CDC, 2015; EU, 2016; World Health Organization, 2015)

A recent and exciting finding is the black soldier fly larvae-mediated degradation of pharmaceuticals. A study showed black soldier fly larvae degraded three pharmaceuticals (e.g. carbamazepine, roxithromycin and trimethoprim) and two pesticides (e.g. azoxystrobin and propiconazole) in a composting system, with no bioaccumulation within the larvae detected (Lalander *et al.*, 2016). Recently, Cai *et al.* (2018b) showed that black soldier fly larvae can degrade tetracycline. Further, the authors found that intestinal bacterial and fungal communities were significantly modified. They also detected an increase in *tet* gene copy number (tetracycline resistance gene) in the gut microbial community, suggesting gut microbes increased tetracycline tolerance by black soldier fly and facilitated tetracycline biodegradation. Their analyses suggested synergistic mechanisms of tetracycline degradation between intestinal microbes and black soldier fly larvae including hydrolysis, oxygenation, deamination, demethylation, ring-cleavage and modification, where degradation products could serve as nutritional resources (Cai *et al.*, 2018b). As a follow-up to this work, Cai *et al.* (2018a) investigated the impact of larval sterility on the bioconversion of poultry manure and the persistence of associated antibiotic resistance genes (ARGs). The authors used qPCR targeting 11 tetracycline resistance genes and 2 integrase genes and found that non-sterile black soldier fly larvae reduced ARGs by 95%, versus 88.7 and 48.4% in sterile larvae and controls, respectively. This significant difference between treatments suggested a synergistic effect between intestinal microbes and black soldier fly in ARG reduction mechanisms. They also showed a shift in the microbial community, with decreases in human pathogens within the waste by 70-92%. Furthermore, they found that environmental factors that were altered by larval activity, such as pH and nitrogen content, influenced bacterial community composition, and thus the persistence of ARGs (Cai *et al.*, 2018a).

Overall, these results are promising for developing strategies for accelerating the degradation of antibiotics by adjusting the intestinal microbiota of black soldier fly larvae. But, though encouraging, these studies only targeted a few ARGs for detection, limiting elucidation of the impact of black soldier fly larvae and associated microbes on a broad range of ARGs and mechanisms. And recently, Gao *et al.* (2019) measured life histories of 4-day old black soldier fly larvae fed with diet supplemented with sulphonamides (e.g. sulphamonomethoxine, sulfamethoxazole, sulfamethazine, and sulfadiazine) in 0.1, 1 to 10 mg/kg concentrations, comparing to unsupplemented controls. Results showed that 0.1 and 1 mg/kg concentrations had no effects on black soldier fly larval survival, pupation, or eclosion. However, 10 mg/kg sulphonamide concentrations had a significantly negative impact on larval and pupal survival, body weight, pupation, and eclosion. Sulfadiazine was also detected in prepupae originating from larvae with feed supplements of 1 and 10 mg/kg, though at lower than input supplement concentrations (0.4663 and 0.7814 mg/100 larvae, respectively). These data indicate that black soldier fly larvae can reduce some sulphonamides, but depending on the concentration, may be detrimental to black soldier fly growth and development, while harbouring residues through the prepupal stage. These data underscore the importance of caution when generalising data across all life stages and antibiotic classes (GAO *et al.*, 2019).

Research on larvae raised on a variety of animal manures with or without antibiotic supplementation, and identifying key pathogens (both naturally occurring and intentionally inoculated), their frequency of occurrence, and expression of ARGs and virulence factors in both the digested manure and black soldier fly larvae are valuable. Such approaches will allow for an assessment of the utility of black soldier fly as a method to reduce antimicrobial agents and related compounds. Data may also allow for industrialisation of such processes that mitigate emergence, spread, or persistence of antibiotic resistant pathogens in the animal waste residue produced. Furthermore, these data would be useful for the federal government when reviewing current regulations on the use of insect protein as feed.

4   Black soldier fly bioproducts for the future: an example of the potential for other insect species

The black soldier fly system offers an innovative option for managing animal wastes that can be translated into widespread industrial use, transforming low value waste into high value, upcycled products (Figure 1), while also producing protein for use as livestock feed. Black soldier fly products for animal feeds (as whole insects, protein meals, fats, etc.) are currently the main (and most well-known) outputs when the black soldier fly is used for waste management. However, we recognise that until the biosafety of products generated from manure management can be ensured through continued research on the topic, in addition to the stigma associated with such a process, products generated from manure management should not enter the food stream and should be as purified as possible. This system would be useful worldwide to support circular economies in rural and urban locations worldwide, and even low-earth-orbit environments.

4.1   *Biodiesel*
Black soldier fly larvae are rich in many nutrients, especially fats and fatty acids, making them ideal candidates for renewable sources of biodiesel (Zheng *et al.*, 2012) The high ratio of medium chain saturated to polyunsaturated fatty acids makes larval black soldier fly fats ideal for producing biodiesel (Surendra *et al.* 2016). Li *et al.* (2011) were the first to investigate the conversion of dairy manure into biodiesel. At their laboratory scale, approximately 1.25 kg of manure yielded approximately 16 g of biodiesel. Although this resulting amount (1.26%) might seem small, at an industrial scale, this translate to a massive amount of biodiesel. For example, the average dairy cow produces approximately 36 kg of manure per day per 450 kg of weight (United States Department of Agriculture, 1995). Consider, an average weight (680 kg) mature Holstein cow produced approximately 54 kg of manure per day. If industrialised, such waste could be transformed to 0.68 kg of biodiesel. A black soldier fly facility co-located with a dairy farm with 2,000 head of cattle could produce 1,360 kg of biodiesel/day. At a density of 0.88 kg/l (Alleman *et al.*, 2016), this equates to approximately 1,550 l of biodiesel/day. At Huazhong Agricultural University in Wuhan, China, a biogas plant mixes agricultural waste streams (i.e. corncobs and

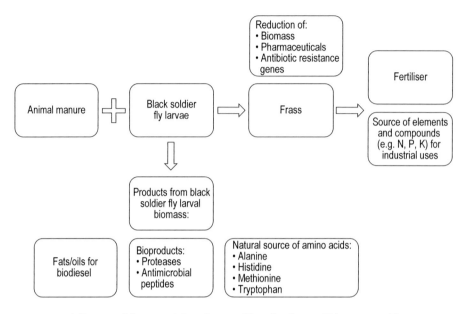

FIGURE 1   A diagram of the potential products and benefits that could be generated by processing animal manure with black soldier fly larvae

swine manure) to optimise energy yield (Li *et al.*, 2015). A 3:2 ratio of volatile solids was found to yield the most biogas, and the resulting residue investigated for biodiesel production. Bioconversion of this residue by black soldier fly larvae yielded 0.79 g biodiesel per 100 g of residue. The authors extrapolated from their study that the amount of corncob and manure needed to produce 1 metric ton of biodiesel could come from 16.78 hectares of corn and 387 head of swine per year. As another example, in June 2019, an estimated 75.5 million pigs were being raised in the USA (Barrett, 2019). If pig wastes were co-digested with corn residue for biogas production, and the biogas residue subsequently digested by black soldier fly larvae, the manure generated from US swine production could yield 195,090 metric tons (or approximately 172 million litres) of biodiesel annually. As of April 2020, China was estimated to produce approximately 2.6 times more pork than the USA, and 1.4 times more than the EU in 2020 (Shahbandeh, 2020). The amount of biodiesel that could result from swine production alone could be massive; the aforementioned volume of 172 million litres could replace approximately 1% of the total diesel used in the USA in 2019 (United States Energy Information Administration, 2020).

4.2   *Microbiological products*
Products from the black soldier fly larvae could result in a thriving industry to supply broad needs in waste management, agriculture and home garden protection, pharmaceuticals, and remediation of industrial biofouling. Resulting black soldier

fly-derived products, such as antimicrobial peptides, probiotics and prebiotics, represent promising strategies for selective microbial control in manure management. But these products may also offer additional practical benefits to society, such as antimicrobials in treating infections or in reducing biofouling. Some of these alternatives have been reported to embody most or all of the essential antibiotics' functions, while also being less likely to induce bacterial resistance (Alvarez *et al.*, 2019; Elhag *et al.*, 2017; Vogel *et al.*, 2018; Wu *et al.*, 2018). As discussed above, research has showed that a black soldier fly diet and associated bioproducts modify gut microbial communities, modulate immune systems, reduce or inhibit pathogens, and establish a favourable gut microbiome. Thus, the continued identification of novel black soldier fly-associated microbial strains could lead to the production of a microbial cocktail for use as a manure pretreatment for more efficient waste conversion to black soldier fly protein and lipids. And, as microbial amendment to manure reduced black soldier fly development time (Yu *et al.*, 2011), this application would have a direct impact on the productivity of black soldier fly in industrial applications of waste management. Furthermore, antimicrobial peptide testing will determine if any of the isolated AMPs have activity against other pathogens, or with microbial signalling or biofilm development. Indeed, research conducted thus far has yielded encouraging data that compel continuation into the investigation of black soldier fly bioconversion of animal manure, for optimisation aiming to ensure animal, food and environmental safety, while producing products of quality and environmental value.

The antimicrobial properties of insect haemolymph have been known and used for many years in traditional medicine (Seabrooks and Hu, 2017). It is not surprising, then, that the AMPs produced in black soldier fly larval haemolymph, excretions, and secretions are attractive options for antimicrobial therapeutics; in fact, they have been shown to have comparable, synergistic, or improved activity over traditional antibiotics against important human pathogens, as described above (Seabrooks and Hu, 2017; Wu *et al.*, 2018). Antimicrobial effects by black soldier fly larval extracts have also been demonstrated toward *E. coli*, *Micrococcus luteus*, *Pseudomonas fluorescens* and *B. subtilis* biofilms (Muller *et al.*, 2017; O'Toole and Kolter, 1998), and also against plant pathogens (Park *et al.*, 2015). These data are encouraging in that black soldier fly AMPs could be suitable as alternative or a combined therapy against a broad array of plant and animal infections, with potential to reduce antimicrobial resistance, with additional application for bioremediation and biofouling control. But, the concentrations of AMPs needed for clinical trials and other large-scale applications may be an obstacle for bringing black soldier fly AMPs to market. Other obstacles, such as potential cytotoxic, immunogenic, and allergenic reactions must also be considered. Nevertheless, continued investigations are merited in that black soldier fly mass production may be beneficial to produce desired AMP concentrations, as would be synthesising AMPs and optimising AMP properties.

4.3   Fertiliser

Once digested by black soldier fly larvae, the material remaining (frass) is lower in moisture content, mass, and nutrient content, making it more suitable (both economically and environmentally) for land applications as a fertiliser than the raw manure itself. Black soldier fly larvae are capable of reducing manure mass and nutrients in manure (Matos *et al.*, 2020). In dairy, poultry, and swine manure, black soldier flies reduce dry matter by 26-78% (Li *et al.*, 2011; Myers *et al.*, 2014; ur Rehman *et al.*, 2017b), 36-60% (Lalander *et al.*, 2019; Xiao *et al.*, 2018), and 33-56% (Miranda *et al.*, 2019; Newton *et al.*, 2005), respectively. Furthermore, black soldier fly larvae also reduce numerous nutrients that could otherwise be dangerous to the environment if overapplied as fertiliser. In dairy, poultry, and swine manure, nitrogen is reduced by 6-50% (Liu *et al.*, 2019; Myers *et al.*, 2014), 14-82% (Liu *et al.*, 2019; Oonincx *et al.*, 2015a), and 13-37% (Lalander *et al.*, 2015; Liu *et al.*, 2019), respectively. In dairy manure, phosphorus content is reduced by 61-70% (Myers *et al.*, 2014), and when reared on swine manure, phosphorus is reduced by 44%, respectively (Newton *et al.*, 2005). Though, Oonincx *et al.* (2015a) and Liu *et al.* (2019) found phosphorous increased over time for all three manure types; therefore, variation across studies may be due to differences in methodologies. Metals such as zinc, iron, and copper are also reduced by 26-85% in dairy, swine, and poultry manure (C.D. Miranda., T.L. Crippen, J.A. Cammack, and J.K. Tomberlin, unpublished data). Unfortunately, few studies exist on the utility of black soldier fly larval frass as a fertiliser, and even fewer studies used animal manure as the starting material. Zhu *et al.* (2012) found that black soldier fly larval frass produced from swine manure resulted in faster germination of cucumber and Chinese cabbage seeds in comparison to traditionally composted swine manure. Although the data are promising, more research is needed to understand the full potential and limitations of black soldier fly digested animal manure as a fertiliser.

4.4   *Other direct black soldier fly products*

Bioactive products from the black soldier fly could also be used as a green technology in pest management. As many organisations and individuals are turning more toward integrated pest management and practices, and organic farming, the black soldier fly and its products may be an attractive resource. For instance, black soldier fly larvae produce allomones that are repellent to house flies, resulting in the reduction in house fly occurrence and subsequent oviposition (Bradley and Sheppard, 1984; Sheppard, 1983). Farming the insect for integrated pest management would provide a means to aid in controlling house fly populations. However, the studies showed that the time black soldier fly larvae occupied the manure influenced house fly ovipositional responses (Bradley and Sheppard, 1984), suggesting that a constant black soldier fly population might need to be in place for efficient housefly reduction. Further, geographical ranges of black soldier flies and systematics of adaptation and genetics of the pests need to be considered for targeted or broad use.

For these reasons, isolation of the allomones would seem to be a more beneficial approach.

Black soldier fly larvae also produce trypsin and chymotrypsin, digestive enzymes responsible for protein break down through interactions with targeted amino acids (Kim *et al.*, 2011a,b; Park *et al.*, 2012). These proteases are utilised commercially for scientific research to understand protein structure and function, but also in broad food processing to improve the functional and nutritional quality of food proteins, as ingredients in detergents, in the pharmaceutical industry as therapeutic agents, and for use in bioremediation (Shah and Mital, 2018; Yu and Ahmedna, 2012; Zhou *et al.*, 2011). Commercial sources for proteases, such as trypsin and chymotrypsin, include plants, animals, and microbes. But several factors such as the availability of plants and animals for protease retrieval, land for cultivation and growth, climatic conditions, and agricultural policies, could restrict commercial production of these resources. Furthermore, isolation from microbes would also require sophisticated equipment and skills to prevent contamination. Mass production of insects, such as the black soldier fly, that would result in protease isolation is an interesting alternative to the previously mentioned sources, due to the ease of rearing and reduced space and nutritional needs. Additionally, the wide range of substrates used by black soldier fly could be exploited for those diverse protease commercial applications. However, the full scope of proteases and other biologically active enzymes in black soldier fly larvae, their activities, substrate specificity, pH tolerance, and thermotolerance is lacking. These, along with bioengineering for a robust yield, will need to be solved before these black soldier fly enzymes could be utilised commercially.

4.5   *Individual components*

In contrast to the old adage, we speculate the sum of the parts is greater than the whole especially as related to the black soldier fly. Black soldier fly larvae are quite nutritionally rich, especially at the amino acid level. Compared to soybean meal, black soldier flies contain higher levels of alanine, methionine, histidine, and tryptophan (Barragan-Fonseca *et al.*, 2017). Black soldier fly larvae are also high in fatty acids (Oonincx *et al.*, 2015b) particularly lauric acid, which has known antiviral and antibacterial properties (Gasco *et al.*, 2018). However, as mentioned above, techniques for the isolation and extraction of these compounds must be available and affordable before the full potential of insects and the associated products can be fully appreciated. Further research is needed to identify the processes (both the order/sequence and extraction/purification techniques) that could be implemented to yield the most benefits from the insect biomass. For example, once the insect biomass is dried, oil can be extracted for use as biodiesel, or further fractionated into specific fatty acids. Additionally, provided that the drying or subsequent extraction techniques do not damage the bioproducts of interest, amino acids, chitin, antimicrobial peptides, and/or enzymes could be subsequently extracted from the defatted biomass.

5   Conclusions

Using black soldier fly larvae for manure management is a win-win across the board. Their digestion reduces biomass, nutrients, odorous volatile organic compounds, and pathogens, while creating numerous products of value that can have both direct and indirect positive impacts on agriculture. Imagine a society where we could truly come full circle: brewery waste, post-consumer food waste, or food processing by products, are diverted to a black soldier fly farm, where it is upcycled by the larvae into feed for chickens. The resulting poultry manure is diverted to yet another black soldier fly farm, where it is upcycled by the larvae into purified proteins and amino acids for aquaculture feeds, and oil for biodiesel that powers the trucks transporting the materials. The frass from both black soldier fly farms is used as a soil amendment in crop production, and used to grow more corn and soybean, which can now be diverted to produce food for people. The possibilities of such a circular economy are endless. Regardless, black soldier fly data to date demonstrate what can be done with insects that digest manure. The question remains: will the lesser mealworm, house fly, and other insects be optimised and embraced for such abilities?

Conflict of interest

The authors declare no conflict of interest.

References

Adhikari, M., Paudel, K.P., Martin Jr, N.R. and Gauthier, W.M., 2005. Economics of dairy waste use as fertilizer in central Texas. Waste Management 25: 1067-1074. https://doi.org/10.1016/j.wasman.2005.06.012

Alleman, T.L., McCormick, R.L., Christensen, E.D., Fioroni, G. and Moriarty, K., 2016. Biodiesel handling and use guide. US Department of Energy, Washington, DC, USA.

Alvarez, D., Wilkinson, K.A., Treilhou, M., Tene, N., Castillo, D. and Sauvain, M., 2019. Prospecting peptides isolated from black soldier fly (Diptera: Stratiomyidae) with antimicrobial activity against *Helicobacter pylori* (Campylobacterales: Helicobacteraceae). Journal of Insect Science 19(6): 17. https://doi.org/10.1093/jisesa/iez120

Alyokhin, A., Buzza, A. and Beaulieu, J., 2019. Effects of food substrates and moxidectin on development of black soldier fly, *Hermetia illucens*. Journal of Applied Entomology 143: 137-143.

Augére-Granier, M.-L. 2019. The EU poultry meat and egg sector: main features, challenges and prospects. European Parliamentary Research Service, Brussels, Belgium.

Axtell, R.C. and Arends, J.J., 1990. Ecology and management of arthropod pests of poultry. Annual Review of Entomology 35: 101-126.

Axtell, R.C. and Edwards, T.D., 1970. *Hermetia illucens* control in poultry manure by larviciding. Journal of Economic Entomology 63: 1786-1787.

Axtell, R.C., 1986. Fly management in poultry production: cultural, biological, and chemical. Poultry Science 65: 657-667.

Azzollini, D., Wibisaphira, T., Lakemond, C.M.M. and Fogliano, V., 2019. Toward the design of insect-based meat analogue: the role of calcium and temperature in coagulation behavior of *Alphitobius diaperinus* proteins. LWT 100: 75-82.

Barragan-Fonseca, K.B., Dicke, M. and Van Loon, J.J.A., 2017. Nutritional value of the black soldier fly (*Hermetia illucens* L.) and its suitability as animal feed – a review. Journal of Insects as Food and Feed 3: 105-120. https://doi.org/10.3920/JIFF2016.0055

Barrett, J., 2019. United States hog inventory up 4 percent. United States Department of Agriculture, National Agricultural Statistics Service, Washington, DC, USA.

Barth, C.L., 1986. Fly control through manure management. Poultry Science 65: 668-674.

Beard, R.L. and Sands, D.C., 1973. Factors affecting degradation of poultry manure by flies. Environmental Entomology 2: 801-806.

Bertone, M.A., Green, J.T., Washburn, S.P., Poore, M.H. and Watson, D.W., 2006. The contribution of tunneling dung beetles to pasture soil nutrition. Forage and Grazinglands 4: 1-12.

Beskin, K.V., Holcomb, C.D., Cammack, J.A., Crippen, T.L., Knap, A.H., Sweet, S.T. and Tomberlin, J.K., 2018. Larval digestion of different manure types by the black soldier fly (Diptera: Stratiomyidae) impacts associated volatile emissions. Waste Management 74: 213-220. https://doi.org/10.1016/j.wasman.2018.01.019

Borrelli, L., Coretti, L., Dipineto, L., Bovera, F., Menna, F., Chiariotti, L., Nizza, A., Lembo, F. and Fioretti, A., 2017. Insect-based diet, a promising nutritional source, modulates gut microbiota composition and SCFAs production in laying hens. Scientific Reports 7: 16269. https://doi.org/10.1038/s41598-017-16560-6

Bradley, S.W. and Sheppard, D.C., 1984. House fly oviposition inhibition by larvae of *Hermetia illucens*, the black soldier fly. Journal of Chemical Ecology 10: 853-859.

Cai, M., Ma, S., Hu, R., Tomberlin, J.K., Thomashow, L.S., Zheng, L., Li, W., Yu, Z. and Zhang, J., 2018a. Rapidly mitigating antibiotic resistant risks in chicken manure by *Hermetia illucens* bioconversion with intestinal microflora. Environmental Microbiology 20: 4051-4062. https://doi.org/10.1111/1462-2920.14450

Cai, M., Ma, S., Hu, R., Tomberlin, J.K., Yu, C., Huang, Y., Zhan, S., Li, W., Zheng, L., Yu, Z. and Zhang, J., 2018b. Systematic characterization and proposed pathway of tetracycline degradation in solid waste treatment by *Hermetia illucens* with intestinal microbiota. Environmental Pollution 242: 634-642. https://doi.org/10.1016/j.envpol.2018.06.105

Cammack, J.A. and Tomberlin, J.K., 2017. The impact of diet protein and carbohydrate on select life-history traits of the black soldier fly *Hermetia illucens* (L.) (Diptera: Stratiomyidae). Insects 8: 56. https://doi.org/10.3390/insects8020056

Carbajal, C., Picazo, A., Gutierrez, S., Blackmon, S., Perez, G., Gonzalez, A., Fuentes, E., Wickham, C., Lopez, C., Johnson, D., and Kannan, S. 2017. Genesis of antibiotic resistance XXVII: action plan for Global Union for Antibiotics Research and Development (GUARD) to mitigate AR pandemic (ARP). FASEB Journal 31: 777.

Centers for Disease Control and Prevention (CDC), 2015. National action plan for combating antibiotic-resistant bacteria (national action plan). CDC, Atlanta, GA, USA.

Chang, C.C.Y. and Janzen, H.H., 1996. Long-term fate of nitrogen from annual feedlot manure applications. Journal of Environmental Quality 25: 785-790.

Chee-Sanford, J.C., Mackie, R.I., Koike, S., Krapac, I.G., Lin, Y.-F., Yannarell, A.C., Maxwell, S. and Aminov, R.I., 2009. Fate and transport of antibiotic residues and antibiotic resistance genes following land application of manure waste. Journal of Environmental Quality 38(3): 1086-1108.

Chen, S., Liao, W., Liu, C., Elliott, D.C., Brown, M.D. and Solana, A.E., 2003. Value-added chemicals from animal manure. US Department of Energy, Oak Ridge, TX, USA. Available at: https://www.osti.gov/biblio/15009485/

Čičková, H., Newton, G.L., Lacy, R.C. and Kozanek, M., 2015. The use of fly larvae for organic waste treatment. Waste Management 35: 68-80. https://doi.org/10.1016/j.wasman.2014.09.026

Čičková, H., Pastor, B., Kozanek, M., Martinez-Sanchez, A., Rojo, S. and Takac, P., 2012. Biodegradation of pig manure by the housefly, *Musca domestica*: a viable ecological strategy for pig manure management. PLoS ONE 7: e32798. https://doi.org/10.1371/journal.pone.0032798

Cosse, A.A. and Baker, T.C., 1996. House flies and pig manure volatiles: wind tunnel behavioral studies and electrophysiological evaluations. Journal of Agricultural Entomology 13: 301-317.

Crippen, T.L., Sheffield, C.L., Esquivel, S.V., Droleskey, R.E. and Esquivel, J.F., 2009. The acquisition and internalization of *Salmonella* by the lesser mealworm, *Alphitobius diaperinus* (Coleoptera: Tenebrionidae). Vector Borne Zoonotic Diseases 9: 65-72. https://doi.org/10.1089/vbz.2008.0103

Crippen, T.L., Zheng, L., Sheffield, C.L., Tomberlin, J.K., Beier, R.C. and Yu, Z., 2012. Transient gut retention and persistence of *Salmonella* through metamorphosis in the lesser mealworm, *Alphitobius diaperinus* (Coleoptera: Tenebrionidae). Journal of Applied Microbiology 112: 920-926. https://doi.org/10.1111/j.1365-2672.2012.05265.x

Davis, C.G., Harvey, D., Zahniser, S., Gale, F. and Liefert, W., 2013. Assessing the growth of US broiler and poultry meat exports. A report from the Economic Research Service. USDA, Washington, DC, USA, pp. 1-28.

Diener, S., Zurbruegg, C. and Tockner, K., 2009. Conversion of organic material by black soldier fly larvae: establishing optimal feeding rates. Waste Management & Research 27: 603-610. https://doi.org/10.1177/0734242X09103838

Edmonds, L., Gollehon, N., Kellogg, R.L., Knight, L., Lander, C., Lemunyon, J., Meyer, D., Moffitt, D.C. and Schaefer, J., 2003. Costs associated with development and implementation of comprehensive nutrient management plans, part I – nutrient management, land treatment, manure and wastewater handling and storage, and recordkeeping. United States Department of Resources Natural Resources Conservation Service, Washington, DC, USA.

Eleftherianos, I., Atri, J., Accetta, J. and Castillo, J.C., 2013. Endosymbiotic bacteria in insects:

guardians of the immune system? Frontiers in Physiology 4: 46. https://doi.org/10.3389/fphys.2013.00046

Elhag, O., Zhou, D., Song, Q., Soomro, A.A., Cai, M., Zheng, L., Yu, Z. and Zhang, J., 2017. Screening, expression, purification and functional characterization of novel antimicrobial peptide genes from *Hermetia illucens* (L.). PLoS ONE 12: e0169582. https://doi.org/10.1371/journal.pone.0169582

Erickson, M.C., Islam, M., Sheppard, C., Liao, J. and Doyle, M.P., 2004. Reduction of *Escherichia coli* O157:H7 and *Salmonella enterica* serovar Enteritidis in chicken manure by larvae of the black soldier fly. Journal of Food Protection 67: 685-690. https://doi.org/10.4315/0362-028x-67.4.685

European Union (EU), 2016. Action on antimicrobial resistance. EU, Brussels, Belgium. Available at: https://ec.europa.eu/health/antimicrobial-resistance/eu-action-on-antimicrobial-resistance_en

Food and Agriculture Organisation (FAO), 2006. Livestock's long shadow: environmental issues and options. FAO, Rome, Italy.

Food and Agriculture Organisation (FAO), 2008. Poultry in the 21st century: avian influenza and beyond. In: Thieme, O. and Pilling, D. (eds.) Proceedings of the International Poultry Conference. November 5-7, 2007. FAO, Bangkok, Thailand.

Gao, Q., Deng, W., Gao, Z., Li, M., Liu, W., Wang, X. and Zhu., F., 2019. Effect of sulfonamide pollution on the growth of manure management candidate *Hermetia illucens*. PLOS ONE. 14: e0216086. https://doi.org/10.1371/journal.pone.0216086

Garcia, R., Baelum, J., Fredslund, L., Santorum, P. and Jacobsen, C.S., 2010. Influence of temperature and predation on survival of *Salmonella enterica* serovar Typhimurium and expression of invA in soil and manure-amended soil. Applied and Environmental Microbiology 76: 5025-5031. https://doi.org/10.1128/AEM.00628-10

Gasco, L., Finke, M. and Van Huis, A., 2018. Can diets containing insects promote animal health? Journal of Insects as Food and Feed 4: 1-4. https://doi.org/10.3920/JIFF2018.x001

Gold, M., Cassar, C.M., Zurbrügg, C., Kreuzer, M., Boulos, S., Diener, S. and Mathys, A., 2020. Biowaste treatment with black soldier fly larvae: increasing performance through the formulation of biowastes based on protein and carbohydrates. Waste Management 102: 319-329.

Grossman, E., 2014. As dairy farms grow bigger, new concerns about pollution. Yale Environment 360. Available at: https://e360.yale.edu/features/as_dairy_farms_grow_bigger_new_concerns_about_pollution

Hall, R.D. and Foehse, M.C., 1980. Laboratory and field tests of CGA-72662 for control of the house fly and face fly in poultry, bovine, or swine manure. Journal of Economic Entomology 73: 564-569.

Haruta, S. and Kanno, N., 2015. Survivability of microbes in natural environments and their ecological impacts. Microbes and Environments 30: 123-125. https://doi.org/10.1264/jsme2.ME3002rh

Hussein, M., Pillai, V.V., Goddard, J.M., Park, H.G., Kothapalli, K.S., Ross, D.A., Ketterings, Q.M., Brenna, J.T., Milstein, M.B. and Marquis, H., 2017. Sustainable production of housefly (*Musca domestica*) larvae as a protein-rich feed ingredient by utilizing cattle manure. PLoS ONE 12: e0171708.

Jiang, X., Islam, M., Morgan, J. and Doyle, M.P., 2004. Fate of *Listeria monocytogenes* in bovine manure-amended soil. Journal of Food Protection 67: 1676-1681. https://doi.org/10.4315/0362-028x-67.8.1676

Kaltenpoth, M., 2009. Actinobacteria as mutualists: general healthcare for insects? Trends in Microbiology 17: 529-535. https://doi.org/10.1016/j.tim.2009.09.006

Kaltenpoth, M., Gottler, W., Herzner, G. and Strohm, E., 2005. Symbiotic bacteria protect wasp larvae from fungal infestation. Current Biology 15: 475-479. https://doi.org/10.1016/j.cub.2004.12.084

Keiding, J., 1986. The house-fly: biology and control. World Health Organization, Geneva, Switzerland.

Khan, H.A.A., Shad, S.A. and Akram, W., 2012. Effect of livestock manures on the fitness of house fly, *Musca domestica* L. (Diptera: Muscidae). Parasitology Research 111: 1165-1171. https://doi.org/10.1007/s00436-012-2947-1

Kim, W., Bae, S., Kim, A., Park, K., Lee, S., Choi, Y., Han, S., Park, Y. and Koh, Y., 2011a. Characterization of the molecular features and expression patterns of two serine proteases in *Hermetia illucens* (Diptera: Stratiomyidae) larvae. BMB Reports 44: 387-392.

Kim, W., Bae, S., Park, K., Lee, S., Choi, Y., Han, S. and Koh, Y., 2011b. Biochemical characterization of digestive enzymes in the black soldier fly, *Hermetia illucens* (Diptera: Stratiomyidae). Journal of Asia-Pacific Entomology 14: 11-14. https://doi.org/10.1016/j.aspen.2010.11.003

Kim, Y.B., Kim, D., Jeong, S., Lee, J.K., T., Lee, H. and Lee, K., 2020. Black soldier fly larvae oil as an alternative fat source in broiler nutrition. Poultry Science 99: 3133-3143.

Kudva, I.T., Blanch, K. and Hovde, C.J., 1998. Analysis of *Escherichia coli* O157:H7 survival in ovine or bovine manure and manure slurry. Applied and Environmental Microbiology 64: 3166-3174. https://doi.org/10.1128/AEM.64.9.3166-3174.1998

Kyakuwaire, M., Olupot, G., Amoding, A., Nkedi-kizza, P. and Ateenyi, B.T., 2019. How safe is chicken litter for land application as an organic fertilizer? A review. International Journal of Environmental Research and Public Health 16: 3521.

Lalander, C., Diener, S., Zurbrügg, C. and Vinnerås, B., 2019. Effects of feedstock on larval development and process efficiency in waste treatment with black soldier fly (*Hermetia illucens*). Journal of Cleaner Production 208: 211-219.

Lalander, C., Senecal, J., Gros Calvo, M., Ahrens, L., Josefsson, S., Wiberg, K. and Vinneras, B., 2016. Fate of pharmaceuticals and pesticides in fly larvae composting. Science of the Total Environment 565: 279-286. https://doi.org/10.1016/j.scitotenv.2016.04.147

Lalander, C.H., Fidjeland, J., Diener, S., Eriksson, S. and Vinnerås, B., 2015. High waste-to-biomass conversion and efficient *Salmonella* spp. reduction using black soldier fly for waste recycling. Agronomy for Sustainable Development 35: 261-271. https://doi.org/10.1007/s13593-014-0235-4

Larrain, P.S. and Salas, C.F., 2008. House fly (*Musca domestica* L.) (Diptera: Muscidae) development in different types of manure. Chilean Journal of Agricultural Research 68: 192-197.

Leni, G., Soetemans, L., Jacobs, J., Depraetere, S., Ganotten, N., Bastiaens, L., Calgiani, A. and Sforza, S., 2020. Protein hydrolysates from *Alphitobius diaperinus* and *Hermetia illucens* larvae treated with commercial proteases. Journal of Insects as Food and Feed 6: 393-404. https://doi.org/10.3920/JIFF2019.0037

Li, Q., Zheng, L., Qiu, N., Cai, H., Tomberlin, J.K. and Yu, Z., 2011. Bioconversion of dairy manure by black soldier fly (Diptera: Stratiomyidae) for biodiesel and sugar production. Waste Management 31: 1316-1320.

Li, W., Zheng, L., Wang, Y., Zhang, J., Yu, X. and Zhang, Y., 2015. Potential biodiesel and biogas production from corncob by anaerobic fermentation and black soldier fly. Biosource Technology 194: 276-282.

Liu, Q., Tomberlin, J.K., Brady, J.A., Sanford, M.R. and Yu, Z., 2008. Black soldier fly (Diptera: Stratiomyidae) larvae reduce *Escherichia coli* in dairy manure. Environmental Entomology 37: 1525-1530. https://doi.org/10.1603/0046-225x-37.6.1525

Liu, T., Awasthi, M.K., Chen, H., Duan, Y., Awasthi, S.K. and Zhang, Z., 2019. Performance of black soldier fly larvae (Diptera: Stratiomyidae) for manure composting and production of cleaner compost. Journal of Environmental Management 251: 109593.

Lyons, B.N., Crippen, T.L., Zheng, L., Teel, P.D., Swiger, S.L. and Tomberlin, J.K., 2017. Susceptibility of *Alphitobius diaperinus* in Texas to permethrin- and beta-cyfluthrin-treated surfaces. Pest Management Science 73: 562-567. https://doi.org/10.1002/ps.4327

MacDonald, J.M., O'Donoghue, E., McBride, W.D., Nehring, R., Sandretto, C. and Mosheim, R., 2007. Profits, costs, and the changing structure of dairy farming. USDA-ERS, Washington, DC, USA.

Manyi-Loh, C.E., Mamphweli, S.N., Meyer, E.L., Makaka, G., Simon, M. and Okoh, A.I., 2016. An overview of the control of bacterial pathogens in cattle manure. International Journal of Environmental Research and Public Health 13(9): 843. https://doi.org/10.3390/ijerph13090843

Matos, J.S., Barberino, A.T.M.S., De Araujo, L.P., Lôbo, I.P. and De Almeida Neto, J.A., 2020. Potentials and limitations of the bioconversion of animal manure using fly larvae. Waste and Biomass Valorization: 1-24.

Mazza, L., Xiao, X., Ur Rehman, K., Cai, M., Zhang, D., Fasulo, S., Tomberlin, J.K., Zheng, L., Soomro, A.A., Yu, Z. and Zhang, J., 2020. Management of chicken manure using black soldier fly (Diptera: Stratiomyidae) larvae assisted by companion bacteria. Waste Management 102: 312-318. https://doi.org/10.1016/j.wasman.2019.10.055

Miller, A., 1954. Dung beetles (Coleoptera, Scarabaeidae) and other insects in relation to human feces in a hookworm area of southern Georgia. American Journal of Tropical Medicine and Hygiene 3: 372-389. https://doi.org/10.4269/ajtmh.1954.3.372

Miller, B., Teotia, J. and Thatcher, T., 1974. Digestion of poultry manure by *Musca domestica*. Journal of British Poultry Science 15: 231-234.

Miranda, C., Cammack, J. and Tomberlin, J., 2020a. Life-history traits of house fly, *Musca domestica* L. (Diptera: Muscidae), reared on three manure types. Journal of Insects as Food and Feed 6: 81-90. https://doi.org/10.3920/JIFF2019.0001

Miranda, C.D., Cammack, J.A. and Tomberlin, J.K., 2019. Life-history traits of the black soldier fly, *Hermetia illucens* (L.) (Diptera: Stratiomyidae), reared on three manure types. Animals 9: 281.

Miranda, C.D., Cammack, J.A. and Tomberlin, J.K., 2020b. Mass production of the black soldier fly, *Hermetia illucens* (L.), (Diptera: Stratiomyidae) reared on three manure types. Animals 20: 1243.

Mohr, C.O., 1943. Cattle droppings as ecological units. Ecological Monographs 13: 275-298.

Morrison, P.E. and Davies, D.M., 1964. Feeding of dry chemically defined diets + egg production in adult house-fly. Nature 201: 104-105. https://doi.org/10.1038/201104a0

Muller, A., Wolf, D. and Gutzeit, H., 2017. The black soldier fly, *Hermetia illucens*-a promising source for sustainable production of proteins, lipids, and bioactive substances. Zeitschrif fur Naturforschung C 72: 351-363.

Myers, H.M., Tomberlin, J.K., Lambert, B.D. and Kattes, D., 2014. Development of black soldier fly (Diptera: Stratiomyidae) larvae fed dairy manure. Environmental Entomology 37: 11-15. https://doi.org/10.1093/ee/37.1.11

National Chicken Council, 2019a. Broiler chicken industry key facts 2019. NCC, Washington, DC, USA. Available at: https://www.nationalchickencouncil.org/about-the-industry/statistics/broiler-chicken-industry-key-facts/

National Chicken Council, 2019b. U.S. broiler production. NCC, Washington, DC, USA. Available at: https://www.nationalchickencouncil.org/about-the-industry/statistics/u-s-broiler-production/

Newton, L., Sheppard, C., Watson, D., Burtle, G. and Dove, R., 2005. Using the black soldier fly, *Hermetia illucens*, as a value-added tool for the management of swine manure. North Carolina State University, Raleigh, NC, USA.

Nguyen, T.T.X., Tomberlin, J.K. and Vanlaerhoven, S., 2013. Influence of resources on *Hermetia illucens* (Diptera: Stratiomyidae) larval development. Journal of Medical Entomology 50: 898-906. https://doi.org/10.1603/me12260

Nichols, E., Alarcon, V., Forgie, S., Gomez-Puerta, L.A. and Jones, M.S., 2017. Coprophagous insects and the ecology of infectious diseases of wildlife. ILAR Journal 58: 336-342. https://doi.org/10.1093/ilar/ilx022

Nichols, E., Spector, S., Louzada, J., Larsen, T., Amezquita, S., Favila, M.E. and The Scarabaeinae Research Network, 2008. Ecological functons and ecosystem services providwed by Scarabaeinae dung beetles. Biological Conservaion 161: 1461-1474.

Olsen, M.E., Goh, J., Phillips, M., Guselle, N. and McAllister, T.A., 1999. *Giardia* cyst and Cryptosporidium oocyst survival in water, soil, and cattle feces. Journal of Environmental Quality 28: 1991-1996.

Olson, M., 2020. Human and animal pathogens in manure. Department of Microbiology and Infectious Diseases, University of Calgary, Calgary, Canada.

Oonincx, D.G.A.B., Van Huis, A. and Van Loon, J.J.A., 2015a. Nutrient utilisation by black soldier flies fed with chicken, pig, or cow manure. Journal of Insects as Food and Feed 1: 131-139. https://doi.org/10.3920/JIFF2014.0023

Oonincx, D.G.A.B., Van Broekhoven, S., Van Huis, A. and Van Loon, J.J.A., 2015b. Feed conversion, survival and development, and composition of four insect species on diets composed of food by-products. PLoS ONE 10: e0144601. https://doi.org/10.1371/journal.pone.0144601

O'Toole, G.A. and Kolter, R., 1998. Initiation of biofilm formation in *Pseudomonas fluorescens* WCS365 proceeds via multiple, convergent signalling pathways: a genetic analysis. Molecular Microbiology 28: 449-461. https://doi.org/10.1046/j.1365-2958.1998.00797.x

Paniagua Voirol, L.R., Frago, E., Kaltenpoth, M., Hilker, M. and Fatouros, N.E., 2018. Bacterial symbionts in lepidoptera: their diversity, transmission, and impact on the host. Frontiers in Microbiology 9: 556. https://doi.org/10.3389/fmicb.2018.00556

Park, B., Um, K., Choi, W. and Park, S., 2017. Effect of feeding black soldier fly pupa meal in the diet on egg production, egg quality, blood lipid profiles and faecal bacteria in laying hens. European Poultry Science 81.

Park, K.H., Choi, Y.C., Nam, S.H. and Kim, W., 2012. Recombinant expression and enzyme activity of chumotrypsin-like protease from black soldier fly, *Hermetia illucens* (Diptera: Stratiomyidae). International Journal of Industrial Entomology 25(2): 181-185.

Park, K.H., Kwak, K.W., Nam, S.H., Choi, J.Y., Lee, S.H., Kim, H.G. and Kim, S.H., 2015. Antibacterial activity of larval extract from the black soldier fly *Hermetia illucens* (Diptera: Stratiomyidae) against plant pahtogens. Journal of Entomology and Zoology Studies 3: 176-179.

Prosser, R.S. and Sibley, P.K., 2015. Human health risk assessment of pharmaceuticals and personal care products in plant tissue due to biosolids and manure amendments, and wastewater irrigation. Environment International 75: 223-233.

Roncolini, A., Milanovic, V., Aquilanti, L., Cardinali, F., Garofalo, C., Sabbatini, R., Clementi, F., Belleggia, L., Pasquini, M., Mozzon, M., Foligni, R., Federica Trombetta, M., Haouet, M.N., Serena Altissimi, M., Di Bella, S., Piersanti, A., Griffoni, F., Reale, A., Niro, S. and Osimani, A., 2020. Lesser mealworm (*Alphitobius diaperinus*) powder as a novel baking ingredient for manufacturing high-protein, mineral-dense snacks. Food Research International 131: 109031. https://doi.org/10.1016/j.foodres.2020.109031

Rumbos, C.I., Karapanagiotidis, I.T., Mente, E. and Athanassiou, C.G., 2019. The lesser mealworm *Alphitobius diaperinus*; a noxious pest or a promising nutrient source? Reviews in Aquaculture 11: 1418-1437.

Sarpong, D., Oduru-Kwarteng, S., Gyasi, S.F., Buamah, R., Donkor, E., Awuah, E. and Baah, M.K., 2019. Biodegradation by composting of municipal organic solid waste into organic fertilizer using the black soldier fly (*Hermetia illucens*) (Diptera: Stratiomyidae) larvae. Internationa Journal of Recycling of Organic Waste in Agriculture 8: 45-54.

Sasaki, T., Kobayashi, M. and Agui, N., 2000. Epidemiological potential of excretion and regurgitation by *Musca domestica* (Diptera: Muscidae) in the dissemination of *Escherichia coli* O157: H7 to food. Journal of Medical Entomology 37: 945-949. https://doi.org/10.1603/0022-2585-37.6.945

Seabrooks, L. and Hu, L., 2017. Insects: an underrepresented resource for the discovery of biologically active natural products. Acta Pharmaceutica Sinica B 7: 409-426. https://doi.org/10.1016/j.apsb.2017.05.001

Shah, D. and Mital, K., 2018. The role of trypsin:chymotrypsin in tissue repair. Advances in Therapy 35: 31-42. https://doi.org/10.1007/s12325-017-0648-y

Shahbandeh, M., 2020. Global pork production, by country. Statista, New York, NY, USA. Availableat:https://www.statista.com/statistics/273232/net-pork-production-worldwide-by-country

Sheng, L., Shen, X., Benedict, C., Su, Y., Tsai, H.C., Schacht, E., Kruger, C.E., Drennan, M. and Zhu, M.J., 2019. Microbial safety of dairy manure fertilizer application in raspberry production. Frontiers in Microbiology 10: 2276. https://doi.org/10.3389/fmicb.2019.02276

Sheppard, C., 1983. House fly and lesser fly control utilizing the black soldier fly in manure management systems for caged laying hens. Environmental Entomology 12: 1439-1442.

Sheppard, D.C., Newton, G.L. and Savage, S., 1994. A value added manure management system using the black soldier fly. Bioresource Technology 50: 275-279.

Sierra, J.M., Fuste, E., Rabanal, F., Vinuesa, T. and Vinas, M., 2017. An overview of antimicrobial peptides and the latest advances in their development. Expert Opinion on Biological Therapy 17: 663-676. https://doi.org/10.1080/14712598.2017.1315402

Sim, W.J., Lee, J.W., Lee, E.S., Shin, S.K., Hwang, S.R. and Oh, J.E., 2011. Occurrence and distribution of pharmaceuticals in wastewater from households, livestock farms, hospitals and pharmaceutical manufactures. Chemosphere 82(2): 179-186.

Sobsey, M.D., Khatib, L.A., Hill, V.R., Alocilja, E. and Pillai, S., 2006. Pathogens in animal wastes and the impacts of waste management practices on their survival, transport and fate. In: Rice, J.M., Caldwell, D.F. and Humenik, F.J. (eds.) Animal agriculture and the environment: national center for manure and animal waste management white papers. ASABE, St. Joseph, MI, USA.

Stanford, K., Reuter, T., Gilroyed, B.H. and McAllister, T.A., 2015. Impacts of sporulation temperature, exposure to compost matrix and temperature on survival of *Bacillus cereus* spores during livestock mortality composting. Journal of Applied Microbiology 118: 989-997. https://doi.org/10.1111/jam.12749

St-Hilaire, S., Cranfill, K., McGuire, M.A., Mosley, E.E., Tomberlin, J.K., Newton, G.L., Sealey, W., Sheppard, C. and Irving, S., 2007. Fish offal recycling by the black soldier fly produces a foodstuff high in omega-3 fatty acids. Journal of the World Aquaculture Society 38: 309-313.

Sun, Y. and O'Riordan, M.X., 2013. Regulation of bacterial pathogenesis by intestinal short-chain fatty acids. Advances in Applied Microbiology 85: 93-118. https://doi.org/10.1016/B978-0-12-407672-3.00003-4

Surendra, K.C., Olivier, R., Tomberlin, J.K., Jha, R. and Khanal, S.K., 2016. Bioconversion of organic wastes into biodiesel and animal feed via insect farming. Renewable Energy 98: 197-202. https://doi.org/10.1016/j.renene.2016.03.022.

Teotia, J.S. and Miller, B.F., 1974. Nutritive content of house fly pupae and manure residue. British Poultry Science 15: 177-182. https://doi.org/10.1080/00071667408416093

Tomberlin, J.K. and Van Huis, A., 2020. Black soldier fly from pest to 'crown jewel' of the insects as feed industry: an historical perspective. Journal of Insects as Food and Feed 6: 1-4. https://doi.org/10.3920/JIFF2020.0003

Tomberlin, J.K., Richman, D. and Myers, H.M., 2008. Susceptibility of *Alphitobius diaperinus* (Coleoptera: Tenebrionidae) from broiler facilities in Texas to four insecticides. Journal of Economic Entomology 101: 480-483.

Tomberlin, J.K., Sheppard, D.C. and Joyce, J.A., 2002. Selected life-history traits of black soldier flies (Diptera: Stratiomyidae) reared on three artificial diets. Annals of the Entomological Society of America 95: 379-386. https://doi.org/10.1603/0013-8746(2002)095[0379
:slhtob]2.0.co;2

United States Department of Agriculture, 2015. Overview of the United States hog industry:
breeding herd efficiency continues to increase. USDA, Washington, DC, USA.

United States Department of Agriculture, 1995. Animal manure management. USDA,
Washington, DC, USA. Available at: https://www.nrcs.usda.gov/wps/portal/nrcs/detail/
null/?cid=nrcs143_014211.

United States Energy Information Administration, 2020. Diesel fuel explained. US Energy
Information Administration, Washington, DC, USA. Available at: https://www.eia.gov/
energyexplained/diesel-fuel/use-of-diesel.php

ur Rehman, K., Cai, M., Xiao, X., Zheng, L., Wang, H., Soomro, A.A., Zhou, Y., Li, W., Yu, Z.
and Zhang, J., 2017a. Cellulose decomposition and larval biomass production from the
co-digestion of dairy manure and chicken manure by mini-livestock (*Hermetia illucens*
L.). Journal of Environmental Management 196: 458-465.

ur Rehman, K., Rehman, A., Cai, M., Zheng, L., Xiao, X., Somroo, A.A., Wang, H., Li, W., Yu, Z.
and Zhang, J., 2017b. Conversion of mixtures of dairy manure and soybean curd residue
by black soldier fly larvae (*Hermetia illucens* L.). Journal of Cleaner Production 154: 366-
373.

Usui, M., Kawakura, M., Yoshizawa, N., San, L.L., Nakajima, C., Suzuki, Y. and Tamura, Y., 2017.
Survival and prevalence of *Clostridium difficile* in manure compost derived from pigs.
Anaerobe 43: 15-20. https://doi.org/10.1016/j.anaerobe.2016.11.004

Vogel, H., Muller, A., Heckel, D.G., Gutzeit, H. and Vilcinskas, A., 2018. Nutritional immunology: diversification and diet-dependent expression of antimicrobial peptides in the black
soldier fly *Hermetia illucens*. Developmental and Comparative Immunology 78: 141-148.
https://doi.org/10.1016/j.dci.2017.09.008

Wallace, J.S., Garner, E., Pruden, A. and Aga, D.S., 2018. Occurrence and transformation of
veterinary antibiotics and antibiotic resistance genes in dairy manure treated by advanced anaerobic digestion and conventional treatment methods. Environmental Pollution 236: 764-772

Wang, H., Li, H., Gilbert, J.A., Li, H., Wu, L., Liu, M., Wang, L., Zhou, Q., Yuan, J. and Zhang, Z.,
2015. Housefly larva vermicomposting efficiently attenuates antibiotic resistance genes
in swine manure, with concomitant bacterial population changes. Applied and Environmental Microbiology 81: 7668-7679.

Wang, H., Sangwan, N., Li, H.-Y., Su, J.-Q., Oyang, W.-Y., Zhang, Z.-J., Gilbert, J.A., Zhu, Y.-G.,
Ping, F. and Zhang, H.-L., 2017. The antibiotic resistome of swine manure is significantly
altered by association with the *Musca domestica* larvae gut microbiome. The ISME Journal 11: 100-111.

Wang, H., Zhang, Z., Czapar, G.F., Winkler, M.K. and Zheng, J., 2013. A full-scale house fly
(Diptera: Muscidae) larvae bioconversion system for value-added swine manure reduction. Waste Management & Research 31: 223-231.

Wasala, L., Talley, J.L., Desilva, U., Fletcher, J. and Wayadande, A., 2013. Transfer of *Escherichia coli* O157:H7 to spinach by house flies, *Musca domestica* (Diptera: Muscidae). Phytopathology 103: 373-380. https://doi.org/10.1094/PHYTO-09-12-0217-FI

Welch, C. and O'Hagan, M., 2010. Officials seek cause of Snohomish dairy-manure spill. The Seattle Times, April 14. Available at: https://www.seattletimes.com/seattle-news/officials-seek-cause-of-snohomish-dairy-manure-spill/

World Health Organization, 2015. Global action plan on antimicrobial resistance. WHO, Geneva, Switzerland.

Wu, Q., Patocka, J. and Kuca, K., 2018. Insect antimicrobial peptides, a mini review. Toxins 10: 461. https://doi.org/10.3390/toxins10110461

Xiao, X., Mazza, L., Yu, Y., Cai, M., Zheng, L., Tomberlin, J.K., Yu, J., Van Huis, A., Yu, Z. and Fasulo, S., 2018. Efficient co-conversion process of chicken manure into protein feed and organic fertilizer by *Hermetia illucens* L.(Diptera: Stratiomyidae) larvae and functional bacteria. Journal of Environmental Management 217: 668-676.

Yu, G., Cheng, P., Chen, Y., Li, Y., Yang, Z., Chen, Y. and Tomberlin, J.K., 2011. Inoculating poultry manure with companion bacteria influences growth and development of black soldier fly (Diptera: Stratiomyidae) larvae. Environmental Entomology 40: 30-35. https://doi.org/10.1603/EN10126

Yu, J. and Ahmedna, M., 2012. Functions/applications of trypsin in food processing and food science research. In: Weaver, K. and Kelley, C. (eds.) Trypsin: structure, biosynthesis and functions. Nova Science Publishers, Hauppauge, NY, USA, pp. 75-96.

Yuan, Q.B., Zhai, Y.F., Mao, B.Y., Schwarz, C. and Hu, N., 2019. Fates of antibiotic resistance genes in a distributed swine wastewater treatment plant. Water Environmental Research 91: 1565-1575.

Zhang, Z., Wang, H., Zhu, J., Suneethi, S. and Zheng, J., 2012. Swine manure vermicomposting via housefly larvae (*Musca domestica*): the dynamics of biochemical and microbial features. Bioresource Technology 118: 563-571. https://doi.org/10.1016/j.biortech.2012.05.048

Zheng, L., Hou, Y., Li, W., Yang, S., Li, Q. and Yu, Z., 2012. Biodiesel production from rice straw and restaurant waste employing black soldier fly assisted by microbes. Energy 47: 225-229.

Zhou, F., Tomberlin, J.K., Zheng, L., Yu, Z. and Zhang, J., 2013. Developmental and waste reduction plasticity of three black soldier fly strains (Diptera: Stratiomyidae) raised on different livestock manures. Journal of Medical Entomology 50: 1224-1230.

Zhou, L., Budge, S.M., Ghaly, A.E., Brooks, M.S. and Dave, D., 2011. Extraction, purification and characterization of fish chymotrypsin: a review. American Journal of Biochemistry and Biotechnology 7: 104-123.

Zhu, F.X., Wang, W.P., Hong, C.L., Feng, M.G., Xue, Z.Y., Chen, Y., Yao, Y.L. and Yu, M., 2012. Rapid production of maggots as feed supplement and organic fertilizer by the two-stage composting of pig manure. Bioresource Technology 116: 485-491.

Zi-zhe, C., De-po, Y., Sheng-qing, W., Yong, W., Reaney, M.J., Zhi-min, Z., Long-ping, Z., Guo, S., Yi, N. and Dong, Z., 2017. Conversion of poultry manure to biodiesel, a practical method of producing fatty acid methyl esters via housefly (*Musca domestica* L.) larval lipid. Fuel 210: 463-471.

Zurek, K. and Nayduch, D., 2016. Bacterial associations across house fly life history: evidence for transstadial carriage from managed manure. Journal of Insect Science 16: 2.

Zurek, L., Schal, C. and Watson, D., 2000. Diversity and contribution of the intestinal bacterial community to the development of *Musca domestica* (Diptera: Muscidae) larvae. Journal of Medical Entomology 37: 924-928.

Diseases in edible insect rearing systems

G. Maciel-Vergara[1,2,3]*, A.B. Jensen[1], A. Lecocq[1] and J. Eilenberg[1]

[1]Department of Plant and Environmental Sciences, Faculty of Science, University of Copenhagen, Thorvaldsensvej 40, 1871 Frederiksberg C, Denmark; [2]Laboratory of Entomology, Wageningen University, Droevendaalsesteeg 1, 6708 PB Wageningen, The Netherlands; [3]Laboratory of Virology, Wageningen University, Droevendaalsesteeg 1, 6708 PB Wageningen, The Netherlands; *gabriela.macielvergara@wur.nl, gmv@plen.ku.dk

Abstract

Due to a swift and continuous growth of the insect rearing industry during the last two decades, there is a need for a better understanding of insect diseases (caused by insect pathogens). In the insect production sector, insect diseases are a bottleneck for every type and scale of rearing system with different degrees of technology investment (i.e. semi-open rearing, closed rearing, industrial production, small-scale farming). In this paper, we provide an overview of insect pathogens that are causing disease in the most common insect species reared or collected for use in food and feed. We also include a few examples of diseases of insect species, which are not (yet) reported to be used as food or feed; those examples may increase our understanding of insect diseases in general and for the development of disease prevention and control measures. We pay special attention to the effect of selected biotic and abiotic factors as potential triggers of insect diseases. We discuss the effect of such factors in combination with other production variables on disease development and insect immunocompetence. Additionally, we touch upon prevention and control measures that have been carried out and suggested up to now for insect production systems. Finally, we point towards possible future research directions with possibilities to enhance the resilience of insect production to insect disease outbreaks.

Keywords

edible insects – insect rearing systems – insect diseases – epizootics – stress factors

1   **Introduction**

A large body of our current knowledge on taxonomic, behavioural and pathobiological aspects of insect host-pathogens interactions is based on a limited number of studies on insect pathogens causing disease outbreaks in insects, either in wild or in captive insect populations (Boucias and Pendland, 1998; Onstad and Carruthers, 1990; Steinhaus, 1963; Weiser, 1977). Usually in the past, the discovery and description of pathogens took place because of striking epidemic disease outbreaks in insect populations or they were based on observations on a few diseased individuals (Andreadis and Weseloh, 1990; Becnel and Andreadis, 2014; Brun, 1984; Majumdar *et al.*, 2008; Valles and Chen, 2006). Historically, biological control of agricultural insect pests using microorganisms (Lacey *et al.*, 2001; Sanchis, 2011; Van Lenteren *et al.*, 2018), diseases in honey bees (Bailey, 1968) and in silkworms (Samson *et al.*, 1990) have been the focus of many studies of insect diseases. Furthermore, insect-microbe interactions have also been studied as models to understand epidemiological aspects of human diseases (Scully and Bidochka, 2006). Insect pathogens have also proven to be beneficial for humans in other ways; baculoviruses for example, are used for biotechnological applications (i.e. for vaccines, and oncological treatments) (Felberbaum, 2015; Hofmann *et al.*, 1995; Van Oers, 2006).

   The presence of insect diseases in rearing facilities is definitely not new. Indeed, the most ancient insect husbandry systems developed by humans, apiculture (bee keeping) and sericulture (silk farming), have long suffered from the effects of diseases (Eilenberg and Jensen, 2018a; James and Li, 2012). Nevertheless, given the vast amount of insect and pathogen species in the world and the many different ways in which insects can be useful for humanity, there is still a lot to learn about insects and their pathogens. This is underlined by the challenge posed by the development of infectious diseases in rearing systems of insects produced for food and feed (further referred to as edible insects). On the bright side, the widespread use of molecular techniques, has increased the discovery of (insect) pathogens, especially of viruses (De Miranda *et al.*, 2021; Junglen and Drosten, 2013; Liu *et al.*, 2015), and the understanding of the microbiome of several insect species, including that of a number of edible insects (Vandeweyer *et al.*, 2017). At the same time, new knowledge is continuously being gathered as more research is conducted on the impact of known (Lecocq *et al.*, 2021) and understudied pathogens (G. Maciel-Vergara *et al.*, unpublished data) on insect health in species commonly reared as food and/or feed.

2   **Pathogens of insects collected from nature or reared as food and feed**

Insects form a diverse class of arthropods harbouring a high diversity of pathogens associated with individual species. Viral, fungal, bacterial, and microsporidian pathogens are frequently found to infect insects or in association with diseased insects (Supplementary Material Table S1). Insect pathogens can be specialists, only infecting one or a few taxonomically closely related species like the fungal genus *Strongwellsea* (Eilenberg and Jensen, 2018b), or they can be generalists infecting a variety of insect species which may not be taxonomically related, which is the case for many hypocrealean fungi (Hajek, 1997). Furthermore, some insect pathogens are known to be opportunistic or facultative. Opportunistic pathogens have a broad host range and are often ubiquitous as they can survive and proliferate on a range of substrates other than the main host (Brodeur, 2012); on the other hand, obligate pathogens need their host to fulfil their life cycle (Han and Weiss, 2017). Normally, opportunistic pathogens only cause disease when insects are subjected to stressful conditions (Jurat-Fuentes and Jackson, 2012; Pagnocca *et al.*, 2012; Sikorowski and Lawrence, 1994).

Viruses infecting insects and causing concern in mass production facilities comprise RNA as well as DNA viruses belong to different virus families (reviewed by Maciel-Vergara and Ros, 2017). Among these viruses, many are host-specific. An exception is the invertebrate iridescent virus 6 (IIV-6), known to infect several hosts in the orders Orthoptera and Blattodea (Just and Essbauer, 2001; Kleespies *et al.*, 1999) including gryllids, locusts, and cockroaches. In addition, larvae of the great wax moth, *Galleria mellonella* have shown susceptibility to IIV-6 under experimental conditions (Jakob *et al.*, 2002) as well as lepidopteran and dipteran cultured cell lines (Bronkhorst *et al.*, 2014; Williams *et al.*, 2009). Most entomopathogenic viruses known up to date are taxonomically distant from vertebrate viruses (Miller and Ball, 1998).

Viruses have a high potential to cause epizootics in insect rearing systems and in some cases they pose a threat to whole production stocks. Acheta domesticus densovirus (AdDV), an important pathogen of the European house cricket *A. domesticus*, is well known to cause disease outbreaks, which in the worst case could lead to major losses and to bankruptcy of cricket rearing companies (Szelei *et al.*, 2011; Weissman *et al.*, 2012).

Overt viral infections are initially identified by the symptoms displayed by infected insect hosts. For example, a disruption in moulting, reduced oviposition, or a reduced weight gain may be symptoms. Other symptoms may be a translucent exoskeleton, swollen and/or translucent abdomen (Figure 1C), enlarged brownish or milky midgut, or hindgut, watery faeces, and paralysis (reviewed by Maciel-Vergara and Ros, 2017). The particular symptoms depend on the virus and the host. Viruses can be transmitted through horizontal transmission (between conspecifics), vertical transmission (from parent to offspring), and sexual transmission. Often, viruses are transmitted through more than one of these transmission routes. Methods for the detection of a virus, include molecular techniques, virus isolation, serological studies, histopathology, and electron microscopy (Eberle *et al.*, 2012; Harrison and Hoover, 2012). However, there is a need for guidelines for standardised methods to

FIGURE 1   Clinical signs of infections in selected insects produced for food and feed. (A) Adult of the
            cricket, *Teleogryllus* sp. with inner organs showing a massive cell growth of *Rickettsiella gryl-
            li*. (B) Adult of black soldier fly, *Hermetia illucens* showing advanced mycosis due to an infec-
            tion with *Beauveria bassiana*. (C) Nymph of the cricket, *Acheta domesticus* with swollen ab-
            domen and liquified inner tissue (arrows) due to an infection with A. domesticus densovirus
            (AdDV). (D) Adult of the cricket, *Gryllus bimaculatus* with a strongly swollen abdomen due
            to an infection with *R. grylli*. (E) Adult of the cricket, *Modycogryllus* sp. showing advanced
            mycosis due to an infection with *Metarhizium* sp. (F) Larvae of the giant mealworm, *Zopho-
            bas morio* showing flaccidity and a dark coloration due to advanced septicaemia caused by
            an infection with *Pseudomonas aeruginosa*. Photos: A, C, D, E and F by Gabriela Maciel-Ver-
            gara, and B by Antoine Lecocq.

increase the reproducibility (including quality control) for validation of these diagnostic methods (Maciel-Vergara and Ros, 2017).

Entomopathogenic bacteria belong to various groups, which differ in biology.
They can belong to spore forming (genus *Bacillus*) or non-spore forming bacterial
(genera *Pseudomonas*, *Serratia* and *Rickettsiella*) groups and they can be generalists
or specialists. In most cases, they infect their hosts orally (Jurat-Fuentes and Jackson, 2012). For example, a specialist bacterium, *Bacillus popilliae* is infectious to few
selected species in the order Coleoptera (Rippere *et al.*, 1998). On the other hand,
strains of *Bacillus thuringiensis* var. *kurstaki* have a broader host spectrum within
the order Lepidoptera and can infect many species. Some generalist and opportunistic bacteria, such as non-spore forming bacteria from the genera *Pseudomonas*
and *Serratia*, can cause problems in insect colonies subjected to stress. They can
also multiply rapidly in hosts that are wounded and cannibalised by conspecific insects (Maciel-Vergara *et al.*, 2018). As tested by artificially induced infection, a strain
of the bacterium *Aeromonas hydrophila* has been reported to be pathogenic to the
yellow mealworm *Tenebrio molitor* (Noonin *et al.*, 2011).

A change in coloration, flaccidity, bad odour, and a cease of (usual) movement of
infected hosts are often first signs of bacterial diseases (Figure 1F). However, bacte-

rial pathogens like *Rickettsiella grylli* cause characteristic symptoms in their hosts such as a swollen abdomen and liquified viscous inner organs (Figure 1A and D). Diagnosis has to be followed by microscopy and molecular methods (Fisher and Garczynski, 2012; Tedersoo *et al.*, 2019).

Insect pathogenic fungi can be specialists or generalists. Entomophthorales, an ancient order of fungi, is mostly comprised of specialists (Boomsma *et al.*, 2014; Vega *et al.*, 2012). The species *Entomophthora muscae* infects the house fly *Musca domestica*. The fungus discharges conidia from dead hosts, which increases the likelihood of the conidia to be spread effectively to new hosts (Bellini *et al.*, 1992). Hypocreales (Ascomycota) is another order of fungi that includes genera like *Metarhizium* and *Beauveria*; species in these genera are mostly generalists and can cause diseases in a wide range of insect species. Fungal species belonging to the two genera can infect mealworms (*T. molitor*, a coleopteran species), silkworms *Bombyx mori*, (a lepidopteran species), *M. domestica* (a dipteran species), and *Locusta migratoria* (an orthopteran species) (see references in Supplementary Material Table S1). A recent study found *Beauveria bassiana* to be pathogenic to adults of the black soldier fly (*Hermetia illucens*) in laboratory infection trials (Lecocq *et al.*, 2021) (Figure 1B). Most fungi infect via penetration of the insect cuticle followed by growth in the haemolymph, and they sporulate externally upon host death. The first diagnosis of a fungal infection can be done by observing conidia or other external features on dead insects (Figure 1B and E) and by subsequent analysis using a microscope to identify the fungal genus. Molecular methods such as DNA sequencing help to identify the fungal species in most of the cases (Castrillo and Humber, 2009; Hajek *et al.*, 2012; Humber, 2012; Inglis *et al.*, 2012).

Microsporidia are unicellular parasitic organisms closely related to fungi. In order to infect their hosts the spores must be orally ingested (Solter *et al.*, 2012a). Most known microsporidian species are specialists, although some species have been reported to 'jump' to another host. Microsporidian infections are classified as chronic and rarely as acute (Becnel and Andreadis, 2014). Their presence is not necessarily immediately lethal to an insect population, although they can cause harm upon reaching a critical mass. The most studied microsporidian species have been found in honey bees and locusts.

Another group of unicellular insect pathogens are gregarines (Lange and Lord, 2012), which occur in the insect gut. Gregarines are only known to be parasitic to insects and mostly non-lethal, but can anyway lower the insects' fitness. They can be present in insect populations without being immediately noticed. The reported effects of gregarines in adult fall field crickets (*Gryllus pennsylvanicus*) are decreased longevity and weight loss under nutritional stress (Zuk, 1987). In addition, a *Gregarina* sp. isolated from the German cockroach *Blattella germanica* was reported as being highly pathogenic, and furthermore as being able to increase the susceptibility of its host to microbial and chemical challenges (Lopes and Alves, 2005). High prevalence of gregarines was found in a survey of protozoan parasites in edible insect species including *Gromphadorhina portentosa* (Madagascar hissing cockroach), *T. molitor*, *A. domesticus*, and *L. migratoria* (Gałęcki and Sokół, 2019). Gregarines have also been reported to occur in tenebrionids *Zophobas morio* (Jahnke, 2005) and *Alphitobius diaperinus* (Bala *et al.*, 1990) (Devetak *et al.*, 2013; Steinkraus *et al.*, 1992). To our knowledge, there is very limited information on the effect of gregarines to edible insects in rearing systems. Conducting more comprehensive research might give insight into the role of gregarines in insect production. Insects that are heavily infected with gregarines can exhibit symptoms such as a swollen abdomens and lethargy (Lopes and Alves, 2005). As for microsporidia, gregarines can be detected by examination of gut samples under the microscope, and quantification can be achieved by staining gut fluid (Solter *et al.*, 2012b).

3 Triggering factors for disease development

In insect rearing systems, the development of insect diseases caused by pathogens is determined by several factors (biotic and abiotic) related to the host and to the pathogen. Such factors are interconnected and largely determined by the production conditions inherent to insect mass rearing. Often, disease outbreaks occur when stressful conditions for an insect population which may converge with favourable conditions for a pathogen. Potential triggers that generate stressful conditions in insect colonies include changes in temperature and/or relative humidity, dietary changes and nutrient deficiency, overcrowding, infection with multiple natural enemies (i.e. pathogens and/or parasitoids), and toxic compounds (Figure 2).

3.1 *Temperature and relative humidity*

Insects are poikilothermic animals; their body temperature vary in line with the environmental temperature. Temperature and relative humidity have a substantial influence on the growth, development and survival of insects and microbes alike (Brindley, 1930; Holmes *et al.*, 2012; Ment *et al.*, 2017; Ratte, 1985). Insects and their pathogens have each an optimal temperature range that overlap to a certain extent. The optimal temperature range for pathogens can be similar among species within a taxon at genus or species level (i.e. bacteria, fungi, protozoa), although in some cases, the optimal temperature range for a pathogen in a certain host-pathogen interaction is pathogen-specific. Nevertheless, temperature has a direct effect on insect mortality and on the speed at which infected insects become symptomatic (Blanford and Thomas, 1999; Hurpin, 1968; Inglis *et al.*, 1997).

Four isolates of *Metarhizium flavoviridae*, a pathogenic fungus of the desert locust *Schistocerca gregaria*, caused nearly 100% mortality in 8 days regardless of the incubation temperature (25 and 30 °C), but the higher temperature (30 °C) increased the pathogen's growth and significantly reduced the time to death (Fargues *et al.*, 1997). Likewise two strains of the pathogenic bacteria *Serratia* sp. showed a

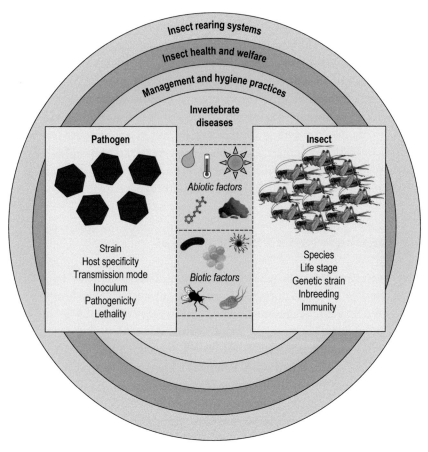

FIGURE 2   Schematic view of the interrelation of aspects inherent to pathogens and insect hosts
             with factors (i.e. biotic and abiotic) that trigger disease outbreaks and that concern in-
             sect health and welfare in insect rearing systems

dose and temperature dependent effect on the mortality and LT_{50} values when infecting the tobacco hornworm (*Manduca sexta*) (Petersen and Tisa, 2012).

At normal hive temperatures, 33 °C for the European honey bee (*Apis mellifera*),
the two common microsporidian pathogens *Nosema apis* and *Nosema ceranae* were
equally virulent, however *N. apis* was less infectious than *N. ceranae* at extreme
temperatures (below 25 and above 37 °C) (Martín-Hernández *et al.*, 2009). Temperature can also influence transmission capacity, illustrated by the duration and
yield of the conidial discharge from insect cadavers for the fungus *E. muscae* after
infecting its natural host, the common house fly (*M. domestica*), with higher conidial yield at lower temperatures (10 and 20 °C, compared to 30 and 38 °C) (Watson
and Petersen, 1993).

Temperature also has a significant effect on the ability of insects to overcome or slow down the infection by pathogens in various ways. A well-known example is the thermoregulatory behaviour displayed in most species of grasshoppers, locusts, and crickets (Order: Orthoptera) when infected with fungi (Blanford and Thomas, 1999; Carruthers *et al.*, 1992; Inglis *et al.*, 1996), bacteria (Louis *et al.*, 1986), and microsporidia (Boorstein and Ewald, 1987). The migratory locust *L. migratoria* raises its body temperature (behavioural fever) in order to suppress or slow down the infection time of fungal diseases caused by *B. bassiana* and *Metarhizium anisopliae* (Ouedraogo *et al.*, 2003; Sangbaramou *et al.*, 2018). Moreover, Mediterranean crickets (*Gryllus bimaculatus*) reared in a temperature gradient, were able to clear the pathogenic form of *R. grylli* off their bodies by rising their body temperature (due to actively moving to a higher temperature zone). However, the effectiveness of behavioural fever is dose- and species-specific, and therefore in some cases, it does not prevent pathogens killing their host (Adamo, 1999; Clancy *et al.*, 2018; Stahlschmidt and Adamo, 2013). Very importantly, thermal behaviour is heavily influenced by the intricate effects of relative humidity and temperature combined, all together determining the dynamics of host-pathogen interactions in insect species that are able to thermoregulate.

Overall, relative humidity and moisture have an effect on the development of disease outbreaks in insect colonies (Benz, 1987; Chakrabarti and Manna, 2008; Fuxa *et al.*, 1999; Mostafa *et al.*, 2005). Relative humidity has been studied more extensively as a key factor for infections caused by fungal pathogens than for pathogens from other taxa (Hajek, 1997; Hall and Papierok, 1982). The effects of relative humidity on the virulence, conidial germination and other aspects related to the infectivity of fungi such as *B. bassiana* and *M. anisopliae* on grasshoppers and locusts are well documented (Arthurs and Thomas, 2001) (Fargues *et al.*, 1997). In another host-pathogen system, mortality of *Tribolium confusum* larvae caused by *M. anisopliae* was negatively correlated with the tested levels of relative humidity (55% and 75%) (Michalaki *et al.*, 2006). For *E. muscae*, relative humidity values did not have any effect on the infection rate of house flies at a constant temperature of 25 °C. However, the effect of the relative humidity on the germination rate was isolate-specific (Watson and Petersen, 1993). Relative humidity did not have a significant effect on the efficacy (measured as the median lethal time, LT_{50}) of two *M. anisopliae* strains to infect the red palm weevil, *Rhynchophorus ferrugineus* (Cheong and Azmi, 2020).

3.2   *Dietary changes and nutrient deficiencies*

Diet composition and nutritional stress play an important role in the insects' immune response to pathogens and their ability to cope with diseases (Alaux *et al.*, 2010b; Ayres and Schneider, 2009; Ponton *et al.*, 2013; Srygley *et al.*, 2009). The protein and carbohydrate contents are especially important for the immune response and survival of insects (Cotter *et al.*, 2011; Ponton *et al.*, 2020). Larvae of the Egyptian cotton ball armyworm *Spodoptera littoralis* (potential feed for quail chicks, (Sayed *et al.*, 2019) challenged with a baculovirus, showed higher immune response and survival when fed on a diet with a high protein content relative to carbohydrate content (P:C ratio) (Lee *et al.*, 2006). In the same study, a group of larvae were allowed to select among diets with varying P:C ratio after being challenged with the virus; those larvae who survived the infection showed a preference for the diet with higher P:C ratio, in comparison to control and dying larvae, suggesting a purposeful change in their feeding behaviour to compensate for the protein costs of building up immunity (Lee *et al.*, 2008).

Similar research on other insect species underlines the dynamics of host feeding behaviour in relation to immunity and survival (prophylactic and therapeutic effects), and adds to the notion that the balance between protein and carbohydrates in the diet varies among insect-host systems and is key for mounting immunity and overcoming infection (Brunner *et al.*, 2014; Povey *et al.*, 2014; Wilson *et al.*, 2019), unless another challenge comes along (see section 'Infection with multiple natural enemies').

Nutritional stress related to food availability or nutrient content has also been connected to cannibalism, which is known as an important route of transmission for pathogens including viruses and bacteria, when healthy individuals feed upon heavily infected (or dead) conspecific insects that are too weak to avoid being preyed on. Baculoviruses have been reported to be transmitted by cannibalism in larvae of the corn earworm *Helicoverpa armigera* (Dhandapani *et al.*, 1993), the beet armyworm *S. exigua* (Elvira *et al.*, 2010), and the fall armyworm *Spodoptera frugiperda* (Chapman *et al.*, 1999). Viruses that are also transmitted due to cannibalism are densoviruses in crickets (Weissmann *et al.*, 2012), entomopoxviruses in grasshoppers (Streett and McGuire, 1990), and iridoviruses in a wide range of hosts (Williams and Hernández, 2006).

3.3   *Population density*

When the population density reaches levels beyond a certain threshold which may be different for each species (overcrowding), an insect colony is in theory at high risk for diseases to develop, due to an increased transmission rate, physiological stress, nutritional stress, and reduced immune response (Anderson and May, 1979; May and Anderson, 1979). Crowding is a stress factor that may be influenced or have an influence on other stressors like temperature, relative humidity, and CO_2 levels. Additionally, in crowded insect populations, increased chances for horizontal pathogen transmission occur as large numbers of seemingly healthy individuals feed on a big supply of food contaminated by the faeces and saliva (i.e. in dipteran production systems) of diseased individuals.

Cannibalism is usually observed in crowded populations as well, increasing the risks for the entry and spread of pathogens through the open wounds that the insects inflict on conspecifics (Steinhaus, 1958). Cannibalism and scavenging were more prevalent in groups of the giant mealworm (*Z. morio*) larvae, when exposed to the opportunistic bacterium *P. aeruginosa* compared to non-exposed larvae. Individual larvae that were artificially injured prior to exposure to *P. aeruginosa*, suffered from higher mortality rates in comparison to non-exposed larvae (Maciel-Vergara *et al.*, 2018). Another opportunistic bacterial pathogen, *Serratia marcescens*, has a higher chance to develop in insect colonies (i.e. silkworms) and mite colonies, when the hosts were subjected to crowding stress (Doane, 1960; Lighthart *et al.*, 1988; Vasantharajan and Munirathnamma, 2013). Solitude can on the other hand lead to a decrease in the melanisation which is part of the immune response as shown in *S. exempta* larvae infected with the virus Spodoptera exigua nucleopolyhedrovirus, SpexNPV or in *T. molitor* infected with the fungus *M. anisopliae* (Reeson *et al.*, 1998; Barnes and Siva-Jothy, 2000).

Nevertheless, the effects of crowding on insect health are not always negative as such effects also depend on behavioural and physiological aspects of specific insects.

3.4   *Infection with multiple natural enemies*

In insects, immune response and disease resistance vary when challenged by multiple pathogens/parasites/parasitoids (simultaneously or sequentially) compared to a challenge by only one pathogen (Malakar *et al.*, 1999; Martin *et al.*, 2012). In nature, mixed infections are fairly common (Virto *et al.*, 2014) and in insect rearing systems, such kind of infections may be more common than we may think (Maciel-Vergara *et al.*, in preparation). Mixed infections can become a stress factor by boosting the pathogenicity of one or more other types of pathogens prevalent in the same host (Hughes *et al.*, 1993). The dynamics between such pathogens in the whole disease process can be synergistic, additive, antagonistic, or independent (Carballo *et al.*, 2017). For instance, research on the effects of a mixed infection by entomopathogenic fungi, showed an additive effect of a low virulent *B. bassiana* strain on the effectiveness of a highly virulent strain of *Metarhizium acridum* when infecting *S. gregaria* (Thomas *et al.*, 2003). Other studies on competition among (viral, microsporidian, bacterial, and fungal) pathogens to thrive in the same host have been conducted using and observing different insect species, although most of the knowledge has been generated for bees and locusts (Evans and Armstrong, 2006; Tounou *et al.*, 2008).

Usually, the shift of a pathogen from being almost innocuous to becoming a threat for its host is related to the suppression of the immune system by competition among various organisms. Similarly, covert viruses can turn overt if their host becomes infected with another viral pathogen or gets challenged by a parasitoid or a parasite. Often, the virus becomes infectious, hosts develop disease symptoms, and mortality increases. Examples of a covert virus becoming overt due to a secondary infection with a non-homologous virus or pathogen are described (Hughes *et al.*, 1993), but maybe the most remarkable is the activation of a number of naturally present viruses in honey bees due to the prevalence of the *Varroa* mite in bee colonies (Alaux *et al.*, 2011; DeGrandi-Hoffman and Chen, 2015; Tritschler *et al.*, 2017).

3.5   *Other stressors and factors related to disease development*

Although vast knowledge on the effects of CO_2 on insect development has been collected (reviewed by Guerenstein and Hildebrand, 2008; Nicolas and Sillans, 1989; Sage, 2002), there is limited evidence of CO_2 as stressor to account for the development of insect diseases. The effects of high levels of CO_2 (either as high – and pure – to induce anaesthesia, or high in proportion to other gases in a mixture) on the insect's physiology and behaviour are described for the house cricket *A. domesticus* (Edwards and Patton, 1965), the German cockroach *B. germanica* (Tanaka, 1982) and other insect species (Bartholomew *et al.*, 2015; Brooks, 1957; Gunasekaran and Rajendran, 2005; Krishnamurthy *et al.*, 1986). Nonetheless, the effects of CO_2 seem to vary greatly in solitary insects compared to social insects, not only in relation to physiological and behavioural aspects but to their immune response as well. A positive correlation between CO_2 anaesthesia and enhanced immunocompetence was found for the common eastern bumble bee, *Bombus impatiens* (Amsalem and Grozinger, 2017) and leaf-cutting ants (Römer *et al.*, 2018).

In the context of host-pathogen interactions, a unique scenario of hyper reactivity to CO_2 and associated high mortality at high concentrations of CO_2 has been registered for *Drosophila melanogaster* infected with the rhabdovirus Drosophila melanogaster sigma virus (reviewed by (L'Héritier, 1948). Moreover, other rhabdoviruses cause hyper reactivity to CO_2 in other dipteran species (Rosen, 1980). Additionally, in a multifactorial set-up where various stressors were tested, reduced virulence of entomopathogenic fungi on *S. gregaria* and *A. domesticus* at increasing CO_2 concentration was observed (2015). An interesting fact relevant for large scale insect rearing systems, is that the effects of CO_2 may vary greatly depending on the insect's developmental stage. In this regard and for future studies, an analogy to the results found by Callier *et al.* (2015) could apply in the sense that while dipteran larvae can thrive in highly hypoxic conditions, these same conditions can severely affect individuals in the adult stage.

Other stressors to take into consideration are heavy metals, toxins and pesticides; chemicals that are known for their diverse effects on insect behaviour (Burden *et al.*, 2019; Chicas-Mosier *et al.*, 2017; Guo *et al.*, 2014; Hladun *et al.*, 2015) and host-pathogen interactions (Jiang *et al.* 2021; Odemer *et al.*, 2018), especially with regards to the immunocompetence of insect hosts (Mir *et al.*, 2020; Shaurub, 2003; Van Ooik *et al.*, 2008). Some studies have evaluated the positive effects of specific chemicals (i.e. silver nanoparticles, silica nanoparticles) on the survival of insects challenged by pathogens (*B. mori* infected with B. mori nucleopolyhedrovirus, BmNPV), however more research is needed to evaluate the effectiveness of using these and other chemicals to manage disease outbreaks in the insect rearing industry (Das *et al.*, 2013; Govindaraju *et al.*, 2011). On the contrary, there is ample evidence of the detrimental effects that chemical exposure has on insect health (particularly pesticide-related chemicals and heavy metals). An example of such negative effects is the increased prevalence and mortality caused by the microsporidian pathogen *N. ceranae* in honey bees and stingless bees exposed to neonicotinoid pesticides (Alaux *et al.*, 2010a; Macías-Macías *et al.*, 2020; Tesovnik *et al.*, 2020). Honey bees exposed to neonicotinoid pesticides, have also been reported to have reduced immunocompetence and increased replication of the deformed wing virus (Di Prisco *et al.*, 2013).

A couple of factors that are not stressors *per se* but that have a crucial effect on the development of insect diseases are the insect developmental stage (Blaser and Schmid-Hempel, 2005; Briggs and Godfray, 1995), and the prevalence of endosymbionts. The effect of endosymbionts (Chrostek *et al.*, 2020; Martinez *et al.*, 2014; Rottschaefer and Lazzaro, 2012; Zug and Hammerstein, 2015) on the insects' health has been explored in the last two decades, although limited knowledge is available for most insect species reared as food or feed (Dillon *et al.*, 2005; Muhammad *et al.*, 2019).

The insects' life stage is one more factor that plays a key role on the disease dynamics in insect colonies. Usually, one or few of the life stages of an insect host are (highly) susceptible to specific pathogens while the other life stages are less susceptible or not susceptible at all (Engelhard and Volkman, 1995; Goulson *et al.*, 1995).

4 Measures to control diseases and pests on rearing systems

Disease outbreaks in farmed insects are inevitable and unfortunately, most diseases are discovered when there is already a significant damage to the insect colony. Depending on the severity of each case, the best solution in many cases has been to perform a thorough inspection, cleaning and eventual disinfection of the production facilities and to start the production over again. A routine inspection for pathogens should be implemented in every insect rearing system. Diagnostics, as suggested by Eilenberg *et al.* (2018), are to be done in collaboration with experts on invertebrate diseases. Diagnostic protocols are available for a handful of insect pathogens, but the most challenging scenarios are posed by the presence of covert infections (i.e. viruses) and other chronic diseases (i.e. protozoa and obligate bacterial pathogens).

Covert viral infections can be detected before a disease outbreak occurs but their detection does not necessarily means that their presence will cause a severe disease outbreak in a rearing system, since such epizootics depend on many trigger factors (Section 3). Some preventive and corrective measures have been used in laboratories, insectaries and in insect rearing systems (i.e. sericulture, apiculture, sterile insect technique facilities), and have helped on the mitigation of insect pathogens (Bindroo and Verma, 2014; Formato and Smulders, 2011; Kariithi, 2013). Such measures are related to the implementation of hygiene at different levels of the production facilities, to the modification of specific steps in the rearing process and to the application of immune-intervention strategies. The application and effectiveness of these measures vary depending on the type of production system (e.g. open, semi-open, closed), on the biology of the insect species, and on the pathogens present in each production system, as well as, on the legislation in place in each region/country.

Discussions on the risks posed by various insect pathogens to different rearing systems have been published, as well as general recommendations on how to try keeping insects healthy (Eilenberg and Jensen, 2018b; Eilenberg *et al.*, 2015, 2018). A guide on good hygiene practices has been made available by the International Platform of Insects for Food and Feed (IPIFF), covering aspects of the insect production and the processing of insect-derived products. The advice in this guide is related to the general hygiene mainly to avoid food-borne pathogens (yet most procedures would also be effective for several insect pathogens) (https://ipiff.org/wp-content/uploads/2019/12/ipiff-guide-on-good-hygiene-practices.pdf). Lately, advances in methods and equipment have been made for the design of a more hygienic, and easy-to-handle insect production; these advances focus on closed high-tech insect production (i.e. crickets and black soldier fly – BSF – production) (Joosten *et al.*, 2020; Mellberg and Wirtanen, 2018;). In addition, a manual for semi-open production of crickets has been recently released. It provides an overview of the good practices advised for the entire rearing process and a guide on how to inspect the cricket rearing process and facilities (Hanboonsong and Durst, 2020).

4.1   *Hygiene and good practices*

Hygiene is without doubt an essential component of any husbandry system and the production of edible insects for feed or food is not an exception. It is important to keep in mind that an integral approach of the hygiene measures and the good production practices should be part of the entire rearing process, concerning: the physical structure (i.e. building, pens, containers, equipment), the feed, the personnel, the insects (i.e. eggs, parent stock), the frass, etc. Hygiene and good production practices are basic aspects for the prevention of food-borne diseases and insect diseases, and are key for starting to engage in the dialogue on insect welfare within the edible insect industry. In our view, and in agreement with the logic of the Brambell's five freedoms (Van Huis, 2019), insect welfare relates to (among other aspects) the ability of captive insect populations to thrive, and to experience less the effects of disease outbreaks by being reared in *ad hoc* conditions. In summary, advices that reinforce the available general recommendations on hygiene and good practices include:

– Cleaning and disinfection agents should be used but they should be approved disinfectants by the corresponding agencies in charge of the regulation of such substances (i.e. EPA, ECHA) and especially, in the production of insects for food, disinfectants should be approved for use in the food industry.

- All the equipment and every surface that is in contact with the insects should be thoroughly washed, disinfected and rinsed every time a new batch of insects is reared.
- If available, steam may be used to disinfect rearing rooms, equipment, oviposition substrate, etc.
- Feed should be inspected (visually) and treated prior to use if needed (i.e. heat). It should be stored in proper conditions, depending on the nature of the feed.
- Fresh feed should be provided regularly to insects (depending on each species need), avoiding the formation and accumulation of moulds.
- Water stagnation and formation of moulds in drinking systems (i.e. for crickets) should be avoided by providing fresh water regularly and by using/designing devices that can be cleaned easily and preferably with materials where microorganisms are not able to thrive.
- Insect frass should be treated prior to disposal, irrespectively if the insect colony was healthy or not, by heating up or fermenting (i.e. compost/silage).
- If applicable to the rearing system, air filtration equipment should be put in place and maintained in appropriate functioning conditions.

4.2   Differentiated breeding (parent stock and 'the rest of the population')

Parent stocks may be reared separately from the rearing of all other instars (i.e. other isolated room in closed containers), as a measure to prevent diseases. Also, more selective and nutritious diet and care may be provided to parent stocks to ensure the quality of egg production. Keeping the parent stock separated from the main production (physically and in terms of nutrition, and care) also ensures a higher biological quality for the parents and a backup solution if the entire production needs to be eventually re-started.

4.3   Mechanical control of pests

Ants, flies, parasitic wasps and mites are the most common insect pests for insect production. Different methods in specific rearing systems are used to keep pests at bay. For example, placing cages or crates for the production of dipteran species (i.e. house fly and BSF) and crickets on elevated platforms with stands submerged in oil or molasses, have been effective to deter ants from entering cages in Ghana, Kenya and Uganda. In closed production systems, double doors prevent the entrance of pests and the escape of insects in rearing systems. Sticky traps and UV-lamp traps are also useful to prevent insect pests to remain inside production facilities. Mites are a major problem, especially in insect rearing systems where substrates have high moisture contents and /or high relative humidity prevails. The most efficient way to control mites is by cleaning the facilities on a regular basis, lowering the relative humidity, keeping the trays/pens free of debris, and preventing the feed from getting too wet and mouldy. A possibly effective but expensive method that might be used to combat mites is the use of the predatory mites e.g. *Stratiolaelaps scim-*

itus, Cheyletus eruditus, and *Cheyletus malaccensis* (Cabrera *et al.*, 2005; Cebolla *et al.*, 2009; Pulpan and Verner, 1965; Rangel and Ward, 2018), however more research should be done to prove their efficacy.

5   Prospects on the control of insect diseases in rearing systems

In the future, novel control strategies can be inspired by methods from other life stock production systems or developed from a deeper understanding of the biology and physiology of host-microbe interactions within the context of insect mass-rearing. Practical constraints for the control of insect diseases in insect rearing systems are especially related to: the insect species, the pathogen species, the size and structure of the facility, the technological investment, the availability of reliable prevention methods, diagnostic tools and direct control methods, and the risk of toxic residues if chemical treatment is pursued (i.e. antibiotics or antivirals).

5.1   *Breeding of disease-resistant/ tolerant strains*
Selective breeding to improve desirable traits in animals and plants has been used by humans for many years and breeding for disease resistance is a classical discipline found within all production systems e.g. crop production (Nelson *et al.*, 2018), aquaculture (Gjedrem, 2015), poultry, pigs, (Proudfoot *et al.*, 2019) and honey bees (Guichard *et al.*, 2019). One of the challenges in resistance breeding is the trade-off with other important traits, which includes responses to abiotic factors, nutritional uptake, growth, and other fitness traits. In addition, resistance to one pathogen might induce susceptibility to another.

In the late 20th century, genomic selection was added to the livestock breeding toolbox; by reading specific locations in the genome and assigning them to measurable production traits, faster improvement in livestock production efficiency has been achieved and the novel CRISPR/Cas technology even allows for genome editing. The CRISPR/Cas gene-editing technique has shown promising results as an antiviral therapy in silkworms (Wei *et al.*, 2017).

Taking the ethical considerations around genome editing into account (i.e. by CRISPR/Cas) (Charo and Greely, 2015; Gjerris *et al.*, 2018), and it will be interesting to see if and how this technology will be used for disease resistance or other functional traits within insects used for mass rearing.

5.2   *Heat shock/thermal therapy*
Temperature plays a key role on the different immune responses of insects against pathogens (5.3.1). The severity of a heat shock (thermal stress) may impact the duration of the immune responses, which varies among insect-pathogen systems. For instance, subjecting *G. mellonella* to a short heat shock (38 °C, 30 min) prior to infection with *B. bassiana* blastospores reduced the infection rate of the fungus,

prolonging the host lifetime (Wojda *et al.*, 2009). Conversely, a prolonged thermal stress (30 and 37 °C, 24 h), provided *G. mellonella* only temporary resistance against *Aspergillus fumigatus* (Browne *et al.*, 2014). Thermal therapy of honey bees at 42 °C for 4 h and back to the normal 32 °C have shown to reduce the viral load of green fluorescent tagged SINV-GFP Sindbis virus in honey bees (McMenamin *et al.*, 2020)

5.3   *Gut microbiota/probiotics*

Gut microbiota modulate insect immune response, enhancing the resiliency of insects against pathogens (Muhammad *et al.*, 2019) or assisting the pathogens to overcome the immune system of their host (Jakubowska *et al.*, 2013). A comprehensive work on this regard has focused on honey bee immunity and its response to bacterial, fungal, and viral pathogens (Evans and Armstrong, 2006; Moran, 2015; Reynaldi *et al.*, 2004).

On the other hand, composition of microbial gut communities in insects (and other animals as well) (Krams *et al.*, 2017; Martínez-Solís *et al.*, 2020; Ponton *et al.*, 2013, 2015) can vary depending on the insect diet. From the perspective of insect rearing, modifying the diet would also modify the microbial composition of insect guts, a feature that could promote higher disease resistance of insects reared under mass-production schemes. As an example, an indigenous gut bacterial strain *Pediococcus pentosaceus* showed increased growth and survival of *T. molitor* larvae (Lecocq *et al.*, in press), and an isolate of the bacterium *Enterococcus mundtii* offered the model insect *Tribolium castaneum* protection towards the bacterial pathogen *B. thuringiensis* (Grau *et al.*, 2017).

5.4   *Biological control*

To our knowledge, very limited information exists on the utilisation of microorganisms to control insect pathogens in insect rearing systems. As mentioned earlier in this paper, virus discovery has increased over the last decade and generally speaking, new viruses that are found by New Generation Sequencing (NGS) technology (Datta *et al.*, 2015) in otherwise healthy hosts, are referred to as insect-specific viruses (ISV's). ISV's are not able to replicate in vertebrate hosts and it is suggested that they persist in insect populations through vertical (transovarial) transmission. Although, we do not exclude the possibility that some newly discovered (covert) viruses may end up being pathogenic to insects reared under stressful conditions in insect rearing systems, the antagonistic interaction between (engineered and wild-type) insect-specific viruses (ISV's) and arboviruses vectored by insects (Adelman *et al.*, 2001; Airs and Bartholomay, 2017; Bolling *et al.*, 2015; Powers *et al.*, 1996), is a starting point to evaluate the trade-offs if ISV's were to be used to increase pathogen resistance in edible insect species. Nouri *et al.* (2018) reviewed the potential applications that ISV's may have for different purposes. An additional alternative to investigate is the use of bacteriophages for the control of bacterial diseases in insects. Bacteriophages have the ability to alter the bacterial genomic material, and might thus disrupt the infection process (Li *et al.*, 2016b; Zimmer *et al.*, 2013).

5.5 *RNA interference*

RNA interference (RNAi) is a technology used for the inhibition of virus replication based on gene-expression regulation, by the neutralisation of targeted mRNA molecules (Aguiar *et al.*, 2016). This biological process is known to protect vertebrate, invertebrate, and plant hosts from virus attacks (Burand and Hunter, 2013; La Fauce and Owens, 2013; Li *et al.*, 2016a; Sidahmed and Wilkie, 2010). RNAi has been used to control (to a low extent), the prevalence of *Glossina pallidipes* salivary gland hypertrophy virus in tsetse fly rearing systems (Abd-Alla *et al.*, 2011a,b). More promising results were seen in reducing the prevalence of the Israeli acute paralysis virus and deformed wing virus in honey bees using RNAi (Brutscher and Flenniken, 2015; Burand and Hunter, 2013; Desai *et al.*, 2012; Hunter *et al.*, 2010).

6   Concluding remarks

Although pathogens and beneficial insects have coexisted in insect rearing systems since ancient times, the recent fast growth of the insect rearing industry (for protein production) has exposed the need for a better understanding of insect diseases that develop in production facilities. Notably, there is more research to be done on the biology of insect pathogens and the interactions they have with their insect hosts (for the subject of this paper focusing on insects produced as food and feed). At the same time, more knowledge is needed on the correlation and/or interaction between production variables and host-pathogen dynamics. Such multifactorial relations are rather complex, with stress factors being critical for the development of disease outbreaks, with often more than one pathogen involved, and several trade-offs that challenge the management of insect diseases in insect production processes.

Additionally, and since many aspects of insect production have an implication on disease development (from insect physiology to in-house hygienic measures), old and novel techniques and possibilities should be extensively explored as preventative and corrective measures. Needless to say, no single solution can address all problems when it comes to the management of diseases. Rearing practices should be continuously revised and changed accordingly. Doing so will allow to find a better balance between enhancing productivity (by optimizing the production) and avoiding insect disease outbreaks while at the same time, taking into account the insects' health.

Ultimately, a holistic approach in understanding the various aspects related to insect diseases in connection with the production process should be taken. Such approach is relevant for the ongoing development of protocols for the management, prevention, and control of diseases.

Acknowledgements

This review was written as a part of the PhD project of Gabriela Maciel-Vergara, supported by the Agricultural Transformation by Innovation, Erasmus Mundus Joint Doctorate Program – (AgTrain – EMJMD; project 512095), funded by the EACEA (Education, Audiovisual and Culture Executive Agency) of the European Commission. Antoine Lecocq was supported by the EU grant SUSINCHAIN (H2020-LC-SFS-17-2019 #861976). The authors like to acknowledge Monique van Oers and Vera Ros for critically reading this manuscript throughout the submission and Joop van Loon for critically reading the last version of it. We thank two anonymous reviewers for their comments and suggestions on this manuscript.

Figure 2 was designed including the use of selected free available graphic resources at www.freepik.com.

Conflict of interest

The authors declare no conflict of interest.

Supplementary material

Table S1. Literature review of pathogens of insects produced or collected in nature as food and feed.

Supplementary material can be found online at https://doi.org/10.3920/JIFF2021.0024.

References

Abd-Alla, A.M.M., Parker, A.G., Vreysen, M.J.B. and Bergoin, M., 2011a. Tsetse salivary gland hypertrophy virus: hope or hindrance for tsetse control? PLoS Neglected Tropical Diseases 5: e1220. http://doi.org/10.1371/journal.pntd.0001220

Abd-Alla, A.M.M., Salem, T.Z., Parker, A.G., Wang, Y., Jehle, J.A., Vreysen, M.J.B. and Boucias, D., 2011b. Universal primers for rapid detection of hytrosaviruses. Journal of Virological Methods 171: 280-283. https://doi.org/10.1016/j.jviromet.2010.09.025

Adamo, S.A., 1999. Evidence for adaptive changes in egg laying in crickets exposed to bacteria and parasites. Animal Behaviour 57: 117-124. https://doi.org/10.1006/anbe.1998.0999

Adelman, Z.N., Blair, C.D., Carlson, J.O., Beaty, B.J. and Olson, K.E., 2001. Sindbis virus-induced silencing of dengue viruses in mosquitoes. Insect Molecular Biology 10: 265-273. https://doi.org/10.1046/j.1365-2583.2001.00267.x

Aguiar, E.R.G.R., Olmo, R.P. and Marques, J.T., 2016. Virus-derived small RNAs: molecular footprints of host-pathogen interactions. WIREs RNA 7: 824-837. https://doi.org/10.1002/wrna.1361

Airs, P.M. and Bartholomay, L.C., 2017. RNA Interference for mosquito and mosquito-borne disease control. Insects 8: 4.

Alaux, C., Brunet, J.-L., Dussaubat, C., Mondet, F., Tchamitchan, S., Cousin, M., Brillard, J., Baldy, A., Belzunces, L.P. and Le Conte, Y., 2010a. Interactions between *Nosema* microspores and a neonicotinoid weaken honeybees (*Apis mellifera*). Environmental Microbiology 12: 774-782. http://doi.org/10.1111/j.1462-2920.2009.02123.x

Alaux, C., Dantec, C., Parrinello, H. and Le Conte, Y., 2011. Nutrigenomics in honey bees: digital gene expression analysis of pollen's nutritive effects on healthy and varroa-parasitized bees. BMC Genomics 12: 496. http://doi.org/10.1186/1471-2164-12-496

Alaux, C., Ducloz, F., Crauser, D. and Conte, Y.L., 2010b. Diet effects on honeybee immuno-competence. Biology Letters 6: 562-565. http://doi.org/10.1098/rsbl.2009.0986

Amsalem, E. and Grozinger, C.M., 2017. Evaluating the molecular, physiological and behavioural impacts of CO_2 narcosis in bumble bees (*Bombus impatiens*). Journal of Insect Physiology 101: 57-65. https://doi.org/10.1016/j.jinsphys.2017.06.014

Anderson, R.M and May, R.M., 1979. Population biology of infectious diseases: Part I. Nature 280: 361-367. https://doi.org/10.1038/280361a0

Andreadis, T.G. and Weseloh, R.M., 1990. Discovery of *Entomophaga maimaiga* in North American gypsy moth, *Lymantria dispar*. Proceedings of the National Academy of Sciences 87: 2461-2465.

Arthurs, S. and Thomas, M.B., 2001. Effects of temperature and relative humidity on sporulation of *Metarhizium anisopliae* var. *acridum* in mycosed cadavers of *Schistocerca gregaria*. Journal of Invertebrate Pathology 78: 59-65. https://doi.org/10.1006/jipa.2001.5050

Ayres, J.S. and Schneider, D.S., 2009. The role of anorexia in resistance and tolerance to infections in *Drosophila*. PLoS Biology 7: e1000150. https://doi.org/10.1371/journal.pbio.1000150

Bailey, L., 1968. Honey bee pathology. Annual Review of Entomology 13: 191-212.

Bala, P., Kaur, D., Lipa, J. and Bhagat, R., 1990. *Gregarina alphitobii sp.* n. and *Mattesia alphitobii sp. n.*, parasitizing *Alphitobius diaperinus* Panz.(Tenebrionidae, Coleoptera). Acta Protozool 29: 245-256.

Barnes, A.I. and Siva-Jothy, M.T., 2000. Density-dependent prophylaxis in the mealworm beetle *Tenebrio molitor* L. (Coleoptera: Tenebrionidae): cuticular melanization is an indicator of investment in immunity. Proceedings of the Royal Society of London. Series B: Biological Sciences 267: 177-182. http://doi.org/10.1098/rspb.2000.0984

Bartholomew, N.R., Burdett, J.M., VandenBrooks, J.M., Quinlan, M.C. and Call, G.B., 2015. Impaired climbing and flight behaviour in *Drosophila melanogaster* following carbon dioxide anaesthesia. Scientific Reports 5: 15298. https://doi.org/10.1038/srep15298

Becnel, J.J. and Andreadis, T.G., 2014. Microsporidia in insects. In: Weiss, L.M. and Becnel, J.J. (eds.) Microsporidia: pathogens of opportunity, 1st edition. John Wiley and Sons, Hoboken, NJ, USA, pp. 521-570. https://doi.org/10.1002/9781118395264.ch21

Bellini, R., Mullens, B. and Jespersen, J., 1992. Infectivity of two members of the *Entomophthora muscae* complex [Zygomycetes: Entomophthorales] for *Musca domestica* [Dipt.: Muscidae]. Entomophaga 37: 11-19.

Benz, G., 1987. Environment. In: Fuxa, J.R. and Tanada, Y. (eds.) Epizootiology of insect diseases. John Wiley and Sons, Hoboken, NJ, USA, pp. 177-214.

Bindroo, B.B. and Verma, S., 2014. Sericulture technologies developed by CSRTI MYSORE. Central Sericultural Research and Training Institute, Mysore, India, 54 pp.

Blanford, S. and Thomas, M.B., 1999. Host thermal biology: the key to understanding host-pathogen interactions and microbial pest control? Agricultural and Forest Entomology 1: 195-202. https://doi.org/10.1046/j.1461-9563.1999.00027.x

Blaser, M. and Schmid-Hempel, P., 2005. Determinants of virulence for the parasite *Nosema whitei* in its host *Tribolium castaneum*. Journal of Invertebrate Pathology 89: 251-257. https://doi.org/10.1016/j.jip.2005.04.004

Bolling, B.G., Weaver, S.C., Tesh, R.B. and Vasilakis, N., 2015. Insect-specific virus discovery: significance for the arbovirus community. Viruses 7: 4911-4928.

Boomsma, J.J., Jensen, A.B., Meyling, N.V. and Eilenberg, J., 2014. Evolutionary interaction networks of insect pathogenic fungi. Annual Review of Entomology 59: 467-485. https://doi.org/10.1146/annurev-ento-011613-162054

Boorstein, S.M. and Ewald, P.W., 1987. Costs and benefits of behavioural fever in *Melanoplus sanguinipes* infected by *Nosema acridophagus*. Physiological Zoology 60: 586-595. http://doi.org/10.1086/physzool.60.5.30156132

Boucias, D.G. and Pendland, J.C., 1998. Principles of insect pathology. Springer Science & Business Media, Berlin, Germany.

Briggs, C.J. and Godfray, H.C.J., 1995. The dynamics of insect-pathogen interactions in stage-structured populations. The American Naturalist 145: 855-887. http://doi.org/10.1086/285774

Brindley, T.A., 1930. The growth and development of *Ephestia kuehniella* Zeller (Lepidoptera) and *Tribolium confusum* Duval (Coleoptera) under controlled conditions of temperature and relative humidity. Annals of the Entomological Society of America 23: 741-757. http://doi.org/10.1093/aesa/23.4.741

Brodeur, J., 2012. Host specificity in biological control: insights from opportunistic pathogens. Evolutionary Applications 5: 470-480. https://doi.org/10.1111/j.1752-4571.2012.00273.x

Bronkhorst, A.W., Van Cleef, K.W.R., Venselaar, H. and Van Rij, R.P., 2014. A dsRNA-binding protein of a complex invertebrate DNA virus suppresses the Drosophila RNAi response. Nucleic Acids Research 19: 12237-12248. https://doi.org/10.1093/nar/gku910

Brooks, M.A., 1957. Growth-retarding effect of carbon-dioxide anaesthesia on the German cockroach. Journal of Insect Physiology 1: 76-84. https://doi.org/10.1016/0022-1910(57)90024-0

Browne, N., Surlis, C. and Kavanagh, K., 2014. Thermal and physical stresses induce a short-term immune priming effect in *Galleria mellonella* larvae. Journal of Insect Physiology 63: 21-26. https://doi.org/10.1016/j.jinsphys.2014.02.006

Brun, G., 1984. Le virus sigma de la Drosophile. Bulletin de la Société Entomologique de France 89: 674-680.

Brunner, F.S., Schmid-Hempel, P. and Barribeau, S.M., 2014. Protein-poor diet reduces host-specific immune gene expression in *Bombus terrestris*. Proceedings. Biological Sciences 281: 20140128. http://doi.org/10.1098/rspb.2014.0128

Brutscher, L.M. and Flenniken, M.L., 2015. RNAi and antiviral defense in the honey bee. Journal of Immunology Research, Article ID: 941897. https://doi.org/10.1155/2015/941897

Burand, J.P. and Hunter, W.B., 2013. RNAi: future in insect management. Journal of Invertebrate Pathology 112: S68-S74. https://doi.org/10.1016/j.jip.2012.07.012

Burden, C.M., Morgan, M.O., Hladun, K.R., Amdam, G.V., Trumble, J.J. and Smith, B.H., 2019. Acute sublethal exposure to toxic heavy metals alters honey bee (*Apis mellifera*) feeding behaviour. Scientific Reports 9: 4253. http://doi.org/10.1038/s41598-019-40396-x

Cabrera, A., Cloyd, R. and Zaborski, E., 2005. Development and reproduction of *Stratiolaelaps scimitus* (Acari: Laelapidae) with fungus gnat larvae (Diptera: Sciaridae), potworms (Oligochaeta: Enchytraeidae) or *Sancassania aff. sphaerogaster* (Acari: Acaridae) as the sole food source. Experimental and Applied Acarology 36: 71-81. https://doi.org/10.1007/s10493-005-0242-x

Callier, V., Hand, S.C., Campbell, J.B., Biddulph, T. and Harrison, J.F., 2015. Developmental changes in hypoxic exposure and responses to anoxia in *Drosophila melanogaster*. The Journal of Experimental Biology 218: 2927-2934. http://doi.org/10.1242/jeb.125849

Carballo, A., Murillo, R., Jakubowska, A., Herrero, S., Williams, T. and Caballero, P., 2017. Co-infection with iflaviruses influences the insecticidal properties of *Spodoptera exigua* multiple nucleopolyhedrovirus occlusion bodies: Implications for the production and biosecurity of baculovirus insecticides. PLoS ONE 12: e0177301. http://doi.org/10.1371/journal.pone.0177301

Carruthers, R.I., Larkin, T.S., Firstencel, H. and Feng, Z., 1992. Influence of thermal ecology on the mycosis of a rangeland grasshopper. Ecology 73: 190-204. http://doi.org/10.2307/1938731

Castrillo, L. and Humber, R., 2009. Molecular methods for identification and diagnosis of fungi. In: Stock, P., Vandenberg, J., Glazer, I., Boemare, N. (eds.) Insect pathogens molecular approaches and techniques. CABI International, Wallingford, UK, pp.50-70.

Cebolla, R., Pekár, S. and Hubert, J., 2009. Prey range of the predatory mite *Cheyletus malaccensis* (Acari: Cheyletidae) and its efficacy in the control of seven stored-product pests. Biological Control 50(1): 1-6. https://doi.org/10.1016/j.biocontrol.2009.03.008

Chakrabarti, S. and Manna, B., 2008. Effect of microsporidian infection on reproductive potentiality on mulberry Silkworm, *Bombyx mori* L. (Lepidoptera: Bombycidae) in different seasons. International Journal of Industrial Entomology 17: 157-163.

Chapman, J.W., Williams, T., Escribano, A., Caballero, P., Cave, R.D. and Goulson, D., 1999. Age-related cannibalism and horizontal transmission of a nuclear polyhedrosis virus in larval *Spodoptera frugiperda*. Ecological Entomology 24: 268-275. https://doi.org/10.1046/j.1365-2311.1999.00224.x

Charo, R.A. and Greely, H.T., 2015. CRISPR critters and CRISPR cracks. The American Journal of Bioethics 15: 11-17. http://doi.org/10.1080/15265161.2015.1104138

Cheong, J.L. and Azmi, W.A., 2020. Dataset on the influence of relative humidity on the pathogenicity of *Metarhizium anisopliae* isolates from Thailand and Malaysia against red palm weevil (*Rhynchophorus ferrugineus*, Olivier) adult. Data in Brief 30: 105482. http://doi.org/10.1016/j.dib.2020.105482

Chicas-Mosier, A.M., Cooper, B.A., Melendez, A.M., Pérez, M., Oskay, D. and Abramson, C.I., 2017. The effects of ingested aqueous aluminum on floral fidelity and foraging strategy in honey bees (*Apis mellifera*). Ecotoxicology and Environmental Safety 143: 80-86. https://doi.org/10.1016/j.ecoenv.2017.05.008

Chrostek, E., Martins, N., Marialva, M.S. and Teixeira, L., 2020. *Wolbachia*-conferred antiviral protection is determined by developmental temperature. bioRxiv: 2020.2006.2024.169169. http://doi.org/10.1101/2020.06.24.169169

Clancy, L.M., Jones, R., Cooper, A.L., Griffith, G.W. and Santer, R.D., 2018. Dose-dependent behavioural fever responses in desert locusts challenged with the entomopathogenic fungus *Metarhizium acridum*. Scientific Reports 8: 1-8.

Cotter, S.C., Simpson, S.J., Raubenheimer, D. and Wilson, K., 2011. Macronutrient balance mediates trade-offs between immune function and life history traits. Functional Ecology 25: 186-198. https://doi.org/10.1111/j.1365-2435.2010.01766.x

Das, S., Bhattacharya, A., Debnath, N., Datta, A. and Goswami, A., 2013. Nanoparticle induced morphological transition of *Bombyx mori* nucleopolyhedrovirus: a novel method to treat silkworm grasserie disease. Applied Microbiology and Biotechnology 97: 6019-6030.

Datta, S., Budhauliya, R., Das, B. and Chatterjee, S., 2015. Next-generation sequencing in clinical virology: discovery of new viruses. World Journal of Virology 4: 265.

De Miranda, J.R., Granberg, F., Onorati, P., Jansson, A. and Berggren, Å., 2021. Virus prospecting in crickets – discovery and strain divergence of a novel Iflavirus in wild and cultivated *Acheta domesticus*. Viruses 13(3): 364. https://doi.org/10.3390/v13030364

DeGrandi-Hoffman, G. and Chen, Y., 2015. Nutrition, immunity and viral infections in honey bees. Current Opinion in Insect Science 10: 170-176. https://doi.org/10.1016/j.cois.2015.05.007

Desai, S.D., Eu, Y.-J., Whyard, S. and Currie, R.W., 2012. Reduction in deformed wing virus infection in larval and adult honey bees (*Apis mellifera* L.) by double-stranded RNA ingestion. Insect Molecular Biology 21: 446-455. https://doi.org/10.1111/j.1365-2583.2012.01150.x

Devetak, D., Omerzu, M. and Clopton, R.E., 2013. Notes on the gregarines (Protozoa: Apicomplexa: Eugregarinorida) of insects in Slovenia. Anali za Istrske in Mediteranske študije Series Historia Naturalis [Annals for Istrian and Mediterranean Studies] 23: 73-90.

Dhandapani, N., Jayaraj, S. and Rabindra, R., 1993. Cannibalism on nuclear polyhedrosis virus infected larvae by *Heliothis armigera* (Hubn.) and its effect on viral infection. International Journal of Tropical Insect Science 14: 427-430.

Di Prisco, G., Cavaliere, V., Annoscia, D., Varricchio, P., Caprio, E., Nazzi, F., Gargiulo, G. and Pennacchio, F., 2013. Neonicotinoid clothianidin adversely affects insect immunity and promotes replication of a viral pathogen in honey bees. Proceedings of the National Academy of Sciences 110: 18466-18471. https://doi.org/10.1073/pnas.1314923110

Dillon, R.J., Vennard, C.T., Buckling, A. and Charnley, A.K., 2005. Diversity of locust gut bacteria protects against pathogen invasion. Ecology Letters 8: 1291-1298. https://doi.org/10.1111/j.1461-0248.2005.00828.x

Doane, C., 1960. Bacterial pathogens of *Scolytus multistriatus* as related to crowding. Journal of Insect Pathology 2: 24-29.

Eberle, K.E., Wennmann, J.T., Kleespies, R.G. and Jehle, J.A., 2012. Basic techniques in insect virology. In: Lacey, L.A. (ed.) Manual of techniques in invertebrate pathology, 2nd edition. Academic Press, San Diego, CA, USA, pp. 15-74. https://doi.org/10.1016/B978-0-12-386899-2.00002-6

Edwards, L.J. and Patton, R.L., 1965. Effects of carbon dioxide anesthesia on the house cricket, *Acheta domesticus* (Orthoptera: Gryllidae). Annals of the Entomological Society of America 58: 828-832. https://doi.org/10.1093/aesa/58.6.828

Eilenberg, J. and Jensen, A.B., 2018a. Prevention and management of diseases in terrestrial invertebrates. In: Hajek, A.E. and Shapiro-Ilan, D.I. (ed.) Ecology of invertebrate diseases. Wiley & Sons, Chennai, India, pp. 495-526.

Eilenberg, J. and Jensen, A.B., 2018b. Strong host specialization in fungus genus *Strongwellsea* (Entomophthorales). Journal of Invertebrate Pathology 157: 112-116. https://doi.org/10.1016/j.jip.2018.08.007

Eilenberg, J., Van Oers, M.M., Jensen, A.B., Lecocq, A., Maciel-Vergara, G., Santacoloma, L.P.A., Van Loon, J.J.A. and Hesketh, H., 2018. Towards a coordination of European activities to diagnose and manage insect diseases in production facilities. Journal of Insects as Food and Feed 4(3): 157-166.

Eilenberg, J., Vlak, J.M., Nielsen-LeRoux, C., Cappellozza, S. and Jensen, A.B., 2015. Diseases in insects produced for food and feed. Journal of Insects as Food and Feed 1(2): 87-102.

Elvira, S., Williams, T. and Caballero, P., 2010. Juvenile hormone analog technology: effects on larval cannibalism and the production of *Spodoptera exigua* (Lepidoptera: Noctuidae) nucleopolyhedrovirus. Journal of Economic Entomology 103: 577-582. https://doi.org/10.1603/ec09325

Engelhard, E.K. and Volkman, L.E., 1995. Developmental resistance in fourth instar *Trichoplusia ni* orally inoculated with Autographa californica M nuclear polyhedrosis virus. Virology 2: 384-389. https://doi.org/10.1006/viro.1995.1270

Evans, J.D. and Armstrong, T.-N., 2006. Antagonistic interactions between honey bee bacterial symbionts and implications for disease. BMC Ecology 6: 4.

Fargues, J., Ouedraogo, A., Goettel, M.S. and Lomer, C.J., 1997. Effects of temperature, humidity and inoculation method on susceptibility of *Schistocerca gregaria* to *Metarhizium flavoviride*. Biocontrol Science and Technology 7: 345-356. https://doi.org/10.1080/09583159730758

Felberbaum, R.S., 2015. The baculovirus expression vector system: a commercial manufacturing platform for viral vaccines and gene therapy vectors. Biotechnology Journal 10: 702-714. https://doi.org/10.1002/biot.201400438

Fisher, T.W. and Garczynski, S.F., 2012. Isolation, culture, preservation, and identification of entomopathogenic bacteria of the Bacilli. In: Lacey, L.A. (ed.) Manual of techniques in invertebrate pathology, 2nd edition. Academic Press, San Diego, CA, USA, pp. 75-99. https://doi.org/10.1016/B978-0-12-386899-2.00003-8

Formato, G. and Smulders, F.J.M., 2011. Risk management in primary apicultural production.

Part 1: bee health and disease prevention and associated best practices. Veterinary Quarterly 31(1): 29-47. http://doi.org/10.1080/01652176.2011.565913

Francardi, V., Benvenuti, C., Roversi, P.F., Rumine, P. and Barzanti, G., 2012. Entomopathogenicity of *Beauveria bassiana* (Bals.) Vuill. and *Metarhizium anisopliae* (Metsch.) Sorokin isolated from different sources in the control of *Rhynchophorus ferrugineus* (Olivier) (Coleoptera Curculionidae). Redia 95: 49-55.

Fuxa, J.R., Sun, J.Z., Weidner, E.H. and LaMotte, L.R., 1999. Stressors and rearing diseases of *Trichoplusia ni:* evidence of vertical transmission of NPV and CPV. Journal of Invertebrate Pathology 74: 149-155. https://doi.org/10.1006/jipa.1999.4869

Gałęcki, R. and Sokół, R., 2019. A parasitological evaluation of edible insects and their role in the transmission of parasitic diseases to humans and animals. PLoS ONE 14: e0219303-e0219303. https://doi.org/10.1371/journal.pone.0219303

Gjedrem, T., 2015. Disease resistant fish and shellfish are within reach: a review. Journal of Marine Science and Engineering 3: 146-153.

Gjerris, M., Gamborg, C. and Röcklinsberg, H., 2018. Could crispy crickets be CRISPR-Cas9 crickets-ethical aspects of using new breeding technologies in intensive insectproduction. Professionals in Food Chains. Wageningen Academic Publishers, Wageningen, The Netherlands, pp. 3-5.

Goulson, D., Hails, R.S., Williams, T., Hirst, M.L., Vasconcelos, S.D., Green, B.M., Carty, T.M. and Cory, J.S., 1995. Transmission dynamics of a virus in a stage-structured insect population. Ecology 76: 392-401. https://doi.org/10.2307/1941198

Govindaraju, K., Tamilselvan, S., Kiruthiga, V. and Singaravelu, G., 2011. Silvernanotherapy on the viral borne disease of silkworm *Bombyx mori* L. Journal of Nanoparticle Research 13: 6377-6388. https://doi.org/10.1007/s11051-011-0390-3

Grau, T., Vilcinskas, A. and Joop, G., 2017. Probiotic *Enterococcus mundtii* isolate protects the model insect *Tribolium castaneum* against *Bacillus thuringiensis*. Frontiers in Microbiology 8: 1261. https://doi.org/10.3389/fmicb.2017.01261

Guerenstein, P.G. and Hildebrand, J.G., 2008. Roles and effects of environmental carbon dioxide in insect life. Annual Review of Entomology 53: 161-178. https://doi.org/10.1146/annurev.ento.53.103106.093402

Guichard, M., Neuditschko, M., Fried, P., Soland, G. and Dainat, B., 2019. A future resistance breeding strategy against *Varroa destructor* in a small population of the dark honey bee. Journal of Apicultural Research 58: 814-823. https://doi.org/10.1080/00218839.2019.1654966

Gunasekaran, N. and Rajendran, S., 2005. Toxicity of carbon dioxide to drugstore beetle *Stegobium paniceum* and cigarette beetle *Lasioderma serricorne*. Journal of Stored Products Research 41: 283-294. https://doi.org/10.1016/j.jspr.2004.04.001

Guo, Z., Döll, K., Dastjerdi, R., Karlovsky, P., Dehne, H.-W. and Altincicek, B., 2014. Effect of fungal colonization of wheat grains with *Fusarium spp.* on food choice, weight gain and mortality of meal beetle larvae (*Tenebrio molitor*). PLoS ONE 9: e100112. https://doi.org/10.1371/journal.pone.0100112

Hajek, A.E., 1997. Ecology of terrestrial fungal entomopathogens. In: Jones, J.G. (ed.) Advances in microbial ecology. Springer, Philadelphia, PA, USA, pp. 193-249.

Hajek, A.E., Papierok, B. and Eilenberg, J., 2012. Methods for study of the Entomophthorales. In: Lacey, L.A. (ed.) Manual of techniques in invertebrate pathology, 2nd edition. Academic Press, San Diego, CA, USA, pp. 285-316. https://doi.org/10.1016/B978-0-12-386899-2.00009-9

Hall, R. and Papierok, B., 1982. Fungi as biological control agents of arthropods of agricultural and medical importance. Parasitology 84: 205-240.

Han, B. and Weiss, L.M., 2017. Microsporidia: obligate intracellular pathogens within the fungal kingdom. In: Heitman, J., Howlett, B.J., Crous, P.W., Stukenbrock, E.H., James, T.Y. and Gow, N.A.R. (eds.) The fungal kingdom. Wiley and Sons, Hoboken, NJ, USA, pp. 97-113. https://doi.org/10.1128/9781555819583.ch5

Hanboonsong, A. and Durst, P., 2020. Guidance on sustainable cricket farming – a practical manual for farmers and inspectors. Food and Agriculture Organization, Rome, Italy.

Harrison, R. and Hoover, K., 2012. Baculoviruses and other occluded insect viruses. In: Vega, F.E. and Kaya, H.K. (eds.) Insect pathology, 2nd edition. Academic Press, San Diego, CA, USA, pp. 73-131. https://doi.org/10.1016/B978-0-12-384984-7.00004-X

Hladun, K.R., Parker, D.R. and Trumble, J.T., 2015. Cadmium, copper, and lead accumulation and bioconcentration in the vegetative and reproductive organs of *Raphanus sativus*: implications for plant performance and pollination. Journal of Chemical Ecology 41: 386-395. https://doi.org/10.1007/s10886-015-0569-7

Hofmann, C., Sandig, V., Jennings, G., Rudolph, M., Schlag, P. and Strauss, M., 1995. Efficient gene transfer into human hepatocytes by baculovirus vectors. Proceedings of the National Academy of Sciences 92: 10099-10103. https://doi.org/10.1073/pnas.92.22.10099

Holmes, L.A., Vanlaerhoven, S.L. and Tomberlin, J.K., 2012. Relative humidity effects on the life history of *Hermetia illucens* (Diptera: Stratiomyidae). Environmental Entomology 41: 971-978. https://doi.org/10.1603/en12054

Hughes, D.S., Possee, R.D. and King, L.A., 1993. Activation and detection of a latent baculovirus resembling Mamestra brassicae Nuclear Polyhedrosis Virus in *M. brassicae* insects. Virology 194: 608-615. https://doi.org/10.1006/viro.1993.1300

Humber, R.A., 2012. Identification of entomopathogenic fungi. In: Lacey, L.A. (ed.) Manual of techniques in invertebrate pathology, 2nd edition. Academic Press, San Diego, CA, USA, pp. 151-187. https://doi.org/10.1016/B978-0-12-386899-2.00006-3

Hunter, W., Ellis, J., Van Engelsdorp, D., Hayes, J., Westervelt, D., Glick, E., Williams, M., Sela, I., Maori, E., Pettis, J., Cox-Foster, D. and Paldi, N., 2010. Large-scale field application of RNAi technology reducing Israeli Acute Paralysis Virus disease in honey bees (*Apis mellifera*, Hymenoptera: Apidae). PLoS Pathogens 6: e1001160. https://doi.org/10.1371/journal.ppat.1001160

Hurpin, B., 1968. The influence of temperature and larval stage on certain diseases of *Melolontha melolontha*. Journal of Invertebrate Pathology 10: 252-262. https://doi.org/10.1016/0022-2011(68)90082-7

Inglis, G., Johnson, D., Cheng, K. and Goettel, M., 1997. Use of pathogen combinations to overcome the constraints of temperature on entomopathogenic hyphomycetes against grasshoppers. Biological Control 8: 143-152.

Inglis, G.D., Enkerli, J. and Goettel, M.S., 2012. Laboratory techniques used for entomopathogenic fungi: hypocreales. In: Lacey, L.A. (ed.) Manual of techniques in invertebrate pathology, 2nd edition. Academic Press, San Diego, CA, USA, pp. 189-253. https://doi.org/10.1016/B978-0-12-386899-2.00007-5

Inglis, G.D., Johnson, D.L. and Goettel, M.S., 1996. Effects of temperature and thermoregulation on mycosis by *Beauveria bassiana* in grasshoppers. Biological Control 7: 131-139. https://doi.org/10.1006/bcon.1996.0076

Jahnke, M., 2005. *Gregarina tibengae* spn. (Apicomplexa: Eugregarinida) described from *Zophobas atratus* Fabricius, 1775 (Coleoptera: Tenebrionidae). Acta Protozool 44: 67-74.

Jakob, N.J., Kleespies, R.G., Tidona, C.A., Müller, K., Gelderblom, H.R. and Darai, G., 2002. Comparative analysis of the genome and host range characteristics of two insect iridoviruses: Chilo iridescent virus and a cricket iridovirus isolate. Journal of General Virology 83: 463-470. https://doi.org/10.1099/0022-1317-83-2-463

Jakubowska, A.K., Vogel, H. and Herrero, S., 2013. Increase in gut microbiota after immune suppression in Baculovirus-infected larvae. PLoS Pathogens 9: e1003379. https://doi.org/10.1371/journal.ppat.1003379

James, R.R. and Li, Z., 2012. From silkworms to bees: diseases of beneficial insects. In: Vega, F.E. and Kaya, H.K. (eds.) Insect pathology, 2nd edition. Academic Press, San Diego, CA, USA, pp. 425-459. https://doi.org/10.1016/B978-0-12-384984-7.00012-9

Jiang, D., Tan, M., Guo, Q. and Yan, S., 2021. Transfer of heavy metal along food chain: a mini-review on insect susceptibility to entomopathogenic microorganisms under heavy metal stress. Pest Management Science 77: 1115-1120. https://doi.org/10.1002/ps.6103

Joosten, L., Lecocq, A., Jensen, A.B., Haenen, O., Schmitt, E. and Eilenberg, J., 2020. Review of insect pathogen risks for the black soldier fly (*Hermetia illucens*) and guidelines for reliable production. Entomologia Experimentalis et Applicata 168: 432-447. https://doi.org/10.1111/eea.12916

Junglen, S. and Drosten, C., 2013. Virus discovery and recent insights into virus diversity in arthropods. Current Opinion in Microbiology 16: 507-513. https://doi.org/10.1016/j.mib.2013.06.005

Jurat-Fuentes, J.L. and Jackson, T.A., 2012. Bacterial entomopathogens. In: Vega, F.E. and Kaya, H.K. (eds.) Insect pathology, 2nd edition. Academic Press, San Diego, CA, USA, pp. 265-349. https://doi.org/10.1016/B978-0-12-384984-7.00008-7

Just, F.T. and Essbauer, S.S., 2001. Characterization of an iridescent virus isolated from *Gryllus bimaculatus* (Orthoptera: Gryllidae). Journal of Invertebrate Pathology 77: 51-61. https://doi.org/10.1006/jipa.2000.4985

Kariithi, H.M., 2013. Glossina hytrosavirus control strategies in tsetse fly factories: application of infectomics in virus management. PhD-thesis, Wageningen University, Wageningen, The Netherlands.

Kleespies, R.G., Tidona, C.A. and Darai, G., 1999. Characterization of a new iridovirus isolated from crickets and investigations on the host range. Journal of Invertebrate Pathology 73: 84-90. https://doi.org/10.1006/jipa.1998.4821

Krams, I.A., Kecko, S., Jõers, P., Trakimas, G., Elferts, D., Krams, R., Luoto, S., Rantala, M.J.,

Inashkina, I., Gudrā, D., Fridmanis, D., Contreras-Garduño, J., Grantiņa-Ieviņa, L. and Krama, T., 2017. Microbiome symbionts and diet diversity incur costs on the immune system of insect larvae. The Journal of Experimental Biology 220: 4204-4212. https://doi.org/10.1242/jeb.169227

Krishnamurthy, T.S., Spratt, E.C. and Bell, C.H., 1986. The toxicity of carbon dioxide to adult beetles in low oxygen atmospheres. Journal of Stored Products Research 22: 145-151. https://doi.org/10.1016/0022-474X(86)90008-1

La Fauce, K. and Owens, L., 2013. Suppression of Penaeus merguiensis densovirus following oral delivery of live bacteria expressing dsRNA in the house cricket (*Acheta domesticus*) model. Journal of Invertebrate Pathology 112: 162-165. https://doi.org/10.1016/j.jip.2012.11.006

Lacey, L.A., Frutos, R., Kaya, H. and Vail, P., 2001. Insect pathogens as biological control agents: do they have a future? Biological Control 21: 230-248.

Lange, C.E. and Lord, J.C., 2012. Protistan entomopathogens. In: Vega, F.E. and Kaya, H.K. (eds.) Insect pathology, 2nd edition. Academic Press, San Diego, CA, USA, pp. 367-394. https://doi.org/10.1016/B978-0-12-384984-7.00010-5

Lecocq, A., Joosten, L., Schmitt, E., Eilenberg, J. and Jensen, A.B., 2021. *Hermetia illucens* adults are susceptible to infection by the fungus *Beauveria bassiana* in laboratory experiments. Journal of Insects as Food and Feed 7: 63-68. https://doi.org/10.3920/jiff2020.0042

Lecocq, A., Natsopolou, M.E., Berggreen, I.E., Eilenberg, J., Heckmann, L.H.L., Nielsen, H.V., Stensvold, C.R. and Jensen, A.B., in press. Probiotic properties of an indigenous *Pediococcus pentosaceus* strain on *Tenebrio molitor* larval growth and survival. Journal of Insects as Food and Feed. https://doi.org/10.3920/JIFF2020.0156

Lee, K.P., Cory, J.S., Wilson, K., Raubenheimer, D. and Simpson, S.J., 2006. Flexible diet choice offsets protein costs of pathogen resistance in a caterpillar. Proceedings. Biological Sciences 273: 823-829. https://doi.org/10.1098/rspb.2005.3385

Lee, K.P., Simpson, S.J. and Wilson, K., 2008. Dietary protein-quality influences melanization and immune function in an insect. Functional Ecology 22: 1052-1061. https://doi.org/10.1111/j.1365-2435.2008.01459.x

L'Héritier, P.H., 1948. Sensitivity to CO_2 in *Drosophila* – a review. Heredity 2: 325-348. https://doi.org/10.1038/hdy.1948.20

Li, M.-L., Weng, K.-F., Shih, S.-R. and Brewer, G., 2016a. The evolving world of small RNAs from RNA viruses. WIREs RNA 7: 575-588. https://doi.org/10.1002/wrna.1351

Li, Z., Li, X., Zhang, J., Wang, X., Wang, L., Cao, Z. and Xu, Y., 2016b. Use of phages to control *Vibrio splendidus* infection in the juvenile sea cucumber *Apostichopus japonicus*. Fish & Shellfish Immunology 54: 302-311. https://doi.org/10.1016/j.fsi.2016.04.026

Lighthart, B., Sewall, D. and Thomas, D.R., 1988. Effect of several stress factors on the susceptibility of the predatory mite, *Metaseiulus occidentalis* (Acari: Phytoseiidae), to the weak bacterial pathogen *Serratia marcescens*. Journal of Invertebrate Pathology 52: 33-42. https://doi.org/10.1016/0022-2011(88)90099-7

Liu, S., Chen, Y. and Bonning, B.C., 2015. RNA virus discovery in insects. Current Opinion in Insect Science 8: 54-61. https://doi.org/10.1016/j.cois.2014.12.005

Lopes, R.B. and Alves, S.B., 2005. Effect of *Gregarina sp.* parasitism on the susceptibility of *Blattella germanica* to some control agents. Journal of Invertebrate Pathology 88: 261-264. https://doi.org/10.1016/j.jip.2005.01.010

Louis, C., Jourdan, M. and Cabanac, M., 1986. Behavioural fever and therapy in a rickettsia-infected Orthoptera. American Journal of Physiology 250: R991-995. https://doi.org/10.1152/ajpregu.1986.250.6.R991

Macías-Macías, J.O., Tapia-Rivera, J.C., De la Mora, A., Tapia-González, J.M., Contreras-Escareño, F., Petukhova, T., Morfin, N. and Guzman-Novoa, E., 2020. *Nosema ceranae* causes cellular immunosuppression and interacts with thiamethoxam to increase mortality in the stingless bee *Melipona colimana*. Scientific Reports 10: 17021. https://doi.org/10.1038/s41598-020-74209-3

Maciel-Vergara, G. and Ros, V.I.D., 2017. Viruses of insects reared for food and feed. Journal of Invertebrate Pathology 147: 60-75. https://doi.org/10.1016/j.jip.2017.01.013

Maciel-Vergara, G., Jensen, A.B. and Eilenberg, J., 2018. Cannibalism as a possible entry route for opportunistic pathogenic bacteria to insect hosts, exemplified by *Pseudomonas aeruginosa*, a pathogen of the giant mealworm *Zophobas morio*. Insects 9: 88.

Majumdar, A., Boetel, M.A. and Jaronski, S.T., 2008. Discovery of *Fusarium solani* as a naturally occurring pathogen of sugarbeet root maggot (Diptera: Ulidiidae) pupae: prevalence and baseline susceptibility. Journal of Invertebrate Pathology 97: 1-8. https://doi.org/10.1016/j.jip.2007.05.003

Malakar, R., Elkinton, J.S., Hajek, A.E. and Burand, J.P., 1999. Within-host interactions of *Lymantria dispar* (Lepidoptera: Lymantriidae) nucleopolyhedrosis virus and *Entomophaga maimaiga* (Zygomycetes: Entomophthorales). Journal of Invertebrate Pathology 73: 91-100. https://doi.org/10.1006/jipa.1998.4806

Martin, S.J., Highfield, A.C., Brettell, L., Villalobos, E.M., Budge, G.E., Powell, M., Nikaido, S. and Schroeder, D.C., 2012. Global honey bee viral landscape altered by a parasitic mite. Science 336: 1304-1306. https://doi.org/10.1126/science.1220941

Martinez, J., Longdon, B., Bauer, S., Chan, Y.S., Miller, W.J., Bourtzis, K., Teixeira, L. and Jiggins, F.M., 2014. Symbionts commonly provide broad spectrum resistance to viruses in insects: a comparative analysis of *Wolbachia* strains. PLoS Pathogens 10: e1004369. https://doi.org/10.1371/journal.ppat.1004369

Martínez-Solís, M., Collado, M.C. and Herrero, S., 2020. Influence of diet, sex, and viral infections on the gut microbiota composition of *Spodoptera exigua* Ccterpillars. Frontiers in Microbiology 11: 753. https://doi.org/10.3389/fmicb.2020.00753

Martín-Hernández, R., Meana, A., García-Palencia, P., Marín, P., Botías, C., Garrido-Bailón, E., Barrios, L. and Higes, M., 2009. Effect of temperature on the biotic potential of honeybee Microsporidia. Applied and Environmental Microbiology 75: 2554-2557. https://doi.org/10.1128/aem.02908-08

May, R.M. and Anderson, R.M., 1979. Population biology of infectious diseases: Part II. Nature 280: 455-461. https://doi.org/10.1038/280455a0

McMenamin, A.J., Daughenbaugh, K.F. and Flenniken, M.L., 2020. The heat shock response in the Western honey bee (*Apis mellifera*) is antiviral. Viruses 12: 245.

Mellberg, S. and Wirtanen, G., 2018. Clean and easy cricket rearing: a guide on hygienic building design in rearing facilities. Helsingin yliopisto Ruralia-instituutti, Helsinki, Finland. https://helda.helsinki.fi/handle/10138/259036

Ment, D., Shikano, I. and Glazer, I., 2017. Abiotic factors, ecology of invertebrate diseases. In: Hajek, A.E. (ed.) Ecology of invertebrate diseases. Wiley and Sons, Hoboken, NJ, USA, pp. 143-186. https://doi.org/10.1002/9781119256106.ch5

Michalaki, M.P., Athanassiou, C.G., Kavallieratos, N.G., Batta, Y.A. and Balotis, G.N., 2006. Effectiveness of *Metarhizium anisopliae* (Metschinkoff) Sorokin applied alone or in combination with diatomaceous earth against *Tribolium confusum* Du Val larvae: influence of temperature, relative humidity and type of commodity. Crop Protection 25: 418-425. https://doi.org/10.1016/j.cropro.2005.07.003

Miller, L.K. and Ball, L.A., 1998. The insect viruses. Springer Science & Business Media, New York, NY, USA, 416 pp.

Mir, A.H., Qamar, A., Qadir, I., Naqvi, A.H. and Begum, R., 2020. Accumulation and trafficking of zinc oxide nanoparticles in an invertebrate model, *Bombyx mori*, with insights on their effects on immuno-competent cells. Scientific Reports 10: 1617. https://doi.org/10.1038/s41598-020-58526-1

Moran, N.A., 2015. Genomics of the honey bee microbiome. Current Opinion in Insect Science 10: 22-28.

Mostafa, A.M., Fields, P.G. and Holliday, N.J., 2005. Effect of temperature and relative humidity on the cellular defense response of *Ephestia kuehniella* larvae fed *Bacillus thuringiensis*. Journal of Invertebrate Pathology 90: 79-84. https://doi.org/10.1016/j.jip.2005.08.007

Muhammad, A., Habineza, P., Ji, T., Hou, Y. and Shi, Z., 2019. Intestinal microbiota confer protection by priming the immune system of red palm weevil *Rhynchophorus ferrugineus* Olivier (Coleoptera: Dryophthoridae). Frontiers in Physiology 10: 1303. https://doi.org/10.3389/fphys.2019.01303

Nelson, R., Wiesner-Hanks, T., Wisser, R. and Balint-Kurti, P., 2018. Navigating complexity to breed disease-resistant crops. Nature Reviews Genetics 19: 21-33. https://doi.org/10.1038/nrg.2017.82

Nicolas, G. and Sillans, D., 1989. Immediate and latent effects of carbon dioxide on insects. Annual Review of Entomology 34: 97-116.

Noonin, C., Jiravanichpaisal, P., Söderhäll, I., Merino, S., Tomás, J.M. and Söderhäll, K., 2011. Melanization and pathogenicity in the insect, *Tenebrio molitor*, and the Crustacean, *Pacifastacus leniusculus*, by *Aeromonas hydrophila* AH-3. PLoS ONE 5: e15728. https://doi.org/10.1371/journal.pone.0015728

Nouri, S., Matsumura, E.E., Kuo, Y.-W. and Falk, B.W., 2018. Insect-specific viruses: from discovery to potential translational applications. Current Opinion in Virology 33: 33-41. https://doi.org/10.1016/j.coviro.2018.07.006

Odemer, R., Nilles, L., Linder, N. and Rosenkranz, P., 2018. Sublethal effects of clothianidin and *Nosema spp.* on the longevity and foraging activity of free flying honey bees. Ecotoxicology 27: 527-538. https://doi.org/10.1007/s10646-018-1925-5

Onstad, D. and Carruthers, R., 1990. Epizootiological models of insect diseases. Annual Review of Entomology 35: 399-419.

Ouedraogo, R.M., Cusson, M., Goettel, M.S. and Brodeur, J., 2003. Inhibition of fungal growth in thermoregulating locusts, *Locusta migratoria*, infected by the fungus *Metarhizium anisopliae var acridum*. Journal of Invertebrate Pathology 82: 103-109. https://doi.org/10.1016/s0022-2011(02)00185-4

Pagnocca, F.C., Masiulionis, V.E. and Rodrigues, A., 2012. Specialized fungal parasites and opportunistic fungi in gardens of attine ants. Psyche, Article ID: 905109.

Petersen, L.M. and Tisa, L.S., 2012. Influence of temperature on the physiology and virulence of the insect pathogen *Serratia sp.* strain SCBI. Applied and Environmental Microbiology 78: 8840-8844. https://doi.org/10.1128/aem.02580-12

Ponton, F., Morimoto, J., Robinson, K., Kumar, S.S., Cotter, S.C., Wilson, K. and Simpson, S.J., 2020. Macronutrients modulate survival to infection and immunity in *Drosophila*. Journal of Animal Ecology 89: 460-470. https://doi.org/10.1111/1365-2656.13126

Ponton, F., Wilson, K., Holmes, A., Raubenheimer, D., Robinson, K.L. and Simpson, S.J., 2015. Macronutrients mediate the functional relationship between *Drosophila* and *Wolbachia*. Proceedings. Biological Sciences 282: 20142029. https://doi.org/10.1098/rspb.2014.2029

Ponton, F., Wilson, K., Holmes, A.J., Cotter, S.C., Raubenheimer, D. and Simpson, S.J., 2013. Integrating nutrition and immunology: a new frontier. Journal of Insect Physiology 59: 130-137. https://doi.org/10.1016/j.jinsphys.2012.10.011

Povey, S., Cotter, S.C., Simpson, S.J. and Wilson, K., 2014. Dynamics of macronutrient self-medication and illness-induced anorexia in virally infected insects. Journal of Animal Ecology 83: 245-255. https://doi.org/10.1111/1365-2656.12127

Powers, A.M., Kamrud, K.I., Olson, K.E., Higgs, S., Carlson, J.O. and Beaty, B.J., 1996. Molecularly engineered resistance to California serogroup virus replication in mosquito cells and mosquitoes. Proceedings of the National Academy of Sciences 93: 4187-4191. https://doi.org/10.1073/pnas.93.9.4187

Proudfoot, C., Lillico, S. and Tait-Burkard, C., 2019. Genome editing for disease resistance in pigs and chickens. Animal Frontiers 9: 6-12. https://doi.org/10.1093/af/vfz013

Pulpán, J. and Verner, P.H., 1965. Control of the tyroglyphoid mites in stored grain by the predatory mite *Cheyletus eruditus* (Schrank). Canadian Journal of Zoology 43(3): 417-432. https://doi.org/10.1139/z65-042

Rangel, J. and Ward, L., 2018. Evaluation of the predatory mite *Stratiolaelaps scimitus* for the biological control of the honey bee ectoparasitic mite *Varroa destructor*. Journal of Apicultural Research 57(3): 425-432. https://doi.org/10.1080/00218839.2018.1457864

Ratte, H.T., 1985. Temperature and insect development. In: Hoffmann, K.H. (ed.) Environmental physiology and biochemistry of insects. Springer Berlin Heidelberg, Berlin, Heidelberg, Germany, pp. 33-66. https://doi.org/10.1007/978-3-642-70020-0_2

Reeson, A.F., Wilson, K., Gunn, A., Hails, R.S. and Goulson, D., 1998. Baculovirus resistance in the noctuid Spodoptera exempta is phenotypically plastic and responds to population density. Proceedings of the Royal Society of London. Series B: Biological Sciences 265: 1787-1791. https://doi.org/10.1098/rspb.1998.0503

Reynaldi, F.J., De Giusti, M.R. and Alippi, A.M., 2004. Inhibition of the growth of *Ascosphaera apis* by selected strains of *Bacillus* and *Paenibacillus* species isolated from honey. Revista Argentina de Microbiología 36: 52-55.

Rippere, K.E., Tran, M.T., Yousten, A.A., Hilu, K.H. and Klein, M.G., 1998. *Bacillus popilliae* and *Bacillus lentimorbus*, bacteria causing milky disease in Japanese beetles and related scarab larvae. International Journal of Systematic and Evolutionary Microbiology 48: 395-402. https://doi.org/10.1099/00207713-48-2-395

Römer, D., Bollazzi, M. and Roces, F., 2018. Carbon dioxide sensing in the social context: leaf-cutting ants prefer elevated CO_2 levels to tend their brood. Journal of Insect Physiology 108: 40-47. https://doi.org/10.1016/j.jinsphys.2018.05.007

Rosen, L., 1980. Carbon dioxide sensitivity in mosquitoes infected with Sigma, vesicular stomatitis, and other rhabdoviruses. Science 207: 989-991. https://doi.org/ 10.1126/science.6101512

Rottschaefer, S.M. and Lazzaro, B.P., 2012. No effect of *Wolbachia* on resistance to intracellular infection by pathogenic bacteria in *Drosophila melanogaster*. PLoS ONE 7: e40500. https://doi.org/10.1371/journal.pone.0040500

Sage, R.F., 2002. How terrestrial organisms sense, signal, and respond to carbon dioxide 1. Integrative and Comparative Biology 42: 469-480. https://doi.org/10.1093/icb/42.3.469

Samson, M., Baig, M., Sharma, S., Balavenkatasubbaiah, M., Sasidharan, T. and Jolly, M., 1990. Survey on the relative incidence of silkworm diseases in Karnataka, India. Indian Journal of Sericulture 29: 248-254.

Sanchis, V., 2011. From microbial sprays to insect-resistant transgenic plants: history of the biospesticide *Bacillus thuringiensis*. A review. Agronomy for Sustainable Development 31: 217-231. https://doi.org/10.1051/agro/2010027

Sangbaramou, R., Camara, I., Huang, X.-Z., Shen, J., Tan, S.-Q. and Shi, W.-P., 2018. Behavioural thermoregulation in *Locusta migratoria manilensis* (Orthoptera: Acrididae) in response to the entomopathogenic fungus, *Beauveria bassiana*. PLoS ONE 13: e0206816. https://doi.org/10.1371/journal.pone.0206816

Sayed, W.A., Ibrahim, N.S., Hatab, M.H., Zhu, F. and Rumpold, B.A., 2019. Comparative study of the use of insect meal from *Spodoptera littoralis* and *Bactrocera zonata* for feeding Japanese quail chicks. Animals 9: 136.

Scully, L.R. and Bidochka, M.J., 2006. Developing insectmodels for the study of current and emerging human pathogens. FEMS Microbiology Letters 263: 1-9. https://doi.org/10.1111/j.1574-6968.2006.00388.x

Shaurub, E.-S.H., 2003. Immune response of insects to abiotic agents: a review of current prospectives. International Journal of Tropical Insect Science 23: 273-279. https://doi.org/10.1017/S1742758400012327

Sidahmed, A.M.E. and Wilkie, B., 2010. Endogenous antiviral mechanisms of RNA interference: a comparative biology perspective. In: Min, W.-P. and Ichim, T. (eds.) RNA interference: from biology to clinical applications. Humana Press, Totowa, NJ, USA, pp. 3-19. https://doi.org/10.1007/978-1-60761-588-0_1

Sikorowski, P.P. and Lawrence, A.M., 1994. Microbial contamination and insect rearing. American Entomologist 40: 240-253. https://doi.org/10.1093/ae/40.4.240

Solter, L.F., Becnel, J.J. and Oi, D.H., 2012a. Microsporidian entomopathogens. In: Vega, F.E. and Kaya, H.K. (eds.) Insect pathology, 2nd edition. Academic Press, San Diego, CA, USA, pp. 221-263. https://doi.org/10.1016/B978-0-12-384984-7.00007-5

Solter, L.F., Becnel, J.J. and Vávra, J., 2012b. Research methods for entomopathogenic microsporidia and other protists. In: Lacey, L.A. (d.) Manual of techniques in invertebrate pathology, 2nd edition. Academic Press, San Diego, CA, USA, pp. 329-371. https://doi.org/10.1016/B978-0-12-386899-2.00011-7

Srygley, R.B., Lorch, P.D., Simpson, S.J. and Sword, G.A., 2009. Immediate protein dietary effects on movement and the generalised immunocompetence of migrating mormon crickets *Anabrus simplex* (Orthoptera: Tettigoniidae). Ecological Entomology 34: 663-668. https://doi.org/10.1111/j.1365-2311.2009.01117.x

Stahlschmidt, Z. and Adamo, S., 2013. Context dependency and generality of fever in insects. Naturwissenschaften 100: 691-696.

Steinhaus, E.A., 1958. Crowding as a Possible Stress Factor in Insect Disease. Ecology, 39: 503-514. https://doi.org/10.2307/1931761

Steinhaus, E.A., 1963. Insect pathology. Vol. 1. An advanced treatise. Academic Press, New York, NY, USA, 663 pp.

Steinkraus, D.C., Brooks, W.M. and Geden, C.G., 1992. Discovery of the neogregarine *Farinocystis tribolii* and an eugregarine in the lesser mealworm, *Alphitobius diaperinus*. Journal of Invertebrate Pathology 59: 203-205. https://doi.org/10.1016/0022-2011(92)90035-3

Streett, D. and McGuire, M., 1990. Pathogenic diseases of grasshoppers. In: Chapman, R.F. (ed.) Biology of grasshoppers. Wiley and Sons, Hoboken, NJ, USA, pp. 483-516.

Szelei, J., Woodring, J., Goettel, M.S., Duke, G., Jousset, F.X., Liu, K.Y., Zadori, Z., Li, Y., Styer, E., Boucias, D.G., Kleespies, R.G., Bergoin, M. and Tijssen, P., 2011. Susceptibility of North-American and European crickets to Acheta domesticus densovirus (AdDNV) and associated epizootics. Journal of Invertebrate Pathology 106: 394-399. https://doi.org/10.1016/j.jip.2010.12.009

Tanaka, A., 1982. Effects of carbon-dioxide anaesthesia on the number of instars, larval duration and adult body size of the German cockroach, *Blattella germanica*. Journal of Insect Physiology 28: 813-821. https://doi.org/10.1016/0022-1910(82)90092-0

Tedersoo, L., Drenkhan, R., Anslan, S., Morales-Rodriguez, C. and Cleary, M., 2019. High-throughput identification and diagnostics of pathogens and pests: overview and practical recommendations. Molecular Ecology Resources 19: 47-76. https://doi.org/10.1111/1755-0998.12959

Tesovnik, T., Zorc, M., Ristanić, M., Glavinić, U., Stevanović, J., Narat, M. and Stanimirović, Z., 2020. Exposure of honey bee larvae to thiamethoxam and its interaction with *Nosema ceranae* infection in adult honey bees. Environmental Pollution 256: 113443. https://doi.org/10.1016/j.envpol.2019.113443

Thomas, M.B., Watson, E.L. and Valverde-Garcia, P., 2003. Mixed infections and insect-pathogen interactions. Ecology Letters 6: 183-188. https://doi.org/10.1046/j.1461-0248.2003.00414.x

Tounou, A.K., Kooyman, C., Douro-Kpindou, O.K. and Poehling, H.M., 2008. Interaction between *Paranosema locustae* and *Metarhizium anisopliae* var. acridum, two pathogens of the desert locust, *Schistocerca gregaria* under laboratory conditions. Journal of Invertebrate Pathology 97: 203-210. https://doi.org/10.1016/j.jip.2007.10.002

Tritschler, M., Vollmann, J.J., Yañez, O., Chejanovsky, N., Crailsheim, K. and Neumann, P.,

2017. Protein nutrition governs within-host race of honey bee pathogens. Scientific Reports 7: 14988. https://doi.org/10.1038/s41598-017-15358-w

Valles, S.M. and Chen, Y., 2006. Serendipitous discovery of an RNA virus from the cricket, *Acheta domesticus*. Florida Entomologist 89: 282-283.

Van Huis, A., 2019. Welfare of farmed insects. Journal of Insects as Food and Feed 5(3): 159-162. https://doi.org/10.3920/JIFF2019.x004

Van Lenteren, J.C., Bolckmans, K., Köhl, J., Ravensberg, W. and Urbaneja, A., 2018. Biological control using invertebrates and microorganisms: plenty of new opportunities. BioControl 63: 39-59. https://doi.org/10.1007/s10526-017-9801-4

Van Oers, M.M., 2006. Vaccines for viral and parasitic diseases produced with baculovirus vectors. Advances in Virus Research. Academic Press, San Diego, CA, USA, pp. 193-253. https://doi.org/10.1016/S0065-3527(06)68006-8

Van Ooik, T., Pausio, S. and Rantala, M.J., 2008. Direct effects of heavy metal pollution on the immune function of a geometrid moth, *Epirrita autumnata*. Chemosphere 71: 1840-1844. https://doi.org/10.1016/j.chemosphere.2008.02.014

Vandeweyer, D., Crauwels, S., Lievens, B. and Van Campenhout, L., 2017. Metagenetic analysis of the bacterial communities of edible insects from diverse production cycles at industrial rearing companies. International Journal of Food Microbiology 261: 11-18. https://doi.org/10.1016/j.ijfoodmicro.2017.08.018

Vasantharajan, V. and Munirathnamma, N., 2013. Studies on silkworm diseases. III. Epizootiology of a septicemic disease of silkworms caused by *Serratia marcesens*. Journal of the Indian Institute of Science 60: 33.

Vega, F.E., Meyling, N.V., Luangsa-ard, J.J. and Blackwell, M., 2012. Fungal entomopathogens. In: Vega, F.E. and Kaya, H.K. (eds.) Insect pathology, 2nd edition. Academic Press, San Diego, CA, USA, pp. 171-220. https://doi.org/10.1016/B978-0-12-384984-7.00006-3

Virto, C., Navarro, D., Tellez, M.M., Herrero, S., Williams, T., Murillo, R. and Caballero, P., 2014. Natural populations of *Spodoptera exigua* are infected by multiple viruses that are transmitted to their offspring. Journal of Invertebrate Pathology 122: 22-27. https://doi.org/10.1016/j.jip.2014.07.007

Watson, D.W. and Petersen, J.J., 1993. Seasonal Activity of *Entomophthora muscae* (Zygomycetes: Entomophthorales) in *Musca domestica* L (Diptera: Muscidae) with reference to temperature and relative humidity. Biological Control 3: 182-190. https://doi.org/10.1006/bcon.1993.1026

Wei, G., Lai, Y., Wang, G., Chen, H., Li, F. and Wang, S., 2017. Insect pathogenic fungus interacts with the gut microbiota to accelerate mosquito mortality. Proceedings of the National Academy of Sciences 114: 5994-5999. https://doi.org/10.1073/pnas.1703546114

Weiser, J., 1977. An atlas of insect diseases. Springer Netherlands, Dordrecht, The Netherlands.

Weissman, D.B., Gray, D.A., Pham, H.T. and Tijssen, P., 2012. Billions and billions sold: pet-feeder crickets (Orthoptera: Gryllidae), commercial cricket farms, an epizootic densovirus, and government regulations make for a potential disaster. Zootaxa 3504: 67-88.

Williams, T. and Hernández, O., 2006. Costs of cannibalism in the presence of an iridovirus pathogen of *Spodoptera frugiperda*. Ecological Entomology 31: 106-113. https://doi.org/10.1111/j.0307-6946.2006.00771.x

Williams, T., Chitnis, N.S. and Bilimoria, S.L., 2009. Invertebrate iridovirus modulation of apoptosis. Virologica Sinica 24(4): 295-304. https://doi.org/10.1007/s12250-009-3060-1

Wilson, J.K., Ruiz, L. and Davidowitz, G., 2019. Dietary protein and carbohydrates affect immune function and performance in a specialist herbivore insect (*Manduca sexta*). Physiological and Biochemical Zoology 92: 58-70. https://doi.org/10.1086/701196

Wojda, I., Kowalski, P. and Jakubowicz, T., 2009. Humoral immune response of *Galleria mellonella* larvae after infection by *Beauveria bassiana* under optimal and heat-shock conditions. Journal of Insect Physiology 55: 525-531.

Zimmer, C.R., Dias de Castro, L.L., Pires, S.M., Delgado Menezes, A.M., Ribeiro, P.B. and Leivas Leite, F.P., 2013. Efficacy of entomopathogenic bacteria for control of *Musca domestica*. Journal of Invertebrate Pathology 114: 241-244. https://doi.org/10.1016/j.jip.2013.08.011

Zug, R. and Hammerstein, P., 2015. *Wolbachia* and the insect immune system: what reactive oxygen species can tell us about the mechanisms of *Wolbachia*-host interactions. Frontiers in Microbiology 6: 1201. https://doi.org/10.3389/fmicb.2015.01201

Zuk, M., 1987. The effects of gregarine parasites on longevity, weight loss, fecundity and developmental time in the field crickets *Gryllus veletis* and *G. pennsylvanicus*. Ecological Entomology 12: 349-354. https://doi.org/10.1111/j.1365-2311.1987.tb01014.x

Nutritional value of insects and ways to manipulate their composition

D.G.A.B. Oonincx[1]* and M.D. Finke[2]

*Animal Nutrition Group, Wageningen University and Research Centre,
De Elst 1, 6708 WD Wageningen, The Netherlands; Mark Finke LLC,
17028 E Wildcat Dr., Rio Verde, AZ 85263, USA; *dennis.oonincx@wur.nl*

Abstract

This article reports on the nutrients present in insects and factors affecting their variability. Data on protein content and amino acid profiles of a variety of insect species are discussed and their amino acid profiles compared to nutrient requirements of growing broiler chicks, catfish, trout, swine, and human adults and young children. Both *in vitro* and *in vivo* protein digestibility data for a variety of insect species is presented and factors affecting these data are discussed. Furthermore, the fat content and fatty acid profiles of a variety of insect species is reviewed, with special attention on omega-6 and omega-3 fatty acids. Information on carbohydrates, fibre and chitin in insects is shown along with potential effects on nutrient availability. This is followed by a discussion of essential minerals in insects with an emphasis on calcium and phosphorus. Data on insect vitamin content is shown along with a discussion of antinutritional factors such as phytate and thiaminase, which can adversely affect their nutritional value. Dietary effects on insect nutrient composition are reviewed with an emphasis on essential minerals, heavy metals, vitamin E, and carotenoids. Lastly, the effects of processing, including protein extraction and various cooking methods on insect composition are discussed. In summary, this article provides an overview of the nutrient content of insects, and how select nutrients can be altered.

Keywords

amino acids – fatty acids – vitamins – minerals – digestibility – nutrient – manipulation

1   Introduction

Insects are important sources of nutrients for humans and a wide variety of other animal species. Hence studies on insect nutrient composition can be found in disciplines ranging from anthropology to zoology. Comprehensive literature reviews of insect nutrient content have been published (Bukkens, 1997; Finke, 2004; Payne *et al.*, 2016; Raubenheimer and Rothman, 2013; Rumpold and Schluter, 2013a). While the first mention of using insects to feed production animals was in 1919 (Linder, 1919), it was not until the 1960s and 1970s that research started in earnest (Calvert *et al.*, 1969; Hale, 1973; Teotia and Miller, 1973, 1974; Ueckert *et al.*, 1972). The renewed interest in this area of study in the last decade yields numerous papers and comprehensive reviews regarding the safety of insects as food and feed, as well as several overviews of feeding trials for production animals (Gasco *et al.*, 2019; Henry *et al.*, 2015; Makkar *et al.*, 2014; Riddick, 2014; Rumpold and Schluter, 2013b; Sanchez-Muros *et al.*, 2014, 2016). Given the comprehensive nature of these recent reviews, this article will summarise this compositional data, focus on nutrients that have received little attention and nutrient manipulation. Data on both wild and produced insects is discussed.

2   Nutrient content of insects

2.1   *Protein and amino acids*

The protein content of insects varies between 25 and 75% on a dry matter (DM) basis (Barker *et al.*, 1998; Bukkens, 1997; Cerda *et al.*, 2001; Finke, 2002, 2013, 2015a; Oonincx and Dierenfeld, 2012; Oonincx and Van der Poel, 2011). Proteins are composed of amino acids and the true protein content equals the sum of amino acids. However, protein content is generally estimated by multiplying nitrogen content with a protein factor (Kp) of 6.25. This results in the so-called crude protein content. This factor is underestimated if not all amino acids are quantified, or due to methodological issues (Oonincx *et al.*, 2019b), such as losses of amino acids during hydrolysis. Conversely, the presence of nonprotein nitrogen from compounds, such as chitin, uric acid, and β-alanine leads to overestimations of true protein content when using this factor (Janssen *et al.*, 2017). An alternative Kp of 4.76 for insects has been suggested based on amino acid data on larvae of yellow mealworms (*Tenebrio molitor*), lesser mealworms (*Alphitobius diaperinus*) and black soldier flies (*Hermetia illucens*) (Janssen *et al.*, 2017). Recalculating the data from 20 insect samples including 13 species and different developmental stages, results in an average Kp of 5.81; range 4.56 to 6.45 (Finke, 2002, 2007, 2013, 2015a,b). This data confirms that a Kp of 6.25 is often a slight overestimate. However, until data for more species and at different life stages are accumulated, retaining a Kp of 6.25 can be beneficial to facilitate comparisons between studies.

The amino acids that make up true protein, are generally grouped as either nutritionally indispensable (essential) amino acids or nutritionally dispensable (non-essential) amino acids. While all amino acids are required, nutritionally indispensable amino acids cannot be synthesised by most animal species and must be

provided in the diet. Variation in amino acid patterns between life stages of a species partially depends on whether that species undergoes complete metamorphosis
(holometabolous) or incomplete metamorphosis (hemimetabolous) (Finke, 2002;
Pieterse and Pretorius, 2014). Comparison of the amino acid pattern (mg amino
acid/g crude protein) for the hemimetabolous house cricket (*Acheta domesticus*)
suggests that amino acid composition is fairly constant and unaffected by diet or
life stage (Table 1). Similarly, black soldier fly prepupae raised on eight different
diets had similar amino acid patterns (Spranghers *et al.*, 2017; Wang *et al.*, 2020), as
did tobacco hornworm larvae (*Manduca sexta*) fed two different diets (Landry *et al.*,
1986). This suggests that for holometabolous species amino acid patterns are fixed
within a specific life stage. Body parts, such as wings, legs or mandibles, have specific physical requirements to function properly; therefore, it is unlikely that their
amino acid composition can be altered by diet. Amino acid patterns between different life stages of holometabolous insects would however be expected to differ because larvae and adults are morphologically dissimilar. This was confirmed for yellow mealworms, where the amino acid pattern for larvae differed from the harder
bodied adults (Finke, 2002). The adults contain more glycine (52%) and tryptophan
(37%) than mealworm larvae, while larvae contain more leucine (29%), phenylalanine (35%) and tyrosine (120%) than adults (Finke, 2002). The amino acid patterns
of larvae and pupae meal of the house fly (*Musca domestica*) raised on the same
diet also differ (Pieterse and Pretorius, 2014). In this case differences were seen for
the amino acids alanine, arginine, aspartic acid, methionine, serine, and tyrosine.

Amino acid patterns are important because they partially determine the suitability of dietary protein sources. This suitability also depends on the amino acid
requirements of the consuming animal. The amino acid with the lowest concentration relative to the requirement of the animal is called the first limiting amino
acid. Table 2 shows the amino acid pattern of four commonly raised insect species and compares them to the amino acid requirements of broiler chicks, catfish,
trout, swine, and adult humans and young children (NRC, 1994, 2011, 2012; WHO/
FAO/UNU, 2007). Methionine and cystine are usually the first limiting amino acids in most insect species when fed to production animals or humans. The exception appears to be house crickets for catfish and swine, where threonine and/or
tryptophan are first limiting. These calculations are supported by data from animal
feeding trials. Methionine and arginine are the first limiting amino acids for broiler
chicks when fed Mormon cricket meal (*Anabrus simplex*) as the sole source of dietary protein in purified diets (Finke *et al.*, 1985), or when house cricket meal is the
source of protein in a corn based diet (Nakagaki *et al.*, 1987). Similarly, methionine
is the first limiting amino acid in a corn-soy-meat meal diet for growing chickens
when dried maggot meal substitutes for meat meal (Bamgbose, 1999). Methionine
is also the first limiting amino acid when adult Mormon cricket meal (Finke *et al.*,
1987), yellow mealworm larvae meal (Goulet *et al.*, 1978), or house fly larvae meal
are fed to growing rats (Onifade *et al.*, 2001). The high amino acid scores of all four

TABLE 1   Amino acid patterns (mg/g crude protein) of house crickets (*Acheta domesticus*)

Amino acid	Finke (2002) nymphs	Finke (2007) nymphs	Finke (2015b) nymphs	Finke (2002) adults	Finke (2007) adults	Yi et al. (2013) adults	Bosch et al. (2014) adults	Nakagaki et al. (1987) adults	Poelaert et al. (2018) adults	Köhler et al. (2019) adults
Alanine	89.0	101.1	90.9	87.8	76.9	81.0	83.6	95.0	65.7	82.6
Arginine	61.0	70.9	78.8	61.0	57.3	65.0	56.8	60.0	63.2	64.4
Aspartic acid	70.8	79.4	82.4	83.9	84.9	76.0	76.4	88.0	71.0	103.5
Cystine	8.4	9.1	9.8	8.3	9.8	NR	11.2	NR	12.9	9.1
Glutamic acid	103.9	117.1	114.5	104.9	104.4	110.0	105.9	117.0	98.1	94.8
Glycine	52.6	60.6	53.5	50.7	45.3	51.0	51.3	59.0	45.5	48.1
Histidine	22.1	25.7	22.1	23.4	22.7	21.0	33.9	26.0	31.3	21.7
Isoleucine	42.9	40.6	40.3	45.9	36.4	36.0	40.0	42.0	39.3	38.6
Leucine	95.5	72.6	70.9	100.0	66.7	66.0	66.4	73.0	63.1	75.7
Lysine	53.9	62.3	57.9	53.7	51.1	53.0	58.0	56.0	57.1	55.8
Methionine	13.0	15.4	16.6	14.6	19.6	NR	15.8	15.0	18.7	15.2
Phenylalanine	27.9	32.0	35.6	31.7	30.2	NR	31.8	22.0	31.7	36.0
Proline	55.2	61.1	59.8	56.1	54.2	54.0	54.4	62.0	51.6	50.8
Serine	41.6	42.9	40.4	49.8	52.0	38.0	37.1	49.0	46.2	59.4
Threonine	35.7	38.9	37.6	36.1	31.1	35.0	36.2	35.0	38.5	39.7
Tryptophan	5.2	6.3	8.7	6.3	7.6	9.0	NR	6.0	8.5	9.8
Tyrosine	55.2	62.9	64.8	48.8	44.0	NR	62.1	41.0	47.4	95.7
Valine	49.4	60.0	59.6	52.2	48.4	55.0	57.2	60.0	52.0	64.0

NR = not reported.

TABLE 2   Amino acid patterns (mg/g crude protein) of four commonly raised insects and amino acids
             scores and first limiting amino acid for various species

Amino acid	*Acheta domesticus* adults/nymphs	*Tenebrio molitor* larvae	*Zophobas morio* larvae	*Hermetia illucens* larvae/prepupae
Alanine	87.8	80.2	72.7	62.7
Arginine	65.7	60.0	57.4	52.8
Aspartic acid	79.1	81.0	83.1	88.3
Glutamic acid	109.2	112.1	127.0	103.7
Glycine	52.3	53.1	48.7	55.0
Histidine	22.8	30.2	31.1	32.4
Isoleucine	40.3	46.1	46.9	43.3
Leucine	78.6	84.9	80.4	69.9
Lysine	55.3	55.4	54.6	59.1
Methionine	15.8	13.3	12.2	18.8
Methionine + cystine	24.9	23.3	21.8	24.4
Phenylalanine	31.5	35.2	37.2	41.5
Phenylalanine + tyrosine	87.5	102.6	108.4	112.3
Proline	56.7	68.1	55.9	55.4
Serine	44.1	47.2	44.1	38.3
Threonine	35.7	40.3	39.9	39.0
Tryptophan	7.2	10.5	11.4	15.1
Valine	54.1	62.9	60.5	63.8

Amino acid score/first limiting amino acid:

Humans				
Children	96/Met+Cys	90/Met+Cys	84/Met+Cys	94/Met+Cys
Adults	113/Met+Cys	106/Met+Cys	99/Met+Cys	111/Met+Cys
Livestock				
Poultry	64/Met+Cys	60/Met+Cys	56/Met+Cys	62/Met+Cys
Catfish	55/Thr+Try	54/Met+Cys	50/Met+Cys	56/Met+Cys
Trout	52/Met+Cys	49/Met+Cys	46/Met+Cys	51/Met+Cys
Swine	59/Try	56/Met+Cys	52/Met+Cys	58/Met+Cys

The amino acid data shown for insects are an average of data from published sources (Finke, 2002, 2007, 2015b; Spranghers *et al.*, 2017; Yi *et al.*, 2013). The insect amino acid scores equal the concentration of their first limiting essential amino acid divided by the concentration of that amino acid in the reference pattern (NRC, (1994, 2011, 2012) for animals and WHO/FAO/UNU (2007) for humans).

insect species for both children and adults suggests they are a high-quality protein source for humans.

Protein quality is also determined by digestibility and hence amino acid availability. Amino acids from insect meals are readily available when fed to poultry with values equal to, or higher than, those from conventional protein sources, such as soybean meal or fish meal (Table 3). The only exception is black solder fly larvae meal which has lower amino acid digestibilities especially for the sulphur containing amino acids, methionine and cystine (DeMarco *et al.*, 2015; Schiavone *et al.*, 2017). Larvae (Liland *et al.*, 2017; Schmitt *et al.*, 2019; Tschirner *et al.*, 2015) and prepupae (Spranghers *et al.*, 2017; Wang *et al.*, 2020) of this species have a highly variable mineral content, which is influenced by their diet. As some protein is bound to their mineralised exoskeleton, an increased mineral content might decrease their digestibility. Therefore, amino acid and protein digestibility in poultry might be increased by raising this species on diets with a lower mineral content. Animal performance metrics and insect meal digestibility in shrimp, poultry, ducks, quail, pigs, rabbits, and various species of fish has been elaborately reviewed by Gasco *et al.* (2019).

In a monogastric digestibility model the *in vitro* protein digestibility for dried (conditions unknown) and milled larvae of the yellow mealworm and black soldier fly was only 65.5-68.7% (Marono *et al.*, 2015). Similarly low values were obtained for raw house crickets (65.5%) and yellow mealworms (72.5%) in a porcine model, these were, however, similar to those for beef muscle and chicken breast (Poelaert *et al.*, 2017). Oven cooking or autoclaving further decreased the *in vitro* digestibility of both house crickets and yellow mealworms and also increased neutral detergent fibre, but not acid detergent fibre (Poelaert *et al.*, 2017). Kiiru *et al.* (2020) also reports very low *in vitro* protein digestibility values for raw (29-30.0%) and extruded (38-50%) cricket flours using a static method and quantifying the free amino acids. The latter quantification method is likely to underestimate digestibility as di- and

TABLE 3 Low, high, and mean amino acid digestibilities (%) of various insect meals when fed to poultry

Species and life stage	Low	High	Mean	Study
Gryllus testaceus adults	85 (Cys)	96 (Ala)	93	Wang *et al.* (2005)
Acrida cinerea adults	85 (Cys)	99 (Thr)	94	Wang *et al.* (2007)
Musca domestica larvae	92 (Ile)	98 (Lys)	95	Hwangbo *et al.* (2009)
M. domestica larvae	83 (Gly)	96 (Tyr)	91	Pieterse and Pretorius (2014)
M. domestica pupae	86 (Ala)	100 (Ser+Asp)	95	Pieterse and Pretorius (2014)
M. domestica larvae	77 (Gly)	91 (Tyr)	83	Hall *et al.* (2018)
Tenebrio molitor larvae	80 (Met)	93 (Ala)	86	DeMarco *et al.* (2015)
Hermetia illucens pupae	42 (Met)	89 (Pro)	68	DeMarco *et al.* (2015)
H. illucens larvae	44 (Cys)	92 (Ala+Tyr)	77	Schiavone *et al.* (2017)
H. illucens larvae	45 (Cys)	99 (Ala)	80	Schiavone *et al.* (2017)

tri-peptides would incorrectly be deemed indigestible. Conversely, higher *in vitro* protein digestibilities (76.4 to 93.3%) were reported for 11 species of freeze-dried insects in a canine model (Bosch *et al.*, 2014, 2016). This included house crickets (91.7%), yellow mealworms (91.3-92.5%), black soldier fly larvae (89.7%) and black soldier fly pupae (77.7%). The *in vitro* protein digestibility of raw yellow mealworms (85.0%) was increased by vacuum cooking (90.5%), boiling (90.1%), oven cooking for 15 and 30 minutes (91.5 and 90.4%, respectively), while frying (87.2%) had no effect (Megido *et al.*, 2018). Kinyuru *et al.* (2010a) also obtained high *in vitro* protein digestibilites (82.3 to 90.5%) for freshly caught termites (*Macrotermes subhylanus*) and grasshoppers (*Ruspolia differens*) that declined slightly after solar drying at approximately 30 °C. Using two different methods, high *in vitro* protein digestibilities were also reported for yellow mealworm meal (90.2%), and black soldier fly larvae meal (93-94%), when extruded with wheat flour (Azzollini *et al.*, 2018; Ottoboni *et al.*, 2018). The high variability between the aforementioned studies suggest that *in vitro* protein digestibility data should be viewed with caution until it is correlated to *in vivo* studies. Processing and drying methods, as well as the used digestibility model, can have a large impact on data obtained *in vitro*.

Most amino acid availability data of insects pertain to poultry. However, numerous protein digestibility studies have been conducted in rats (Table 4). Typically, the milk protein casein is used as a benchmark in such studies. The digestibility of insect protein was in most cases equal to, or slightly lower, than that for casein. The lower values for protein digestibility in Table 4 for freeze dried yellow mealworms (Goulet *et al.*, 1978), dried termites (*Macrotermes falciger*) (Phelps *et al.*, 1975), and silkworm (*Bombyx mori*) pupae (Rao, 1994) are for apparent digestibility. True digestibility values are higher because they are corrected for endogenous protein/nitrogen losses. True protein digestibility is 7-15% higher than apparent protein digestibility in dried honey bees (*Apis melifera*) (Ozimek *et al.*, 1985), crickets (Poelaert *et al.*, 2018), yellow mealworms (Jensen *et al.*, 2019; Poelaert *et al.*, 2018) and lesser mealworms (Jensen *et al.*, 2019). Rats fed silkworm pupae consumed 33% less feed than those fed casein, which would negatively affect apparent protein digestibility (Rao, 1994). Other studies using silkworm meal obtained much higher protein digestibility values. This suggests these low values are not representative. The wide range of protein digestibilities of insects when fed to rats is likely a function of species and life stage, but also how the material was processed prior to use in feeding trials.

Protein digestibility of insects is generally high and influenced by the following factors: (1) if a larger proportion of amino acids is present in cuticular proteins complexed with chitin, or is highly sclerotised protein digestibility is likely decreased (Finke, 2007; Ozimek *et al.*, 1985); (2) removing certain body parts (typically wings, or heads) before further processing and feeding to rats increases between study variability; (3) processing methods such as drying can decrease protein digestibility depending on time and temperature (Dreyer and Wehmeyer, 1982; Pieterse and Pretorius, 2014).

2.2 *Fats and fatty acids*

The fat content of insects varies between 10 and 70% on a dry matter basis (Bukkens, 1997; Finke, 2013; Yang *et al.*, 2006). Fat content is typically estimated by an extraction which determines the total weight of all fat-soluble molecules (crude

TABLE 4 True protein digestibility (%) of various insect species, stages and process methods when fed to rats, compared to benchmark values for protein digestibility of casein

Order	Species and life stage	Protein digestibility	Processing method	Casein benchmark	Study
Coleoptera	*Tenebrio molitor* larvae meal*	75.1	freeze dried	88.4	Goulet *et al.* (1978)
	T. molitor larvae meal + methionine*	78.9	freeze dried	88.4	Goulet *et al.* (1978)
	T. molitor larvae meal	91.9	freeze dried	98.8	Poelaert *et al.* (2018)
	T. molitor larvae meal	92.0	freeze dried	ND	Jensen *et al.* (2019)
	Alphitobius diaperinus larvae meal	93.7	freeze dried	ND	Jensen *et al.* (2019)
	A. diaperinus larvae meal	92.5	dried at 120-160 °C	ND	Jensen *et al.* (2019)
	A. diaperinus larvae meal	91.5	vacuum dried at 40 °C	ND	Jensen *et al.* (2019)
	A. diaperinus larvae meal	94.2	defatted then dried	ND	Jensen *et al.* (2019)
	A. diaperinus larvae meal	91.4	extruded then dried at 120 °C	ND	Jensen *et al.* (2019)
	A. diaperinus larvae meal	92.4	freeze dried with enzymes added	ND	Jensen *et al.* (2019)
	Rhynchophorus phoenicis larvae	91.8	raw	92.6	Ekpo (2011)
	R. phoenicis larvae	92.6	boiled	92.6	Ekpo (2011)
	R. phoenicis larvae	92.1	fried	92.6	Ekpo (2011)
	Oryctes rhinoceros larvae	89.3	raw	92.6	Ekpo (2011)
	O. rhinoceros larvae	90.1	boiled	92.6	Ekpo (2011)
	O. rhinoceros larvae	89.7	fried	92.6	Ekpo (2011)
Diptera	*Musca domestica* larvae meal*	90.6	dried at 65 °C	93.3	Iñiguez-Covarrubias *et al.* (1994)
	M. domestica larvae meal	89.0-91.0	unknown	89.0-91.0	Kouamé *et al.* (2011)

TABLE 4   True protein digestibility (%) of various insect species when fed to rats, compared to benchmark
values (*cont.*)

Order	Species and life stage	Protein digestibility	Processing method	Casein benchmark	Study
Hymenoptera	dried *Apis mellifera* adults	71.5	dried at 70 °C	96.8	Ozimek *et al.* (1985)
	A. mellifera adult concentrate	94.3	alkaline extraction of bees dried at 70 °C	96.8	Ozimek *et al.* (1985)
Isoptera	*Macrotermes falciger* meal*	45.3-50.7	dewinged, dried and lightly fried	84.2	Phelps *et al.* (1975)
	Macrotermes bellicosus adults	90.2	dewinged, raw	92.6	Ekpo (2011)
	M. bellicosus adults	91.0	dewinged, boiled	92.6	Ekpo (2011)
	M. bellicosus adults	90.9	dewinged, fried	92.6	Ekpo (2011)
Lepidoptera	*Bombyx mori* chrysalid meal	88.9	dried, conditions unknown	99.8	Lin *et al.* (1983)
	B. mori chrysalid meal	89.4-89.8	dried, conditions unknown, water extracted	99.8	Lin *et al.* (1983)
	B. mori pupae meal *	67.0	unknown	84.0	Rao (1994)
	Samia ricinii pupae meal	87.0	dried at 60-70 °C and defatted	92.0	Longvah *et al.* (2011)
	S. ricinii pupae meal	87.0	dried at 60-70 °C and defatted	92.0	Longvah *et al.* (2011)
	Imbrasia belina larvae meal	85.5	gastrointestinal contents removed then dried, drying conditions unknown	98.0	Dreyer and Wehmeyer (1982)
	I. belina larvae meal	83.9	gastrointestinal contents removed then dried, drying conditions unknown then soaked in water and canned	98.0	Dreyer and Wehmeyer (1982)
	I. belina larvae	86.0	raw	92.6	Ekpo (2011)
	I. belina larvae	88.4	boiled	92.6	Ekpo (2011)
	I. belina larvae	87.8	fried	92.6	Ekpo (2011)
	Clanis bilineata larvae meal	95.8	heads removed and dried at 60 °C	96.1	Xia *et al.* (2012)
Orthoptera	*Acheta domesticus* meal	83.9	freeze dried	98.8	Poelaert *et al.* (2018)

* = apparent protein digestibility; ND = not determined.

fat). This includes glycerides but also waxes, sterols, fat soluble vitamins and other fat-soluble compounds.

Fat is composed of fatty acids. Two or three fatty acids are coupled to glycerol and form diglycerides and triglycerides, respectively. These fatty acids are divided into saturated, mono-unsaturated, and poly-unsaturated fatty acids based on their degree of saturation. Poly-unsaturated fatty acids are subdivided into omega 3, 6 or 9 poly-unsaturated fatty acids, based on the relative position of the first double bond.

The fatty acid composition of insects depends on the species and life stage as well as environmental factors such as diet, temperature, and light (Finke and Oonincx, 2017). Males of most species have smaller fat reserves than females (Kulma *et al.*, 2019; Lease and Wolf, 2011; Liu *et al.*, 2017; Nestel *et al.*, 2005; Rho and Lee., 2014; Zhou *et al.*, 1995). Commercially produced insects seem to have a higher fat content than those collected from the wild (Finke, 2002, 2013; Lease and Wolf, 2011; Oonincx and Dierenfeld, 2012; Yang *et al.*, 2006). This might be due to decreased energy expenditure in captivity, ready access to diets with a high energy content, or both.

In general, wild caught insects contain relatively high amounts of linoleic acid (18:2 n-6) and linolenic acid (18:3 n-3) (Fast, 1970; Thompson, 1973). Commercially raised insects also contain high levels of linoleic acid but much lower levels of linolenic acid than wild caught insects, because their diet often contains large amounts of grains and grain by-products, which have low levels of linolenic acid (Dreassi *et al.*, 2017; Finke, 2002, 2013; Jones *et al.*, 1972; Oonincx *et al.*, 2015b, 2019a; Paul *et al.*, 2017). Larvae of the black soldier fly have an unusual fatty acid profile rich in lauric acid (C12:0) irrespective of the diet (Finke 2013; Oonincx *et al.*, 2015a; Spranghers *et al.*, 2017; St-Hilaire *et al.*, 2007; Surendra *et al.*, 2016). Like vertebrates, most insects can synthesise saturated and monounsaturated fatty acids (Beenakkers *et al.*, 1985, Tietz and Stern, 1969). However, most species are unable to synthesise linoleic acid and linolenic acid, which makes those essential nutrients. There are exceptions such as the American cockroach (*Periplaneta americana*) and the house cricket, which apparently can synthesise these fatty acids (Borgeson and Blomquist, 1993; Borgeson *et al.*, 1991).

In three insect species with an aquatic larval stage, the concentration of linolenic acid increased, while arachidonic acid (20:4 n-6), and eicosapentaenoic acid (20:5 n-3) decreased in the adults after they emerged from their aquatic habitat (Hanson *et al.*, 1985). This is likely due to differences in the fatty acid content of their diets. Several studies reporting on the fatty acid content of field collected insects find higher levels of eicosapentaenoic acid in aquatic species than in terrestrial species, which rarely contain these fatty acids (Fontaneto *et al.*, 2011; Ghioni *et al.*, 1996; Sushchik *et al.*, 2003, 2013;). In aquatic ecosystems certain microalgae produce eicosapentaenoic acid, which is transferred to higher trophic levels including insects (Gladyshev *et al.*, 2011). Eicosapentaenoic acid is not produced by higher plants in terrestrial ecosystems, therefore herbivorous terrestrial insects are less likely to obtain this fatty acid.

2.3   *Carbohydrates*

In general, carbohydrates calculated as nitrogen free extract, are present in small amounts in insects (Barker *et al.*, 1998; Finke, 2002, 2013, Oonincx and Dierenfeld, 2011; Pennino *et al.*, 1991). The carbohydrate content of yellow mealworm larvae can vary between one and seven percent (Ramos-Elorduy *et al.*, 2002). However, these differences are likely the result of the food remaining in the gastro-intestinal tract.

2.4   *Fibre and chitin*

Insects contain significant amounts of fibre as measured by crude fibre, acid detergent fibre, or neutral detergent fibre (Barker *et al.*, 1998; Finke, 2002, 2007, 2013; Lease and Wolf, 2010; Madibela *et al.*, 2007; Marono *et al.*, 2015; Oonincx and Dierenfeld, 2011; Pennino *et al.*, 1991). The components in these fibre fractions are not well known, although in whole insects they include sclerotised proteins and proteins, minerals and other compounds bound to chitin (Finke, 2007; Kramer *et al.*, 1995; Madibela *et al.*, 2007; Marono *et al.*, 2015). Chitin is a N-acetyl-β-D-glucosamine polymer, which provides rigidity to the insect's exoskeleton.

The outmost part of insects, the cuticle, is a matrix of proteins, lipids, minerals and other compounds (Kramer *et al.*, 1995). Chitin is only present in the procuticle, the two innermost layers of the cuticle (Moussian, 2010), and therefore makes up a small part of an insect's weight. Quantitative data on the chitin content of whole insects is limited and comparisons between studies are difficult, due to the use of various analytical methods (Cauchie, 2002; Finke, 2007; Henriques *et al.*, 2020; Kaspari, 1991; Lease and Wolf, 2010; Woods *et al.*, 2020). Traditional fibre assays (crude, neutral detergent or acid detergent) are sometimes used to approximate chitin, even though these methods overestimate true chitin levels (Finke, 2007, 2013; Madibela *et al.*, 2007; Marono *et al.*, 2015). Many animal species, including humans, have chitinases, which can digest chitin (Fujimoto *et al.*, 2002; Lindsay *et al.*, 1984; Paoletti *et al.*, 2007; Strobel *et al.*, 2013; Tabata *et al.*, 2018; Whitaker *et al.*, 2004). To what extent these enzymes are effective *in vivo* is currently unknown.

The predominant compound in the cuticle of most insects is not chitin, but protein (Kramer *et al.*, 1995). 'Harder bodied' insects such as adult yellow mealworm beetles contain higher fibre levels than softer bodied insects such as silkworms and house cricket nymphs. This is due to higher levels of amino acids in the acid detergent fibre fraction (Finke, 2007). The amino acid pattern of whole insects differs from the pattern in the acid detergent fibre fraction, and differs between species (Finke, 2007). Particularly valine, histidine, and glycine concentrations are higher in the acid detergent fibre fraction than in whole insects. These amino acids likely contribute to the relative strength, stiffness, elasticity and other physical properties of sclerotised and cuticular proteins. Insects with 'harder' cuticles do not necessarily contain more chitin, but contain more sclerotised protein and proteins cross-linked to chitin than softer bodied insects. However, the chitin content of adult beetles does seem higher than in their larvae (Finke, 2007; Kaya *et al.*, 2014; Shin *et al.*, 2019).

2.5   *Minerals*

Minerals are classified as macro-minerals (calcium, phosphorus, magnesium, sodium, potassium and chloride) and micro or trace minerals (iron, zinc, copper, manganese, iodine and selenium). This classification is based on the amount needed to meet dietary requirements. For most species, requirements for macro-minerals are measured in g per kg and micro-mineral requirements in mg per kg.

Insects generally contain low levels of calcium because they lack a mineralised skeleton. Calcium levels are typically less than 0.3% DM although higher levels have been reported for stoneflies (Plecoptera) and some other species (Barker *et al.*, 1998; Finke, 2002, 2013; Oonincx and Dierenfeld, 2012; Oonincx and Van der Poel, 2011; Punzo, 2003; Studier and Sevick, 1992). Higher levels of calcium occasionally reported for feeder crickets are likely due to dietary calcium remaining in the gut (Barker *et al.*, 1998; Finke, 2002; Frye and Calvert, 1989; Hatt *et al.*, 2003; Oonincx and Dierenfeld, 2011; Oonincx and Van der Poel, 2011; Punzo, 2003). While the exoskeleton of most insects is primarily composed of protein and chitin, black solder fly larvae (Finke, 2013; Spranghers *et al.*, 2017), and face fly larvae (*Musca autumnalis*) (Dashefsky *et al.*, 1976; Koo *et al.*, 1980; Roseland *et al.*, 1985), have a mineralised exoskeleton in which calcium and other minerals are incorporated into the cuticle. Therefore, they can contain high levels of calcium.

Most insects contain more phosphorus than calcium, except for species with a mineralised exoskeleton (i.e. face fly and black soldier fly larvae). Significant amounts of phosphorus are present in plants as phytate and are hence unavailable for digestion. In contrast, phosphorus in insects is likely to be bioavailable as was reported when face fly pupa were fed to poultry (Dashefsky *et al.*, 1976).

Magnesium levels in insects likely meet the dietary requirements of most species (Barker *et al.*, 1998; Finke, 2002, 2013; Oonincx and Dierenfeld, 2012; Studier and Sevick, 1992). Black soldier fly larvae contain 3 to 10 times more magnesium than most other insects (Finke, 2013; Spranghers *et al.*, 2017). This is likely the result of their mineralised exoskeleton in which minerals such as calcium and magnesium form a complex with chitin (Diener *et al.*, 2015).

Only a few studies report the sodium and potassium content of insects (Finke, 2002, 2013; Oonincx and Dierenfeld, 2011; Oonincx and Van der Poel, 2011; Reichle *et al.*, 1969; Studier and Sevick, 1992). Insects typically contain more potassium than sodium whereas chloride levels are intermediate. Most insect species likely contain adequate amounts of these three minerals to meet the dietary requirements of most animal species.

Most insects contain sufficiently high levels of the trace minerals iron, zinc, copper, manganese and selenium, which would meet the dietary requirements for most animals (Barker *et al.*, 1998; Finke, 2002, 2013; Oonincx and Dierenfeld, 2012; Oonincx and Van der Poel, 2011; Punzo, 2003; Studier and Sevick, 1992). However, studies suggest possible species-specific differences for certain trace minerals. For example, the manganese concentrations in the alates of five termite species (2,710-

5,150 mg/kg DM) were extremely high compared to mopane worms (*Gonimbrasia belina*; 39 mg/kg DM), house crickets (38 mg/kg DM), yellow mealworms (5 mg/kg DM), or migratory locusts (10 mg/kg DM) determined in the same study (Verspoor *et al.*, 2020). However, other termite species from other regions, for instance *Nasutitermes* spp., from Venezuela, reportedly have lower manganese concentrations (32-115 mg/kg DM) (Oyarzun *et al.*, 1996). Caste differences were apparent in the latter study; alates had a far lower manganese concentration than workers (37 vs 115 mg/kg DM).

Adult fruit flies (*Drosophila melanogaster*) (Barker *et al.*, 1998; Oonincx and Dierenfeld, 2012), and house flies (Finke, 2013), contain relatively high levels of iron (125 to 454 mg/kg DM) compared to most other insect species. Iron and zinc concentrations in insects seem positively correlated (Spearman's $\rho = 0.592$), that is higher iron concentrations often coincide with higher zinc concentrations (Mwangi *et al.*, 2018). The contents of the gastrointestinal tract can be a significant percentage of the total weight of an insect (4-7% of the live weight) (Finke, 2003); and therefore diet also directly influences the (trace) mineral content of fully fed insects.

Mineral availability can be inhibited by so-called anti-nutritional factors such as phytate, oxalate, and tannins. For example, caterpillars of the Pallid emperor moth (*Cirina forda*) contain both phytic acid (10 mg/kg) and oxalate (40 mg/kg), but no tannins (Omotoso, 2006). Tannins (10.4 mg/kg), phytic acid (13.5 mg/kg) and oxalates (0.8 mg/kg) were found in larvae of the African palm weevil (*Rhynchophorus phoenicis*) (Ekpo, 2011). The source of these anti-nutritional factors is likely from the food present in the gastrointestinal tract. Evidence of accumulation of these substances in insects is lacking. The concentrations found seem to be relatively low; in seeds and cereal grains concentrations up to 7% DM have been found (Zhou and Erdman, 1995).

2.6   *Vitamins and carotenoids*

2.6.1   Vitamin A
Vitamin A is a group of compounds composed of retinoids and carotenoids. Like most vertebrates, insects obtain retinoids via the cleavage of various carotenoids (Von Lintig, 2012). However, carotenoid cleavage in vertebrates primarily takes place in the intestine, and the resulting retinoids are stored in the liver (Olson, 1989), whereas insects convert carotenoids to retinoids only in the compound eye (Von Lintig, 2012). In fruit flies and honey bees (*Apis mellifera*) retinoids are only found in the compound eye while other parts of the adult insect do not contain retinoids nor do fruit fly larvae (Giovannucci and Stephenson, 1999; Goldsmith and Warner, 1964). This explains why holometabolous adults contain very low levels of vitamin A/retinoids (Barker *et al.*, 1998; Finke, 2002, 2013; Oonincx and Dierenfeld, 2012; Oonincx and Van der Poel, 2011; Pennino *et al.*, 1991), and holometabolous larvae, which lack compound eyes, do not contain preformed retinoids (retinal or 3-hydroxyretinal).

In many animal species certain carotenoids can be converted to vitamin A/retinal (Bender, 2002; Levi *et al.*, 2012; McComb, 2010; Olson, 1989). High levels of carotenoids, including those that can be converted to retinal, are found in various wild insect species (Arnold *et al.*, 2010; Cerda *et al.*, 2001; Eeva *et al.*, 2010; Isaksson and Andersson, 2007; Newbrey *et al.*, 2013; Seki *et al.*, 1998; Ssepuuya *et al.*, 2017) whereas commercially produced insects contain far lower quantities (Finke, 2002, 2013, 2015a; Oonincx and Van der Poel, 2011). This difference is likely a result of dietary carotenoid intake as was shown for both silkworm larvae (Chieco *et al.*, 2019) and fruit flies (Giovannucci and Stephenson, 1999).

2.6.2   Vitamin D

For a long time, insects were considered to contain low levels of vitamin D (typically < 400 IU/kg DM) (Finke, 2002, 2013, Oonincx *et al.*, 2010). However, data from three wild caught species indicated a high variability in its concentrations with values ranging from below the detection limit (100 IU vitamin D3/kg DM) up to 1,288 IU vitamin D3/kg DM (Finke, 2015b). More recently it was discovered that certain insects, like vertebrate animals, can synthesise vitamin D3 *de novo* when exposed to UV-B (Oonincx *et al.*, 2018). This capacity varies greatly between species; no evidence of *de novo* synthesis is found in black soldier fly larvae, but yellow mealworm larvae can reach over 6,000 IU/kg DM (Oonincx *et al.*, 2018).

2.6.3   Vitamin E

Vitamin E content of insects, including house crickets, yellow mealworm larvae and black soldier fly larvae, varies. Values for house crickets range from 8 to 195 IU vitamin E/kg DM, yellow mealworm larvae range from < 22 to 116 IU vitamin E/kg DM and black soldier fly larvae range from 10 to 235 IU vitamin E/kg DM (Finke, 2002; Pennino *et al.*, 1991). This variation is likely due to dietary differences resulting in different amounts being incorporated in the insect's tissues and dietary vitamin E in the insect's gastrointestinal tract. While variable between species, the vitamin E content of most commercially raised insects is below 37 IU/kg DM (Barker *et al.*, 1998; Finke, 2002, 2013; Oonincx and Dierenfeld, 2011). Wild insects appear to contain higher levels of vitamin E than commercially raised insects, which are often provided with diets containing low levels of vitamin E (Pennino *et al.*, 1991; Punzo, 2003).

2.6.4   B-vitamins

Several studies report on the B-vitamin content of commercially raised insects (Bawa *et al.*, 2020; Finke, 2002, 2013, 2015b; Jones *et al.*, 1972). However, the B-vitamin content of unprocessed wild-caught insects is limited (Finke, 2015b; Igwe *et al.*, 2011; Kinyuru *et al.*, 2010a,b). Some information on B-vitamin content of wild-caught insects consumed by humans is available (Banjo *et al.*, 2006; Dreyer and Wehmeyer, 1982; Igwe *et al.*, 2011; Kinyuru *et al.*, 2010a; Kodondi *et al.*, 1987; Santos Oliveira *et al.*,

1976; Teffo *et al.*, 2007). These, however, are usually dried, boiled, fried, or roasted and specific body parts are often removed. During this processing some B-vitamins can be destroyed by exposure to heat, light or oxygen. Furthermore, differences in sample preparation and analytical methods (microbiological versus chemical techniques) complicate direct comparisons.

Despite these issues some general observation regarding B-vitamins in insects can be made. Most studies show unprocessed insects are a very good source of riboflavin (vitamin B2), niacin, pantothenic acid, pyridoxine (vitamin B6), biotin, folic acid, and cyanocobalamin (vitamin B12). Low levels for several of these vitamins occasionally reported in the literature are likely due to losses during processing and storage (Kinyuru *et al.*, 2010a).

One B-vitamin that appears to be low in many species of insects is thiamine (vitamin B1). Many species of commercially raised or wild caught insects including house crickets, adult yellow mealworms, superworms (*Zophobas morio*), butterworms, (*Chilecomadia moorei*) Turkistan roaches (*Blatta lateralis*), pallid wing grasshoppers (*Trimerotropis pallidipennis*), and rhinoceros beetles (*Oxygrylius ruginasus*) contain low levels of thiamine (< 3.0 mg/kg DM) although others (black soldier fly larvae, adult house flies, silkworms, yellow mealworm larvae, waxworms (*Galleria mellonella*), and white-lined sphinx moths (*Hyles lineata*) contain much higher levels (5.0 to 45.0 mg thiamine/kg DM) (Bawa *et al.*, 2020; Finke, 2002, 2013, 2015b). This variation is similar to that observed for 14 insect species collected in Nigeria (0.3 to 32.4 mg thiamine/kg dry mater) (Banjo *et al.*, 2006) and three species from Angola (1.3 to 36.7 mg thiamine/kg DM) (Santos Oliveira *et al.*, 1976). Dried and smoked Attacidae caterpillars (*Nudaurelia melanops, Imbrasia truncata, Imbrasia epimethea*) from Zaire contain low thiamine levels (1.5 to 2.7 mg/kg dry product) (Kodondi *et al.*, 1987) and unprocessed longhorn grasshoppers from East Africa contained no detectable thiamine (Kinyuru *et al.*, 2010b). Because thiamine is relatively unstable and most of the values reported are for insects processed (i.e. body parts removed and dried, smoked or fried), it is unclear how representative these values are for unprocessed insects. That said, the range of values for thiamine in processed insects is similar to that observed for raw insects.

Two species of edible insects (domesticated silkworm larvae and African silkworm pupae (*Anaphe* spp.) contain the enzyme thiaminase (Nishimune *et al.*, 2000). Unless inactivated during processing, thiaminase destroys thiamine. Thiamine deficiency due to the consumption of *Anaphe* pupae is suggested as a cause of a seasonal ataxia in local populations in Nigeria (Adamolekun, 1993; Adamolekun *et al.*, 1997). Most enzymes are inactivated by heat, but some thiaminase activity remains after brief exposure of the extract of *Anaphe* pupae to 100 °C, suggesting it is heat tolerant. This thiaminase had an optimal activity at 70 °C and only after 15 minutes exposure to 100 °C was it largely deactivated (Nishimune *et al.*, 2000). The gut of the variegated grasshopper (*Zonocerus variegatus*) also contains a thiaminase (Ehigie *et al.*, 2013). This, however, has its optimal activity at a lower temperature (50 °C) indicating it is a different enzyme. It is currently unknown how widespread thiaminases are in insects.

2.7   *Choline*

Choline is a component of both lecithin and the neurotransmitter acetylcholine. It also plays an important role in one-carbon metabolism and as such can spare the need for dietary methionine (Pesti *et al.*, 1979). The data on choline in insects, albeit limited, indicate insects contain high levels of choline (1,570 to 7,258 mg/kg DM) (Finke, 2002, 2013, 2015a,b; Fogang *et al.*, 2017; Noland and Baumann, 1949). Because methionine is typically the first-limiting amino acid when fed to most species, choline content is important when using insects as food or feed.

2.8   *Taurine*

Most animal species can synthesise taurine from its precursor methionine. For some species, such as cats and foxes, this amino sulfonic acid is a required nutrient. In general insect larvae contain little, if any, taurine (Bodnaryk, 1981; Finke, 2002, 2013, 2015a; Massie *et al.*, 1989; Ramsey and Houston, 2003). Larvae of both the bertha armyworm (*Mamestra configurata*) and fruit flies contain little (< 200 μg/g fresh weight), or undetectable taurine but levels increase in pupae and peak in adults (Bodnaryk, 1981; Massie *et al.*, 1989) with values ranging from 500 to 1,100 μg/g fresh weight, likely because insect flight muscles contain high concentrations of taurine (Whitton *et al.*, 1987). Adult house crickets (Finke, 2002, 2015a), adult pallid wing grasshoppers (Finke, 2015b), adult fruit flies (Massie *et al.*, 1989), house flies (Finke, 2013), and adult moths (Bodnaryk, 1981; Finke, 2015b) are rich sources of taurine, while three species of adult beetles (*T. molitor*, *O. ruginasus* and an unreported species) contain only low levels (Finke, 2002, 2015b; Ramsey and Houston, 2003).

2.9   *Sterols*

Unlike most animals, insects cannot synthesise the characteristic ring structure of sterols and hence, require a dietary source (Jing and Behmer, 2020). The sterol form therefore depends on the diet and the insect species. Phytosterols such as stigmasterol, sitosterol, and campesterol, or zoosterols such as cholesterol and 7-dehydrocholesterol, are enzymatically converted for their required function. Typically they form structural components of phospholipid bilayers in cell membranes, but are also used as precursors for vitamin D or moulting hormone. Total sterol content and the variation seems a function of species, diet, life stage and season (Connor *et al.*, 2006; Koštál *et al.*, 2013; Liland *et al.*, 2017). While insects contain a variety of sterols often β-sitosterol, cholesterol, or 7-dehydrocholesterol are the most abundant (Cerda *et al.*, 2001; Cheseto *et al.* 2015; Connor *et al.*, 2006; Jing and Behmer, 2020; Koštál *et al.*, 2013; Liland *et al.*, 2017; Mlček *et al.*, 2019; Sabolová *et al.*, 2016; Svoboda *et al.*, 1995). High dietary levels of sterols, such as cholesterol, can increase the risk of atherosclerosis in humans and some species of monkeys (Hopkins, 1992; Rudel *et al.*, 1998).

3   Life stage effects on fat content

Fat stores are usually highest in the final larval or nymphal stage (Fast, 1970). This pattern depends to some extent on whether the species is holometabolous or hemimetabolous. Larvae of holometabolous species have a higher fat content than adults (Lease and Wolf, 2011; Punzo, 2003). For instance, the fat content of the larvae of fruit flies (Church and Robertson 1966), house flies (Pearincott, 1960), black soldier flies (Liu *et al.*, 2017) and yellow mealworms (Finkel, 1948) increases as they develop but drops significantly in the adult because fat is used as an energy source during pupation. A similar decrease is observed in insects undergoing diapause or hibernation (Ali and Ewiess, 1977; Downer and Matthews, 1976).

   Similar to holometabolous insects, nymphs of hemimetabolous insects slowly increase their fat content as they develop as was shown for house crickets (Hutchins and Martin, 1968; Lipsitz and McFarlane, 1971). However, unlike holometaboulous insects, the fat contents of newly emerged hemimetabolous adults are similar to late stage nymphs and then slowly decline (Lipsitz and McFarlane, 1971).

4   Dietary effects on insect nutrient composition

Diet can affect the composition of fats/fatty acids, vitamins, carotenoids, and minerals in insects. The amount and the fatty acid composition of insect fat is highly variable and affected by both life stage and diet. This subsequently affects the levels of other nutrients, most notably protein and moisture. An increase in fat content, with stable amounts of moisture and protein, dilutes the concentrations of the latter two as was shown for house fly larvae (Pearincott, 1960).

4.1   *Fat*

The extent to which dietary changes in crude fat content and fatty composition of insects are possible depends strongly on the species. For instance, the fat content of yellow mealworms fed four different diets ranged from 23 to 29% DM, while Argentinean cockroaches (*Blaptica dubia*) fed these diets ranged from 16 to 40% DM (Oonincx *et al.*, 2015a). Similarly, the fat content of house crickets fed five different diets ranged from 9 to 44% (Bawa *et al.*, 2020). The fat content of velvet bean caterpillars (*Anticarsia gennatalis*) is 40 to 65% higher on an artificial diet compared to leaves from three plants (Cookman *et al.*, 1984). This increase in fat content carried through to the adult stage; moths from larvae fed the artificial diet contained 71 to 105% more fat than those fed plant leaves (Cookman *et al.*, 1984). Similarly, higher fat contents were found in dried insect meals from larvae of the tobacco hornworm (21 vs 17% dry DM) and the fall armyworm (*Spodoptera frugiperda*) (21 vs 12% DM) fed artificial diets, compared to those fed leaves or grasses (Landry *et al.*, 1986).

   Besides the variability in total fat content, large differences in fatty acid composition of insects when fed different diets are apparent, which is similar to most vertebrates. Numerous studies indicate that the fatty acid composition in both larval and adult insects partially reflects the fatty acid composition of the diet (Cookman *et al.*, 1984; Madariaga *et al.*, 1971; Schaefer, 1968; Spranghers *et al.*, 2017; St-Hilaire *et al.*, 2007; Van Broekhoven *et al.*, 2015). However, due to selective accumulation, catabolism and/or fatty acid synthesis these fatty acid compositions are not identical (Cookman *et al.*, 1984; Madariaga *et al.*, 1971; Oonincx *et al.*, 2015a; Schaefer, 1968).

The omega-3 fatty acid content of house fly larvae (Hussein *et al.*, 2017), black soldier fly larvae (Erbland *et al.*, 2020; Oonincx *et al.*, 2019a; St-Hilaire *et al.*, 2007), Jamaican field crickets (Komprda *et al.*, 2013; Starčević *et al.*, 2017), house crickets (Finke 2015b; Oonincx *et al.*, 2019a), yellow mealworms (Fasel *et al.*, 2017; Finke 2015b; Oonincx *et al.*, 2019a), lesser mealworms (Oonincx *et al.*, 2019a), superworms, and waxworms (Finke, 2015b; Komprda *et al.*, 2013), are increased by providing a diet enriched with these fatty acids. Black soldier fly larvae fed diets containing fish oil or fish by-products accumulate both eicosapentaenoic acid and docosahexaenoic acid (Erbland *et al.*, 2020; St-Hilaire *et al.*, 2007). In contrast house crickets, superworms and waxworms fed diets containing fish oil accumulate eicosapentaenoic acid but not docosahexaenoic acid, even though both were present in the diets (Finke, 2015b). To which extent this is due to species, diet, or experimental setup requires further studies. When, and for how long these fatty acids should be provided will partially depend on the further use of the insect. Addition at the final developmental stage (i.e. gut-loading) might be suitable if whole insects, including the gastrointestinal contents, are used.

4.2   *Vitamins*
There are little data regarding the effect of diet on the vitamin content in insects. Supplementing the diets of house crickets, yellow mealworms, superworms, and waxworms with high levels of vitamin E during growth, resulted in high levels in the insects (116-440 IU/kg DM) (Finke, 2015a). Because the insects were fasted prior to analysis, these elevated levels are mainly due to incorporation of vitamin E into the tissue of the insect. This is similar to swine, where elevated levels of dietary vitamin E increase the vitamin E concentration of various tissues (Asghar *et al.*, 1991). Carotenoid content can also be increased via the insect's diet. Feeding a diet containing β-carotene to species that use retinal as their chromophore such as house crickets, yellow mealworm larvae, and superworm larvae increases β-carotene concentration in their tissues (Finke, 2015a). Providing a diet containing β-carotene to species that use 3-hydroxyretinal as their chromophore such as Diptera and Lepidoptera typically does not increase β-carotene but elevates zeaxanthin levels as was shown for blow flies (*Calliphora*) (Vogt and Kirschfeld, 1984), fruit flies (Giovannucci and Stephenson, 1999), and waxworm larvae (Finke, 2015b). However, pupae from two silkworm strains did contain some β-carotene when the larvae were fed diets containing β-carotene (Chieco *et al.*, 2019). Fruit fly larvae accumulate carotenoids and

convert these to retinoids when they form compound eyes during pupation (Seki *et al.*, 1998; Von Lintig, 2012).

4.3   *Minerals and heavy metals*

To some extent mineral composition can be altered by diet, depending on the mineral and the species of insect. Across three different studies the calcium content of black soldier fly prepupae fed nine different diets ranged from 1 to 66 g/kg DM (Proc *et al.*, 2020; Spranghers *et al.*, 2017; Wang *et al.*, 2020). Iron, zinc, and manganese also varied while phosphorus and potassium were similar across dietary treatments. The mineral content of black soldier fly larvae fed sixteen different diets also varied greatly (8 to 83 g/kg DM) although in this case part of the effects are likely a result of the diet remaining in the gastrointestinal tract of the larvae (Liland *et al.*, 2017; Schmitt *et al.*, 2019; Tschirner *et al.*, 2015). The mineralised exoskeleton of black soldier fly larvae might be the reason for these extreme variations. Zinc also accumulates in black soldier fly larvae, prepupae and adults when incorporated in the larval diet (Diener *et al.*, 2015). Zinc accumulates in larval and prepupal exuvia, further indicating that the variability in mineral concentrations is due to the mineralised exoskeleton of this species. In housefly pupae, reared on diet spiked with zinc chloride, the zinc concentration increased by approximately 10% of the dietary concentration, which might indicate effects from gut loading, rather than selective accumulation and storage (Maryanski *et al.*, 2002). Like zinc, the heavy metals copper, cadmium and lead accumulate in black soldier fly larvae (Diener *et al.*, 2015; Purschke *et al.*, 2017; Van der Fels-Klerx *et al.*, 2016; Wu *et al.*, 2020). A measure for the efficiency of accumulation is the bio accumulation factor (BAF). It is calculated by dividing the concentration in the insect, by the concentration in the diet. Hence, a BAF above 1 indicates selective accumulation and a BAF below 1 indicates selective excretion. The aforementioned studies indicate that the BAF for arsenic, mercury, chrome, and nickel are below 1 in the black soldier fly. However, contrasting data has been reported by Schmitt *et al.* (2019) confirming accumulation of cadmium and mercury and selective excretion for lead and copper. The accumulation of cadmium seems to occur in several fly species, including houseflies, fruit flies, several midge species, and the flesh fly *Boettcherisca peregrina* (Charlton *et al.*, 2015; Gao *et al.*, 2017; Kazimírová and Ortel, 2000; Maroni and Watson, 1985; Maryanski *et al.*, 2002; Postma *et al.*, 1996; Purschke *et al.*, 2017; Timmermans and Walker, 1989; Wu *et al.*, 2006).

   In contrast to these Diptera, larvae of the yellow mealworm larvae accumulate arsenic, but not copper, zinc, lead or cadmium (Van der Fels-Klerx *et al.*, 2016; Vijver *et al.*, 2003). The concentrations of lead and cadmium in this species can, however, increase partially due to the presence of these elements in the gastro-intestinal tract (Vijver *et al.*, 2003). This was for instance shown for the predatory beetle (*Notiophilus biguttatus*), which efficiently excreted retained cadmium when it was removed from the diet (Jansen *et al.*, 1991). Similarly, the predatory carabid beetle (*Poecilus cupreus*) had increased cadmium levels (approximately 10% of the dietary concentration), and zinc (less than 10% of the dietary concentration) when provided with spiked diets (Maryanski *et al.*, 2002). However, when provided with a cadmium free diet, concentrations fell rapidly (Kramarz, 1999). The cadmium concentration in the sago grub was below 10% of the dietary concentration (Köhler *et al.*, 2020) and no evidence of cadmium or zinc accumulation was found in the beetle *Neochetina eichhorniae* (Jamil and Hussain, 1992).

Several studies on heavy metals in Orthopterans, either taken from contaminated sites or experimentally exposed, have been conducted. Higher levels in the environment lead to higher levels of lead, cadmium, and mercury in four species of grasshoppers (Devkota and Schmidt, 2000). The BAF for cadmium was above 1, whereas it was below 1 for lead. Interestingly, the BAF for mercury was below 1 for three of the species, but > 2 for the fourth. A higher BAF for cadmium than for lead was also reported for the Oriental longheaded grasshopper (*Acrida chinensis*) and migratory locusts (Zhang *et al.*, 2012). Both cadmium and copper concentrations increased in migratory locusts with increased dietary levels (Crawford *et al.*, 1996). The same was also found in the common field grasshopper (*Chorthippus brunneus*), and it was shown that most of the cadmium and copper (85% for both) accumulated in the integument (Hunter *et al.*, 1987). This might explain why lower concentrations are often found in later instar nymphs and adults which have a lower surface to volume ratio.

Jamaican field crickets (*Gryllus assimilis*) also accumulate dietary cadmium (Bednarska *et al.*, 2015). However, their zinc levels remain constant at different dietary levels indicating that zinc concentrations are more tightly regulated. Tight regulation of heavy metals that are of nutritional relevance, such as iron, zinc, and copper, seems common over a variety of insect species. This is plausible, as a variety of selective transporters are present to regulate their concentrations within the insect (Mwangi *et al.*, 2018).

Tobacco cutworms (*Spodoptera litura*) also concentrate copper and zinc from their diet in amounts that seem to fit physiological levels (7 and 70 mg/kg fresh weight, respectively) (Zhuang *et al.*, 2009). Lead, and especially cadmium, are largely excreted by this lepidopteran. A noteworthy ecological study on the Bogong moth (*Agrotis infusa*) indicates that they store arsenic and can effectively transport this pollutant during migration (Green, 2008).

Based on the information above, it seems that accumulation and excretion patterns show similarities at the order level. However, these patterns differ between life stages (Lindqvist, 1992; Timmermans and Walker, 1989), and species-specific exceptions likely exist. Therefore, if a new species is to be commercially produced, controlled studies on accumulation and excretion patterns are advisable. Variation in dietary intake levels will to some extent affect insect concentrations. For species collected from the field, seasonal differences in host plants and host plant concentrations will affect insect concentrations (Ortiz *et al.*, 2015; Zhang *et al.*, 2009).

Furthermore, insects from contaminated areas can have elevated concentrations
due to contaminated diets, but might also take up these minerals directly from the
environment.

5   Environmental effects on insect composition

Environmental factors such as temperature, light, and humidity are known to affect
growth and development (Akman Gündüz and Gülel 2002; Ali and Ewiess, 1977; Ali
et al., 2011; Han *et al.*, 2008; Roe *et al.*, 1980, 1985). Some effects on chemical composition are also known to occur. For instance, bean beetles (*Acanthoscelides obtectus*)
raised at 30 °C contain more protein and the same amount of fat as their counterparts raised at 20 °C (Sönmez and Gülel, 2008). An increase in rearing temperature
from 20 to 27 °C in two-spotted field crickets (*Gryllus bimaculatus*) decreases their
protein concentration and increases the concentration of fat (Hoffmann, 1973).
Their fatty acid profile also changes with higher temperatures leading to a higher
degree of saturation. This coincides with an increased fresh weight per individual,
possibly indicating an increased synthesis of saturated fatty acids at 27 °C. Similarly, females of the mosquito, *Culex tarsalis*, contain more unsaturated fatty acids
(C16:1, C18:1 and C18:2) when raised at 22 °C, compared to counterparts raised at
30 °C which contain more short chain fatty acids (Harwood and Takata, 1965). This
accumulation of unsaturated fatty acids is more pronounced when this species is
raised under a short photoperiod, which would increase cold survival. Indeed, photoperiodic effects on insect nutrient content are likely indirect, acting through other processes such as preparing for diapause.
    Besides photoperiod, the emission spectrum of the light source can influence
composition. If UV-B is emitted, this will lead to vitamin D3 synthesis in certain
insects such as the yellow mealworm (Oonincx *et al.*, 2018). The vitamin D3 concentration increases over time with prolonged exposure until it reaches a plateau.
A higher plateau is reached with a higher UV-B intensity. A higher abundance of
short wavelengths within the UV-B spectrum is also likely to lead to a higher vitamin D3 content in insects, as was shown in the vertebrate *Pogona vitticeps* (Diehl
et al., 2018).

6   Processing effects on insect nutrient composition

Although specific information on the effects of various processing methods on the
nutrient content and availability of insect products is limited, it is likely similar
to other human foods or animal feeds (Ssepuuya *et al.*, 2017). Heat and processing
effects on protein/amino acid availability, effects of traditional cooking methods
on trace mineral, fatty acid and fat content and processing effects on B-vitamins in
insect and insect products have been reported.

Processing using high pressures and temperatures decreases amino acid availability and protein quality (Batterham *et al.*, 1986). As such, the protein quality of insect meals is likely decreased by processing conditions, especially drying temperatures. The lower average amino acid availability of house fly larvae meal (91%), compared to house fly pupae meal (95%), might partially be due to longer drying times (Pieterse and Pretorius, 2014). Likewise, dried Mopane caterpillars exposed to high temperatures during canning have a slightly reduced apparent protein digestibility compared to those that are not canned (83.9 and 85.8%, respectively) (Dreyer and Wehmeyer, 1982). In contrast, no effect of processing (either boiling or frying) on protein digestibility or biological value was seen for four species of insects when fed to rats (Ekpo, 2011). Similarly, Jensen *et al.* (2019) report that neither protein digestibility, nor the biological value of lesser mealworm larvae is altered by freeze drying, vacuum drying at low (40 °C), or higher (120-160 °C) temperatures, industrial drying, defatting, extrusion, or by adding a blend of enzymes to the diet when fed to rats.

In addition to processing insects into meals for animal feed, protein extraction from raw insects for use in human food has also been evaluated (Lin *et al.*, 1983; Ozimek *et al.*, 1985, Yi *et al.*, 2013). Dried and ground silkworm chrysalid protein was compared to an aqueous extract of ground silkworm chrysalid protein with and without pretreatment with hydrogen peroxide (Lin *et al.*, 1983). Amino acid patterns were similar for the three treatments; however, sodium hydroxide treatment prior to extraction decreased cysteine levels. Whereas protein digestibility was similar between treatments, net protein utilisation was lower for the protein extract pre-treated with sodium hydroxide, probably due to cysteine destruction.

While some processing methods negatively affect protein/amino acid availability, some can improve protein quality. Ozimek *et al.* (1985) evaluated the effect on amino acid pattern and availability and the net protein value of an alkaline extraction of dried adult honeybees fed to rats. This extraction increased protein quality of the concentrate compared to dried honeybees by elevating the relative abundance (amino acid/protein) of leucine (54%) and methionine (59%). Amino acid availability was also improved by extraction (90%), compared to dried honeybees (65.4%), and was similar to casein (92.2%). The improved amino acid availability may partially be due to the removal of cuticular proteins complexed with chitin; the alkaline extraction decreased chitin content from 11.1% DM to below the detection limit. The improved amino acid profile and increased amino acid availability in the extract improved net protein utilisation from 41.5 to 60.6% compared to dried adult honeybees.

The nutrient content of insects prepared for human consumption may be altered by traditional cooking methods in two ways. Depending on the type of cookware used, iron, zinc, copper, and nickel can leach from the cookware and into the insect during preparation. The high values for copper, iron and zinc reported for some insects prepared using traditional methods may be partially due to the cookware used (Payne *et al.*, 2015; Santos Oliveira *et al.*, 1976). Additionally, many insects consumed by humans are either broiled, roasted, or fried prior to consumption. Broiled or roasted insects contain less fat than raw insects as fat is lost during cooking. A comparison of traditional processing methods using four insect species (black soldier fly prepupae, house crickets, longhorn grasshoppers and the Egyptian cotton leafworm (*Spodoptera littoralis*)) showed decreases in fat content (0.8-51%) and consequently increases of the crude protein (1.2-22%) content (Nyangena *et al.*, 2020). Fat content was the highest in the raw product and decreased in the following order: toasting, boiling, oven drying and solar-drying.

For fried insects, the fat content increases as the fat from the cooking oil is absorbed by the insect (Santos Oliveira *et al.*, 1976). This also changes the fatty acid profiles, because this material contains both the fat naturally present in the insect and the fat in the cooking oil (Santos Oliveira *et al.*, 1976).

As previously mentioned, some B-vitamins are relatively unstable when exposed to heat, light, or oxygen. The vitamin content of insects processed for human consumption is lower than raw unprocessed insects, possibly due to losses during cooking and drying. Kinyuru *et al.*, (2010a) evaluated the effect of toasting (frying at approximately 150 °C for 5 minutes) and drying (solar drying at 30 °C) on the niacin, riboflavin, pyridoxine, and folic acid content of Mendi termites and longhorn grasshoppers. Toasting decreased the folic acid content of these species by 37 and 43%, respectively, and riboflavin by 23 and 34%, respectively. In contrast there was little loss of pyridoxine and niacin. Drying of termites and grasshoppers also decreased both folic acid by 47 to 66%, and riboflavin by 29 to 46%. There was little loss of pyridoxine during drying and a significant loss of niacin for termites (26%), but not for grasshoppers. The reason for these differences is unclear. Of note is that thiamine (vitamin B1) which is known to be unstable during heating was not measured in this study.

Dreyer and Wehmeyer (1982) showed that canning reduces thiamine (84%), riboflavin (20%) and niacin (42%) levels in milled Mopane caterpillars, most likely because of the heat used to sterilise the cans. In contrast to the loss of B-vitamins, β-carotene was unaffected by the canning process.

7   Conclusions

Insects have a high protein content and their amino acid profiles are suitable for production animals and humans alike. This protein is highly digestible; hence, insects are generally a good source of dietary protein. In most cases methionine is the first limiting amino acid when insects are used as food or feed. Fat is the next most prominent nutrient in insects. Large variations in fat content and fatty acid composition occur due to species, life stage, diet and gender.

Carbohydrates are virtually absent in the insect body but fibre is present in significant amounts. This fibre consists primarily of chitin and sclerotised proteins, which are components of the insect exoskeleton. Whereas calcium levels in insects are lower than in animals with an endoskeleton, most other minerals are present in adequate concentrations to meet nutrient requirements of animals and humans. Mineral concentrations can be especially high in certain fly larvae. Vitamin A levels in insects are low, however carotenoids can be present in relatively high concentrations, depending on the diet. Vitamin D content is highly variable and largely depends on the availability of UV-B radiation during insect development. Also vitamin E concentrations are highly variable and depend on dietary levels. Most B-vitamins are present in adequate concentrations in most insect species. The taurine content of insects is highly variable and depends on both species and life stage.

Several factors such as species and life stage influence the composition of insect species. Diet can strongly affect the concentration of most nutrients, although some, especially amino acids and minerals of nutritional relevance, tend to be less flexible. Heavy metals can accumulate in insects. However, accumulation patterns vary between species and orders. Insect processing can affect their nutritional value in many ways, but the most common are destruction of vitamins and denaturation of proteins due to heat. Although large variations in insect composition are known, in general they are to be considered a prime source of nutrients.

Conflict of interest

The authors declare no conflict of interest.

References

Adamolekun, B., 1993. *Anaphe venata* entomophagy and seasonal ataxic syndrome in southwest Nigeria. Lancet 341: 629.

Adamolekun, B., McCandless, D.W. and Butterworth, R.F., 1997. Epidemic of seasonal ataxia in Nigeria following ingestion of the African silkworm *Anaphe venata*: role of thiamine deficiency? Metabolic Brain Disease 12: 251-258.

Akman Gündüz, N.E. and Gülel, A., 2002. Effect of temperature on development, sexual maturation time, food consumption and body weight of *Schistocerca gregaria* Forsk. (Orthoptera: Acrididae). Türk Zooloji Dergisi 26: 223-227.

Ali, M. and Ewiess, M.A., 1977. Photoperiodic and temperature effects on rate of development and diapause in the green stink bug, *Nezara viridula* L. (Heteroptera: Pentatomidae). Zeitschrift Fur Angewandte Entomologie 84: 256-264.

Ali, M.F., Mashaly, A.M.A., Mohammed, A.A. and El-Magd Mahmoud Mohammed A., 2011. Effect of temperature and humidity on the biology of *Attagenus fasciatus* (Thunberg) (Coleoptera: Dermestidae). Journal of Stored Product Research 47: 25-31. https://doi.org/10.1016/j.jspr.2010.07.002

Arnold, K.E., Ramsay, S.L., Henderson, L. and Larcombe, S., 2010. Seasonal variation in diet quality: antioxidants, invertebrates and blue tits *Cyanistes caeruleus*. Biological Journal of the Linnean Society 99: 708-717.

Asghar, A., Gray, J.I., Miller, E.R., Ku, P.K., Booren, A.M. and Buckley, D.J., 1991. Influence of supranutritional vitamin E supplementation in the feed on swine growth performance and deposition in different tissues. Journal of the Science of Food and Agriculture 57: 19-29.

Azzollini, D., Derossi, A., Fogliano, V., Lakemond, C.M.M. and Severini, C., 2018. Effects of formulation and process conditions on microstructure, texture and digestibility of extruded insect-riched snacks. Innovative Food Science and Emerging Technologies 45: 344-353.

Bamgbose, A.M., 1999. Utilization of maggot-meal in cockerel diets. Indian Journal of Animal Sciences 69: 1056-1058.

Banjo, A.D., Lawal, O.A. and Songonuga, E.A., 2006. The nutritional value of fourteen species of edible insects in southwestern Nigeria. African Journal of Biotechnology 5: 298-301.

Barker, D., Fitzpatrick, M.P. and Dierenfeld, E.S., 1998. Nutrient composition of selected whole invertebrates. Zoo Biology 17: 123-134.

Batterham, E.S., Darnell, R.E., Herbert, L.S. and Major, E.J., 1986. Effect of pressure and temperature on the availability of lysine in meat and bone meal as determined by slope-ratio assays with growing pigs, rats and chicks and by chemical techniques. British Journal of Nutrition 55: 441-453.

Bawa, M., Songsempong, S., Kaewtapee, C. and Chanput, W., 2020. Effect of diet on the growth performance, feed conversion, and nutrient content of the house cricket. Journal of Insect Science 20: 1-10.

Bednarska, A.J., Opyd, M.O., Zurawicz, E. and Laskowski, R., 2015. Regulation of body metal concentrations: Toxicokinetics of cadmium and zinc in crickets. Ecotoxicology and Environmental Safety 119: 9-14.

Beenakkers, M.T.A., Van der Horst, D.J. and Van Marrewijk, W.J.A., 1985. Insect lipids and lipoproteins, and their role in physiological processes. Progress in Lipid Research 24: 19-67.

Bender, D.A., 2002. Introduction to nutrition and metabolism. CRC Press, London, UK, 450 pp.

Bodnaryk, R.P., 1981. The biosynthesis function, and fate of taurine during the metamorphosis of the Noctuid moth *Mamestra configurata* Wlk. Insect Biochemistry 11: 199-205.

Borgeson, C.E. and Blomquist, G.J., 1993. Subcellular location of the Δ^{12} desaturase rules out bacteriocyte contribution to linoleate biosynthesis in the house cricket and the American cockroach. Insect Biochemistry and Molecular Biology 23: 297-302.

Borgeson, C.E., Kurtti, T.J., Munderloh, U.G. and Blomquist, G.J., 1991. Insect tissues, not microorganisms, produce linoleic-acid in the house cricket and the American cockroach. Experientia 47: 238-241.

Bosch, G., Vervoort, J.J.M and Hendriks, W.H., 2016. *In vitro* digestibility and fermentability of selected insects for dog foods. Animal Feed Science and Technology 221: 174-184.

Bosch, G., Zhang, S., Oonincx, D.G.A.B. and Hendriks, W.H., 2014. Protein quality of insects as potential ingredients for dog and cat foods. Journal of Nutritional Science 3: e29.

Bukkens, S.G.F., 1997. The nutritional value of edible insects. Ecology of Food and Nutrition 36: 287-319.

Calvert, C.C., Martin, R.D. and Morgan, N.O., 1969. House fly pupae as food for poultry. Journal of Economic Entomology 62: 938-939.

Cauchie, H.M., 2002. Chitin production by arthropods in the hydrosphere. Hydrobiologia 470: 63-96.

Cerda, H., Martinez, R., Briceno, N., Pizzoferrato, L., Manzi, P., Tommaseo Ponzetta, M., Marin, O. and Paoletti, M.G., 2001. Palm worm: (*Rhynchophorus palmarum*) traditional food in Amazonas, Venezuela – nutritional composition, small scale production and tourist palatability. Ecology of Food and Nutrition 40: 13-32.

Charlton, A., Dickinson, M., Wakefield, M., Fitches, E., Kenis, M., Han, R., Zhu, F., Kone, N., Grant, M. and Devic, E., 2015. Exploring the chemical safety of fly larvae as a source of protein for animal feed. Journal of Insects as Food and Feed 1: 7-16. https://doi.org/10.3920/JIFF2014.0020

Cheseto, X., Kuate, S.P., Tchouassi, D.P., Ndung'u, M., Teal, P.E. and Torto, B., 2015. Potential of the desert locust *Schistocerca gregaria* (Orthoptera: Acrididae) as an unconventional source of dietary and therapeutic sterols. PLoS ONE 10: e0127171.

Chieco, C., Morrone, L., Bertazza, G., Cappellozza, S., Saviane, A., Gai, F., Di Virgilio, N. and Rossi, F., 2019. The effect of strain and rearing medium on the chemical composition, fatty acid profile and carotenoid content in silkworm (*Bombyx mori*) pupae. Animals 9: 103.

Church, R.B. and Robertson, F.W., 1966. A biochemical study of the growth of *Drosophila melanogaster*. Journal of Experimental Zoology 162: 337-351.

Connor, W.E., Wang, Y., Green, M. and Lin, D.S., 2006. Effects of diet and metamorphosis upon the sterol composition of the butterfly *Morpho peleides*. Journal of Lipid Research 47: 1444-1448.

Cookman, J.E., Angelo, M.J., Slansky Jr, F. and Nation, J.L., 1984. Lipid content and fatty acid composition of larvae and adults of the velvetbean caterpillar, *Anticarsia gemmatalis*, as affected by larval diet. Journal of Insect Physiology 30: 523-527.

Crawford, L.A., Lepp, N.W. and Hodkinson, I.D., 1996. Accumulation and egestion of dietary copper and cadmium by the grasshopper *Locusta migratoria* R and F (Orthoptera: Acrididae). Environmental Pollution 92: 241-246. https://doi.org/10.1016/0269-7491(96)00004-8

Dashefsky, H.S., Anderson, D.L., Tobin, E.N. and Peters, T.M., 1976. Face fly pupae: a potential feed supplement for poultry. Environmental Entomology 5: 680-682.

DeMarco, M., Martinez, S., Hernandez, F., Madrid, J., Gai, F., Rotolo, L., Belforti, M., Bergero, D., Katz, H., Dabbou, S., Kovitvadhi, A., Zoccarato, I., Gasco, L. and Schiavone, A., 2015. Nutritional value of two insect larval meals (*Tenebrio molitor* and *Hermetia illucens*) for broiler chickens: apparent nutrient digestibility, apparent ileal amino acid digestibility and apparent metabolizable energy. Animal Feed Science and Technology 209: 211-218.

Devkota, B. and Schmidt, G.H., 2000. Accumulation of heavy metals in food plants and grasshoppers from the Taigetos Mountains, Greece. Agriculture, Ecosystems and Environment 78: 85-91.

Diehl, J.J.E, Baines, F.M., Heijboer, A.C., Van Leeuwen, J.P., Kik, M., Hendriks, WH. and Oonincx, D.G.A.B., 2018. A comparison of UV b compact lamps in enabling cutaneous vitamin D synthesis in growing bearded dragons. Journal of Animal Physiology and Animal Nutrition 102: 308-316.

Diener, S., Zurbrugg, C. and Tockner, K., 2015. Bioaccumulation of heavy metals in the black soldier fly, *Hermetia illucens* and effects on its life cycle. Journal of insects and Food and Feed 1: 261-270. https://doi.org/10.3920/JIFF2015.0030

Downer, R.G.H. and Matthews, J.R., 1976. Patterns of lipid distribution and utilisation in insects. American Zoologist 16: 733-745.

Dreassi, E., Cito, A., Zanfini, A., Materozzi, L., Botta, M. and Francardi, V., 2017. Dietary fatty acids influence the growth and fatty acid composition of the yellow mealworm *Tenebrio molitor* (Coleoptera: Tenebrionidae). Lipids 52: 285-294.

Dreyer, J.J. and Wehmeyer, A.S., 1982. On the nutritive value of mopanie worms. South African Journal of Science 78: 33-35.

Eeva, T., Helle, S. and Salminen, J.P., 2010. Carotenoid composition of invertebrates consumed by two insectivorous bird species. Journal Chemical Ecology 36: 608-613.

Ehigie, L.O., Emuebie, O. and Ehigie, F.A., 2013. Biochemical properties of thiaminase, a toxic enzyme in the gut of grasshoppers (*Zonocerus variegatus* Linn). Cameroon Journal of Experimental Biology 9: 9-16.

Ekpo, K.E. 2011. Effect of processing on the protein quality of four popular insects consumed in Southern Nigeria. Archives of Applied Science Research 3: 307-326.

Erbland, P., Alyokhin, A., Perkins, L.B. and Peterson, M., 2020. Dose-dependent retention of omega-3 fatty acids by black soldier fly larvae (Diptera: Stratiomyidae). Journal of Economic Entomology 113: 1221-1226. https://doi.org/10.1093/jee/toaa045

Fasel, N.J., Mène-Safrané, L., Ruczynski, I., Komar, E. and Christe, P. 2017. Diet induced modifications of fatty acid composition in mealworm larvae (*Tenebrio molitor*). Journal of Food Research 6: 5. https://doi.org/10.5539/jfr.v6n5p22

Fast, P.G., 1970. Insect lipids. Progress in the Chemistry of Fats and other Lipids 11: 181-242.

Finke M.D., 2002. Complete nutrient composition of commercially raised invertebrates used as food for insectivores. Zoo Biology 21: 269-285.

Finke, M.D., 2003. Gut loading to enhance the nutrient content of insects as food for reptiles: a mathematical approach. Zoo Biology 22: 147-162.

Finke, M.D. 2004. The nutrient content of insects. In: Capinara, J.L. (ed.) Encyclopedia of entomology. Vol 2. Kluwer Academic Publishers, Dordrecht, The Netherlands, p. 1562-1575.

Finke, M.D., 2007. Estimate of chitin in raw whole insects. Zoo Biology 26: 105-115.

Finke, M.D., 2013. Complete nutrient content of four species of feeder insects. Zoo Biology 32: 27-36.

Finke, M.D., 2015a. Complete nutrient content of three species of wild caught insects, pallid-winged grasshopper, rhinoceros beetles and white-lined sphinx moth. Journal of Insects as Food and Feed 1: 281-292. https://doi.org/10.3920/JIFF2015.0033

Finke, M.D., 2015b. Complete nutrient content of four species of commercially available feeder insects fed enhanced diets during growth. Zoo Biology 34: 554-564.

Finke, M.D. and Oonincx, D.G.A.B., 2017. Nutrient content of insects. In: Van Huis, A. and Tomberlin, J.K. (eds) Insects as food and feed: from production to consumption. Wageningen Academic Publishers, Wageningen, The Netherlands, pp. 290-317.

Finke, M.D., DeFoliart, G.R. and Benevenga, N.J., 1987. Use of a four-parameter logistic model to evaluate the protein quality of mixtures of Mormon cricket meal and corn gluten meal in rats. Journal of Nutrition 117: 1740-1750.

Finke, M.D., Sunde, M.L. and DeFoliart, G.R., 1985. An evaluation of the protein quality of Mormon crickets (*Anabrus simplex* Haldeman) when used as a high protein feedstuff for poultry. Poultry Science 64: 708-712

Finkel, A.J., 1948. The lipid composition of *Tenebrio molitor* larvae. Physiological Zoology 21: 111-133.

Fogang, A.R., Kansci, G., Viau, M., Hafnaoui, N., Meynier, A., Demmano, G. and Genot, C., 2017. Lipid and amino acids profiles support the potential of *Rhynchophorus phoenicis* larvae for human nutrition. Journal of Food Composition and Analysis 60: 64-73.

Fontaneto, D., Tommaseo-Ponzetta, M., Galli, C., Risé, P., Glew, R.H. and Paoletti, M.G., 2011. Differences in fatty acid composition between aquatic and terrestrial insects used as food in human nutrition. Ecology of Food and Nutrition 50: 351-367.

Frye, F.L. and Calvert, C.C., 1989. Preliminary information on the nutritional content of mulberry silk moth (*Bombyx mori*) larvae. Journal of Zoo and Wildlife Medicine 20: 73-75.

Fujimoto, W., Suzuki, M., Kimura, K. and Iwanaga, T., 2002. Cellular expression of the gut chitinase in the stomach of frogs *Xenopus laevis* and *Rana catesbeiana*. Biomedical Research-Tokyo 23: 91-99.

Gao, Q., Wang, X., Wang, W., Lei, C. and Zhu, F., 2017. Influences of chromium and cadmium on the development of black soldier fly larvae. Environmental Science and Pollution Research 24: 8637-8644.

Gasco, L., Biasato, I., Dabbou, S., Schiavone, A. and Gai, F., 2019. Animals fed insect-based diets: State-of-the-art on digestibility, performance and product quality. Animals 9: 170. https://doi.org/10.3390/ani9040170

Ghioni, C., Bell, J.G. and Sargent, J.R., 1996. Polyunsaturated fatty acids in neutral lipids and phospholipids of some freshwater insects. Comparative Biochemistry and Physiology B 114: 161-170.

Giovannucci, D.R. and Stephenson, R.S., 1999. Identification and distribution of dietary precursors of the *Drosophila* visual pigment chromophore: analysis of carotenoids in wild type and ninaD mutants by HPLC. Vision Research 39: 219-229.

Gladyshev, M.I., Kharitonov, A.Y., Popova, O.N., Sushchik, N.N., Makhutova, O.N. and Kalacheva, G.S., 2011. Quantitative estimation of dragonfly role in transfer of essential polyunsaturated fatty acids from aquatic to terrestrial ecosystems. Doklady Biochemistry and Biophysics 438: 141-143.

Goldsmith, T.H. and Warner, L.T., 1964. Vitamin A in the vision of insects. Journal of General Physiology 47: 433-441.

Goulet, G., Mullier, P., Sinave, P. and Brisson, G.J., 1978. Nutritional evaluation of dried *Tenebrio molitor* larvae in the rat. Nutrition Reports International 18: 11-15.

Green, K., 2008. Migratory bogong moths (*Agrotis infusa*) transport arsenic and concentrate it to lethal effect by estivating gregariously in alpine regions of the Snowy Mountains of Australia. Arctic, Antarctic, and Alpine Research 40: 74-80.

Hale, O.M., 1973. Dried *Hermetia illucens* (Stratiomyidae) as a feed additive for poultry. Journal of the Georgia Entomological Society 8: 16-20.

Hall, H.N., Masey O'Neill H.V., Scholey D., Burton E., Dickinson M. and Fitches E.C. 2018. Amino acid digestibility of larval meal (*Musca domestica*) for broiler chickens. Poultry Science 97: 1290-1297.

Han, R.D., Parajulee, M., Zhong, H. and Feng, G., 2008. Effects of environmental humidity on the survival and development of pine caterpillars, *Dendrolimus tabulaeformis* (Lepidoptera: Lasiocampidae). Insect Science 15: 147-152.

Hanson, B.J., Cummins, K.W., Cargill, A.S. and Lowry, R.R., 1985. Lipid content, fatty acid composition, and the effect of diet on fats of aquatic insects. Comparative Biochemistry and Physiology B 80: 257-276.

Harwood, R.F. and Takata, N., 1965. Effect of photoperiod and temperature on fatty acid composition of the mosquito *Culex tarsalis*. Journal of Insect Physiology 11: 711-716

Hatt, J.M., Hung, E. and Wanner, M., 2003. The influence of diet on the body composition of the house cricket (*Acheta domesticus*) and consequences for their use in zoo animal nutrition. Der Zoologische Garten 73: 238-244.

Henriques, B.S., Garcia, E.S., Azambuja, P. and Genta, F.A., 2020. Determination of chitin content in insects: an alternative method based on calcofluor staining. Frontiers in Physiology 11: 117. https://doi.org/10.3389/fphys.2020.00117

Henry, M., Gascob, L., Piccoloc, G. and Fountoulakia, E., 2015. Review on the use of insects in the diet of farmed fish: past and future. Animal Feed Science and Technology 203: 1-22.

Hoffmann, K.H., 1973. Effects of temperature on chemical composition of crickets (*Gryllus*, orthopt.) (In German). Oecologia 13: 147-175.

Hopkins, P.N. 1992. Effects of dietary cholesterol on serum cholesterol: a meta-analysis and review. American Journal of Clinical Nutrition 55: 1060-1070.

Hunter, B.A., Hunter, L.M., Johnson, M.S. and Thompson, D.J., 1987. Dynamics of metal accumulation in the grasshopper *Chorthippus brunneus* in contaminated grasslands. Archives of Environmental Contamination and Toxicology 16: 711-716.

Hussein, M., Pillai, V.V., Goddard, J.M., Park, H.G., Kothapalli, K.S., Ross, D.A., Ketterings, Q.M., Brenna, J.T., Milstein, M.B., Marquis, H., Johnson, P.A., Nyrop, J.P. and Selvaraj, V., 2017. Sustainable production of housefly (*Musca domestica*) larvae as a protein-rich feed ingredient by utilizing cattle manure. PLoS ONE 12: e0171708.

Hutchins, R.F.N. and Martin, M.M., 1968. The lipids of the common house cricket, *Acheta domesticus* L. I. Lipid classes and fatty acid distribution. Lipids 3: 247-249.

Hwangbo, J., Hong, E.C., Jang, A., Kang, H.K., Oh, J.S., Kim, B.W. and Park, B.S., 2009. Utilization of house fly-maggots, a feed supplement in the production of broiler chickens. Journal of Environmental Biology 30: 609-614.

Igwe, C.U., Ujowundu, C.O., Nwaogu, L.A. and Okwu, G.N., 2011. Chemical analysis of an edible African termite, *Macrotermes nigeriensis*, a potential antidote to food security problem. Biochemistry and Analytical Biochemistry 1: 1-4.

Iñiguez-Covarrubias, G., De Franco-Gómez, M.J. and Del R Andrade-Maldonado, G., 1994. Biodegradation of swine waste by house-fly larvae and evaluation of their protein quality in rats. Journal of Applied Animal Research 5: 65-74.

Isaksson, C. and Andersson, S., 2007. Carotenoid diet and nestling provisioning in urban and rural great tits *Parus major*. Journal of Avian Biology 38: 564-572.

Jamil, K. and Hussain, S., 1992. Biotransfer of metals to the insect *Neochetina eichhornae* via aquatic plants. Archives of Environmental Contamination and Toxicology 22: 459-463.

Janssen, M., Bruins, A., De Vries, T. and Van Straalen, N., 1991. Comparison of cadmium kinetics in four soil arthropod species. Archives of Environmental Contamination and Toxicology 20: 305-312.

Janssen, R.H., Vincken, J.P., Van den Broek, L.A.M., Fogliano, V. and Lakemond, C.M.M., 2017. Nitrogen-to-protein conversion factors for three edible insects: *Tenebrio molitor, Alphitobius diaperinus*, and *Hermetia illucens*. Journal of Agricultural and Food Chemistry. 65: 2275-2278.

Jensen, L.D., Miklos, R., Dalsgaard, T.K., Heckmann, L.H. and Nørgaard, J.V., 2019. Nutritional evaluation of common (*Tenebrio molitor*) and lesser (*Alphitobius diaperinus*) mealworms in rats and processing effect on the lesser mealworm. Journal of Insects as Food and Feed 5: 257-266. https://doi.org/10.3920/JIFF2018.0048

Jing, X. and Behmer, S.T., 2020. Insect sterol nutrition: physiological mechanisms, ecology, and applications. Annual Review of Entomology 65: 251-271.

Jones, L.D., Cooper, R.W. and Harding, R.S., 1972. Composition of mealworm *Tenebrio molitor* larvae. The Journal of Zoo Animal Medicine 3: 34-41.

Kaspari, M., 1991. Prey preparation as a way that grasshopper sparrows (*Ammodramus savannarum*) increase the nutrient composition of their prey. Behavioral Ecology 2: 234-241.

Kaya, M., Baran, T., Erdoğan, S., Menteş, A., Aşan Özüsağlam, M. and Çakmak, Y.S., 2014. Physicochemical comparison of chitin and chitosan obtained from larvae and adult Colorado potato beetle (*Leptinotarsa decemlineata*). Materials Science and Engineering: C 45: 72-81.

Kazimírová, M. and Ortel, J., 2000. Metal accumulation by *Ceratitis capitata* (Diptera) and transfer to the parasitic wasp *Coptera occidentalis* (Hymenoptera). Environmental Toxicology and Chemistry 19: 1822-1829.

Kiiru, S.M., Kinyuru, J.N., Kiage, B.N. and Marel, A.K. 2020. Partial substitution of soy protein isolates with cricket flour during extrusion affects firmness and *in vitro* protein digestibility. Journal of Insects as Food and Feed 6: 169-177. https://doi.org/10.3920/JIFF2019.0024

Kinyuru, J.N., Kenji, G.M., Muhoho, S.N. and Ayieko, M. 2010b. Nutritional potential of longhorn grasshopper (*Ruspolia differens*) consumed in Siaya district, Kenya. Journal of Agriculture, Science and Technology 12: 32-46.

Kinyuru, J.N., Kenji, G.M., Njoroge, S.M. and Ayieko, M. 2010a. Effect of processing methods on the *in vitro* protein digestibility and vitamin content of edible winged termite (*Macrotermes subhylanus*) and grasshopper (*Ruspolia differens*). Food Bioprocess Technology 3: 778-782.

Kodondi, K.K., Leclercq, M. and Gaudin-Harding, F., 1987. Vitamin estimations of three edible species of Attacidae caterpillars from Zaire. International Journal of Vitamin and Nutrition Research 57: 333-334.

Köhler, R., Irias-Mata, A., Ramandey, E., Purwestri, R. and Biesalski, H.K., 2020. Nutrient composition of the Indonesian sago grub (*Rhynchophorus bilineatus*). International Journal of Tropical Insect Science 40: 677-686. https://doi.org/10.1007/s42690-020-00120-z

Köhler, R., Kariuki, L., Lambert, C. and Biesalski, H.K., 2019. Protein, amino acid and mineral composition of some edible insects from Thailand. Journal of Asia-Pacific Entomology 22: 372-378.

Komprda, T., Zornikova, G., Rozikova, V., Borkovcova, M. and Przywarova, A., 2013. The effect of dietary *Salvia hispanica* seed on the content of n-3 long-chain polyunsaturated fatty acids in tissues of selected animal species, including edible insects. Journal of Food Composition and Analysis 32: 36-43.

Koo, S.I., Currin, T.A., Johnson, M.G., King, E.W. and Turk, D.E., 1980. The nutritional value and microbial content of dried face fly pupae (*Musca autumnalis* (De Geer)) when fed to chicks. Poultry Science 59: 2514-2518.

Koštál, V., Urban, T., Rimnáčová, L., Berková, P., and Simek,, P., 2013. Seasonal changes in minor membrane phospholipid classes, sterols and tocopherols in overwintering insect, *Pyrrhocoris apterus*. Journal of Insect Physiology 59: 934-41.

Kouamé, B., Marcel, G., Brou André, K., Alassane, M., Koffi Gabouet, K. and Coulibally Séraphi, K., 2011. Détermination du taux optimal de farine d'asticots séchés dans le régime du rat en croissance. Journal of Animal and Plant Sciences 12: 1553-1559.

Kramarz, P., 1999. Dynamics of accumulation and decontamination of cadmium and zinc in carnivorous invertebrates. 1. The ground beetle, *Poecilus cupreus* L. Bulletin of Environmental Contamination and Toxicology 63: 531-537.

Kramer, K.J., Hopkins, T.L. and Schaefer, J., 1995. Applications of solids NMR to the analysis of insect sclerotized structures. Insect Biochemistry and Molecular Biology 25: 1067-1080.

Kulma, M., Kouřimská, L., Plachý, V., Božik, M., Adámková, A. and Vrabec, V., 2019. Effect of sex on the nutritional value of house cricket, *Acheta domestica* L. Food Chemistry 272: 267-272.

Landry, S., DeFoliart, G.R. and Sunde, M.L., 1986. Larval protein quality of six species of Lepidoptera (Saturniidae, Sphingidae, Noctuidae). Journal of Economic Entomology 79: 600-604.

Lease, H.M. and Wolf, B.O., 2010. Exoskeletal chitin scales isometrically with body size in terrestrial insects. Journal of Morphology 271: 759-768.

Lease, H.M. and Wolf, B.O., 2011. Lipid content of terrestrial arthropods in relation to body size, phylogeny, ontogeny and sex. Physiological Entomology 36: 29-38.

Levi, L., Ziv, T., Admon, A., Levavi-Sivan, B. and Lubzens, E., 2012. Insight into molecular pathways of retinal metabolism, associated with vitellogenesis in zebrafish. American Journal of Physiology – Endocrinology and Metabolism 302: 626-644.

Liland, N.S., Biancarosa, I., Araujo, P., Biemans, D., Bruckner, C.G., Waagbø, R., Bente, E., Torstensen, B.E., and Lock, E.J., 2017. Modulation of nutrient composition of black soldier fly (*Hermetia illucens*) larvae by feeding seaweed-enriched media. PLoS ONE 12: e0183188.

Lin, S., Njaa, L.R., Eggum, B.O. and Shen, H., 1983. Chemical and biological evaluation of silk worm chrysalid protein. Journal of the Science of Food and Agriculture 34: 896-900.

Linder, P., 1919. Extraction of fat from small animals. Zoological Technology and Biology 7: 213-220.

Lindqvist, L., 1992. Accumulation of cadmium, copper, and zinc in five species of phytopha-
gous insects. Environmental entomology 21: 160-163.

Lindsay, G.J.H., Walton, M.J., Adron, J.W., Fletcher, T.C., Cho, C.Y. and Cowey, C.B., 1984. The
growth of rainbow trout (*Salmo gairdneri*) given diets containing chitin and its relationship to chitinolytic enzymes and chitin digestibility. Aquaculture 37: 315-334.

Lipsitz, E.Y. and McFarlane, J.E., 1971. Analysis of lipid during the life cycle of the house
cricket, *Acheta domesticus*. Insect Biochemistry 1: 446-460.

Liu, X., Chen, X., Wang, H., Yang, Q., Ur Rehman, K., Li, W., Cai, M., Li, Q., Mazza, L., Zhang,
J., Yu, Z., and Zheng, L., 2017. Dynamic changes of nutrient composition throughout the
entire life cycle of black soldier fly. PLoS ONE 12: e0182601.

Longvah, T., Mangthya, K. and Ramulu, P., 2011. Nutrient composition and protein quality evaluation of eri silkworm (*Samia racinii*) prepupae and pupae. Food Chemistry 128: 400-403.

Madariaga, M.A., Mata, F., Municio, A.M. and Ribera, A. 1971. Effect of the lipid composition
of the larval diet on the fatty acid composition during development of *Ceratitis capitata*.
Insect Biochemistry 2: 249-256.

Madibela, O.R., Seitiso, T.K., Thema, T.F. and Letso, M., 2007. Effect of traditional processing
methods on chemical composition and *in vitro* true dry matter digestibility of the mopanie worm (*Imbrasia belina*). Journal of Arid Environments 68: 492-500.

Makkar, H.P.S., Tran, G., Heuze, V. and Ankers, P., 2014: State-of-the-art on use of insects as
animal feed. Animal Feed Science and Technology 197: 1-33.

Maroni, G. and Watson, D., 1985. Uptake and binding of cadmium, copper and zinc by *Drosophila melanogaster* larvae. Insect Biochemistry 15: 55-63.

Marono, S., Piccolo, G., Loponte, R., Di Meo, C., Attia, Y.A., Nizza, A. and Bovera, F., 2015. *In
vitro* crude protein digestibility of *Tenebrio molitor* and *Hermetia illucens* insect meals
and its correlation with chemical composition traits. Italian Journal of Animal Sciences
14: 338-343.

Maryanski, M., Kramarz, P., Laskowski, R. and Niklinska, M., 2002. Decreased energetic reserves, morphological changes and accumulation of metals in carabid beetles (*Poecilus
cupreus* L.) exposed to zinc-or cadmium-contaminated food. Ecotoxicology 11: 127-139.

Massie, H.R., Williams, T.R. and DeWolfe, L.K., 1989. Changes in taurine in aging fruit flies
and mice. Experimental Gerontology 24: 57-65.

McComb, A., 2010. Evaluation of vitamin A supplementations for captive amphibian species. MSc thesis, North Carolina State University, Raleigh, North Carolina, 129 pp.

Megido, R.C., Poelaert, C., Ernens, M., Liotta, M., Blecker, C., Danthine, S., Tyteca, E., Haubruge, E., Alabi, T., Bindelle, J. and Francis, F., 2018. Effect of household cooking techniques on the microbiological load and the nutritional quality of mealworms (*Tenebrio
molitor* L. 1758). Food Research International 106: 503-508.

Mlček, J., Adámková, A., Adámek, M., Borkovcová, M., Bednářová, M. and Knížková, I., 2019.
Fat from Tenebrionidae bugs – sterols content, fatty acid profiles, and cardiovascular risk
indexes. Polish Journal of Food Nutrition Sciences 69: 247-254

Moussian, B., 2010. Recent advances in understanding mechanisms of insect cuticle differentiation. Insect Biochemistry and Molecular Biology 40: 363-375.

Mwangi, M.N., Oonincx, D.G.A.B., Stouten, T., Veenenbos, M., Melse-Boonstra, A., Dicke, M. and Van Loon, J.J., 2018. Insects as sources of iron and zinc in human nutrition. Nutrition Research Reviews 31: 248-255.

Nakagaki, B.J., Sunde, M.L. and Defoliart, G.R., 1987. Protein quality of the house cricket, *Acheta domesticus*, when fed to broiler chicks. Poultry Science 66: 1367-1371.

National Research Council (NRC), 1994. Nutrient requirements of poultry. Ninth Edition. National Academy Press, Washington, DC, USA, 157 pp.

National Research Council (NRC), 2011. Nutrient requirements of fish and shrimp. National Academy Press, Washington, DC, USA, 376 pp.

National Research Council (NRC), 2012. Nutrient requirements of Swine. Eleventh Edition. National Academy Press, Washington, DC, USA, 420 pp.

Nestel, D., Papadopoulos, N.T., Liedo, P., Gonzales-Ceron, L. and Carey, J.R., 2005. Trends in lipid and protein contents during medfly aging: An harmonic path to death. Archives of Insect Biochemistry and Physiology 60: 130-139.

Newbrey, J.L., Paszkowski, C.A. and Dumenko, E.D., 2013. A comparison of natural and restored wetlands as breeding bird habitat using a novel yolk carotenoid approach. Wetlands 33: 471-482.

Nishimune, T., Watanabe, Y., Okazaki, H. and Akai, H., 2000. Thiamin is decomposed due to *Anaphe* spp. entomophagy in seasonal ataxia patients in Nigeria. Journal of Nutrition 130: 1625-1628.

Noland, J.L. and Baumann, C.A. 1949. Requirement of the German cockroach for choline and related compounds. Experimental Biology and Medicine 70: 198-201.

Nyangena, D.N., Mutungi, C., Imathiu, S., Kinyuru, J., Affognon, H., Ekesi, S., Nakimbugwe, D. and Fiaboe, K.K., 2020. Effects of traditional processing techniques on the nutritional and microbiological quality of four edible insect species used for food and feed in East Africa. Foods 9: 574.

Olson, J.A., 1989. Provitamin A function of carotenoids: the conversion of beta-carotene into vitamin A. Journal of Nutrition 119: 105-108.

Omotoso, O.T. 2006. Nutritional quality, functional properties and anti-nutrient compositions of the larva of *Cirina forda* (Westwood) (Lepidoptera: Saturniidae). Journal of Zhejiang University Science B 7: 51-55.

Onifade, A.A., Oduguwa, O.O., Fanimo, A.O., Abu, A.O., Olutunde, T.O., Arije, A. and Babatunde, G.M., 2001. Effects of supplemental methionine and lysine on the nutritional value of housefly larvae meal (*Musca domestica*) fed to rats. Bioresource Technology 78: 191-194.

Oonincx, D.G.A.B and Dierenfeld, E.S., 2012. An investigation into the chemical composition of alternative invertebrate prey. Zoo Biology 31: 40-54.

Oonincx, D.G.A.B and Van der Poel, A.F., 2011. Effects of diet on the chemical composition of migratory locusts (*Locusta migratoria*). Zoo Biology 30: 9-16

Oonincx, D., Bosch, G., Van Der Borght, M., Smets, R., Gasco, L., Fascetti, A., Yu, Z., Johnson, V., Tomberlin, J.K. and Finke, M., 2019b. A cross-laboratory study on analytical variability of amino acid content in three insect species. In: Book of Abstracts of the 70th Annual

Meeting of the European Federation of Animal Science. Wageningen Academic Publishers, Wageningen, The Netherlands, p. 327.

Oonincx, D.G.A.B, Laurent, S., Veenenbos, M.E. and Van Loon, J.J A., 2019a. Dietary enrichment of edible insects with omega 3 fatty acids. Insect Science 27: 500-509. https://doi.org/10.1111/1744-7917.12669

Oonincx, D.G.A.B., Stevens, Y., Van den Borne, J.J.G.C., Van Leeuwen, J.P.T.M. and Hendriks, W.H., 2010. Effects of vitamin D3 supplementation and UVb exposure on the growth and plasma concentration of vitamin D3 metabolites in juvenile bearded dragons (*Pogona vitticeps*). Comparative Biochemistry and Physiology B 156: 122-128.

Oonincx, D.G.A.B., Van Broekhoven, S., Van Huis, A. and Van Loon, J.J.A., 2015a. Feed conversion, survival and development, and composition of four insect species on diets composed of food by-products. PLoS ONE 10: e0144601.

Oonincx, D.G.A.B., Van Keulen, P., Finke, M.D., Baines, F.M., Vermeulen, M. and Bosch, G., 2018. Evidence of vitamin D synthesis in insects exposed to UVb light. Scientific Reports 8 10807.

Oonincx, D.G.A.B., Van Leeuwen, J.P., Hendriks, W.H. and Van der Poel, A.F.B., 2015b. The diet of free-roaming Australian central bearded dragons (*Pogona vitticeps*). Zoo Biology 34: 271-277.

Ortiz, C., Weiss-Penzias, P.S., Fork, S. and Flegal, A.R., 2015. Total and monomethyl mercury in terrestrial arthropods from the central California coast. Bulletin of environmental contamination and toxicology 94: 425-430.

Ottoboni, M., Spranghers, T., Pinotti, L., Baldi, A., De Jaeghere, W. and Eeckhout, M., 2018. Inclusion of *Hermetia illucens* larvae or prepupae in an experimental extruded feed: process optimisation and impact on *in vitro* digestibility. Italian Journal of Animal Science 17: 418-427.

Oyarzun, S.E., Crawshaw, G.J. and Valdes, E.V., 1996. Nutrition of the tamandua: I. Nutrient composition of termites (*Nasutitermes spp.*) and stomach contents from wild tamanduas (*Tamandua tetradactyla*). Zoo Biology 15: 509-524.

Ozimek, L., Sauer, W.C., Kozikowski, V., Ryan, J.K., Jorgensen, H. and Jelen, P., 1985. Nutritive value of protein extracted from honey bees. Journal of Food Science 50: 1327-1329.

Paoletti, M.G., Norberto, L., Damini, R. and Musumeci, S., 2007. Human gastric juice contains chitinase that can degrade chitin. Annals of Nutrition and Metabolism 51: 244-251.

Paul, A., Frederich, M., Megido, R.C., Alabi, T., Malik, P., Uyttenbroeck, R., Francis, F., Blecker, C., Haubruge, E., Lognay, G. and Danthine, S., 2017. Insect fatty acids: a comparison from three Orthopteran and *Tenebrio molitor* L. larvae. Journal of Asia-Pacific Entomology 20: 337-240.

Payne, C.L.R., Scarborough, P., Rayner, M. and Nonaka, K., 2016. A systematic review of nutrient composition data available for twelve commercially available edible insects and comparison with reference values. Trends in Food Science and Technology 47: 69-77.

Payne, C.L.R., Umemura, M., Dube, S., Azuma, A., Takenaka, C. and Nonaka, K., 2015. The mineral composition of five insects as sold for human consumption in Southern Africa. African Journal of Biotechnology 14: 2443-2448.

Pearincott, J.V., 1960. Changes in lipid content during growth and metamorphosis of the house fly *Musca domestica* Linnaeus. Journal of Cellular and Comparative Physiology 55: 167-174.

Pennino, M., Dierenfeld, E.S. and Behler, J.L., 1991. Retinol, alpha-tocopherol, and proximate nutrient composition of invertebrates used as feed. International Zoo Yearbook 30:143-149.

Pesti, G.M., Harper, A.E. and Sunde, M.L., 1979. Sulfur amino acid and methyl donor status of corn-soy diets fed to starting broiler chicks and turkey poults. Poultry Science 58: 1541-1547.

Phelps, R.J., Struthers, J.K. and Moyo, S.J., 1975. Investigations into the nutritive value of *Macrotermes falciger* (Isoptera: Termitidae). Zoologica Africana 10: 123-132.

Pieterse, E. and Pretorius, Q., 2014. Nutritional evaluation of dried larvae and pupae meal of the housefly (*Musca domestica*) using chemical and broiler-based biological assays. Animal Production Science 54: 347-355.

Poelaert, C., Despret, X., Sindic, M., Beckers, Y., Francis, F., Portetelle, D., Soyeurt, H., Thévis, A. and Bindelle, J., 2017. Cooking has variable effects on the fermentability in the large intestine of the fraction of meats, grain legumes, and insects that is resistant to digestion in the small intestine in an *in vitro* model of the pig's gastrointestinal tract. Journal of Agricultural and Food Chemistry 65: 435-444.

Poelaert, C., Francis, F., Alabi, T., Caparros Megido, R., Crahay, B., Bindelle, J. and Beckers, Y., 2018. Protein value of two insects, subjected to various heat treatments, using growing rats and the protein digestibility-corrected amino acid score. Journal of Insects as Food and Feed 4: 77-87. https://doi.org/10.3920/JIFF2017.0003

Postma, J.F., VanNugteren, P. and De Jong, M.B.B., 1996. Increased cadmium excretion in metal-adapted populations of the midge *Chironomus riparius* (Diptera). Environmental Toxicology and Chemistry: An International Journal 15: 332-339.

Proc, K., Bulak, P., Wiącek, D. and Bieganowski, A. 2020. *Hermetia illucens* exhibits bioaccumulative potential for 15 different elements – implications for feed and food production. Science of the Total Environment 723: 138125.

Punzo, F., 2003. Nutrient composition of some insects and arachnids. Florida Scientist 66: 84-98.

Purschke, B., Scheibelberger, R., Axmann, S., Adler, A. and Jäger, H., 2017. Impact of substrate contamination with mycotoxins, heavy metals and pesticides on growth performance and composition of black soldier fly larvae (*Hermetia illucens*) for use in the feed and food value chain. Food Additives and Contaminants: Part A 34: 1410-1420.

Ramos-Elorduy, J., Gonzalez, E.A., Hernandez, A.R. and Pino, J.M., 2002. Use of *Tenebrio molitor* (Coleoptera: Tenebrionidae) to recycle organic wastes and as feed for broiler chickens. Journal of Economic Entomology 95: 214-220.

Ramsay, S.L. and Houston, D.C., 2003. Amino acid composition of some woodland arthropods and its implications for breeding tits and other passerines. Ibis 145: 227-232.

Rao, P.U., 1994. Chemical composition and nutritional evaluation of spent silk worm pupae. Journal Agriculture and Food Chemistry 42: 2201-2203.

Raubenheimer, D. and Rothman, J.M. 2013. Nutritional ecology of entomophagy in humans and other primates. Annual Review of Entomology 58: 141-160.

Reichle, D.E., Shanks, M.H. and Crossley, D.A., 1969. Calcium, potassium, and sodium content of forest floor arthropods. Annals of the Entomological Society of America 62: 57-62.

Rho, M.S. and Lee, K.P., 2014. Geometric analysis of nutrient balancing in the mealworm beetle, *Tenebrio molitor* L. (Coleoptera: Tenebrionidae). Journal of Insect Physiology 71: 37-45.

Riddick, E.W., 2014. Insect protein as a partial replacement for fishmeal in the diets of juvenile fish and crustaceans. In: Morales-Ramos, J.A., Rojas, M.G. and Shapiro, D.A. (eds) Mass production of beneficial organisms invertebrates and entomopathogens. Elsevier, Waltham, MA, USA, pp 565-582.

Roe, R.M., Clifford, C.W. and Woodring, J.P., 1980. The effect of temperature on feeding, growth, and metabolism during the last larval stadium of the female house cricket, *Acheta domesticus*. Journal of Insect Physiology 26: 639-644.

Roe, R.M., Clifford, C.W. and Woodring, J.P., 1985. The effect of temperature on energy distribution during the last-larval stadium of the female house cricket, *Acheta domesticus*. Journal of Insect Physiology 31: 371-378.

Roseland, C.R., Grodowitz, M.J., Kramer, K.J., Hopkins, T.L. and Broce, A.B., 1985. Stabilization of mineralized and sclerotized puparial cuticle of muscid flies. Insect Biochemistry 15: 521-528.

Rudel, L.L., Parks, J.S., Hedrick, L., Thomas, M. and Willifors, K. 1998. Lipoprotein and cholesterol metabolism in diet-induced coronary artery atherosclerosis in primates. Role of cholesterol and fatty acids. Progress in Lipid Research 37: 353-370.

Rumpold, B.A. and Schluter, O.K., 2013a. Nutritional composition and safety aspects of edible insects. Molecular Nutrition Food Research 57: 802-823.

Rumpold, B.A. and Schluter, O.K., 2013b. Potential and challenges of insects as an innovative source for food and feed production. Innovative Food Science and Emerging Technologies 17: 1-11.

Sabolová, M., Adámková, A., Kouřimská, L., Chrpová, D. and Pánek, J., 2016. Minor lipophilic compounds in edible insects. Potravinarstvo 10: 400-406.

Sanchez-Muros, M.J., Barroso, F.G. and De Haro, C., 2016. Brief summary of insect usage as an industrial animal feed/feed ingredient. In: Dorsey, A.T., Morales-Ramos, J.A. and Rojas, M.G. (ed) Insects as sustainable food ingredients production, processing and food applications. Elsevier, Waltham, MA, USA, pp 273-309.

Sanchez-Muros, M.J., Barroso, F.G. and Manzano-Agugliaro, F., 2014. Insect meal as renewable source of food for animal feeding: a review. Journal of Cleaner Production 65: 16-27.

Santos Oliveira, J.F., Passos De Carvalho, J., Bruno De Sousa, R.F.X. and Simao, M.M., 1976. The nutritional value of four species of insects consumed in Angola. Ecology of Food and Nutrition 5: 91-97.

Schaefer, C.H., 1968. The relationship of the fatty acid composition of *Heliothis zea* larvae to that of its diet. Journal of Insect Physiology 14: 171-178.

Schiavone, A., De Marco, M., Martínez, S., Dabbou, S., Renna, M., Madrid, J., Hernandez, F., Rotolo, L., Costa, P., Gai, F. and Gasco, L., 2017. Nutritional value of a partially defatted and a highly defatted black soldier fly larvae (*Hermetia illucens* L.) meal for broiler chickens: apparent nutrient digestibility, apparent metabolizable energy and apparent ileal amino acid digestibility. Journal of Animal Science and Biotechnology 8: 51. https://doi. org/10.1186/s40104-017-0181-5

Schmitt, E., Belghit, I., Johansen, J., Leushuis, R., Lock, E.J., Melsen, D., Shanmugam, R.K.R., Van Loon, J. and Paul, A., 2019. Growth and safety assessment of feed streams for black soldier fly larvae: a case study with aquaculture sludge. Animals 9: 189. https://doi. org/10.3390/ani9040189

Seki, T., Isono, K., Ozaki, K., Tsukahara, Y., Shibata-Katsuta, Y., Ito, M., Irie, T. and Katagiri, M., 1998. The metabolic pathway of visual pigment chromophore formation in *Drosophila melanogaster* all-trans (3S)-3-hydroxyretinal is formed from all-trans retinal via (3R)-3-hydroxyretinal in the dark. European Journal of Biochemistry 257: 522-527.

Shin, C.S., Kim, D.Y. and Shin, W.S., 2019. Characterization of chitosan extracted from mealworm beetle (*Tenebrio molitor, Zophobas morio*) and rhinoceros beetle (*Allomyrina dichotoma*) and their antibacterial activities. International Journal of Biological Macromolecules 125: 72-77.

Sonmez, E. and Gulel, A., 2008. Effects of different temperatures on the total carbohydrate, lipid and protein amounts of the bean beetle, *Acanthoscelides obtectus* Say (Coleoptera: Bruchidae). Pakistan Journal of Biological Sciences 11: 1803-1808.

Spranghers, T., Ottoboni, M., Klootwijk, C., Ovyn, A., Deboosere, S., De Meulenaer, B., Michiels, J., Eeckhout, M., De Clercq, P. and De Smet, S., 2017. Nutritional composition of black soldier fly (*Hermetia illucens*) prepupae reared on different organic waste substrates. Journal of the Science of Food and Agriculture 97: 2594-2600.

Ssepuuya, G., Mukisa, I.M. and Nakimbugwe, D., 2017. Nutritional composition, quality, and shelf stability of processed *Ruspolia nitidula* (edible grasshoppers). Food Science & Nutrition 5: 103-112.

St-Hilaire, S., Cranfill, K., McGuire, M.A., Mosley, E.E., Tomberlin, J.K., Newton, L., Sealey, W., Sheppard, C. and Irving, S., 2007. Fish offal recycling by the black soldier fly produces a foodstuff high in omega-3 fatty acids. Journal of the World Aquaculture Society 38: 309-313.

Starčević, K., Gavrilović, A., Gottstein, Z. and Mašek, T., 2017. Influence of substitution of sunflower oil by different oils on the growth, survival rate and fatty acid composition of Jamaican field cricket (*Gryllus assimilis*). Animal Feed Science and Technology 228: 66-71.

Strobel, S., Roswag, A., Becker, N.I., Trenczek, T.E. and Encarnação, J.A., 2013. Insectivorous bats digest chitin in the stomach using acidic mammalian chitinase. PLoS ONE 8: e72770.

Studier, E.H. and Sevick, S.H., 1992. Live mass, water content, nitrogen and mineral levels in some insects from south-central lower Michigan. Comparative Biochemistry and Physiology A 103: 579-595.

Surendra, K.C., Olivier, R., Tomberlin, J.K., Jha, R. and Khanal, S.K. 2016. Bioconversion of organic wastes into biodiesel and animal feed via insect farming. Renewable Energy 98: 197-202.

Sushchik, N.N., Gladyshev, M.I., Moskvichova, A.V., Makhutova, O.N. and Kalachova, G.S. 2003. Comparison of fatty acid composition in major lipid classes of the dominant benthic invertebrates of the Yenisei river. Comparative Biochemistry and Physiology B 134: 111-122.

Sushchik, N.N., Yurchenko, Y.A., Gladyshev, M.I., Belevich, O.E., Kalachova, G.S. and Kolma-
kova, A.A., 2013. Comparison of fatty acid contents and composition in major lipid classes of larvae and adults of mosquitoes (Diptera: Culicidae) from a steppe region. Insect
Science 20: 585-600.

Svoboda, J., Schiff, N. and Feldlaufer, M., 1995. Sterol composition of three species of sawflies
(Hymenoptera: Symphyta) and their dietary plant material. Experientia 51: 150-152.

Tabata, E., Kashimura, A., Kikuchi, A. Masuda, H., Miyahara, R., Hiruma, Y., Wakita, S., Ohno,
M., Sakaguchi, M., Sugahara, Y., Matoska, V., Bauer, P.O. and Oyama, F., 2018. Chitin digestibility is dependent on feeding behaviors, which determine acidic chitinase mRNA levels
in mammalian and poultry stomachs. Scientific Reports 8: 1461. https://doi.org/10.1038/
s41598-018-19940-8

Teffo, L.S., Toms, R.B. and Eloff, J.N., 2007. Preliminary data on the nutritional composition
of the edible stink-bug, *Encosternum delegorguei* Spinola, consumed in Limpopo prov-
ince, South Africa. South African Journal of Science 103: 434-436

Teotia, J.S. and Miller, B.F., 1973. Fly pupae as a dietary ingredient for starting chicks. Poultry
Science 52: 1830-1835.

Teotia, J.S. and Miller, B.F., 1974. Nutritive content of house fly pupae and manure residue.
British Poultry Science 15: 177-182.

Thompson, S.N., 1973. Review and comparative characterization of fatty-acid compositions
of seven insect orders. Comparative Biochemistry and Physiology 45: 467-482.

Tietz, A. and Stern, N., 1969. Stearate desaturation by microsomes on the locust fat-body.
FEBS Letters 2: 286-288.

Timmermans, K.R. and Walker, P.A., 1989. The fate of trace metals during the metamorphosis of chironomids (Diptera, Chironomidae). Environmental pollution 62: 73-85.

Tschirner, M. and Simon, A., 2015. Influence of different growing substrates and processing
on the nutrient composition of black soldier fly larvae destined for animal feed. Journal
of Insects as Food and Feed 1: 249-259. https://doi.org/10.3920/JIFF2014.0008

Ueckert, D.N., Yang, S.P. and Albin, R.C. 1972. Biological value of rangeland grasshoppers as a
protein concentrate. Journal of Economic Entomology 65: 1286-1288.

Van Broekhoven, S., Oonincx, D.G.A.B., Van Huis, A. and Van Loon, J.J.A., 2015. Growth performance and feed efficiency of three edible mealworm species (Coleoptera: Tenebrionidae) on diets composed of organic by-products. Journal of Insect Physiology 73: 1-10.

Van der Fels-Klerx, H.J., Camenzuli, L., Van der Lee, M.K. and Oonincx, D.G.A.B., 2016. Uptake of cadmium, lead and arsenic by *Tenebrio molitor* and *Hermetia illucens* from contaminated substrates. PLoS ONE 11: e0166186.

Vijver, M., Jager, T., Posthuma, L. and Peijnenburg, W., 2003. Metal uptake from soils and
soil-sediment mixtures by larvae of *Tenebrio molitor* (L.)(Coleoptera). Ecotoxicology and
Environmental Safety 54: 277-289.

Vogt, K. and Kirschfeld, K., 1984. Chemical identity of the chromophore of fly visual pigment. Naturwissenschaften 71: 211-213.

Von Lintig, J., 2012. Metabolism of carotenoids and retinoids related to vision. Journal of
Biological Chemistry 287: 1627-1634.

Wang, D., Zhai, S.W., Zhang, C.X., Bai, Y.Y., An, S.H. and Xu, Y.N., 2005. Evaluation on nutritional value of field crickets as a poultry feedstuff. Asian-Australasian Journal of Animal Sciences 18: 667-670.

Wang, D., Zhai, S.W., Zhang, C.X., Zhang, Q. and Chen, H., 2007. Nutrition value of the Chinese grasshopper *Acrida cinerea* (Thunberg) for broilers. Animal Feed Science and Technology 135: 66-74.

Wang, S. Y., Wu, L., Li, B. and Zhang, D., 2020. Reproductive potential and nutritional composition of *Hermetia illucens* (Diptera: Stratiomyidae) prepupae reared on different organic wastes. Journal of Economic Entomology 113: 527-537.

Whitaker, Jr., J.O., Dannelly, H.K. and Prentice, D.A., 2004. Chitinase in insectivorous bats. Journal of Mammalogy 85: 15-18.

Whitton, P.S., Strang, R.H.C. and Nicholson, R.A., 1987. The distribution of taurine in the tissues of some species of insects. Insect Biochemistry 17: 573-577.

WHO/FAO/UNU Expert Consultation, 2007. Protein and amino acid requirements in human nutrition. WHO Technical Reports Series 935. World Health Organization, Geneva, Switzerland.

Woods, M.J., Goosen, N.J., Hoffman, L.C. and Pieterse, E., 2020. A simple and rapid protocol for measuring the chitin content of *Hermetia illucens* (L.) (Diptera: Stratiomyidae) larvae. Journal of Insects as Food and Feed 6: 285-290. https://doi.org/10.3920/JIFF2019.0030

Wu, G.X., Ye, G.E., Hu, C. and Cheng, J.A., 2006. Accumulation of cadmium and its effects on growth, development and haemolymph biochemical compositions in *Boettcherisca peregrina* larvae (Diptera: Sarcophagidae). Insect Science 13: 31-39.

Wu, N., Wang, X., Xu, X., Cai, R. and Xie, S., 2020. Effects of heavy metals on the bioaccumulation, excretion and gut microbiome of black soldier fly larvae (*Hermetia illucens*). Ecotoxicology and Environmental Safety 192: 110323.

Xia, Z., Wu, S., Pan, S. and Kim, J.M., 2012. Nutritional evaluation of protein from *Clanis bilineata* (Lepidoptera), an edible insect. Journal of the Science of Food and Agriculture 92: 1479-1482.

Yang, L.F., Siriamornpun, S. and Li, D., 2006. Polyunsaturated fatty acid content of edible insects in Thailand. Journal of Food Lipids 13: 277-285.

Yi, L., Lakemond, C.M.M., Sagis, L.M.C., Eisner-Schadler, V., Van Huis, A. and Van Boekel, M.A.J.S., 2013. Extraction and characterisation of protein fractions from five insect species. Food Chemistry 141: 3341-3348.

Zhang, Z., Song, X., Wang, Q. and Lu, X., 2012. Cd and Pb contents in soil, plants, and grasshoppers along a pollution gradient in Huludao City, Northeast China. Biological Trace Element Research 145: 403-410.

Zhang, Z.-S., Lu, X.-G., Wang, Q.-C. and Zheng, D.-M., 2009. Mercury, cadmium and lead biogeochemistry in the soil-plant-insect system in Huludao City. Bulletin of environmental contamination and toxicology 83: 255.

Zhou, J.R. and Erdman, J.W., 1995. Phytic acid in health and disease. Critical Reviews in Food Science and Nutrition 35: 495-508.

Zhou, X., Honek, A., Powell, W. and Carter, N., 1995. Variations in body length, weight, fat

content and survival in *Coccinella septempunctata* at different hibernation sites. Entomologia Experimentalis et Applicata 75: 99-107.

Zhuang, P., Huiling, Z. and Wensheng, S., 2009. Biotransfer of heavy metals along a soil-
plant-insect-chicken food chain: field study. Journal of Environmental Sciences 21: 849-
853.

Genetic and genomic selection in insects as food and feed

T. Eriksson[1,2] and C.J. Picard[1]*

[1]Department of Biology, Indiana University, Purdue University Indianapolis, 723 W. Michigan Street, Indianapolis, IN 46202, USA; [2]Beta Hatch Inc., 200 Titchenal Road, Cashmere, WA 98815, USA; *cpicard@iupui.edu

Abstract

This review will summarise existing tools and resources and highlight areas of focus for the insects as food and feed industry for the production of insects as alternative protein sources. By applying knowledge gained from other agricultural organisms coupled with the ease of insect population growth and rearing capabilities, and the increase in biotechnological advances, strains optimised for various economic and biological traits should be one of the most attainable goals for researchers and insect farmers alike. We have reviewed strengths (and weaknesses) of various genetic and genomic approaches, and consider the future of insect farming in the context of genetic and genomic selection of insects.

Keywords

yellow mealworm – black soldier fly – BSF – honey bee – silkworm – crickets – genetic selection – genomics

1   Introduction

The first animal genome belongs to the fruit fly, *Drosophila melanogaster,* published in 2000 (Adams *et al.,* 2000), and human genome drafts followed suite in 2001 after nearly 2 decades of sequencing (Lander *et al.,* 2001; Venter *et al.,* 2001). Sequencing technologies have advanced rapidly since, significantly lowering cost and increasing coverage. Indeed, here we are in 2020 and one can generate an animal genome of sufficient quality that fits on a USB flash drive plugged into a standard laptop in less than a week (Fologea *et al.,* 2005). Furthermore, along with the progress in sequencing technology, advances in software for the biologists have enabled those without an informatics background to assemble, annotate and analyse genomes to answer basic biological questions (www.digitalinsights.qiagen.com). Together these innovations have revealed new avenues for industries whose business models benefit substantially from optimising traits of organisms (such as those in the agri- and aquacultural spaces), by increasing the speed and precision of optimisation. In some cases this means creating transgenic organisms by introducing the DNA sequence of one species into another to confer, for example, insect resistance on *Bt* corn (EPA, 1995). A more common, tried-and-true method of trait enhancement relies on selectively breeding conspecific individuals carrying desirable genetic and phenotypic profiles to produce offspring with desirable traits. This practice has been carried out for a very long time using traditional breeding strategies (e.g. breed males to multiple females and observe phenotypes, and reciprocal cross- ing), but can be time-consuming, labour-intensive, and sometimes unsuccess- ful without knowledge of the underlying traits' genetics. With next-generation biotechnologies, the process can be accelerated and more efficient, since geno- type-phenotype relationships can be more readily identified and can be used to formulate more optimal artificial selection and crossing schemes, reducing false leads and the number of generations required to increase the frequency of opti- mal phenotypes.

Selective breeding is not only useful to traditional crops, livestock and other ag- ricultural products, but will also be an important tool used for the optimisation of insects as alternative protein sources (Van Huis, 2013; Van Huis *et al.*, 2015). In this context, the main differences between traditional agri- and aquacultural animals, most of which are vertebrates, and insects, lie in the mode of reproduction. Most insects evolved to produce many offspring quickly but each with little chance of survival (i.e. the r-selected strategy) whereas many vertebrates produce small num- bers of more expensive offspring with typically some parental care ensuring their increased survival (i.e. the K-selected strategy) (Pianka, 1970). Therefore, artificial selection and selective breeding efforts can be done more efficiently in insects. However, it should be noted that the diversity of insect sex determination mecha- nisms is far more diverse than across vertebrate systems (Bachtrog *et al.*, 2014), and may have an influence on insect management.

The modern trait-optimisation workflow may involve genome-wide association studies (GWAS) in which loci are identified as being linked to economically im- portant traits (e.g. fecundity, stress resiliency, or growth rates), followed by genetic modifications that introduce existing DNA sequences from conspecifics (or het- erospecifics) into a controlled baseline genetic background, with the overall goal of reducing the number of undesirable effects (e.g. reduced fitness) while optimising the desirable ones (Box 1). This review will examine known cases of artificial se- lection and selective breeding in insects, outline strengths (and weaknesses) sur- rounding these methods, and discuss what we see as the future for insects as food and feed.

2   Optimisation via artificial selection and selective breeding

Similar to its natural counterpart, artificial selection operates on standing heritable genotypic variation in a population in favour of individuals with phenotypes deemed desirable by the human selector. Once identified, the selected individuals are allowed to mate, over successive generations fixing the associated alleles and resulting in strains with particular phenotypes (Box 2). In free-ranging insects such as the honey bee *Apis mellifera*, natural populations serve as reservoirs for variation. Regional variability is well described in many subspecies (also called races or ecotypes) native to different parts of the world (Alqarni *et al.*, 2011; Engel, 1999) which exhibit distinct characteristics including coloration, morphometry (Alpatov, 1929; Alqarni *et al.*, 2011), venom biochemistry (Palma and Brochetto-Braga, 1993), and behaviour (Alqarni *et al.*, 2011; Winston and Katz, 1982), with new variation still being discovered (Sheppard and Meixner, 2003). Managed colonies require open ranges to forage and therefore can interact and intermix with feral hives, and may lead to a reduction in the genetic diversity of wild populations (Meixner *et al.*, 2010).

Box 1: Genetically modified organisms

It is important to note some important differences in some terms used when discussing genetic selection. Genetically modified organisms (GMOs) is a broad term that describes an organism that has been genetically modified using technology in which would not have happened naturally. This broad term represents a wide variety of modifications, some of which are related to genetic selection. More specific are the terms transgenesis and cisgenesis. Transgenesis refers to the practice of using genes or other genetic loci from another organism, and this could not happen naturally as the organisms being considered are heterospecific with divergent evolutionary histories and therefore not sexually compatible. An example of transgenesis is the engineering of foods we eat such as corn (maize). The classic example is *Bt* corn, a strain of corn that contains a bacterial gene that confers insecticide resistance, or corn with increased nutritional value by expressing a common bacterial gene that produces methionine, an essential amino acid.

The alternative to this is cisgenesis, in which a gene is introduced using genetic engineering methods into an organism, but that gene comes from a sexually compatible organism (for example, the same species or a closely-related species). In this situation, the gene, in theory, could be introduced via mating and introgression, but it is unlikely to happen naturally perhaps due to geographical barriers preventing reproductively compatible populations from intermixing. Examples of this include taking a gene or a non-coding regulatory element (e.g. promoter and enhancer) from the same organism by using genetic engineering to introduce it by itself into a background which contains a variety of desirable traits. Currently, a lot of controversy surrounds the use of genetically modified organisms, mostly related to transgenetics, and many of the regulations are based on the process (how genes are engineered) rather than the type of gene used (gene function and species of origin).

In each and every case, careful considerations of local regulations, the potential for environmental spread through the sale of live insects, and consumer acceptance will play a role in the selection of technologies and strategies outlined in this review. In many cases, genetic modifications can play an useful role in functional characterisations important for an overall understanding of the insects' biology.

Box 2: Strains

Historically in entomology, a strain generally refers to populations, wild or captive, that possess some stable, consistent characteristic over multiple generations that enable distinguishing one population from another (Carpenter and Bloem, 2002; Russo *et al.*, 2001; Tabashnik *et al.*, 2009; Zatsepina *et al.*, 2001), such as susceptibility to insecticides (Huang *et al.*, 2004), stress resistance (Force *et al.*, 1995), aggression (Alaux *et al.*, 2009) and sex-specific sterility (Heinrich and Scott, 2000). This usage is common, for example, in the literature of insect pests (Argentine *et al.*, 1992; Kuno, 2010; Liao *et al.*, 2019; Liu *et al.*, 2004; Mertz, 1975; Pashley, 1988), *Drosophila* fruit flies (Arking *et al.*, 2002; Cingolani *et al.*, 2012; Seong *et al.*, 2019; Zatsepina *et al.*, 2001), and the silkworm *Bombyx mori* (Murakami and Ohtsuki, 1989; Ruiz and Almanza, 2018; Zhou *et al.*, 2008). Often strain, line, and breed are interchangeably used (Furdui *et al.*, 2014; Seong *et al.*, 2019) but occasionally distinctions are made. Another commonly used convention relates to colony, which can be a strain, line or breed, but is typically formed as an off-shoot from a wild or field-collected sample or from other colonies already established. For the purposes of this review, we will use strain as the dominant term referring to unique population, naturally or artificially selected for.

On the other hand, mating of parasitoids, the black soldier fly *Hermetia illucens*, and the house cricket *Acheta domesticus* can be done in closed spaces, thus preventing outcrossing with wild populations. This capability minimises pathogen exposure and enables controlled breeding, but promotes mating among close kin which can lead to inbreeding depression. Nonetheless, genetic depauperation may be ameliorated by starting a colony with sufficiently high effective population size, and occasional outcrossing with wildtypes or stocks from different sources. For instance in *H. illucens*, human-assisted dispersal has broaden the natural distribution significantly (Marshall *et al.*, 2015), with cultivated populations reared in factories and laboratories worldwide (Dzepe *et al.*, 2020; Kenis *et al.*, 2018; Sheppard *et al.*, 2002; Zhou *et al.*, 2013). Phenotypic divergence has been observed between laboratory strains (Zhou *et al.*, 2013), likely driven by unintentional selection and local adaptation. Although challenging, it is critical to strike a balance between minimising the genetic load and preserving selected phenotypes in established strains, while also preventing modified organisms from escaping into the local ecosystem. The latter point is an obvious concern with small, mobile, and highly fecund insects such as those mentioned above.

The domesticated silkworm *Bombyx mori*, arguably the only truly domesticated insect, originated from the wild silkworm *Bombyx mandarina* through to a single domestication event in China with little subsequent gene flow between wild ancestors and the domesticated stock (Arunkumar *et al.*, 2006; Cheng *et al.*, 2015; Xia *et al.*, 2009). Therefore, wild *B. mori* does not exist, and consequently standing genetic variation exists solely among commercial and laboratory strains. Interestingly, the initial domestication bottleneck did not significantly reduce standing genetic variation in *B. mori* (Xia *et al.*, 2009), enabling subsequent successful artificial selection. Today, existing strains number in the thousands (Jingade *et al.*, 2011). A similar situation exists for the yellow mealworm *Tenebrio molitor*. A known pest of cereal stores, this mealworm species has evolved in close association with humans and presumably no truly wild population currently exists. It has been a popular feed species mostly for exotic pets and therefore produced in large number by the pet feed industry. *T. molitor* is also widely reared with populations maintained by hobbyists and at many laboratories (Morales-Ramos *et al.*, 2019).

3   Improving traits for the production of insects: the old way

Principles of natural selection can be applied to artificially change phenotypic frequencies in insect populations. In the context of insect production, first, we may wish to make a distinction between phenotypes associated with economic and fitness traits. Economic traits are defined as those directly linked to the intrinsic commercial value of the insect (e.g. production rate of silk and honey). Fitness traits such as fecundity, immunity, and environmental tolerance affect how robust an insect is at surviving and reproducing under artificial conditions. Optimising an economic trait may be at the expense of a fitness trait, and vice versa.

Even without knowledge of the underlying genetic landscape, desirable traits can be intentionally propagated by selective breeding of individuals with desirable traits, such as what was done in early agricultural systems with livestock, poultry and crops. Most traits under selection are continuous (e.g. weight, number of eggs produced, offspring size) with a range of possible phenotypes expressed in a population. Hybridisation, the mating of individuals from distinct genetic backgrounds, is often used to bring together naturally arisen variation from reproductively isolated populations, to create novel genetic landscapes and interactions that may extend the range of selectable phenotypic variation. As above, this method (introducing genetically variable individuals to induce hybrid vigour) is also at the mercy of the unintended consequences for the possibility of the introduction of potentially positive, but also, detrimental phenotypes.

The typical aim of selection programs is to increase the prevalence of phenotypes that would increase the quantity or quality of the commercial product (e.g. amount of silk per cocoon or honey per hive, neatness and reelability of silk threads, etc.), decrease in the amount time to harvest (e.g. growth rate, development time, voltinism), or promote the stability and health in artificial rearing environments (e.g. disease and parasite resistance, temperature tolerance, cannibalism reduction). In other words, it is desirable to optimise economic traits as much as possible without sacrificing fitness. Principles of genetic improvements and domestication of cultivated insects have been reviewed elsewhere (Gregory, 2009; Hoy, 1976). Below we highlight outstanding examples in insect-rearing history.

For millennia, humans have cultivated insects in large numbers for a wide range of valuable products and services they provide. The best documented and most widespread practices are the sericulture of *B. mori* for silk threads and the apiculture of *A. mellifera* for honey and crop pollination, both of which date back to about 5,000 years ago (Alqarni *et al.*, 2011; Ruiz and Almanza, 2018). Also significant is the century-old biological pest-control industry (Mackauer, 1972; McGugan and Coppel, 1962; Turnbull and Chant, 1961; Van Den Bosch and Messenger, 1973) which grow natural enemies (e.g. the parasitoid wasps *Trichogramma* sp.) to control populations of pest insects (Greany *et al.*, 1984). More recently, the insect as feed and food industry has emerged and is rapidly growing, producing insects for livestock, poultry, aquaculture and pets, and to some extent, human food (Makkar *et al.*, 2014; Van Huis, 2013). The three insects most widely produced and commercialised for these purposes are *H. illucens*, *T. molitor*, and *A. domesticus*.

3.1   Sericulture

In *B. mori*, the economic value of silk threads, spun as cocoons by larvae in preparation for pupation, has long directed artificial selection efforts. As previously mentioned, the balance between economic and fitness traits is important, and both were optimised. For example, while economic traits important for silk productions were under extensive selection (weight of cocoons, cocoon's raw silk content, reelability, and length, size, and neatness of silk filament (Datta *et al.*, 2001; Kumar *et al.*, 1995; Shekar and Basavaraja, 2008)), traits related to domestication and fitness were also in focus (e.g. flightlessness, synchronous hatching, number of annual generations, larval duration, survival and hardiness, pupation rate, egg production (Cheng *et* al., 2015; Moorthy et al., 2007; Murakami and Ohtsuki, 1989; Pradeep et al., 2005; Shekar and Basavaraja, 2008)). Because many of these traits are polygenic, quantitative and interact via epistasis, completely disentangling one from the others is often challenging. Furthermore, the environment can impact the economics of cultivating silkworm. Environmental tolerance is an emerging need to expand the reach of this valuable insect to communities in climatically diverse localities and to cope with global climate change, possibly resulting in improved hybrids (Datta et al., 2001; Shekar and Basavaraja, 2008). In modern sericulture, selective breeding and assessment of outcomes are highly quantitative, informed by both molecular and phenotypic data combined with multivariate statistical analyses and indices (Datta et al., 2001; Hasan et al., 2011; Mano, 1993; Mirhosseini et al., 2005; Moorthy et al., 2007; Shekar and Basavaraja, 2008). In fact, many reviews have been published related to modern methods for optimising silkworm breeding and production including artificial selection history (Nagaraju et al., 1996; Neshagaran Hemmatabadi et al., 2016), genetics (Jingade et al., 2011; Mizoguchi and Okamoto, 2013; Tanaka, 1953), silk properties (Mondal et al., 2007), diseases and immunity (Bhat et al., 2009; Jing et al., 2013), trait heritability (Singh et al., 2011; Tanaka, 1953), rearing conditions (Rahmathulla, 2012), diet (Kanafi et al., 2007), and molecular resources (Goldsmith et al., 2005; Zhou et al., 2008).

3.2   *Apiculture*

In addition to generating valuable hive products including honey and wax, *A. mellifera* plays a critical role in crop pollination and food security (Carreck and Williams, 1988; Klein et al., 2007; Southwick and Southwick Jr, 1992). The honey bee's economic importance and unusual genetic sex determination system have spurred basic research on many traits, many of which have been shown to be variable and heritable including parasite defence (Moritz, 1985), queen mating frequency (Kraus et al., 2005), foraging behaviour (Page Jr et al., 2000), honey production and alarm response (Collins et al., 1984), body size (Oldroyd et al., 1991), and pollen preference (Basualdo et al., 2007). A variety of genes and their functions and correlations to traits are characterised (Alaux et al., 2009; Kerr et al., 2010; Scott Schneider et al., 2004), including those associated with pollen hoarding (Hunt et al., 1995), hygiene (Lapidge et al., 2002), parasite defence (Spötter et al., 2016), and alarm behaviour (Moritz and Southwick, 1987). To improve their colonies, apiculturists and breeders have imposed selection on both economic traits including honey production (Guzmán-Novoa and Page Jr, 1999) and pollen preference (Basualdo et al., 2007; Nye and Mackensen, 1968; 1970) as well as vitality traits including climate tolerance (Alqarni et al., 2011), hygiene behaviours (Perez-Sato et al., 2009), parasite tolerance (Huang et al., 2014), pollen hoarding (Page and Fondrk, 1995). Comprehensive, standardised protocols for queen care, trait selection, and selective breeding are well established (Büchler et al., 2013; Uzunov et al., 2017).

3.3   *Biological pest control*

Populations of pest insects can be controlled with systematic releases of their natural enemies or via the release of sterile individuals (Greany *et al.*, 1984; Sørensen *et
al.*, 2012; Van Lenteren and Nicoli, 2004). Examples of natural enemies include the
chalcid wasps *Dahlbominus fuscipennis* (Wilkes, 1942), *Aphytis lingnanensis* (White
et al., 1970), and *Trichogramma* sp. (Sorati *et al.*, 1996), all of which are parasitoids.
Insects reared for sterile release include the melon fly *Bactrocera cucurbitae* (Miyatake, 2006) and the New World screwworm *Cochliomyia hominivorax* (Krafsur,
1998). The economic value of these insects is tied with quantity which is strongly influenced by their ability to survive and reproduce in artificial rearing environment.
Therefore, artificial selection has focused on improving life-history traits including
development rate (Miyatake, 2006), sex ratio (Simmonds, 1947; Wilkes, 1947), and
female fecundity (Wilkes, 1947). For example, a laboratory selection for productive
females resulted in a 40% increase in female productivity, and thus ensuing in a
much greater number of offspring generated without having to invest more on materials and labour, just by removing less productive females (Wilkes, 1947). Similar
to the silkworm, environmental tolerance is important for pest-control insects, resulted in selection for temperature tolerance (White *et al.*, 1970; Wilkes, 1942; 1947)
and insecticide resistance (Wilkes *et al.*, 1952). Cultivated strains were also under
selection to maintain normal foraging behaviours and vitality upon field release
(Boller, 1972; Van Lenteren and Nicoli, 2004).

3.4   *Entomoculture*

Unlike the aforementioned established industries, growing insects as feed and
food alternatives is still in its infancy. Recent focused reviews have evaluated the
nutritional content and suitability for animal feed (Makkar *et al.*, 2014) as well as
challenges to mass-rearing, genetic improvements and management practices (Jensen *et al.*, 2017). Though many insects are reared worldwide for these purposes, the
more promising species include the black soldier fly *H. illucens*, the yellow mealworm *T. molitor*, and the house cricket *A. domesticus* have been attracting the most
attention and interest. Other noteworthy species include the oak moth *Antherae
pernyi* (Li *et al.*, 2017, 2020), the house fly *Musca domestica* (Elahi *et al.*, 2020; Hall
et al., 2018), and the palm weevil *Rhynchophorus ferrugineus* (Chinarak *et al.*, 2020).

In the last few decades, the black soldier fly's potential as a feed alternative for
livestock and aquaculture, and a processor of organic waste has been recognised
and promoted (Newton *et al.*, 1977; Sheppard, 1983; Sheppard *et al.*, 1994; Wang and
Shelomi, 2017). The latter is being actively explored for manure management (Rehman *et al.*, 2017) and biofuel conversion (Elsayed *et al.*, 2020). Life history (Booth
and Sheppard, 1984; Cammack and Tomberlin, 2017; Furman *et al.*, 1959; Zhou *et
al.*, 2013), distribution and diversity (Park *et al.*, 2017; Stahls *et al.*, 2020), mating
behaviours (Giunti *et al.*, 2018), and response to variable rearing conditions (Dzepe
et al., 2020; Hoc *et al.*, 2019; Lalander *et al.*, 2019; Meneguz *et al.*, 2018; Rhode *et al.*,

2020; Tschirner and Simon, 2015; Zhou *et al.*, 2013), gut endosymbiont (Bruno *et al.*, 2019; Cifuentes *et al.*, 2020; Varotto Boccazzi *et al.*, 2017), nutritional values (Makkar *et al.*, 2014) and livestock feeding on these flies (Newton *et al.*, 1977; Rimoldi *et al.*, 2019; Schiavone *et al.*, 2017) have all been investigated. Innovations enabling scaling up and industrialising *H. illucens* rearing are being rapidly developed (Liu *et al.*, 2020; Marien *et al.*, 2018; Zhan *et al.*, 2020). Although wild founders can start captive colonies, rapid collapse occurred within several generations, an outcome attributed to inbreeding (Rhode *et al.*, 2020). Rapid inbreeding depression can greatly hinder selective mating following artificial selection, and consequently genetic improvement of traits, and although improvement efforts are undoubted underway, so far none has been published.

Another species being mass produced as feed is the yellow mealworm *T. molitor*. Its usage began with the pet feed industry as early as late 1900's (Martin *et al.*, 1976), and more recently being seriously considered as a viable alternative source of protein and lipid for livestock and aquaculture (Choi *et al.*, 2018; Henry *et al.*, 2018; Oonincx and De Boer, 2012; Veldkamp and Bosch, 2015) and for food as it is currently a part of the human diet in Africa, Asia, Australia and the Americas. In addition, yellow mealworms can also degrade mycotoxins present in crops (Van Broekhoven *et al.*, 2017) and plastics (Yang *et al.*, 2018a,b), and produce nitrogen-rich frass (Poveda *et al.*, 2019), greatly increasing their versatility and commercial value. Life cycle, effects of environmental conditions and diet on development and fitness traits including reproduction and immunity (McConnell and Judge, 2018; Ribeiro *et al.*, 2018; Vigneron *et al.*, 2019), and mating behaviours (Worden and Parker, 2001) have been studied. Experimental evolution experiments successfully used artificial selection and selective breeding to improve important economic and vitality traits including body size, body weight, growth rate (Leclercq, 1963; Morales-Ramos *et al.*, 2019), fecundity and food conversion efficiency (Morales-Ramos *et al.*, 2019), and improved immunity (Armitage and Siva-Jothy, 2005). Hybridisation technique was successfully used to study segmentation patterns (Hein, 1924).

Like the yellow mealworm, the house cricket *A. domesticus* has been sold commercially as pet feed (Nakagaki *et al.*, 1987). *A. domesticus* is also produced for human consumption in some parts of the world (Nakagaki and DeFoliart, 1991), and maintained in laboratories as a research model (Wilson *et al.*, 2010). Along with other edible crickets, this species is being produced by tens of thousands of growers around the world (Halloran *et al.*, 2016; Wilkie, 2018). Risk profile has been investigated (Fernandez-Cassi *et al.*, 2019). Mass rearing protocols are developed, some capable of producing 6,000 individuals daily (Parajulee *et al.*, 1993). Some cooperatives can produce upward of 700-800 kg daily (Halloran *et al.*, 2016). Studies have explored the effects of diet and abiotic conditions on the efficiency of farming *A. domesticus* at different scales (Clifford and Woodring, 1990; Collavo *et al.*, 2005; Orinda *et al.*, 2017). Experimental selection has produced crickets with larger body sizes, a heritable trait positively correlated with immune function (Ryder and Siva-

Jothy, 2001), by manipulating size of food parcels for 10 generations (Tennis, 1985). Body size was studied in great detail in the context of mate choice and sexual selection (Castillo, 2005). Strains with increased longevity can also be selectively bred, resulting in long-living individuals more resilient to oxidative stresses induced by environmental toxins (Flasz *et al.*, 2020). Natural trade-offs between immunity and fitness traits have been characterised in *A. domesticus* (Bascunan-Garcia *et al.*, 2010) and other members of the subfamily Gryllinae (Kerr *et al.*, 2010). However, we found no published record of trait improvements in the context of mass production.

3.5   *Trait trade-offs*

Generations of selection can lead to negative correlations between traits, or trade-offs, such as among life history characteristics (Stearns, 2000), and between immunity and reproduction (Schwenke *et al.*, 2016). Evolutionary trade-offs in natural populations may be viewed as a complex interplay of reproductive success optimisation, lineage-specific constraints, density and frequency dependent selection, and short-term population-specific dynamics (Stearns, 2000). These forces are also present in production populations of cultivated insects. Selective sweeps may be a possible mechanism that generate negative trait correlations, where strong directional selection on a desirable trait brings along linked, deleterious alleles (Berry *et al.*, 1991; Smith and Haigh, 1974) that in a closed, inbreeding populations manifest as undesirable phenotypes in other traits.

In mass-produced insects, trade-offs have been observed between economic traits and fitness traits, as well as among them (Neumann and Blacquiere, 2017). In *B. mori*, strains with better survival tend to be less productive (Datta *et al.*, 2001; Shekar and Basavaraja, 2008) and there is a negative correlation between cocoon weight and shell percentage (Mirhosseini *et al.*, 2005). In *A. mellifera*, selection for increased pollen hoarding resulted in workers with smaller bodies (Page and Fondrk, 1995) and larvae with increased sensitivity to the nutritional environment during development (Linksvayer *et al.*, 2011). Furthermore, there is evidence that hybrid bees are less fit than their pure-bred parents, afflicted with reduced foraging efficiency (Quezada-Euán *et al.*, 1996), lower metabolic rates, and increased wing shape asymmetry (Scott Schneider *et al.*, 2004). The importance of considering traits influencing bee colony fitness and population stability, and not just those immediately linked to economic values, has been advocated (Meixner *et al.*, 2010). In fact, commercial values are very much dependent on long term stability of managed populations. For parasites reared for biological control, trade-offs can occur between longevity and fecundity (Miyatake, 2006; Nagarkatti and Nagaraja, 1978). Mass-rearing conditions may lead to degradation of host acceptance (Van Bergeijk *et al.*, 1989) perhaps due to inadvertent adaptation to rearing diets and conditions. However, host acceptance can be rescued by means of artificial selection (Kölliker-Ott *et al.*, 2003). In *T. molitor*, selection for increased pupal weight unintentionally resulted in reduced larval survival in one example, ultimately impacting total biomass production (Leclercq, 1963; Morales-Ramos *et al.*, 2019). In the absence of directed artificial selection, local adaptation by natural selection can maintain life history variation and phenotypic plasticity between strains (Urs and Hopkins, 1973).

3.6   *Inbreeding*

Inbreeding depression is a longstanding problem for many closed populations of insects, a consequence of mating among close relatives that is exacerbated in captive populations founded from a small effective population of founders. Prolonged inbreeding in just about any species quickly results in decreased genetic variation among the progeny (Jingade *et al.*, 2011; Rhode *et al.*, 2020).

In the honey bee, the negative impacts of management on colony fitness are of top concerns among apiculturists, and the positive effect of genetic diversity on colony wellbeing is well documented (Seeley and Tarpy, 2007). In some cases, loss of genetic variation is exacerbated by commercial queen-breeding practices that produce large number of offspring from relatively few matrilines, and the persistent non-random mating to maintain pure lines (Meixner *et al.*, 2010), however, in this particular case, it may not be due entirely to management but rather natural bottlenecks caused by out-of-Africa expansions (Harpur *et al.*, 2012). Therefore, protocols for queen selection and mating techniques have been refined to minimise depletion of genetic variation (Büchler *et al.*, 2013).

A similar situation is observed among biological control insects. Inbred lines of a chalcid parasite were less long-lived, likely due to inbreeding depression, but survival in females could be improved with selection on field stock (Wilkes, 1947). Inbreeding depression can take time to manifest, for instance, no negative changes in female fecundity, male mating success, and morphometry (Sorati *et al.*, 1996) were observed over four generations of sibling mating in a *Trichogramma* parasitic wasp.

4   Improving traits for the production of insects: the new way

4.1   *Genetic selection*

Selection has long been used in the agriculture space. In the early days, this was based on the inference of heritability based on pedigrees and breeding programs (for example, Mendel's peas; Mendel, 1865), in which the actual genes responsible for the phenotypes were not known. This process works well, albeit, it is very slow to optimise, especially when the traits can only be measured after several months or years, especially true in domesticated agricultural animals. Furthermore, due to the possible presence of other genes which may not be entirely desirable to transmit, the animal may optimise in one phenotype to the detriment of another.

To complicate matters further for selection methods, many traits are not monogenetic, expressed as the result of more than one gene. Monogenetic traits are probably uncommon in nature, but some have been discovered. A clear example is hair-lessness in certain dog breeds, attributed to the transcription factor FOXI3 which is involved in the expression of hair and teeth (Drogemuller *et al.*, 2008). More commonly are quantitative or complex traits, in which many genes are involved, and each individual gene has a small effect on the phenotype. For example, milk protein yield in a breed of cattle is associated with 144 single nucleotide polymorphisms (SNPs) (Daetwyler *et al.*, 2008), and multiple quantitative trait loci (QTL) affecting cocoon traits and ectoparasite defence have been identified in the silkworm and honey bee, respectively (Behrens *et al.*, 2011; Lu *et al.*, 2004). Additionally, the environment can have an impact, with a continuum of phenotypic responses across individuals.

Genetic selection is using genetic tools to identify genes or genetic loci that are associated with traits, then selectively cross those individuals with the appropriate genetics to produce subsequent generations. However, most traits are quantitative or complex traits, meaning multiple genes/genetic loci are responsible for producing the continuous variability we see (e.g. milk yield in dairy cows or clutch size in birds). The ability to identify/characterise the genes that produce phenotypic changes and that are heritable forms the basis of genetic selection. Unfortunately, there are many challenges to identifying these genes that would allow us to make inferences about the heritability of a particular trait (i.e. how much of the variation is attributable to the genes being studied), in part, many of the effects that are seen in phenotypes due to a single gene is a small effect, it is only the additive genetic effect that produces a large effect.

4.2   *Tools and resources*

Next-generation sequencing technologies have been leveraged for many cultivated insects to assemble nuclear genomes (The Honeybee Genome Sequencing Consortium, 2006; The International Silkworm Genome Consortium, 2008; Ferguson *et al.*, 2020; Xia *et al.*, 2009; Zhan *et al.*, 2020), transcriptomes (Cheng *et al.*, 2015; Liu *et al.*, 2015; Oppert *et al.*, 2020; Zhan *et al.*, 2020; Zhu *et al.*, 2019), and microbiomes (Jung *et al.*, 2014; Zhan *et al.*, 2020).

In addition, an increasing number of individual genomic loci have been characterised, including genetic markers (Kumar *et al.*, 2009; Solignac *et al.*, 2003), coding genes and proteins (Gao *et al.*, 2019; Giannetto *et al.*, 2017; Lee *et al.*, 2016; Park *et al.*, 2015). As an example, for *T. molitor*, much research has been done in characterising antimicrobial peptides, immunity-related genes, and their regulation (Jo *et al.*, 2017; Johnston *et al.*, 2014; Keshavarz *et al.*, 2020; Kim *et al.*, 1998; Lee *et al.*, 1996; Moon *et al.*, 1994). Other subjects of investigations include peptidases (Oppert *et al.*, 2012; Prabhakar *et al.*, 2007), insulin (Sevala *et al.*, 1993), and microsatellites (Petitpierre *et al.*, 1988).

Multiple strategies exist for applying next-generation sequencing technologies in different contexts. For example, a GWAS as discussed above is typically done using pre-designed arrays that allow for the quick and efficient genotyping of millions of SNPs, but this only works if there is an existing array. An alternative to discovering genome-wide SNPs across the genome is to genotyping by sequencing, wherein the genomic DNA is typically fragmented using a restriction enzyme, and then sequenced using massively parallel sequencing technologies (Elshire *et al.*, 2011). Where genome-wide expression patterns are of interest, transcriptomes are a valuable resource. Transcriptomics refers to the generation and assembly of transcriptomes, or the coding part of the genome, that allow for the analysis of variation in expression across cell types and tissues or among organisms and strains (e.g. see if multiple genes are changing in response to a stimulus, how gene expression varies between selected strains, etc.). Transcriptomes tend to be less expensive and easier to assemble relative to genomes, as they are typically less complex, are less repetitive, and reference genes typically exist. In some situations, metagenomics, a growing field that studies the environments' genetic material most often attributed to microbial and other microbiological material, is also relevant.

These resources are modernising genetic improvements, where genomic DNA can be modified precisely to bring together desirable genetic variances while preventing hitchhiking of deleterious alleles, or to design transgenic organisms more efficiently (Heinrich and Scott, 2000; Wen *et al.*, 2010). Editing methods relying on the CRISPR/Cas nuclease technology (Ran *et al.*, 2013) are being successfully developed for many arthropods (Sun *et al.*, 2017) including *A. mellifera* (Hu *et al.*, 2019; Kohno *et al.*, 2016), *B. mori* (Ma *et al.*, 2014; Wei *et al.*, 2014; Xu *et al.*, 2019), *H. illucens* (Zhan *et al.*, 2020), and *Tribolium casteneum*, a close relative of *T. molitor* (Adrianos *et al.*, 2018; Gilles *et al.*, 2015). Some examples of traits manipulated successfully in CRISPR/Cas in experiments include silk proteins (Kojima *et al.*, 2007; Xu *et al.*, 2018), sensory systems and behaviours (Liu *et al.*, 2017), and expression of recombinant proteins (Acharya *et al.*, 2002; Wen *et al.*, 2010) in *B. mori,* and larval duration, body size, and flightlessness in *H. illucens* (Zhan *et al.*, 2020).

5   Methods for genetic selection and engineering

There are generally two approaches to identifying genomic loci for genetic engineering and subsequent selection. They may be labelled inside-out and outside-in. In the former, you look at a targeted list of loci that may have some functional association already established in other, often model, species. For example, in the insect world, the *D. melanogaster* database (www.flybase.org) has a wealth of information on well-characterised genes that have been experimentally associated with a wide range of biological and molecular processes with homology in other insects. Researchers can leverage such a database as a starting point to identify loci that affect a trait of interest in their focal organisms. While relevant genes may actually be identified from knowledge of other organisms, the amount of phenotypic variation observed in the focal species that can be attributed to those identified genes may be small, because traits of interest are often polygenic and may have divergent underlying genetics between different species, therefore the accounting of effector loci is often incomplete.

Instead of focusing on a handful of genes with known functions, the alternative outside-in approach takes a bird-eye view of genomic variation to establish genotype-phenotype correlations in a large number of individuals displaying a range of phenotypes. Without a reference genome, researchers may choose to carry out a GWAS (see Figure 1 for a schematic diagram) to pinpoint alleles associate with desirable phenotypes. For example, there may be a variable position that is located in or near an effector locus, be it a gene or a non-coding regulatory element, and due to its close proximity and the lack of recombination it is an informative marker of an uncharacterised chromosomal location that contains a significant allele. Currently for the common insects reared as feed and food, nuclear genome assemblies are only available for the black soldier fly (Zhan *et al.*, 2020) and the yellow mealworm (Eriksson *et al.*, 2020). While this resource has yet to exist for the other important feed and food insects such as crickets and the palm weevil, basic research goals can still be accomplished using published genomes of closely related species. For example, prior to the completion of a *T. molitor* reference genome assembly in 2020, the *Tribolium castenum* genome (*Tribolium* Genome Sequencing Consortium *et al.*, 2008) was used to study the yellow mealworm in our laboratory.

Re-sequencing experiments become possible with a reference genome, where it can be used to map whole genome reads sequenced from multiple individuals exhibiting different phenotypes. Variances can then be called from the mappings, and their significance to traits can be statistically established from known phenotypic scores. Albeit more prone to false positives, the outside-in strategy enables the discovery of non-coding loci with regulatory functions (e.g. transcription start sites, enhancers, promoters, silencers) and novel taxon-specific genes (i.e. genes only found in a single species or group). This method would instead allow you to discover new genetic loci, especially those that strongly affect phenotypes, and gain a better understanding of the underlying genetic variation attributed to a given phenotype, and allow you to screen a large number of individuals for these specific alleles, and allowing for selective mating to occur.

A genome-editing project may benefit from a combination of both of the approaches discussed above. Armed with an understanding of individual genes' molecular functions and genotype-phenotype correlations among multiple strains, one can more accurately predict candidate loci linked to important traits that would cause appreciable phenotypic changes following genome editing.

5.1   *Challenges and drawbacks*
As with nearly all scientific innovation, there is a balance needed between the cost and the value. Spending money upfront to generate a large dataset could be advan-

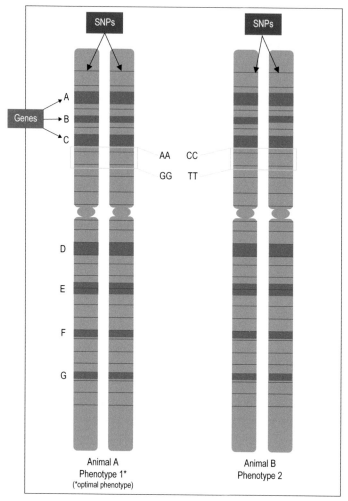

FIGURE 1   Simplified design of a genome association study, in which multiple locations (red lines:
single nucleotide polymorphisms (SNPs), represented here as a single chromosome)
randomly span the length of the chromosome. A genome-wide association study would
include multiple individuals of varying phenotypic levels, and then the association of
the genotypes per phenotype can be assigned. Often the SNPs are not located in genes,
but are physically close to genes, and inferences are possible.

tageous in the long run, as there would likely be little need to run additional samples if your existing dataset contains a lot of data. For re-sequencing and GWAS,
the number of samples is highly important. Although more costly, the greater the
number of unrelated individuals across the phenotypes of interest, the more power the method has to detect actual associated genotypes. Advances in sequencing
technology have made running larger number of samples more affordable, but the
DNA library preparation for sequencing is still currently costly.

The basic design of a GWAS is to sample animals with a range of phenotypes of interest, and then genotype those animals using SNP arrays or panels. Then, complex statistical analyses are performed to identify SNPs linked to the trait while reducing the false positive rate. Admixture is a source of bias in many GWAS, thus it is important to account for in your statistical model and to select appropriate samples. In correlational analyses, including individuals sharing recent ancestry (i.e. non-independent samples, such as siblings, parents and offspring) risks statistically biasing an association between a trait and a marker. For insect populations, sampling a group of sufficiently unrelated individuals can be challenging given managed insects' high reproductive outputs and lack of outcrossing opportunities, with some insects more challenging than others. For example, all extant yellow mealworm populations are in captivity, and to avoid sampling closely related individuals a GWAS would need to obtain specimens from different source populations (e.g. commercial suppliers, laboratories). This challenge also highlights the necessity of keeping detailed pedigrees and estimating relatedness within and among strains. Other insects such as the black soldier fly have robust wild, geographical populations with different evolutionary histories, and therefore less likely to suffer from sampling bias. Another drawback of large-scale association studies is that they are often labour intensive and time consuming. In each case, researchers would need to establish separate laboratory populations using founders from different sources, and maintain them in conditions that would not skew the expression of the traits in question, measure a suite of phenotypes for several generations, and then genotype.

When designing an experiment, SNP density (i.e. how close the SNPs are to one another?) and location should be considered. If a SNP is too far away from the quantitative-trait locus of interest, then it would not be detected as 'linked' by the analysis. A dense SNP array usually improves the precision in detecting a linked SNP. This feature is sensitive to linkage disequilibrium (LD), the magnitude of which varies between species and even across regions based on genomic features as seen in the honey bee (Wallberg *et al.*, 2015). The sample size of a GWAS is determined in part by the effect size or, the proportion of the phenotype that can be explained by the genotype. A single gene with large, obvious effects on phenotypes is easily detectable in a small number of animals, but those with smaller effects require a larger sample to detect. This in turn is all dependent on heritability of the trait, the proportion of the variation explained by the QTL, and LD. In the end, if you have many markers, and many unrelated samples, then the discovery of informative SNPs is likely. Since the false discovery rate when using a large number of SNPs is very high and that the SNP showed up as significant is expected by chance, it is often recommended to validate the detected variants in additional, independent sets of samples.

Relative to vertebrates, there are inherent advantages to using r-selected insects for selection experiments, namely short generation time and high fecundity. Indeed, within a few years of artificial selection and selective breeding, a population

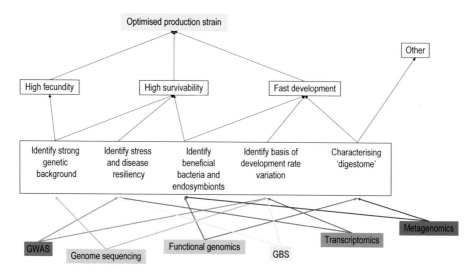

FIGURE 2   In this simplified and non-exhaustive example, the production of optimised strains can
encompass multiple areas (including some not included here), and can be determined
using many different genetic and genomic tools (GBS = genotype by sequencing; GWAS
= genome wide association study).

with phenotypic improvements could be generated. These conventional means
work but are vulnerable to the hitch-hiking of deleterious or undesirable alleles in
LD with desirable ones.

And finally, there are copy number and structural variants (large insertions or de-
letions, or inversions) to consider. These variants are much more difficult to detect
using traditional massively parallel sequencing platforms. However, technology has
improved to be able to address this. Long-read sequencing technology, offers by
company such as PacBio, can be used to bridge these gaps, but suffers from higher
error rates relative to short-read Illumina sequencing and requires computationally
expensive error corrections. Other companies like 10x Genomics takes advantage of
less error prone short reads and proprietary partitioning methods to assemble longer scaffolds, albeit they have discontinued this technology. And Hi-C makes use of
library preparation protocols that link long-range chromosomal segments together
to understand spatial organisation of a chromosome. Generally, 10X outperforms
(Srikanth *et al.*, 2020; Srivastava *et al.*, 2020) other currently available long read sequencing platforms. The first insect to be sequenced exclusively using this platform
includes the mealworm (Eriksson *et al.*, 2020).

5.2   *Future directions*

There is little doubt alternatives to conventional ranching and farming will soon be
needed to meet growing global demand for protein. Climate change, increased hu-

man population and antibiotic resistance are shifting the economics of raising traditional livestock and poultry with large carbon footprints. In response, the emerging insect as feed and food industry is rapidly scaling up capacity, taking advantage
of insects' natural abilities to mature rapidly, produce large number of offspring,
tolerate crowded rearing conditions, and consume a wide range of feed, including
waste. We predict that many innovations will be fuelled by next-generation molecular technologies, particularly with the CRISPR/Cas genome editing toolkit (Ran *et
al.*, 2013). This trend is already in motion in many of the insects discussed, and has
yielded promising results.

With generally two approaches available for the isolation and characterisation
of genetic variants linked to particular (desirable or not) phenotypes, one method
holds promise over the other. In selective breeding strategies that focus on improving specific phenotypes, linked genotypes may be quickly identified in subsequent
genomic analyses. The main issue with this method is that many genomic changes
could have taken place during selection, and it may be not able to isolate the causative genotypes (lots of genetic hitchhiking). An alternative approach is to use population genetics to identify adaptive genes that can then be correlated to the phenotypes. This requires large samples sizes across many phenotypes, and requires
native or wild populations to be variable across the phenotypic landscape, but it is
more likely to yield successful outcomes (Ross-Ibarra *et al.*, 2007).

Another aspect for consideration when assessing the functional characterisation
of any significant variants correlated to a phenotype would include an understanding of the impacts of the microbial communities associated with these insects. Endosymbionts, such as *Wolbachia*, are well-known to manipulate biological processes from cytoplasmic incompatibilities (Werren and Windsor, 2000; Werren *et al.*,
2008), contributions to degradation of waste (e.g. (Przemieniecki *et al.*, 2020), to
the potential nutritional mutualism (Nikoh *et al.*, 2014). There is very little research
of *Wolbachia* and other endosymbionts on the majority of the proposed insects
discussed here (crickets, mealworms and back soldier flies), albeit well-known in
other insects used for biocontrol, possibly an indication of its inexistence, or more
likely, this is an area of needed research for the continuation of an understanding
of the biology of the insects, from wild to farmed populations.

As an example of what insect farmers and researchers can look forward to, the
first reference cattle genome was available in 2009 (The Bovine Genome Sequencing and Analysis Consortium *et al.*, 2009), launching the practice of linking traits to
genotypes with SNPs identified that were unique to different breeds of cattle. The
1000 bull genome project (Hayes and Daetwyler, 2019) began as a means to quantify
the variation present in various cattle breeds (n = 121), and to date, there are 2,703
cattle genomes. With each subsequent analyses the group does by adding new individuals, the number of SNPs increases, and for species with a lot of variability that
number increases with each subsequent analysis. As opposed to species with little
variation, that number plateaus despite the addition of new genomes. Most SNPs

are intergenic variants (i.e. not in a gene) with intronic SNPs the second most common (Hayes and Daetwyler, 2019). This particular dataset demonstrates the need for additional genetic and genomic data to better understand the insects we wish to commercialise, while highlighting limitations with current agricultural processes. Genome editing may be able to address this problem, by selectively changing the genomic DNA sequence at single-nucleotide level, but without the basic research and knowledge to start from, it will remain to be seen. However, academic and the industry are armed with the best tools and resources to disrupt traditional agricultural processes in producing sustainable alternative protein sources.

Conflicts of interest

The authors declare no conflicts of interest.

References

Acharya, A., Sriram, S., Sehrawat, S., Rahman, M., Sehgal, D. and Gopinathan, K., 2002. *Bombyx mori* nucleopolyhedrovirus: molecular biology and biotechnological applications for large-scale synthesis of recombinant proteins. Current Science 83: 455-465.

Adams, M.D., Celniker, S.E., Holt, R.A., Evans, C.A., Gocayne, J.D., Amanatides, P.G., Scherer, S.E., Li, P.W., Hoskins, R.A., Galle, R.F., George, R.A., Lewis, S.E., Richards, S., Ashburner, *et al.*, 2000. The genome sequence of *Drosophila melanogaster*. Science 287: 2185-2195.

Adrianos, S., Lorenzen, M. and Oppert, B., 2018. Metabolic pathway interruption: CRISPR/Cas9-mediated knockout of tryptophan 2, 3-dioxygenase in *Tribolium castaneum*. Journal of Insect Physiology 107: 104-109.

Alaux, C., Sinha, S., Hasadsri, L., Hunt, G.J., Guzmán-Novoa, E., DeGrandi-Hoffman, G., Uribe-Rubio, J.L., Southey, B.R., Rodriguez-Zas, S. and Robinson, G.E., 2009. Honey bee aggression supports a link between gene regulation and behavioral evolution. Proceedings of the National Academy of Sciences 106: 15400-15405.

Alpatov, W., 1929. Biometrical studies on variation and races of the honey bee (*Apis mellifera* L.). The Quarterly Review of Biology 4: 1-58.

Alqarni, A.S., Hannan, M.A., Owayss, A.A. and Engel, M.S., 2011. The indigenous honey bees of Saudi Arabia (Hymenoptera, Apidae, *Apis mellifera jemenitica* Ruttner): their natural history and role in beekeeping. Zookeys 134: 83-98. https://doi.org/10.3897/zookeys.134.1677

Argentine, J.A., Clark, J.M. and Lin, H., 1992. Genetics and biochemical mechanisms of abamectin resistance in two isogenic strains of Colorado potato beetle. Pesticide biochemistry and physiology 44: 191-207.

Arking, R., Novoseltseva, J., Hwangbo, D.-S., Novoseltsev, V. and Lane, M., 2002. Different age-specific demographic profiles are generated in the same normal-lived *Drosophila* strain by different longevity stimuli. The Journals of Gerontology Series A: Biological Sciences and Medical Sciences 57: B390-B398.

Armitage, S.A. and Siva-Jothy, M.T., 2005. Immune function responds to selection for cuticular colour in *Tenebrio molitor*. Heredity 94: 650-656. https://doi.org/10.1038/sj.hdy.6800675

Arunkumar, K.P., Metta, M. and Nagaraju, J., 2006. Molecular phylogeny of silkmoths reveals the origin of domesticated silkmoth, *Bombyx mori* from Chinese *Bombyx mandarina* and paternal inheritance of *Antheraea proylei* mitochondrial DNA. Molecular Phylogenetics and Evolution 40: 419-427. https://doi.org/10.1016/j.ympev.2006.02.023

Bachtrog, D., Mank, J.E., Peichel, C.L., Kirkpatrick, M., Otto, S.P., Ashman, T.L., Hahn, M.W., Kitano, J., Mayrose, I., Ming, R., Perrin, N., Ross, L., Valenzuela, N. and Vamosi, J.C., 2014. Sex determination: why so many ways of doing it? PLoS Biology 12: e1001899. https://doi.org/10.1371/journal.pbio.1001899

Bascunan-Garcia, A.P., Lara, C. and Cordoba-Aguilar, A., 2010. Immune investment impairs growth, female reproduction and survival in the house cricket, *Acheta domesticus*. Journal of Insect Physiology 56: 204-211. https://doi.org/10.1016/j.jinsphys.2009.10.005

Basualdo, M., Rodríguez, E., Bedascarrasbure, E. and De Jong, D., 2007. Selection and estimation of the heritability of sunflower (*Helianthus annuus*) pollen collection behavior in *Apis mellifera* colonies. Genetics and Molecular Research 6: 374-381.

Behrens, D., Huang, Q., Gessner, C., Rosenkranz, P., Frey, E., Locke, B., Moritz, R.F. and Kraus, F.B., 2011. Three QTL in the honey bee *Apis mellifera* L. suppress reproduction of the parasitic mite *Varroa destructor*. Ecology and Evolution 1: 451-458. https://doi.org/10.1002/ece3.17

Berry, A.J., Ajioka, J. and Kreitman, M., 1991. Lack of polymorphism on the *Drosophila* fourth chromosome resulting from selection. Genetics 129: 1111-1117.

Bhat, S.A., Bashir, I. and Kamili, A.S., 2009. Microsporidiosis of silkworm, *Bombyx mori* L. (Lepidoptera-Bombycidae): a review. African Journal of Agricultural Research 4: 1519-1523.

Boller, E., 1972. Behavioral aspects of mass-rearing of insects. Entomophaga 17: 9-25.

Booth, D.C. and Sheppard, C., 1984. Oviposition of the black soldier fly, *Hermetia illucens* (Diptera: Stratiomyidae): eggs, masses, timing, and site characteristics. Environmental entomology 13: 421-423.

Bruno, D., Bonelli, M., De Filippis, F., Di Lelio, I., Tettamanti, G., Casartelli, M., Ercolini, D. and Caccia, S., 2019. The intestinal microbiota of *Hermetia illucens* larvae is affected by diet and shows a diverse composition in the different midgut regions. Applied and environmental microbiology 85: e01864-18. https://doi.org/10.1128/AEM.01864-18

Büchler, R., Andonov, S., Bienefeld, K., Costa, C., Hatjina, F., Kezic, N., Kryger, P., Spivak, M., Uzunov, A. and Wilde, J., 2013. Standard methods for rearing and selection of *Apis mellifera* queens. Journal of Apicultural Research 52: 1-30. https://doi.org/10.3896/ibra.1.52.1.07

Cammack, J.A. and Tomberlin, J.K., 2017. The impact of diet protein and carbohydrate on select life-history traits of the black soldier fly *Hermetia illucens* (L.) (Diptera: Stratiomyidae). Insects 8: 56. https://doi.org/10.3390/insects8020056

Carpenter, J.E. and Bloem, S., 2002. Interaction between insect strain and artificial diet in diamondback moth development and reproduction. Entomologia Experimentalis et Applicata 102: 283-294.

Carreck, N. and Williams, I., 1988. The economic value of bees in the UK. Bee World 79: 115-123. https://doi.org/10.1080/0005772x.1998.11099393

Castillo, R.C.d., 2005. The quantitative genetic basis of female and male body size and their implications on the evolution of body size dimorphism in the house cricket *Acheta domesticus* (Gryllidae). Genetics and Molecular Biology 28: 843-848.

Cheng, T., Fu, B., Wu, Y., Long, R., Liu, C. and Xia, Q., 2015. Transcriptome sequencing and positive selected genes analysis of *Bombyx mandarina*. PLoS ONE 10: e0122837. https://doi.org/10.1371/journal.pone.0122837

Chinarak, K., Chaijan, M. and Panpipat, W., 2020. Farm-raised sago palm weevil (*Rhynchophorus ferrugineus*) larvae: Potential and challenges for promising source of nutrients. Journal of Food Composition and Analysis 92: 103542.

Choi, I.-H., Kim, J.-M., Kim, N.-J., Kim, J.-D., Park, C., Park, J.-H. and Chung, T.-h., 2018. Replacing fish meal by mealworm (*Tenebrio molitor*) on the growth performance and immunologic responses of white shrimp (*Litopenaeus vannamei*). Acta Scientiarum. Animal Sciences 40: e39077. https://doi.org/10.4025/actascianimsci.v40i1.39077.

Cifuentes, Y., Glaeser, S.P., Mvie, J., Bartz, J.-O., Müller, A., Gutzeit, H.O., Vilcinskas, A. and Kämpfer, P., 2020. The gut and feed residue microbiota changing during the rearing of *Hermetia illucens* larvae. Antonie van Leeuwenhoek 113: 1323-1344. https://doi.org/10.1007/s10482-020-01443-0

Cingolani, P., Platts, A., Wang le, L., Coon, M., Nguyen, T., Wang, L., Land, S.J., Lu, X. and Ruden, D.M., 2012. A program for annotating and predicting the effects of single nucleotide polymorphisms, SnpEff: SNPs in the genome of *Drosophila melanogaster* strain w1118; iso-2; iso-3. Fly (Austin) 6: 80-92. https://doi.org/10.4161/fly.19695

Clifford, C.W. and Woodring, J., 1990. Methods for rearing the house cricket, *Acheta domesticus* (L.), along with baseline values for feeding rates, growth rates, development times, and blood composition. Journal of Applied Entomology 109: 1-14.

Collavo, A., Glew, R.H., Huang, Y.-S., Chuang, L.-T., Bosse, R. and Paoletti, M.G., 2005. House cricket small-scale farming. Ecological implications of minilivestock: potential of insects, rodents, frogs and snails 27: 515-540.

Collins, A.M., Rinderer, T.E., Harbo, J.R. and Brown, M.A., 1984. Heritabilities and correlations for several characters in the honey bee. Journal of Heredity 75: 135-140. https://doi.org/10.1093/oxfordjournals.jhered.a109888

Daetwyler, H.D., Schenkel, F.S., Sargolzaei, M. and Robinson, J.A., 2008. A genome scan to detect quantitative trait loci for economically important traits in Holstein cattle using two methods and a dense single nucleotide polymorphism map. Journal of Dairy Science 91: 3225-3236. https://doi.org/10.3168/jds.2007-0333

Datta, R., Basavaraja, H., Reddy, N.M., Kumar, S.N., Kumar, N.S., Babu, M.R., Ahsan, M. and Jayaswal, K., 2001. Breeding of new productive bivoltine hybrid, CSR12×CSR6 of silkworm *Bombyx mori* L. International Journal of Industrial Entomology 3: 127-133.

Drogemuller, C., Karlsson, E.K., Hytonen, M.K., Perloski, M., Dolf, G., Sainio, K., Lohi, H., Lindblad-Toh, K. and Leeb, T., 2008. A mutation in hairless dogs implicates FOXI3 in ectodermal development. Science 321: 1462. https://doi.org/10.1126/science.1162525

Dzepe, D., Nana, P., Kuietche, H.M., Kuate, A.F., Tchuinkam, T. and Djouaka, R., 2020. Role of pupation substrate on post-feeding development of black soldier fly larvae, *Hermetia illucens* (Diptera: Stratiomyidae). 8: 760-764.

Elahi, U., Ma, Y.B., Wu, S.G., Wang, J., Zhang, H.J. and Qi, G.H., 2020. Growth performance, carcass characteristics, meat quality and serum profile of broiler chicks fed on housefly maggot meal as a replacement of soybean meal. Journal of Animal Physiology and Animal Nutrition 104: 1075-1084. https://doi.org/10.1111/jpn.13265

Elsayed, M., Ran, Y., Ai, P., Azab, M., Mansour, A., Jin, K., Zhang, Y. and Abomohra, A.E.-F., 2020. Innovative integrated approach of biofuel production from agricultural wastes by anaerobic digestion and black soldier fly larvae. Journal of Cleaner Production 263: 121495. https://doi.org/10.1016/j.jclepro.2020.121495

Elshire, R.J., Glaubitz, J.C., Sun, Q., Poland, J.A., Kawamoto, K., Buckler, E.S. and Mitchell, S.E., 2011. A robust, simple genotyping-by-sequencing (GBS) approach for high diversity species. PLoS ONE 6: e19379. https://doi.org/10.1371/journal.pone.0019379

Engel, M.S., 1999. The taxonomy of recent and fossil honey bees (Hymenoptera: Apidae; *Apis*). Journal of Hymenoptera Research 8: 165-196.

Environmental Protection Agency (EPA), 1995. Pesticide fact sheet. Name of chemical(s): *Bacillus thuringiensus* CrylA9(b) d-endotoxin and the genetic material necessary for its production (plasmid vector pCIB4431) in corn. Environmental Protection Agency, Washington, DC, USA.

Eriksson, T., Andere, A.A., Kelstrup, H., Emery, V.J. and Picard, C.J., 2020. The yellow mealworm (*Tenebrio molitor*) genome: a resource for the emerging insect as food and feed industry. Journal of Insects as Food and Feed 6: 445-455. https://doi.org/10.3920/jiff2019.0057

Ferguson, K.B., Kursch-Metz, T., Verhulst, E.C. and Pannebakker, B.A., 2020. Hybrid genome assembly and evidence-based annotation of the egg parasitoid and biological control agent *Trichogramma brassicae*. G3: Genes, Genomes, Genetics 10: 3533-3540. https://doi.org/10.1534/g3.120.401344

Fernandez-Cassi, X., Supeanu, A., Vaga, M., Jansson, A., Boqvist, S. and Vagsholm, I., 2019. The house cricket (*Acheta domesticus*) as a novel food: a risk profile. Journal of Insects as Food and Feed 5: 137-157. https://doi.org/10.3920/jiff2018.0021

Flasz, B., Dziewięcka, M., Kędziorski, A., Tarnawska, M. and Augustyniak, M., 2020. Vitellogenin expression, DNA damage, health status of cells and catalase activity in *Acheta domesticus* selected according to their longevity after graphene oxide treatment. Science of The Total Environment 737. https://doi.org/10.1016/j.scitotenv.2020.140274

Fologea, D., Gershow, M., Ledden, B., McNabb, D.S., Golovchenko, J.A. and Li, J., 2005. Detecting single stranded DNA with a solid state nanopore. Nano Letters 5: 1905-1909. https://doi.org/10.1021/nl051199m

Force, A.G., Staples, T., Soliman, S. and Arking, R., 1995. Comparative biochemical and stress analysis of genetically selected *Drosophila* strains with different longevities. Developmental genetics 17: 340-351.

Furdui, E.M., Marghitas, L.A., Dezmirean, D.S., Pasca, I., Pop, I.F., Erler, S. and Schluns, E.A.,

2014. Genetic characterization of *Bombyx mori* (Lepidoptera: Bombycidae) breeding and hybrid lines with different geographic origins. Journal of Insect Science 14: 211. https://doi.org/10.1093/jisesa/ieu073

Furman, D.P., Young, R.D. and Catts, P.E., 1959. *Hermetia illucens*(Linnaeus) as a factor in the natural control of Musca domestica Linnaeus. Journal of economic entomology 52: 917-921.

Gao, Z., Deng, W. and Zhu, F., 2019. Reference gene selection for quantitative gene expression analysis in black soldier fly (*Hermetia illucens*). PLoS ONE 14: e0221420. https://doi.org/10.1371/journal.pone.0221420

Giannetto, A., Oliva, S., Mazza, L., Mondello, G., Savastano, D., Mauceri, A. and Fasulo, S., 2017. Molecular characterization and expression analysis of heat shock protein 70 and 90 from *Hermetia illucens*reared in a food waste bioconversion pilot plant. Gene 627: 15-25. https://doi.org/10.1016/j.gene.2017.06.006

Gilles, A.F., Schinko, J.B. and Averof, M., 2015. Efficient CRISPR-mediated gene targeting and transgene replacement in the beetle *Tribolium castaneum*. Development 142: 2832-2839.

Giunti, G., Campolo, O., Laudani, F. and Palmeri, V., 2018. Male courtship behaviour and potential for female mate choice in the black soldier fly *Hermetia illucens* L. (Diptera: Stratiomyidae). Entomologia Generalis 38: 29-46. https://doi.org/10.1127/entomologia/2018/0657

Goldsmith, M.R., Shimada, T. and Abe, H., 2005. The genetics and genomics of the silkworm, *Bombyx mori*. Annual Review of Entomology 50: 71-100. https://doi.org/10.1146/annurev.ento.50.071803.130456

Greany, P., Vinson, S. and Lewis, W., 1984. Insect parasitoids: finding new opportunities for biological control. American Institute of Biological Sciences Circulation, McLean, VA, USA.

Gregory, T.R., 2009. Artificial selection and domestication: modern lessons from Darwin's enduring analogy. Evolution: Education and Outreach 2: 5-27. https://doi.org/10.1007/s12052-008-0114-z

Guzmán-Novoa, E. and Page Jr, R.E., 1999. Selective breeding of honey bees (Hymenoptera: Apidae) in Africanized areas. Journal of economic entomology 92: 521-525.

Hall, H.N., Masey O'Neill, H.V., Scholey, D., Burton, E., Dickinson, M. and Fitches, E.C., 2018. Amino acid digestibility of larval meal (*Musca domestica*) for broiler chickens. Poultry Science 97: 1290-1297. https://doi.org/10.3382/ps/pex433

Halloran, A., Roos, N., Flore, R. and Hanboonsong, Y., 2016. The development of the edible cricket industry in Thailand. Journal of Insects as Food and Feed 2: 91-100. https://doi.org/10.3920/jiff2015.0091

Harpur, B.A., Minaei, S., Kent, C.F. and Zayed, A., 2012. Management increases genetic diversity of honey bees via admixture. Molecular Ecology 21: 4414-4421. https://doi.org/10.1111/j.1365-294X.2012.05614.x

Hasan, M.A., Rahman, S.M. and Ahsan, M.K., 2011. Genetic variability, correlation, path analysis and construction of selection index in mulberry silkworm, *Bombyx mori* L. I Genetic variability. University Journal of Zoology, Rajshahi University 30: 33-36.

Hayes, B.J. and Daetwyler, H.D., 2019. 1000 bull genomes project to map simple and complex genetic traits in cattle: applications and outcomes. Annual Review of Animal Biosciences 7: 89-102. https://doi.org/10.1146/annurev-animal-020518-115024

Hein, S.A., 1924. Studies on variation in the mealworm, *Tenebrio molitor*. Journal of Genetics 14: 1-38.

Heinrich, J.C. and Scott, M.J., 2000. A repressible female-specific lethal genetic system for making transgenic insect strains suitable for a sterile-release program. Proceedings of the National Academy of Sciences 97: 8229-8232.

Henry, M., Gasco, L., Chatzifotis, S. and Piccolo, G., 2018. Does dietary insect meal affect the fish immune system? The case of mealworm, *Tenebrio molitor* on European sea bass, *Dicentrarchus labrax*. Developmental & Comparative Immunology 81: 204-209.

Hoc, B., Noel, G., Carpentier, J., Francis, F. and Caparros Megido, R., 2019. Optimization of black soldier fly (*Hermetia illucens*) artificial reproduction. PLoS ONE 14: e0216160. https://doi.org/10.1371/journal.pone.0216160

Hoy, M.A., 1976. Genetic improvement of insects: fact or fantasy. Environmental entomology 5: 833-839.

Hu, X.F., Zhang, B., Liao, C.H. and Zeng, Z.J., 2019. High-Efficiency CRISPR/Cas9-Mediated Gene Editing in Honeybee (*Apis mellifera*) Embryos. G3: Genes, Genomes, Genetics 9: 1759-1766. https://doi.org/10.1534/g3.119.400130

Huang, F., Subramanyam, B. and Toews, M.D., 2004. Susceptibility of laboratory and field strains of four stored-product insect species to spinosad. Journal of Economic Entomology 97: 2154-2159.

Huang, Q., Lattorff, H.M., Kryger, P., Le Conte, Y. and Moritz, R.F., 2014. A selective sweep in a microsporidian parasite Nosema-tolerant honeybee population, *Apis mellifera*. Animal Genetics 45: 267-273. https://doi.org/10.1111/age.12114

Hunt, G.J., Page, R., Fondrk, M.K. and Dullum, C.J., 1995. Major quantitative trait loci affecting honey bee foraging behavior. Genetics 141: 1537-1545.

Jensen, K., Kristensen, T.N., Heckmann, L.-H. and Sørensen, J.G., 2017. Breeding and maintaining high-quality insects. In: Van Huis, A. and Tomberlin, J.K. (eds) Insects as food and feed: from production to consumption. Wageningen Academic Publishers, Wageningen, The Netherlands, pp. 174-198. https://doi.org/10.3920/978-90-8686-849-0

Jing, W., Xiangyang, Z. and Xiujin, Z., 2013. A review of innate immunity of silkworm, *Bombyx mori*. African Journal of Agricultural Research 8: 2319-2325. https://doi.org/10.5897/AJARx11.079

Jingade, A., Vijayan, K., Somasundaram, P., Srinivasababu, G. and Kamble, C., 2011. A review of the implications of heterozygosity and inbreeding on germplasm biodiversity and its conservation in the silkworm, *Bombyx mori*. Journal of Insect Science 11(1): 8. https://doi.org/10.1673/031.011.0108

Jo, Y.H., Kim, Y.J., Park, K.B., Seong, J.H., Kim, S.G., Park, S., Noh, M.Y., Lee, Y.S. and Han, Y.S., 2017. TmCactin plays an important role in Gram-negative and -positive bacterial infection by regulating expression of 7 AMP genes in *Tenebrio molitor*. Scientific reports 7: 46459.

Johnston, P.R., Makarova, O. and Rolff, J., 2014. Inducible defenses stay up late: temporal patterns of immune gene expression in *Tenebrio molitor*. G3: Genes, Genomes, Genetics 4: 947-955.

Jung, J., Heo, A., Park, Y.W., Kim, Y.J., Koh, H. and Park, W., 2014. Gut microbiota of *Tenebrio molitor* and their response to environmental change. Journal of Microbiology and Biotechnology 24: 888-897.

Kanafi, R.R., Ebadi, R., Mirhosseini, S., Seidavi, A., Zolfaghari, M. and Etebari, K., 2007. A review on nutritive effect of mulberry leaves enrichment with vitamins on economic traits and biological parameters of silkworm *Bombyx mori* L. Invertebrate Survival Journal 4: 86-91.

Kenis, M., Bouwassi, B., Boafo, H., Devic, E., Han, R., Koko, G., Koné, N.G., Maciel-Vergara, G., Nacambo, S., Pomalegni, S.C.B., Roffeis, M., Wakefield, M., Zhu, F. and Fitches, E., 2018. Small-scale fly larvae production for animal feed. In: Halloran, A., Flore, R., Vantomme, P. and Roos, N. (eds) Edible insects in sustainable food systems. Springer, Cham, Switzerland, pp. 239-261. https://doi.org/10.1007/978-3-319-74011-9_15

Kerr, A.M., Gershman, S.N. and Sakaluk, S.K., 2010. Experimentally induced spermatophore production and immune responses reveal a trade-off in crickets. Behavioral Ecology 21: 647-654. https://doi.org/10.1093/beheco/arq035

Keshavarz, M., Jo, Y.H., Edosa, T.T., Bae, Y.M. and Han, Y.S., 2020. TmPGRP-SA regulates antimicrobial response to bacteria and fungi in the fat body and gut of *Tenebrio molitor*. International Journal of Molecular Sciences 21: 2113.

Kim, D.-H., Lee, Y.T., Lee, Y.J., Chung, J.H., Lee, B.L., Choi, B.S. and Lee, Y., 1998. Bacterial expression of tenecin 3, an insect antifungal protein isolated from *Tenebrio molitor*, and its efficient purification. Molecules & Cells 8: 786-789.

Klein, A.M., Vaissiere, B.E., Cane, J.H., Steffan-Dewenter, I., Cunningham, S.A., Kremen, C. and Tscharntke, T., 2007. Importance of pollinators in changing landscapes for world crops. Proceedings of the Royal Society B: Biological Sciences 274: 303-313. https://doi.org/10.1098/rspb.2006.3721

Kohno, H., Suenami, S., Takeuchi, H., Sasaki, T. and Kubo, T., 2016. Production of knockout mutants by CRISPR/Cas9 in the European honeybee, *Apis mellifera* L. Zoological Science 33: 505-512. https://doi.org/10.2108/zs160043

Kojima, K., Kuwana, Y., Sezutsu, H., Kobayashi, I., Uchino, K., Tamura, T. and Tamada, Y., 2007. A new method for the modification of fibroin heavy chain protein in the transgenic silkworm. Bioscience, Biotechnology, and Biochemistry 71: 2943-2951. https://doi.org/10.1271/bbb.70353

Kölliker-Ott, U.M., Bigler, F. and Hoffmann, A.A., 2003. Does mass rearing of field collected *Trichogramma brassicae* wasps influence acceptance of European corn borer eggs? Entomologia Experimentalis et Applicata 109: 197-203.

Krafsur, E.S., 1998. Sterile insect technique for suppressing and eradicating insect populations: 55 years and counting. Journal of Agricultural Entomology 15: 303-317.

Kraus, F.B., Neumann, P. and Moritz, R.F.A., 2005. Genetic variance of mating frequency in the honeybee (*Apis mellifera* L.). Insectes Sociaux 52: 1-5. https://doi.org/10.1007/s00040-004-0766-9

Kumar, G.A., Jalali, S., Nagesh, M., Venkatesan, T. and Niranjana, P., 2009. Genetic variation in artificially selected strains of the egg parasitoid, *Trichogramma chilonis* Ishii (Hymenoptera: Trichogrammatidae) using RAPD analysis. Journal of Biological Control 23: 353-359.

Kumar, P., Bhutia, R. and Ahsan, M., 1995. Estimates of genetic variability for commercial quantitative traits and selection indices in bivoltine races of mulberry silkworm (*Bombyx mori* L.). Indian Journal of Genetics and Plant Breeding 55: 109-116.

Kuno, G., 2010. Early history of laboratory breeding of *Aedes aegypti* (Diptera: Culicidae) focusing on the origins and use of selected strains. Journal of Medical Entomology 47: 957-971. 10.1603/me10152

Lalander, C., Diener, S., Zurbrügg, C. and Vinnerås, B., 2019. Effects of feedstock on larval development and process efficiency in waste treatment with black soldier fly (*Hermetia illucens*). Journal of Cleaner Production 208: 211-219. https://doi.org/10.1016/j.jclepro.2018.10.017

Lander, E.S., Linton, L.M., Birren, B., Nusbaum, C., Zody, M.C., Baldwin, J., Devon, K., Dewar, K., Doyle, M., FitzHugh, W., Funke, R., Gage, D., Harris, K., Heaford, A., Howland, J., Kann, L., *et al.*, 2001. Initial sequencing and analysis of the human genome. Nature 409: 860-921. https://doi.org/10.1038/35057062

Lapidge, K.L., Oldroyd, B.P. and Spivak, M., 2002. Seven suggestive quantitative trait loci influence hygienic behavior of honey bees. Naturwissenschaften 89: 565-568.

Leclercq, J., 1963. Artificial selection for weight and its consequences in *Tenebrio molitor* L. Nature 198: 106-107.

Lee, Y.-S., Seo, S.-H., Yoon, S.-H., Kim, S.-Y., Hahn, B.-S., Sim, J.-S., Koo, B.-S. and Lee, C.-M., 2016. Identification of a novel alkaline amylopullulanase from a gut metagenome of *Hermetia illucens*. International journal of biological macromolecules 82: 514-521.

Lee, Y.J., Chung, T.J., Park, C.W., Hahn, Y., Chung, J.H., Lee, B.L., Han, D.M., Jung, Y.H., Kim, S. and Lee, Y., 1996. Structure and expression of the tenecin 3 gene in *Tenebrio molitor*. Biochemical and Biophysical Research Communications 218: 6-11.

Li, Q., Li, Y.-P., Ambuhl, D., Liu, Y.-P., Li, M.-W. and Qin, L., 2020. Nutrient composition of Chinese oak silkworm, *Antheraea pernyi*, a traditional edible insect in China: a review. Journal of Insects as Food and Feed 6: 355-369. https://doi.org/10.3920/JIFF2019.0059

Li, W., Zhang, Z., Lin, L. and Terenius, O., 2017. *Antheraea pernyi* (Lepidoptera: Saturniidae) and its importance in sericulture, food consumption, and traditional Chinese medicine. Journal of Economic Entomology 110: 1404-1411. https://doi.org/10.1093/jee/tox140

Liao, J., Xue, Y., Xiao, G., Xie, M., Huang, S., You, S., Wyckhuys, K.A.G. and You, M., 2019. Inheritance and fitness costs of resistance to *Bacillus thuringiensis* toxin Cry2Ad in laboratory strains of the diamondback moth, *Plutella xylostella* (L.). Scientific Reports 9: 6113. https://doi.org/10.1038/s41598-019-42559-2

Linksvayer, T.A., Kaftanoglu, O., Akyol, E., Blatch, S., Amdam, G.V. and Page, R.E., Jr., 2011. Larval and nurse worker control of developmental plasticity and the evolution of honey bee queen-worker dimorphism. Journal of Evolutionary Biology 24: 1939-1948. https://doi.org/10.1111/j.1420-9101.2011.02331.x

Liu, H., Cupp, E.W., Guo, A. and Liu, N., 2004. Insecticide resistance in Alabama and Florida mosquito strains of *Aedes albopictus*. Journal of Medical Entomology 41: 946-952.

Liu, Q., Liu, W., Zeng, B., Wang, G., Hao, D. and Huang, Y., 2017. Deletion of the *Bombyx mori* odorant receptor co-receptor (BmOrco) impairs olfactory sensitivity in silkworms. Insect Biochemistry and Molecular Biology 86: 58-67. https://doi.org/10.1016/j.ibmb.2017.05.007

Liu, S., Shi, X.-X., Jiang, Y.-D., Zhu, Z.-J., Qian, P., Zhang, M.-J., Yu, H., Zhu, Q.-Z., Gong, Z.-J. and Zhu, Z.-R., 2015. *De novo* analysis of the *Tenebrio molitor* (Coleoptera: Tenebrionidae) transcriptome and identification of putative glutathione S-transferase genes. Applied entomology and zoology 50: 63-71.

Liu, Z., Najar-Rodriguez, A.J., Minor, M.A., Hedderley, D.I. and Morel, P.C.H., 2020. Mating success of the black soldier fly, *Hermetia illucens* (Diptera: Stratiomyidae), under four artificial light sources. Journal of Photochemistry and Photobiology B 205: 111815. https://doi.org/10.1016/j.jphotobiol.2020.111815

Lu, C., Li, B., Zhao, A. and Xiang, Z., 2004. QTL mapping of economically important traits in silkworm (*Bombyx mori*). Science in China Series C: Life Sciences 47: 477-484.

Ma, S., Chang, J., Wang, X., Liu, Y., Zhang, J., Lu, W., Gao, J., Shi, R., Zhao, P. and Xia, Q., 2014. CRISPR/Cas9 mediated multiplex genome editing and heritable mutagenesis of BmKu70 in *Bombyx mori*. Scientific Reports 4: 4489. https://doi.org/10.1038/srep04489

Mackauer, M., 1972. Genetic aspects of insect production. Entomophaga 17: 27-48.

Makkar, H.P., Tran, G., Heuzé, V. and Ankers, P., 2014. State-of-the-art on use of insects as animal feed. Animal Feed Science and Technology 197: 1-33.

Mano, Y., 1993. A new method to select promising silkworm breeds/combinations. Indian Silk 31: 53.

Marien, A., Debode, F., Aerts, C., Ancion, C., Francis, F. and Berben, G., 2018. Detection of *Hermetia illucens*by real-time PCR. Journal of Insects as Food and Feed 4: 115-122. https://doi.org/10.3920/jiff2017.0069

Marshall, S., Woodley, N. and Hauser, M., 2015. The historical spread of the Black Soldier Fly, *Hermetia illucens*(L.)(Diptera, Stratiomyidae, Hermetiinae), and its establishment in Canada. The Journal of the Entomological Society of Ontario 146: 51-54.

Martin, R., Rivers, J. and Cowgill, U., 1976. Culturing mealworms as food for animals in captivity. International Zoo Yearbook 16: 63-70.

McConnell, M.W. and Judge, K.A., 2018. Body size and lifespan are condition dependent in the mealworm beetle, *Tenebrio molitor*, but not sexually selected traits. Behavioral Ecology and Sociobiology 72: 32. https://doi.org/10.1007/s00265-018-2444-3

McGugan, B. and Coppel, H., 1962. Part II. Biological control of forest insects, 1910-1958. In: McLeod, J.H., Coppel, H.C. and McGugan, B.M. A review of the biological control attempts against insects and weeds in Canada. Commonwealth Agricultural Bureau, Franham Royal, UK, pp. 35-127.

Meixner, M.D., Costa, C., Kryger, P., Hatjina, F., Bouga, M., Ivanova, E. and Büchler, R., 2010. Conserving diversity and vitality for honey bee breeding. Journal of Apicultural Research 49: 85-92. https://doi.org/10.3896/ibra.1.49.1.12

Mendel, G., 1865. Versuche uber Pflanzen-Hybriden. Abhandl. d. Naturf. Vereins in Brunn 4: 3-47.

Meneguz, M., Gasco, L. and Tomberlin, J.K., 2018. Impact of pH and feeding system on black soldier fly (*Hermetia illucens*, L; Diptera: Stratiomyidae) larval development. PLoS ONE 13: e0202591. https://doi.org/10.1371/journal.pone.0202591

Mertz, D.B., 1975. Senescent decline in flour beetle strains selected for early adult fitness. Physiological Zoology 48: 1-23.

Mirhosseini, S., Ghanipoor, M., Shadparvar, A. and Etebari, K., 2005. Selection indices for cocoon traits in six commercial silkworm (*Bombyx mori* L.) lines. The Philippine Agricultural Scientist 88: 328-336.

Miyatake, T., 2006. Quantitative genetic aspects of the quality control of mass-reared insects: the case of the melon fly (*Bactrocera cucurbitae*). Formosan Entomolist 26: 307-318.

Mizoguchi, A. and Okamoto, N., 2013. Insulin-like and IGF-like peptides in the silkmoth *Bombyx mori*: discovery, structure, secretion, and function. Frontiers in Physiology 4: 217. https://doi.org/10.3389/fphys.2013.00217

Mondal, M., Trivedy, K. and Nirmal Kumar, S., 2007. The silk proteins, sericin and fibroin in silkworm, *Bombyx mori* Linn. – a review. Caspian Journal of Environmental Sciences 5: 63-76.

Moon, H.J., Lee, S.Y., Kurata, S., Natori, S. and Lee, B.L., 1994. Purification and molecular cloning of cDNA for an inducible antibacterial protein from larvae of the coleopteran, *Tenebrio molitor*. The Journal of Biochemistry 116: 53-58.

Moorthy, S., Das, S., Kar, N. and Urs, S.R., 2007. Breeding of bivoltine breeds of *Bombyx mori* suitable for variable climatic conditions of tropics. International Journal of Industrial Entomology 14: 99-105.

Morales-Ramos, J.A., Kelstrup, H.C., Rojas, M.G. and Emery, V., 2019. Body mass increase induced by eight years of artificial selection in the yellow mealworm (Coleoptera: Tenebrionidae) and life history trade-offs. Journal of Insect Science 19: 4. https://doi.org/10.1093/jisesa/iey110

Moritz, R.F., 1985. Heritability of the postcapping stage in *Apis mellifera* and its relation to varroatosis resistance. Journal of Heredity 76: 267-270.

Moritz, R.F. and Southwick, E.E., 1987. Phenotype interactions in group behavior of honey bee workers (*Apis mellifera* L.). Behavioral Ecology and Sociobiology 21: 53-57.

Murakami, A. and Ohtsuki, Y., 1989. Genetic studies on tropical races of silkworm (*Bombyx mori*), with special reference to cross breeding strategy between tropical and temperate races. 2. Multivoltine silkworms in Japan and their origin. Journal of the Association for Research in Otolaryngology 23: 123-127.

Nagaraju, J., Raje, U. and Datta, R., 1996. Crossbreeding and heterosis in the silkworm, *Bombyx mori*: a review. Sericologia 36: 1-20.

Nagarkatti, S. and Nagaraja, H., 1978. Experimental comparison of laboratory reared vs. wild-type *Trichogramma confusum* [Hym.: Trichogrammatidae] I. Fertility, fecundity and longevity. Entomophaga 23: 129-136.

Nakagaki, B. and DeFoliart, G., 1991. Comparison of diets for mass-rearing *Acheta domesticus* (Orthoptera: Gryllidae) as of food conversion efficiency with values reported for livestock. Journal of Economic Entomology 84: 891-896.

Nakagaki, B.J., Sunde, M.L. and Defoliart, G.R., 1987. Protein quality of the house cricket, *Acheta domesticus*, when fed to broiler chicks. Poultry Science 66: 1367-1371. https://doi.org/10.3382/ps.0661367

Neshagaran Hemmatabadi, R., Seidavi, A. and Gharahveysi, S., 2016. A review on correlation, heritability and selection in silkworm breeding. Journal of Applied Animal Research 44: 9-23. https://doi.org/10.1080/09712119.2014.987289

Neumann, P. and Blacquiere, T., 2017. The Darwin cure for apiculture? Natural selection and managed honeybee health. Evolutionary Applications 10: 226-230. https://doi.org/10.1111/eva.12448

Newton, G., Booram, C., Barker, R. and Hale, O., 1977. Dried *Hermetia illucens* larvae meal as a supplement for swine. Journal of Animal Science 44: 395-400.

Nikoh, N., Hosokawa, T., Moriyama, M., Oshima, K., Hattori, M. and Fukatsu, T., 2014. Evolutionary origin of insect-*Wolbachia* nutritional mutualism. Proceedings of the National Academy of Sciences 111: 10257-10262. https://doi.org/10.1073/pnas.1409284111

Nye, W.P. and Mackensen, O., 1968. Selective breeding of honeybees for alfalfa pollen: fifth generation and backcrosses. Journal of Apicultural Research 7: 21-27.

Nye, W.P. and Mackensen, O., 1970. Selective breeding of honeybees for alfalfa pollen collection: with tests in high and low alfalfa pollen collection regions. Journal of Apicultural Research 9: 61-64.

Oldroyd, B., Rinderer, T. and Buco, S., 1991. Heritability of morphological characters used to distinguish European and Africanized honeybees. Theoretical and Applied Genetics 82: 499-504.

Oonincx, D.G. and De Boer, I.J., 2012. Environmental impact of the production of mealworms as a protein source for humans – a life cycle assessment. PLoS ONE 7: e51145. https://doi.org/10.1371/journal.pone.0051145

Oppert, B., Martynov, A.G. and Elpidina, E.N., 2012. *Bacillus thuringiensis* Cry3Aa protoxin intoxication of *Tenebrio molitor* induces widespread changes in the expression of serine peptidase transcripts. Comparative Biochemistry and Physiology Part D: Genomics and Proteomics 7: 233-242.

Oppert, B., Perkin, L.C., Lorenzen, M. and Dossey, A.T., 2020. Transcriptome analysis of life stages of the house cricket, *Acheta domesticus*, to improve insect crop production. Scientific Reports 10: 3471. https://doi.org/10.1038/s41598-020-59087-z

Orinda, M.A., Mosi, R.O., Ayieko, M.A. and Amimo, F.A., 2017. Growth performance of Common house cricket (*Acheta domesticus*) and field cricket (*Gryllus bimaculatus*) crickets fed on agro-byproducts. Journal of Entomology and Zoology Studies 5: 1664-1668.

Page Jr, R., Fondrk, M., Hunt, G., Guzman-Novoa, E., Humphries, M., Nguyen, K. and Greene, A., 2000. Genetic dissection of honeybee (*Apis mellifera* L.) foraging behavior. Journal of Heredity 91: 474-479.

Page, R.E. and Fondrk, M.K., 1995. The effects of colony-level selection on the social organization of honey bee (*Apis mellifera* L.) colonies: colony-level components of pollen hoarding. Behavioral Ecology and Sociobiology 36: 135-144.

Palma, M.S. and Brochetto-Braga, M., 1993. Biochemical variability between venoms from different honey-bee (*Apis mellifera*) races. Comparative Biochemistry and Physiology Part C: Pharmacology, Toxicology and Endocrinology 106: 423-427.

Parajulee, M.N., Defoliart, G.R. and Hogg, D.B., 1993. Model for use in mass-production of *Acheta domesticus* (Orthoptera: Gryllidae) as food. Journal of Economic Entomology 86: 1424-1428.

Park, S., Choi, H., Choi, J.-Y. and Jeong, G., 2017. Population structure of the exotic black soldier fly, *Hermetia illucens* (Diptera: Stratiomyidae) in Korea. Korean Journal of Environment and Ecology 31: 520-528. https://doi.org/10.13047/kjee.2017.31.6.520

Park, S.I., Kim, J.W. and Yoe, S.M., 2015. Purification and characterization of a novel antibacterial peptide from black soldier fly (*Hermetia illucens*) larvae. Developmental and Comparative Immunology 52: 98-106. https://doi.org/10.1016/j.dci.2015.04.018

Pashley, D.P., 1988. Current status of fall armyworm host strains. Florida Entomologist 71: 227-234.

Perez-Sato, J.A., Chaline, N., Martin, S.J., Hughes, W.O. and Ratnieks, F.L., 2009. Multi-level selection for hygienic behaviour in honeybees. Heredity 102: 609-615. https://doi.org/10.1038/hdy.2009.20

Petitpierre, E., Gatewood, J. and Schmid, C., 1988. Satellite DNA from the beetle *Tenebrio molitor*. Experientia 44: 498-499.

Pianka, E.R., 1970. On R- and K-selection. American Naturalist 104: 592-597.

Poveda, J., Jiménez-Gómez, A., Saati-Santamaría, Z., Usategui-Martín, R., Rivas, R. and García-Fraile, P., 2019. Mealworm frass as a potential biofertilizer and abiotic stress tolerance-inductor in plants. Applied Soil Ecology 142: 110-122.

Prabhakar, S., Chen, M.S., Elpidina, E., Vinokurov, K., Smith, C., Marshall, J. and Oppert, B., 2007. Sequence analysis and molecular characterization of larval midgut cDNA transcripts encoding peptidases from the yellow mealworm, *Tenebrio molitor* L. Insect Molecular Biology 16: 455-468.

Pradeep, A.R., Chatterjee, S.N. and Nair, C.V., 2005. Genetic differentiation induced by selection in an inbred population of the silkworm *Bombyx mori*, revealed by RAPD and ISSR marker systems. Journal of Applied Genetics 46: 291.

Przemieniecki, S.W., Kosewska, A., Ciesielski, S. and Kosewska, O., 2020. Changes in the gut microbiome and enzymatic profile of *Tenebrio molitor* larvae biodegrading cellulose, polyethylene and polystyrene waste. Environmental Pollution 256: 113265. https://doi.org/10.1016/j.envpol.2019.113265

Quezada-Euán, J.J.G., Echazarreta, C.M. and Paxton, R.J., 1996. The distribution and range expansion of Africanized honey bees (*Apis mellifera*) in the state of Yucatan, Mexico. Journal of Apicultural Research 35: 85-95. https://doi.org/10.1080/00218839.1996.11100917

Rahmathulla, V.K., 2012. Management of climatic factors for successful silkworm (*Bombyx mori* L.) crop and higher silk production: a review. Psyche: A Journal of Entomology 2012: Article ID 121234. https://doi.org/10.1155/2012/121234

Ran, F.A., Hsu, P.D., Wright, J., Agarwala, V., Scott, D.A. and Zhang, F., 2013. Genome engineering using the CRISPR-Cas9 system. Nature Protocols 8: 2281-2308. https://doi.org/10.1038/nprot.2013.143

Rehman, K.u., Rehman, A., Cai, M., Zheng, L., Xiao, X., Somroo, A.A., Wang, H., Li, W., Yu, Z. and Zhang, J., 2017. Conversion of mixtures of dairy manure and soybean curd residue by black soldier fly larvae (*Hermetia illucens* L.). Journal of Cleaner Production 154: 366-373. https://doi.org/10.1016/j.jclepro.2017.04.019

Rhode, C., Badenhorst, R., Hull, K.L., Greenwood, M.P., Bester-van der Merwe, A.E., Andere, A.A., Picard, C.J. and Richards, C., 2020. Genetic and phenotypic consequences of early domestication in black soldier flies (*Hermetia illucens*). Animal Genetics 51: 752-762. https://doi.org/10.1111/age.12961

Ribeiro, N., Abelho, M. and Costa, R., 2018. A review of the scientific literature for optimal conditions for mass rearing *Tenebrio molitor* (Coleoptera: Tenebrionidae). Journal of Entomological Science 53: 434-454.

Rimoldi, S., Gini, E., Iannini, F., Gasco, L. and Terova, G., 2019. The effects of dietary insect meal from *Hermetia illucens* prepupae on autochthonous gut microbiota of rainbow trout (*Oncorhynchus mykiss*). Animals 9(4): 143. https://doi.org/10.3390/ani9040143

Ross-Ibarra, J., Morrell, P.L. and Gaut, B.S., 2007. Plant domestication, a unique opportunity to identify the genetic basis of adaptation. Proceedings of the National Academy of Sciences 104 Suppl 1: 8641-8648. https://doi.org/10.1073/pnas.0700643104

Ruiz, X. and Almanza, M., 2018. Implications of genetic diversity in the improvement of silkworm *Bombyx mori* L. Chilean journal of agricultural research 78: 569-579. https://doi.org/10.4067/s0718-58392018000400569

Russo, J., Brehelin, M. and Carton, Y., 2001. Haemocyte changes in resistant and susceptible strains of *D. melanogaster* caused by virulent and avirulent strains of the parasitic wasp *Leptopilina boulardi*. Journal of Insect Physiology 47: 167-172.

Ryder, J. and Siva-Jothy, M., 2001. Quantitative genetics of immune function and body size in the house cricket, *Acheta domesticus*. Journal of Evolutionary Biology 14: 646-653.

Schiavone, A., Cullere, M., De Marco, M., Meneguz, M., Biasato, I., Bergagna, S., Dezzutto, D., Gai, F., Dabbou, S., Gasco, L. and Dalle Zotte, A., 2017. Partial or total replacement of soybean oil by black soldier fly larvae (*Hermetia illucens* L.) fat in broiler diets: effect on growth performances, feed-choice, blood traits, carcass characteristics and meat quality. Italian Journal of Animal Science 16: 93-100. https://doi.org/10.1080/182805 1X.2016.1249968

Schwenke, R.A., Lazzaro, B.P. and Wolfner, M.F., 2016. Reproduction-immunity trade-offs in insects. Annual Review of Entomology 61: 239-256. https://doi.org/10.1146/annurev-ento-010715-023924

Scott Schneider, S., DeGrandi-Hoffman, G. and Smith, D.R., 2004. The African honey bee: factors contributing to a successful biological invasion. Annual Review of Entomology 49: 351-376. https://doi.org/10.1146/annurev.ento.49.061802.123359

Seeley, T.D. and Tarpy, D.R., 2007. Queen promiscuity lowers disease within honeybee colonies. Proceedings of the Royal Society B: Biological Sciences 274: 67-72.

Seong, K.M., Mittapalli, O., Clark, J.M. and Pittendrigh, B.R., 2019. A review of DDT resistance as it pertains to the 91-C and 91-R strains in *Drosophila melanogaster*. Pesticide Biochemistry and Physiology 161: 86-94. https://doi.org/10.1016/j.pestbp.2019.06.003

Sevala, V., Sevala, V. and Loughton, B., 1993. Insulin-like molecules in the beetle *Tenebrio molitor*. Cell and tissue research 273: 71-77.

Shekar, K.C. and Basavaraja, H., 2008. Changes in qualitative and quantitative characters in bivoltine silkworm breeds of *Bombyx mori* L. under different selection methods. Indian Journal of Sericulture 47: 175-182.

Sheppard, C., 1983. House fly and lesser fly control utilizing the black soldier fly in manure management systems for caged laying hens. Environmental entomology 12: 1439-1442.

Sheppard, D.C., Newton, G.L., Thompson, S.A. and Savage, S., 1994. A value added manure management system using the black soldier fly. Bioresource technology 50: 275-279.

Sheppard, D.C., Tomberlin, J.K., Joyce, J.A., Kiser, B.C. and Sumner, S.M., 2002. Rearing methods for the black soldier fly (Diptera: Stratiomyidae). Journal of Medical Entomology 39: 695-698.

Sheppard, W.S. and Meixner, M.D., 2003. *Apis mellifera pomonella*, a new honey bee subspecies from Central Asia. Apidologie 34: 367-375. https://doi.org/10.1051/apido:2003037

Simmonds, F., 1947. Improvement of the sex-ratio of a parasite by selection. The Canadian Entomologist 79: 41-44.

Singh, T., Bhat, M.M. and Khan, M.A., 2011. Critical analysis of correlation and heritability phenomenon in the silkworm, *Bombyx mori* (Lepidoptera: Bombycidae). Advances in Bioscience and Biotechnology 02: 347-353. https://doi.org/10.4236/abb.2011.25051

Smith, J.M. and Haigh, J., 1974. The hitch-hiking effect of a favourable gene. Genetics Research 23: 23-35.

Solignac, M., Vautrin, D., Loiseau, A., Mougel, F., Baudry, E., Estoup, A., Garnery, L., Haberl, M. and Cornuet, J.-M., 2003. Five hundred and fifty microsatellite markers for the study of the honeybee (*Apis mellifera* L.) genome. Molecular Ecology Notes 3: 307-311. https://doi.org/10.1046/j.1471-8286.2003.00436.x

Sorati, M., Newman, M. and Hoffman, A.A., 1996. Inbreeding and incompatibility in *Trichogramma* nr. *brassicae*: evidence and implications for quality control. Entomologia experimentalis et applicata 78: 283-290.

Sørensen, J.G., Addison, M.F. and Terblanche, J.S., 2012. Mass-rearing of insects for pest management: challenges, synergies and advances from evolutionary physiology. Crop Protection 38: 87-94.

Southwick, E.E. and Southwick Jr, L., 1992. Estimating the economic value of honey bees (Hymenoptera: Apidae) as agricultural pollinators in the United States. Journal of economic entomology 85: 621-633.

Spötter, A., Gupta, P., Mayer, M., Reinsch, N. and Bienefeld, K., 2016. Genome-wide association study of a *Varroa*-specific defense behavior in honeybees (*Apis mellifera*). Journal of Heredity 107: 220-227.

Srikanth, K., Park, J.E., Lim, D., Cha, J., Cho, S.R., Cho, I.C. and Park, W., 2020. A comparison between hi-C and 10X genomics linked read sequencing for whole genome phasing in Hanwoo cattle. Genes 11: 332. https://doi.org/10.3390/genes11030332

Srivastava, S., Srikanth, K., Won, S., Son, J.H., Park, J.E., Park, W., Chai, H.H. and Lim, D., 2020. Haplotype-based genome-wide association study and identification of candidate genes associated with carcass traits in Hanwoo cattle. Genes 11: 551. https://doi.org/10.3390/genes11050551

Stahls, G., Meier, R., Sandrock, C., Hauser, M., Sasic Zoric, L., Laiho, E., Aracil, A., Doderovic, J., Badenhorst, R., Unadirekkul, P., Mohd Adom, N.A.B., Wein, L., Richards, C., Tomberlin, J.K., Rojo, S., Veselic, S. and Parviainen, T., 2020. The puzzling mitochondrial phylogeography of the black soldier fly (*Hermetia illucens*), the commercially most important insect protein species. BMC Evolutionary Biology 20: 60. https://doi.org/10.1186/s12862-020-01627-2

Stearns, S.C., 2000. Life history evolution: successes, limitations, and prospects. Naturwissenschaften 87: 476-486. https://doi.org/10.1007/s001140050763

Sun, D., Guo, Z., Liu, Y. and Zhang, Y., 2017. Progress and prospects of CRISPR/Cas systems in insects and other arthropods. Frontiers in physiology 8: 608.

Tabashnik, B.E., Van Rensburg, J. and Carrière, Y., 2009. Field-evolved insect resistance to *Bt* crops: definition, theory, and data. Journal of economic entomology 102: 2011-2025.

Tanaka, Y., 1953. Genetics of the silkworm, *Bombyx mori*. Advances in Genetics 5: 239-317. https://doi.org/10.1016/S0065-2660(08)60409-5

Tennis, P., 1985. Long-term divergence in body size produced by food size in laboratory populations of *Acheta domesticus* (Orthoptera: Gryllidae). Canadian Journal of Zoology 63: 1395-1401.

The Bovine Genome Sequencing and Analysis Consortium, Elsik, C.G., Tellam, R.L., Worley, K.C., Gibbs, R.A., Muzny, D.M., Weinstock, G.M., Adelson, D.L., Eichler, E.E., Elnitski, L., *et al.*, 2009. The genome sequence of taurine cattle: a window to ruminant biology and evolution. Science 324: 522-528. https://doi.org/10.1126/science.1169588

The Honeybee Genome Sequencing Consortium, 2006. Insights into social insects from the genome of the honeybee *Apis mellifera*. Nature 443: 931.

The International Silkworm Genome Consortium, 2008. The genome of a lepidopteran model insect, the silkworm *Bombyx mori*. Insect biochemistry and molecular biology 38: 1036-1045.

Tribolium Genome Sequencing Consortium, Richards, S., Gibbs, R.A., Weinstock, G.M., Brown, S.J., Denell, R., Beeman, R.W., Gibbs, R., Beeman, R.W., Brown, S.J., Bucher, G., *et al.*, 2008. The genome of the model beetle and pest *Tribolium castaneum*. Nature 452: 949-955. https://doi.org/10.1038/nature06784

Tschirner, M. and Simon, A., 2015. Influence of different growing substrates and processing on the nutrient composition of black soldier fly larvae destined for animal feed. Journal of Insects as Food and Feed 1: 249-259. https://doi.org/10.3920/jiff2014.0008

Turnbull, A.L. and Chant, D.A., 1961. The practice and theory of biological control of insects in Canada. Canadian Journal of Zoology 39: 697-753.

Urs, K. and Hopkins, T., 1973. Effect of moisture on growth rate and development of two strains of *Tenebrio molitor* L.(Coleoptera, Tenebrionidae). Journal of Stored Products Research 8: 291-297.

Uzunov, A., Brascamp, E.W. and Büchler, R., 2017. The basic concept of honey bee breeding programs. Bee World 94: 84-87. https://doi.org/10.1080/0005772x.2017.1345427

Van Bergeijk, K., Bigler, F., Kaashock, N. and Pak, G., 1989. Changes in host acceptance and host suitability as an effect of rearing *Trichogramma maidis* on a factitious host. Entomologia Experimentalis et Applicata 52: 229-238.

Van Broekhoven, S., Gutierrez, J.M., De Rijk, T., De Nijs, W. and Van Loon, J.J.A., 2017. Degradation and excretion of the *Fusarium* toxin deoxynivalenol by an edible insect, the yellow mealworm (*Tenebrio molitor* L.). World Mycotoxin Journal 10: 163-169. https://doi.org/10.3920/WMJ2016.2102

Van Den Bosch, R. and Messenger, P., 1973. Entomophagous insects. Biological control. Intext Press, Inc., New York, NY, USA, pp. 35-47.

Van Huis, A., 2013. Potential of insects as food and feed in assuring food security. Annual Review of Entomology 58: 563-583. https://doi.org/10.1146/annurev-ento-120811-153704

Van Huis, A., Dicke, M. and Van Loon, J.J.A., 2015. Insects to feed the world. Journal of Insects as Food and Feed 1: 3-5. https://doi.org/10.3920/JIFF2015.x002

Van Lenteren, J.C. and Nicoli, G., 2004. Quality control of mass produced beneficial insects. In: Heinz, R., Van Driesche, R. and Parrella, M.P. (eds) Biological control of arthropod pests in protected cultures. Ball Publishing, Batavia, IL, USA, pp 503-526.

Varotto Boccazzi, I., Ottoboni, M., Martin, E., Comandatore, F., Vallone, L., Spranghers, T., Eeckhout, M., Mereghetti, V., Pinotti, L. and Epis, S., 2017. A survey of the mycobiota associated with larvae of the black soldier fly (*Hermetia illucens*) reared for feed production. PLoS ONE 12: e0182533. https://doi.org/10.1371/journal.pone.0182533

Veldkamp, T. and Bosch, G., 2015. Insects: a protein-rich feed ingredient in pig and poultry diets. Animal Frontiers 5: 45-50.

Venter, J.C., Adams, M.D., Myers, E.W., Li, P.W., Mural, R.J., Sutton, G.G., Smith, H.O., Yandell, M., Evans, C.A., Holt, R.A., Gocayne, J.D., Amanatides, P., Ballew, R.M., Huson, D.H., Wortman, J.R., *et al.*, 2001. The sequence of the human genome. Science 291: 1304-1351. https://doi.org/10.1126/science.1058040

Vigneron, A., Jehan, C., Rigaud, T. and Moret, Y., 2019. Immune defenses of a beneficial pest: the mealworm beetle, *Tenebrio molitor*. Frontiers in Physiology 10: 138. https://doi.org/10.3389/fphys.2019.00138

Wallberg, A., Glemin, S. and Webster, M.T., 2015. Extreme recombination frequencies shape genome variation and evolution in the honeybee, *Apis mellifera*. PLoS Genetics 11: e1005189. https://doi.org/10.1371/journal.pgen.1005189

Wang, Y.S. and Shelomi, M., 2017. Review of black soldier fly (*Hermetia illucens*) as animal feed and human food. Foods 6(10): 91. https://doi.org/10.3390/foods6100091

Wei, W., Xin, H., Roy, B., Dai, J., Miao, Y. and Gao, G., 2014. Heritable genome editing with CRISPR/Cas9 in the silkworm, *Bombyx mori*. PLoS ONE 9: e101210. https://doi.org/10.1371/journal.pone.0101210

Wen, H., Lan, X., Zhang, Y., Zhao, T., Wang, Y., Kajiura, Z. and Nakagaki, M., 2010. Transgenic silkworms (*Bombyx mori*) produce recombinant spider dragline silk in cocoons. Molecular biology reports 37: 1815-1821.

Werren, J.H., Baldo, L. and Clark, M.E., 2008. *Wolbachia*: master manipulators of invertebrate biology. Nature Reviews Microbiology 6: 741-751. https://doi.org/10.1038/nrmicro1969

Werren, J.H. and Windsor, D.M., 2000. *Wolbachia* infection frequencies in insects: evidence of a global equilibrium? Proceedings of the Royal Society B: Biological Sciences 267: 1277-1285. https://doi.org/10.1098/rspb.2000.1139

White, E.B., Debach, P. and Garber, M.J., 1970. Artificial selection for genetic adaptation to temperature extremes in *Aphytis lingnanensis* Compere (Hymenoptera: Aphelinidae). Hilgardia 40: 161-192.

Wilkes, A., 1942. The influence of selection on the preferendum of a chalcid (*Microplectron fuscipennis* Zett.) and its significance in the biological control of an insect pest. Proceedings of the Royal Society of London. Series B-Biological Sciences 130: 400-415.

Wilkes, A., 1947. The effects of selective breeding on the laboratory propagation of insect parasites. Proceedings of the Royal Society of London. Series B-Biological Sciences 134: 227-245.

Wilkes, A., Pielou, D. and Glasser, R., 1952. Selection for DDT tolerance in a beneficial insect. In: Conference on Insecticide Resistance and Insect Physiology, December 8-9, 1951, University of Cincinnati. National Academy of Sciences, National Research Council, Vol. 219, pp. 78-81.

Wilkie, R.M., 2018. 'Minilivestock' farming: who is farming edible insects in Europe and North America? Journal of Sociology 54: 520-537.

Wilson, A.D.M., Whattam, E.M., Bennett, R., Visanuvimol, L., Lauzon, C. and Bertram, S.M., 2010. Behavioral correlations across activity, mating, exploration, aggression, and antipredator contexts in the European house cricket, *Acheta domesticus*. Behavioral Ecology and Sociobiology 64: 703-715. https://doi.org/10.1007/s00265-009-0888-1

Winston, M.L. and Katz, S.J., 1982. Foraging differences between cross-fostered honeybee workers (*Apis mellifera*) of European and Africanized races. Behavioral Ecology and Sociobiology 10: 125-129.

Worden, B.D. and Parker, P.G., 2001. Polyandry in grain beetles, *Tenebrio molitor*, leads to greater reproductive success: material or genetic benefits? Behavioral Ecology 12: 761-767.

Xia, Q., Guo, Y., Zhang, Z., Li, D., Xuan, Z., Li, Z., Dai, F., Li, Y., Cheng, D. and Li, R., 2009. Complete resequencing of 40 genomes reveals domestication events and genes in silkworm (*Bombyx*). Science 326: 433-436.

Xu, J., Chen, R.M., Chen, S.Q., Chen, K., Tang, L.M., Yang, D.H., Yang, X., Zhang, Y., Song, H.S. and Huang, Y.P., 2019. Identification of a germline-expression promoter for genome editing in *Bombyx mori*. Insect Science 26: 991-999. https://doi.org/10.1111/1744-7917.12657

Xu, J., Dong, Q., Yu, Y., Niu, B., Ji, D., Li, M., Huang, Y., Chen, X. and Tan, A., 2018. Mass spider silk production through targeted gene replacement in *Bombyx mori*. Proceedings of the National Academy of Sciences 115: 8757-8762. https://doi.org/10.1073/pnas.1806805115

Yang, S.S., Brandon, A.M., Andrew Flanagan, J.C., Yang, J., Ning, D., Cai, S.Y., Fan, H.Q., Wang, Z.Y., Ren, J., Benbow, E., Ren, N.Q., Waymouth, R.M., Zhou, J., Criddle, C.S. and Wu, W.M., 2018a. Biodegradation of polystyrene wastes in yellow mealworms (larvae of *Tenebrio molitor* Linnaeus): factors affecting biodegradation rates and the ability of polystyrene-fed larvae to complete their life cycle. Chemosphere 191: 979-989. https://doi.org/10.1016/j.chemosphere.2017.10.117

Yang, S.S., Wu, W.M., Brandon, A.M., Fan, H.Q., Receveur, J.P., Li, Y., Wang, Z.Y., Fan, R., Mc-

Clellan, R.L., Gao, S.H., Ning, D., Phillips, D.H., Peng, B.Y., Wang, H., Cai, S.Y., Li, P., *et al.*, 2018b. Ubiquity of polystyrene digestion and biodegradation within yellow mealworms, larvae of *Tenebrio molitor* Linnaeus (Coleoptera: Tenebrionidae). Chemosphere 212: 262-271. https://doi.org/10.1016/j.chemosphere.2018.08.078

Zatsepina, O.G., Velikodvorskaia, V.V., Molodtsov, V.B., Garbuz, D., Lerman, D.N., Bettencourt, B.R., Feder, M.E. and Evgenev, M.B., 2001. A *Drosophila melanogaster* strain from sub-equatorial Africa has exceptional thermotolerance but decreased Hsp70 expression. Journal of Experimental Biology 204: 1869-1881.

Zhan, S., Fang, G., Cai, M., Kou, Z., Xu, J., Cao, Y., Bai, L., Zhang, Y., Jiang, Y., Luo, X., Xu, J., Xu, X., Zheng, L., Yu, Z., Yang, H., Zhang, Z., Wang, S., Tomberlin, J.K., Zhang, J. and Huang, Y., 2020. Genomic landscape and genetic manipulation of the black soldier fly *Hermetia illucens*, a natural waste recycler. Cell Research 30: 50-60. https://doi.org/10.1038/s41422-019-0252-6

Zhou, F., Tomberlin, J.K., Zheng, L., Yu, Z. and Zhang, J., 2013. Developmental and waste reduction plasticity of three black soldier fly strains (Diptera: Stratiomyidae) raised on different livestock manures. Journal of Medical Entomology 50: 1224-1230. https://doi.org/10.1603/me13021

Zhou, Z., Yang, H. and Zhong, B., 2008. From genome to proteome: great progress in the domesticated silkworm (*Bombyx mori* L.). Acta Biochimica et Biophysica Sinica 40: 601-611. https://doi.org/10.1111/j.1745-7270.2008.00432.x

Zhu, Z., Rehman, K.U., Yu, Y., Liu, X., Wang, H., Tomberlin, J.K., Sze, S.H., Cai, M., Zhang, J., Yu, Z., Zheng, J. and Zheng, L., 2019. De novo transcriptome sequencing and analysis revealed the molecular basis of rapid fat accumulation by black soldier fly (*Hermetia illucens*, L.) for development of insectival biodiesel. Biotechnology for Biofuels 12: 194. https://doi.org/10.1186/s13068-019-1531-7

Insect left-over substrate as plant fertiliser

M. Chavez[1,2]* and M. Uchanski[1,2]

[1]Department of Horticulture and Landscape Architecture, Colorado State University, 1173 campus delivery, Fort Collins, CO 80523, USA; [2]Graduate Degree Program in Ecology, Colorado State University, 102 Johnson Hall, Fort Collins, CO 80523, USA; *maria.chavez@colostate.edu

Abstract

The production of insect protein as human food and livestock feed (entomophagy) may provide a more environmentally beneficial alternative to traditional animal agriculture. However, the resulting waste product from insect production has resulted in large accumulations of left-over substrate and frass. Due to its nutrient and microbial profile, this left-over product has the potential to be utilised as a biofertiliser for high value crop production. Studies have been conducted using the frass of various insects (e.g. black soldier flies, houseflies, and mealworms) to monitor its impact on crop productivity. Overall, frass tends to have similar or better results when compared to inorganic fertilisers, especially when combined with them. Aside from productivity and growth, frass may also preserve soil fertility by decreasing leaching and infiltration, and reducing the prevalence of disease and pathogens. In addition, chitin found in frass also has beneficial properties for plant/crop growth and disease resistance. Monitoring the dietary inputs of industrially reared insects may be the best way of mitigating the potential negative impacts of frass application, such as increased electrical conductivity and heavy metal toxicity. No single study confirms all of these benefits at once. Future studies should focus building onto these results by demonstrating systems levels benefits.

Keywords

biofertiliser – chitin – decomposition – entomophagy – fertility amendment

1 Introduction

Entomophagy is defined as the consumption of insect protein for human and animal nutrition (Table 1). The practice of eating insects is a well-known tradition in many areas of the world and is beginning to become more common in the United States (Durst *et al.*, 2010). Currently the main source of protein in the modern American diet comes from livestock, such as cattle, poultry, and swine. Animal agriculture is responsible for the largest extractions and loss of water in the environment, occupies large amounts of land, and is one of the highest producers of atmospheric methane, which is a potent greenhouse gas (Sims *et al.*, 2005). Additionally, accumulated manure may result in environmental degradation, known as eutrophication, when bacteria and nutrients runoff into surrounding streams and lakes. Infiltration into the groundwater from these same sources can also occur, causing complications for the industry (Gay and Knowlton, 2005). Insects can provide an alternative protein source that has the potential to improve consumer health and reduce the negative impacts on the environment compared to other sources of fertility (Chia *et al.*, 2019).

Though the entomophagy industry provides a possible solution to many agricultural problems, it has also resulted in a few of its own. One of particular concern is the waste created during the process of insect rearing. Currently, this digested product does not have any widely adopted practices of application. As a result, the digestive residue, also known as frass or digestate, is mainly discarded as a waste product. Due to its concentration of nitrogen and phosphorus, this product has the potential to utilise existing waste streams to create a fertility amendment for food crop production (Schmitt, 2020). These additional nutrients from frass may serve as a supplement or substitute for the inorganic fertilisers that are applied during typical crop production (Figure 1). Decreases in inorganic fertiliser use can reduce the subsequent environmental problems, such as eutrophication.

Crops require a combination of resources for survival such as carbon dioxide, mineral nutrients, heat, and light (Swiader and Ware, 2002). The most important mineral nutrients include nitrogen, phosphorus, and potassium. These are abun-

TABLE 1 Definition of key terms

Term	Definition	Reference
Bioconversion	Digestion of organic material into a more enriched material via animal intake and excrement.	Putra *et al.* (2017)
Biofertiliser	Bioconverted waste material via insect digestion.	Xiao *et al.* (2018)
Chitin	Polysaccharide found in insect exoskeletons.	Sharp (2013)
Decomposition	Modification of detritus in the soil.	Chapin *et al.* (2011)
Digestate	Type of frass produced during insect production, often black soldier flies.	Temple *et al.* (2013)
Entomophagy	The practice of eating insects, especially by people.	Durst *et al.* (2010)
Frass	Insect excrement (solid).	Kagata and Ohgushi (2012a)
Fertility amendment	Additional supplement to provide nutrients for crops.	Schmitt (2020)
Herbivory	Consumption of plant material by animals.	Chapin *et al.* (2011)

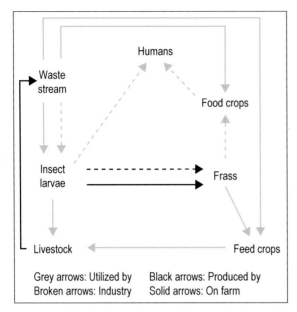

FIGURE 1   Production and entomophagy: rearing insects for human and livestock consumption

dant on earth in their elemental forms, but must be present in bioavailable forms in order to be absorbed and utilised during the vegetative growth of plants. Unfortunately, bioavailable nutrients are considerably more limited and sparse within our biosphere. Because of this cultivated food crop production requires special soil fertility management in order to meet nutritional needs (Swiader and Ware, 2002).

There are several large challenges with the conventional system of fertility management. Many of the resources required are limited and non-renewable. For example, phosphorus is mined in several areas throughout the world for inorganic liquid fertigation. However, these sources are being depleted at extremely high rates, complicating the likelihood that our current reserves can sustain future production (Koppelaar and Weikard, 2013). Peat is also in high demand by agricultural producers and is found only in limited and sensitive wetland areas. Mining peatlands causes dramatic losses in biodiversity and cultural services. Additionally, drained peatlands are large producers of greenhouse gas emissions (Couwenberg et al., 2011).

Others concerns for agricultural production include nutrient infiltration and runoff into surrounding landscapes. This occurs because farmers apply nutrients in higher dosages than is actually accessible to the plant (Chapin et al., 2011). When nutrients are not absorbed by the plant, they leach into the soil or across land surfaces and are carried into the surrounding environments. The nutrient pollution to the system increases algae, plant growth, and oxygen levels. This results in environmental degradation, known as eutrophication, and leads to large decreases in biodiversity (Chapin et al., 2011). Though nitrogen is the most common problem due to its frequent over application in fertiliser regimes, phosphorus run-off is a large detriment to natural systems as well (Eghball and Gilley, 1999). Phosphorus run-off increases with soil erosion, a problem that is common on agricultural land. (Daverede *et al.*, 2004).

Additionally, ammonia emissions from wet manure and compost applications to production fields have considerable impacts on the environment (Fuchs *et al.*, 2008). Volatised ammonia can travel large distances and be deposited as nitrogen pollution in surrounding lakes and oceans, escalating eutrophication in those systems. It can also be deposited in soil increasing acidity and reducing plant growth (Gay and Knowlton, 2005). All of these reasons provide further justification for the increasing demand and interest in alternative amendments for crop fertility, or fertility amendments (Table 1).

Frass is solid insect waste material that has been converted to a microbially rich substance through digestion. This results in a product higher in organic matter and lower in nitrogen and phosphorus content than its' parent source material (Kagata and Ohgushi, 2012a). Once digested, nutrients such as nitrogen, phosphorus, and potassium become more bioavailable for plants (Table 1). The total nutrients are reduced and the leaching of these nutrients outside of the system decrease as well. The ratio of nitrogen to phosphorus decreases after biological digestion, or bioconversion (Table 1), and provides a more balanced product for plants than the traditional amendments. This is valuable because most composts are very high in nitrogen and low in phosphorus (Swiader and Ware, 2002).

Chitin is the predominant substance in the shed exoskeletons of insects, and is often found mixed in their left-over substrate, such as waste material and digestate. It naturally occurs in large abundances, making it ideal for application in resource-intensive crop production. As such, over the past few decades chitin has gained interest as a potential biofertiliser, and is mostly sourced from salt water crustacean exoskeletons (Sharp, 2013). For example, shrimp waste that is high in nitrogen and phosphorus, also contains abundant chitin (Fatima *et al.*, 2018).

Studies have demonstrated that crop production increases in the presence of chitin. Lettuce leaf number, leaf area per plant, leaf dry weight, leaf fresh weight, and chlorophyll index increased in the presence of chitosan applications that were 0.05, 0.10, 0.15 and 0.02% of soil mixtures (Xu and Mou, 2018). Another study by Spiegel *et al.* (1987) observed that chitin applications increased fresh root weight of bean, corn and tomato plants. However, more observations in economic yield improvement will be important for this field going forward.

Chitin may also improve pathogen and disease resistance in plants. A study by Bell *et al.* (1998) observed that 3 mg/ml of chitin soil applications reduced both incidence and severity of fusarium yellows (*Fusarium oxysporum*) in celery petioles, while celery with chitosan root dips experienced less severe incidences of yellows. Chitin applications in 0.5% of soil mixtures have been shown to reduce incidence of the root-knot nematode, *Meloidogyne fava*, in bean, corn, and tomato plants (Spiegel *et al.*, 1987). Chitin can enhance the antifungal action of bacterial isolates in root rot disease (Ahmed *et al.*, 2003). Chitin that has been deproteinised and demineralised from shrimp samples ranges from 5 to 32% of their dry weight. Chitin from insect samples range from 2 to 36% (Hahn *et al.*, 2020). Since shrimp chitin has proved to produce plant growth changes related to fertility and pathogen protection, the similarities in the chitin concentration of insects reared for entomophagy may drive similar processes as well.

Insects contribute to nutrient diversion through decomposition and herbivory (Table 1) (Chapin *et al.*, 2011). Though 95% of decomposition occurs by microbes, such as bacteria and fungi, insects and other arthropods can also influence decomposition in several different ways, both directly and indirectly. Insects can also influence the chemical alteration stage of decomposition. Certain species act directly as decomposers by eating dead organic matter that is coated in bacteria. Insects ingest the bacterial coated material and recycle the organic matter as a waste product that is higher in carbon and nitrogen (Chapin *et al.*, 2011). Decomposers may also have specialised mechanisms for consuming and processing plant material. This can include microbial populations in their gut to digest recalcitrant lignin and cellulose compounds (Kagata and Ohgushi, 2012a).

Herbivory can help increase decomposition rates (Chapman *et al.*, 2003). Specifically, insect herbivores contribute to decomposition by consuming live plant material and converting it into frass. Frass is rich in bioavailable compounds that are beneficial to plants and other smaller decomposers. Additionally this process may injure leaves that are still attached to the stem, forcing early abscission of damaged leaves and their deposition on the soil floor (Chapin *et al.*, 2011). Frass has been shown to decompose more quickly than leaf litter resulting in large releases of total nitrogen (Kagata and Ohgushi, 2012b). The impact of insect solid waste excrement (frass) on soil nutrient dynamics has been well studied over the course of the past few decades. Frost and Hunter (2004) analysed how frass deposition influenced soil carbon and nitrogen. Northern red oak (*Quercus rubra*) plants were used to create an artificial system to observe the herbivory of the eastern tent caterpillar (*Malacosoma americanum*). Herbivory of *Q. rubra* decreased nitrogen in the soil, while additions of *M. americanum* frass increased inorganic soil nitrogen and carbon.

Insect herbivore diet selection can affect litter quality and subsequently nutrient cycling. Changes in nutrient cycling alter composition of plant species and communities. Therefore, nitrogen availability and plant abundance increases with the presence of insects, such as grasshoppers (Belovsky and Slade, 2000). Entomophagy seeks to produce insect-based protein and take advantage of these processes to reduce environmental impacts for production.

2   Entomophagy

Entomophagy is the animal consumption of insects (Durst *et al.*, 2010). A variety of insects are grown for the purpose of animal consumption and each has their own benefits and challenges. Insect production requires minimal land, supports a closed nutrient cycle, benefits a more circular economy, and provides alternative sources for feed stock (Figure 1) (Chia *et al.*, 2019). We will briefly examine a few insects that are of interest for frass utilisation.

Black soldier fly (*Hermetia illucens*) larvae have become a species of particular interest for research in the past few decades. Black soldier fly larvae are efficient in reducing and metabolising waste, such as animal manure. Studies have shown they can reduce manure dry matter by 30 to 50% (Miranda *et al.*, 2019) and metabolise approximately 17.6 to 32.5% of poultry waste fed (Diener *et al.*, 2009). Additionally, the protein and fat content of the larvae is high and depends on the insects' diet (Table 2) (Miranda *et al.*, 2019). They are a valuable resource to livestock farmers who accumulate considerable amounts of wet manure. Wet manure is problematic for farmers as it is expensive and labour intensive to manage.

Farmers can use black soldier flies to reduce accumulated manure moisture and dry matter, while also harvesting the larvae themselves as a protein rich food source for other livestock, such as poultry. Black soldier fly larvae are also known to provide additional benefits such as reducing the oviposition of house flies (Bradley and Sheppard, 1984), *Escherichia coli* populations (Liu *et al.*, 2008), and odour causing compounds in manure (Beskin *et al.*, 2018). Black soldier fly larvae can be reared on already existing waste streams and can possibly decrease the ecological consequences of food production (Bosch *et al.*, 2019).

A less adopted decomposer of interest are beetle larvae, or mealworms. Mealworms are either fed to livestock, mostly poultry, or processed as a flour for human consumption. There are three species of edible meal worms: *Tenebrio molitor*, *Zophobas atratus*, and *Alphitobius diaperinus*. Conversion efficiency of feed by larvae depends on species and diet, ranging from 6.36 to 34.37% of dry matter (Van Broekhoven *et al.*, 2015). Similar to black soldier fly larvae, meal worms are also high in protein and fat (Table 2). Like black soldier fly larvae, meal worms also produce less

TABLE 2   Protein and fat content of insect larvae reared for entomophagy

Insect	Crude protein (%)	Crude fat (%)	Reference
Hermetia illucens (black soldier fly)	42–48	28	Miranda *et al.* (2019)
Tenebrio molitor (mealworm)	47	25	Van Broekhoven *et al.* (2015)
Zophobas atratus (mealworm)	40	38	Van Broekhoven *et al.* (2015)
Alphitobius diaperinu (mealworm)	64	19	Van Broekhoven *et al.* (2015)
Musca domestica (housefly)	56	22	Yang *et al.* (2015a)

greenhouse gas emissions and require less land than other animal livestock (Oonincx and De Boer, 2012).

Houseflies (*Musca domestica*) have similar fat and protein content to mealworms. They are typically acknowledged as a nuisance to livestock farmers, but efforts have been made to demonstrate that their larvae could be useful as livestock feed as well. Houseflies reared on gibberellin fermentation produce similar protein and fat concentrations to meal worms and black soldier fly larvae (Yang *et al.*, 2015b). Poultry waste bioconverted by housefly larvae have resulted in large reductions of moisture content (75 to 50%). The final products were also greatly reduced in biomass (80%) and become an odourless, granular substance (El Boushy, 1991). Improvements in livestock body conditions have also been observed. Broiler chickens raised on diets supplemented with housefly larvae were significantly heavier than control groups that were not fed larvae (Hwangbo *et al.*, 2009).

All three of these insects provide promise for altering the typical framework of protein production and introducing novel food sources into human and livestock diets.

3   Use of insect frass as fertiliser

Horticultural food production often demands large quantities of inorganic fertilisers to promote plant growth (Swiader and Ware, 2002). The value of insect production for protein sources is recognised, but what is less understood is how nutrient diversion during this process could be utilised as potential plant biofertiliser. The bioconversion of waste streams by insects in the entomophagy industry is an exciting opportunity for food crop production and biofertiliser utilisation. Biofertilisers are an interesting fertility option in alternative agriculture and may provide a more sustainable option for farmers concerned with eutrophication and soil degradation (Xiao *et al.*, 2018). Waste from insect production could also potentially serve as a peat replacement substrate or as an additional amendment to a reduced inorganic fertiliser regime. Several studies have been conducted to investigate uses of insect frass for application in horticultural production.

The frass of various insects have been analysed to determine their nutrient profile (Table 3). These profiles are useful as a comparison to conventional manures and composts. Manure that has not been composted tends to have the highest carbon, nitrogen, and phosphorus concentrations (Table 3). Insect digested manure and traditionally composted manure tend to have similar carbon, nitrogen, and phosphorus concentrations (Table 3). For example, a study by Zhang *et al.* (2012) monitored the impacts of manure quality on bioconversion by houseflies. Manure treated with houseflies saw decreases in total nitrogen and phosphorus, while bioavailable nitrogen and phosphorus increased. Temple *et al.* (2013) conducted an incubation trial to assess the nutrient profile and availability of black soldier fly

TABLE 3   Insect larvae frass composition

Waste treatment	pH	EC (dS/m)	Moisture (%)	C (%)	N (%)	P (%)	K (%)	S (%)	Ca (%)	Mg (%)	Reference
Mealworm	–	–	–	38.9	2.92	1.53	1.86	0.18	0.1	0.54	Poveda et al. (2019)
Mealworm	–	–	–	38.8	2.67	1.44	1.97	0.17	0.09	0.52	Poveda et al. (2019)
Mealworm larvae	–	–	–	42.44	7.75	1.02	1.15	0.28	0.11	0.34	Poveda et al. (2019)
Black soldier fly	5.5	44	–	42.9	4.54	1.23	2.44	0.49	0.64	0.13	Temple et al. (2013)
Black soldier fly	–	–	–	31.1	1.27	0.46	2.79	–	–	–	Rosmiati et al. (2017)
Black soldier fly	8.84	8.5	51.4	35.2	4.4	5.2	4.1	–	4.5	0.8	Setti et al. (2019)
Housefly	7.78	–	18.55	3.36	4.66	2.7	1.3	–	–	10.55	Zhu et al. (2015)
Housefly	8.5	–	29.8	78.23	3.2	2	–	–	–	0	Yang et al. (2015a)
Traditional compost	7.3	11	–	40.7	2.8	1.81	2.24	0.65	3.69	0.66	Temple et al. (2013)
Untreated manure	6.59	–	72.42	84.8	6.23	3.72	2.4	–	–	10.55	Zhu et al. (2015)
Peat	6.1	1.3	–		0.15						Setti et al. (2019)

larvae digestate compared to other organic amendments commonly applied in the vegetable crops produced. Zhu *et al.* (2015) used houseflies to compost swine manure and analysed the left over substrates. Larvae compost was highest in total phosphorus and potassium concentrations and lowest in total nitrogen, creating a more balanced nutrient regime for plant growth. Larvae treated compost also had the highest levels of micro nutrients such as zinc, copper, iron, and cadmium. Zhu *et al.* (2012) observed decreases in total nitrogen and water soluble carbon when manure was treated with houseflies as compared to naturally composted manure. Detoxification and pH were also higher in fly treated manure.

Plant utilisation of nutrients may also be improved by frass application. Putra *et al.* (2017) observed differences in nutrient utilisation depending on treatment. Nitrogen utilisation was highest in raw manure treatments. However, phosphorus and potassium utilisation were highest in in the black soldier fly digestate treatments. Improving nutrient utilisation of a quickly depleting resource, such as phosphorus, may provide one solution to reducing phosphorus inputs and improving plant growth. Companies that produce insect for food and feed are also highly interested in utilising the waste stream they create during this process. Enterra Feed Corporations (Maple Ridge, BC, Canada) conducted a study to test their black soldier fly larvae digestate product in field, seedling, and incubation trials (Temple *et al.*, 2013).

Frass may also improve soil properties, improving the overall environment for crop production. Houben *et al.* (2020) assessed the nutrient profile of mealworm frass and revealed similar concentrations of nitrogen, phosphorus, and potassium to raw manure. The frass in this experiment mineralised quickly, resulting in increased rates of decomposition. This can reduce nutrient leaching into surrounding environments. Aligned with these observations, soluble phosphorus concentrations were significantly lower in frass treatments, continuing to decrease the likelihood of leaching and infiltration. Klammsteiner *et al.* (2019) fed black soldier fly larvae different diets and utilised their resulting frass as separate treatments. They monitored the impacts of each frass treatment on ryegrass productivity. They did not see any differences in soil total carbon and nitrogen or biomass. However, overall in the black soldier fly larvae treatments there was a significant increase in phosphorous bioavailability, similar to what had been seen by Temple *et al.* (2013). Zhan and Quilliam (2017) observed improvements in soil organic matter from all treatments, but most consistent improvements occurred in the black soldier fly larvae digestate treatments. It seems that at lower levels black soldier fly larvae frass can produce comparable results to traditional compost and conventional inorganic fertilisers, while also improving soil organic matter. Increases in organic carbon were also observed in house fly treatments. Gibberellin fermentation residue digested by houseflies contained 1.3% potassium, 3.2% total nitrogen, 2.0% inorganic phosphorus, and 91.5% organic matter (Yang *et al.*, 2015b). These concentrations are appropriate proportions to what is often required for crop productivity and further indicate housefly waste as a potential biofertiliser.

Other characteristics of waste material can be improved through bioconversion. The mechanical conversion of waste could also be helpful in improving the waste sources as amendments in crop production. Housefly (*M. domestica*) larvae feed on a variety of waste materials and can efficiently convert animal manures into odourless, coarse substrates (Čičková *et al.*, 2015). Zhang *et al.* (2012) saw similar decreases in moisture, faecal coliforms, and odour causing compounds. Zhu *et al.* (2012) also observed increases in compost maturation time, detoxification, and pH and decreases in percent moisture when manure was treated with house flies as compared to naturally composted manure.

Frass can serve various purposes in crop production systems (Table 4). Most studies have been conducted in pots within a controlled environment setting, such as a greenhouse or growth chamber. Some studies have considered it as a growth medium replacement for peat. Setti *et al.* (2019) applied black soldier fly larvae digestate in lettuce, tomato, and basil crops. They analysed the chemical characteristics of the black soldier fly larvae digestate (Table 3). From these results, we can determine that a modest decrease in peat utilisation, supplemented with black soldier fly larvae digestate, can sustain the current conventional growth parameters of crop production. Over time this 10% reduction in peat extraction, if widely adopted, could result in some conservation of peat resources.

Frass can also be used to reduce existing waste streams and convert those streams into useful plant amendments. Coffee husks, a coffee waste product, are abundant in tropical production areas around the world. The organic waste material created from this accumulation has little value, but can possibly be improved by bioconversion, or the intake and excrement of organic material by animals. Putra *et al.* (2017) and Rosmiati *et al.* (2017) analysed coffee husks digested by black soldier fly larvae for their nutrient profile, effectiveness as a fertiliser treatment, and nutrient utility. The black soldier fly larvae digestate had lower carbon, nitrogen and potassium compared to the undigested coffee husks, but higher amounts of phosphorus. How-

TABLE 4 Utilisation of insect frass in crop production

Use	Reference
Replacement for peat	Setti *et al.* (2019)
Improved waste streams for fertility amendment	Putra *et al.* (2017); Rosmiati *et al.* (2017)
Reduced plant pest, pathogens, and disease	Choi and Hassanzadeh (2019); Klammsteiner *et al.* (2019); Vickerson *et al.* (2016)
Improved resistance to stress	Poveda *et al.* (2019)
Increased crop productivity and yield	Beesigamukama *et al.* (2020); Buenvinida and Tamban (2016); Choi and Hassanzadeh (2019); Houben *et al.* (2020); Poveda *et al.* (2019); Putra *et al.* (2017); Rosmiati *et al.* (2017); Setti *et al.* (2019); Temple *et al.* (2013); Xiao *et al.* (2018); Zhan and Quilliam (2017); Zhu *et al.* (2012)

ever, based on these results, major modifications need to be made in coffee husk bioconversion before it can be considered a valuable replacement for raw manure.

Insect frass may also reduce incidences of plant pathogen and disease. Several studies show the relationship between the use of insect frass as promoters of resistance to biotic and abiotic stressors. For example, black soldier fly larvae frass has been proposed to have insecticidal qualities. Insecticides are natural compounds that can decrease pest insect populations. It was shown that when applied as an insecticide through this method, the frass reduced wireworm populations (Vickerson *et al.*, 2016).

Choi and Hassanzadeh, 2019 also demonstrated that black soldier fly larvae frass inhibited the growth of *F. oxysporum* and *Rhizoctonia solani*. Additionally treatments including the frass also reduced disease prevalence in bean plants inoculated with *Trichoderma*. A study by Klammsteiner *et al.* (2019) revealed black soldier fly larvae frass produced from chicken feed, fresh produce, and grass cutting diets entirely eliminated *Salmonella* population in the treatments. *E. coli* populations were reduced in the fresh produce and grass cutting diets, but not the chicken manure. This indicates that bioconversion of animal wastes could decrease risks of human pathogen transmission as well as improve crop production.

Poveda *et al.* (2019) was able to demonstrate the role of meal worm frass (*T. molitor*) in lettuce resistance to abiotic stresses under field conditions. Three types of meal worm frass were produced from different diets. These frass treatments were exposed to salinity, drought, and flooding stress. Typically fertilisation improved survival and performance in stress conditions. Two of the three diets saw improved dry weight, root length, and aerial part length compared to the negative unfertilised control and equal to the fertilised positive control. This indicates that meal worm frass can provide equivalent support for crops under stress conditions as compared to a conventional fertiliser.

Various crops have been studied to determine the impact of frass on productivity (Table 5). Most of these studies have been pot studies in controlled environments. Lettuce amended with black soldier fly larvae digestate at a 90:10 ratio (peat:digestate) produced increases of height, stem diameter, leaf dry weight, and leaf area that were statistically comparable, or higher than the control (Setti *et al.*, 2019). Putra *et al.* (2017) also treated lettuce with black soldier fly larvae digestate. The digestate treatments achieved significantly higher wet and dry weights than the undigested coffee husks, though the highest dry and wet weights of lettuce vegetative growth were seen in the raw manure treatments. A related study conducted by Rosmiati *et al.* (2017) showed similar results with applications of black soldier fly larvae bioconverted coffee husks on lettuce production. Plant height, number of leaves, leaf area, and chlorophyll content of lettuce were all highest in the raw manure treatments, similar to Putra *et al.* (2017).

A study by Poveda *et al.* (2019) revealed that mealworm (*T. molitor*) frass may have some success with improving chard yield. The fresh weight, length aerial part,

TABLE 5   Influence of insect larvae frass and other amendments on crop productivity in controlled environment studies

Crop	Amendment	Application (% media)	Germination index (%)	Height (cm)	Weight (g)	Reference
Baby lettuce	black soldier fly	10	95.4	11.7	0.9	Setti et al. (2019)
Baby lettuce	black soldier fly	20	70.7	10.3	0.7	Setti et al. (2019)
Lettuce	black soldier fly	50	–	–	21.00	Putra et al. (2017)
Lettuce	black soldier fly	33.33	–	–	11.78	Putra et al. (2017)
Lettuce	black soldier fly	50	–	14,860	–	Rosmiati et al. (2017)
Lettuce	black soldier fly	33.33	–	12,390	–	Rosmiati et al. (2017)
Chard	mealworm	2	–	70	70	Poveda et al. (2019)
Basil	black soldier fly	10	120.1	11.8	0.15	Setti et al. (2019)
Basil	black soldier fly	20	143.4	8.3	0.09	Setti et al. (2019)
Tomato	black soldier fly	10	122.0	18.1	0.7	Setti et al. (2019)
Tomato	black soldier fly	20	113.3	18.6	0.5	Setti et al. (2019)
Spring onion	black soldier fly	5	–	359.7	1.0	Zhan and Quilliam (2017)
Spring onion	black soldier fly	10	–	250.6	1.1	Zhan and Quilliam (2017)
Spring onion	black soldier fly	15	–	66.7	1.1	Zhan and Quilliam (2017)
Green bush bean	black soldier fly	1.5	–	56	1.83	Choi and Hassanzadeh (2019)
Barley	mealworm	50	–	–	3	Houben et al. (2020)
Barley	mealworm	100	–	–	3.5	Houben et al. (2020)
Maize	black soldier fly	0.186	–	95.4	6.83	Gärttling et al. (2020)
Maize	black soldier fly	1.09	–	96.7	6.61	Gärttling et al. (2020)
Baby lettuce	peat	100	78.4	8.1	0.4	Setti et al. (2019)
Baby lettuce	peat and fertiliser	100	94.9	11.7	0.7	Setti et al. (2019)
Basil	peat	100	97.5	5.3	0.11	Setti et al. (2019)
Basil	peat and fertiliser	100	100.0	9.8	0.16	Setti et al. (2019)
Tomato	peat	100	98.7	14.9	0.2	Setti et al. (2019)
Tomato	peat and fertiliser	100	91.3	18.3	0.6	Setti et al. (2019)

width of basal stem, and chlorophyll content of the leaves all indicated significant increases with application of meal worm frass when compared to the unfertilised negative control. Statistically the fertilised positive control and the frass treatments produced comparable measurements indicating that frass could be a substitute for inorganic commercial fertiliser.

Basil experienced increases in height, stem diameter, leaf number, leaf dry weight, stem dry weight, total dry weight, and leaf area that were statistically comparable or higher to the control when black soldier fly larvae digestate was applied in a 90:10 (peat:digestate) mixture (Setti *et al.*, 2019). One study collected lab raised caterpillar frass and added the frass to basil plants in different quantities. The greatest plant heights occurred in the treatments with the largest amount of frass, 15 and 10 gram applications. Leaf width was greatest in the two higher frass treatments as well (Buenvinida and Tamban, 2016).

Setti *et al.* (2019) applied black soldier fly larvae digestate to tomato plants. Height, leaf number, leaf dry weight, stem dry weight, total dry weight, and leaf area were highest in the 90:10 mixture and statistically comparable, or higher than the control.

Zhan and Quilliam (2017) investigated the influence of black soldier fly larvae frass on spring onion plants. Plant height and weight were similar across treatments. The lowest heights and weights were observed in the highest level frass treatment. Choi and Hassanzadeh demonstrated the use of black soldier fly larvae digestate as a fertiliser for green bush beans. They found that digestate in combination with humic acid produced the statistically highest height, biomass, and nitrate concentrations.

Houben *et al.* (2020) used mealworm frass to investigate its impact on barley productivity. Barley treated with mealworm frass applications had similar biomass, nitrogen, phosphorus, and potassium concentrations in comparison to inorganic fertiliser treatments. Maize treated with black soldier fly larvae digestate produced statistically comparable yields to other organic fertilisers, though these yields were significantly lower than the standard inorganic fertiliser (Gärttling *et al.*, 2020). These studies provide further evidence that application in a controlled environment can provide similar benefits to currently adopted fertiliser regime.

Studies conducted in the field indicate that bioconversion of waste material can improve its ability to benefit plant production, regardless of the production setting (Table 6). Enterra Feed Corporation applied black soldier fly larvae digestate at zero, five, and ten tons per hectare to lettuce crops in a field setting. Fresh weight and mortality of lettuce was statistically higher than the controls and improved by a second application of digestate (Temple *et al.*, 2013). They applied these same condition to bok choi crops in the field. Bok choi fresh weight was highest and morality was the lowest in the ten t/ha treatment (Temple *et al.*, 2013). Black soldier fly larvae digestate was applied to potato fields. Potato tubers produced the highest yield at five t/ha digestate compared to the other treatments. From these results,

TABLE 6   Influence of black soldier fly larvae frass on crop productivity in field studies

Crop	Waste	Application rate (kg/m^2)	Yield (kg/m^2)	Reference
Lettuce	Enterra Feed Corp	0.5	2.7	Temple *et al.* (2013)
Lettuce	Enterra Feed Corp	1.0	3.3	Temple *et al.* (2013)
Bok choi	Enterra Feed Corp	0.5	5.9	Temple *et al.* (2013)
Bok choi	Enterra Feed Corp	1.0	7.3	Temple *et al.* (2013)
Bean	Enterra Feed Corp	0.5	5.1	Temple *et al.* (2013)
Bean	Enterra Feed Corp	1.0	5.9	Temple *et al.* (2013)
Potato	Enterra Feed Corp	0.5	3.4	Temple *et al.* (2013)
Potato	Enterra Feed Corp	1.0	3.1	Temple *et al.* (2013)
Chinese Cabbage	–	1.21	0.279	Choi *et al.* (2009)
Maize	brewery	3.0	0.520	Beesigamukama *et al.* (2020)
Maize	brewery	6.0	0.595	Beesigamukama *et al.* (2020)
Maize	brewery	10.0	0.615	Beesigamukama *et al.* (2020)
Shallots	poultry	0.25	0.21	Quilliam *et al.* (2020)
Shallots	brewery	0.25	0.22	Quilliam *et al.* (2020)

they recommend that digestate should be applied to each crop at a rate of five tons per hectare, twice per season (Temple *et al.*, 2013).

Additional studies have been done that support these findings (Table 6). Choi *et al.* (2009) applied black soldier fly digestate to Chinese cabbage and saw statistically similar results to the standard fertiliser control. Beesigamukama *et al.* (2020) applied black soldier fly frass in three quantities of nitrogen content (30, 60, and 100 kg) and compared their impact on maize yield when a standard fertiliser was applied at the same rates. Thirty and 60 kg of black soldier fly larvae frass produced significantly higher maize yields than their corresponding standard fertiliser. The highest yields were produced at 100 kg and were statistically comparable to the corresponding standard fertiliser. Quilliam *et al.* (2020) utilised two different origins of waste on shallot crops. The statistically highest yields were produced when treated with brewery biofertiliser. Poultry waste fertiliser was also statistically comparable to a standard fertiliser treatment. These results indicate that black soldier fly larvae left-over substrates can sufficiently support yields comparable to a standard fertiliser regime. These adaptations to conventional practices in the field may also lead to less negative impacts that are associated with commonly used organic amendments, such as raw manure.

Different stages of crop growth can be useful in determining the impact of frass applications. Temple *et al.* (2013) conducted seedling trials with several different crops to assess black soldier fly larvae digestate as a growing medium for seeds in the field. Bok choi, onion, bean, and tomato seedlings saw highest dry yield when applied at 15 t/ha. Lettuce and squash saw highest dry yield at 10 t/ha. These results and their value for commercial production are highly dependent on crop and adaptation practice. Zhu *et al.* (2012) measured the germination index of Chinese cabbage and cucumber seeds treated with fly digested composts. The composts exhibited higher percentages of germination than those treated with naturally composted manure. Setti *et al.* (2019) also saw increases in germination index of lettuce when black soldier fly larvae digestate was applied to the crop. In the same study, basil experienced increased rates of emergence after digestate applications. This indicates that fly bioconversion of manure can improve emergence time in crop production.

Certain considerations may be required to improve the efficacy of frass application. A study by Xiao *et al.* (2018) combined the effects of a naturally found digestive bacteria from within black soldier fly larvae gut contents and black soldier fly larvae frass to analyse their synergistic impact on plant productivity. The bacteria and larvae treated manure saw higher reductions in manure biomass and increases in rate of reduction compared to the non-bacterial treated control. There were also lower levels of moisture and nutrients in the digested manure, which is desirable to livestock farmers. The bacteria treatments increased the germination index of Chinese cabbage and rape seeds compared to the control. Though this process is not entirely realistic for commercial production, gut isolated bacteria could provide further insights in improving bioconversion efficiency for biofertiliser production.

Vermicomposting is a common practice in small scale, organic production and the literature supporting its usefulness as a fertiliser for high value specialty crops is voluminous. Earthworms are added to manure in order to break down carbon-based materials, and this biological conversion creates several positive impacts. Compared to other fertilisers, vermicompost results in lower runoff and infiltration rates (Elliot *et al.*, 2007), and may improve yield, organic matter content, and soil fertility (Paul and Metzger, 2005). Though earthworms are not insects the process of bioconversion undergone by the worms is similar to the digestive processes of insects (Chapin *et al.*, 2011). Because of these similarities, it is reasonable to anticipate that they are capable of decomposing similar food sources and that insect frass may provide many of the same benefits as worm castings to soil fertility and crop production.

Similar to insect frass, several studies have produced significant increases in vegetable crops total yields when vermicompost was applied in combination with inorganic fertilisers (Alam *et al.*, 2007; Ansari, 2008; Arancon *et al.*, 2003; Atiyeh *et al.*, 2000; Gutiérrez-Miceli *et al.*, 2007; Narayan *et al.*, 2013; Nongmaithme and Pal, 2001; Yourtchi *et al.*, 2013). Within these studies vermicompost treatments also produced consistent increases in seedling growth, plant height, stem diameter, leaf area, germination, leaf number, shoot length, emergence, branch number, total biomass, and fruit, leaf, shoot, and root weight for many species of vegetable crops. Typically, the most effective treatments were supplemented with 10 to 50% vermicompost applications by volume, or 10:90 and 50:50 vermicompost to peat mixtures.

Vermicompost studies have also shown that the substitution of biofertilisers can decrease pathogens and disease in crop production. Vermicompost application can reduce the occurrence of blossom end rot of tomatoes (Surrage *et al.*, 2010). Decreases in pathogens such as nematodes (Renčo and Kováčik, 2015), pea leafminers (Suryawan and Reyes, 2016), and scab (Singhai *et al.*, 2011) have occurred when applied to potatoes.

4   Conclusions, special considerations, and future directions

Insect waste streams present a product with possible applications to food crop production. Many of these studies show that insect frass in combination with inorganic fertiliser actually produce the best results concerning crop productivity, and pathogen and disease resistance. The most effective treatments typically include 10 to 40 by volume percent frass applications; this varies greatly by crop, insect, waste source, treatment, and environment (e.g. field vs greenhouse). The addition of non-traditional amendments, such as humic acid (Choi and Hassanzadeh, 2019) and bacterial isolates (Xiao *et al.*, 2018) may serve as a further supplement of conventional fertilisers during the production of food crops.

One of the consistent concerns mentioned in this research refers to the sodium content and electrical conductivity levels of the soil after application. This may require that the industry monitors the diets of industrially produced insects so that their frass does not have negative repercussions on plant productivity and soil health (Temple *et al.*, 2013). For example, compared to inorganic conventional fertilisers, black soldier fly larvae frass has a much higher electrical conductivity (Zhan and Quilliam, 2017) than traditional composts. Additionally, there is some evidence that starting diet composition affects the bacterial populations present in fly larvae frass, which further complicates the system (Klammsteiner *et al.*, 2019). House fly frass also showed higher levels of heavy metals such as lead and arsenic (Zhu *et al.*, 2015). Each of these issues are important considerations for larval production and biofertiliser utilisation in the future.

Studies that show high levels of bioavailable phosphorus in insect frass are especially promising for agricultural areas limited in phosphorus moving forward. As described, phosphorus is a non-renewable resource and prospects to replace it should be of high priority for future research and adoption of practices. Lastly, common units should be adopted to express relative quantities of insect digestate applied across different environments. For example, kg/ha for field applications and litre:litre for greenhouse use. These factors should be noted and considered as this body of literature is expanded through additional studies.

Conflict of interest

The authors declare no conflict of interest.

References

Ahmed, A.S., Ezziyyani, M., Sánchez, C.P. and Candela, M.E., 2003. Effect of chitin on biological control activity of *Bacillus* spp. and *Trichoderma harzianum* against root rot disease
in pepper (*Capsicum annuum*) plants. European Journal of Plant Pathology 109: 633-637.

Alam, M.N., Jahan, M.S., Ali, M.K., Ashraf, M.A. and Islam, M.K., 2007. Effect of vermicompost and chemical fertilizers on growth, yield and yield components of potato in barind
soils of Bangladesh. Journal of Applied Sciences Research 3: 1879-1888.

Ansari, A.A., 2008. Effect of vermicompost and vermiwash on the productivity of spinach
(*Spinacia oleracea*), onion (*Allium cepa*) and potato (*Solanum tuberosum*). World Journal
of Agricultural Sciences 4: 554-557.

Arancon, N.Q., Edwards, C.A., Bierman, P., Metzger, J.D., Lee, S. and Welch, C., 2003. Effects
of vermicomposts on growth and marketable fruits of field-grown tomatoes, peppers and
strawberries: the 7th International Symposium on Earthworm Ecology, Cardiff, Wales,
2002. Pedobiologia 47: 731-735.

Atiyeh, R.M., Arancon, N., Edwards, C.A. and Metzger, J.D., 2000. Influence of earth-
worm-processed pig manure on the growth and yield of greenhouse tomatoes. Bioresource Technology 75: 175-180.

Beesigamukama, D., Mochoge, B., Korir, N., Musyoka, M.W., Fiaboe, K.K., Nakimbugwe, D.,
Khamis, F.M., Subramanian, S., Dubois, T., Ekesi, S. and Tanga, C.M., 2020. Nitrogen Fertilizer equivalence of black soldier fly frass fertilizer and synchrony of nitrogen mineralization for maize production. Agronomy 10: 1-19.

Bell, A.A., Hubbard, J.C., Liu, L., Davis, R.M. and Subbarao, K.V., 1998. Effects of chitin and
chitosan on the incidence and severity of *Fusarium* yellows of celery. Plant Disease 82:
322-328.

Belovsky, G.E. and Slade, J.B., 2000. Insect herbivory accelerates nutrient cycling and increases plant production. Proceedings of the National Academy of Sciences 97: 14412-
14417.

Beskin, K.V., Holcomb, C.D., Cammack, J.A., Crippen, T.L., Knap, A.H., Sweet, S.T. and Tomberlin, J.K., 2018. Larval digestion of different manure types by the black soldier fly (Diptera:
Stratiomyidae) impacts associated volatile emissions. Waste Management 74: 213-220.

Bosch, G., Van Zanten, H.H.E., Zamprogna, A., Veenenbos, M., Meijer, N.P., Van der Fels-
Klerx, H.J. and Van Loon, J.J.A., 2019. Conversion of organic resources by black soldier fly
larvae: egislation, efficiency and environmental impact. Journal of Cleaner Production
222: 355-363.

Bradley, S.W. and Sheppard, D.C., 1984. House fly oviposition inhibition by larvae of *Hermetia illucens*, the black soldier fly. Journal of Chemical Ecology 10: 853-859.

Buenvinida, L.P. and Tamban, V.E., 2016. Effectiveness of caterpillar frass as fertilizer on the growth of basil plants (*Ocimum basilicum*). 4th Global Summit on Education, March 14-15, 2016, Kuala Lumpur, Malaysia.

Chapin III, F.S., Matson, P.A. and Vitousek, P., 2011. Principles of terrestrial ecosystem ecology. Springer Science & Business Media, Berlin, Germany.

Chapman, S.K., Hart, S.C., Cobb, N.S., Whitham, T.G. and Koch, G.W., 2003. Insect herbivory increases litter quality and decomposition: an extension of the acceleration hypothesis. Ecology 84: 2867-2876.

Chia, S.Y., Tanga, C.M., Van Loon, J.J. and Dicke, M., 2019. Insects for sustainable animal feed: inclusive business models involving smallholder farmers. Current Opinion in Environmental Sustainability 41: 23-30.

Choi, S. and Hassanzadeh, N., 2019. BSFL Frass: a novel biofertilizer for improving plant health while minimizing environmental impact. The Canadian Science Fair Journal 2: 41-46.

Choi, Y.C., Choi, J.Y., Kim, J.G., Kim, M.S., Kim, W.T., Park, K.H., Bae, S.W. and Jeong, G.S., 2009. Potential usage of food waste as a natural fertilizer after digestion by *Hermetia illucens* (Diptera: Stratiomyidae). International Journal of Industrial Entomology 19: 171-174.

Čičková, H., Newton, G.L., Lacy, R.C. and Kozánek, M., 2015. The use of fly larvae for organic waste treatment. Waste management 35: 68-80.

Couwenberg, J., Thiele, A., Tanneberger, F., Augustin, J., Bärisch, S., Dubovik, D., Liashchynskaya, N., Michaelis, D., Minke, M., Skuratovich, A. and Joosten, H., 2011. Assessing greenhouse gas emissions from peatlands using vegetation as a proxy. Hydrobiologia 674: 67-89.

Daverede, I.C., Kravchenko, A.N., Hoeft, R.G., Nafziger, E.D., Bullock, D.G., Warren, J.J. and Gonzini, L.C., 2004. Phosphorus runoff from incorporated and surface-applied liquid swine manure and phosphorus fertilizer. Journal of Environmental Quality 33: 1535-1544.

Diener, S., Zurbrügg, C. and Tockner, K., 2009. Conversion of organic material by black soldier fly larvae: establishing optimal feeding rates. Waste Management & Research 27: 603-610.

Durst, P.B., Johnson, D.V., Leslie, R.N. and Shono, K. (eds.), 2010. Forest insects as food: humans bite back. In: Proceedings of a workshop on Asia-Pacific resources and their potential for development, 19-21 February 2008, Chiang Mai, Thailand. Food and Agriculture Organization of the United Nations RAP Publication 2010/02. Available at: http://www.fao.org/docrep/012/i1380e/i1380e00.pdf.

Eghball, B. and Gilley, J.E., 1999. Phosphorus and nitrogen in runoff following beef cattle manure or compost application. Journal of Environmental Quality 28: 1201-1210.

El Boushy, A.R., 1991. House-fly pupae as poultry manure converters for animal feed: a review. Bioresource Technology 38: 45-49.

Elliott, A.L., Davis, J.G., Waskom, R.M., Self, J.R. and Christensen, D.K., 2007. Phosphorus fertilizers for organic farming systems. Colorado State University, Fort Collins, CO, USA. Available at: https://extension.colostate.edu/topic-areas/agriculture/phosphorus-fertilizers-for-organic-farming-systems-0-569/

Fatima, B., Zahrae, M.F. and Razouk, R., 2018. Chitin/chitosan's bio-fertilizer: usage in vegetative growth of wheat and potato crops. Chitin-Chitosan: Myriad Functionalities in Science and Technology. InTech Open, London, UK, 321 pp.

Frost, C.J. and Hunter, M.D., 2004. Insect canopy herbivory and frass deposition affect soil nutrient dynamics and export in oak mesocosms. Ecology 85: 3335-3347.

Fuchs, J.G., Kupper, T., Tamm, L. and Schenk, M., 2008. Compost and digestate: sustainability, benefits, impacts for the environment and for plant production. Research Institute of Organic Agriculture (FiBL), Frick, Switzerland.

Gärttling, D., Kirchner, S.M. and Schulz, H., 2020. Assessment of the N-and P-fertilization effect of black soldier fly (diptera: stratiomyidae) by-products on maize. Journal of Insect Science 20: 1-11.

Gay, S.W. and Knowlton, K.F., 2005. Ammonia emissions and animal agriculture. Virginia Cooperative Extension, Blacksburg, VA, USA.

Gutiérrez-Miceli, F.A., Santiago-Borraz, J., Molina, J.A.M., Nafate, C.C., Abud-Archila, M., Llaven, M.A.O., and Dendooven, L., 2007. Vermicompost as a soil supplement to improve growth, yield and fruit quality of tomato (*Lycopersicum esculentum*). Bioresource Technology 98: 2781-2786.

Hahn, T., Tafi, E., Paul, A., Salvia, R., Falabella, P. and Zibek, S., 2020. Current state of chitin purification and chitosan production from insects. Journal of Chemical Technology & Biotechnology 95(11): 2775-2795.

Houben, D., Daoulas, G., Faucon, M.P. and Dulaurent, A.M., 2020. Potential use of mealworm frass as a fertilizer: impact on crop growth and soil properties. Scientific Reports 10: 1-9.

Hwangbo, J., Hong, E.C., Jang, A., Kang, H.K., Oh, J.S., Kim, B.W. and Park, B.S., 2009. Utilization of house fly-maggots, a feed supplement in the production of broiler chickens. Journal of Environmental Biology 30: 609-614.

Kagata, H. and Ohgushi, T., 2012a. Non-additive effects of leaf litter and insect frass mixture on decomposition processes. Ecological Research 27: 69-75.

Kagata, H. and Ohgushi, T., 2012b. Positive and negative impacts of insect frass quality on soil nitrogen availability and plant growth. Population Ecology 54: 75-82.

Klammsteiner, T., Turan, V., Oberegger, S., Insam, H. and Juárez, M.F.D., 2019. Black soldier fly (*Hermetia illucens*) frass as plant fertilizer. 7th International Conference on Sustainable Solid Waste Management. June 26-29, 2019. Heraklion, Crete.

Koppelaar, R.H.E.M. and Weikard, H.P., 2013. Assessing phosphate rock depletion and phosphorus recycling options. Global Environmental Change 23: 1454-1466.

Liu, Q., Tomberlin, J.K., Brady, J.A., Sanford, M.R. and Yu, Z., 2008. Black soldier fly (Diptera: Stratiomyidae) larvae reduce *Escherichia coli* in dairy manure. Environmental Entomology 37: 1525-1530.

Miranda, C.D., Cammack, J.A. and Tomberlin, J.K., 2019. Life-history traits of the black soldier fly, *Hermetia illucens* (L.) (Diptera: Stratiomyidae), reared on three manure types. Animals 9: 281.

Narayan, S., Kanth, R.H., Narayan, R., Khan, F.A., Singh, P. and Rehman, S.U., 2013. Effect of integrated nutrient management practices on yield of potato. Potato Journal 40: 84-86.

Nongmaithem, D. and Pal, D., 2011. The effect of organic sources of nutrients on the growth attributes and yields of potato (*Solanum tuberosum* L.). Journal of Crop and Weed 7: 67-69.

Oonincx, D.G.A.B. and De Boer, I.J.M., 2012. Environmental impact of the production of meal-worms – as a protein source for humans – a life cycle assessment. PLoS ONE 7: e51145.

Paul, L.C. and Metzger, J.D., 2005. Impact of vermicompost on vegetable transplant quality. HortScience 40: 2020-2023.

Poveda, J., Jiménez-Gómez, A., Saati-Santamaría, Z., Usategui-Martín, R., Rivas, R. and García-Fraile, P., 2019. Mealworm frass as a potential biofertilizer and abiotic stress toler-ance-inductor in plants. Applied Soil Ecology 142: 110-122.

Putra, R.E., Hutami, R., Suantika, G. and Rosmiati, M., 2017. Application of compost pro-duced by bioconversion of coffee husk by black soldier fly larvae (*Hermetia illucens*) as solid fertilizer to lettuce (*Lactuca sativa* var. crispa): impact to harvested biomass and utilization of nitrogen, phosphor, and potassium. Proceedings of the International Con-ference on Green Technology 8: 466-472.

Quilliam, R.S., Nuku-Adeku, C., Maquart, P., Little, D., Newton, R. and Murray, F., 2020. Inte-grating insect frass biofertilisers into sustainable peri-urban agro-food systems. Journal of Insects as Food and Feed 6: 315-322. https://doi.org/10.3920/JIFF2019.0049

Renčo, M. and Kováčik, P., 2015. Assessment of the nematicidal potential of vermicompost, vermicompost tea, and urea application on the potato-cyst nematodes *Globodera rosto-chiensis* and *Globodera pallida*. Journal of Plant Protection Research 55: 187-192.

Rosmiati, M., Nurjanah, K.A., Suantika, G. and Putra, R.E., 2017. Application of compost pro-duced by bioconversion of coffee husk by black soldier fly larvae (*Hermetia illucens*) as solid fertilizer to lettuce (Lactuca sativa var. crispa): impact to growth. Proceedings of the International Conference on Green Technology 8: 38-44.

Schmitt, E., 2020. Potential benefits of using *Hermetia illucens* frass as a soil amendement on food production and for environmental impact reduction. Current Opinion in Green and Sustainable Chemistry 25: 100335.

Setti, L., Francia, E., Pulvirenti, A., Gigliano, S., Zaccardelli, M., Pane, C. and Ronga, D., 2019. Use of black soldier fly (*Hermetia illucens* (L.), Diptera: Stratiomyidae) larvae processing residue in peat-based growing media. Waste Management 95: 278-288.

Sharp, R.G., 2013. A review of the applications of chitin and its derivatives in agriculture to modify plant-microbial interactions and improve crop yields. Agronomy 3: 757-793.

Sims, J.T., Bergström, L., Bowman, B.T. and Oenema, O., 2005. Nutrient management for in-tensive animal agriculture: policies and practices for sustainability. Soil Use and Manage-ment 21: 141-151.

Singhai, P.K., Sarma, B.K. and Srivastava, J.S., 2011. Biological management of common scab of potato through *Pseudomonas* species and vermicompost. Biological Control 57: 150-157.

Spiegel, Y., Chet, I. and Cohn, E., 1987. Use of chitin for controlling plant plant-parasitic nem-atodes. Plant and Soil 98: 337-345.

Surrage, V.A., Lafreniere, C., Dixon, M. and Zheng, Y., 2010. Benefits of vermicompost as a constituent of growing substrates used in the production of organic greenhouse toma-toes. HortScience 45: 1510-1515.

Suryawan, I.B.G. and Reyes, S.G., 2016. The influence of cultural practice on population of pea leafminer (*Liriomyza huidobrensis*) and its parasitoids in potato. Indonesian Journal of Agricultural Science 7: 35-42.

Swiader, J.M. and Ware, G.W., 2002. Producing vegetable crops. Interstate Publisher, Inc., Danville, IL, USA.

Temple, W.D., Radley, R., Baker-French, J. and Richardson, F., 2013. Use of Enterra natural fertilizer (black soldier fly larvae digestate) as a soil amendment. Enterra Feed Corporation, Vancouver, British Columbia, Canada.

Van Broekhoven, S., Oonincx, D.G., Van Huis, A. and Van Loon, J.J., 2015. Growth performance and feed conversion efficiency of three edible mealworm species (Coleoptera: Tenebrionidae) on diets composed of organic by-products. Journal of Insect Physiology 73: 1-10.

Vickerson, A., Radley, R., Marchant, B., Kaulfuss, O. and Kabaluk, T., Enterra Feed Corp, 2016. *Hermetia illucens* frass production and use in plant nutrition and pest management. US Patent: WO2015013826A1. Available at: https://patents.google.com/patent/WO2015013826A1/en

Xiao, X., Mazza, L., Yu, Y., Cai, M., Zheng, L., Tomberlin, J.K., Yu, J., Van Huis, A., Yu, Z., Fasulo, S. and Zhang, J., 2018. Efficient co-conversion process of chicken manure into protein feed and organic fertilizer by *Hermetia illucens* L. (Diptera: Stratiomyidae) larvae and functional bacteria. Journal of Environmental Management 217: 668-676.

Xu, C. and Mou, B., 2018. Chitosan as soil amendment affects lettuce growth, photochemical efficiency, and gas exchange. HortTechnology 28: 476-480.

Yang, L., Zhao, F., Chang, Q., Li, T. and Li, F., 2015a. Effects of vermicomposts on tomato yield and quality and soil fertility in greenhouse under different soil water regimes. Agricultural Water Management 160: 98-105.

Yang, S., Xie, J., Hu, N., Liu, Y., Zhang, J., Ye, X. and Liu, Z., 2015b. Bioconversion of gibberellin fermentation residue into feed supplement and organic fertilizer employing housefly (*Musca domestica* L.) assisted by *Corynebacterium variabile*. PLoS ONE 10(5): e0110809.

Yourtchi, M.S., Hadi, M.H.S. and Darzi, M.T., 2013. Effect of nitrogen fertilizer and vermicompost on vegetative growth, yield and NPK uptake by tuber of potato (Agria CV.). International Journal of Agriculture and Crop Sciences 5: 2033-2040.

Zahn, N.H. and Quilliam, R., 2017. The effects of insect frass created by *Hermetia illucens* on spring onion growth and soil fertility. BSc-thesis, University of Stirling, Stirling, UK.

Zhang, Z., Wang, H., Zhu, J., Suneethi, S. and Zheng, J., 2012. Swine manure vermicomposting via housefly larvae (*Musca domestica*): the dynamics of biochemical and microbial features. Bioresource Technology 118: 563-571.

Zhu, F.X., Wang, W.P., Hong, C.L., Feng, M.G., Xue, Z.Y., Chen, X.Y., Yao, Y.L. and Yu, M., 2012. Rapid production of maggots as feed supplement and organic fertilizer by the two-stage composting of pig manure. Bioresource Technology 116: 485-491.

Zhu, F.X., Yao, Y.L., Wang, S.J., Du, R.G., Wang, W.P., Chen, X.Y., Hong, C.L., Qi, B., Xue, Z.Y. and Yang, H.Q., 2015. Housefly maggot-treated composting as sustainable option for pig manure management. Waste Management 35: 62-67.

Impacts of insect consumption on human health

V.J. Stull

University of Wisconsin-Madison, Global Health Institute, 1050 Medical Sciences Center, 1300 University Avenue, Madison, WI 53706, USA; vstull@wisc.edu

Abstract

Edible insects represent an understudied food resource that may promote human health. They characteristically contain ample protein, healthy fatty acids, minerals, and vitamins, and have been touted for their environmental benefits given their efficient resource use. While numerous *in vitro,* animal, and nutrient quantification studies have elucidated a framework of potential health impacts of entomophagy, few have measured direct health outcomes. This review investigates and summarises existing evidence on health impacts derived exclusively from human interventions. A systematic literature search was conducted in three databases: SCOPUS, Web of Science, and PubMed. Out of 1,691 initial results, only nine studies met the inclusion criteria. In these limited studies, insects were shown to have potential to: (1) promote growth and influence iron status when added to complementary foods; (2) modulate gut microbiota with some prebiotic effects; and (3) provide amino acids similar to soya protein. One study also provided isolated evidence that an insect-herb mixture could possibly reduce symptoms of chronic obstructive pulmonary disease when added to routine treatment. Importantly, results reveal a significant lack of human subjects research directly measuring health outcomes from insect consumption. Findings from the included studies indicate that insects are generally safe and offer both beneficial and neutral outcomes compared to other foods. These discoveries, in tandem with extensive evidence from non-human studies, support claims that insect consumption could further enhance health by addressing micronutrient deficiencies or promoting gut health. There are also other plausible health promoting properties of insects that could help ameliorate complications with hypertension and other non-communicable disease. More rigorous and better controlled human intervention trials are fundamental to confirm health benefits and better assess risks associated with entomophagy, while also addressing unanswered questions regarding nutrient bioavailability, the fate of dietary chitin, and *in vivo* activity of bioactive peptides.

Keywords

entomophagy – edible insects – health impacts – nutrition – clinical trial – dietary intervention

1   Introduction

The global food system faces an unprecedented array of challenges, which will have
ramifications for human health. Farmers and communities are pressured by climate change, inequitable trade, food waste, supply chain disruptions, and the need
to support a growing population while demand for resource-intensive animal products rises (Ranganathan *et al.*, 2016). Animals are an important protein source for
many, but livestock production, particularly beef, is one of the largest contributors
of greenhouse gas emissions and has also been blamed for other harms to the environment (Poore and Nemecek, 2018). Consequently, novel and sustainable protein
production will be required in the decades to come.

Global food security has improved in recent decades, but advances have not been
uniform and challenges remain. An estimated 690 million people faced chronic un-
dernourishment in 2019 (FAO *et al.*, 2020), despite adequate global food supplies.
Malnutrition kills more people than AIDS, malaria, and tuberculosis combined
(WFP, 2015). Undernutrition weakens the immune system and is responsible for
more than 45% (> 3 million) of annual deaths among children under five in developing countries (Black *et al.*, 2013). Severe protein energy malnutrition (PEM) can
manifest in muscle atrophy, wasting syndrome, or an increased risk of tuberculosis
and gastroenteritis. Anaemia resulting from iron or vitamin B_{12} deficiency is a major
global health risk (Ritchie and Roser, 2017), and low haemoglobin during pregnancy
contributes to the between 2.5 and 3.4 million maternal and neonatal deaths that
occur each year (Liu *et al.*, 2012; Stevens *et al.*, 2013). In other contexts, overweight
and obesity stress healthcare systems as about 650 million people experienced obesity in 2016 (WHO, 2020). The double burden of simultaneous under- and over-
nutrition also challenges many governments. Increasing insect consumption could
help address nutrition-related health problems by safely supplying micro and macronutrients.

While largely overlooked in Western cuisine today, insects are frequently included in traditional dishes across continents. Entomophagy – the practice of eating
insects – is not a new phenomenon. Many species have served as an important part
of the diet throughout history (Bodenheimer, 1951; DeFoliart, 1995). The practice is
highlighted in ancient religious texts, including the Bible (see Leviticus 11:22), and
more than 2,100 known edible insects have been identified to-date (Jongema, 2017).

A recent research boom has begun to unpack the many prospects of insects as
both food and feed, where previously a paucity of studies existed (Baiano, 2020).
The nutritional value of insects has been widely explored, though these data are
still sparse relative to the surfeit of edible species and myriad variables impacting
nutrient composition. Recent anthropological research has explored insects in human evolution (Lesnik, 2018) and entomologists are even unpacking edible insect
genomes (Ylla *et al.*, 2020). Other studies have focused on purported environmental
benefits of insect agriculture. There is evidence they require less water, land, and

feed than conventional livestock while emitting fewer greenhouse gases (Oonincx *et al.*, 2010; Van Huis *et al.*, 2013).

Generalising nutrient composition across species and growth parameters is difficult, but insects characteristically have nutrient-dense profiles akin to meat – rich in protein and fat, low in carbohydrates. Many species contain ample energy, protein, and healthy fatty acids, while also providing essential minerals, and even some vitamins (Rumpold and Schlüter, 2013). Insects are considered a good source of protein with crude levels comparable to and sometimes even higher than other conventional meat sources (Belluco *et al.*, 2013; Finke, 2002; Melo *et al.*, 2011). The quality of insect protein is said to be high; it is broadly considered bioavailable and relatively digestible (Belluco *et al.*, 2013; Finke, 2015; Verkerk *et al.*, 2007). Insects also provide essential amino acids for human nutrition (Bukkens, 2005; Collavo *et al.*, 2005; Rumpold and Schlüter, 2013), which are particularly valuable in contexts where malnutrition is problematic. Edible insects are also often a good source of fat with numerous species providing healthy polyunsaturated fatty acids (Womeni *et al.*, 2009) including omega-3s (Paul *et al.*, 2017; Wathne *et al.*, 2018). There is some evidence that modifying insect diets can improve fat composition and ratios (Oonincx *et al.*, 2019). Notably, data related to the nutrient content of insects are often low-quality and limited to within-species variation in proportions of micro- and macronutrients (Payne *et al.*, 2016).

Insects offer myriad potential health benefits that have yet to be fully explored, including those stemming from bioactive compounds with antioxidant or other properties and dietary fibre (Roos and Van Huis, 2017). Insect consumption is not without risks, however. Food safety considerations are important when consuming insects, as they are also potential sources of allergenic, toxic, or antinutrient substances (Rumpold and Schlüter, 2013).

Numerous studies have investigated *potential* health impacts of insect consumption – from nutritional analyses (Belluco *et al.*, 2013; Rumpold and Schlüter, 2013; Womeni *et al.*, 2009) and hypotheses as to impacts on deficient populations (Nadeau *et al.*, 2015), to laboratory studies where insect components are tested on cell cultures *in vitro* (Latunde-Dada *et al.*, 2016) or fed to animals (Bergmans *et al.*, 2020). These studies provide a foundational understanding of how entomophagy *could* influence human health. However, they do not demonstrate causality, nor do they fully explore other key variables driving health outcomes from entomophagy, such as nutritional/immune status of the consumer, presence of antinutrients in insects, processing methods, nutrient bioavailability, allergens, or contaminants.

It remains relatively unclear if and how eating insects improves human health. Are insects unique when compared to other animal foods? Do they contain nutrients or properties that correlate with benefits or detriments to human health? This purpose of this review is to summarise existing evidence from human intervention studies that measure human health outcomes resulting from entomophagy.

2   Methods

2.1   *Search strategy*

To define the scope of this broad topic and investigate health impacts of insect consumption, a systematic review was conducted. The search strategy was developed in consultation with a health sciences academic librarian. Three primary databases were searched, including Web of Science, PubMed, and SCOPUS. The final searches were completed on July 10, 2020. To standardise, two 'AND' categories were applied to basic title, abstract, and keyword searches in all three databases. An example of a primary search syntax is included in Appendix A. No limits for language, year of publication, author affiliation, or geographic region were applied.

Following the database searches, a secondary search using Google Scholar was conducted to find articles cited in found manuscripts with relevant titles not identified by preliminary searches. Additionally, a repeat of the key search strategy was implemented in Google Scholar with one modification. The names of specific edible insect types (e.g. 'cricket,' 'caterpillar,' 'termite,' 'beetle,' 'weevil,' 'stinkbug,' 'grasshopper,' 'ant,' 'wasp,' 'bee,' 'dragonfly') were added to the first search category to identify articles lacking generic terms such as 'edible insects' or 'entomophagy' in the title or abstract.

2.2   *Inclusion and exclusion criteria*

Following the searches, duplicate results were removed. Next, all titles, abstracts, and keywords were examined for eligibility via the primary inclusion and exclusion criteria. For studies that appeared to meet the inclusion criteria or where no determination could be made based on abstract and title alone, full text articles were reviewed.

All types of peer-reviewed clinical trials, dietary interventions, and acceptability studies whereby human subjects consumed edible insects and health outcomes were measured were included. Studies using insects in various states (raw, whole, processed, modified, added to other foods, or deconstructed into component parts) with or without a control group were included. Studies on insect products (e.g. silk proteins, honey) were excluded. For the purposes of this review, the category 'edible insects' included non-insects arthropods, such as spiders, that are widely grouped with insects in this field.

Search results that were not peer reviewed, such as books and reports, were excluded. Additionally, surveys, animal feeding trials, systematic and narrative reviews, commentaries, consumer taste trials, basic nutrient analyses, and studies not involving direct and monitored human consumption of insects with measured health impacts were excluded. Retrospective case reports outlining health outcomes associated with self-reported insect consumption were excluded.

3 Results

A total of 2,668 citations were collected in the initial searches of Web of Science, SCOPUS, and PubMed. Titles were screened, and a secondary search using Google Scholar for references in key manuscripts and insect-specific studies identified 15 additional papers, making the total 2,681. Systematic deduplication yielded 1,691 citations and 48 manuscripts were pulled for thorough review after screening title and abstract. Of these, nine papers met the inclusion criteria. Figure 1 highlights the selection process.

The nine human intervention studies included in this review were published between 2009 and 2020. They fell into four broad categories: (1) acceptability and nutritional implications of using insects as a component of complementary foods (CFs) (n = 5); (2) impacts of insect consumption on gut microbiota (n = 1); (3) quality and impact of insects as a protein supplement (n = 2); (4) clinical implications of adding an insect granule supplement to routine treatment for chronic obstructive pulmonary disease (COPD) (n = 1) (Table 1).

Eight of the studies involved randomised dietary interventions where insect consumption was compared to other treatments, a control, or a placebo. The remaining study was not randomised, but did involve feeding insect foods to human subjects to ascertain acceptability and tolerability (Bauserman *et al.*, 2015b). None of the studies controlled or tracked the entire diet of study participants. Eight studies utilised processed insects integrated into another food – a complementary cereal (Bauserman *et al.*, 2015a,b; Konyole *et al.*, 2012, 2019; Skau *et al.*, 2015), smoothie and muffin (Stull *et al.*, 2018), protein bar (Vangsoe *et al.*, 2018a), or drink (Vangsoe *et al.*, 2018b). One study provided 30 g of silkworm 'granules' daily, but processing methods for this product were not specified (Hu *et al.*, 2020). The primary subjects were children (n = 1,170).

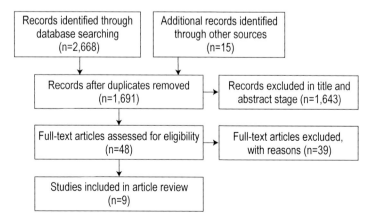

FIGURE 1   PRISMA flow diagram outlining screening process for literature review (modified from Moher *et al.*, 2009)

TABLE 1   Articles included for final review

Authors	Konyole *et al.* (2012)
Full title	Acceptability of Amaranth grain-based nutritious complementary foods with Dagaa fish (*Rastrineobola argentea*) and edible termites (*Macrotermes subhylanus*) compared to corn soya blend plus among young children/mothers dyads in western Kenya
Insect type	Termite: *Macrotermes subhylanus* (Isoptera: Termitidae)
Study design	Randomised crossover clinical trial + at home dietary intervention (acceptability study)
Location	Makunga Sub-District Hospital, Kakamega County, Western Kenya
Population	Healthy children (6-24 months),
Intervention	(1) 150 ml of WinFood Classic Complementary Food Mix provided to infants for 1 day: Amaranth Grain + 10% edible termites + 3% dagaa (small fish); (2) 100 g flour mix given to mothers to take home. [Compared to two other foods without insects or fish]
Sample	n = 57 pairs (child and mother); n = 57 consuming insects
Duration	1 day with 1 day washout; 4 week follow-up
Outcomes measured	Tolerance via morbidity data; Maternal and Child Acceptability (ranking & child consumed > 75% of serving).
Results	Complementary food with termites was acceptable by the target population with no adverse effects found.
Authors	Skau *et al.* (2015)
Full title	Effects of animal source food and micronutrient fortification in complementary food products on body composition, iron status, and linear growth: a randomised trial in Cambodia
Insect type	Edible spider: *Haplopelma* (species unspecified)
Study design	Randomised and single-blinded, community-based dietary intervention
Location	PeaRieng and Sithor Kandal Districts, Prey Veng province, Cambodia
Population	Infants (6 months old)
Intervention	Age-based provision of rice-based complementary food products: WinFood (WF) with small fish and edible spiders. (WF contained 1.8 g spiders per 100 g dry weight.) [Compared to three other formulations without spider]
Sample	n = 419; n = 106 consuming insects
Duration	9 months
Outcomes measured	Changes in fat-free mass (FFM); change in plasma ferritin and soluble transferrin receptors; changes in anthropometric variables, including knee-heel length.
Results	No difference in fat-free mass increment in insect porridge (WF) compared with control [WF: +0.04 kg (95% CI: 20.20, 0.28 kg); No effect on iron status across the treatments, however iron status deteriorated for all infants over study period for all treatments. No significant differences in anthropometric measurements across treatments.

TABLE 1   Articles included for final review (*cont.*)

Authors	Bauserman *et al.* (2015a)
Full title	Caterpillar cereal as a potential complementary feeding product for infants and young children: nutritional content and acceptability
Insect type	Caterpillar (species unspecified)
Study design	Acceptability study
Location	Rural Equateur Province of the Democratic Republic of Congo
Population	Healthy infants (8-10 months) and their mothers
Intervention	30 g sachet of complementary cereal with dried, ground caterpillar fed to infants daily (actual amount of caterpillar unclear). [Not randomised; no control]
Sample	n = 20 pairs (infant and mother); n = 20 consuming insects
Duration	7 days
Outcomes measured	Child illness or symptoms as reported by the mother; Maternal acceptability (by rating); Child acceptability (by % consumed); Nutrient content of cereal; Microbiologic assays for food safety.
Results	Caterpillar cereal is safe for consumption (no serious side-effects), has appropriate macro- and micronutrient contents for complementary feeding, and is acceptable to mothers and infants.
Authors	Bauserman *et al.* (2015b)
Full title	A cluster-randomised trial determining the efficacy of caterpillar cereal as a locally available and sustainable complementary food to prevent stunting and anaemia
Insect type	Caterpillar (species unspecified)
Study design	Cluster-randomised controlled feeding trial
Location	Rural Equateur Province of the Democratic Republic of Congo
Population	Infants (6 months old)
Intervention	30 or 45 g sachet of complementary cereal with dried, ground caterpillar fed to infants daily depending on age (actual amount of caterpillar unclear). [Compared with control that did not receive food]
Sample	n = 222; n = 111 consuming insects
Duration	12 months
Outcomes measured	Compliance with feeding; 24 hour dietary recall; Feeding practices via Infant and Child Feeding Index (ICFI); anthropometric measurements on infants at 6, 9, 12 and 18 months of age; blood indices of Fe status at 18 months.
Results	Supplementation with caterpillar cereal did not reduce prevalence of stunting at 18 months; Infants who consumed caterpillar cereal had higher Hb concentrations than the control group (10.7 vs 10.1 g/dl, $P = 0.03$) and fewer were anaemic (26 vs 50%, $P = 0.006$); no difference in estimates of body Fe stores (6.7 vs 7.2 mg/kg body weight, $P = 0.44$).

TABLE 1   Articles included for final review (*cont.*)

Authors	Stull *et al.* (2018)
Full title	Impact of edible cricket consumption on gut microbiota in healthy adults, a double-blind, randomised crossover trial
Insect type	Tropical house cricket: *Gryllodes sigillatus* F. Walker (Orthoptera: Gryllidae)
Study design	Crossover clinical trial, double blind, randomised
Location	Colorado, USA
Population	Healthy adults (> 18 years)
Intervention	25 g whole cricket powder added to diet daily via muffin and shake. [Compared with control foods without insects]
Sample	n = 20; n = 20 consuming insects
Duration	2 week intervention (6 weeks total with washout and control)
Outcomes measured	Digestive health questionnaire (self-report); Comprehensive Metabolic Panel; Microbiome changes via faecal DNA; Changes in Microbial Metabolism; Faecal Triglycerides; Measures of Inflammation.
Results	Cricket consumption is tolerable and non-toxic at the studied dose; no changes in excreted bile acids or faecal triglycerides; increase in abundance of the probiotic bacterium *Bifidobacterium animalis* (5.7 fold); associated with decrease of other bacterium. Cricket consumption also associated with reduced plasma TNF-α. Suggestive of improved gut health and reduced systemic inflammation.
Authors	Vangsoe *et al.* (2018a)
Full title	Effects of insect protein supplementation during resistance training on changes in muscle mass and strength in young men
Insect type	Lesser mealworm: *Alphitobius diaperinus* Panzer (Coleoptera: Tenebrionidae)
Study design	Clinical trial, randomised, controlled, single-blinded trial
Location	Aarhus University, Denmark
Population	Healthy young men (18 to 30 years)
Intervention	Ingested an insect-protein bar containing 0.4 g protein/kg (Pro, n = 9) within 1 h after each training and 1 h before night sleep on training days. [Compared to isocaloric carbohydrate control bar]
Sample	n = 18; n = 9 consuming insects
Duration	8 weeks (resistance training 4/week)
Outcomes measured	Strength (leg press and bench press); body composition (via whole-body dual-energy X-ray absorptiometry); dietary records were obtained (three-day registration periods) before and during the intervention period.
Results	No significant differences in body composition and muscle strength improvements observed; compliance indicative of tolerability.

TABLE 1   Articles included for final review (*cont.*)

Authors	Vangsoe *et al.* (2018b)
Full title	Ingestion of insect protein isolate enhances blood amino acid concentrations similar to soya protein in a human trial
Insect type	Lesser mealworm: *A. diaperinus* Panzer (Coleoptera: Tenebrionidae)
Study design	Randomised cross-over study
Location	Aarhus University, Denmark
Population	Healthy young men (18 to 30 years)
Intervention	25 g of crude insect protein ingested postprandially. [Compared to whey, soya, or water placebo]
Sample	n = 6; n = 6 consuming insects
Duration	4 days
Outcomes measured	Amino acid concentration in blood before and 0 min, 20 min, 40 min, 60 min, 90 min, and 120 min determined using 1H NMR spectroscopy.
Results	Ingestion of whey, soya, and insect protein isolates increased blood concentrations of amino acids (AA) over a 120 min period (whey > insect = soya). Insect protein induced blood AA concentrations similar to soya protein. Higher blood AA concentrations at 120 min post ingestion was observed for insect protein, suggesting it is a 'slow' digestible protein source.
Authors	Konyole *et al.* (2019)
Full title	Effect of locally produced complementary foods on fat-free mass, linear growth, and iron status among Kenyan infants: a randomised controlled trial
Insect type	Termites: *Macrotermes subhylanus* (Isoptera: Termitidae)
Study design	Randomised and double-blind, controlled trial
Location	Mumias Sub-County in Kakamega County, Western Kenya
Population	Infants (6 months old)
Intervention	Provision of: WinFood Classic (WFC) (71% grain amaranth, 10.4% corn, 0.6% soyabean oil, 5% sugar, 10% edible termites, and 3% small fish dry weight); given in age-adjusted daily rations. [Compared to two other foods without insects]
Sample	n = 499 pairs (mother and infants); n = 165 consuming insects
Duration	9 months
Outcomes measured	FFM accrual, linear growth, and iron status; Changes in FFM, length, plasma ferritin, plasma transferrin receptors, and haemoglobin from 6 to 15 months of age. Secondary outcomes: change in weight, MUAC, head circumference, skinfolds, weight.
Results	WFC insect food was associated with weight gain over the 9 months, predominantly fat free mass. Fat mass was also preserved during the 9-month intervention on all diets. WFC did not differ from the other complementary foods or controls in terms of weight gain or mass. Plasma ferritin decreased, whereas plasma transferrin receptors increased in all three groups over the 9 months indicating an overall deterioration in iron status.

TABLE 1   Articles included for final review (*cont.*)

Authors	Hu *et al.* (2020)
Full title	Effects of compound Caoshi silkworm granules on stable COPD patients and their relationship with gut microbiota: a randomised controlled trial
Insect type	Caoshi silkworm (species unspecified)
Study design	Randomised controlled trial
Location	Jinhua Hospital of Zhejiang University, China
Population	Adults 40-80 with COPD grades II and III and history of 2 or more prior exacerbations
Intervention	Routine COPD treatment plus Compound Caoshi Silkworm Granules (CCSGs) containing 30 g silkworm and 5 g Astragalus, once per day. These granules were given orally, one bag at a time and once a day. [Compared to a control without CCSGs]
Sample	n = 40; n = 20 consuming insects
Duration	3 months
Outcomes measured	St. George's Respiratory Questionnaire (SGRQ) score and lung function. Microbial abundance from stool samples analysed using 16s rRNA sequencing before treatment.
Results	Safe treatment. Individuals treated with CCSGs showed improved SGRQ scores. No difference in forced expiratory volume / forced vital capacity. Baseline abundance of gut microbiota in patients with the top 10 SGRQ scores differed from the abundance of gut microbiota in those with the lowest 10 SGRQ scores. CCSGs can be considered effective for patients with stable COPD, and gut microbiota may be a new target.

Healthy adults were the next most represented group (n = 44) followed by adults with confirmed COPD (n = 40). Children and individuals with severe health complications that could compromise the result of these studies were excluded.

None of the studies reported severe adverse health outcomes or a significant increase in morbidity or mortality with insect consumption, even for vulnerable populations. Two studies measured only basic health outcomes, such as tolerance via reported morbidity, which are difficult to interpret and generalise (Bauserman *et al.*, 2015b; Konyole *et al.*, 2012). However, all studies on CFs confirmed that insects are a safe addition to the diet for children. Overall, insect consumption was associated with predominantly positive or neutral results.

3.1   *Insects in complementary feeding and impacts on nutritional status*

The majority (5 out of 9) of human intervention studies evaluated insects as a novel supplement to CF for infants (6 months) and young children (up to 24 months). Two studies took place in Kenya, two in the Democratic Republic of Congo (DRC), and one in Cambodia. Outcomes included anthropometric measurements, acceptability and tolerance, and iron status. All of these studies assessed health impacts of adding insects, along with other locally available ingredients, to CFs. Overall, results suggest insects are an acceptable ingredient in CFs; insect-fortified CFs are comparable to existing CFs. Other observed improvements were not consistent across studies.

Bauserman *et al.* (2015a) conducted a cluster-randomised trial in DRC to determine the effectiveness of complementary cereal including caterpillars (species unspecified) to reduce stunting and anaemia in 175 infants (treatment group n = 91). While caterpillar-supplemented cereal did not reduce the prevalence of stunting in children over 12 months relative to a control group (who ate regular diet), caterpillar consumption was associated with an increase in haemoglobin concentrations (Hb levels 10.7 g/dl in treatment, 10.1 g/dl in control, $P = 0.03$) and lower anaemia rates (26% in treatment versus 50% control) (Bauserman *et al.*, 2015a).

Building from an acceptability study in 2012, Konyole *et al.* (2019) compared three CFs: (1) WinFood Classic (WFC), a germinated amaranth, maize, and fish product fortified with 10% termites by dry weight, *Macrotermes subhylanus* (Isoptera: Termitidae); (2) WinFood Lite (WFL), another germinated amaranth and corn-based food not containing insects but supplemented with micronutrients; and (3) corn-soya blend (CSB+), a control corn and soya blend with a vitamin premix. Infants eating all three foods gained weight and there were no differences in body composition (fat free body mass or fat mass) over the 9-month intervention. Weight gained across all groups was primarily fat free body mass. Compared with the CSB++ control, both WFC (with termites) and WFL (with micronutrients) were associated with a decrease in plasma ferritin (WFC 0.6 µg/l (95% CI [0.4, 0.8]) and an increased in plasma transferrin receptors (WFC 3.3 mg/l (95% CI [1.7, 4.9]). The deterioration in iron status was slightly more pronounced among children who ate the termite-fortified food (WFC). Including insects and fish did not impact body composition or promote growth differently than the micronutrient fortified WFL or CSB+ control. Given that none of the CFs improved iron nutritional status, the authors speculate that environmental factors, such as exposure to hazards, enteric disfunction, malaria, and/or breastfeeding may be playing a role in iron absorption (Konyole *et al.*, 2019).

In a similar randomised, single-blinded, community-based trial, Skau *et al.* (2015) demonstrated weight gain in infants who consumed pre-cooked semi-instant porridge with edible spiders (*Haplopelma* species unspecified). Here, infants were fed one of four treatments: WinFood (rice fortified with fish and edible spiders), WFL (rice fortified with small fish only), CSB+ (fortified corn-soya blend), and CSB++ (fortified corn-soya blend with 8% dried skimmed milk). No differences in growth across the four treatments were observed, including changes in fat free body mass and fat mass. No differences in plasma ferritin and soluble transferrin receptor were found between the interventions. However, the WF group had a marginally smaller change in haemoglobin than CSB++ control, and a higher prevalence of anaemia in infants who received the WinFood spider cereal compared to the WFL and the CSB++ corn-soya control (fortified with skim milk) was observed. Anaemia in the insect treatment group was 53.7%, whereas the WinFood Lite was 35.2% and the control was 35.2%. However, anaemia increased across all intervention groups during the study period, likely due to inadequate initial supply of iron in all treatment foods. Overall, locally produced CFs containing edible spiders did not differ from the existing fortified corn-soya blend products.

It is difficult to conduct dietary interventions in food insecure contexts. Not surprisingly, the quality of these complementary feeding trials varied. Some began with heterogeneous baseline participant populations and the full sample of nine studies utilised varying measures of anthropometry. Moreover, environmental factors and other variables that could impact iron status were not measured. However, Konyole *et al.* (2019) and Skau *et al.* (2015) recruited children of similar ages and included insect-free comparison treatment arms.

3.2   *Medicinal properties of edible insects*

One study explored potential medicinal effects of edible insects. Silkworms have been used historically in traditional Chinese medicine to treat a variety of health conditions, including COPD (Hu *et al.*, 2020; Li *et al.*, 2017). Hu *et al.* (2020) investigated the impacts of taking compound Caoshi silkworm granules (CCSGs), in addition to routine treatment, on COPD patient symptoms and lung function for 3 months. Results were compared with patients who received only routine care. Each CCSG packet given to patients contained 30 g Caoshi silkworm (species unspecified) along with 5 g Astragalus medicinal herb. There was no significant difference in the incidence of adverse events between the two groups ($P > 0.05$). Although ingesting silkworm granules did not change measures of lung function, results suggest that it may have improved symptomology, measured via St. George's respiratory questionnaire (SGRQ) ($P = 0.02$). The mechanism for action is not clear but may be linked to baseline microbiota among patients. Importantly, it is not feasible to differentiate effects of the silkworm from effects of the Astragalus herb. Additionally, it is likely that other studies reporting medicinal efficacy of silkworms or edible insects are reported in the literature but may not appear in English language journals.

3.3   *Insects and the microbiome*

Only one study in this review assessed the potential for insects to serve as modulators of gut microbiota. In a double-blind, crossover trial, Stull *et al.* (2018) measured the health impacts of consuming 25/g of tropical house cricket, *Gryllodes sigillatus* F. Walker (Orthoptera: Gryllidae), per day on healthy adults. Whole ground cricket consumption did not result in toxicity, cause gastrointestinal distress, or yield global disruptions to gut microbiota. Only one small metabolic difference was detected: a slight increase of alkaline phosphatase (ALP) with cricket consumption (95% CI (0.11-0.399) $P < 0.05$). Low ALP levels are sometimes associated with malnutrition or malabsorption. However, all participants stayed within normal ALP ranges. No changes were observed across treatment groups in lipid metabolism or phyla-level microbiota composition. Only one significant change in inflammation was observed: tumour necrosis factor alpha was lower after cricket consumption relative to the control (-0.525 (95% CI (-0.93--0.12; $P < 0.05$) (Stull *et al.*, 2018).

Five bacterial taxa were significantly increased after cricket consumption and ten were decreased. Little is known regarding all of these taxa, however cricket supplementation (containing ~2 g/day chitin) was associated with a 5.7 fold increase in faecal levels of the probiotic species, *Bifidobacterium animalis,* a 4-fold decrease in faecal *Lactobacillus,* and a 3.5-fold decrease in genus *Acidaminococcus* relative to the control (Stull *et al.*, 2018). Increased levels of *B. animalis* could benefit health, especially through pathogen inhibition. A reduction in probiotic *Lactobacillus reuteri*, is not desirable, but the authors speculate that chitooligosaccharides may inhibit lactic acid producing bacteria (LAB) and the dietary fibre present in the control diet may have promoted LAB growth. Reductions in *Acidaminococcus* may be desirable, given its association with growth deficits in infants (Gough *et al.*, 2015).

These results are suggestive of potential gastrointestinal and systemic health benefits associated with consuming crickets. However, additional placebo-controlled human feeding trials, along with more *in vitro* studies, are needed to determine mechanisms and long-term health impacts of insects and their component parts on gut microbiota (specifically chitin, specific amino acids, and unsaturated fatty acids).

Interestingly, Hu *et al.* (2020) also considered that microbiota could play a role in how insects impact health via the gut-lung axis. The authors divided participants who consumed silkworm into two groups according to their SGRQ scores (top 10 and bottom 10) and compared their microbiota components at baseline. Significant differences across species value, Shannon value, top 10 levels of phylum, and heatmap of functional predictions were observed. Faecal microbiota may play a regulatory role during administration of silkworm granules, but this is not yet understood.

3.4   *Insect protein quality*

Two studies investigated the quality and utility of insect protein isolate in healthy men. Neither reported unique or novel properties of insect protein. In the first, Vangsoe *et al.* (2018a) investigated the effect of a dietary supplement of insect protein from the lesser mealworm, *Alphitobius diaperinus* Panzer (Coleoptera: Tenebrionidae), on muscle hypertrophy and strength gains during prolonged resistance training compared to an isocaloric carbohydrate supplement. No dietary restrictions were placed on participants. Fat- and bone-free mass improved with both treatments across the study period. No differences in body composition or muscle strength gains were observed between groups. Although the insect protein treatment group consumed more protein on training days than the carbohydrate group, participants in both groups ate adequate protein outside the intervention. Energy balance variation might explain why no differences in hypertrophy or strength were detected between groups. Ultimately, the authors attribute the change in fat- and bone-free mass to the training program itself and energy consumption, rather than insect protein-supplementation.

In the second study, Vangsoe *et al.* (2018b) investigated the quality of insect protein from the lesser mealworm by comparing postprandial amino acid availability and blood amino acid profiles after participants consumed protein isolate from whey, soya, or insects. Six healthy men were randomised to consume 25 g of protein treatment or a water placebo on four different days. Blood samples were taken before ingesting protein, at the time of ingestion, and then 20, 40, 60, 90, and 120 minutes after. Results indicated a significant rise in blood concentrations of essential and branched-chain amino acids, as well as leucine, over the time period for all protein supplements. This change was greatest for whey, with soya and insects comparable to one another ($P < 0.05$). Insect protein behaved similar to soya protein. Higher amino acid concentrations beyond the 120 min period were observed with insect protein consumption, indicating that insects may be considered a slow digestible protein. Insect protein isolate did not exhibit superior or unique characteristics compared to plant-based soya protein.

4   Discussion

The results of this review indicate a dearth of published human intervention studies investigating edible insects. From those published, entomophagy appears to be safe at the studied doses and offer several potential benefits to human health. Results should not be extrapolated given the number of species studied, small sample sizes, and limited health outcomes measured. Due to the small number of studies per species, no valuable comparisons across insect species can be made at this time. Only one study analysed health outcomes from insect consumption alone and not in combination with other foods (Vangsoe *et al.*, 2018b). Since none of the studies controlled or monitored the complete diet of participants, it is difficult to determine which outcomes resulted from the insects versus other food ingredients or the combination thereof. There is a true need for additional human intervention studies and clinical trials measuring the health impacts of entomophagy.

It would be short-sighted not to put the findings of this review in context with the other extensive research on edible insects from animal studies, *in vitro* experiments, and nutrient analyses. This discussion contextualises the results of the systematic review and highlights other potential health impacts not yet explored through human trials.

5 Improving nutritional status

5.1 *Micronutrient deficiencies*

Insects can be a good source of several essential micronutrients, particularly minerals like potassium, calcium, copper, magnesium, manganese, phosphorous, and selenium (Finke, 2002; Rumpold and Schlüter, 2013; Schabel, 2010). Notably, they also often contain relevant concentrations of zinc and iron (Christensen *et al.*, 2006), with several reporting iron concentrations greater than ground beef or chicken (Finke, 2002; Mwangi *et al.*, 2018; Oonincx and Van der Poel, 2011; USDA, 2018).

The rich micronutrient content of edible insects is particularly relevant in food insecure contexts, where iron-deficiency anaemia is prevalent. As this review demonstrates, the impact of dietary insect iron on iron status is not well understood and outcomes have been inconsistent. In one study, infants that consumed caterpillar-enriched cereal for 12 months had higher iron concentrations and lower rates of anaemia than those that did not (Bauserman *et al.*, 2015a). In two other studies, insect-fortified CFs were not associated with improvements in iron status (Konyole *et al.*, 2019; Skau *et al.*, 2015); however, other ingredients in these foods and/or non-dietary confounders could have influenced iron status.

It remains unclear exactly how well insect iron is absorbed. It is uniquely packaged compared to iron in vertebrates or plants, containing both haem and non-haem molecules. Predominantly, insect iron is found in the non-haem form (ferritin and holoferritin), but insects also contain smaller quantities of haem iron (found in cytochromes) (Mwangi *et al.*, 2018). Several *in vitro* and animal studies suggest that bioavailability of insect iron is good. A recent simulated digestion study assessed the solubility and bioavailability of both iron and zinc from several insect species and compared results with sirloin beef (Latunde-Dada *et al.*, 2016). Iron solubility for insects was greater than steak or whole wheat flour, but bioaccessibility (ferritin concentration in Caco-2 cells) varied by species and was not uniform across foods. Adding insects to wheat flour (1:1) demonstrated better iron and zinc solubility than beef in composite mixtures. It is therefore plausible that adding insects to starch-based diets could help address deficiencies and even increase the bioavailability of native or fortified iron in the starchy foods (Mwangi *et al.*, 2018). In a study of malnourished rats, researchers compared iron bioavailability from two insects, the house cricket, *Acheta domesticus* Linnaeus (Orthoptera: Gryllidae), and the palm weevil, *Rhynchphorus phoenicis* Fabricius (Coleoptera: Curculionidae), a positive control casein, and ferrous sulphate ($FeSO_4 \cdot 7H_2O$) diet. After inducing malnutrition, rats were fed supplemented diets for 14 days. During this time, an increase in haemoglobin relative biological value was observed with both insects, comparable to the control, indicating that these insects are an excellent source of bioavailable iron (Agbemafle *et al.*, 2019). Another study fed palm weevil larvae enriched biscuits to female albino rats for 28 days. Rats fed palm weevil had haematological parameters similar to rats fed the control diet, but their cholesterol concentrations were slightly higher than the control group. No inflammation was observed (Ayensu *et al.*, 2020).

Additional human and animal studies are needed to untangle these contradictory findings and further explore iron bioavailability in humans. Analyses of other variables that impact insect iron absorption, such as its release from the food matrix, effect of food preparation methods, presence of antinutrients, iron form, and systemic factors are needed.

5.2   *Vitamin B12*

There is some evidence that certain insects contain B vitamins such as biotin, pyridoxine, riboflavin, pantothenic acid, folate, and niacin, and B12 (cobalamin) (Nowak *et al.*, 2014; Rumpold and Schlüter, 2013; Schabel, 2010). Major dietary sources of vitamin B12 include meat, milk, animal-sourced foods, and to a lesser extent nori, nutritional yeast, and some bacteria or algae. People eating a majority plant-based diet are often at risk of deficiency. Only a handful of insects have been assessed for B12 content, and most demonstrated negligible quantities. However, the domestic house cricket has been shown to contain an impressive 8 mcg per 100 g fresh weight (FAO, 2017), relevant given the recommended dietary allowance for adults older than 14 is 2.4 mcg per day (NIH, 2018). Some B vitamins are not heat stable and may decrease through processing. The content, stability, and bioavailability of insect B12 requires more research.

5.3   *Protein and amino acid deficiencies*

The quality of insect protein is thought to be good, but there are discrepancies in the literature regarding its digestibility. The two relevant studies in this review did not demonstrate that insect protein from the lesser mealworm is unique or superior to other proteins. Vangsoe *et al.* (2018b) found that postprandial amino acid availability from insect protein isolate is comparable to soya, a high-quality plant protein. But since insects provide animal-based protein, it was expected that they would be more digestible than plant-based protein sources. A previous study in rats showed that termite protein was about 61% as digestible as casein (Phelps *et al.*, 1975); moreover, chitin could modulate protein digestibility (Churchward-Venne *et al.*, 2017). It is possible that processing insects to remove chitin may improve digestibility.

On the contrary, other studies have found variations in insect protein digestibility by species, but relatively high overall levels. In a study of 87 species, protein digestibility ranged from 76 to 96%, higher than most plant proteins but lower than eggs (95%) or beef (98%) (Ramos-Elorduy *et al.*, 1997). Another study measured protein digestibility of the farmed yellow mealworm, *Tenebrio molitor* Linnaeus (Coleoptera: Tenebrionidae), and house crickets fed to rats. Digestibility for both raw (84-92%) and heat-treated insects (84-90%) was high (Poelaert *et al.*, 2018). Others have reported insects having superior or equivalent protein quality compared with soya (Finke *et al.*, 1989; Yi *et al.*, 2013).

Regardless of overall protein digestibility, supplementing inadequate diets with specific amino acids that are deficient could be one means to combat PEM. For example, lysine is the most limiting essential amino acid in corn grain (Alan, 2009). People that get more than 50% of their daily calories from eating corn, such as many communities across Southern Africa, are at risk of developing PEM (Nuss and Tanumihardjo, 2011). Numerous species of insects contain relevant quantities of lysine (Oibiokpa *et al.*, 2018; Sogbesan and Ugwumba, 2008; Stull *et al.*, 2019). Insects are also thought to supply important amino acids that are deficient in other cereal, tuber, and legume-based diets (Bukkens, 2005; Manditsera *et al.*, 2019; Nadeau *et al.*, 2015; Van Huis *et al.*, 2013). A recent study of malnutrition recovery in mice indicated that a tropical house cricket-based diet was comparable to a both a peanut- and milk-based diet in terms of body weight recovery, but differed in impacts on immune and metabolic markers (Bergmans *et al.*, 2020). Parker *et al.* (2020) assessed nutrient content of raw, roasted, and groundnut mixed palm weevil larvae in Ghana to supplement CFs. Mixing larvae in all forms with other local foods (e.g. potatoes) was found to supply otherwise deficient lysine to the diet and generate more complete amino acid profiles, which could enhance CFs to help meet nutritional requirements for children (Parker *et al.*, 2020).

It should be noted that insects are not homogenous, and protein and amino acid values in insects do differ by order, species, and rearing conditions. Processing and cooking methods could damage amino acids; Sulphur-containing amino acids such as methionine are particularly susceptible to damage via oxidation depending on temperature and duration of exposure (Hendriks, 2018). More research is needed to understand the quality of insect proteins, the impact of processing on amino acids, and the bioavailability of proteins from insects in the diet.

6   Insect chitin and derivatives

Insects may have health impacts beyond simply providing nutrients; one plausible mechanism stems from the utilisation of insect fibre – a non-nutrient – in the gut. The relationship between insects and the microbiome is not well understood. Only one study in this review assessed impacts of insect consumption on the microbiome (Stull *et al.*, 2018). Another speculated as to interactions between the microbiome and health effects from silkworm consumption on COPD patients (Hu *et al.*, 2020). However, there are several reasons to hypothesise that insects may modulate gut microbiota and therefore indirectly influence health.

Unlike all other animal foods, edible insects contain meaningful levels of dietary fibre, predominantly in the form of chitin, an insoluble fibre. Chitin, a modified polysaccharide (poly-beta-1,4-*N*-acetylglucosamine) with a structure analogous to indigestible cellulose, is the primary component of the insect exoskeleton, respiratory linings, digestive and excretory systems (Clark and Smith, 1935). Through deacetylation, chitin can change structure into more soluble forms, namely chitosan or chitooligosaccharides (COS). COS from other sources are known prebiotics, so it is possible that insect chitin and derivatives could affect human health by selectively promoting the growth of beneficial bacterial species in the intestines.

In the first clinical trial to assess the impact of whole insect consumption on gut microbiota, daily consumption was associated with an increase in in the abundance of a beneficial probiotic bacterial species, *B. animalis* (Stull *et al.*, 2018). This particular bacterium has been studied extensively. It helps to inhibit pathogens such as *Escherichia coli* and *Salmonella* in the gut (Collado *et al.*, 2007; Martins *et al.*, 2009). The mechanism of action remains unclear, however. Recent studies demonstrate that several human cell types do produce chitinases (including gastric chitinases) and chitinase-like proteins (Mack *et al.*, 2015; Paoletti *et al.*, 2007) that could break down chitin. In primates, chitinase evolution has been linked to insectivory (Janiak *et al.*, 2017). No evidence clearly explains how chitin is broken down (deacetylated or otherwise) during human digestion, however.

Laboratory and animal studies have confirmed the potential prebiotic effect of insects and indicated that insects can modify the microbiome. Using an *in vitro* simulation model with *T. molitor* insect flour, De Carvalho *et al.* (2019) showed that *T. molitor* insect flour enhanced growth of almost all studied probiotic bacteria, increased production of short chain fatty acids, and helped maintain culture viability even while under nutritional stress (De Carvalho *et al.*, 2019). In another *in vitro* study, digestion resistant constituents from grass grub larvae (predominantly chitin) were fermented in faecal batch cultures and were associated with an increase in *Faecalibacterium*, which is thought to be beneficial for its anti-inflammatory properties and potential to improve gut health (Young *et al.*, 2020). Animal feeding trials have exhibited similar and more pronounced effects, as in these studies insects can make up a large portion of a controlled diet. Laying hens fed black soldier fly, *Hermetia illucens* Linnaeus (Diptera: Stratiomyidae) larvae meal (in place of soyabean meal) experienced dramatic changes in caecal microbiota, pointing to a prebiotic effect (Borrelli *et al.*, 2017).

Conversely, insect chitin could also inhibit the growth of bacteria, fungi, or other organisms in the gut. For example, a decrease in faecal probiotic *Lactobacillus* was also observed in healthy adults after 2 weeks of consuming cricket (Stull *et al.*, 2018). This is not altogether surprising, given that in laboratory studies, COS was shown to inhibit the growth of *Lactobacillus* spp. (Jeon *et al.*, 2001). In industry, chitosan has been used to prevent beer spoilage by preventing *Lactobacillus* growth (Gil *et al.*, 2004). Antibiotic and antimicrobial properties could also be positive, however. For example, chitin derivatives from various sources have been shown to inhibit growth of pathogens and other less desirable enteric organisms, such as *Clostridium perfringens* (Tsai and Hwang, 2004) and *E. coli* (Chien *et al.*, 2016; Selenius *et al.*, 2018) and even *Vibrio cholera*, and *Salmonella typhimurium* (Fernandes *et al.*, 2008)

Additionally, some evidence suggests that insect chitin could also stimulate activation of innate and adaptive immune cells (Komi *et al.*, 2018), improve glucose tolerance and insulin secretion (Zheng *et al.*, 2018), and exhibit anticancer, antiviral, and antifungal activity (Piccolo *et al.*, 2017). Animal studies indicate that chitin could help control lipid absorption in the intestines, lower total plasma cholesterol and low-density lipoprotein (LDL)-cholesterol, while increasing excretion of triglycerides in faeces (Zacour *et al.*, 1992). Many of these studies are exploratory and more research is needed to understand the properties of insect chitin and all possible health impacts from consumption. Additionally, we need to better appreciate how insects interact with the complex gut microbiome.

7   Other potential benefits

7.1   *Bioactive compounds*

Insects may offer other benefits due to the presence of bioactive compounds. Although none of the articles included in this review specifically investigated bioactive compounds, there is some evidence that these compounds could reduce health risks and boost immune function. For example, insects may contain bioactive compounds that could strengthen immune system function or reduce health risks (Roos and Van Huis, 2017). Specifically, several publications have confirmed that insects are a good source of biologically active peptides (Nongonierma and FitzGerald, 2017; Wu *et al.*, 2015; Zielińska *et al.*, 2017a,b), which have possible impacts discussed below.

7.2   *Hypertension and cardiovascular disease*

High blood pressure is a key risk factor for stroke, heart attack, chronic kidney disease, disability, and premature death; however it is largely considered preventable (Mills *et al.*, 2016). The angiotensin-converting enzyme (ACE) causes vasoconstriction and subsequently increases blood pressure. Excessive activity of ACE plays a role in regulating electrolyte and water balance, as well as blood pressure (Brunner *et al.*, 1972). Thus, inhibiting ACE activity using drugs (or nutraceuticals) is a common treatment for cardiovascular disease.

ACE inhibitory peptides have been identified in in handful of edible insects, and activity via insect peptides is comparable to some bioactive peptides found in other animal protein sources (Cito *et al.*, 2017a,b). *In vitro* studies have shown that peptides from the silkworm, *Bombyx mori* Linnaeus (Lepidoptera: Bombycidae), have inhibitory activity against ACE as well as the enzyme α-glucosidase (Vercruysse *et al.*, 2005; Wu *et al.*, 2015; Zhang *et al.*, 2016). These studies suggest that insect protein or its hydrolysates could serve as a functional ingredient in supplementary therapeutic foods to treat hypertension or control cholesterol, but human studies are needed.

7.3   *Metabolic syndrome*

Bioactive peptides derived from insects may inhibit enzyme activity associated with metabolic syndrome (Zielinska *et al.*, 2020). Antioxidant and anti-inflammatory peptides in several insects have been observed (Zielińska *et al.*, 2017a), and these peptides have demonstrated antiradical, chelating ions, lipoxygenase, and cyclooxygenase-2 activity. A follow-on study demonstrated that thermal processing the desert locust, *Schistocerca gregaria* Forskål. (Orthoptera: Acrididae), may boost peptide properties (Zielinska *et al.*, 2020).

7.4   *Type 2 diabetes and obesity*

Type 2 diabetes is often complicated by hypertension, which is also a risk factor for stroke, heart attack, heart failure, and kidney disease (Sowers and Epstein, 1995). There are several ways that insects could help manage type 2 diabetes. Synthetic ACE inhibitors are often used to treat hypertension *with* type 2 diabetes, but their side effects are unpleasant and natural product alternatives are desired (Wu *et al.*, 2015). ACE inhibitors from insects may be useful. Inhibiting enzyme α-glucosidase is another therapeutic target for suppressing hyperglycaemia in type 2 diabetes (Zhang *et al.*, 2016), and only a handful of α-glucosidase inhibitors for diabetes have been identified. Recent research illustrates that peptide fractions from desert locusts could inhibit α-glucosidase to help treat type 2 diabetes (Zielinska *et al.*, 2020). Moreover, glycosaminoglycan derived from the two-spotted cricket *Gryllus bimaculatus* De Geer (Orthoptera: Gryllidae) has been shown to decrease blood glucose, LDL-cholesterol, and alkaline phosphatase levels while reducing oxidative damage in diabetic mice (Ahn *et al.*, 2019).

A few studies have investigated the impact of insects and their bioactive compounds on weight control in mice (Kim *et al.*, 2016; Seo *et al.*, 2017). Results have been favourable, as daily intake of yellow mealworm larvae by obese mice mitigated body weight gain. By reducing fat accumulation in the body and the concentration of triglycerides in fat cells (adipocytes), the insect powder may have contained a bioactive compound that could induce weight loss (Seo *et al.*, 2017). Another study found that extracts of Korean horn beetle, *Allomyrina dichotoma* Linnaeus (Coleoptera: Scarabaeidae), can reduce endoplasmic reticulum stress when injected into the brain tissue of obese mice (Kim *et al.*, 2016).

Insect chitin or chitosan may also impact weight. The influence of dietary chitosan supplements on body weight and fat metabolism is poorly understood. In a recent study, chitosan from fungus was associated with reductions in mean body weight and blood sugar (Trivedi *et al.*, 2016), and a meta-analysis concluded that chitosan supplementation can yield a short- and medium-term effect on weight loss and improvement of serum lipid profiles, as well as cardiovascular factors (Moraru *et al.*, 2018). Human studies are needed to improve our understanding in each of these three dimensions.

7.5   *Antioxidants*

Edible insects are a dietary source of antioxidants. In a comparison of antioxidant capacity of aqueous and liposoluble extracts from insects with olive oil and fresh orange juice, Di Mattia *et al.* (2019) reported that numerous insect species displayed antioxidant capacity values two- or three-fold higher. African caterpillars, grasshoppers, and crickets exhibited the highest levels of reducing power, whereas grasshoppers, silkworms, and crickets demonstrated the highest antioxidant capacity values (Di Mattia *et al.*, 2019).

Processing insects may impact antioxidants. In a study of three species (yellow mealworm, desert locust, and tropical house cricket), processing methods impacted antioxidant and anti-inflammatory activities of the insect peptides. All insects in the study were a good source of bioactive peptides with antioxidant activity, and after digestion and absorption processes they exhibited high antiradical activity. Moreover, heat treatment positively affected antioxidant properties of the insect peptides (Zielińska *et al.*, 2017a).

Insect antioxidants could also mediate oxidative stress. A study of the free radical scavenging activities of the aqueous extract of lesser banded hornet, *Vespa affinis* Linnaeus (*Hymenoptera*: Vespidae), demonstrated antioxidant potential, which could mediate therapeutic activities in conditions that are linked to oxidative stress (Dutta *et al.*, 2016). Additionally, edible insect hydrolysates from digested dubia roach, *Blaptica dubia* Serville (*Blattodea*: Blaberidae), Madagascar hissing cockroach, *Gromphadorhina portentosa* Schaum (Blattodea: Blaberidae), migratory locust, *Locusta migratoria* Linnaeus (Orthoptera: Acrididae), superworm larvae, *Zophobas morio* Fabricius (Coleoptera: Tenebrionidae), and a cricket, *Amphiacusta annulipes* Serville (Orthoptera; Gryllidae), were assessed for free radical-scavenging activity and ability to chelate metal ions. Results demonstrated that these edible insects were a rich source of bioactive peptides with both antioxidant activity and the ability to chelate metal ions (Zielińska *et al.*, 2017b). Reported antioxidants found in insects could help prevent molecular damage, cardiovascular disease, and oxidative stress, but human studies are needed.

8   Health risks

Although the studies included in this review report mostly positive or neutral health outcomes, potential health risks from insect consumption have also been identified. Allergies, food safety concerns associated with microbiological or chemical contamination, and the presence of antinutrients that could impede nutrient absorption are the primary risks currently associated with entomophagy. There is little evidence, however, of disease or parasitoid transmission to humans when insects are handled under sanitary conditions (Van Huis *et al.*, 2013). In general, insects known to be edible are considered safe to eat if they are properly processed and handled.

8.1   *Allergies*

Although excluded from analyses in this review for not meeting the inclusion criteria, more than 250 articles were found that discussed or investigated insect allergens. These studies were not controlled; most were case reports of people eating insects and then falling ill, reviews, or laboratory studies assessing allergens in
sera. A review by De Gier and Verhoeckx (2018) provides a comprehensive overview
of case reports of allergy following insect consumption, cross-reactive and other
proteins involved in insect allergy, and possible alteration of allergenic potential
through processing and digestion. Testa *et al.* (2017) also discuss allergies in a review
of possible health risks associated with entomophagy. Allergenic effects from edible
insects are difficult to study (Yates-Doerr, 2014), but they have been identified to
varying degrees. Insects, like other arthropods, can produce mild allergic reactions
in consumers (Ayuso, 2011). There are several known antigenic determinants con-
served across arthropods. Cross-reactivity between edible insects and crustaceans
has been established (Belluco *et al.*, 2013), and current data highlight the role played
by arthropod pan-allergens including arginine kinase and tropomyosin (Ribeiro *et
al.*, 2018). Individuals allergic to shrimp may be more likely to be allergic to yellow
mealworm than other insects (Broekman *et al.*, 2017), and sera from people with
dust mite and crustacean allergies cross-reacted with mealworm proteins (Verhoeckx *et al.*, 2014) indicating that people with dust mite allergies may be allergic to insect proteins. Researchers also found a strong correlation between specific Immunoglobulin E (IgE) levels in the two-spotted cricket, shrimp-specific IgE
levels, and responses, indicating that crickets may induce a reaction in individuals
with a crustacean allergy (Kamemura *et al.*, 2018). Although uncommon, extreme
reactions to edible insects can cause dangerous or fatal anaphylaxis. Anaphylactic
shock has been observed with entomophagy in rare cases (Ji *et al.*, 2008; Kung *et al.*,
2011; Okezie *et al.*, 2010).

It is reasonable to recommend that individuals with known shellfish or mollusc
allergies avoid entomophagy (Testa *et al.*, 2017). Occupational exposures leading to
allergies also pose a threat to people who work closely with edible insects (Mlcek *et
al.*, 2014). Double-blind and placebo-controlled trials, food challenges, and research
on molecular mechanisms are needed to confirm and further explore this topic.

8.2   *Food safety*

There is no evidence that insect eating is accompanied by more health hazards than
eating other animal products (Mezes, 2018). Like all meat products, insects should
be properly processed to ensure food safety. Heat treatments are advised to reduce
risks (Vandeweyer *et al.*, 2018). Microbial contamination is of particular concern at
specific stages of the value chain, including processing, storage, and transportation.
Under certain conditions, insect microflora can facilitate the growth of hazardous
microorganisms including *Enterobacteriaceae* (Klunder *et al.*, 2012) known to cause
food borne illness. Insects can be contaminated via improper handling and process-

ing (Opara *et al.*, 2012) and bacterial spores may survive some cooking methods (Ter Beek and Brul, 2010). Therefore, proper processing, handling, and storage at all times is required to destroy pathogenic microorganisms present on or inside edible insects to ensure food safety (Klunder *et al.*, 2012; Simpanya *et al.*, 2000; Testa *et al.*, 2017).

Chemical contamination and the presence of toxins can threaten food safety. Edible insects can be contaminated by pesticides in their feed source, and some species contain defence-related toxins or repellents (Rumpold and Schlüter, 2013). For example, histamine toxicity and poisoning can occur with eating fried insects. Histidine, found in high concentrations in grasshoppers and silkworm pupae, is decarboxylated by bacteria to histamine, a heat stable toxin that can cause rashes, nausea, vomiting, diarrhoea, headaches, dyspnoea, chest tightness and other complications (Chomchai and Chomchai, 2018). Caution should be followed when consuming the scarab beetle, *Eulepida mashona* Arrow (Coleoptera: Scarabaeidae), in Zimbabwe, which may contain cyanogenic compounds even after traditional cooking (Musundire *et al.*, 2016b). The bioaccumulation of heavy metals in insects, such as lead in grasshoppers (Handley *et al.*, 2007), or presence of mycotoxins such as aflatoxin may pose additional threats (Banjoy *et al.*, 2010; Musundire *et al.*, 2016a); however, more research is needed to quantify these risks.

8,3 *Malabsorption*

Antinutrients identified in some insects, such as phytate, tannin, oxalate, hydrocyanide, saponins, and alkaloids may affect and inhibit the availability of dietary nutrients (Chakravorty *et al.*, 2016; Longvah *et al.*, 2012; Musundire *et al.*, 2014; Omotoso, 2006; Zhou and Han, 2006). Phytic acids, for example, may act as chelators to reduce the bioavailability of minerals. Antinutrients in insects may be absorbed from the feed source or synthesised directly by the insect; thus, rearing conditions could influence this aspect of food safety. Heating and processing may also impact anti-nutrient concentration and nutrient digestibility (El Hassan *et al.*, 2008), and vitamins that are not heat stable (Williams *et al.*, 2016). More research to understand what antinutrients are present in insects and investigate ways to eliminate them before consumption or mediate risk is needed.

9 Conclusions

Edible insects represent a diverse and relatively underexplored food group. There is ample evidence that insects are nutrient-dense, providing valuable macro- and micronutrients for human diets. However, more standardised research, particularly *in vivo* human trials, is needed to ascertain direct health impacts of consumption. Only nine human intervention studies met inclusion criteria for this review, highlighting a dearth of available research, especially given the thousands of edible species to consider. The evidence to-date demonstrates that insects can be a valuable addition to the diet, particularly in food insecure contexts where they may supplement nutrients that are otherwise deficient. They may also serve as a valuable, lower-resource protein in wealthier contexts. Bourgeoning research on the health impacts of edible insects beyond basic nutrient provision also suggests that insect fibre and bioactive compounds may provide health benefits that warrant further investigation.

Conflict of interest

The author declares no conflict of interest.

Appendix A

Example primary search syntax used in SCOPUS
TITLE-ABS-KEY ('edible insects' OR 'entomophagy' OR 'edible insect' OR 'minilivestock' OR 'microlivestock' OR 'mini-livestock' OR 'micro-livestock' OR 'farmed insect' OR 'insect cultivation' OR 'insect farming' OR 'insect agriculture' OR 'insect protein' OR 'insect oil' OR 'insect fat' OR 'insect protein isolate' OR 'insect meal' OR 'insect powder') AND TITLE-ABS-KEY ('human trial' OR 'intervention' OR 'clinical trial' OR 'crossover' OR 'health study' OR 'dietary' OR 'dietary intervention' OR 'feeding study' OR 'case study' OR 'epidemiological study' OR 'supplementation' OR 'experimental group' OR 'experiment' OR 'trial' OR 'case control' OR 'case-control' OR 'case study').

References

Agbemafle, I., Hanson, N., Bries, A.E. and Reddy, M.B., 2019. Alternative protein and iron sources from edible insects but not *Solanum torvum* improved body composition and iron status in malnourished rats. Nutrients 11: 2481.

Ahn, E.-M., Myung, N.-Y., Jung, H.-A. and Kim, S.-J., 2019. The ameliorative effect of *Protaetia brevitarsis* larvae in HFD-induced obese mice. Food Science and Biotechnology 28: 1177-1186.

Alan, L.K., 2009. Enhancement of amino acid availability in corn grain. In: Kriz, P.D.A.L. and Larkins, P.D.B.A. (eds.) Molecular genetic approaches to maize improvement. Springer, Berlin Heidelberg, Germany, pp. 79-89.

Ayensu, J., Larbie, C., Annan, R.A., Lutterodt, H., Edusei, A., Loh, S.P. and Asiamah, E.A., 2020. Palm weevil larvae (*Rhynchophorus phoenicis* Fabricius) and orange-fleshed sweet potato-enriched biscuits improved nutritional status in female Wistar albino rats. Journal of Nutrition and Metabolism, Article ID: 8061365. https://doi.org/10.1155/2020/8061365

Ayuso, R., 2011. Update on the diagnosis and treatment of shellfish allergy. Current Allergy and Asthma Reports 11: 309-316.

Baiano, A., 2020. Edible insects: an overview on nutritional characteristics, safety, farming, production technologies, regulatory framework, and socio-economic and ethical implications. Trends in Food Science & Technology 100: 35-50.

Banjoy, A.D., Lawal, O.A., Fasunwon, B.T. and Alimi, G.O., 2010. Alkali and heavy metal contaminants of some selected edible arthropods in South Western Nigeria. American-Eurasian Journal of Toxicological Sciences 2: 25-29.

Bauserman, M., Lokangaka, A., Gado, J., Close, K., Wallace, D., Kodondi, K.-K., Tshefu, A. and Bose, C., 2015a. A cluster-randomized trial determining the efficacy of caterpillar cereal as a locally available and sustainable complementary food to prevent stunting and anaemia. Public Health Nutrition 18: 1785-1792.

Bauserman, M., Lokangaka, A., Kodondi, K., Gado, J., Viera, A.J., Bentley, M.E., Engmann, C., Tshefu, A. and Bose, C., 2015b. Caterpillar cereal as a potential complementary feeding product for infants and young children: nutritional content and acceptability. Maternal & Child Nutrition 11: 214-220.

Belluco, S., Losasso, C., Maggioletti, M., Alonzi, C.C., Paoletti, M.G. and Ricci, A., 2013. Edible insects in a food safety and nutritional perspective: a critical review. Comprehensive Reviews in Food Science and Food Safety 12: 296-313.

Bergmans, R.S., Nikodemova, M., Stull, V.J., Rapp, A. and Malecki, K.M.C., 2020. Comparison of cricket diet with peanut-based and milk-based diets in the recovery from protein malnutrition in mice and the impact on growth, metabolism and immune function. PLoS ONE 15: e0234559.

Black, R.E., Victora, C.G., Walker, S.P., Bhutta, Z.A., Christian, P., De Onis, M., Ezzati, M., Grantham-McGregor, S., Katz, J., Martorell, R. and Uauy, R., 2013. Maternal and child undernutrition and overweight in low-income and middle-income countries. The Lancet 382: 427-451.

Bodenheimer, F.S., 1951. Insects as human food: a chapter of the ecology of man. W. Junk, The Hague, The Netherlands, 356 pp.

Borrelli, L., Coretti, L., Dipineto, L., Bovera, F., Menna, F., Chiariotti, L., Nizza, A., Lembo, F. and Fioretti, A., 2017. Insect-based diet, a promising nutritional source, modulates gut microbiota composition and SCFAs production in laying hens. Scientific Reports 7: 16269.

Broekman, H., Knulst, A.C., De Jong, G., Gaspari, M., Den Hartog Jager, C.F., Houben, G.F. and Verhoeckx, K.C.M., 2017. Is mealworm or shrimp allergy indicative for food allergy to insects? Molecular Nutrition & Food Research 61(9): 1601061. https://doi.org/10.1002/mnfr.201601061

Brunner, H.R., Laragh, J.H., Baer, L., Newton, M.A., Goodwin, F.T., Krakoff, L.R., Bard, R.H. and Bühler, F.R., 1972. Essential hypertension: renin and aldosterone, heart attack and stroke. New England Journal of Medicine 286: 441-449.

Bukkens, S., 2005. Insects in the human diet. In: Paoletti, M.G. (ed.) Ecological implications of minilivestock. Potential of insects, rodents, frogs and snails. Science Publishers, Enfield, NH, USA, pp. 545-577.

Chakravorty, J., Ghosh, S., Megu, K., Jung, C. and Meyer-Rochow, V.B., 2016. Nutritional and anti-nutritional composition of *Oecophylla smaragdina* (Hymenoptera: Formicidae) and *Odontotermes* sp. (Isoptera: Termitidae): two preferred edible insects of Arunachal Pradesh, India. Journal of Asia-Pacific Entomology 19: 711-720.

Chien, R.-C., Yen, M.-T. and Mau, J.-L., 2016. Antimicrobial and antitumor activities of chitosan from shiitake stipes, compared to commercial chitosan from crab shells. Carbohydrate Polymers 138: 259-264.

Chomchai, S. and Chomchai, C., 2018. Histamine poisoning from insect consumption: an outbreak investigation from Thailand. Clinical Toxicology 56: 126-131.

Christensen, D.L., Orech, F.O., Mungai, M.N., Larsen, T., Friis, H. and Aagaard-Hansen, J., 2006. Entomophagy among the Luo of Kenya: a potential mineral source? International Journal of Food Sciences & Nutrition 57: 198-203.

Churchward-Venne, T.A., Pinckaers, P.J.M., Van Loon, J.J.A. and Van Loon, L.J.C., 2017. Consideration of insects as a source of dietary protein for human consumption. Nutrition Reviews 75: 1035-1045.

Cito, A., Botta, M., Francardi, V. and Dreassi, E., 2017a. Insects as source of angiotensin converting enzyme inhibitory peptides. Journal of Insects as Food and Feed 3: 231-240. https://doi.org/10.3920/JIFF2017.0017

Cito, A., Dreassi, E., Frosinini, R., Zanfini, A., Pianigiani, C., Botta, M. and Francardi, V., 2017b. The potential beneficial effects of *Tenebrio molitor* (Coleoptera Tenebrionidae) and *Galleria mellonella* (Lepidoptera Pyralidae) on human health. Redia 100: 125-133.

Clark, G.L. and Smith, A.F., 1935. X-ray diffraction studies of chitin, chitosan, and derivatives. The Journal of Physical Chemistry 40: 863-879.

Collado, M.C., Grześkowiak, Ł. and Salminen, S., 2007. Probiotic strains and their combination inhibit *in vitro* adhesion of pathogens to pig intestinal mucosa. Current Microbiology 55: 260-265.

Collavo, A., Glew, R.H., Huang, Y.S., Chuang, L.T., Bosse, R. and Paoletti, M.G., 2005. House cricket small-scale farming. In: Paoletti, M.G. (ed.) Ecological implications of minilivestock. Potential of insects, rodents, frogs and snails. Science Publishers, Enfield, NH, USA, pp. 519-544.

De Carvalho, N.M., Walton, G.E., Poveda, C.G., Silva, S.N., Amorim, M., Madureira, A.R., Pintado, M.E., Gibson, G.R. and Jauregi, P., 2019. Study of *in vitro* digestion of *Tenebrio molitor* flour for evaluation of its heck impact on the human gut microbiota. Journal of Functional Foods 59: 101-109.

De Gier, S. and Verhoeckx, K., 2018. Insect (food) allergy and allergens. Molecular Immunology 100: 82-106.

DeFoliart, G.R., 1995. Edible insects as minilivestock. Biodiversity & Conservation 4: 306-321.

Di Mattia, C., Battista, N., Sacchetti, G. and Serafini, M., 2019. Antioxidant activities *in vitro* of water and liposoluble extracts obtained by different species of edible insects and invertebrates. Frontiers in Nutrition 6: 106. https://doi.org/10.3389/fnut.2019.00106

Dutta, P., Dey, T., Manna, P. and Kalita, J., 2016. Antioxidant potential of *Vespa affinis* L., a traditional edible insect species of North East India. PLoS ONE 11: e0156107.

El Hassan, N.M., Hamed, S.Y., Hassan, A.B., Eltayet, M.M. and Baiker, E., 2008. Nutritional evaluation and physiochemical properties of boiled and fried tree locust. Pakistan Journal of Nutrition 7: 325-329.

Fernandes, J.C., Tavaria, F.K., Soares, J.C., Ramos, Ó.S., João Monteiro, M., Pintado, M.E. and Xavier Malcata, F., 2008. Antimicrobial effects of chitosans and chitooligosaccharides, upon *Staphylococcus aureus* and *Escherichia coli*, in food model systems. Food Microbiology 25: 922-928.

Finke, M.D., 2002. Complete nutrient composition of commercially raised invertebrates used as food for insectivores. Zoo Biology 21: 269-285.

Finke, M.D., 2015. Complete nutrient content of four species of commercially available feeder insects fed enhanced diets during growth. Zoo Biology 34: 554-564.

Finke, M.D., DeFoliart, G.R. and Benevenga, N.J., 1989. Use of a four-parameter logistic model to evaluate the quality of the protein from three insect species when fed to rats. The Journal of Nutrition 119: 864-871.

Food and Agriculture Organisation (FAO), 2017. FAO/INFOODS food composition database for biodiversity. Version 4.0. FAO, Rome, Italy. Available at: http://www.fao.org/infoods/infoods/tables-and-databases/faoinfoods-databases/en/

Food and Agriculture Organisation / International Fund for Agricultural Development / United Nations International Children's Emergency Fund / World Food Programme / World Health Organisation (FAO/IFAD/UNICEF/WFP/WHO), 2020. The state of food security and nutrition in the world 2020. Transforming food systems for affordable healthy diets. FAO, Rome, Italy.

Gil, G., Mónaco, S., Cerrutti, P. and Galvagno, M., 2004. Selective antimicrobial activity of chitosan on beer spoilage bacteria and brewing yeasts. Biotechnology Letters 26: 569-574.

Gough, E.K., Stephens, D.A., Moodie, E.E.M., Prendergast, A.J., Stoltzfus, R.J., Humphrey, J.H. and Manges, A.R., 2015. Linear growth faltering in infants is associated with *Acidaminococcus* sp. and community-level changes in the gut microbiota. Microbiome 3: 24.

Handley, M.A., Hall, C., Sanford, E., Diaz, E., Gonzalez-Mendez, E., Drace, K., Wilson, R., Villalobos, M. and Croughan, M., 2007. Globalization, binational communities, and imported food risks: results of an outbreak investigation of lead poisoning in Monterey County, California. American Journal of Public Health 97: 900-906.

Hendriks, W.H., 2018. 46 amino acid availability in heat-damaged ingredients. Journal of Animal Science 96: 25.

Hu, Y., Shi, Q., Ying, S., Zhu, D., Chen, H., Yang, X., Xu, J., Xu, F., Tao, F. and Xu, B., 2020. Effects of compound Caoshi silkworm granules on stable COPD patients and their relationship with gut microbiota: a randomized controlled trial. Medicine 99: e20511.

Janiak, M.C., Chaney, M.E. and Tosi, A.J., 2017. Evolution of acidic mammalian chitinase genes (CHIA) is related to body mass and insectivory in primates. Molecular Biology and Evolution 35: 607-622. https://doi.org/10.1093/molbev/msx312

Jeon, Y.-J., Park, P.-J. and Kim, S.-K., 2001. Antimicrobial effect of chitooligosaccharides produced by bioreactor. Carbohydrate Polymers 44: 71-76.

Ji, K.-M., Zhan, Z.-K., Chen, J.-J. and Liu, Z.-G., 2008. Anaphylactic shock caused by silkworm pupa consumption in China. Allergy 63: 1407-1408.

Jongema, Y., 2017. List of edible insects of the world. Wageningen University & Research, Wageningen, The Netherlands. Available at: http://tinyurl.com/mestm6p.

Kamemura, N., Sugimoto, M., Tamehiro, N., Adachi, R., Tomonari, S., Watanabe, T. and Mito, T., 2018. Cross-allergenicity of crustacean and the edible insect *Gryllus bimaculatus* in patients with shrimp allergy. Molecular Immunology 106: 127-134.

Kim, J., Yun, E.-Y., Park, S.-W., Goo, T.-W. and Seo, M., 2016. *Allomyrina dichotoma* larvae regulate food intake and body weight in high fat diet-induced obese mice through mTOR and Mapk signaling pathways. Nutrients 8: 100.

Klunder, H.C., Wolkers-Rooijackers, J., Korpela, J.M. and Nout, M.J.R., 2012. Microbiological aspects of processing and storage of edible insects. Food Control 26: 628-631.

Komi, D.E.A., Sharma, L. and Dela Cruz, C.S., 2018. Chitin and its effects on inflammatory and immune responses. Clinical Reviews in Allergy & Immunology 54: 213-223.

Konyole, S., Kinyuru, J., Owuor, B., Kenji, G., Onyango, C., Estambale, B., Friis, H., Roos, N. and Owino, V., 2012. Acceptability of amaranth grain-based nutritious complementary foods with Dagaa fish (*Rastrineobola argentea*) and edible termites (*Macrotermes subhylanus*) compared to corn soy blend plus among young children/mothers dyads in Western Kenya. Journal of Food Research 1: 111.

Konyole, S.O., Omollo, S.A., Kinyuru, J.N., Skau, J.K.H., Owuor, B.O., Estambale, B.B., Filteau, S.M., Michaelsen, K.F., Friis, H., Roos, N. and Owino, V.O., 2019. Effect of locally produced complementary foods on fat-free mass, linear growth, and iron status among Kenyan infants: a randomized controlled trial. Maternal & Child Nutrition 15: e12836.

Kung, S.-J., Fenemore, B. and Potter, P.C., 2011. Anaphylaxis to Mopane worms (*Imbrasia belina*). Annals of Allergy, Asthma & Immunology 106: 538-540.

Latunde-Dada, G.O., Yang, W. and Vera Aviles, M., 2016. *In vitro* iron availability from insects and Sirloin beef. Journal of Agricultural and Food Chemistry 64: 8420-8424.

Lesnik, J.J., 2018. Edible insects and human evolution. University Press of Florida, Gainesville, FL, USA.

Li, W., Zhang, Z., Lin, L. and Terenius, O., 2017. *Antheraea pernyi* (Lepidoptera: Saturniidae) and its importance in sericulture, food consumption, and traditional Chinese medicine. Journal of Economic Entomology 110: 1404-1411.

Liu, L., Johnson, H.L., Cousens, S., Perin, J., Scott, S., Lawn, J.E., Rudan, I., Campbell, H., Cibulskis, R., Li, M., Mathers, C. and Black, R.E., 2012. Global, regional, and national causes of child mortality: an updated systematic analysis for 2010 with time trends since 2000. The Lancet 379: 2151-2161.

Longvah, T., Manghtya, K. and Qadri, S.S.Y.H., 2012. Eri silkworm: a source of edible oil with a high content of α-linolenic acid and of significant nutritional value. Journal of the Science of Food and Agriculture 92: 1988-1993.

Mack, I., Hector, A., Ballbach, M., Kohlhäufl, J., Fuchs, K.J., Weber, A., Mall, M.A. and Hartl, D., 2015. The role of chitin, chitinases, and chitinase-like proteins in pediatric lung diseases. Molecular and Cellular Pediatrics 2: 3.

Manditsera, F.A., Luning, P.A., Fogliano, V. and Lakemond, C.M.M., 2019. The contribution of wild harvested edible insects (*Eulepida mashona* and *Henicus whelluni*) to nutrition security in Zimbabwe. Journal of Food Composition and Analysis 75: 17-25.

Martins, F.S., Silva, A.A., Vieira, A.T., Barbosa, F.H.F., Arantes, R.M.E., Teixeira, M.M. and Nicoli, J.R., 2009. Comparative study of *Bifidobacterium animalis, Escherichia coli, Lactobacillus casei* and *Saccharomyces boulardii* probiotic properties. Archives of Microbiology 191: 623-630.

Melo, V., Garcia, M., Sandoval, H., Jiménez, H.D. and Calvo, C., 2011. Quality proteins from edible indigenous insect food of Latin America and Asia. Emirates Journal of Food and Agriculture 23: 283-289.

Mezes, M., 2018. Food safety aspect of insects: a review. Acta Alimentaria 47: 513-522.

Mills, K.T., Bundy, J.D., Kelly, T.N., Reed, J.E., Kearney, P.M., Reynolds, K., Chen, J. and He, J., 2016. Global disparities of hypertension prevalence and control: a systematic analysis of population-based studies from 90 countries. Circulation 134: 441-450.

Mlcek, J., Rop, O., Borkovcova, M. and Bednarova, M., 2014. A comprehensive look at the possibilities of edible insects as food in Europe – a review. Polish Journal of Food and Nutrition Sciences 64: 147-157.

Moher, D., Liberati, A., Tetzlaff, J., Altman, D.G. and The PRISMA Group, 2009. Preferred reporting items for systematic reviews and meta-analyses: the PRISMA statement. PLOS Medicine 6: e1000097. 10.1371/journal.pmed.1000097

Moraru, C., Mincea, M.M., Frandes, M., Timar, B. and Ostafe, V., 2018. A meta-analysis on randomised controlled clinical trials evaluating the effect of the dietary supplement chitosan on weight loss, lipid parameters and blood pressure. Medicina 54: 109. https://doi.org/10.3390/medicina54060109

Musundire, R., Osuga, I.M., Cheseto, X., Irungu, J. and Torto, B., 2016a. Aflatoxin contamination detected in nutrient and anti-oxidant rich edible stink bug stored in recycled grain containers. PLoS ONE 11: e0145914.

Musundire, R., Zvidzai, C.J., Chidewe, C., Ngadze, R.T., Macheka, L., Manditsera, F.A., Mubaiwa, J. and Masheka, A., 2016b. Nutritional and bioactive compounds composition of *Eulepida mashona*, an edible beetle in Zimbabwe. Journal of Insects as Food and Feed 2: 179-187. https://doi.org/10.3920/JIFF2015.0050

Musundire, R., Zvidzai, C.J., Chidewe, C., Samende, B.K. and Manditsera, F.A., 2014. Nutrient and anti-nutrient composition of *Henicus whellani* (Orthoptera: Stenopelmatidae), an edible ground cricket, in south-eastern Zimbabwe. International Journal of Tropical Insect Science 34: 223-231.

Mwangi, M.N., Oonincx, D.G.A.B., Stouten, T., Veenenbos, M., Melse-Boonstra, A., Dicke, M. and Van Loon, J.J.A., 2018. Insects as sources of iron and zinc in human nutrition. Nutrition Research Reviews 2: 248-255.

Nadeau, L., Nadeau, I., Franklin, F. and Dunkel, F., 2015. The potential for entomophagy to address undernutrition. Ecology of Food and Nutrition 54: 200-208.

National Institutes of Health (NIH), 2018. Vitamin B12 – factsheet for health professionals. NIH, Bethesda, MD, USA. Available at: https://ods.od.nih.gov/factsheets/VitaminB12-HealthProfessional/

Nongonierma, A.B. and FitzGerald, R.J., 2017. Unlocking the biological potential of proteins from edible insects through enzymatic hydrolysis: a review. Innovative Food Science & Emerging Technologies 43: 239-252.

Nowak, V., Persijn, D., Rittenschober, D. and Charrondiere, U.R., 2014. Review of food composition data for edible insects. Food Chemistry 193: 39-46. https://doi.org/10.1016/j.foodchem.2014.10.114

Nuss, E.T. and Tanumihardjo, S.A., 2011. Quality protein maize for Africa: closing the protein inadequacy gap in vulnerable populations. Advances in Nutrition 2: 217-224.

Oibiokpa, F.I., Akanya, H.O., Jigam, A.A., Saidu, A.N. and Egwim, E.C., 2018. Protein quality of four indigenous edible insect species in Nigeria. Food Science and Human Wellness 7: 175-183.

Okezie, O.A., Kgomotso, K.K. and Letswiti, M.M., 2010. Mopane worm allergy in a 36-year-old woman: a case report. Journal of Medical Case Reports 4: 42.

Omotoso, O.T., 2006. Nutritional quality, functional properties and anti-nutrient compositions of the larva of *Cirina forda* (Westwood) (Lepidoptera: Saturniidae). Journal of Zhejiang University Science B 7: 51-55.

Oonincx, D.G.A.B. and Van der Poel, A.F.B., 2011. Effects of diet on the chemical composition of migratory locusts (*Locusta migratoria*). Zoo Biology 30: 9-16.

Oonincx, D.G.A.B., Laurent, S., Veenenbos, M.E. and Van Loon, J.J.A., 2019. Dietary enrichment of edible insects with omega 3 fatty acids. Insect Science 27(3): 500-509. https://doi.org/10.1111/1744-7917.12669

Oonincx, D.G.A.B., Van Itterbeeck, J., Heetkamp, M.J.W., Van den Brand, H., Van Loon, J.J.A. and Van Huis, A., 2010. An exploration on greenhouse gas and ammonia production by insect species suitable for animal or human consumption. PLoS ONE 5: e14445.

Opara, M., Sanyigha, F.T., Ogbuewu, I.P. and Okoli, I.C., 2012. Studies on the production trend and quality characteristics of palm grubs in the tropical rainforest zone of Nigeria. Journal of Agricultural Technology 8: 851-860.

Paoletti, M.G., Norberto, L., Damini, R. and Musumeci, S., 2007. Human gastric juice contains chitinase that can degrade chitin. Annals of Nutrition and Metabolism 51: 244-251.

Parker, M.E., Zobrist, S., Lutterodt, H.E., Asiedu, C.R., Donahue, C., Edick, C., Mansen, K., Pelto, G., Milani, P., Soor, S., Laar, A. and Engmann, C.M., 2020. Evaluating the nutritional content of an insect-fortified food for the child complementary diet in Ghana. BMC Nutrition 6: 7.

Paul, A., Frederich, M., Megido, R.C., Alabi, T., Malik, P., Uyttenbroeck, R., Francis, F., Blecker, C., Haubruge, E., Lognay, G. and Danthine, S., 2017. Insect fatty acids: a comparison of lipids from three Orthopterans and *Tenebrio molitor* L. larvae. Journal of Asia-Pacific Entomology 20: 337-340.

Payne, C.L.R., Scarborough, P., Rayner, M. and Nonaka, K., 2016. A systematic review of nutrient composition data available for twelve commercially available edible insects, and comparison with reference values. Trends in Food Science & Technology 47: 69-77.

Phelps, R.J., Struthers, J.K. and Moyo, S.J.L., 1975. Investigations into the nutritive value of *Macrotermes falciger* (Isoptera: Termitidae). Zoologica Africana 10: 123-132.

Piccolo, G., Iaconisi, V., Marono, S., Gasco, L., Loponte, R., Nizza, S., Bovera, F. and Parisi, G., 2017. Effect of *Tenebrio molitor* larvae meal on growth performance, *in vivo* nutrients digestibility, somatic and marketable indexes of gilthead sea bream (*Sparus aurata*). Animal Feed Science and Technology 226: 12-20.

Poelaert, C., Francis, F., Alabi, T., Megido, R.C., Crahay, B., Bindelle, J. and Beckers, Y., 2018. Protein value of two insects, subjected to various heat treatments, using growing rats and the protein digestibility-corrected amino acid score. Journal of Insects as Food and Feed 4: 77-87. https://doi.org/10.3920/JIFF2017.0003

Poore, J. and Nemecek, T., 2018. Reducing food's environmental impacts through producers and consumers. Science 360: 987-992.

Ramos-Elorduy, J., Moreno, J.M.P., Prado, E.E., Perez, M.A., Otero, J.L. and De Guevara, O.L., 1997. Nutritional value of edible insects from the State of Oaxaca, Mexico. Journal of Food Composition and Analysis 10: 142-157.

Ranganathan, J., Vennard, D., White, R., Dumas, P., Lipinski, B. and Searchinger, T., 2016. Shifting diets for a sustainable food future. World Resources Institute, Washington, DC, USA.

Ribeiro, J.C., Cunha, L.M., Sousa-Pinto, B. and Fonseca, J., 2018. Allergic risks of consuming edible insects: a systematic review. Molecular Nutrition & Food Research 62: 1700030. https://doi.org/10.1002/mnfr.201700030

Ritchie, H. and Roser, M., 2017. Micronutrient deficiency. Our World in Data. Available at: https://ourworldindata.org/micronutrient-deficiency

Roos, N. and Van Huis, A., 2017. Consuming insects: are there health benefits? Journal of Insects as Food and Feed 3: 225-229. https://doi.org/10.3920/JIFF2017.x007

Rumpold, B.A. and Schlüter, O.K., 2013. Nutritional composition and safety aspects of edible insects. Molecular Nutrition & Food Research 57: 802-823.

Schabel, H.G., 2010. Forest insects as food: a global review. Food and Agriculture Organization of the United Nations, Rome, Italy, pp. 37-64.

Selenius, O., Korpela, J., Salminen, S. and Gallego, C.G., 2018. Effect of chitin and chitooligosaccharide on *in vitro* growth of *Lactobacillus rhamnosus* GG and *Escherichia coli* TG. Applied Food Biotechnology 5: 163-172.

Seo, M., Goo, T.-W., Chung, M.Y., Baek, M., Hwang, J.-S., Kim, M.-A. and Yun, E.-Y., 2017. *Tenebrio molitor* larvae inhibit adipogenesis through AMPK and MAPKs signaling in 3T3-L1 adipocytes and obesity in high-fat diet-induced obese mice. International Journal of Molecular Sciences 18: 518. https://doi.org/10.3390/ijms18030518

Simpanya, M.F., Allotey, J. and Mpuchane, S.F., 2000. A mycological investigation of phane, an edible caterpillar of an emperor moth, *Imbrasia belina*. Journal of Food Protection 63: 137-140.

Skau, J., Touch, B., Chhoun, C., Chea, M., Umni, U., Makurat, J., Filteau, S., Wieringa, F., Dijkhuizen, M., Ritz, C., Wells, J., Berger, J., Friis, H., Michaelsen, K. and Roos, N., 2015. Effects of animal source food and micronutrient fortification in complementary food products on body composition, iron status, and linear growth: a randomized trial in Cambodia. The American Journal of Clinical Nutrition 101: 742-751.

Sogbesan, A.O. and Ugwumba, A.A.A., 2008. Nutritional evaluation of termite (*Macrotermes subhyalinus*) meal as animal protein supplements in the diets of *Heterobranchus longifilis* (Valenciennes, 1840) fingerlings. Turkish Journal of Fisheries and Aquatic Sciences 8: 149-158.

Sowers, J.R. and Epstein, M., 1995. Diabetes mellitus and associated hypertension, vascular disease, and nephropathy. Hypertension 26: 869-879.

Stevens, G.A., Finucane, M.M., De-Regil, L.M., Paciorek, C.J., Flaxman, S.R., Branca, F., Peña-Rosas, J.P., Bhutta, Z.A. and Ezzati, M., 2013. Global, regional, and national trends in haemoglobin concentration and prevalence of total and severe anaemia in children and pregnant and non-pregnant women for 1995-2011: a systematic analysis of population-representative data. The Lancet Global Health 1: e16-e25.

Stull, V.J., Finer, E., Bergmans, R.S., Febvre, H.P., Longhurst, C., Manter, D.K., Patz, J.A. and Weir, T.L., 2018. Impact of edible cricket consumption on gut microbiota in healthy adults, a double-blind, randomized crossover trial. Scientific Reports 8: 10762.

Stull, V.J., Kersten, M., Bergmans, R.S., Patz, J.A. and Paskewitz, S., 2019. Crude protein, amino acid, and iron content of Tenebrio molitor (Coleoptera, Tenebrionidae) reared on an agricultural byproduct from maize production: an exploratory study. Annals of the Entomological Society of America 112: 533-543. https://doi.org/10.1093/aesa/saz024

Ter Beek, A. and Brul, S., 2010. To kill or not to kill bacilli: opportunities for food biotechnology. Current Opinion in Biotechnology 21: 168-174.

Testa, M., Stillo, M., Maffei, G., Andriolo, V., Gardois, P. and Zotti, C.M., 2017. Ugly but tasty: a systematic review of possible human and animal health risks related to entomophagy. Critical Reviews in Food Science and Nutrition 57: 3747-3759.

Trivedi, V., Satia, M., Deschamps, A., Maquet, V., Shah, R., Zinzuwadia, P. and Trivedi, J., 2016. Single-blind, placebo controlled randomised clinical study of chitosan for body weight reduction. Nutrition Journal 15: 3.

Tsai, G. and Hwang, S., 2004. In vitro and in vivo antibacterial activity of shrimp chitosan against some intestinal bacteria. Fisheries Science 70: 675-681.

United States Department of Agriculture (USDA), 2018. National nutrient database for standard reference 1, release April, 2018. USDA Agricultural Research Service, Washington, DC, USA. Available at: http://www.ars.usda.gov/ba/bhnrc/ndl

Van Huis, A., Van Itterbeeck, J., Klunder, H., Mertens, E., Halloran, A., Muir, G. and Vantomme, P., 2013. Edible insects: future prospects for food and feed security. FAO Forestry Paper no. 171. Food and Agriculture Organization of the United Nations, Rome, Italy. Available at: http://www.fao.org/docrep/018/i3253e/i3253e00.htm

Vandeweyer, D., Wynants, E., Crauwels, S., Verreth, C., Viaene, N., Claes, J., Lievens, B. and Campenhout, L.V., 2018. Microbial dynamics during industrial rearing, processing, and storage of tropical house crickets (Gryllodes sigillatus) for human consumption. Applied and Environmental Microbiology 84: e00255-18.

Vangsoe, M.T., Joergensen, M.S., Heckmann, L.-H.L. and Hansen, M., 2018a. Effects of insect protein supplementation during resistance training on changes in muscle mass and strength in young men. Nutrients 10: 335. https://doi.org/10.3390/nu10030335

Vangsoe, M.T., Thogersen, R., Bertram, H.C., Heckmann, L.-H.L. and Hansen, M., 2018b. Ingestion of insect protein isolate enhances blood amino acid concentrations similar to soy protein in a human trial. Nutrients 10: 1357. https://doi.org/10.3390/nu10101357

Vercruysse, L., Smagghe, G., Herregods, G. and Van Camp, J., 2005. ACE inhibitory activity in enzymatic hydrolysates of insect protein. Journal of Agricultural and Food Chemistry 53: 5207-5211.

Verhoeckx, K.C.M., Van Broekhoven, S., Den Hartog-Jager, C.F., Gaspari, M., De Jong, G.A.H., Wichers, H.J., Van Hoffen, E., Houben, G.F. and Knulst, A.C., 2014. House dust mite (Der p 10) and crustacean allergic patients may react to food containing yellow mealworm proteins. Food and Chemical Toxicology 65: 364-373.

Verkerk, M.C., Tramper, J., Van Trijp, J.C.M. and Martens, D.E., 2007. Insect cells for human food. Biotechnology Advances 25: 198-202.

Wathne, A.M., Devle, H., Naess-Andresen, C.F. and Ekeberg, D., 2018. Identification and quantification of fatty acids in *T. viridissima*, *C. biguttulus*, and *C. brunneus* by GC-MS. Journal of Lipids, Article ID: 3679247. https://doi.org/10.1155/2018/3679247

Williams, J.P., Williams, J.R., Kirabo, A., Chester, D. and Peterson, M., 2016. Nutrient content and health benefits of insects. In: Dossey, A.T., Morales-Ramos, J.A. and Rojas, M.G. (eds.) Insects as sustainable food ingredients. Academic Press, San Diego, CA, USA, pp. 61-84.

Womeni, H.M., Linder, M., Tiencheu, B., Mbiapo, F.T., Villeneuve, P., Fanni, J. and Parmentier, M., 2009. Oils of insects and larvae consumed in Africa: potential sources of polyunsaturated fatty acids. Oilseeds and fats, Crops and Lipids 16: 230-235.

World Food Programme (WFP), 2015. Hunger statistics. Fighting hunger worldwide. World Food Programme, Rome, Italy. Available at: http://www.wfp.org/hunger/stats

World Health Organisation (WHO), 2020. Obesity and overweight. WHO, Geneva, Switzerland. Available at: https://www.who.int/news-room/fact-sheets/detail/obesity-and-overweight

Wu, Q., Jia, J., Yan, H., Du, J. and Gui, Z., 2015. A novel angiotensin-I converting enzyme (ACE) inhibitory peptide from gastrointestinal protease hydrolysate of silkworm pupa (*Bombyx mori*) protein: biochemical characterization and molecular docking study. Peptides 68: 17-24.

Yates-Doerr, E., 2014. The world in a box? Food security, edible insects, and 'One World, One Health' collaboration. Social Science & Medicine 129: 106-112. https://doi.org/10.1016/j.socscimed.2014.06.020

Yi, L., Lakemond, C.M.M., Sagis, L.M.C., Eisner-Schadler, V., Van Huis, A. and Van Boekel, M.A.J.S., 2013. Extraction and characterisation of protein fractions from five insect species. Food Chemistry 141: 3341-3348.

Ylla, G., Nakamura, T., Itoh, T., Kajitani, R., Toyoda, A., Tomonari, S., Bando, T., Ishimaru, Y., Watanabe, T., Fuketa, M., Matsuoka, Y., Noji, S., Mito, T. and Extavour, C.G., 2020. Cricket genomes: the genomes of future food. bioRxiv, https://doi.org/10.1101/2020.07.07.191841

Young, W., Arojju, S.K., McNeill, M.R., Rettedal, E., Gathercole, J., Bell, N. and Payne, P., 2020. Feeding bugs to bugs: edible insects modify the human gut microbiome in an *in vitro* fermentation model. Frontiers in Microbiology Frontiers 11: 1763. https://doi.org/10.3389/fmicb.2020.01763

Zacour, A.C., Silva, M.E., Cecon, P.R., Bambirra, E.A. and Vieira, E.C., 1992. Effect of dietary chitin on cholesterol absorption and metabolism in rats. Journal of Nutritional Science and Vitaminology 38: 609-613.

Zhang, Y., Wang, N., Wang, W., Wang, J., Zhu, Z. and Li, X., 2016. Molecular mechanisms of novel peptides from silkworm pupae that inhibit α-glucosidase. Peptides 76: 45-50.

Zheng, J., Yuan, X., Cheng, G., Jiao, S., Feng, C., Zhao, X., Yin, H., Du, Y. and Liu, H., 2018. Chitosan oligosaccharides improve the disturbance in glucose metabolism and reverse the dysbiosis of gut microbiota in diabetic mice. Carbohydrate Polymers 190: 77-86.

Zhou, J. and Han, D., 2006. Safety evaluation of protein of silkworm (*Antheraea pernyi*) pupae. Food and Chemical Toxicology 44: 1123-1130.

Zielińska, E., Baraniak, B. and Karaś, M., 2017a. Antioxidant and anti-inflammatory activities of hydrolysates and peptide fractions obtained by enzymatic hydrolysis of selected heat-treated edible insects. Nutrients 9: 970. https://doi.org/10.3390/nu9090970

Zielińska, E., Karaś, M. and Jakubczyk, A., 2017b. Antioxidant activity of predigested protein obtained from a range of farmed edible insects. International Journal of Food Science & Technology 52: 306-312.

Zielinska, E., Karas, M., Baroniak, B. and Jakubczyk, A., 2020. Evaluation of ACE, alpha-glucosidase, and lipase inhibitory activities of peptides obtained by *in vitro* digestion of selected species of edible insects. European Food Research and Technology 246: 1361-1369.

Beyond the protein concept: health aspects of using edible insects on animals

L. Gasco[1]*, A. Józefiak[2] and M. Henry[3]

[1]Department of Agricultural, Forest and Food Sciences, University of Torino, Largo P. Braccini 2, 10095 Grugliasco, Italy; [2]Department of Preclinical Sciences and Infectious Diseases, Poznan University of Life Sciences, Wołyńska 35, 60-637 Poznań, Poland; [3]Institute of Marine Biology, Biotechnology and Aquaculture, Hellenic Centre for Marine Research, 46.7 Athinon – Souniouave, 19013 Anavissos, Attiki, Greece; *laura.gasco@unito.it

Abstract

There is an increasing interest in the use of insects in animal feed since they contain high proteins levels, lipids, vitamins and minerals. In particular, insect-derived proteins are seen as one of the potential solution to face the increasing protein shortage and are able to fully substitute soybean meal or fishmeal in aquaculture or livestock feeds. However, beside their interesting nutritional composition, insects are also rich in bioactive compounds such as chitin, antimicrobial peptides or specific fatty acids with immunostimulating, antimicrobial and/or anti-inflammatory properties able to sustain animal health, increase their resistance to diseases. Further studies will also have to investigate whether insects share similarities with bacterial or parasitical pathogens and may act as immunostimulants. These recent findings may launch insects beyond the protein concept into healthy animal feeds. This review presents the effects of insects and their bioactive compounds on fish and crustaceans, poultry, pigs and rabbits immune system, gut health, microbiota and resistance to diseases.

Keywords

gut health – microbiota – antioxidant enzymes – immunity – disease resistance

1   Introduction

Since the dawn of time, insects have been part of our life, being considered as a pest, a resource, food and, more recently, as a feed for intensive livestock production. Pest insects are known to have detrimental impacts on agricultural and food production as they damage crops and parasitise livestock (Bradshaw *et al.*, 2016; FAO, 2020; Paini *et al.*, 2016), and they can also be a great nuisance and health hazard for human (WHO, 2015).

Nevertheless, apart from being considered a great nuisance, insects are responsible for the production of about one third of our food through the pollination process and, as such, they are a great asset playing a key role in all terrestrial and freshwater ecosystems. If all insects were to disappear, human food supply would run out in about four years (Goulson, 2019; Noriega *et al.*, 2018).

Insects were already a major food source 1.2 million years ago (Hardy *et al.*, 2016; Van Huis, 2017) and they are still today part of the modern diet of more than two billion people, in particular in non-Western countries (Kouřimská and Adámková, 2016; van Huis, 2020).

The interest in insects has recently turned to their use in animal nutrition. Insect have a high nutritional value and species that undergo a non-feeding phase (pupae), store energy as fat and thus contain large quantities of triacylglycerol which are recalled in periods of high-energy demand. The fatty acid (FA) profile of insects is species-specific and related to the sex, life stage and environmental conditions of the insects (Oonincx *et al.* 2015), and most interestingly, it can be modulated by the rearing substrate to ameliorate the FA profile of the fed animals (Danieli *et al.*, 2019; Liland *et al.*, 2017; St Hilaire *et al.*, 2007). A fraction of this fat is usually extracted during the insect meal production process, and the resulting defatted raw material is a powder which vary in protein content and can exceed 70% (on a dry matter (DM) basis) (Figure 1). (Gasco *et al.*, 2020a; Józefiak *et al.*, 2016). As a result of their high nutritional value, and in particular their high protein content, they represent excellent alternatives to conventional protein sources, such as soybean meal (SBM) or fishmeal (FM) in animal feed (De Souza *et al.*, 2019; Gasco *et al.*, 2018a, 2019a, 2020a,b; Henry *et al.*, 2015; Józefiak and Engberg 2017; Khan, 2018; Koutsos *et al.*, 2019; Lock *et al.*, 2018; Nogales-Mérida *et al.* 2019; Sogari *et al.*, 2019). Moreover, many insect species bio-convert organic substrates into valuable protein- and energy-rich products thus contributing to the Circular Economy principle (Gasco *et al.*, 2020b). In fact, the use of otherwise non-utilised side-streams as substrates for insects enables the production of high-value products with low environmental impacts (Bosch *et al.*, 2019; Smetana *et al.*, 2019).

So far, only a few insect species are being mass reared for feed purposes, the most studied being the black soldier fly (*Hermetia illucens*; HI), the common housefly (*Musca domestica*; MD), and the yellow mealworm (*Tenebrio molitor*; TM). They have a valuable amino acid profile similar to those of SBM and FM partially or totally covering the requirements of fish and livestock species (Figure 2).

Although, at the beginning, insects were mainly appreciated for their protein and energy supply, their newly discovered bioactive compounds may promote animal health and launch insects beyond the 'simple' protein concept (Gasco *et al.*, 2018b; Józefiak and Engberg 2017; Lee *et al.*, 2018; Wu *et al.*, 2018). This review presents

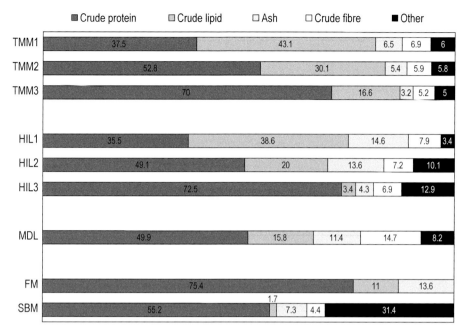

FIGURE 1   Nutrient composition of *Tenebrio molitor, Hermetia illucens* and *Musca domestica* larvae meals compared to fishmeal and soybean meal (% of dry matter) (elaborated from Gasco *et al.*, 2018a). FM = fishmeal; HIL1, HIL2, HIL3 = *Hermetia illucens* larvae meals obtained under different production process; MDL = *Musca domestica* larvae meal; SBM = soybean meal; TMM1, TMM2, TMM3 = *Tenebrio molitor* larvae meals obtained under different production process

the main effects of insects and their bioactive compounds on fish and crustaceans, poultry, pigs and rabbits.

Preparing this article, an accurate bibliographical search was performed to collect all relevant references on the effects of dietary insects on animal health. The interest on this subject is recent and the search returned increasing numbers of references from 2011 to 2020. In addition to personal databases of articles collected since 2011 by the 3 authors and shared between them, academic articles were sourced from Google, Web of Science, Scopus and Science direct. Different key words were used, alone or in association with each others, such as 'insects', 'edible insects', 'insect feed', 'poultry', 'fish', 'teleost', 'pigs', 'piglets', 'rabbit', 'health effect', 'gut health', 'microbiota' 'immunology', 'antimicrobial peptides'. The search terms also consisted of the insect common names in English (e.g. 'yellow mealworm', 'black soldier fly', 'lesser meal worm' or 'common housefly') or the Latin names (e.g. '*Tenebrio molitor*', '*Hermetia illucens*', '*Zophobas morio*' or '*Musca domestica*'). The full text of each article was downloaded, read and when pertinent, used for the current review. All newly found articles were added to the common database. In all, about 950 articles were found of which 130 on fish, 113 on poultry, 36 on pigs, 7 on crusta-

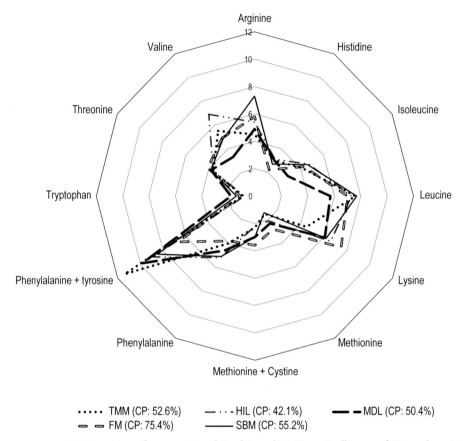

FIGURE 2   Main amino acid composition of *Tenebrio molitor, Hermetia illucens* and *Musca domesti-*
            ca larvae meals compared to fishmeal and soybean meal (% of protein dry matter) (Gas-
            co *et al.*, 2018a and Feedipedia.org). FM = fishmeal; HIL = *Hermetia illucens* larvae meal;
            MDL = *Musca domestica* larvae meal; SBM = soybean meal; TMM = *Tenebrio molitor*
            larvae meal

ceans, 4 on rabbits and more than 93 concerning antimicrobial peptides, chitin or
microbiome.

Considering insect products and their effects on animal health, only peer-reviewed articles written in English language and published during the last decade
were considered for further evaluation to ensure contemporary scientific quality
and timeliness of the review. At the end, without considering more general articles
used in the introduction part, a total of 129 articles were included in this review.

2 Microbiota modulation and gut health

The importance of gut health in general health has recently been underlined for both mammals and other animals. Until recently, gut health was mainly assessed through histological observations. The development of 16S RNA based on high throughput sequencing techniques has enabled the precise detection of complex microbial communities in animals gastrointestinal tract (GIT), named microbiota, which has been closely linked not only to digestion but also to immunity and resistance of the animals to diseases (Egerton *et al.*, 2018; Kayama *et al.*, 2020; McCarville *et al.*, 2020). Diet composition is recognised as one of the major factors affecting this gut microbiota and dietary insects may subsequently modulate the gut microbiota of the animals feeding on these insects (Vogel *et al.*, 2018).

2.1 *Fish and crustaceans*

Crustaceans and fish gut health and microbiota are influenced by dietary insects (HI, listed in Table 1; TM, listed in Table 2; and other insects listed in Table 3).

2.2 *Fish and crustaceans gut morphology and physiology*

Insect meal did not generally show any adverse effect on the gut morphology of fish (Caimi *et al.*, 2020; Elia *et al.*, 2018; Lock *et al.*, 2016; Wang *et al.*, 2019) or on their gut digestive enzymes (Belghit *et al.*, 2019; Li *et al.*, 2017a; Wang *et al.*, 2019). Some exceptions concerned dietary HI (Table 1) found to induce some villus irregularities in Jian carp (Li *et al.*, 2017a), villus shortening in clown fish (Vargas-Abundez *et al.*, 2019) and rainbow trout (Cardinaletti *et al.*, 2019), increased gut lipase activity in Japanese sea bass (Wang *et al.*, 2019) but decreased in European sea bass (Magalhaes *et al.*, 2017), increased weight of the distal intestine of Atlantic salmon (Li *et al.*, 2019, 2020) and reduced steatosis (i.e. abnormal fat retention) of the proximal intestine in Atlantic salmon (Li *et al.*, 2020). The dietary inclusion of HI also increased submuscosa cellularity and production of neutral mucin by the goblet cells in the proximal intestine of rainbow trout (Cardinaletti *et al.*, 2019; Elia *et al.*, 2018). Mucosal and muscular layer thickening of the GIT was also shown in sturgeon fed TM (Józefiak *et al.*, 2019a; Table 1 and 2). This may be related to the production of short-chain fatty acids (SCFA; such as butyrate), which are metabolites of the microbiota known to stimulate the proliferation of mucosal cells in the colon of humans (Mortensen *et al.*, 1999). Microbiota studies on crustaceans and fish have been performed on rainbow trout and few other freshwater and marine fish species fed mainly on HI (Table 1) and TM (Table 2), but recent studies have also investigated the effects of MD, *Gryllus sigillatus*, *Blatta lateralis* and *Zophoba morio* (ZM) (Table 3).

TABLE 1 Effects of *Hermetia illucens* products on gut health and microbiota in fish and crustacean

Animal [IBW-FBW, g] {no. days}	Insect [form]	Insect composition (% DM)		% insect inclusion	Type and % of main protein in CTRL diet	% of substitution	Gut health and microbiota	Reference
		CP	CL					
Marron (*Cherax cainii*) [65-89] {60}	HIL	–	–	12, 11	FM: 41% or PBP: 39%	22, 19	Increased richness and diversity of microbiota; Bacterioidetes, Proteobacteria and Firmicutes LAB; decrease of *Vibrio* and Enterobacter	Foysal *et al.* (2019)
Zebrafish (*Danio rerio*) [2-21 dps] {19}	HIP [FF]			10.5, 21	FM: 42%	25, 50	No intestinal inflammation; lipid accumulation in liver of fish fed 21% HIL; higher chitinase expression at 21% HI	Zarantoniello *et al.* (2018)
Jian carp (*Cyprinus carpio* var. Jian) [35-110] {59}	HIL [DF]	–	–	2.6, 5.3, 7.9, 10.6	FM: 10%	25, 50, 75, 100	At HIL > 75%, irregularities of gut microvilli shape, no effect on gut digestive enzymes	Li *et al.* (2017a,b)
Siberian sturgeon (*Acipenser baerii*) [24-159] {118}	HIL [DF]	65.8	4.24	18.5, 37.5	FM: 70%	25, 50	No effect on histology of liver or gut	Caimi *et al.* (2020)
Siberian sturgeon (*A. baerii*) [640-1,200] {60}	HIL [FF]	40.4	33.5	15	FM: 26%	30	Reduced thickness of mucosa, no difference of villi height; increased richness of microbiota, increased numbers of *Clostridium*, Enterobacteriaceae, *Aeromonas, Bacillus, Carnobacterium, Enterococcus, Lactobacillus*	Józefiak *et al.* (2019a)
Rainbow trout, (*Oncorhynchus mykiss*) [53-166] {71}	HIP [FF]	56.3	23.5	20	FM: 34.8%	30	Slight decrease of villi height in the proximal intestine; no change of mucosal thickness; increased richness of microbiota; increased *Clostridium* and LAB	Józefiak *et al.* (2019b)

Species	Insect meal [processing]	IBW	FBW	Inclusion levels (%)	FM	Effects	Reference
Rainbow trout (*O. mykiss*) [66-223] {84}	HIL [DF]	48.6	20.6	10, 20, 30	FM: 60%	Increased richness and diversity of microbiota in faecal samples, increased lactic acid and butyrate-producing bacteria; Actinomycetaceae, Bacillaceae, Lactobacillaceae, Staphylococcaceae, reduced Vibrionacae, Pseudomonaceae and Enterobacteriaceae; Firmicutes rich in faeces of HIL-fed trout, not in gut mucosa. Gut mucosa (adhered intestinal microbiota) of HIL-fed fish showed increased diversity but not richness; reduction of Proteobacteria, increased abundance of Mycoplasma in gut mucosa.	Terova *et al.* (2019) (gut content); Rimoldi *et al.* (2019) (gut mucosa)
Rainbow trout (*O. mykiss*) [137-300] {98}	HIP [FF]	39.0	41.9	10.5, 21	FM: 42%	Increased golblet cells (hyperplasia) producing neutral mucins in hind intestine; shortening of mid intestine villi; no intestinal inflammation; increased lipid accumulation in liver at 21% HIP	Cardinaletti *et al.* (2019)
Rainbow trout (*O. mykiss*) [179-540] {78}	HIL [FF]	55.3	18	20, 40	FM: 60%	No change of gut morphology; no effect on histology of liver, spleen or gut, acidic and neutral mucin; increased diversity of microbiota especially at 20%, Proteobacter and Tenericutes most abundant in HIL-fed trout	Elia *et al.* (2018); Bruni *et al.* (2018)
European sea bass (*Dicentrarchus labrax*) [50-129] {62}	HIP [DF]	55.8	5.5	6.5, 13, 19.5	FM: 32.4%	No effect on gut digestive enzymes except for lipase lower at 6.5 compared to FM ctrl and 19.5% HIP in both anterior and posterior intestine	Magalhaes *et al.* (2017)
Clownfish (*Amphiprion ocellaris*) [juveniles] {106}	HIL [DF]	–	–	20, 40, 60	FM: 60%	No intestinal inflammation, reduced villi at HIL > 40%	Vargas-Abundez *et al.* (2019)

CL = crude lipid; CP = crude protein; CTRL = control diet; dps = days post-spawning; DF = defatted; DM = dry matter; FBW = final body weight; FF = full fat; FM = fishmeal; HIL = *Hermetia illucens* larvae meal; HIP = *Hermetia illucens* prepupae meal; IBW = initial body weight; LAB = lactic-acid producing bacteria; PBP = poultry by-product meal; SBM = soybean meal; SPC = soy protein concentrate.

TABLE 1   Effects of *Hermetia illucens* products on gut health and microbiota in fish and crustacean (*cont.*)

Animal [IBW-FBW, g] {no. days}	Insect [form]	Insect composition (% DM)		% insect inclusion	Type and % of main protein in CTRL diet	% of substitution	Gut health and microbiota	Reference
		CP	CL					
Rainbow trout (*O. mykiss*) [202-276] {35}	HIP [FF]	–	–	30	FM: 50%	30	Best diversity and richness of microbiota; rich in Actinobacteria	Huyben *et al.* (2019)
	HIL [FF]	–	–				Intermediate diversity and lowest richness of microbiota between 3 HIL but still richer than FM-fed fish; high firmicutes; high LAB	
	HIL [DF]	–	–				Lowest diversity and intermediate richness of microbiota still better than FM; very high Firmicutes (Bacillaceae)	
Atlantic salmon (*Salmo salar*) [49-143] {56}	HIL [DF]		–	60	FM: 35% + SPC: 29.6%	85	Decreased hyper-vacuolisation of proximal intestinal enterocytes; normal histological structure of mid- and distal intestine but of increased weight compared to ctrl fish, maybe related to increased butyrate	Li *et al.* (2019); Belghit *et al.* (2018)
Atlantic salmon (*S. salar*) [247-from 359 to 575] {105}	HIL [FF]	54.1	26.4	5, 10, 25	FM: 20%	25, 50, 100	No effect on gut histology	Lock *et al.* (2016)
		61.0	17.8	5, 25		25, 100		
Atlantic salmon (*S. salar*) [1,400-3,702] {112}	HIL [DF]	52	18	5, 10, 15	FM: 10%	33, 66, 100	At 15% HIL, lower enterocytes steatosis in proximal intestine, increased distal intestine weight. At 5, 10 and 15% HIL, no effect on digestive brush border enzymes or total bile acids in the digesta	Belghit *et al.* (2019); Li *et al.* (2020)
European sea bass (*Dicentrarchus labrax*) [50-129] {62}	HIP [DF]	55.8	5.5	6.5, 13, 19.5	FM: 32.4%	15, 30, 45	No effect on gut digestive enzymes except for lipase lower at 6.5 compared to FM ctrl and 19.5% HIP in both anterior and posterior intestine	Magalhaes *et al.* (2017)
Clownfish (*Amphiprion ocellaris*) [juveniles] {106}	HIL [DF]	–	–	20, 40, 60	FM: 60%	25, 50, 75	No intestinal inflammation, reduced villi at HIL >40%	Vargas-Abundez *et al.* (2019)

CL = crude lipid; CP = crude protein; CTRL = control diet; dps = days post-spawning; DF = defatted; DM = dry matter; FBW = final body weight; FF = full fat; FM = fishmeal; HIL = *Hermetia illucens* larvae meal; HIP = *Hermetia illucens* prepupae meal; IBW = initial body weight; LAB = lactic-acid producing bacteria; PBP = poultry by-product meal; SBM = soybean meal; SPC = soy protein concentrate.

TABLE 2   Effects of *Tenebrio molitor* products on gut health and microbiota in fish and crustacean

Animal [IBW-FBW, g] {no. days}	Insect [form]	Insect composition (% DM)		% insect inclusion	Type and % of main protein in CTRL diet	% of substitution	Gut health and microbiota	Reference
		CP	CL					
Siberian sturgeon (*Acipenser baerii*) [640-1,200] {60}	TMM [FF]	56.3	25.3	15	FM: 26%	40	Increased thickness of muscular layer, no difference of villi height; reduced richness of microbiota; no effect on intestinal *Clostridium*, Enterobacteriaceae, *Lactobacillus* or *Enterococcus*. Increased *Bacillus*, *Carnobacterium* and *Enterococcus*	Józefiak *et al.* (2019a)
Rainbow trout (*Oncorhynchus mykiss*) [53-166] {71}	TMM [FF]	56.3	25.3	20	FM: 34.8%	40	Decrease of villi height in the proximal intestine; no change of mucosal thickness; strongly increased richness of microbiota; strongly increased *Clostridium*; Enterobacteriaceae and LAB	Józefiak *et al.* (2019b)
Rainbow trout (*O. mykiss*) [115-no data] {90}	TMM [FF]	51.9	23.6	60	FM: 70%	86	Decreased diversity of microbiota	Antonopoulou *et al.* (2019)
Sea trout (*Salmo trutta* m. *trutta*) [5-21] {56}	TMM [hydrolysed]	47	29.6	10	FM: 25%	58	Slight increase of villi height; no change in muscular layer thickness of the anterior intestine; no change in microbiota richness; reduction of *Carnobacterium* and *Lactobacillus*	Mikolajczak *et al.* (2020)
European sea bass (*Dicentrarchus labrax*) [5-from 17 to 22] {70}	TMM [FF]	51.9	23.6	50	FM: 70%	71	Increased diversity of microbiota, increased abundance of Tenericutes	Antonopoulou *et al.* (2019)
Gilthead seabream (*Sparus aurata*) [105-240] {163}	TMM [FF]	51.9	23.6	50	FM: 50%	74	increased diversity of microbiota, increased abundance of Tenericutes	Antonopoulou *et al.* (2019)

CL = crude lipid; CP = crude protein; CTRL = control diet; dps = days post-spawning; DM = dry matter; FBW = final body weight; FF = full fat; FM = fishmeal; IBW = initial body weight; LAB = lactic-acid producing bacteria; TMM = *Tenebrio molitor* larvae meal.

TABLE 3   Effects of *Musca domestica*, Chironomidae, *Gryllus sigillatus*, *Blatta lateralis* and *Zophoba morio* on gut health and microbiota in fish and crustacean

Animal [IBW-FBW, g] {no. days}	Insect [form]	Insect composition (% DM)		% insect inclusion	Type and % of main protein in CTRL diet	% of substitution	Gut health and microbiota	Reference
		CP	CL					
Swamp eel (*Monopterus albus*) [27-31] {56}	MDL [whole]	64	24.3	1 day every 3 (1d/3) or 7 days (1d/7)	FM: 45%		At 1d/7, poorer microbiota and less diverse, decreased *Pseudomonas*. At 1d/3, more Bacterioidaceae, decreased *Pseudomonas*	Xiang *et al.* (2019)
Rainbow trout (*Oncorhynchus mykiss*) [0.13-0.65] {28}	ChL [whole frozen]	66	9	100	FM: 12.4%		Distal intestine more folded, with higher brush border, less goblet cells	Ostaszewska *et al.* (2011)
Rainbow trout (*O. mykiss*) [53-166] {71}	GSA [FF]	61.3	19.5	20	FM: 34.8%	48	Decrease of villi height in the proximal intestine; decrease of mucosal thickness; increased richness of microbiota; increased *Clostridium* and LAB	Józefiak *et al.* (2019b)
Rainbow trout (*O. mykiss*) [53-166] {71}	BLA [FF]	54.6	26.1	20	FM: 34.8%	42	Decreased growth; Increase of villi height in the proximal intestine; increase of mucosal thickness; increased richness of microbiota; increased *Clostridium* and LAB	Józefiak *et al.* (2019b)
Sea trout (*Salmo trutta m. trutta*) [5-21] {56}	ZMM [hydrolysed]	49.3	33.6	10	FM: 25%	56	No change in villi height or muscular layer thickness of the anterior intestine; no change in richness of microbiota; reduction of *Aeromonas*, *Carnobacterium* and *Enterococcus*	Mikolajczak *et al.* (2020)

BLA = *Blatta lateralis* adults; ChL = Chironomidae larvae; CL = crude lipid; CP = crude protein; CTRL = control diet; dps = days post-spawning; DM = dry matter; FBW = final body weight; FF = full fat; FM = fishmeal; GSA = *Gryllus sigillatus* adults; IBW = initial body weight; LAB = lactic-acid producing bacteria; MDL = *Musca domestica* larvae meal; ZMM = *Zophoba morio* larval meal.

2.3   *Fish and crustaceans gut microbiota*

The microbiota of fish is closely correlated to the host diet and dietary insects have been suggested to act as prebiotics through both an improvement of the gut absorptive area and the stimulation of the host immunity to fight pathogenic bacteria, thus favouring beneficial commensal bacteria (Guerrerio *et al.*, 2018; Józefiak *et al.*, 2019a). In crustaceans, a dietary supplementation of HI increased the richness and diversity of the gut microbiota of marron (*Cherax cainii*) (Foysal *et al.*, 2019). The opposite was true in swamp eel where the replacement of a control diet with whole MD larva every 3 or 7 days reduced the gut microbiota richness and diversity. However, the fish did not grow well in this unique study showing adverse effect of dietary insects on fish microbiota (Xiang *et al.*, 2019). Indeed, although dietary TM or ZM did not affect the richness of sea trout's microbiota, they significantly reduced the occurrence of potential pathogens (Mikołajczak *et al.*, 2020). Moreover, several studies have shown that both HI and TM increased the gut microbiota diversity of rainbow trout (Antonopoulou *et al.*, 2019; Huyben *et al.*, 2019; Józefiak *et al.*, 2019b). Dietary crustacean chitin and chitosan have been shown to enrich and diversify fish microbiota (Askarian *et al.*, 2012; Ringø *et al.*, 2012; Udayangani *et al.*, 2017), acting as prebiotics and contributing to a healthy gut, decreasing the abundance of fish pathogenic bacteria and thus increasing disease resistance (Qin *et al.*, 2014; Terova *et al.*, 2019; Zhou *et al.*, 2013). Insect chitin may have a similar prebiotic effect (Huyben *et al.*, 2019), although this has not been demonstrated as yet. Nevertheless, dietary insects have been shown to cause a shift of microbiota of crustaceans and fish towards an increased importance of Tenericutes, Actinomycetes, Clostridiales and/or Firmicutes suggested to be chitinolytic bacteria facilitating the digestion of chitin-rich nutrients (Elia *et al.*, 2018; Bruni *et al.*, 2018; Antonopoulou *et al.*, 2019; Foysal *et al.*, 2019; Huyben *et al.*, 2019; Józefiak *et al.*, 2019b; Terova *et al.*, 2019). In particular, Firmicutes lactic acid producing bacteria (LAB), often used as probiotics and known to improve mucosal activity and to produce antimicrobial bacteriocins, were more abundant in the intestines of rainbow trout and Siberian sturgeon fed with HI (Bruni *et al.*, 2018; Huyben *et al.*, 2019; Józefiak *et al.*, 2019a,b, Terova *et al.*, 2019). However, this was not the case of sea trout fed TM, where *Lactobacillus* counts reduced compared to FM-fed fish (Mikołajczak *et al.*, 2020). Other commensal bacteria with antibacterial or antiviral activity (e.g. *Pseudomonas stutzeri*, *Carnobacterium divergens*, *Acinetobacter* spp.) and with the potential to help the host fight or compete with pathogens were observed to be particularly abundant in HI-fed rainbow trout (Bruni *et al.*, 2018). Moreover, bacterial genera associated with common fish or human pathogens (*Pseudomonas*, *Vibrio*, *Aeromonas* and *Enterococcus*) decreased in HI-fed rainbow trout (Rimoldi *et al.*, 2019; Terova *et al.*, 2019), TM- and ZM-fed sea trout (Mikołajczak *et al.*, 2020) as well as in MD-fed swamp eel (Xiang *et al.*, 2019). Insect meals are usually rich in saturated SCFA such as lauric acid, propionate and butyrate, which are antibacterial and antiviral, and may thus be responsible for some of the modulating activity of dietary insects on the gut microbiota of fish described above (Terova *et al.*, 2019). In turn, dietary insects were shown to favour commensal bacteria of the Clostridiales order, which also produce SCFA reinforcing their contribution to the gut microbiomic homeostasis of fish (Józefiak *et al.*, 2019 a,b; Terova *et al.*, 2019). Moreover, SCFA are also known to be anti-inflammatory (Vargas-Abundez *et al.*, 2019) and may be involved in the absence of gut inflammation observed in insect meal-fed fish (Cardinaletti *et al.*, 2019; Li *et al.*, 2019; Zaratoniello *et al.*, 2018).

2.4   *Poultry*

In nature, most birds are primarily insectivorous. For example, wild red jungle fowl (*Gallus gallus*) is known to eat ants and termites and juvenile wild turkeys eat approximately 60% of insects (Klasing, 2005). When farmed poultry have access to outdoor areas, they dedicate part of their time to searching for and ingesting live preys including insects (Józefiak *et al.*, 2016; Schiavone *et al.*, 2019). Table 4 and 5 report the main literature concerning the effects of insect-derived products inclusion in poultry diets.

On the one hand, the effects of dietary insect meals inclusion on GIT morphometry in poultry, the results are often inconsistent. For instance, a decrease in the villi height, crypt depth and villi/crypt ratio was reported in the GIT segments (duodenal, jejunal and ileum) of laying hens (Cutrignelli *et al.*, 2018; Moniello *et al.*, 2019) and broilers (Dabbou *et al.*, 2018) fed with HI especially when the inclusion levels surpassed 10%. On the contrary, no histopathological changes were reported for Muscovy ducks (Gariglio *et al.*, 2019) or broiler chickens (Biasato *et al.*, 2018a) fed HI larva meal or for chickens fed TM meal (Biasato *et al.*, 2016, 2017, 2018b). The use of insect-derived oils/fats did not seem to affect the GIT morphology (Schiavone *et al.*, 2018; Sypniewski *et al.*, 2020).

On the other hand, dietary insects consistently modulate the microbiota populations in the GIT of poultry, but it may vary depending on the considered gut tract, the insect species or the level of insect inclusion (Benzeriha *et al.*, 2019; Józefiak *et al.*, 2018, 2020). Low levels of insect meals in poultry diets diversified the GIT microflora and increased the bacteria with positive effects on gut health (Józefiak *et al.*, 2019a). However, levels higher than 10% (and in particular 15%) of insect meal inclusion seemed to reduce microbial complexity and richness in beneficial bacteria including those with mucolytic activity (Biasato *et al.*, 2019a, 2020a). Following the ingestion of 0.2% of defatted TM, potentially pathogenic bacteria, such as the *Bacteroides-Prevotella* cluster, reduced in crops and caeca, while the ileum showed an increase in *Lactobacillus* spp./*Enterococcus* spp. (Benzertiha *et al.*, 2019; Józefiak *et al.*, 2018). An increase in butyrate producing bacteria, such as the *Clostridium coccoides – Eubacterium rectale* cluster and the *Clostridium leptum* subgroup, was observed in broiler fed 0.2% of full fat HI in all the GIT and ileum, respectively (Józefiak *et al.*, 2018).

TABLE 4   Effects of *Tenebrio molitor*, *Shelfordella lateralis* and *Zophobas morio* meals in gut health and microbiota modulation in poultry

Poultry species [IBW-FBW, g] {no. days}	Insect [form]	Insect composition (% DM)		% insect inclusion	Gut health and microbiota	Reference
		CP	CL			
Broiler chickens (Ross 308) [42-2,122] {35}	TMM [FF]	47.0	29.6	0.2, 0.3	Caeca: TML 0.2% tended to show highest relative abundance of the family Ruminococcaceae and *Lactobacillus reuteri*. Significant increase in level of Clostridia.	Józefiak *et al.* (2020)
Broiler chickens (Ross 308) [42-2,127] {35}	TMM [FF]	56.3	25.3	0.2	Crop: decrease in *Bacteroides-Prevotella* cluster. Ileum: increase in *Clostridium coccoides – Eubacterium rectale* and *Lactobacillus* spp./ *Enterococcus* spp. Caeca: increase in *C. coccoides – E. rectale*.	Józefiak *et al.* (2018)
Broiler chickens (Ross 708) [no data-2,309] {40}	TMM [FF]	55.3	25.1	5, 10, 15	No effect on gut morphology.	Biasato *et al.* (2017)
Free-range broiler chickens [716-no data] {53}	TMM [FF]	55.3	25.1	7.5	Significant increase of the relative abundances of *Firmicutes* and decrease of *Bacteroidetes* phyla. No effect on gut morphology. Higher relative abundance of *Clostridium, Oscillospira, Ruminococcus, Coprococcus* and *Sutterella* genera compared to CTRL.	Biasato *et al.* (2018a)
Broiler chickens (Ross 708) [no data-3,641] {53}	TMM [FF]	57.5	30.7	5, 10, 15	TMM 15% showed lower villus height, higher crypt depth, and lower villus height to crypt depth ratio compared to CLRL and TMM 5%.	Biasato *et al.* (2018b)
Broiler chickens (Ross 308) [42-2,076] {35}	SLI [FF]	54.6	26.1	0.05, 0.1, 0.2	Crop: increase in *Clostridium leptum* subgroup. Ileum: increase in *C. coccoides – E. rectale* cluster and *Lactobacillus* spp./*Enterococcus* spp. Ceca: increase in *Bacteroides-Prevotella* cluster.	Józefiak *et al.* (2018)
Broiler chickens (Ross 308) [42-2,122] {35}	ZMM [FF]	49.3	33.6	0.2, 0.3	Caeca: ZML 0.2% resulted in an increase in relative abundance of the Actinobacteria, including the family Bifidobacteriaceae, with the highest relative abundance of genus *Bifidobacterium pseudolongum*. ZML 0.2% resulted in an increase in the number of *Lactobacillus agilis*.	Józefiak *et al.* (2020)

CL = crude lipid; CP = crude protein; CTRL = control diet; DM = dry matter; FBW = final body weight; FF = full fat; IBW = initial body weight; SLI = *Shelfordella lateralis* imago meal; TMM = *Tenebrio molitor* larvae meal; ZMM = *Zophobas morio* larvae meal.

TABLE 5 Effects of *Hermetia illucens* products in gut health and microbiota modulation in poultry

Poultry species [IBW-FBW, g] {no. days}	Insect [form]	Insect composition (% DM)		% insect inclusion	Gut health and microbiota	Reference
		CP	CL			
Laying hens (Lohmann Brown Classic) [1,410-no data] {168}	HIL [DF]	60.0	9.0	7.3, 14.6	In the duodenum, villi height and villi/crypt ratio were lower in both HIL groups than in control. In ileum, only the villi height was reduced in birds fed the highest level of HIL.	Moniello et al. (2019)
Laying hens (Lohmann Brown Classic) [1,780-2,009] {147}	HIL [DF]	62.7	4.7	17	Increased relative abundance of Elusimicrobia, Lentisphaerae and Cyanobacteria and decreased Fusobacteria. The HIL group showed a higher villi height in the duodenum, while the opposite happened in the jejunum and the ileum. Only in the ileum, higher crypt depth in the HIL group than in the SBM.	Borrelli et al. (2017); Cutrignelli et al. (2018)
Broiler chickens (Ross 308) [40.2-2,278] {35}	HIL [DF]	55.3	18.0	5, 10, 15	The HIL 15% showed a lower villus height, a higher crypt depth and a lower villus height-to-crypt depth ratio than the other groups. Caecal digesta analysis of HIL 15% birds displayed significant increase of the relative abundance of Proteobacteria phylum than HIL 10%. Characteristic microbiota for 5% HIL: L-Ruminococcus (Lachnospiraceae family), Faecalibacterium, Blautia and Clostridium genera were found to be characteristic in caecal microbiota. Characteristic microbiota for 10% HIL: Lactobacillus and Ruminococcus and Bacteroides, Roseburia. Characteristic microbiota for 15% HIL: Helicobacter genera. Lower mucin in small and large intestine of chicken fed 10 and 15% addition of HIL compared to CTRL.	Biasato et al. (2020a); Dabbou et al. (2018)

Species	Insect				Effect	Reference
Broiler chickens (Ross 308) [42–2,101] {35}	HIL [FF]	40.4	33.5	0.2	Crop: increase in *Clostridium coccoides – Eubacterium rectale* cluster and *Lactobacillus* spp./*Enterococcus* spp. Ileum: increase in *C. coccoides – E. rectale* cluster and decrease in *Lactobacillus* spp./*Enterococcus* spp. Caeca: increase in *Bacteroides-Prevotella* cluster, *Streptococcus* spp./*Lactococcus* spp., *C. coccoides – E. rectale* cluster and *Lactobacillus* spp./*Enterococcus* spp.	Józefiak et al. (2018)
Broiler chicken (Ross 308) [817.8–3,751.3] {27}	HIF	–		3.43, 6.87	No influence on gut morphometric indexes	Schiavone et al. (2018)
Muscovy ducks (*Cairina moschata domestica*) [704–2,554] {47}	HIL [DF]	56.7	10.7	3, 6, 9	No effect on gut morphology	Gariglio et al. (2019)
Turkey (B.U.T. 6, AVIAGEN) [179–1,396] {35}	HIF	–	–	2.5, 5	Decrease of Enterobacteriaceae counts	Sypniewski et al. (2020)

CL = crude lipid; CP = crude protein; CTRL = control diet; DF = defatted; DM = dry matter; FBW = final body weight; FF = full fat; HIL = *Hermetia illucens* larvae meal; HIF = *Hermetia illucens* fat; IBW = initial body weight; SBM = soybean meal.

Similar results were found in broilers fed 0.2 and 0.3% of TM or ZM full fat meals where Bacteroides-Prevotella cluster decreased in the caeca and a decrease in *Clostridium perfringens* counts decreased in groups treated with TM (0.2 and 0.3%) and ZM (0.2%). In contrast, *C. perfringens* counts were increased in the birds fed 0.3% ZM resulting in a negative effect as this very pathogenic bacteria causes necrotic enteritis in poultry (Benzertiha *et al.*, 2019).

Similarly, low inclusion levels (0.2, 0.3%) of TM and ZM in broiler chicken also improved caecal commensal microbiota with an increase in Clostridiales and particularly Lactobacillales and Bifidobacteriaceae with *Bifidobacterium pseudolongum* showing the highest abundance in the caecum digesta of birds fed 0.2% ZM (Józefiak *et al.*, 2020).

The dietary inclusion of 7.5% of full fat TM larva meal positively modulated the gut microbiota of free-range broiler chickens, inducing an increase in Firmicutes and a decrease in Bacteroidetes, but did not influence the mucin composition (Biasato *et al.*, 2018a).

Insect meal was also shown to positively modulate the gut microbiota composition of laying hens. Borrelli *et al* (2017), for instance, indicated that the total substitution of SBM by defatted HI larvae increased the relative abundance of Elusimicrobia, Lentisphaerae and Cyanobacteria and decreased Fusobacteria, compared with SBM-fed laying hens. These changes were correlated with the production of butyric acid in the caeca, potentially linked to chitin degradation.

Dietary oil replacement by insect fat also influenced the microbiota as shown in turkey where total substitution of soybean oil (SBO) by HI fat reduced the proliferation of potentially pathogenic bacteria (i.e. Enterobacteriaceae spp.) (Sypniewski *et al.*, 2020).

The modulation of the animal microbiota may be due to antimicrobial peptides (AMPs) produced by the insects introduced in the animal diet (Józefiak and Engberg, 2017). The addition of 2 AMPs (A3 and P5) to the chicken diet showed beneficial effects not only on growth performance and nutrient retention, (Choi *et al.*, 2013a,b) but also on the histomorphology (dose-dependent increase in villi height and depth) and intestinal microflora through a decrease in the population of anaerobic bacteria in the ileal and caecal digesta. The same was true of chicken fed with a cecropin hybrid which showed increased villus height and villus height: crypt depth ratio, but showed a negative effect on the crypt depth of the duodenum and ileum and decreased total aerobic bacteria in the jejunal and caecal digesta (Wen and He, 2012).

2.5   *Pigs and rabbits*

Literature on the influence of insect products on the microbiota and GIT health of pigs and rabbits is scarce as evident from Table 6. The somehow controversial results may be due to differences in composition of the used insect products.

Spranghers *et al.* (2018) did not observe any difference in the histo-morphologi-

TABLE 6   Effects of insect products in gut health and microbiota modulation in pigs and rabbits

Animal [IBW-FBW, kg] {no. days}	Insect [form]	Insect composition (% DM)		% insect inclusion	Type and % of main protein in CTRL	% substitution	Gut health and microbiota	Reference
		CP	CL					
Weaning pigs [6.1-no data] {15}	HIL [FF]	42.01	42.13	4, 8	Toasted soy-beans: 12%	50 and 100%	No effects on gut morphology. Increase in lauric acid in gut segment of pig fed 8% FF HIL	Spranghers et al. (2018)
	HIL [DF]	62.84	8.06	5.42		72%		
Weaning piglets [7.9-17.6] {28}	HIL [FF]	37.92	38.48	1, 2, 4	FM: 4%	25, 50 and 100%	Increase of villus height in the jejunum of piglets fed 2% HIL	Yu et al. (2020a,b)
Weaning pigs [6.1-32.4] {61}	HIL [DF]	59.0	8.97	5, 10	SBM: 20% (phase I); 18.5% (phase II)	Phase I: 30 and 60%; Phase II: 32 and 65%	No effects on gut morphology. Increase of β diversity in HIL fed groups. Increase in SCFAs producing bacteria. Decrease in jejunal inflammation. Increase in neutral mucins	Biasato et al. (2019b, 2020b)
Finishing pigs [76.0-116] {46}	HIL [FF]	37.92	38.48	4, 8	SBM: 16.98%	18.4 and 36.7%	Increase in Lactobacillus, Pseudobutyrivibrio, Roseburia, and Faecalibacterium. Decrease in Streptococcus	Yu et al. (2019)
Rabbits [1.05-2.9] {41}	HIF and TMO			0.75, 1.5	SBO: 1.5%	50 and 100%	No effects on gut morphology	Gasco et al. (2019b)

CL = crude lipid; CP = crude protein; CTRL = control diet; DF = defatted; DM = dry matter; FBW = final body weight; FF = full fat; HIL = *Hermetia illucens* larvae meal; HIF = *Hermetia illucens* fat; IBW = initial body weight; SBM = soybean meal; SBO = soybean oil; SCFAs = short chain fatty acids; TMO = *Tenebrio molitor* oil.

cal traits or in the bacterial count in the intestinal segments of weaned pigs fed full fat or defatted HI meal. Nevertheless, as an increase in lauric acid was found in all the gut segments of piglets fed HI meal, with significant differences between pigs fed 8% full fat HI meal compared to all the other treatments in the digestive tract compartments, the authors suggested a possible dose-dependent antibacterial effect (Spranghers et al., 2018).

Yu et al. (2019, 2020a) highlighted that the inclusion of a full fat HI larvae meal in finishing diets for pigs induced changes in the bacterial composition of the gut that were correlated to an enhanced colonic mucosal immune homeostasis. They suggested a prebiotic effect of low levels (2%) of HI meal inclusion, likely due to chitin. Their results showed an increase in the abundance of health-promoting bacteria, such as Lactobacillus and Bifidobacterium, which are known to preserve the integrity of the intestine lining. At the same time, they also reported an increase in the abundance of butyrate-producing bacteria (Roseburia, Pseudobutyrivibrio and Faecalibacterium) and in the number of Clostridium XIVa clusters in the colon, with a consequent significant increase in the production of butyrate and SCFAs. As far as bacteria or bacterial metabolites with potential negative effects are concerned, their results showed a decrease in Streptococcus sp., Escherichia coli and in the metabolites involved in amino acids metabolism. The same authors (Yu et al., 2020b) also reported an increase in the relative weight of the small intestine of piglets fed 2 and 4% of HI meal and in the villus height in the jejunum of piglets fed 2% of HI meal, thus suggesting a promoting effect on gut development and metabolism. In contrast, Biasato et al. (2019b) reported no effect on the intestinal morphology of pigs fed HI defatted meal.

Feeding weaning piglets with an inclusion of up to 10% of defatted HI larvae meal had a positive effect on the caecal microbiota and the small intestine mucin composition, showing a modulation of the potentially beneficial bacteria and a preservation of the mature mucin secretory architecture together with an attenuation of jejunal inflammation associated with an increased proportion of SCFA-producing bacteria in the caecal microbiota (Biasato et al., 2020b).

The dietary ingestion of HI and TM fats in substitution of 50 and 100% of soybean oil in the diets of growing rabbits did not influence the gut morphometry or alter the liver functions (Gasco et al., 2019b), although a positive effect on microbiota has been suggested (Dabbou et al., 2020). Indeed, an increase in the relative abundance of Bacteroides, Clostridium, Akkermansia and Ruminococcus has been observed. Akkermansia has been indicated as being able to degrade the mucin in the intestine, with a consequent production of SCFAs and consequent positive effects on gut health. Ruminococcus are also butyrate producing bacteria that contribute to the positive effect on the caecal microbiota of rabbits (Dabbou et al., 2020).

3   Immunological status

Beside their nutritional interest, insects contain biological active components such as chitin, which is a structural polysaccharide forming the outer shell of crustaceans and the exoskeleton of insects. It has been shown that crustacean chitin and chitin derivatives activate and boost the innate immune response of animals (Ringø *et al.*, 2012) and its seems that insect chitin may play a similar role (Józefiak and Engberg, 2017; Lee *et al.*, 2008).

3.1   *Fish and crustaceans*

In fish, dietary crustacean chitin and chitosan were potent immunostimulants and improved bacterial and parasitical disease resistance (Esteban *et al.*, 2001; Lee *et al.*, 2008; Shanthi Mari *et al.*, 2014). However, few studies have investigated the effects of insects on the crustaceans and fish immune system (Table 7, 8 and 9). In crustaceans, dietary TM significantly increased both the total hemocyte counts and phenoloxidase activity of the Pacific white shrimp, *Litopenaeus vannamei* (Motte *et al.*, 2019), and dietary HI increased lysozyme, total haemocyte counts and cytokine expression in the intestine of crayfish marron (*C. cainii*) (Foysal *et al.*, 2019). In fish, only two studies have shown no effect of dietary insects on the immune system of fish (Mikołajczak *et al.*, 2020; Wang *et al.*, 2019). All other studies, concerning dietary inclusion of whole chironomid larvae or meals from HI, TM or MD, showed modulation of the fish immune system through a more rapid or increased antibacterial complement activity (Henry *et al.*, 2018b; Ming *et al.*, 2013), increased lysozyme (Ming *et al.*, 2013; Pei *et al.*, 2019; Sankian *et al.*, 2018; Su *et al.*, 2017), increased phagocytic activity (Ido *et al.*, 2015) and increased myeloperoxidase (MPO) activity in rainbow trout (Henry *et al.*, 2018b), but decreased MPO in European sea bass (Henry *et al.*, 2018a), increased trypsin-inhibition (Henry *et al.*, 2018a,b), increased alkaline phosphatase (Pei *et al.*, 2019), increased IgM titres (Su *et al.*, 2017), increased proliferation and apoptotic indexes of the proximal intestine (Ostazewska *et al.*, 2011) and stimulation of T-cells (Li *et al.*, 2019). As for the expression of the genes involved in immunity, and cytokines in particular, dietary insects induced an up-regulation of hepcidin, major histocompatibility complex II (Su *et al.*, 2017), HSP70 (Cardinaletti *et al.*, 2019; Li *et al.*, 2017a; Zarantoniello *et al.*, 2018) pro-inflammatory cytokines, such as interleukin (IL)1β, IL17F, tumour necrosis factor (TNF)α and IL6 expression (Foysal *et al.*, 2019; Zarantoniello *et al.*, 2018) anti-inflammatory cytokines such as IL10 (Foysal *et al.*, 2019), IL2, a cytokine which contributes to the differentiation of T-cells (Su *et al.*, 2017), while HSP-70 and the pro-inflammatory cytokine IL1β were down-regulated in swamp eel (Xiang *et al.*, 2019). Some studies have shown no effect of dietary insects on inflammation or on the expression of inflammation-related cytokines (Henry *et al.*, 2018b; Li *et al.*, 2020; Stenberg *et al.*, 2019), but inflammation has been observed to decrease in European sea bass, as suggested by the ceruloplasmin activity (Henry *et al.*, 2018a). Dietary insects certainly affect the cytokines related to the inflammatory response, but this effect was not found to be consistent between fish species, and it is somewhat difficult to define their precise role. It has been suggested that chitin acts as a pathogen-associated molecular patterns by binding to pathogens recognition receptors and stimulating the production of several cytokines and immune mediators (Stenberg *et al.*, 2019). Regarding the anti-oxidant and stress-response, an *in vitro* study showed that HI down-regulated the bacteria triggered in head-kidney cells of Atlantic salmon (Stenberg *et al.*, 2019). Dietary insects have also been shown to modulate the hepatic or intestinal antioxidant response of fish (Caimi *et al.*, 2020; Elia *et al.*, 2018; Henry *et al.*, 2018b; Ming *et al.*, 2013; Pei *et al.*, 2019; Sankian *et al.*, 2018; Song *et al.*, 2018; Su *et al.*, 2017) and to increase the detoxification capacity in the distal intestine of fish, as suggested by the elevated expression of the Cyp1a1 gene in the intestine of Atlantic salmon (Li *et al.*, 2019). However, this may be linked to potentially higher concentrations of heavy metals in IM, and the absence of effects on toxicity indicators (cell apoptosis or tissue regeneration) or on gut histology tended to infirm this hypothesis (Li *et al.*, 2020).

3.2   *Poultry*

Diet is one of the factor affecting the immunological mechanism of poultry (Kaiser and Balic, 2015). Immunological effects of poultry fed insects products are reported in Table 10.

A recent study in broiler chickens showed that small amounts (0.2 and 0.3%) of TM and ZM significantly decreased IgM levels but up-regulated IL-2. Both dietary insects also decreased the weight of Bursa of Fabricius associated with the differentiation of B-lymphocytes. (Benzertiha *et al.*, 2020). Moreover, insect oil extracted from both TM and ZM affected the expression of selected genes involved in immunological homeostasis, such as GIMAP5, a key regulator of hematopoietic integrity and lymphocyte homeostasis, Kierończyk *et al.* (2018).

The ingestion of low levels of HI (1, 2 and 3%) also immunostimulated broilers by increasing the CD4+ T lymphocyte counts, serum lysozyme activity, spleen lymphocyte proliferation and phagocytic activity in a dose-dependent manner (Lee *et al.*, 2018).

Blood analyses of birds fed insect meals have shown that levels corresponding to about 1 g/d of ingested chitin increased the globulin levels and decreased the albumin to globulin ratio indicating a better immune response (Bovera *et al.*, 2015, 2018; Marono *et al.*, 2017). Feeding laying hens with defatted HI also reduced triglycerides and cholesterol levels (Bovera *et al.*, 2018). On the other hand, the inclusion of HI meal in jumbo quail diets increased the albumin/globulin ratio, which may have compromised the birds' immunity and caused plasma disorders (Mbhele *et al.*, 2019). However, other researchers did not show any differences in blood traits following the ingestion of insect meals, thus suggesting the need to further study the mechanism of action (Biasato *et al.*, 2016, 2017, 2018b; Dabbou *et al.*, 2018; Elahi *et al.*, 2020).

The partial substitution of SBO by HIL fat down-regulated the pro-inflamma-tory TNF-α, while the total substitution reduced the proliferation of potentially pathogenic bacteria (i.e. Enterobacteriaceae spp.) and down-regulated the pro-in-flammatory IL-6, thereby supporting the immune response of turkeys (Sypniewski et al., 2020). Dietary HIF however showed no effect on broiler chicken haematolog-ical traits (Schiavone et al., 2018).

3.3   Pigs

Immunological effects of pigs and rabbits fed insects products are reported in Ta-ble 11. Dietary HI larvae meal at levels varying between 3.5 and 10% had no effect on the haematological parameters of piglets (Biasato et al., 2019; Driemeyer, 2016). Indeed, these authors reported no major effects on the blood profile parameters, all of which fell within the physiological range for pigs, although a linear increase in monocytes and a quadratic response of neutrophils were observed (Biasato et al., 2019b). Dietary TM had no effect on the blood profile or immune system of weaned piglets (Ao et al., 2020). At dietary levels up to 6%, TM meal did not affect immu-noglobulin G (IgG) and immunoglobulin A (IgA) titres, but increased the serum insulin-like growth factor (Jin et al., 2016).

An increase in neutrophils was found in growing pigs fed diets with increasing levels of HI full fat larvae meal in partial and total substitution of FM (Chia et al. 2019), thus suggesting an immunostimulating activity of dietary HI. The authors ascribed the antimicrobial response to the high level of lauric acid in HI larva meal and speculated an increased protection effect against pathogens such as bacteria, viruses and fungi.

The beneficial effect of full fat HI meal (2 and 4%) dietary inclusion on the im-mune homeostasis of piglets is linked to the increase in serum anti-inflammatory cytokines IL-10 and IgA and simultaneous decrease in the pro-inflammatory cyto-kine IFN-γ (Yu et al. 2020b).

In vitro MD extracts were shown to have a strong anti-inflammatory effect (in-duced by lipopolysaccharide) on mice macrophage cell line probably through low-molecular extracted components of the insect (Chu et al., 2013).

4   Antimicrobial effects and resistance to diseases

The natural biotope of insects is often infested with hostile microorganisms (Józe-fiak and Engberg, 2017). To protect themselves, insects produce a wide range of bio-active substances with anti-microbial activity. Thus, HI larvae grown on chicken manure have been shown to reduce the load of Gram negative potential pathogens in the substrate (Józefiak and Engberg, 2017). For example, insects produce many AMPs exhibiting activity against bacteria, fungi, parasites and viruses (Alvarez et al., 2019; Dang et al., 2010; Elhag et al., 2017; Faruk et al., 2016; Imamura et al. 1999;

TABLE 7 Effects of *Hermetia illucens* meals on antioxidant enzymes, immune parameters in crustacean and fish

Animal [IBW-FBW, g] {no. days}	Insect [form]	Insect composition (% DM)		% insect inclusion	Type and % of main protein in CTRL diet	% of substitution	Antioxidant enzymes	Immune parameters	Reference
		CP	CL						
Marron (*Cherax cainii*) [65–89] {60}	HIL	–	–	12, 11	FM: 41% or PBP: 39%	22, 19		Increased lysozyme, total haemocyte counts, increased IL1β, IL8, IL10, IL17F and TNFα expression in intestine	Foysal et al. (2019)
Zebrafish (*Danio rerio*) [2–21 dps] {19}	HIP [FF]	–	–	10.5, 21	FM: 42%	25, 50		Transient increase in HSP70 at 14 dps then significantly lower in HIP 10.5 and 21% than in ctrl fish, higher TNFa and IL6 at 21% HIP	Zaranton-iello et al. (2018)
Jian carp (*Cyprinus carpio*, var Jian) [35–10] {59}	HIL [DF]	–	–	2.6, 5.3, 7.9, 10.6%	FM: 10%	25, 50, 75, 100	Reduced cholesterol; no effect on serum SOD or MDA, increased serum CAT; more apoptotic hepatocytes in fish fed HIL100%	At HI > 75%, enhanced hepatic HSP70 gene expression suggesting induced stress	Li et al. (2017a,b)
Siberian sturgeon, (*Acipenser baerii*), [24–159] {118}	HIL [DF]	65.8	4.24	18.5, 37.5	FM: 70%	25, 50	Alteration of hepatic oxidative stress biomarkers (increased SOD, GR, decreased GPx at 37.5%). SOD, GR, EROD and GST increased in kidney of fish fed 37.5% HIL		Caimi et al. (2020)
Yellow catfish (*Pelteobagrus fulvidraco*) [1.5–48.5] {65}	HIL [FF]	47	17.7	5.5, 10.8, 16.5, 22.3, 34.3, 46.2, 58.5	FM: 40%	13, 25, 37, 48, 68, 85, 100	Increased serum SOD at 10.8% inclusion. Hepatopancreas SOD not affected	Tendency for increased lysozyme and phagocytosis activities at 5.5 and 10.8% inclusion levels	Xiao et al. (2018)

Rainbow trout (*Oncorhynchus mykiss*) [137-300] {98}	HIP [FF]	39	41.9	10.5, 21	FM: 42%	25, 50	No difference in cholesterol or hepatic GR	HSP70 increase at 21% HIP; increased IL10 and TNFα in mid-intestine	Cardinaletti *et al.* (2019)
Rainbow trout (*O. mykiss*) [179-540] {78}	HIL [FF]	55.3	18	20, 40	FM: 60%	25, 50	Unchanged liver and kidney MDA, SOD, CAT, GPx, GR, reduced SeGPx in liver and kidney, increased GSH and EROD in liver, increased GST in kidney		Elia *et al.* (2018)
Atlantic salmon (*Salmo salar*) [49-143] {56}	HIL [DF]	–	–	60	FM: 35% + SPC: 29.6%	83	Elevated xenobiotic detoxification response in intestine (*cyp1a1* expression)	Stimulation of regulatory T-cell activity; no difference in anti-IPVN antibodies	Li *et al.* (2019); Belghit *et al.* (2018)
Atlantic salmon (*S. salar*) [1,400-2,550] {56}	HIL [DF]	52	18	10, 15	FM: 10%	66;100	down-regulated SOD and GPx expression in LPS-triggered cells	No effect on inflammation response to bacteria and virus; antioxidant and stress response to bacteria downregulated (SOD, GPx, HSP70, C/EBPβ, p38MAPK and TLR22)	Stenberg *et al.* (2019)
Atlantic salmon (*S. salar*) [1,400-3,702] [112]	HIL [DF]	52	18	5, 10, 15	FM: 10%	33;66;100		No effect on haemoglobin	Belghit *et al.* (2019)
Japanese sea bass (*Lateolabrax japonicus*), [14-58] {56}	HIL [DF]	55.4	1.6	5, 10, 15, 20	FM: 25%	16, 32, 48, 64	no difference in serum CAT, GPx, SOD; lower MDA	No effect on complement, alcaline phosphatase or lysozyme activity or on concentration of intestinal inflammatory cytokines	Wang *et al.* (2019)

CAT = catalase; C/EBPβ = CCAAT-enhancer-binding proteins; CL = crude lipid; CP = crude protein; CTRL = control diet; DF = defatted; EROD = ethoxyresorufin O-deethylase; FBW = final body weight; FF = full fat; FM = fishmeal; GPx = glutathione peroxidase; GR = glutathione reductase; GST = glutathione S-transferase; GSH = glutathione; HIL = *Hermetia illucens* larvae meal; HIP = *Hermetia illucens* prepupae meal; HSP70 = heat shock proteins 70; IBW = initial body weight; IL10 = interleukin-10; IL17F = interleukin-17F; IL1β = interleukin 1 beta; IL8 = interleukin-1; LAB = lactic-acid producing bacteria; LPS = lipopolysaccharide; MDA = malondialdehyde; p38MAPK = phosphokinase p38; PBP = poultry by-products; SBM = soybean meal; SOD = superoxide dismutase; SPC = soy protein concentrate; TNF-α = tumour necrosis factor-alpha; TRL22 = tool-like receptor 22.

TABLE 8   Effects of *Tenebrio molitor* larvae meal on antioxidant enzymes, immune parameters and resistance to bacterial diseases in crustacean and fish

Animal [IBW-FBW, g] {no. days}	Insect [form]	Insect composition (% DM)		% insect inclusion	Type and % of main protein in CTRL diet	% of substitution	Antioxidant enzymes	Immune parameters	Resistance to diseases	Reference
		CP	CL							
Pacific white shrimp (*Litopenaeus vannamei*) [1.5-6] {56}	TMM [DF]	74.8	12.6	5, 10, 15, 20	FM: 25%	25, 50, 75, 100		Increased phenoloxidase activity and total haemocyte counts; decreased bacterial number	Increased resistance to *Vibrio parahaemolyticus* (55; 75; 84; 87; 86)	Motte *et al.* (2019)
Mandarin fish (*Sinperca scherzeri*) [21-36] {56}	TMM [FF]	52.5	34.1	10, 20, 30	FM: 65%	11, 22, 32	GPx increased at 30%; SOD not affected	Reduced cholesterol et 30%; lysozyme increased, significantly at 30%; myeloperoxidase not affected		Sankian *et al.* (2018)
Yellow catfish (*Pelteobagrus fulvidraco*) [-o-no data] {35}	TMM [-]			9, 18, 27	FM: 24%	25, 50, 75	Lower MDA; increased plasma SOD	Enhanced immune (lysozyme, IgM); up-regulation of hepcidin, MHC II and IL2	Enhanced resistance to *Edwardsiella ictaluri* (67; 70; 74; 87%)	Su *et al.* (2017)
Rainbow trout (*Oncorhynchus mykiss*) [116-314] {90}	TMM [FF]	51.9	23.6	25, 50	FM: 75%	35, 66	increase of intestinal SOD, CAT, G6DP; reduced MDA	More rapid antibacterial activity of complement and increased myeloperoxidase and anti-protease activities; no effect on inflammation or lysozyme activity		Henry *et al.* (2018b); Belforti *et al.* (2015)

Species	Insect meal			Inclusion level	FM replacement	Days	Antioxidant	Immune response	Disease challenge	Reference
Sea trout (*Salmo trutta*) [5-21] {56}	TMM [hydrolysed]	47	29.6	10	FM: 25%	58		No change in IgM or lysozyme activity		Mikolajczak et al. (2020)
European sea bass (*Dicentrarchus labrax*) [65] {42}	TMM [FF]	51.9	23.6	25	FM: 70%	35		Decreased inflammatory and stress-related response (myeloperoxidase and nitric oxide); increased anti-protease activity. No effect on lysozyme or complement		Henry et al. (2018a,b); Gasco et al. (2016)
Pearl gentian grouper (*Epinephelus lanceolatus* × *E. fuscoguttatus*) [7-67] {50}	TMM [DF]	65	3	2.5, 5, 7.5, 10, 12.5	FM: 40%	6, 12, 18, 25, 31	decreased SOD and MDA at 7.5%, increased GR at 2.5%		Better survival to *Vibrio harveyi* with 7.5% TMM (35; 33; 30; 50; 10;5%)	Song et al. (2018)
Red seabream (*Pagrus major*) [24-34] {56}	TMM [DF]	76.5	5.3	5, 10	FM: 50%	10, 20	SOD not affected		Increased survival to *Edwarsiella tarda* (21; 55; 67%)	Ido et al. (2019)

CAT = catalase; CL = crude lipid; CP = crude protein; CTRL = control diet; DF = defatted; FBW = final body weight; FF = full fat; FM = fishmeal; GPx = glutathione peroxidase; GR = glutathione reductase; G6DP = anti oxidative enzyme; IBW = initial body weight; IgM = immunoglobulin M; IL2 = interleukin-2; MDA = malondialdehyde; MHC = major histocompatibility complex gene; SOD = superoxide dismutase; TMM = *Tenebrio molitor* larvae meal.

TABLE 9 Effects of *Bombyx mori*, *Musca domestica*, Chironomid meals and *Zophoba morio* on antioxidant enzymes, immune parameters and resistance to bacterial diseases in crustacean and fish

Animal [IBW-FBW, g] {no. days}	Insect [form]	Insect composition (% DM)		% insect inclusion	Type and % of main protein in CTRL diet	% of substitution	Antioxidant enzymes	Immune parameters	Resistance to diseases	Reference
		CP	CL							
Prawns (*Litopenaeus vannamei*) [0.6-2.4] {30}	BM			0.0125, 0.25, 5 µg/g	FM: 24.5%				Increased survival to *Vibrio penaecida* (0; 90; 89.5; 100)	Ali et al. (2018)
Prawns (*Marsupenaeus japonicus*) [0.9-1.4] {15}	BM [DF]			0.001, 0.01, 0.1	FM: 20%				Increased survival to *V. penaecida* (0; 73; 77; 76)	
Asian swamp eel (*Monopterus albus*) [27-31] {56}	MDL [whole]	64	24.3	1 day every 3 (1d/3) or 7 days (1d/7)	FM: 45%			HSP70 and IL1β downregulated, no difference in IgM	Better resistance to *Aeromonas hydrophila* (73; 93 for 1d/3; 83% for 1d/7)	Xiang et al. (2019)
Asian swamp eel (*M. albus*) [30-58] {40}	MDL [DF]	60	2.6	10, 20, 30	70% earthworm	not isonitrogenous and isolipidic	Increased anti-oxidant enzymes (SOD, GPx, CAT)	Increased lysozyme, acid and alkaline phosphatase; lower serum triglycerides	Increased survival to *A. hydrophila* (30; 37.5; 45; 37.5)	Pei et al. (2019)
Black carp (*Mylopharyngodon piceus*) [72-no data] {60}	MDL [-]			2.5	together with 150 mg/kg carnitine		Increased serum GPx and liver SOD,GPx and CAT; reduced MDA	Increased lysozyme and complement	Improved resistance to *A. hydrophila*	Ming et al. (2013)

Species	Insect meal						Effect	Reference
Red seabream (*Pagrus major*) [26-no data] {60}	MD pupae [FF]	50.6	21.9	5	FM: 50%	10	Increased resistance to *Edwardsiella tarda* (0; 100%)	Ido et al. (2015)
Red seabream (*P. major*) [48-no data]{10}				0.75, 7.5		1.5, 15	Increased phagocytic activity	
Rainbow trout (*Oncorhynchus mykiss*) [0.13-0.65] {28}	ChL [whole]	66	9	100	FM: 12.4%		Higher proliferation and apoptotic indexes in the proximal intestine, no difference in distal intestine	Ostaszewska et al. (2011)
Sea trout (*Salmo trutta*) [5-21] {56}	ZMM [hydrolysed]	49.3	33.6	10	FM:25%	56	No change in IgM or lysozyme activity	Mikolajczak et al. (2020)

BM = *Bombyx mori*; CAT = catalase; ChL = Chironomidae frozen larvae; CL = crude lipid; CP = crude protein; CTRL = control diet; DF = defatted; FBW = final body weight; FF = full fat; FM = fishmeal; GPx = glutathione peroxidase; HSP70 = heat shock proteins 70; IBW = initial body weight; IgM = immunoglobulin M; $IL_1\beta$ = interleukin 1 beta; MDA = malondialdehyde; MDL = *Musca domestica* larvae meal; SOD = superoxide dismutase; ZMM = *Zophoba morio* larvae meal.

TABLE 10   Immunological status of poultry fed insect products

Poultry species [IBW-FBW, kg] {no. days}	Insect species and form	Insect composition (% DM)		% insect inclusion	Immunological status	References
		CP	CL			
Jumbo quails [no data-223] {42}	HIL [DF]	55.5	11.1	2.5, 5, 7.5, 10	Increase of albumin/globulin ratio.	Mbhele et al. (2019)
Laying hens (Hy-line Brown) [1,410-1,857] {140}	HIL [DF]	60.0	9.0	7.3, 14.6	Decrease in albumin/globulin ratio. Decrease in cholesterol and triglycerides levels.	Bovera et al. (2018)
Laying hens Lohmann Brown Classic [1,790-1,890] {140}	HIL [DF]	62.7	4.7	17	Increase level of globulin. Decrease albumin/globulin ratio.	Marono et al. (2017)
Broiler chickens (Ross) [no data-1,300] {30}	HIL [FF]	no data	no data	1, 2, 3	Increase in frequency of CD4+ T lymphocyte, serum lysozyme activity, and spleen lymphocyte proliferation.	Lee et al. (2018)
Broiler chickens (Ross 308) [70.7-2,554] {35} [40.2-2,278] {35}	HIL [DF]	55.3	18.0	5, 10, 15	No differences in haematological traits. No differences in histopathological examination.	Dabbou et al. (2018); Biasato et al. (2020a)
Muscovy ducks (Cairina moschata domestica) [70.4-2,554] {47}	HIL [DF]	56.7	10.7	3, 6, 9	No influence on haematological traits.	Gariglio et al. (2019)
Broiler chickens (Ross 308) [42-2,104] {35}	TMM [FF]	47.0	29.6	0.2, 0.3	Significantly decreased IgM levels. Significant increase of IL-2 and TNF-α at 0.3% TMM. No significant changes in level of IgY, IgA, and IL-6. Decreased of Bursa of Fabricius	Benzertiha et al. (2020)
Broiler chickens (Shaver brown) [1,760-3,470] {32}	TMM [FF]	55.3	23.0	30	Decreased albumin/globulin ratio- Increased aspartate aminotransferase and alanine aminotransferase.	Bovera et al. (2015)
Broiler chickens (Ross 708) [no data-2,309] {40}	TMM [FF]	55.3	25.1	5, 10, 15	Correlation between increasing TMM and changes in the blood and serum parameters: increased erythrocytes level and albumin, while the gamma glutamyl transferase decreased.	Biasato et al. (2017)

Broiler chickens (Ross 308) [42-2,253/2,273] {42}	TMM [FF] fresh	20.15	11.49	10.48	No differences in haematological traits.	Elahi *et al.* (2020)
	TMM [FF] dried	52.89	30.05	2, 4, 8		
Broiler chickens (Ross 708) [no data-3,641] {53}	TMM [FF]	57.5	30.7	5, 10, 15	No differences in histopathological examination. Histopathological alterations varied from absent/minimal to severe in spleen, thymus, bursa of Fabricius and liver for each dietary treatment.	Biasato *et al.* (2018b)
Broiler chickens (Ross 308) [42-2,122] {35}	ZMM [FF]	49.3	33.6	0.2, 0.3	Significant decrease of IgY at 0.2% ZMM, IgM at 0.2 and 0.3%ZMM. Significant increase of IL2. No significant changes in level of IgA and TNF-α. Decreased of Bursa of Fabricius.	Benzertiha *et al.* (2020)
Broiler chickens (Ross 308) [42-1,555/1,566] {28}	TMO, ZMO	–	–	5	Liver: ZMO diet: the APOA1 gene was upregulated; TMO diet: the HNF4A gene was downregulated; ZMO and TMO diets: the GIMPA5 gene was significantly downregulated.	Kieronczyk *et al.* (2018)
Broiler chickens (Ross 308) [817.8-3,751.3] {27}	HIF	–	–	3.43, 6.87	No influence in haematological traits.	Schiavone *et al.* (2018)
Turkey (B.U.T. 6, AVIAGEN) [179-1,398] {35}	HIF	–	–	2.5, 5	Partial substitution of SBO: reduction of TNF-α concentration. Total substitution of SBO: reduction of IL-6.	Sypniewski *et al.* (2020)

APOA1 = apolipoprotein A1; CF = crude fat; CP = crude protein; DF = defatted; DM = dry matter; FBW = final body weight; FF = full fat; GIMPA5 = guanosine triphosphatases of the immune-associated protein 5; HIL = *Hermetia illucens* larvae meal; HIF = *Hermetia illucens* fat; HNF4A = hepatocyte nuclear factor 4 alpha; IBW = initial body weight; IgA = immunoglobulin A; IgM = immunoglobulin M; IgY = immunoglobulin Y; IL-2 = interleukin-2; IL-6 = interleukin-6; SBO = soybean oil; TMM = *Tenebrio molitor* larvae meal; TMO = *Tenebrio molitor* oil; TNF = tumour necrosis factor-alpha; ZMM = *Zophobas morio* larvae meal; ZMO = *Zophobas morio* oil.

TABLE 11 Immunological status of pigs and rabbits fed insect products

Animal [IBW-FBW, kg] {no. days}	Insect [form]	Insect composition (% DM)		% insect inclusion	Type and % of main protein in CTRL	% substitution	Immunological status	Reference
		CP	CL					
Piglets [1.4-6.7] {28}	HIL [FF]	35.9	48.1	3-5	SBM FF: 12%; FM: 11.32%	SBM FF: 35%; FM: 35%	No effects on haematological parameters	Driemeyer (2016)
Weaning pigs [6.1-no data] {15}	HIL [FF]	42.01	42.13	4; 8	Toasted soybean: 12	50 and 100%	No effects on bacterial counts (Lactobacilli and D-Streptococci) in stomach, proximal small intestine and distal small intestine.	Spranghers et al. (2018)
	HIL [DF]	62.84	8.06	5-42		72%		
Weaning pigs [6.1-32.4] {61}	HIL [DF]	59.0	8.97	5; 10	SBM: 20% (phase I); 18.5% (phase II)	Phase I: 30 and 60%; Phase II: 32 and 65%	Increase in monocytes (linear response) and neutrophils (quadratic response)	Biasato et al. (2019b)
Weaning piglets [7.9-17.6] {28}	HIL [FF]	37.92	38.48	1, 2, 4	FM: 4%	25, 50 and 100%	Decrease in serum cytokines (IFN-γ) (minimum in pigs fed 4% HIL). Increase in IgA (maximum in pigs fed 2% HIL). Increase in anti-inflammatory cytokine IL-10 in pig fed 2% of HIL	Yu et al. (2020b)
Growing pigs [18.3-53.2] {63}	HIL [FF]	46.6	–	9, 12, 14.5, 18.5	FM: 10%	FM: 25, 50, 75 and 100%	Increase in neutrophils	Chia et al. (2019)
Weaning piglets [6.9-21.4] {35}	TMM [FF]	58.0	31.65	1, 2	FM: 2%	50 and 100%	No effect on red blood cells, white blood cells, lymphocyte, total protein, blood urea nitrogen, insulin-like growth factor and IgG.	Ao et al. (2020)
Weaning pigs [8.0-from 17.8 to 20.2] {35}	TMM [FF]	43.27	32.93	1.5, 3.0, 4.5, 6.0	SMB: 35.1% (phase I); 27.3 (phase II)	Phase I: 5, 10, 15 and 20%; Phase II: 6, 12, 18, 24%	No effect on IgG and IgA. Increase in serum insulin-like growth factor.	Jin et al. (2016)

CL = crude lipid; CP = crude protein; CTRL = control diet; DF = defatted; DM = dry matter; FBW = final body weight; FF = full fat; FM = fishmeal; HIL = Hermetia illucens larvae meal; IBW = initial body weight; IgA = immunoglobulin A; IgG = immunoglobulin G; SBM = soybean meal; TMM = Tenebrio molitor larvae meal.

Józefiak and Engberg, 2017; Li *et al.*, 2017b; Moon *et al.*, 1994; Mylonakis *et al.*, 2016; Ohta *et al.*, 2014, 2016; Park *et al.*, 2014, 2015; Ravi *et al*, 2011; Schuhmann *et al.* 2003; Vogel *et al.*, 2018; Wu *et al.* 2018). AMPs' mode of action is usually based on their cationic nature enabling them to form pores in the microbe cell membranes making them natural alternatives to medicinal treatments reducing the risk of developing microbial resistance (Zhang and Gallo, 2016). Moreover, peptides extracted from the larvae of MD and HI were shown to have an anti-tumoral activity (Sun *et al.*, 2014; Tian *et al.*, 2018).

Beside AMPs, FAs may have antimicrobial activities (Suresh *et al.* 2014). Insects are rich in SCFAs and medium chain FA, such as valeric (5:0), enanthic (7:0), caprylic (8:0), pelargonic (9:0), capric (10:0), myristic (14:0), palmitic (16:0), palmitoleic (16:1), stearic (18:0), oleic (18:1) and linoleic (18:2) which are known to have antifungal and antibacterial activity against both Gram-negative and Gram-positive bacteria (Urbanek *et al.*, 2012). Lauric acid (12:0) in particular has a strong impact on Gram-positive bacteria (Ankaku *et al.*, 2017; Dayrit, 2015).

Chitin and its deacetylated form, chitosan, have also been recognised as having an antimicrobial effect against bacteria, fungi and yeast (Benhabiles *et al.*, 2012; Shin *et al.* 2019). Their antibacterial activity relies on the interaction between the positively charged chitin/chitosan and the negatively charged microbial cell membranes (Goy *et al.*, 2009). Chitin also has a wound-healing action which could complement its immunostimulating and disease resistance promoting effect (Goy *et al.*, 2009).

The GIT of some carnivorous and omnivorous fish species show enzymatic activity. The chitinases produced by gastric glands and the pancreas are able to hydrolyse the glycosidic bonds of polysaccharides in the chitin of insects and crustaceans (Rangaswamy, 2006; Nogales-Mérida *et al.*, 2019). In general, fish and poultry produce chitinases and present GIT genetically adapted to consume insects, which have been part of their natural diet since ancient times and chitinase activity may increase in animals fed a chitin-rich diet (Nogales-Mérida *et al.*, 2019). Apart from the enzymatic hydrolysis of chitin, chitinase may also be antioxidant and immunostimulant with a potential protective role against bacterial and parasitical infections (Di Rosa *et al.*, 2016; Ngo and Kim, 2014) and such an activity could also partly explain the improved resistance of animals to the bacterial diseases listed below.

4.1   *Fish and crustaceans*

The rich intestine microbial diversity shown in the first section of this review increases the competition with pathogens for nutrient and colonisation sites in the intestine, and thus may improve resistance to diseases (Cerezuela *et al.*, 2013). Moreover, the modulation of the immune system of fish and crustaceans, as a result of the inclusion of insects in their diets described in the previous section, has been translated in a remarkably improved resistance to bacterial diseases as shown in the few existing studies on dietary TM, MD or SWP listed in Table 8 and 9. To our knowledge, no study has been done so far on the effects of dietary HI on crustaceans or fish resistance to diseases. In Pacific white shrimps, *L. vannamei*, 10-20% dietary TM replacing 50-100% of FM drastically reduced *Vibrio parahaemolyticus*-induced mortality, from 45 to 13-16% (Motte *et al.*, 2019). Very low dietary doses of SWP, or silkrose purified from SWP, greatly increased the survival of prawns subjected to an immersion challenge with *Vibrio penaecida*, from 0% for control diet-fed prawns to 73-77 and 90-100% for SWP- and silkrose-fed prawns, respectively (Ali *et al.*, 2018). In fish, 5 and 10% of dietary TM increased the survival of red seabream (*Pagrus major*) infected with *Edwardsiella tarda*, from 21% in the control fish to 55-67% in the insect-fed fish (Ido *et al.*, 2019). Moreover, 7.5% of dietary TM improved the resistance of pearl gentian grouper to *Vibrio harveyi*, but other levels of dietary inclusions either had no effect or even reduced the fish survival rate (Song *et al.*, 2018). The resistance of yellow catfish (*Pelteobagrus fulvidraco*) to *Edwardsiella ictaluri* was significantly improved by the addition of 27% of dietary TM (Su *et al.*, 2017). Dietary MD also improved the survival of swamp eel and black carp (*Mylopharyngodon piceus*) to *Aeromonas hydrophila* (Ming *et al.*, 2013; Pei *et al.*, 2019; Xiang *et al.*, 2019) and of red seabream (*P. major*) to *E. tarda* (Ido *et al.*, 2015). Because of the similarities between the exoskeletons of insects and parasites, it can be hypothesised that they share some pathogen-associated molecular patterns which may not only improve fish resistance to bacterial diseases as listed before but also resistance to parasitic or even fungal diseases. Although no such study has been undertaken with insect chitin, the closely related crustacean chitin and/or chitosan, have recently been shown to reduce the prevalence and intensity of monogenean in rohu, *Labeo rohita* (Kumar *et al.*, 2019) and to greatly increased survival of Indian major carp, *Cirrhina mrigala*, to 75-80%, compared to 10% in control-fed fish when infected with the aquatic fungus *Aphanomyces invadans* (Shanthi *et al.*, 2014). These results taken together are very positive and suggest that dietary inclusion of insects may represent a healthy supplement for aquafeed of the future.

4.2   *Poultry*

The immunostimulation of broiler chickens fed 3% HI discussed in the previous section was translated into an improved survival of chicken when experimentally infected with *Salmonella* Gallinarum (Lee *et al.*, 2018) (Table 12).

Islam and Yang (2017) showed that 0.4% of dietary full-fat TM or ZM decreased mortality and increased IgG and IgA levels in broiler chickens challenged with *Salmonella* and *E. coli*. The authors speculated that the chitin in TM and ZM larva meal had a probiotic effect that was able to act as a natural antibiotic (Table 12).

4.3   *Pigs and rabbits*

The antimicrobial effects of HI fat on the microbiota of pigs has been investigated *in vitro* (Spranghers *et al.* 2018). The results showed a strong antibacterial effect against Lactobacilli and in particular against D-Streptococci, a Gram-positive bac-

TABLE 12   Antimicrobial effects and resistance to diseases induced by insect products in poultry

Poultry species [IBW-FBW, kg] {no. days}	Insect [form]	Insect composition (% DM)		% insect inclusion	Antimicrobial and diseases resistance	References
		CP	CL			
Broiler chickens (Ross) [no data-1,300] {30}	HIL [FF]	–	–	1, 2, 3	Reinforced bacterial clearance and increased survival against *Salmonella* Gallinarum. Prophylactic properties: Reduced bacterial burden against *S.* Gallinarum. Increased survival rate of chicken experimentally infected with *S.* Gallinarum in 3% HIL group.	Lee *et al.* (2018)
Broiler chickens (Ross 308) [39.3-74.8] {7}	TMM [FF]	27.26	11.50	0.4%	Insect meals were fermented with *Lactobacillus plantarum* and *Saccharomyces cerevisiae* to form probiotics. Increase in survival rate in birds challenged *with Escherichia coli* and *Salmonella* spp.	Islam and Yang (2017)
	ZMM [FF]	27.15	8.70			

CL = crude lipid; CP = crude protein; DM = dry matter; FBW = final body weight; FF = full fat; HIL = *Hermetia illucens* larvae meal; IBW = initial body weight; TMM = *Tenebrio molitor* meal; ZMM = *Zophoba morio* larvae meal.

teria genus that can cause severe damage in the pig industry. The authors did not report any effect against Gram-negative bacteria. The *in vitro* antibacterial activity of insects fats extracted from both HI and TM larvae were also investigated in rabbit against common Gram-positive and Gram-negative bacterial pathogenic strains (*Salmonella tiphymurium, Salmonella enteritidis, Yersinia enterocolitica, Pasteurella multocida* and *Listeria monocytogenes*) (Dabbou *et al.*, 2020). These bacteria are foodborne pathogens posing public health concern (De Cesare *et al.* 2017; Kylie *et al.* 2017; Massacci *et al.* 2018; Rodriguez-Calleja *et al.* 2006). The growth of *Y. enterocolitica, L. monocytogenes and P. multocida* was impaired by HI fat, while TM fat only inhibited the growth of *P. multocida* (Table 13).

In vivo, Ji *et al.* (2016) suggested a positive effect of AMPs from insects and reported a decrease in the incidence of diarrhoea in weaning piglets fed 5% of TM, MD or ZM in total substitution of plasma protein meal. However, this effect was not found by Yu *et al.* (2020b) who fed piglets with up to 4% of full fat HI meal (Table 13).

TABLE 13   Antimicrobial effects and resistance to diseases induced by insect products in pigs and rabbits

Animal [IBW-FBW, kg] {no. days}	Insect [form]	Insect composition (% DM)		% insect inclusion	Type and % of main protein in CTRL	% substitution	Antimicrobial and diseases resistance	Reference
		CP	CL					
in vivo trial								
Weaning pigs [4·7-no data] {56}	TMM [FF]	50.2	29	5	Plasma protein powder: 5%	100%	Decrease in diarrhoea rate between 15 and 28 days of trial	Ji *et al.* (2016)
	MDL [FF]	45.6	30.1					
	ZMM [FF]	45.1	41.7					
in vitro trials								
Weaning pigs [6.1-no data] {15}	HIF			Addition of 0.20, 0.50, 1.00, or 1.50 g/100 mL of HIF (corresponding to 0.1, 0.29, 0.58 and 0.87% C12:0) to an incubation medium containing synthetic diet + phosphate buffer (pH 5) + a microbial inoculum from one donor piglet			0.58% had a strong *in vitro* effect against D-streptococci	Spranghers *et al.* (2018)
Rabbits [1.05-2.9] {41}	HIF TMO			0.75, 1.5	SBO: 1.5%	50 and 100%	*In vitro* inhibitory effect of HIO against *Yersinia enterocolitica, Pasteurella multocida and Listeria monocytogenes* and of TMO against *P. multocida*	Dabbou *et al.* (2020)

CL = crude lipid; CP = crude protein; CTRL = control diet; DM = dry matter; FBW = final body weight; FF = full fat; HIF = *Hermetia illucens* fat; IBW = initial body weight; MDL = *Musca domestica* larvae meal; TMM = *Tenebrio molitor* larvae meal; SBO = soybean oil; TMO = *Tenebrio molitor* oil; ZMM = *Zophoba morio* larvae meal.

5   Conclusions

Dietary insects can successfully be used to partially replace conventional protein sources (FM, SBM) without affecting the growth of the animals. However, the lack of large-scale production units (with the exception of housefly) and legislation uncertainties have resulted in high prices and limited amounts of available quantities, both of which currently impair their massive use in animal nutrition. However, the exoskeleton of insects resembles that of parasites, and insects may share pathogen-associated molecular patterns with bacterial or parasitical fish pathogens. They are also rich in short- and medium-chain fatty acids and in chitin, and produce polysaccharides and peptides, all of which may have antimicrobial and/or immunostimulating activity. The immunostimulation and drastically improved disease resistance highlighted in the present publication (particularly in crustacean and fish studies) suggest that low dietary levels of insects may represent a potent supplement providing 'healthy animal feeds' or 'natural non-specific oral vaccines' to boost the animal immune system. They could be used as a 'preventive cure' before stressful events, such as transport or seasonal temperature changes, which are usually linked to a rise of infection risk.

Further studies will have to determine for each insect species with a potential for mass production, doses and duration of administration and the range of diseases protection offered to aquatic and terrestrial farmed animals.

Conflict of interest

The authors declare no conflict of interest.

References

Ali, M.F.Z., Yasin, I.A., Ohta, T., Hashizume, A., Ido, A., Takahashi, T., Miura, C. and Miura, T., 2018. The silkrose of *Bombyx mori* effectively prevents vibriosis in penaeid prawns via the activation of innate immunity. Scientific Reports 8. https://doi.org/10.1038/s41598-018-27241-3

Alvarez, D., Wilkinson, K.A., Treilhou, M., Tene, N., Castillo, D. and Sauvain, M., 2019. Prospecting peptides isolated from black soldier fly (Diptera: Stratiomyidae) with antimicrobial activity against *Helicobacter pylori* (Campylobacterales: Helicobacteraceae). Journal of Insect Science (Online) 19: 17. https://doi.org/10.1093/jisesa/iez120

Ankaku, A.A, Akyala, J.I., Juliet, A. and Obianuju, E.C., 2017. Antibacterial activity of lauric acid on some selected clinical isolates. Annals of Clinical and Laboratory Research 5: 2. https://doi.org/10.21767/2386-5180.1000170

Antonopoulou, E., Nikouli, E., Piccolo, G., Gasco, L., Gai, F., Chatzifotis, S., Mente, E. and

Kormas, K.A., 2019. Reshaping gut bacterial communities after dietary *Tenebrio molitor* larvae meal supplementation in three fish species. Aquaculture 503: 628-635. https://doi.org/10.1016/j.aquaculture.2018.12.013

Ao, X., Yoo, J.S., Wu, Z.L. and Kim, I.H., 2020. Can dried mealworm (*Tenebrio molitor*) larvae replace fish meal in weaned pigs? Livestock Science 239: 104103. https://doi.org/10.1016/j.livsci.2020.104103.

Askarian, F., Zhou, Z., Olsen, R.E., Sperstad, S. and Ringo, E., 2012. Culturable autochthonous gut bacteria in Atlantic salmon (*Salmo salar* L.) fed diets with or without chitin. Characterization by 16S rRNA gene sequencing, ability to produce enzymes and *in vitro* growth inhibition of four fish pathogens. Aquaculture 326: 1-8. https://doi.org/10.1016/j.aquaculture.2011.10.016

Belforti, M., Gai, F., Lussiana, C., Renna, M., Malfatto, V., Rotolo, L., De Marco, M., Dabbou, S., Schiavone, A., Zoccarato, I. and Gasco, L., 2015. *Tenebrio molitor* meal in rainbow trout (*Oncorhynchus mykiss*) diets: effects on animal performance, nutrient digestibility and chemical composition of fillets. Italian Journal of Animal Science14: 670-675. https://doi.org/10.4081/ijas.2015.4170

Belghit, I., Liland, N.S., Gjesdal, P., Biancarosa, I., Menchetti, E., Li, Y., Waagbo, R., Krogdahl, A. and Lock, E.-J., 2019. Black soldier fly larvae meal can replace fish meal in diets of sea-water phase Atlantic salmon (*Salmo salar*). Aquaculture 503: 609-619. https://doi.org/10.1016/j.aquaculture.2018.12.032

Belghit, I., Liland, N.S., Waagbo, R., Biancarosa, I., Pelusio, N., Li, Y., Krogdahl, A. and Lock, E.-J., 2018. Potential of insect-based diets for Atlantic salmon (*Salmo salar*). Aquaculture 491, 72-81. https://doi.org/10.1016/j.aquaculture.2018.03.016

Benhabiles, M. S., Salah, R., Lounici, H., Drouiche, N., Goosen, M. F. A. and Mameri, N., 2012. Antibacterial activity of chitin, chitosan and its oligomers prepared from shrimp shell waste. Food Hydrocolloids 29: 48-56. https://doi.org/10.1016/j.foodhyd.2012.02.013

Benzertiha, A., Kierończyk, B., Kołodziejski, P., Pruszyńska-Oszmałek, E., Rawski, M., Józefiak, D. and Józefiak, A., 2020. *Tenebrio molitor* and *Zophobas morio* full-fat meals as functional feed additives affect broiler chickens' growth performance and immune system traits. Poultry Science 99: 196-206. https://doi.org/10.3382/PS/PEZ450

Benzertiha, A., Kierończyk, B., Rawski, M., Józefiak, A., Kozłowski, K., Jankowski, J. and Józefiak, D., 2019. *Tenebrio molitor* and *Zophobas morio* full-fat meals in broiler chicken diets: effects on nutrients digestibility, digestive enzyme activities, and cecal microbiome. Animals 9: 1128. https://doi.org/10.3390/ani9121128

Biasato, I. Ferrocino, I., Biasibetti, E., Grego, E., Dabbou, S., Sereno, A., Gai, F., Schiavone, A. and Capucchio, M.T., 2018a. Modulation of intestinal microbiota, morphology and mucin composition by dietary insect meal inclusion in free-range chickens. BMC Veterinary Research 14: 383. https://doi.org/10.1186/s12917-018-1690-y

Biasato, I., De Marco, M., Rotolo, L., Renna, M., Lussiana, C., Dabbou, S., Capucchio, M.T., Biasibetti, E., Costa, P., Gai, F., Pozzo, L., Dezzutto, D., Bergagna, S., Martínez, S., Tarantola, M., Gasco, L. and Schiavone, A., 2016. Effects of dietary *Tenebrio molitor* meal inclusion in free-range chickens. Journal of Animal Physiology and Animal Nutrition 100: 1104-1112. https://doi.org/10.1111/jpn.12487

Biasato, I., Ferrocino, I., Colombino, E., Gai, F., Schiavone, A., Cocolin, L., Capucchio, M.T. and Gasco, L., 2020b. Effects of dietary *Hermetia illucens* meal inclusion on cecal microbiota and small intestine mucin dynamics and inflammatory status of weaned piglets. Journal of Animal Science and Biotechnology, 11: 64. https://doi.org/10.1186/s40104-020-00466-x

Biasato, I., Ferrocino, I., Dabbou, S., Evangelista, R., Gai, F., Gasco, L., Cocolin, L., Capucchio, M.T. and Schiavone, A., 2020a. Black soldier fly and gut health in broiler chickens: insights into the relationship between cecal microbiota and intestinal mucin composition. Journal of Animal Science and Biotechnology 11: 11. https://doi.org/10.1186/s40104-019-0413-y

Biasato, I., Ferrocino, I., Grego, E., Dabbou, S., Gai, F., Gasco, L., Cocolin,L., Capucchio, M.T. and Schiavone, A., 2019a. Gut Microbiota and Mucin Composition in Female Broiler Chickens Fed Diets including Yellow Mealworm. Animals 9: 213. https://doi.org/10.3390/ani9050213

Biasato, I., Gasco, L., De Marco, M., Renna, M., Rotolo, L., Dabbou, S., Capucchio, M.T., Biasibetti, E., Tarantola, M., Sterpone, L., Cavallarin, L., Gai, F., Pozzo, L., Bergagna, S., Dezzutto, D., Zoccarato, I., Schiavone, A., 2018b. Yellow mealworm larvae (*Tenebrio molitor*) inclusion in diets for male broiler chickens: effects on growth performance, gut morphology, and histological findings. Poultry Science 97: 540-548. https://doi.org/10.3382/ps/pex308

Biasato, I., Gasco, L., De Marco, M., Renna, M., Rotolo, L., Dabbou, S., Capucchio, M.T., Biasibetti, E., Tarantola, M., Bianchi, C., Cavallarin, L., Gai, F., Pozzo, L., Dezzutto, D., Bergagna, S. and Schiavone, A., 2017. Effects of yellow mealworm larvae (*Tenebrio molitor*) inclusion in diets for female broiler chickens: implications for animal health and gut histology. Animal Feed Science and Technology 234: 253-263. https://doi.org/10.1016/j.anifeedsci.2017.09.014

Biasato, I., Renna, M., Gai, F., Dabbou, S., Meneguz, M., Perona, G., Martinez, S., Barroeta-Lajusticia, A.C., Bergagna, S., Sardi, L., Capucchio, M.T., Bressan, E., Dama, A., Schiavone, A. and Gasco, L., 2019b. Partially defatted black soldier fly larva meal inclusion in piglet diets: effects on the growth performance, nutrient digestibility, blood profile, gut morphology and histological features. Journal of Animal Science and Biotechnology 10: 12. https://doi.org/10.1186/s40104-019-0325-x

Borrelli, L., Coretti, L., Dipineto, L., Bovera, F., Menna, F., Chiariotti, L., Nizza, A., Lembo, F. and Fioretti, A., 2017. Insect-based diet, a promising nutritional source, modulates gut microbiota composition and SCFAs production in laying hens. Scientific Reports 7: 16269. https://doi.org/10.1038/s41598-017-16560-6

Bosch, G., van Zanten, H.H.E., Zamprogna, A., Veenenbos, M., Meijer, N.P., van der Fels-Klerx, H.J. and van Loon, J.J.A., 2019. Conversion of organic resources by black soldier fly larvae: Legislation, efficiency and environmental impact. Journal of Cleaner Production 222: 355-363. https://doi.org/10.1016/j.jclepro.2019.02.270

Bovera, F., Loponte, R., Pero, M.E., Cutrignelli, M.I., Calabrò, S., Musco, N., Vassalotti, G., Panettieri, V., Lombardi, P., Piccolo, G., Di Meo, C., Siddi, G., Fliegerova, K. and Moniello, G., 2018. Laying performance, blood profiles, nutrient digestibility and inner organs traits of hens fed an insect meal from *Hermetia illucens* larvae. Research in Veterinary Science 120: 86-93. https://doi.org/10.1016/j.rvsc.2018.09.006

Bovera, F., Piccolo, G., Gasco, L., Marono, S., Loponte, R., Vassalotti, G., Mastellone, V., Lombardi, P., Attia Y.A and Nizza A. 2015. Yellow mealworm larvae (*Tenebrio molitor*, L.) as a possible alternative to soybean meal in broiler diets. British Poultry Science 56: 569-575, https://doi.org/10.1080/00071668.2015.1080815

Bradshaw, C.J.A., Leroy, B., Bellard, C., Roiz, D., Albert, C., Fournier, A., Barbet-Massin, M., Salles, J.M., Simard, F. and Courchamp F., 2016. Massive yet grossly underestimated global costs of invasive insects. Nature Communications 7: 12986. https://doi.org/10.1038/ncomms12986

Bruni, L., Pastorelli, R., Viti, C., Gasco, L. and Parisi, G., 2018. Characterization of the intestinal microbial communities of rainbow trout (*Oncorhynchus mykiss*) fed with *Hermetia illucens* (black soldier fly) partially defatted larva meal as partial dietary protein source. Aquaculture 487: 56-63. https://doi.org/10.1016/j.aquaculture.2018.01.006

Caimi, C., Gasco, L., Biasato, I., Malfatto, V., Varello, K., Prearo, M., Pastorino, P., Bona, M.C., Francese, D.R., Schiavone, A., Elia, A.C., Dörr, A.J.M. and Gai, F., 2020. Could Dietary Black Soldier Fly Meal Inclusion Affect the Liver and Intestinal Histological Traits and the Oxidative Stress Biomarkers of Siberian Sturgeon (*Acipenser baerii*) Juveniles? Animals 10: 155. https://doi.org/10.3390/ani10010155

Cardinaletti, G., Randazzo, B., Messina, M., Zarantoniello, M., Giorgini, E., Zimbelli, A., Bruni, B., Parisi, G., Olivotto, I. and Tulli, F., 2019. Effects of Graded Dietary Inclusion Level of Full-Fat *Hermetia illucens* Prepupae Meal in Practical Diets for Rainbow Trout (*Oncorhynchus mykiss*). Animals 9(5): 251. https://doi.org/10.3390/ani9050251

Cerezuela, R., Meseguer, J. and Esteban, M.A., 2013. Effects of dietary inulin, *Bacillus subtilis* and microalgae on intestinal gene expression in gilthead seabream (*Sparus aurata* L.). Fish & Shellfish Immunology 34: 843-848. https://doi.org/10.1016/j.fsi.2012.12.026

Chia, S.Y., Tanga, C.M., Osuga, I.M., Alaru, A.O., Mwangi, D.M., Githinji, M., Subramanian, S., Fiaboe, K.K.M., Ekesi, S., van Loon, J.J.A. and Dicke, M., 2019. Effect of dietary replacement of fishmeal by insect meal on growth performance, blood profiles and economics of growing pigs in Kenya. Animals 9: 705. https://doi.org/10.3390/ani9100705

Choi, S.C., Ingale, S.L., Kim, J.S., Park, Y.K., Kwon, I.K. and Chae, B.J., 2013a. An antimicrobial peptide-A3: effects on growth performance, nutrient retention, intestinal and faecal microflora and intestinal morphology of broilers. British Poultry Science 54: 738-746. https://doi.org/10.1080/00071668.2013.838746

Choi, S.C., Ingale, S.L., Kim, J.S., Park, Y.K., Kwon, I.K. and Chae, B.J., 2013b. Effects of dietary supplementation with an antimicrobial peptide-P5 on growth performance, nutrient retention, excreta and intestinal microflora and intestinal morphology of broilers. Animal Feed Science and Technology 185: 78-84. https://doi.org/10.1016/j.anifeedsci.2013.07.005

Chu, F.J., Jin, X.B., Xu, Y.Y., Ma, Y., Li, X.B., Lu, X.M., Liu, W.B. and Zhu, J.Y., 2013. Inflammatory regulation effect and action mechanism of anti-inflammatory effective parts of housefly (*Musca domestica*) larvae on atherosclerosis. Evidence Based Complementary and Alternative Medicine: Article ID 340267: 10 https://doi.org/10.1155/2013/340267

Cutrignelli, M.I., Messina, M., Tulli, F., Randazzo, B., Olivotto, I., Gasco, L., Loponte, R. and Bovera, F., 2018. Evaluation of an insect meal of the Black Soldier Fly (*Hermetia illucens*) as soybean substitute: Intestinal morphometry, enzymatic and microbial activity in laying hens. Research in Veterinary Science 117: 209-215. https://doi.org/10.1016/j.rvsc.2017.12.020

Dabbou S., Ferrocin, I., Gasco L., Schiavone A., Trocino, A., Xiccato, G., Barroeta A.C., Maione, S., Soglia, D., Biasato, I., Cocolin, L., Gai, F. and Nucera, D.D., 2020. Antimicrobial effects of black soldier fly and yellow mealworm fats and their impact on gut microbiota of growing rabbit. Animals 10: 1292. https://doi.org/10.3390/ani10081292

Dabbou S., Gai F., Biasato I., Capucchio MT., Biasibetti E., Dezzutto D., Meneguz M., Plachà I., Gasco L. and Schiavone A., 2018. Black soldier fly defatted meal as a dietary protein source for broiler chickens: effects on growth performance, blood traits, gut morphology and histological features. Journal of Animal Science and Biotechnology 9: 49. https://doi.org/10.1186/s40104-018-0266-9

Dang, X. L., Wang, Y. S., Huang, Y. D., Yu, X. Q. and Zhang, W. Q., 2010. Purification and characterization of an antimicrobial peptide, insect defensin, from immunized house fly (Diptera: Muscidae), Journal of Medical Entomology 47: 1141-1145. https://doi.org/10.1603/ME10016

Danieli, P.P., Lussiana, C., Gasco, L., Amici, A. and Ronchi, B., 2019. The effects of diet formulation on the yield, proximate composition, and fatty acid profile of the Black soldier fly (*Hermetia illucens* L.) prepupae intended for animal feed. Animals 9: 178. https://doi.org/10.3390/ani9040178

Dayrit, F.M., 2015. The properties of lauric acid and their significance in coconut oil. Journal of the American Oil Chemists' Society 92: 1-15. https://doi.org/10.1007/s11746-014-2562-7

De Cesare, A., Parisi, A., Mioni, R., Comin, D., Lucchi, A. and Manfreda, G., 2017. *Listeria monocytogenes* circulating in rabbit meat products and slaughterhouses in Italy: prevalence data and comparison among typing results. Foodborne Pathogens and Disease 14: 167-176. https://doi.org/10.1089/fpd.2016.2211.

De Souza-Vilela, J., Andrew, N.R. and Ruhnke, I., 2019. Insect protein in animal nutrition. Animal Production Science 59: 2029-2036. https://doi.org/10.1071/AN19255

Di Rosa, M., Distefano, G., Zorena, K. and Malaguarnera, L., 2016. Chitinases and immunity: ancestral molecules with new functions. Immunobiology 221: 399-411. https://doi.org/10.1016/j.imbio.2015.11.014

Driemeyer, H., 2016. Evaluation of black soldier fly (*Hermetia illucens*) larvae as an alternative protein source in pig creep diets in relation to production, blood and manure microbiology parameters. MSc thesis, University of Stellenbosch, South Africa, p. 114.

Egerton, S., Culloty, S., Whooley, J., Stanton, C. and Ross, R.P., 2018. The gut microbiota of marine fish. Frontiers Microbiology, 9: 873. https://doi.org/10.3389/fmicb.2018.00873.

Elahi, U., Wang, J., Ma, Y.B., Wu, S.G., Wu, J., Qi, G.H. and Zhang, H.J., 2020. Evaluation of yellow mealworm meal as a protein feedstuff in the diet of broiler chicks. Animals 10: 224. https://doi.org/10.3390/ani10020224

Elhag, O., Zhou, D., Song, Q., Soomro, A.A., Cai, M., Zheng, L., Yu, Z. and Zhang, J., 2017.

Screening, expression, purification and functional characterization of novel antimicrobial peptide genes from *Hermetia illucens* (L.). PLoS ONE 12: e0169582. https://doi.org/10.1371/journal.pone.0169582

Elia, A.C., Capucchio, M.T., Caldaroni, B., Magara, G., Dorr, A.J.M., Biasato, I., Biasibetti, E., Righetti, M., Pastorino, P., Prearo, M., Gai, F., Schiavone, A. and Gasco, L., 2018. Influence of *Hermetia illucens* meal dietary inclusion on the histological traits, gut mucin composition and the oxidative stress biomarkers in rainbow trout (*Oncorhynchus mykiss*). Aquaculture 496: 50-57. https://doi.org/10.1016/j.aquaculture.2018.07.009

Esteban, M.A., Cuesta, A., Ortuño, J. and Meseguer, J., 2001. Immunomodulatory effects of dietary intake of chitin on gilthead seabream (*Sparus aurata* L.) innate immune system. Fish & Shellfish Immunology 11: 303-315. https://doi.org/10.1006/fsim.2000.0315

Food and Agriculture Organization of the United Nations (FAO), 2020. Forecasting threats to the food chain affecting food security in countries and regions. Food Chain Crisis Early Warning Bulletin no. 34, January-March 2020. FAO, Rome, Italy. Available at http://www.fao.org/publications/card/en/c/ca7582en/.

Faruk, M.O., Yusof, F and Chowdhury, S., 2016. An overview of antifungal peptides derived from insect. Peptides 80: 80-88. https://doi.org/10.1016/j.peptides.2015.06.001

Foysal, M.J., Fotedar, R., Tay, C. and Gupta, S.K., 2019. Dietary supplementation of black soldier fly (*Hermetica illucens*) meal modulates gut microbiota, innate immune response and health status of marron (*Cherax cainii*, Austin 2002) fed poultry-by-product and fishmeal based diets. PeerJ 7. https://doi.org/10.7717/peerj.6891

Gariglio, M., Dabbou, S., Crispo, M., Biasato, I., Gai, F., Gasco, L., Piacente, F., Odetti, P., Bergagna, S., Plachà, I., Valle, E., Colombino, E., Capucchio, M.T. and Schiavone, A., 2019. Effects of the dietary inclusion of partially defatted black soldier fly (*Hermetia illucens*) meal on the blood chemistry and tissue (spleen, liver, thymus, and bursa of Fabricius) histology of muscovy ducks (*Cairina moschata domestica*). Animals 9: 307. https://doi.org/10.3390/ani9060307

Gasco, L., Acuti, G., Bani, P., Dalle Zotte, A., Danieli, P.P., De Angelis, A., Fortina, R., Marino, R., Parisi, G., Piccolo, G., Pinotti, L., Prandini, A., Schiavone, A., Terova, G., Tulli, F. and Roncarati, A., 2020a. Insects and fish by-products as sustainable alternatives to conventional animal proteins in animal nutrition. Italian Journal of Animal Science 19: 360-372. https://doi.org/10.1080/1828051X.2020.1743209

Gasco, L., Biancarosa, I. and Liland, N.S., 2020b. From waste to feed: a review of recent knowledge on insects as producers of protein and fat for animal feeds. Current Opinion in Green and Sustainable Chemistry 23: 67-79. https://doi.org/10.1016/j.cogsc.2020.03.003

Gasco, L., Biasato, I., Dabbou, S., Schiavone, A. and Gai, F., 2019a. Animals fed insect-based diets: state-of-the-art on digestibility, performance and product quality. Animals 9: 170. https://doi.org/10.3390/ani9040170.

Gasco, L., Dabbou, S., Trocino, A., Xiccato, G., Capucchio, M.T., Biasato, I., Dezzutto, D., Birolo, M., Meneguz, M., Schiavone, A. and Gai, F., 2019b. Effect of dietary supplementation with insect fats on growth performance, digestive efficiency and health of rabbits. Journal of Animal Science and Biotechnology 10: 4. https://doi.org/10.1186/s40104-018-0309-2

Gasco, L., Finke, M. and Van Huis, A., 2018b. Can diets containing insects promote animal health? Journal of Insects as Food and Feed 4: 1-4. https://doi.org/10.3920/jiff2018.x001

Gasco, L., Gai, F., Maricchiolo, G., Genovese, L., Ragonese, S., Bottari, T. and Caruso, G., 2018a. Fish meal alternative protein sources for aquaculture feeds. In: Gasco, L., Gai, F., Maricchiolo, G., Genovese, L., Ragonese, S., Bottari, T., Caruso, G. (eds) Feeds for the aquaculture sector – current situation and alternative sources. Springer Briefs in Molecular Science. Lightning Source UK Ltd, Cham, Switzerland, pp. 1-28. https://doi.org/10.1007/978-3-319-77941-6

Gasco, L., Henry, M., Piccolo, G., Marono, S., Gai, F., Renna, M., Lussiana, C., Antonopoulou, E., Mola, P. and Chatzifotis, S., 2016. *Tenebrio molitor* meal in diets for European sea bass (*Dicentrarchus labrax* L.)juveniles: growth performance, whole body composition and *in vivo* apparent digestibility. Animal Feed Science and Technology 220: 34-45. https://doi.org/10.1016/j.anifeedsci.2016.07.003

Goulson, D., 2019. The insect apocalypse and why it matters. Current Biology 29: R942-R995. https://doi.org/10.1016/j.cub.2019.06.069

Goy, R.C., Debritto, D. and Assis, O.B.G., 2009. A review of the antimicrobial activity of chitosan. Polímeros 19: 241-247. https://doi.org/10.1590/S0104-14282009000300013

Guerrerio, I. Oliva-Teles, A. and Enes, P., 2018. Prebiotics as functional ingredients: focus on Mediterranean fish aquaculture. Reviews in Aquaculture 10: 800-832. https://doi.org/10.1111/raq.12201

Hardy, K., Radini, A., Buckley, S., Blasco, R., Copeland, L., Burjachs, F., Girbal, J., Yll, R., Carbonell, E. and Bermúdez de Castro, J.M., 2016. Diet and environment 1.2 million years ago revealed through analysis of dental calculus from Europe's oldest hominin at Simadel Elefante, Spain. The Science of Nature 104: 2. https://doi.org/10.1007/s00114-016-1420-x

Henry, M., Gasco, L., Piccolo, G. and Fountoulaki, E., 2015. Review on the use of insects in the diet of farmed fish: Past and future. Animal Feed Science and Technology 203: 1-22. https://doi.org/10.1016/j.anifeedsci.2015.03.001

Henry, M.A., Gai, F., Enes, P., Perez-Jimenez, A. and Gasco, L., 2018b. Effect of partial dietary replacement of fishmeal by yellow mealworm (*Tenebrio molitor*) larvae meal on the innate immune response and intestinal antioxidant enzymes of rainbow trout (*Oncorhynchus mykiss*). Fish & Shellfish Immunology 83: 308-313.

Henry, M.A., Gasco, L., Chatzifotis, S. and Piccolo, G., 2018a. Does dietary insect meal affect the fish immune system? The case of mealworm, *Tenebrio molitor* on European sea bass, *Dicentrarchus labrax*. Developmental & Comparative Immunology 81: 204-209. https://doi.org/10.1016/j.dci.2017.12.002

Huyben, D., Vidakovic, A., Werner Hallgren, S. and Langeland, M., 2019. High-throughput sequencing of gut microbiota in rainbow trout (*Oncorhynchus mykiss*) fed larval and pre-pupae stages of black soldier fly (*Hermetia illucens*). Aquaculture 500: 485-491. https://doi.org/10.1016/j.aquaculture.2018.10.034

Ido, A., Hashizume, A., Ohta, T., Takahashi, T., Miura, C. and Miura, T., 2019. Replacement of fish meal by defatted yellow mealworm (*Tenebrio molitor*) larvae in diet improves growth performance and disease resistance in red seabream (*Pagrus major*). Animals 9: 100. https://doi.org/10.3390/ani9030100

Ido, A., Iwai, T., Ito, K., Ohta, T., Mizushige, T., Kishida, T., Miura, C. and Miura, T., 2015. Dietary effects of housefly (*Musca domestica*) (Diptera: Muscidae) pupae on the growth performance and the resistance against bacterial pathogen in red sea bream (*Pagrus major*) (Perciformes: Sparidae). Applied Entomology and Zoology 50: 213-221. https://doi.org/10.1007/s13355-015-0325-z

Imamura, M., Wada, S., Koizumi, N., Kadotani, T., Yaoi, K., Sato, R. and Iwahana, H., 1999. Acaloleptins A: inducible antibacterial peptides from larvae of the beetle, *Acalolepta luxuriosa*. Archives of Insect Biochemistry and Physiology 40: 88-98. https://doi.org/10.1002/(SICI)1520-6327(1999)40:2<88::AID-ARCH3>3.0.CO;2-B

Islam, M.M. and Yang, C.-J., 2017. Efficacy of mealworm and super mealworm larvae probiotics as an alternative to antibiotics challenged orally with *Salmonella* and *E. coli* infection in broiler chicks. Poultry Science 96: 27-34. https://doi.org/10.3382/ps/pew220

Ji, Y.J., Liu, H. N., Kong, X. F., Blachier, F., Geng, M.M., Liu, Y. Y. and Yin Y. L., 2016. Use of insect powder as a source of dietary protein in early-weaned piglets. Journal Animal Science 94: 111-116. https://doi.org/10.2527/jas2015-9555

Jin, X.H., Heo, P.S., Hong, J.S., Kim, N.J. and Kim, Y.Y., 2016. Supplementation of dried mealworm (*Tenebrio molitor* larva) on growth performance, nutrient digestibility and blood profiles in weaning pigs. Asian-Australasian Journal Animal Science 29: 979-86. https://doi.org/10.5713/ajas.15.0535

Józefiak, A. and Engberg, R.M., 2017. Insect proteins as a potential source of antimicrobial peptides in livestock production. A review. Journal of Animal and Feed Sciences 26: 87-99. https://doi.org/10.22358/jafs/69998/2017

Józefiak, A., Benzertiha, A., Kierończyk, B., Łukomska, A., Wesołowska, I. and Rawski, M., 2020. Improvement of cecal commensal microbiome following the insect additive into chicken diet. Animals 10: 577. https://doi.org/10.3390/ani10040577

Józefiak, A., Kierończyk, B., Rawski, M., Mazurkiewicz, J., Benzertiha, A., Gobbi, P., Nogales-Mérida, S., Świątkiewicz, S. and Józefiak, D., 2018. Full-fat insect meals as feed additive – the effect on broiler chicken growth performance and gastrointestinal tract microbiota. Journal of Animal and Feed Sciences 2: 131-139. https://doi.org/10.22358/jafs/91967/2018

Józefiak, A., Nogales-Merida, S., Mikołajczak, Z. and Mazurkiewicz, J., 2019b. The utilization of full-fat insect meal in rainbow trout (*Oncorhynchus mykiss*) nutrition: The effects on growth performance, intestinal microbiota and gastro-intestinal tract histomorphology. Annals of Animal Science 19: 747-765. https://doi.org/10.2478/aoas-2019-0020

Józefiak, A., Nogales-Merida, S., Rawski, M., Kieronczyk, B. and Mazurkiewicz, J., 2019a. Effects of insect diets on the gastrointestinal tract health and growth performance of Siberian sturgeon (*Acipenser baerii* Brandt, 1869). BMC Veterinary Research 15: 348. https://doi.org/10.1186/s12917-019-2070-y

Józefiak, D., Józefiak, A., Kierończyk, B., Rawski, M., Świątkiewicz, S., Długosz, J. and Engberg, R.M., 2016. Insects – a natural nutrient source for poultry – a review. Annals of Animal Science 16. https://doi.org/10.1515/aoas-2016-0010

Kaiser, P. and Balic, A., 2015. The avian immune system. Chapter 17. In: Scanes, C.G (ed.)

Sturkie's avian physiology (6th Ed.). Academic Press, San Diego, CA, USA, pp. 403-418. https://doi.org/10.1016/B978-0-12-407160-5.00017-8

Kayama, H., Okumura, R. and Takeda, K., 2020. Interaction between the microbiota, epithelia, and immune cells in the intestine. Annual Review of Immunology 38: 23-48. https://doi.org/10.1146/annurev-immunol-070119-115104

Khan, S.H., 2018. Recent advances in role of insects as alternative protein source in poultry nutrition. Journal of Applied Animal Research 46: 1144-1157. https://doi.org/10.1080/097 12119.2018.1474743

Kierończyk, B., Rawski, M., Józefiak, A., Mazurkiewicz, J., Świątkiewicz, S., Siwek, M., Bednarczyk, M., Szumacher-Strabel, M., Cieślak, A., Benzertiha, A. and Józefiak, D. 2018. Effects of replacing soybean oil with selected insect fats on broiler. Animal Feed Science and Technology 240: 170-183. https://doi.org/10.1016/j.anifeedsci.2018.04.002

Klasing, K.C., 2005. Poultry Nutrition: A Comparative Approach. Journal of Applied Poultry Research 14: 426-436. https://doi.org/10.1093/japr/14.2.426

Kouřimská, L. and Adámková A., 2016. Nutritional and sensory quality of edible insects. NFS Journal 4: 22-26. https://doi.org/10.1016/j.nfs.2016.07.001

Koutsos, L., McComb A. and Finke, M., 2019. Insect composition and uses in animal feeding applications: A brief review. Annals of the Entomological Society of America 112: 544-551. https://doi.org/10.1093/aesa/saz033

Kumar, R., Kaur, N. and Kamilya, D., 2019. Chitin modulates immunity and resistance of *Labeo rohita* (Hamilton, 1822) against gill monogeneans. Aquaculture 498: 522-527. https://doi.org/10.1016/j.aquaculture.2018.09.013

Kylie, K., McEwen, S.A., Boerlin, P., Reid-Smith, M.J., Weese, J.S. and Turner, P.V., 2017. Prevalence of antimicrobial resistance in fecal *Escherichia coli* and *Salmonella enterica* in Canadian commercial meat, companion, laboratory, and shelter rabbits (*Oryctolagus cuniculus*) and its association with routine antimicrobial use in commercial meat rabbits. Preventive Veterinary Medicine 147: 53-57. https://doi.org/10.1016/j.prevetmed.2017.09.004.

Lee, C.G., Da Silva, C.A., Lee, J.Y., Hartl, D. and Elias, J.A., 2008. Chitin regulation of immune responses: an old molecule with new roles. Current Opinion in Immunology 20: 684-691. https://doi.org/10.1016/j.coi.2008.10.002

Lee, J., Kim, Y-M., Park, Y-K., Yang, Y-C., Jung, B-G. and Lee B-J., 2018. Black soldier fly (*Hermetia illucens*) larvae enhances immune activities and increases survivability of broiler chicks against experimental infection of *Salmonella* Gallinarum. Journal of Veterinaly Medical Science 80: 736-740. https://doi.org/10.1292/jvms.17-0236

Li, S., Ji, H., Zhang, B., Zhou, J. and Yu, H., 2017a. Defatted black soldier fly (*Hermetia illucens*) larvae meal in diets for juvenile Jian carp (*Cyprinus carpio* var. Jian): Growth performance, antioxidant enzyme activities, digestive enzyme activities, intestine and hepatopancreas histological structure. Aquaculture and Fisheries Management 477: 62-70. https://doi.org/10.1016/j.aquaculture.2017.04.015

Li, Y., Kortner, T.M., Chikwati, E.M., Belghit, I., Lock, E.-J. and Krogdahl, A., 2020. Total replacement of fish meal with black soldier fly (*Hermetia illucens*) larvae meal does not compromise the gut health of Atlantic salmon (*Salmo salar*). Aquaculture 520: 734967. https://doi.org/10.1016/j.aquaculture.2020.734967

Li, Y., Kortner, T.M., Chikwati, E.M., Munang'andu, H.M., Lock, E.-J. and Krogdahl, A., 2019. Gut health and vaccination response in pre-smolt Atlantic salmon (*Salmo salar*) fed black soldier fly (*Hermetia illucens*) larvae meal. Fish & Shellfish Immunology 86: 1106-1113. https://doi.org/10.1016/j.fsi.2018.12.057

Li, Z., Mao, R., Teng, D. Hao, Y., Chen, H., Wang, X., Wang, X, Yang, N and Wang, J. 2017b. Antibacterial and immunomodulatory activities of insect defensins-DLP2 and DLP4 against multidrug-resistant *Staphylococcus aureus*. Scientific Reports, 7: 12124. https://doi.org/10.1038/s41598-017-10839-4

Liland, N.S., Biancarosa, I., Araujo, P., Biemans, D., Bruckner, C.G., Waagbø, R., Torstensen, B.E. and Lock, E-J., 2017. Modulation of nutrient composition of black soldier fly (*Hermetia illucens*) larvae by feeding seaweed-enriched media. PLoS ONE 12:e0183188. https://doi.org/10.1371/journal.pone.0183188

Lock, E.R., Arsiwalla, T. and Waagbo, R., 2016. Insect larvae meal as an alternative source of nutrients in the diet of Atlantic salmon (*Salmo salar*) postsmolt. Aquaculture Nutrition 22: 1202-1213. https://doi.org/10.1111/anu.12343

Lock, E-J., Biancarosa, I. and Gasco, L., 2018. Insects as raw materials in compound feed for aquaculture. In: Halloran, A., Flore, R., Vantomme, P., Roos, N. (eds) Edible insects in sustainable food systems. Springer, Cham, Switzerland, pp 263-276. https://doi.org/10.1007/978-3-319-74011-9_16

Magalhaes, R., Sanchez-Lopez, A., Leal, R.S., Martinez-Llorens, S., Oliva-Teles, A. and Peres, H., 2017. Black soldier fly (*Hermetia illucens*) pre-pupae meal as a fish meal replacement in diets for European seabass (*Dicentrarchus labrax*). Aquaculture 476: 79-85. https://doi.org/10.1016/j.aquaculture.2017.04.021

Marono, S., Loponte, R., Lombardi, P., Vassalotti, G., Pero, M.E., Russo, F., Gasco, L., Parisi, G., Piccolo, G., Nizza, S., Di Meo, C., Attia, Y.A. and Bovera, F., 2017. Productive performance and blood profiles of laying hens fed Hermetia illucens larvae meal as total replacement of soybean meal from 24 to 45 weeks of age. Poultry Science 96, 1783-1790. https://doi.org/10.3382/ps/pew461

Massacci, F.R., Magistrali, C.F., Cucco, L., Curcio, L., Bano, L., Mangili, P., Scoccia, E.,Bisgaard, M., Aalbæk, B. and Christensen, H., 2018. Characterization of *Pasteurella multocida* involved in rabbit infections. Veterinary Microbiology 213: 66-72. https://doi.org/10.1016/j.vetmic.2017.11.023

Mbhele, F.G.T., Mnisi, C.M. and Mlambo, V.A., 2019. Nutritional evaluation of insect meal as a sustainable protein source for jumbo quails: physiological and meat quality responses. Sustainability 11: 6592. https://doi.org/10.3390/su11236592

McCarville, J.L., Chen, G.Y., Cuevas, V.D., Troha, K. and Ayres, J.S., 2020. Microbiota Metabolites in Health and Disease. Annual Review of Immunology 38: 147-170. https://doi.org/10.1146/annurev-immunol-071219-125715

Mikołajczak, Z., Rawski, M., Mazurkiewicz, J., Kierończyk, B. and Józefiak, D., 2020. The effect of hydrolyzed insect meals in sea trout fingerling (*Salmo trutta*) diets on growth performance, microbiota and biochemical blood parameters. Animals 10: 1031. https://doi.org/10.3390/ani10061031

Ming, J., Ye, J., Zhang, Y., Yang, X., Wu, C., Shao, X. and Liu, P., 2013. The influence of maggot meal and l-carnitine on growth, immunity, antioxidant indices and disease resistance of black carp (*Mylopharyngodon piceus*). Journal of the Chinese Cereals and Oils Association 28: 80-86

Moniello, G., Ariano, A., Panettieri, V., Tulli, F., Olivotto, I., Messina, M., Randazzo, B., Severino, L., Piccolo, G., Musco, N., Addeo, N.F., Hassoun, G. and Bovera, F., 2019. Intestinal Morphometry, Enzymatic and Microbial Activity in Laying Hens Fed Different Levels of a *Hermetia illucens* Larvae Meal and Toxic Elements Content of the Insect Meal and Diets. Animals 9: 86. https://doi.org/10.3390/ani9030086

Moon, H.J., Lee, S.Y., Kurata, S., Natori, S. and Lee, B.L., 1994. Purification and molecular cloning of cDNA for an inducible antibacterial protein from larvae of the coleopteran, *Tenebrio molitor*. Journal of Biochemistry 116: 53-58. https://doi.org/10.1093/oxfordjournals.jbchem.a124502

Mortensen, F.V., Langkilde, N.C., Joergensen, J.C. and Hessov, I., 1999. Short-chain fatty acids stimulate mucosal cell proliferation in the closed human rectum after Hartmann's procedure. International Journal of Colorectal Disease 14: 150-154. https://doi.org/10.1007/s003840050201

Motte, C., Rios, A., Lefebvre, T., Do, H., Henry, M. and Jintasataporn, O., 2019. Replacing fish meal with defatted insect meal (yellow mealworm *Tenebrio molitor*) improves the growth and immunity of pacific white shrimp (*Litopenaeus vannamei*). Animals 9: 258. https://doi.org/10.3390/ani9050258

Mylonakis, E., Podsiadlowski, L., Muhammed, M. and Vilcinskas, A., 2016. Diversity, evolution and medical application of insect antimicrobial peptides. Philosophical Transactions of the Royal Society B 371: 20150290. https://doi.org/10.1098/rstb.2015.0290

Ngo, D-H. and Kim, S-E., 2014. Antioxidant effects of chitin, chitosan, and their derivatives. Advances in Food and Nutrition Research 73: 15-31. https://doi.org/10.1016/B978-0-12-800268-1.00002-0

Nogales-Mérida, S., Gobbi, P., Józefiak, D., Mazurkiewicz, J., Dudek, K., Rawski, M., Kierończyk, B. and Józefiak, A., 2019. Insect meals in fish nutrition. Reviews in Aquaculture 11: 1080-1103. https://doi.org/10.1111/raq.12281

Noriega, J.A., Hortal, J., Azcárate, F.M., Berg, M., Bonada, N., Briones, M.J., Del Toro, I., Goulson, D., Ibañez, S., Landis, D.A., Moretti, M., Pott, S.G., Slade, E.M., Stout, J.C., Ulyshen, M.D., Wackers, F.L., Woodcock, B.A and Santos, A.M.C., 2018. Research trends in ecosystem services provided by insects. Basic and Applied Ecology 26: 8-23. https://doi.org/10.1016/j.baae.2017.09.006

Ohta, T., Ido, A., Kusano, K., Miura, C. and Miura, T., 2014. A Novel Polysaccharide in Insects Activates the Innate Immune System in Mouse Macrophage RAW264 Cells. PLoS ONE 9: e114823. https://doi.org/10.1371/journal.pone.0114823

Ohta, T., Kusano, K., Ido, A., Miura, C. and Miura, T., 2016. Silkrose: A novel acidic polysaccharide from the silkmoth that can stimulate the innate immune response. Carbohydrate Polymers 136: 995-1001. https://doi.org/10.1016/j.carbpol.2015.09.070

Oonincx, D.G.A.B., Van Broekhoven, S.V., Van Huis, A. and Van Loon, J.J.A., 2015. Feed conversion, survival and development, and composition of four insect species on diets composed of food by-products. PLoS ONE, 10: e0144601. https://doi.org/10.1371/journal.pone.0144601

Ostaszewska, T., Dabrowski, K., Kwasek, K., Verri, T., Kamaszewski, M., Sliwinski, J. and Napora-Rutkowski, L., 2011. Effects of various diet formulations (experimental and commercial) on the morphology of the liver and intestine of rainbow trout (*Oncorhynchus mykiss*) juveniles. Aquaculture Research 42: 1796-1806. https://doi.org/10.1111/j.1365-2109.2010.02779.x

Paini, D.R. Sheppard, A.W., Cook, D.C., De Barro, P.J., Worner, S.P and Thomas, M.B., 2016. Global threat to agriculture from invasive species. Proceedings of the National Academy of Sciences of the United States of America 113: 7575-7579. https://doi.org/10.1073/pnas.1602205113

Park S.I., Kim, J-W. and Yoe, S.M., 2015. Purification and characterization of novel antibacterial peptide from black soldier fly (*Hermetia illucens*) larvae. Developmental and Comparative Immunology 52: 98-106. https://doi.org/10.1016/j.dci.2015.04.018

Park, S.I., Chang, B.S. and Yoe. S.M., 2014. Detection of antimicrobial substances from larvae of the black soldier fly, *Hermetia illucens* (Diptera: Stratiomyidae). Entomological Research. 44: 58-64. https://doi.org/10.1111/1748-5967.12050

Pei, M.T., Yang, C., Yang, D.Q. and Yi, T.L., 2019. Effects of housefly maggot meal and earthworms on growth and immunity of the Asian swamp eel *Monopterus albus* (Zuiew). The Israeli Journal of Aquaculture – Bamidgeh 71: 8. http://hdl.handle.net/10524/62904

Qin, C., Zhang, Y., Liu, W., Xu, L., Yang, Y. and Zhou, Z., 2014. Effects of chito-oligosaccharides supplementation on growth performance, intestinal cytokine expression, autochthonous gut bacteria and disease resistance in hybrid tilapia *Oreochromis niloticus* ♀ x *Oreochromis aureus* ♂. Fish & Shellfish Immunolo!y 40: 267-274. https://doi.org/10.1016/j.fsi.2014.07.010

Rangaswamy, C.P., 2006. Physiology of digestion in fish and shrimp. In: Ali SA (ed.) Training manual on shrimp and fish nutrition and feed management. Central Institute of Brackish Water Aquaculture, Chennai, India, pp. 2-9.

Ravi, C., Jeyashree, A. and Renuka Devi, K., 2011. Antimicrobial peptides from insects: an overview. Research in Biotechnology 2: 1-7.

Rimoldi S., Gini, E., Iannini, F., Gasco, L.and Terova, G., 2019. The effects of dietary insect meal from *Hermetia illucens* prepupae on autochthonous gut microbiota of rainbow trout (*Oncorhynchus mykiss*). Animals 9: 143 https://doi.org/10.3390/ani9040143

Ringø, E., Zhou, Z., Olsen, R.E. and Song, S.K., 2012. Use of chitin and krill in aquaculture – the effect on gut microbiota and the immune system: a review. Aquaculture Nutrition 18: 117-131. https://doi.org/10.1111/j.1365-2095.2011.00919.x

Rodriguez-Calleja, J.M., Garcia-Lopez, I., Garcia Lopez, M.L., Santos, J.A. and Otero, A., 2006. Rabbit meats as a source of bacterial foodborne pathogens. Journal of Food Protection 69: 1106-1112. https://doi.org/10.4315/0362-028x-69.5.1106

Sankian, Z., Khosravi, S., Kim, Y.O, and Lee, S.M., 2018. Effects of dietary inclusion of yellow mealworm (*Tenebrio molitor*) meal on growth performance, feed utilization, body composition, plasma biochemical indices, selected immune parameters and antioxidant enzyme activities of mandarin fish (*Siniperca scherzeri*) juveniles. Aquaculture 496: 79-87. https://doi.org/10.1016/j.aquaculture.2018.07.012

Schiavone, A., Dabbou, S., De Marco, M., Cullere, M., Biasato, I., Biasibetti, E., Capucchio, M.T., Bergagna, S., Dezzutto, D., Meneguz, M., Gai, F., Dalle Zotte A. and Gasco, L., 2018. Black soldier fly larva fat inclusion in finisher broiler chicken diet as an alternative fat source. Animal 12: 2032-2039. https://doi.org/10.1017/S1751731117003743.

Schiavone, A., Dabbou, S., Petracci, M., Zampiga, M., Sirri, F., Biasato, I., Gai, F. and Gasco, L. 2019. Black soldier fly defatted meal as a dietary protein source for broiler chickens: Effects on carcass traits, breast meat quality and safety. Animal 13: 2397-2405. https://doi.org/10.1017/S1751731119000685

Schuhmann, B., Seitz, V., Vilcinskas, A. and Podsiadlowski, L., 2003. Cloning and expression of gallerimycin, an antifungal peptide expressed in immune response of greater wax moth larvae, *Galleria mellonella*. Archives of Insect Biochemistry and Physiology 53: 125-133. https://doi.org/10.1002/arch.10091

Shanthi Mari, L.S., Jagruthi, C., Anbazahan, S.M., Yogeshwari, G., Thirumurugan, R., Arockiaraj, J., Mariappan, P., Balasundaram, C. and Harikrishnan, R., 2014. Protective effect of chitin and chitosan enriched diets on immunity and disease resistance in *Cirrhina mrigala* against *Aphanomyces invadans*. Fish & Shellfish Immunology 39: 378-385. https://doi.org/10.1016/j.fsi.2014.05.027

Shin, C.-S., Kim, D.-Y. and Shin, W.-S., 2019. Characterization of chitosan extracted from mealworm beetle (*Tenebrio molitor*, Zophobasmorio) and rhinoceros beetle (*Allomyrina dichotoma*) and their antibacterial activities. International Journal of Biological Macromolecules 125: 72-77. https://doi.org/10.1016/j.ijbiomac.2018.11.242

Smetana, S., Schmitt, E. and Mathys, A., 2019. Sustainable use of *Hermetia illucens* insect biomass for feed and food: Attributional and consequential life cycle assessment. Resources, Conservation and Recycling 144: 285-296. https://doi.org/10.1016/j.resconrec.2019.01.042

Sogari, G., Amato, M., Biasato, I., Chiesa, S. and Gasco, L. 2019. The potential role of insects as feed: a multi-perspective review. Animals 9: 119. https://doi.org/10.3390/ani9040119

Song, S.G., Chi, S.Y., Tan, B.P., Liang, G.L., Lu, B.Q., Dong, X.H., Yang, Q.H., Liu, H.Y. and Zhang, S., 2018. Effects of fishmeal replacement by *Tenebrio molitor* meal on growth performance, antioxidant enzyme activities and disease resistance of the juvenile pearl gentian grouper (*Epinephelus lanceolatus* male x *Epinephelus fuscoguttatus* female). Aquaculture Research 49: 2210-2217. https://doi.org/10.1111/are.13677

Spranghers, T., Joris, M., Vrancx, J., Ovyn, A., Eeckhout, M., De Clercq, P. and De Smet, S., 2018. Gut antimicrobial effects and nutritional value of black soldier fly (*Hermetia illucens* L.) prepupae for weaned piglets. Animal Feed Science and Technology 235: 33-42. https://doi.org/10.1016/j.anifeedsci.2017.08.012

Stenberg, O.K., Holen, E., Piemontese, L., Liland, N.S., Lock, E.-J., Espe, M. and Belghit, I., 2019. Effect of dietary replacement of fish meal with insect meal on *in vitro* bacterial and viral induced gene response in Atlantic salmon (*Salmo salar*) head kidney leukocytes. Fish & Shellfish Immunology 91: 223-232. https://doi.org/10.1016/j.fsi.2019.05.042

St-Hilaire, S., Cranfill, K., McGuire, M.A., Mosley, E.E., Tomberlin, J.K., Newton, L., Sealey, W., Sheppard, C., Irving, S., 2007. Fish offal recycling by the black soldier fly produces a food-stuff high in omega-3 fatty acids. Journal of the World Aquaculture Society 38: 309-313.

Su, J., Gong, Y., Cao, S., Lu, F., Han, D., Liu, H., Jin, J., Yang, Y., Zhu, X. and Xie, S., 2017. Effects of dietary *Tenebrio molitor* meal on the growth performance, immune response and disease resistance of yellow catfish (*Pelteobagrus fulvidraco*). Fish & Shellfish Immunology 69: 59-66. https://doi.org/10.1016/j.fsi.2017.08.008

Sun, H-X., Chen, L-Q., Zhang, J. and Chen, F-Y., 2014. Anti-tumor and immunomodulatory activity of peptide fraction from *Musca domestica*. Journal of Ethnopharmacology 153: 831-839. https://doi.org/10.1016/j.jep.2014.03.052

Suresh, A., Praveenkumar, R., Thangaraj, R., Oscar, F.L., Baldev, E., Dhanasekaran, D. and Thajuddin, N., 2014. Microalgal fatty acid methyl ester a new source of bioactive compounds with antimicrobial activity. Asian Pacific Journal of Tropical Disease 4, Suppl. 2: S979-S984. https://doi.org/10.1016/S2222-1808(14)60769-6

Sypniewski, J., Kierończyk, B., Benzertiha, A., Mikołajczak, Z., Pruszyńska-Oszmałek, E., Kołodziejski, P., Sassek, M., Rawski, M., Czekała, W. and Józefiak, D., 2020. Replacement of soybean oil by *Hermetia illucens* fat in turkey nutrition: effect on performance, digestibility, microbial community, immune and physiological status and final product quality. British Poultry Science 61: 294-302. https://doi.org/10.1080/00071668.2020.1716302

Terova, G., Rimoldi, S., Ascione, C., Gini, E., Ceccotti, C. and Gasco, L., 2019. Rainbow trout (*Oncorhynchus mykiss*) gut microbiota is modulated by insect meal from *Hermetia illucens* prepupae in the diet. Reviews in Fish Biology and Fisheries 29: 465-486. https://doi.org/10.1007/s11160-019-09558-y

Tian, Z., Feng, Q., Sun, H., Liao, Y., Du, L., Yang, R., Li, X., Yang, Y. and Xia, Q., 2018. Isolation and purification of active antimicrobial peptides from *Hermetia illucens* L., and its effects on CNE2 cells. bioRxiv 353367. https://doi.org/10.1101/353367

Udayangani, R.M.C., Dananjaya, S.H.S., Nikapitiya, C., Heo, G.-J., Lee, J. and De Zoysa, M., 2017. Metagenomics analysis of gut microbiota and immune modulation in zebrafish (*Danio rerio*) fed chitosan silver nanocomposites. Fish & Shellfish Immunoloy 66: 173-184. https://doi.org/10.1016/j.fsi.2017.05.018

Urbanek, A., Szadziewski, R., Stepnowski, P., Boros-Majewska, J., Gabriel, I., Dawgul, M., Kamysz, W., Sonsowaska, D. and Gołębiowski, M., 2012. Composition and antimicrobial activity of fatty acids detected in the hygroscopic secretion collected from the secretory setae of larvae of the biting midge *Forcipomyia nigra* (Diptera: Ceratopogonidae). Journal of Insect Physiology, 58: 1265-1276. https://doi.org/10.1016/j.jinsphys.2012.06.014

Van Huis, A. 2017. Did early humans consume insects? Journal of Insects as Food and Feed 3: 161-163. https://doi.org/10.3920/jiff2017.x006

Van Huis, A., 2020. Insects as food and feed, a new emerging agricultural sector: a review. Journal of Insects as Food and Feed 6: 27-44. https://doi.org/10.3920/jiff2019.0017

Vargas-Abúndez, A., Randazzo, B., Foddai, M., Sanchini, L., Truzzi, C., Giorgini, E., Gasco, L. and Olivotto, I., 2019. Insect meal based diets for clownfish: Biometric, histological, spectroscopic, biochemical and molecular implications. Aquaculture 498: 1-11. https://doi.org/10.1016/J.aquaculture.2018.08.018

Vogel, H., Muller, A., Heckel, D.G., Gutzeit, H. and Vilcinskas, A., 2018. Nutritional immunology: Diversification and diet-dependent expression of antimicrobial peptides in the
black soldier fly *Hermetia illucens*. Developmental & Comparative Immunology 78: 141-
148. https://doi.org/10.1016/j.dci.2017.09.008

Wang, G., Peng, K., Hu, J., Yi, C., Chen, X., Wu, H. and Huang, Y., 2019. Evaluation of defatted
black soldier fly (*Hermetia illucens* L.) larvae meal as an alternative protein ingredient
for juvenile Japanese seabass (*Lateolabrax japonicus*) diets. Aquaculture 507: 144-154.
https://doi.org/10.1016/j.aquaculture.2019.04.023

Wen, L.-F. and He, J.-G., 2012. Dose-response effects of an antimicrobial peptide, a cecropin
hybrid, on growth performance, nutrient utilisation, bacterial counts in the digesta and
intestinal morphology in broilers. British Journal of Nutrition 108: 1756-1763. https://doi.
org/10.1017/S0007114511007240

World Health Organization (WHO), 2015. World Malaria Report 2015. WHO, Geneva, Switzerland, 280 pp. Available at https://www.who.int/malaria/publications/world-malar-
ia-report-2015/report/en/.

Wu, Q., Patočka, J. and Kuča, K., 2018. Insect antimicrobial peptides, a mini review. Toxins 10:
461. https://doi.org/10.3390/toxins10110461

Xiang, J., Qin, L., Zhao, D., Xiong, F., Wang, G., Zou, H., Li, W., Li, M., Song, K. and Wu, S., 2019.
Growth performance, immunity and intestinal microbiota of swamp eel (*Monopterus
albus*) fed a diet supplemented with house fly larvae (*Musca domestica*). Aquaculture
Nutrition 26: 693-704. https://doi.org/10.1111/anu.13029

Xiao, X., Jin, P., Zheng, L., Cai, M., Yu, Z., Yu, J. and Zhang, J., 2018. Effects of black soldier fly
(*Hermetia illucens*) larvae mealprotein as a fishmeal replacement on the growth and immune index of yellow catfish (*Pelteobagrus fulvidraco*). Aquaculture Research 49: 1569-
1577. https://doi.org/10.1111/are.13611

Yu, M., Li, Z., Chen, W., Rong, T., Wang, G. and Ma, X. 2019. *Hermetia illucens* larvae as a potential dietary protein source altered the microbiota and modulated mucosal immune
status in the colon of finishing pigs. Journal of Animal Science and Biotechnology 10: 50.
https://doi.org/10.1186/s40104-019-0358-1

Yu, M., Li, Z., Chen, W., Rong, T., Wang, G., Wang, F. and Ma, X. 2020b. Evaluation of full-
fat *Hermetia illucens* larvae meal as a fishmeal replacement for weanling piglets: Effects
on the growth performance, apparent nutrient digestibility, blood parameters and gut
morphology. Animal Feed Science and Technology 264: 114431. https://doi.org/10.1016/j.
anifeedsci.2020.114431

Yu, M., Li, Z., Chen, W., Wang, G., Rong, T., Liu, Z., Wang, F. and Ma, X. 2020a. *Hermetia illucens* larvae as a fishmeal replacement alters intestinal specific bacterial populations and
immune homeostasis in weanling piglets. Journal of Animal Science 98: akz395. https://
doi.org/10.1093/jas/skz395

Zarantoniello, M., Bruni, L., Randazzo, B., Vargas, A., Gioacchini, G., Truzzi, C., Annibaldi,
A., Riolo, P., Parisi, G., Cardinaletti, F., Tulli, F. and Olivotto, I., 2018. Partial dietary inclusion of *Hermetia illucens* (black soldier fly) full-fat prepupae in zebrafish feed: biometric,
histological, biochemical, and molecular implications. Zebrafish 15: 519-532. https://doi.
org/10.1089/zeb.2018.1596

Zhang, L-J. and Gallo, R.L., 2016. Antimicrobial peptides. Current Biology 26: R14-9. https://
    doi.org/10.1016/j.cub.2015.11.017

Zhou, Z., Karlsen, O., He, S., Olsen, R.E., Yao, B. and Ringo, E., 2013. The effect of dietary chi-
    tin on the autochthonous gut bacteria of Atlantic cod (*Gadus morhua* L.). Aquaculture
    Research 44: 1889-1900. https://doi.org/10.1111/j.1365-2109.2012.03194.x

A meta-analysis on the nutritional value of insects in aquafeeds

N.S. Liland[1]*, P. Araujo[1], X.X. Xu[1,2], E.-J. Lock[1,3], G. Radhakrishnan[1,3], A.J.P. Prabhu[1] and I. Belghit[1]**

[1]Institute of Marine Research, P.O. Box 1870 Nordnes, 5817 Bergen, Norway; [2]College of Animal Science and Technology, Northwest A&F University, Yangling 712100, Shaanxi, China, P.R.; [3]Department of Biology, University of Bergen, Postbox 7803, 5020 Bergen, Norway; *nina.liland@hi.no; **ikram.belghit@hi.no

Abstract

A major challenge for development of sustainable aquafeeds is its dependence on fish meal and fish oil. Similarly, it is unwanted to include more plant ingredients which adds more pressure on resources like arable land, freshwater and fertilisers. New ingredients that do not require these resources but rather refine and valorise organic side streams, like insects, are being developed. Increasing evidence indicates that using insect ingredients in aquafeeds are a sustainable alternative and considerable progress has been made on this topic in the past years. The aim of this chapter is to present a comprehensive and systematic analysis of the data available on the impact of insects in aquafeeds. Systematic search, collection and selection of relevant literature from databases such as Web of Science and NCBI was performed. The literature search enabled 91 scientific papers from peer-reviewed journals, comprising a dataset of 415 experimental diets, including 35 different aquatic species and 14 insect species to be included in this meta-analysis, covering what we consider a close to complete representation of credible publications on this topic. Information on aquatic species, insect species, dietary composition (amino acids, fatty acids, proximate composition) and performance outputs (growth performance indicators and nutrient digestibility) were included in the construction of the dataset. Regression models and principal component analyses were performed on the meta-data. The results from the meta-analysis revealed a great degree of variation in the maximum threshold for insect inclusion in aquafeeds (from 4 to 37%) based on subgroups of trophic level of aquatic species, insect species used, statistical method and the output parameter. Overall, a maximum threshold of 25-30% inclusion of insects in aquafeeds for uncompromised performance is suggested. Reduction in protein digestibility, imbalanced amino acid profile and increasing levels of saturated fatty acid were identified as major factors limiting higher inclusion of insects in aquafeeds.

Keywords

aquafeed – aquaculture – black soldier fly – mealworm – sustainable feed

1   Introduction

By 2050, the world population is expected to reach 9.7 billion (UN, 2019), and the demand for feed and food crops is expected to increase to 25-70% above today's levels. Aquaculture is currently producing more than half of the seafood destined for human consumption and is considered by the FAO as having the capacity to grow and become an even more important source of animal protein for the expanding world population (FAO, 2020). Around 70% of the total aquaculture production today is made up of fed species, and the aquafeed industry must therefore keep up with this growth. Many of the traditional non-fed forms of finfish aquaculture are also being intensified, often resulting in the use of formulated feeds, thus putting a further pressure on the production of aquaculture feeds. With a lot of pressure on land- and water-use by agriculture, and an already maximised output from most fisheries, the aquafeed industry is looking for new sources for feed ingredients with a smaller impact on the global environment and resource use.

Due to a limited availability of fish meal (FM) and fish oil (FO), there has already been a considerable reduction in the inclusion level of such marine ingredients in aquafeeds during the last two decades. A large amount of research lies behind this shift in dietary composition, which has focused on the nutritional requirements of each fish species (Glencross, 2020; Glencross *et al.*, 2007; Turchini *et al.*, 2019). The main ingredients in most aquafeeds nowadays are from terrestrial sources, dominated by plant-derived products such as soy protein concentrates and different vegetables oils, as well as animal by-products, depending on the region and fish species (FAO, 2011). While the replacement of FM and FO with terrestrial ingredients has allowed the sector to grow, some of these aquafeed ingredients could be used directly for human consumption. Additionally, by sourcing feedstuffs from agriculture, there is a risk of contributing to the global deforestation and a non-sustainable resource consumption (Wilfart *et al.*, 2016). The current research is therefore focused on finding novel sources of ingredients to replace both traditional marine and plant-based ingredients. Most research has considered animal by-products, microalgae, single cell proteins, blue mussels, krill and insects (Hua *et al.*, 2019; Sørensen *et al.*, 2012).

The first report of rearing insect dates back to thousands of years ago when the cultivation of silkworms (*Bombyx mori*) for silk production was initiated in China. Insects have also been an important protein source for people in many cultures (Evans *et al.*, 2015). The earliest published research on the use of insects to convert animal waste into high quality protein to then be used in animal feeds, appeared in the 1970s (Hale, 1973; Newton *et al.*, 1977). In these trials, the authors included black soldier fly (BSF) larvae (*Hermetia illucens*), reared on cattle faeces and urine slurry, as a dietary supplement for chicken and swine. Later, the larvae of this fly species was tested as a feed ingredient in the diets of channel catfish and tilapia (Bondari and Sheppard, 1981). A more industrialised farming of insects for feed purposes started in Western countries in the last decade, accompanied by a large amount of research on the use of this feed ingredient in animal diets (Van Huis, 2020a,b). Since then, the interest in using insects in fish nutrition has increased exponentially, leaving a large amount of data from scientific studies (e.g. Arru *et al.*, 2019; Gasco *et al.*, 2019; Henry *et al.*, 2015; Hua, 2021; Nogales-Mérida *et al.*, 2019). Insect meals (IM) have a great potential for supplying the protein required for future aquafeed production. There has therefore been a strong support for research and industries in connection with available insect processing, upscaling production and cost (Hua *et al.*, 2019; IPIFF, 2019a,b, 2020a). There is today an increase in companies focusing on the farming of insect species on a large scale (Cadinu *et al.*, 2020; IPIFF, 2019a,b, 2020b). In 2017, the European Union authorised the use of processed IMs from seven insect species (BSF, common housefly, yellow mealworm, lesser mealworm, house cricket, banded cricket and field cricket – EU, 2017) in aquafeeds. One year later, the first insect-fed rainbow trout was seen in the French market. The aquaculture sector consumed more than 50% of the total European insect protein production in 2019, which was approximatively 5,000 tons (IPIFF, 2019b). Both the insect production and the use of insect products in aquafeeds are predicted to rise (IPIFF, 2019b). However, despite the great progress in research on the topic and large growth in insect production, insects are currently not produced in sufficient volumes to be used more extensively in commercial aquafeed production.

   Scientific literature concerning the nutritional properties of different insect species and the use of insect-based ingredients in aquafeed have been extensively reviewed (Gasco *et al.*, 2019; Henry *et al.*, 2015; Hua, 2021; Nogales-Merida *et al.*, 2019). Dietary inclusion of insect protein meal and/or insect oil in aquaculture diets without negative effect on growth performances have been successfully demonstrated in some feeding trials (Belghit *et al.*, 2019a; Bruni *et al.*, 2020; Fawole *et al.*, 2020; Li *et al.*, 2017; Magalhães *et al.*, 2017; Wang *et al.*, 2019). However, in other studies, growth performance and feed utilisation were repressed by using insects in the diets (Gasco *et al.*, 2016; Kroeckel *et al.*, 2012; Reyes *et al.*, 2020). The variations in growth and feed utilisation seen between the studies might be due to differences in tolerance level of insect ingredients between aquatic species and life stages. Logically, the quality of the insect ingredient utilised will also play a large role, as the quality of this ingredient varies due the insect species, nutritional profile, use of different rearing substrate and the processing methods. A vital part of how well insect ingredients will perform in a compound diet is the production and composition of the diet. The living conditions for the aquatic species are also important, so factors such as water quality and general care for the aquatic species can also affect the results. The large amount of research done on the use of insects as aquafeed ingredients, makes it hard to see the general tendencies concerning what might be more or less successful when using insects as part of a compound feed. One way of achieving a comprehensive inter-study comparison of the available data is to perform a meta-analysis. The quantitative evaluation of input (e.g. diet composition) and output (e.g. growth) by different mathematical and statistical approaches (e.g. univariate/multivariate analysis) might allow for a better understanding of the most influential parameters affecting the performance of aquatic species fed diets containing insects.

Meta-analysis studies have gained recognition in many scientific fields, including the field of fish nutrition, as a useful tool to obtain more knowledge of the existing published data. This approach has been used to analyse data on dietary requirements for amino acids (AA) and minerals in fish (Kaushik and Seiliez, 2010, Prabhu *et al.*, 2013; 2016), and for evaluating the effects of FM replacement with alternative protein sources in fish diets (Hua, 2021; Hua and Bureau, 2012). Recently, Hua (2021) employed a meta-analysis to quantify the relationship between FM replacement with IM in aquafeeds and growth performance (specific growth rate – SGR) of different aquatic species. In this study, based on 33 published trials, the author concluded that low or moderate replacement of FM with IM (below 29% of diet) had no effects on fish growth performance. The author highlights the importance for further research to elucidate the different biological and dietary factors that influence tolerance levels of IM in the diets of farmed aquatic species.

This chapter aims to summarise the current knowledge on the use of insects in aquaculture feeds by performing a meta-analysis. We wanted to search for nutritional factors of diets containing insect ingredients that affect the performance of the aquatic species. In addition, we wanted to give a comprehensive overview over existing studies on the use of insects in aquafeeds. We have gathered detailed information on diet composition (AA, fatty acids (FA), and proximate composition) as well as several growth performance parameters (such as SGR and feed conversion ratio (FCR)). This information was used for a meta-analysis to search for patterns in what factors cause changes in performance. In addition, we provide the complete dataset which can be a useful tool for future studies, both for planning and for comparing results.

2   Methods

2.1   *Literature search*
A systematic literature search was conducted through the electronic databases Web of Science and NCBI in October 2020, using a combination of search terms (insect OR dietary insect OR *H. illucens* OR *Tenebrio molitor*) AND (fish OR fish nutrition OR fish growth). Three peer-reviewed articles in Chinese were also included in the meta-analysis. A first screening step was based on the title and abstract of each

article. This procedure allowed for the exclusion of irrelevant studies according to the criteria listed below: (1) no data reported on the growth or on the feed intake; (2) study on ornamental aquatic species; and (3) using live insects or dietary frass from insect species in fish diets.

The initial search generated 165 articles. Among these, 12 duplicates and 62 studies not fitting the criteria were removed, leaving 91 relevant articles for the meta-analysis.

2.2   *Selection of parameters to include in meta-analyses*

The data selection was based on the most reported parameters in published studies. We also chose parameters that we knew are generally analysed by the same or similar methods, to allow for better comparison. The nature of data collected is summarised in Table 1. It must be mentioned that some data not intended for meta-analysis (e.g. crude protein and lipid of the IMs and their AA composition) were also collected and included as a separate supplementary table (Table S1). The complete data set used in the meta-analysis study is also presented in Table S1.

2.3   *Aquaculture species included in the meta-analysis*

The meta-analysis included 91 studies in peer-reviewed manuscripts where insect meals or oils were used in diets for aquatic species. The studies were performed on 35 different species of fish and shellfish. The main species cited were Rainbow trout (16 papers) and Nile tilapia (9 papers). A complete overview of the species used in the studies is found in Table S2. The feeding ranged in days from 12 to 163 with a median at 56 days (Table S1). The trial lasting 12 days was not a growth trial and only recorded for nutrient digestibility. There was a large range in the size of aquatic species used in the trials (1 g - 3.7 kg final body weight), although most of the studies were done on juvenile aquatic species. The median of final body weights was at 62 g and 68% of the 378 dietary groups were from aquatic species with a final body weight lower than 100 g. We checked the data reported about trial execution in each trial and found that most trials were performed according to physical and nutritional needs of each species. Some exceptions were found, such as a trial with European seabass performed in temperatures in the upper tolerance range of this species (27-28 °C, Abdel-tawwab *et al.*, 2020), and yellow catfish (up to 32 °C, Hu *et al.*, 2017), but none where they exceeded the tolerance range (Table S1).

2.4   *Insect species and products included in the meta-analysis*

Most of the experimental trials used BSF as a feed ingredient, representing 46% of the diet groups of the full data, followed by yellow mealworm (MW, *T. molitor*), including superworm (*Zophobas morio*), with 32% of the data, silkworm (*B. mori*) accounting for 13% of the data, housefly (*Musca domestica*) with 8% and the remaining 1% of the studies being done on tropical house cricket (*Gryllodes sigillatus*), field cricket (*Gryllus assimilis*), Turkestan cockroach (*Blatta lateralis*), speckled

TABLE 1   An overview of the extracted data from the selected 91 papers

	Input data – diets
Input data – aquatic species and trials	– Aquatic species
	– Developmental stage of aquatic species
	– Length of trial (days)
Input data – insect ingredients	– Insect species
	– Insect developmental stage (larvae, pre-pupae, pupae, adult insect)
	– Substrate used to grow insects
	– Type of product/processing
	– Insect proximate composition (crude protein, ether extract, ash, dry matter as % of insect product)
	– Insect content of essential and semi-essential AAs (Arg, His, Ile, Leu, Lys, Met, Phe, Thr, Trp, Val, Ala, Glu, Tau as % of protein)
Input data – dietary composition	– Diet proximate composition (crude protein, ether extract, ash, dry matter as % of diet)
	– Inclusion level of insect product in diet (% of diet)
	– Dietary essential and semi-essential AAs (Arg, His, Ile, Leu, Lys, Met, Phe, Thr, Trp, Val, Ala, Glu, Tau as % of diet)
	– Dietary content of selected FA (lauric acid, EPA, DHA, sum saturated FA, sum polyunsaturated FA as % of total FA and ratio n-3/n-6 FA)
Output data – aquatic species performance	– Initial body weight in grams
	– Final body weight in grams and normalised per study (control set to 100%)
	– Feed conversion ratio[1]
	– Digestibility of crude protein and lipid
	– Specific growth rate[2]

AA = amino acids; DHA = docosahexaenoic acid; EPA = eicosapentaenoic acid; FA = fatty acid.
[1] Feed conversion ratio = feed intake (g) / growth (g).
[2] Specific growth rate = (ln(final weight in grams) − ln(initial weight in grams)) / t (in days) × 100.

cockroach (*Nauphoeta cinerea*), Madagascar hissing cockroach (*Gromphadorhina portentosa*), green bottle fly (*Lucilia sericata* Meigen), shea caterpillar (*Cirina butyrospermi*), nonbiting midges (chironomid) and grasshoppers (locust). Insect meals and oils included were from in total 14 different species (Table S3). The collected data includes studies done with 105 different dried IMs, 7 different insect oils, 3 protein hydrolysates, 1 fermented product and 2 products using wet, ground insect products. Most of the studies are performed on dry insect meals, either used as is or with different degrees of defatting. A complete overview of the insect species and citing papers can be found in Table S3.

2.5   Composition of insect meals included in the meta-analysis

The composition of IMs used for formulating the diets included in this meta-analysis is presented in Table 2. There was quite a large range in protein and lipid content, reflecting the different degrees of defatting performed on the insect ingredients. The average protein content was 52±12% of the meal, ranging from 24 to 83% of dry matter. The content of the typical limiting AAs lysine, threonine, methionine and tryptophane also varied a lot between the different meals (Table S1). The content of AAs in the insect meals is reported as % of protein to reflect the quality of the protein in the meal, rather than the quantity, and to be able to compare between meals with large differences in protein content. Some insect meals contained essential AAs below the requirement of most aquatic species. Arginine was, for example, low in some full-fat BSF meals (Fabrikov *et al.*, 2020; Józefiak *et al.*, 2019b), while the content of this AA was higher in defatted mealworm and aquatic insects (Basto *et al.*, 2020; Roncarati *et al.*, 2019). The variations in AA composition, even within the same insect species and similar processing of the meal, emphasises the importance of analysing AA composition before the production of any insect-containing feed.

2.6   Composition of experimental diets included in the meta-analysis

We collected information on the proximate composition of diets used in the reported trials and compared to requirements of the aquatic species in question, when this was available (NRC, 2011). The average protein content of the experimental diets was 47±5% for salmonids, 46±5% for European seabass/seabream, 42±3% for catfish, 38±2% for shrimp, 36±3% for carp and 35±6% for tilapia. Dietary protein contents were all within protein requirements of respective species. Almost all the included studies fulfilled the requirements for the essential AA, except some studies where the levels of iso-leucine (Muin *et al.*, 2017; Sánchez-Muros *et al.*, 2015) and

TABLE 2   The nutrient profile of the insect meals used for diet formulation in the studies included in the meta-analysis

Parameters	Range values (min-max)	Median	S.D.	Data points
Proximate composition (% of insect meal)				
Dry matter	79-98	93	3.5	58
Crude protein	24-83	52	12	87
Lipid	1.6-35	22	9.4	86
Concentration of selected essential amino acids (% of crude protein)				
Lysine	2.3-9.2	5.2	1.5	47
Threonine	1.5-6.4	3.5	1.0	47
Methionine	0.6-3.9	1.4	0.7	47
Tryptophane	0.7-2.3	1.3	0.4	10

leucine (Devic *et al.*, 2017) were below requirement for Nile tilapia (NRC, 2011). As seen from the references, the dietary deficiencies of AA were mostly found in Nile tilapia (Devic *et al.*, 2017; Muin *et al.*, 2017; Sanchez-Muros *et al.*, 2015). Most trials therefore reported diets with sufficient AA contents.

The dietary lipid content also varied with aquatic species and were in according to general acceptance and requirement of lipids in the respective species. The average lipid contents of the experimental diets were (in order of highest to lowest) 19±5% for salmonids, 15±5% for European seabass/seabream, 13±5% for catfish, 11±5% for tilapia, 9±5% for shrimp and 7±5% for carp. FA composition varied according to the use of FO and plant oils. The inclusion of the fat fraction of BSF meals led to an introduction of the medium chained FA lauric acid ($C_{12:0}$) to the diets, which was found in levels as high as 33% of total FA (Kroeckel *et al.*, 2012).

2.7   *Data analyses and statistics*

2.7.1   Distribution of data
The entries for control and insect inclusion in diets in Table S1 were labelled as -1 and 1, respectively. The distributions in terms of FCR, SGR, digestibility of crude protein (DCP) and digestibility of crude lipid (DCL) were studied by setting the control diet as the reference distribution. The responses (FCR, SGR, DCP and DPL) were standardised by subtracting the mean (μ) from the control diets and dividing by its associated standard deviation (σ).The results (Figure S1) demonstrated that the distributions for the control and insect inclusion diets overlapped according to a normal distribution ($\mu = 0$ and $\sigma = 1$). Few responses for the control diets were outside the 99.9% area of the normal curve. For example, two FCR values and one SGR value were outside the range $\pm 3\sigma$ (Figure S1). Although most of the responses in the insect inclusion diets overlapped with the control diets, the former diets showed that some FCR and SGR responses differed from the latter diet, probably due to the influence of the percentage of insect inclusion in the diets, which is a parameter studied in more detail below.

In this meta-analysis, few studies (~12%) reported four performance parameters (FCR, SGR, DCP and DCL) simultaneously, while the 70% of the studies were focused on reporting two performance parameters simultaneously, for instance FCR/SGR (65%), DCP/DCL (3%) or SGR/DCP (1%).

2.7.2   Regression analysis
The meta-data collected, namely normalised final weight (NFW), FCR, SGR, DCP and DCL, were subjected to regression analyses. Linear regression was performed using the simple model $Y = bX + a$, where Y is the output, x is the insect inclusion level in the diet, b is the slope and a is the intercept. Broken-line analyses with two slopes was used for non-linear regression. The broken-line regression used was: $Y_1 = a_1 + b_1 * X$; $YatXbp = b_1 * Xbp + a_1$; $Y_2 = YatXbp + b_2 * (X - Xbp)$; $Y = IF(X < Xbp, Y_1,$

Y2) as described in Prabhu *et al.* (2013). In the above model, X is the insect inclusion level in the diet as percent; Y1, Y2 are output measures; b1, b2 are slopes of the two regression segments; a1, a2, the intercepts and Xbp is the breakpoint which provides the estimated insect inclusion level at which the inflection in response was observed. The analysis was performed on three different constraint levels: no constraints, medium constraint $-0.2 < b_1 < 0$ and high constraint $-0.1 < b_1 < 0$. In addition to the broken line regression, simple quadratic regression was also tested. Regression analyses was performed on the entire dataset for each of the specified output measures and on sub-grouped datasets based on (1) trophic groups; (2) insect species groups and (3) AA supplemented vs non-supplemented studies. For any dataset, both linear and non-linear regression analyses was performed, and the best fit model was chosen to describe the trend and parameter estimates such as slope, breakpoint or intercept. All the regression analyses were performed using GraphPad Prism® (version 8.03 for Windows GraphPad Software, La Jolla, CA, USA).

Aquatic species group: The trophic level (TL) expresses the interactions of organisms with each other within their food web. The TL estimates of the aquatic species were taken from the Fish Base data (FishBase, www.fishbase.org; Froese and Pauly, 2007). The TL of fish generally range from 2 (e.g. the detritus feeding blue-barred parrotfish) to 4.7 (e.g. the piscivorous striped marlin), cephalopods from 3.2 (e.g. the planktivorous Patagonian squid) to 4.5 (e.g. the piscivorous greater hooked squid). Thus, the different aquatic species in the current study were categorised into two groups according to their TL, TL value above or below 3.5. The first category of aquatic species (TL > 3.5) included blackspot seabream, Chum salmon, tench, seabream, African catfish, common catfish, largemouth bass, Japanese seabass, Asian seabass, rainbow trout, meagre, turbot, Atlantic salmon, red seabream, Olive flounder, gilthead seabream, black sea salmon, Eurasian perch and rockfish. The second category of aquatic species (TL < 3.5) comprised Nile tilapia, rohu carp, gibel carp, field eel, rice field eel, jian carp, mirror carp, Siberian sturgeon, seatrout, yellow catfish, dark barbell catfish, European seabass and climbing perch. The third group included crustaceans which comprised Pacific white shrimp, giant freshwater prawn and marron. The pearl gentian grouper was excluded in this analysis as the TL of this species was not provided in Fish Base data.

Insect species group: The full data set was also divided into two subgroups based on the most studied insect species, being: (1) BSF (46% of the data); and (2) MW (32% of the data).

Amino acid supplementation: Two data sets were compiled using the full data set, each containing closely related aquatic species. The trials were then split into two: (1) trials where diets were supplemented with AAs (lysine and/or methionine); or (2) trials with non-supplemented diets. The only groups that had enough data for a comparison of supplemented vs non-supplemented were trout (sea trout and rainbow trout combined, 9 trials with supplemented diets and 11 trials with non-supplemented diets) and the carnivore marine fish seabream and seabass combined

(5 studies with supplemented diets and 7 trials with non-supplemented diets). The dietary inclusion of IM was plotted against the main output factors: normalised final body weight, SGR and FCR. Regression analyses were performed as described above to find the best fit line and to discover any significant relationships between inclusion level of insects in diets supplemented or not with AA. We also used a t-test to see if the two groups showed any different response. Outliers, clearly a result of either typographical errors or trials not being growth experiments, were not included (SGR in rainbow trout trials, Jeong *et al.*, 2020; Gelincek and Yamaner, 2020).

2.7.3   Principal component analysis

The parameters summarised in Table S1 were arranged as a *m×n* data matrix, where *m* represented the independent variables (e.g. aquatic species, percentage of inclusion, fish subgroups, etc) and *n* represented the measured responses (e.g. FCR, SGR, DCP and DCL), and submitted to principal component analysis (PCA) to reduce the dimensionality of the data set. The matrices submitted to PCA are standardised by subtracting their means and dividing by their standard deviations to construct linear combinations of the predictor variables *n* that contains the greatest variance. Statgraphics 19® Centurion (Version 16.1.11, StatPoint Technologies, Inc., The Plains, VA, USA) was used for statistical analysis of the data set in Table S1.

3   Results and discussion

The aquaculture feed industry needs new sources of feed ingredients, and insects are on their way to become one of these novel sources of protein and fats. When overcoming the obstacle of production quantities, both insect protein and lipid should have a natural place in aquafeeds. This comprehensive overview of work performed on the use of insect ingredients in aquafeeds shows that a massive research effort has been focused on this topic, but that the studies in total have some shortcomings as they are almost all from quite small-size aquatic species. This might partially be due to the smaller cost related to feeding trials with smaller animals. The fact that production quantities are a limiting factor for the use of insects in aquafeeds might also have led to more studies being done on juvenile stages. These stages are less demanding for large quantities of feeds and feed ingredients than the grow-out phases, and, as such, an obvious starting place for implementing commercial insect-containing diets. Also, compared to commercial diets, feed prepared for research scale on small-size aquatic species are generally not extruded at high pressure and temperature. The demonstrated links between specific nutrients in the diets and growth performance shown here can be the basis for future studies. The attached data file containing all collected data from a total of 91 publications using insect ingredients in aquafeeds can also be helpful for planning future studies and for comparing results from future trials.

3.1 *Growth and performance impact of insect ingredients in aquafeeds*

Normalised final body weights were plotted against insect inclusion in diets. There were in total 87 control diets and 285 insect-containing diets included in this analysis. Of the 285, 121 observations were higher, and 164 were lower than their respective controls (Figure 1A). In summary, we found that 58% of the insect-fed groups ended up with a final body weight below the control, and 42% showed a higher final body weight than their respective controls. Quite many of the studies using insect ingredients thus demonstrate a better growth than their respective controls, showing that making well-performing diets containing insect ingredients is quite possible. A quantitative estimation of the threshold inclusion of insect ingredients in the diets at which the growth performance started to decline was performed on the same data set. The breaking point estimated for low or little reduction in normalised final weight was at 25.0% insect inclusion in the diets (Figure 1A). The slope (b2) for the line after 25% insect inclusion was -0.42 (Figure 1A). Little effects on normalised final body weight were therefore found with up to 25% insects in the diets, with a gradual decrease in final weights obtained in the trials using higher dietary inclusion of insects. The same effect of a strong breaking point was not seen on the SGR, which showed a negative regression with increasing dietary insect inclusion starting from the Y intercept ($Y = 2.36 − 0.008627x$, x = % insect in diet, Figure 2A). The FCR plotted against insect inclusion in diets did not have a significant slope or breaking point detectable (Figure 2B). The digestibility of crude protein was also affected by insect inclusion in the diets, with a breaking point at 24.6% dietary insect inclusion (Figure 2C), while the digestibility of lipids was not affected by insects in the diets (Figure 2D). Since 95% of the diets included in this study contained between zero and 40% insect ingredients, the strongest conclusions can be drawn in this area and larger insecurities of effect will be associated with performance of diets with insect ingredients above this level. The spread of data as seen in Figure 1A and in the non-linear regression analyses is an indication of other inherent variations such as aquaculture species, insect species or other criteria from the different studies included in the dataset.

A PCA was then performed to provide an overview of the complete dataset and the relationship between insect inclusion in the diets (spread from 0% to 76% of the diets) and the variable performance responses (FCR, SGR, DCP and DCL). The PCA (Figure 3) revealed that the performance responses in aquatic species fed diets containing between 4% and 30% insect ingredients were grouped together with the control groups (0% inclusion). Similar trends were observed for the data grouped into TL, namely TL > 3.5, TL < 3.5 and crustaceans. The PCA plot (Figure S2) confirmed that the three subgroups are clustered together towards low insect inclusion in the diets, where most of the data is found. The highest FCR and SGR were observed for subgroups having TL above 3.5 and below 3.5, respectively. Overall, the PCA results showed that high levels of IM are not recommended for all aquafeed purposes and that levels in the range of 5-30% might be regarded as appropriate to produce results in the range observed in control diets.

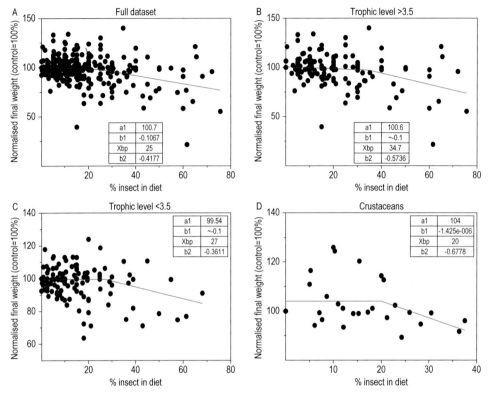

FIGURE 1   Broken line analysis of normalised weight gain against insect inclusion in the diet in (A) the
full dataset and (B) data set divided by trophic levels (TL): TL > 3.5, (C) TL < 3.5 and (D) crustaceans. a1 = intercept; b1 = slope line 1; b2 = slope line 2; Xbp = x at breaking point

3.1.1   Growth and performance of aquatic species subgroups

The performance parameters (NFW, FCR, SGR, DCP and DCL) of the three subgroups (TL > 3.5, TL < 3.5 and crustaceans) were analysed. There were breaking
points detected in the normalised final body weight, at 34.7, 27 and 20% of insect
inclusion in the diets for the aquatic species with TL > 3.5, TL < 3.5 and crustaceans,
respectively (Figures 1B, C and D, respectively). Similar clear breaking points were
not seen for SGR (Figures S3B, S3C and S3D). For FCR, no breaking point was
obtained for TL groups above and below 3.5 (Figures S4B and S4C); however there
was a breaking point at 17.4% of insects in the diets for crustaceans (Figure S4D),
with increasing FCR at higher insect inclusion than this. Further, clear breaking
points were also detected for the digestibility of CP with the full dataset (24.6%
insects in diet, Figure 2C) and for the group with TL above 3.5 (25.6% insects in the
diet, Figure S5C). In general, this comparative study of growth performance based
on the TL of aquatic species and crustaceans, showed that aquatic species with a
TL above 3.5 have a higher level of acceptance for dietary insect (35%), than the

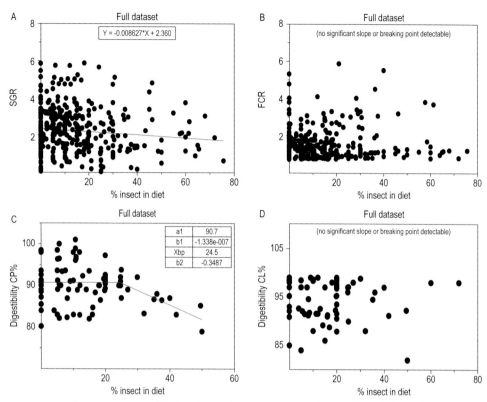

FIGURE 2   Linear regression and broken line analyses was performed on results on (A) specific growth rate (SGR); (B) feed conversion ratio (FCR); (C) digestibility of crude protein (CP); and (D) digestibility of crude lipid (CL), against percentage insect in diet in the full dataset, as well as in data separated into trophic levels. a1 = intercept; b1 = slope line 1; b2 = slope line 2; Xbp = x at breaking point

aquatic species with TL below 3.5. The crustaceans had the lowest tolerance level of insects in the diet (20%).

Of dietary components, saturated fats significantly decreased the normalised final body weights of the full dataset from a breaking point at 39% SFA of total FA in the diets (Figure 4A). A similar trend was observed for lauric acid, with a significant decrease of the normalised final weight of the full data set from the breaking point at 26% C12% of total FA in diet (Figure 4B). The same tendencies in reduction of normalised final body weight was seen in the aquatic species with TL > 3.5 (breaking point at 38.6% SFA of FA in diet, Figure S7B). Normalised final body weight of the same TL group was also decreased by increased lauric acid in the diets (Y = 98.77 − 0.53x, x = % C12% of FA in diet, Figure S8B), while the other performance indicators (FCR, DCP, DCL, SGR) were not affected (data not shown). The relation between saturated fats or lauric acid in diet and decreased final body weight was not seen in the group of aquatic species with TL < 3.5 (Figures S7C and S8C). This

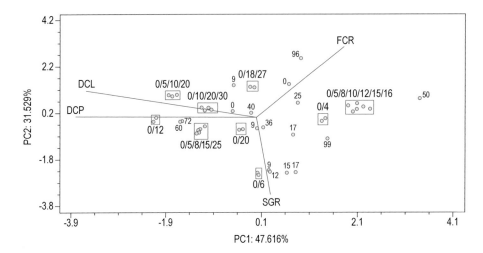

FIGURE 3   Principal component analysis (PCA) showing the relationship between percentage of
             insect inclusion in the diets (ID) and the performance outputs feed conversion ratio
             (FCR), specific growth rate (SGR), digestibility of crude protein (DCP) and digestibil-
             ity of crude lipid (DCL) recorded in published studies. The 91 bibliographic references
             (Table S1) were arranged as a 415×5 matrix where the first column with 415 rows rep-
             resents the ID (ranging from 0 to 76%) and the remaining four columns represent the
             experimental responses (FCR, SGR, DCP and DCL). A total of two components (PC1
             and PC2) were able to explain a 79.145% of the data variability and detect specific insect
             inclusion diets that exhibited a similar performance to control diets. The red rectangles
             indicate those insect inclusion diets that were clustered with the control diets and the
             slash-separated numbers the control (0%) and nominal percentage of ID associated to
             every cluster.

is, however, most likely due to lower levels of SFA and lauric acid used in diets in
this trophic group, with few dietary groups above the breaking point detected for
the group with TL above 3.5 (Figures S7C and S8C). For the subgroup of data from
studies using insect ingredients in trials on crustaceans, there was no data available (Figures S7D and S8D). Increased content of saturated fats in insect-based
diets, especially related to the increased lauric acid content in BSF ingredients, may
therefore reduce growth performance of the aquatic species. This should be given
extra attention in studies where high contents of this insect species is used, especially when full-fat insect products are used.

3.1.2   Impact of insect species subgroups on growth performances
The performance of aquaculture species was divided into two subgroups depending on the insect species they were fed (BSF or MW). For BSF, the main effect of
increased insect inclusion was on NFW (Figure 5A). The calculated breaking point
was located at 40% BSF inclusion in the diets. There was a slow but gradual decrease in NFW from 0 to 40% BSF inclusion, which was changed to a strong decrease after this point. There was also a reduction in SGR with increased BSF in

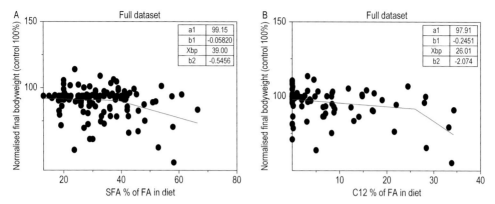

FIGURE 4   Broken line analysis of the normalised final body weight plotted against (A) dietary saturated fats (SFA) and (B) dietary C12:0 lauric acid for the full dataset. Data divided into the three trophic groups is presented in Figures S7 and S8. a1 = intercept; b1 = slope line 1; b2 = slope line 2; Xbp = x at breaking point

diet, but the relation was not very strong and there was not a breaking point detected ($y = 2.469 - 0.015x$, x = % insects in diet, Figure 5C). This decrease in normalised final weight might be due to increased lauric acid content, since a significant linear relationship between final body weight and dietary lauric acid was also found in the BSF-fed aquatic species (data not shown). It is, however, difficult to separate these two factors due to the natural contribution of $C12:0$ in BSF. In aquafeeds containing MW, a similar decrease as in BSF fed aquatic species was seen, but only in SGR ($Y = 2.9 - 0.03x$, x = % insects in diet, Figure 5D), not in final weight (Figure 5B). Since there were only few values with high weights in this dataset on MW, we excluded the data on fish weighing more than 500 g in these analyses. This was to reduce bias due to fish size, which will affect the SGR. There was no effect of BSF or MW inclusion on FCR (data not shown).

Similar results were seen in the recent meta-analysis done by Hua (2021), where inclusion levels of BSF larvae higher than 29% in the diets resulted in a poorer growth (SGR) than aquatic species fed FM control diets, while dietary inclusion of MW at any level did not reduce growth of aquatic species. The current study revealed a reduced performance in both BSF and MW fed aquatic species, but a bit differently. The reduction in SGR in aquatic species fed MW was more immediate than with the introduction of BSF, which kept quite stable till ~40% inclusion levels. However, with no significant reduction in final weights, only in SGR, when increasing MW content in aquafeeds, a more detailed study of individual species etc. would have to be done to elucidate on this topic. BSF larvae and MW contain approximatively the same level of crude protein (47 and 50%, respectively) and fat (24 and 22%, respectively), but the FA profile differs largely between the two species (Nogales-Mérida *et al.*, 2019). Thus, the divergent effects of BSF and MW on animal growth might be due to the FA profile but also related to nutrient digestibility or presence of antinutritional factors.

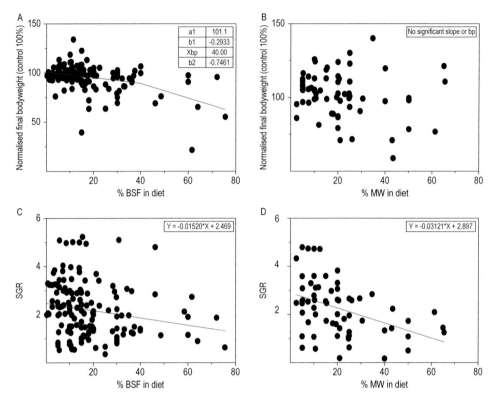

FIGURE 5   Linear regression of normalised final body weight and specific growth rate (SGR) plotted
against percent inclusion of black soldier fly (BSF) (A and C) and mealworm (MW) (B and D)
in aquafeeds. a1 = intercept; b1 = slope line 1; b2 = slope line 2; Bp = breaking point; Xbp = x at
breaking point

3.2 *Making an optimal insect-containing aquafeed*

3.2.1 Estimation of crude protein in insect meals
The protein requirement varies between aquatic species and their life stages. Generally spoken, carnivorous fish and shellfish used in aquaculture, often marine,
usually have higher protein requirements (38-55% protein of diet) than omnivores
(20-40%) (NRC, 2011). The protein content of the diets included in this meta-analysis were mostly covering the requirements of the cultured animal. The dietary protein content is, however, almost exclusively reported as crude protein calculated by
using the standard 6.25 N-to-protein factor. The problem with this is that insects
contain high concentrations of non-protein nitrogen and the protein content is
therefore often overestimated by using this factor, as reported earlier (Janssen *et al.*,
2017; Liland *et al.*, 2017). It has been proposed to use nitrogen-to-protein conversion
factors of 4.76 for BSF larvae, MW and lesser MW, and between 4.53-4.80 for house

cricket meals (Belghit *et al.*, 2019b; Janssen *et al.*, 2017). These factors provide a more correct estimate of IM protein content, which is a vital determinant when insects are to replace other protein sources in aquafeed. Of the 91 articles selected for this meta-analysis, only two used a corrected protein factor. Some trials reporting reduced growth due to the introduction of IMs to the diets, may in fact be reporting a change in performance due to lack of dietary protein. The general trend is, however, an increased crude protein content in the diets with increased insect inclusion (crude protein = 41.9 + 0.07x, x = % insects in the diets, P = 0.006, R^2 = 0.02, data not shown), suggesting that many experiments have, in fact, considered the overestimation of crude protein content when using the standard 6.25 N-to protein factor in their diet formulations. The information from the included studies suggests that future studies should consider the substantial amounts of non-protein nitrogen in IM and use alternative conversion factors more suited for these protein sources.

3.2.2   Effects of supplementing dietary amino acids

When using an alternative protein source in aquafeed, it is essential to have precise and quantitative data on the dietary content of essential AAs, as many protein sources of non-animal origin have an unbalanced composition of these. In a setting of commercial aquaculture, limiting AAs would be added to the aquafeeds and should therefore also be supplemented in experiments to fill the requirements of the species. The general AA profile of IM varies according to species and are taxon dependent, with the Diptera having a similar profile to FM, while the Celeoptera and Orthoptera have a close profile to soybean meal and are generally deficient in methionine and lysine (Henry *et al.*, 2015). The growth and performance of rainbow trout and seatrout were analysed together while seabass and seabream were analysed together and are presented in Figure 6. Dietary supplementation of limiting AAs did not affect the normalised final body weight in either of the two groups of aquatic species (seabass/seabream and trout, Figure 6A and 6B, respectively). The FCR in the supplemented trout did increase with increased insect inclusion in the diets (Figure 6D), but since these diets all clustered around 0-20% inclusion levels, this could level out at higher inclusion levels, as seen in the AA supplemented diets. There was no linear relationship between the SGR of the aquatic species and the dietary inclusion of insects (Figures 6C and D). The AA-supplemented diets led to a higher SGR in both groups of species (Figures 6E and 6F). Supplementing with limiting AAs also led to a lower FCR in seabass/seabream (Figure 6C). It must be noted that all the diets, supplemented or not, fulfilled the requirement of the respective species. The effect of dietary lysine, arginine, leucine and phenylalanine on FCR and SGR were clearly visible in different subset of data (data not presented due to no clear relation between these AA contents in diets and insect inclusion), so being aware of the positive effects of adequate AA contents of the diets is of great importance, as expected. The overall quality of the aquafeed might be the most important factor when planning a trial using insect ingredients, possibly explaining why some

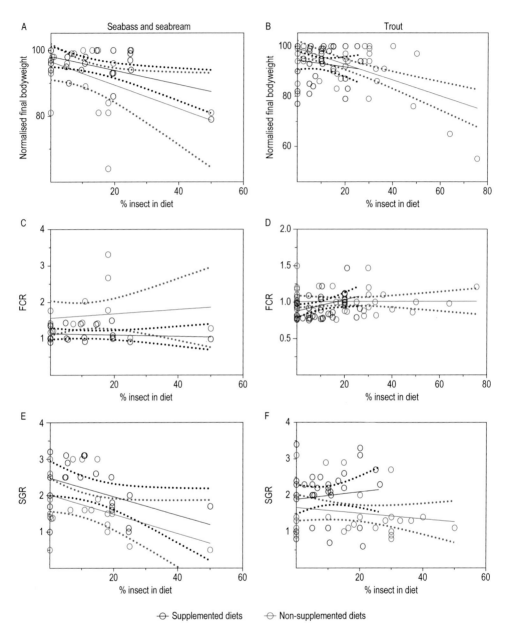

FIGURE 6   Normalised final body weight (control set to 100%), feed conversion ratio (FCR) and specific
            growth rate (SGR) plotted against the percentage inclusion of insects in diet in seabass and
            seabream (A, C and E) and trout (B, D and F). Data from trials using supplemented diets (and
            its best-fit line and 95% confidence interval) are shown in black, while the non-supplemented
            diets are shown in red.

experiments fail to make the animals perform optimally on these feeds, while some experiments succeed.

3.2.3 Estimation of chitin in insect meals

Chitin is the primary structural polysaccharide of the arthropod exoskeleton. Insects are known to contain acid detergent fibre (ADF) and neutral detergent fibre and these fibres represent chitin. Whole insects contain between 5-25% chitin, with 35% of chitin only in the exoskeletons from BSF larvae (Hahn *et al.*, 2018). The content of chitin was reported only in a few studies in the collected data and varied considerably between IMs. For example, chitin was reported to be 2.7-16.6, 2.8-22.6 and 9.2-16.7% of dried IM for BSF, MW and silkworm meals, respectively. In these studies, two main methods were reported for chitin quantification in IM, namely ADF and N- acetyl glucosamine estimation. Finke (2007) demonstrated that the ADF fraction contains a high amount of AA, which leads to an overestimation of insect chitin content as ADF contains both protein and chitin. Furthermore, according to Clark *et al.* (1993) N- acetyl glucosamine estimation (NAG) was used as an indicator of chitin digestion based on the calorimetric method (Reissig *et al.*, 1955). This method was adopted by Alves *et al.* (2020) for estimating the chitin content. Also, studies conducted by Marono *et al.* (2015) calculated chitin considering the protein linked to acid detergent fibre (AOAC, 2003). However, these calculated values could underestimate the true amount of chitin in different IMs. Up to date, it seems that the quantification of chitin is still challenging, since this polymer is always associated with other compounds (protein, carbohydrates, lipids or minerals) in IMs. Furthermore, chitin is a hard, inelastic, N-acetylated amino polysaccharide and it is insoluble in water and most solvents, making direct quantification of chitin very challenging. There are, however, new methods being developed which could be useful for future studies, for example by using calcofluor staining (Henriques *et al.*, 2020). An important note is to find a method that is easily accessible and that could be generally agreed on as the benchmark chitin method for better comparison between trials.

3.2.4 Which factors reduce crude protein digestibility of animals fed insect-based diets?

Meta-analysis of protein digestibility data (by both non-linear regression and PCA) revealed that the digestibility of crude protein declined in diets containing more than 25% insects. Considering the insect inclusion thresholds for uncompromised growth of 10 to 37% (by non-linear regression) and 5 to 30% (by PCA), it can be inferred that protein digestibility is an important factor, influencing the growth performance of aquatic species fed insect-containing diets. Altered growth of aquatic species fed insect-based ingredients related to reduction of the bioavailability and digestibility of nutrients are often attributed to the content of chitin in IMs (Gasco *et al.*, 2019; Kroeckel *et al.*, 2012). Inclusion of dietary chitin in aquafeeds without negative effects on growth performances and digestibility of diets has been success-fully demonstrated in some fish feeding trials (Fines and Holt, 2010; Gopalakannan and Arul, 2006; Karlsen *et al.*, 2017), but not in others (Shiau and Yu, 1999). It has been reported that the structural form of chitin inhibits nutrient absorption from the intestinal tract and therefore reduces protein and lipid bioavailability in mice and poultry (Han *et al.*, 1999; Tanaka *et al.*, 1997). In this meta-analysis, it was, how-ever, not possible to look for a correlation between the content of chitin in the feeds and the digestibility of protein or lipids due to the very limited data and large vari-ations in analytical methods used for chitin measurements. Recently, Fisher *et al.* (2020) showed that the digestibility of dietary BSF was higher than diets using soy protein concentrate, corn protein concentrate or FM as a protein source in Atlantic salmon. However, when comparing the digestibility of BSF and MW in the diets of European seabass, MW-containing feeds had higher protein and AA digestibility than diets with BSF (Basto *et al.*, 2020). The authors speculated that the difference in digestibility between BSF and MW could be caused by the chitin, which was higher in the BSF meal than the MW meal (6.2% and 4.6% of meals, respectively) (Basto *et al.*, 2020). It is, however, also possible that this could have been cause by other factors in the BSF, like lauric acid or other compounds not yet described. Some studies have also shown that the processing of BSF can improve protein di-gestibility, such as BSF hydrolysate protein having higher protein digestibility com-pared to defatted insect protein meal (Roques *et al.*, 2020). Different methods to process BSF meals have also been shown to strongly affect growth in trials with Atlantic salmon (Lock *et al.*, 2016). The processing method may therefore have a great impact on the bioavailability of both AAs and FAs (Dumas *et al.*, 2018; Roques *et al.*, 2020). Based on the limited available data, it is not possible to conclude which factors reduce crude protein digestibility of aquatic species fed insect-based diets. More research is needed to be able to pinpoint which components of the IMs are causing this decline in protein digestibility.

4   Conclusions

The analysis of the compiled information on diet composition and performance of 415 experimental diets, reported in 91 different publications, allows formulating diets containing up to 25-30% insect inclusion without reduced performance of the aquaculture species. The main limitations for the use of insects as feed ingredients might be the high variation in quality and compositions of IMs as well as large changes in dietary fatty acid profiles of the diets when using non-defatted meals and insect oils. There should also be specific use of true protein for the assessment of protein in insect-containing diets. Given the high degree of data heterogeneity, the results have been interpreted cautiously by different mathematical/statistical approaches. For example, there is a general lack of data on aquacultured species up to market size, with much of the data coming from studies done on juvenile stages. However, the information from the 91 analysed references might be regarded as the core evidence on this topic to date, and along with the present meta-analysis may allow for corresponding improvements to future research.

Acknowledgments

This study was supported by Nærings- og fiskeridepartementet, IMR and fellowship provided by the Indian Council of Agricultural Research (ICAR) through the ICAR Netaji Subash International Fellowship.

Conflicts of interest

No potential conflicts of interest were reported by the authors.

Supplementary material

Figure S1. Normal distribution of data from the control group (red circles) and insect-diet groups (black circles) in terms of the performance parameters feed conversion ratio (FCR), specific growth rate (SGR), digestibility of crude protein (DCP) and digestibility of crude lipid (DCL).

Figure S2. Principal component analysis showing the relationship between percentage of insect inclusion in the diets (ID) and (A) FCR and SGR and (B) DCP and DCL.

Figure S3. Relation between specific growth rate (SGR) and % insect in diet in the full data set (A), trophic level > 3.5 (B), trophic level < 3.5 (C) or in crustaceans (D).

Figure S4. Relation between feed conversion ratio (FCR) and the % of insects in the diets in the full dataset (A), trophic level > 3.5 (B), trophic level < 3.5 (C) or in crustaceans (D).

Figure S5. Relation between digestibility of crude protein (CP) and % insect in diet at 24.57 % insect in diet for the full data set (A).

Figure S6. Relation between digestibility of crude lipid % and the % of insects in the diets in the full dataset (A), trophic level > 3.5 (B), or trophic level < 3.5 (C).

Figure S7. Relation between content of saturated fatty acids (SFA) and normalised final body weight.

Figure S8. Relation between content of lauric acid (C12% of fatty acid (FA) in diet and normalised final body weight in the full dataset (26% of total FA) (A) and in the trophic level > 3.5 (B), respectively.

Table S1. Full dataset.
Table S2. Fish species included in studies.
Table S3. Overview of insect species and products used in feeding trials included in meta-analysis.

Supplementary material can be found online at https://doi.org/10.3920/JIFF2020.0147.

References

Abdel-Tawwab, M., Khalil, R.H., Metwally, A.A., Shakweer, M.S., Khallaf, M.A. and Abdel-Latif, H.M.R., 2020. Effects of black soldier fly (*Hermetia illucens* L.) larvae meal on growth performance, organs-somatic indices, body composition, and hemato-biochemical variables of European sea bass, *Dicentrarchus labrax*. Aquaculture 522: 735136. https://doi.org/10.1016/j.aquaculture.2020.735136

Alves, A.P.d.C., Paulino, R.R., Pereira, R.T., Da Costa, D.V. and e Rosa, P.V., 2020. Nile tilapia fed insect meal: growth and innate immune response in different times under lipopolysaccharide challenge. Aquaculture Research 52: 529-540. https://doi.org/10.1111/are.14911

Arru, B., Furesi, R., Gasco, L., Madau, F.A., Pulina, P., 2019. The introduction of insect meal into the fish diet: the first economic analysis on European sea bass farming. Sustainability 11: 1697. https://doi.org/10.3390/su11061697

Association of Official Analytical Chemists (AOAC), 2003. Official methods of analysis, 17th edition. AOAC, Washington, DC, USA.

Basto, A., Matos, E. and Valente, L.M.P., 2020. Nutritional value of different insect larvae meals as protein sources for European sea bass (*Dicentrarchus labrax*) juveniles. Aquaculture 521: 735085. https://doi.org/10.1016/j.aquaculture.2020.735085

Belghit, I., Liland, N.S., Gjesdal, P., Biancarosa, I., Menchetti, E., Li, Y., Waagbø, R., Krogdahl, Å. and Lock, E.-J., 2019a. Black soldier fly larvae meal can replace fish meal in diets of sea-water phase Atlantic salmon (*Salmo salar*). Aquaculture 503: 609-619. https://doi.org/10.1016/j.aquaculture.2018.12.032

Belghit, I., Lock, E.-J., Fumière, O., Lecrenier, M.-C., Renard, P. Dieu, M., Berntssen, M.H.G., Palmblad, M. and Rasinger, J.D., 2019b. Species-specific discrimination of insect meals for aquafeeds by direct comparison of tandem mass spectra. Animals 9: 222. https://doi.org/10.3390/ani9050222

Bondari, K. and Sheppard, D.C., 1981. Soldier fly larvae as feed in commercial fish production. Aquaculture 24: 103-109. https://doi.org/10.1016/0044-8486(81)90047-8

Bruni, L., Randazzo, B., Cardinaletti, G., Zarantoniello, M., Mina, F., Secci, G., Tulli, F., Olivotto, I. and Parisi, G., 2020. Dietary inclusion of full-fat *Hermetia illucens* prepupae meal in practical diets for rainbow trout (*Oncorhynchus mykiss*): lipid metabolism and fillet quality investigations. Aquaculture 529: 735678. https://doi.org/10.1016/j.aquaculture.2020.735678

Cadinu, L.A., Barra, P., Torre, F., Delogu, F. and Madau, F.A., 2020. Insect rearing: potential, challenges, and circularity. Sustainability 12: 4567. https://doi.org/10.3390/su12114567

Clark, D.J., Lawrence, A.L. and Swakon, D.H.D., 1993. Apparent chitin digestibility in penaeid shrimp. Aquaculture 109: 51-57. https://doi.org/10.1016/0044-8486(93)90485-H

Cummins, V.C., Rawles, S.D., Thompson, K.R., Velasquez, A., Kobayashi, Y., Hager, J. and Webster, C.D., 2017. Evaluation of black soldier fly (*Hermetia illucens*) larvae meal as partial or total replacement of marine fish meal in practical diets for Pacific white shrimp (*Litopenaeus vannamei*). Aquaculture 473: 337-344. https://doi.org/10.1016/j.aquaculture.2017.02.022

Devic, E., Leschen, W., Murray, F. and Little, D.C., 2017. Growth performance, feed utilization and body composition of advanced nursing Nile tilapia (*Oreochromis niloticus*) fed diets containing black soldier fly (*Hermetia illucens*) larvae meal. Aquaculture Nutrition 24: 416-423. https://doi.org/10.1111/anu.12573

Dumas, A., Raggi, T., Barkhouse, J., Lewis, E. and Weltzien, E., 2018. The oil fraction and partially defatted meal of black soldier fly larvae (*Hermetia illucens*) affect differently growth performance, feed efficiency, nutrient deposition, blood glucose and lipid digestibility of rainbow trout (*Oncorhynchus mykiss*). Aquaculture 492: 24-34. https://doi.org/10.1016/j.aquaculture.2018.03.038

European Union (EU), 2017. Commission Regulation (EU) 2017/893 of 24 May 2017 amending Annexes I and IV to Regulation (EC) No 999/2001 of the European Parliament and of the Council and Annexes X, XIV and XV to Commission Regulation (EU) No 142/2011 as regards the provisions on processed animal protein. Official Journal of the European Union L 138: 92. Available at: https://tinyurl.com/uzt956aw.

Evans, J., Alemu, M.H., Flore, R., Frøst, M.B., Halloran, A., Jensen, A.B., Maciel-Vergara, G., Meyer-Rochow, V.B., Münke-Svendsen, C., Olsen, S.B., Payne, C., Roos, N., Rozin, P., Tan, H.S.G., Van Huis, A., Vantomme, P. and Eilenberg, J., 2015. 'Entomophagy': an evolving terminology in need of review. Journal of Insects as Food and Feed 1: 293-305. https://doi.org/10.3920/JIFF2015.0074

Fabrikov, D., Sánchez-Muros, M.J., Barroso, F.G., Tomás-Almenar, C., Melenchón, F., Hidalgo, M.C., Morales, A.E., Rodriguez-Rodriguez, M. and Montes-Lopez, J., 2020. Comparative study of growth performance and amino acid catabolism in *Oncorhynchus mykiss*, *Tinca tinca* and *Sparus aurata* and the catabolic changes in response to insect meal inclusion in the diet. Aquaculture 529: 735731. https://doi.org/10.1016/j.aquaculture.2020.735731

Fawole, F.J., Adeoye, A.A., Tiamiyu, L.O., Ajala, K.I., Obadara, S.O. and Ganiyu, I.O., 2020. Substituting fishmeal with *Hermetia illucens* in the diets of African catfish (*Clarias gariepinus*): effects on growth, nutrient utilization, haemato-physiological response, and oxidative stress biomarker. Aquaculture 518: 734849. https://doi.org/10.1016/j.aquaculture.2019.734849

Fines, B.C. and Holt, G.J., 2010. Chitinase and apparent digestibility of chitin in the digestive tract of juvenile cobia, *Rachycentron canadum*. Aquaculture 303: 34-39. https://doi.org/10.1016/j.aquaculture.2010.03.010

Finke, M.D., 2007. Estimate of chitin in raw whole insects. American Zoo and Aquarium Association 26: 105-115. https://doi.org/10.1002/zoo.20123

Fisher, H.J., Collins, S.A., Hanson, C., Mason, B., Colombo, S.M. and Anderson, D.M., 2020. Black soldier fly larvae meal as a protein source in low fish meal diets for Atlantic salmon (*Salmo salar*). Aquaculture 521: 734978. https://doi.org/10.1016/j.aquaculture.2020.734978

Food and Agriculture Organization of the United Nations (FAO), 2011. FAOSTAT statistical database. FAO, Rome, Italy.

Food and Agriculture Organization of the United Nations (FAO), 2020. The state of world fisheries and aquaculture 2020. Sustainability in action. FAO, Rome, Italy.

Froese, R. and Pauly, D. (eds.), 2019. FishBase. Available at: www.fishbase.org.

Gasco, L., Biasato, I., Dabbou, S., Schiavone, A. and Gai, F., 2019. Animals fed insect-based diets: state-of-the-art on digestibility, performance and product quality. Animals 9: 170. https://doi.org/10.3390/ani9040170

Gasco, L., Henry, M., Piccolo, G., Marono, S., Gai, F., Renna, M., Lussiana, C., Antonopoulou, E., Mola, P. and Chatzifotis, S., 2016. *Tenebrio molitor* meal in diets for European sea bass (*Dicentrarchus labrax L.*) juveniles: growth performance, whole body composition and *in vivo* apparent digestibility. Animal Feed Science and Technology 220: 34-45. https://doi.org/10.1016/j.anifeedsci.2016.07.003

Gelinçek, İ. and Yamaner, G., 2020. An investigation on the gamete quality of Black Sea trout (*Salmo trutta labrax*) broodstock fed with mealworm (*Tenebrio molitor*). Aquaculture Research 51: 2379-2388. https://doi.org/10.1111/are.14581

Glencross, B.D., 2020. A feed is still only as good as its ingredients: an update on the nutritional research strategies for the optimal evaluation of ingredients for aquaculture feeds. Aquaculture Nutrition 6: 1871-1883. https://doi.org/10.1111/anu.13138

Glencross, B.D., Booth, M. and Allan, G.L., 2007. A feed is only as good as its ingredients – a review of ingredient evaluation strategies for aquaculture feeds. Aquaculture Nutrition 13: 17-34. https://doi.org/10.1111/j.1365-2095.2007.00450.x

Gopalakannan, A. and Arul, V., 2006. Immunomodulatory effects of dietary intake of chitin, chitosan and levamisole on the immune system of *Cyprinus carpio* and control of *Aeromonas hydrophila* infection in ponds. Aquaculture 255: 179-187. https://doi.org/10.1016/j.aquaculture.2006.01.012

Hahn, T., Roth, A., Febel, E., Fijalkowska, M., Schmitt, E., Arsiwalla, T. and Zibek, S., 2018. New methods for high-accuracy insect chitin measurement. Journal of the Science of Food and Agriculture 98: 5069-5073. https://doi.org/10.1002/jsfa.9044

Hale, O.J.G.E.S.J., 1973. Dried *Hermetia illucens* larvae (Diptera: Stratiomyidae) as a feed additive for poultry. Journal of the Georgia Entomological Society 8: 16-20.

Han, L.K., Kimura, Y. and Okuda, H., 1999. Reduction in fat storage during chitin-chitosan treatment in mice fed a high-fat diet. International Journal of Obesity 23: 174-179. https://doi.org/10.1038/sj.ijo.0800806

Henriques, B.S., Garcia, E.S., Azambuja, P. and Genta, F.A., 2020. Determination of chitin content in insects: an alternate method based on calcofluor staining. Frontiers in Physiology 11: 117. https://doi.org/10.3389/fphys.2020.00117

Henry, M., Gasco, L., Piccolo, G. and Fountoulaki, E., 2015. Review on the use of insects in the diet of farmed fish: Past and future. Animal Feed Science and Technology 203: 1-22. https://doi.org/10.1016/j.anifeedsci.2015.03.001

Hu, J., Wang, G., Huang, Y., Sun, Y., He, F., Zhao, H. and Li, N., 2017. Effects of substitution of fish meal with black soldier fly (*Hermetia illucens*) larvae meal, in yellow catfish (*Pelteobagrus fulvidraco*) diets. Israeli Journal of Aquaculture – Bamidgeh 69.

Hua, K., 2021. A meta-analysis of the effects of replacing fish meals with insect meals on growth performance of fish. Aquaculture 530: 735732. https://doi.org/10.1016/j.aquaculture.2020.735732

Hua, K. and Bureau, D.P., 2012. Exploring the possibility of quantifying the effects of plant protein ingredients in fish feeds using meta-analysis and nutritional model simulation-based approaches. Aquaculture 356-357: 284-301. https://doi.org/10.1016/j.aquaculture.2012.05.003

Hua, K., Cobcroft, J.M., Cole, A., Condon, K., Jerry, D.R., Mangott, A., Praeger, C., Vucko, M.J., Zeng, C., Zenger, K. and Strugnell, J.M., 2019. The future of aquatic protein: implications for protein sources in aquaculture diets. One Earth 1: 316-329. https://doi.org/10.1016/j.oneear.2019.10.018

International Platform of Insects for Food and Feed (IPIFF), 2019a. Building bridges between the insect production chain, research and policymakers. IPIFF, Brussels, Belgium. Available at: https://ipiff.org/wp-content/uploads/2019/12/IPIFF-researchpriorities-HorizonEurope.pdf

International Platform of Insects for Food and Feed (IPIFF), 2019b. The European insect sector today: challenges, opportunities and regulatory landscape. IPIFF, Brussels, Belgium. Available at: https://ipiff.org/wp-content/uploads/2019/12/2019IPIFF_VisionPaper_updated.pdf

International Platform of Insects for Food and Feed (IPIFF), 2020a. Promoting insects for human consumption and animal feed. IPIFF, Brussels, Belgium.

International Platform of Insects for Food and Feed (IPIFF), 2020b. Edible insects on the Euopean market (factsheet). IPIFF, Brussels, Belgium. Available at: https://ipiff.org/wp-content/uploads/2020/06/10-06-2020-IPIFF-edible-insects-market-factsheet.pdf

Janssen, R.H., Vincken, J.-P., Van den Broek, L.A., Fogliano, V., Lakemond, C.M.J.J.o.A. and Chemistry, F., 2017. Nitrogen-to-protein conversion factors for three edible insects: *Tenebrio molitor, Alphitobius diaperinus*, and *Hermetia illucens*. Journal of Agricultural and Food Chemist 65: 2275-2278. https://doi.org/10.1021/acs.jafc.7b00471

Jeong, S.-M., Khosravi, S., Mauliasari, I.R. and Lee, S.-M., 2020. Dietary inclusion of mealworm (*Tenebrio molitor*) meal as an alternative protein source in practical diets for rainbow trout (*Oncorhynchus mykiss*) fry. Fisheries and Aquatic Sciences 23: 1-8. https://doi.org/10.1186/s41240-020-00158-7

Józefiak, A., Nogales-Mérida, S., Mikołajczak, Z., Rawski, M., Kierończyk, B. and Mazurkiewicz, J., 2019a. The utilization of full-fat insect meal in rainbow trout (*Oncorhynchus mykiss*) nutrition: the effects on growth performance, intestinal microbiota and gastrointestinal tract histomorphology. Annals of Animal Science 19: 747-765. http://doi.org/10.2478/aoas-2019-0020

Józefiak, A., Nogales-Merida, S., Rawski, M., Kieronczyk, B. and Mazurkiewicz, J., 2019b. Effects of insect diets on the gastrointestinal tract health and growth performance of Siberian sturgeon (*Acipenser baerii* Brandt, 1869). BMC Veterinary Research 15: 348. https://doi.org/10.1186/s12917-019-2070-y

Karlsen, Ø., Amlund, H., Berg, A. and Olsen, R.E., 2017. The effect of dietary chitin on growth and nutrient digestibility in farmed Atlantic cod, Atlantic salmon and Atlantic halibut. Aquaculture Research 48: 123-133. https://doi.org/10.1111/are.12867

Kaushik, S.J. and Seiliez, I., 2010. Protein and amino acid nutrition and metabolism in fish: current knowledge and future needs. Aquaculture Research 41: 322-332. https://doi.org/10.1111/j.1365-2109.2009.02174.x

Kroeckel, S., Harjes, A.G.E., Roth, I., Katz, H., Wuertz, S., Susenbeth, A. and Schulz, C., 2012. When a turbot catches a fly: evaluation of a pre-pupae meal of the black soldier fly (*Hermetia illucens*) as fish meal substitute – growth performance and chitin degradation in juvenile turbot (*Psetta maxima*). Aquaculture 364-365: 345-352. https://doi.org/10.1016/j.aquaculture.2012.08.041

Li, S., Ji, H., Zhang, B., Zhou, J. and Yu, H., 2017. Defatted black soldier fly (*Hermetia illucens*) larvae meal in diets for juvenile Jian carp (*Cyprinus carpio var. Jian*): growth performance, antioxidant enzyme activities, digestive enzyme activities, intestine and hepatopancreas histological structure. Aquaculture 477: 62-70. https://doi.org/10.1016/j.aquaculture.2017.04.015

Liland, N.S., Biancarosa, I., Araujo, P., Biemans, D., Bruckner, C.G., Waagbø, R., Torstensen, B.E. and Lock, E.-J., 2017. Modulation of nutrient composition of black soldier fly (*Hermetia illucens*) larvae by feeding seaweed-enriched media. PLoS ONE 12: e0183188. https://doi.org/10.1371/journal.pone.0183188

Lock, E.R., Arsiwalla, T. and Waagbø, R., 2016. Insect larvae meal as an alternative source of nutrients in the diet of Atlantic salmon (*Salmo salar*) postsmolt. Aquaculture Nutrition 22: 1202-1213. https://doi.org/10.1111/anu.12343

Magalhães, R., Sánchez-López, A., Leal, R.S., Martínez-Llorens, S., Oliva-Teles, A. and Peres, H., 2017. Black soldier fly (*Hermetia illucens*) pre-pupae meal as a fish meal replacement in diets for European seabass (*Dicentrarchus labrax*). Aquaculture 476: 79-85. https://doi.org/10.1016/j.aquaculture.2017.04.021

Marono, S., Piccolo, G., Loponte, R., Di Meo, C., Attia, Y.A., Nizza, A. and Bovera, F., 2015. *In vitro* crude protein digestibility of *Tenebrio molitor* and *Hermetia illucens* insect meals and its correlation with chemical composition traits. Italian Journal of Animal Science 14: 338-349. https://doi.org/10.4081/ijas.2015.3889

Muin, H., Taufek, N.M., Kamarudin, M.S. and Razak, S.A., 2017. Growth performance, feed utilization and body composition of nile tilapia, *Oreochromis niloticus* (Linnaeus, 1758) fed with different levels of black soldier fly, *Hermetia illucens* (Linnaeus, 1758) maggot meal diet. Iranian Journal of Fisheries Sciences 16: 567-577. Available at: http://jifro.ir/article-1-2721-en.html

National Research Council (NRC), 2011. Nutrient requirements of fish and shrimp. National Academies Press, Washington, DC, USA. https://doi.org/10.17226/13039

Newton, G.L., Booram, C.V., Barker, R.W. and Hale, O.M., 1977. Dried *Hermetia Illucens* larvae meal as a supplement for swine. Journal of Animal Science 44: 395-400. https://doi.org/10.2527/jas1977.443395x

Nogales-Mérida, S., Gobbi, P., Józefiak, D., Mazurkiewicz, J., Dudek, K., Rawski, M., Kierończyk, B. and Józefiak, A., 2019. Insect meals in fish nutrition. Reviews in Aquaculture 11: 1080-1103. https://doi.org/10.1111/raq.12281

Prabhu, A.J.P., Schrama, J.W. and Kaushik, S.J., 2013. Quantifying dietary phosphorus requirement of fish – a meta-analytic approach. Aquaculture Nutrition 19: 233-249. https://doi.org/10.1111/anu.12042

Prabhu, A.J.P., Schrama, J.W. and Kaushik, S.J., 2016. Mineral requirements of fish: a systematic review. Reviews in Aquaculture 8: 172-219. https://doi.org/10.1111/raq.12090

Reissig, J.L., Strominger, J.L. and Leloir, L.F.J.J.o.B.C., 1955. A modified colorimetric method for the estimation of N-acetylamino sugars. Journal of Biological Chemistry 217: 959-966. https://doi.org/10.1016/S0021-9258(18)65959-9

Roncarati, A., Cappuccinelli, R., Meligrana, M.C., Anedda, R., Uzzau, S. and Melotti, P., 2019. Growing trial of gilthead sea bream (*Sparus aurata*) juveniles fed on chironomid meal as a partial substitution for fish meal. Animals 9: 144. https://doi.org/10.3390/ani9040144

Roques, S., Deborde, C., Guimas, L., Marchand, Y., Richard, N., Jacob, D., Skiba-Cassy, S., Moing, A. and Fauconneau, B., 2020. Integrative metabolomics for assessing the effect of insect (*Hermetia illucens*) protein extract on rainbow trout metabolism. Metabolites 10: 83. https://doi.org/10.3390/metabo10030083

Sánchez-Muros, M.J., De Haro, C., Sanz, A., Trenzado, C.E., Villareces, S. and Barroso, F.G., 2015. Nutritional evaluation of *Tenebrio molitor* meal as fishmeal substitute for tilapia (*Oreochromis niloticus*) diet. Aquaculture Nutrition 22: 943-955. https://doi.org/10.1111/anu.12313

Shiau, S.-Y. and Yu, Y.-P., 1999. Dietary supplementation of chitin and chitosan depresses growth in tilapia, *Oreochromis niloticus×O. aureus*. Aquaculture 179: 439-446. https://doi.org/10.1016/S0044-8486(99)00177-5

Sørensen, M., Berge, G.M., Thomassen, M.S., Ruyter, B., Hatlen, B., Ytrestøyl, T., Aas, T.S. and Åsgård, T.E.J.N.R., 2012. Today's and tomorrow's feed ingredients in Norwegian aquaculture *Nofima rapportserie*. Nofima AS. Available at: http://hdl.handle.net/11250/2557672

Tanaka, Y., Tanioka, S.-I., Tanaka, M., Tanigawa, T., Kitamura, Y., Minami, S., Okamoto, Y., Miyashita, M. and Nanno, M., 1997. Effects of chitin and chitosan particles on BALB/c mice by oral and parenteral administration. Biomaterials 18: 591-595. https://doi.org/10.1016/S0142-9612(96)00182-2

Turchini, G.M., Trushenski, J.T. and Glencross, B.D.J.N.A.J.o.A., 2019. Thoughts for the future of aquaculture nutrition: realigning perspectives to reflect contemporary issues related to judicious use of marine resources in aquafeeds. North American Journal of Aquaculture 81: 13-39. https://doi.org/10.1002/naaq.10067

United Nations, 2019. Peace, dignity and equality on a health planet. UN, New York, NY, USA.

Van Huis, A., 2020a. Insects as food and feed, a new emerging agricultural sector: a review. Journal of Insects as Food and Feed 6: 27-44. https://doi.org/10.3920/JIFF2019.0017

Van Huis, A., 2020b. Prospects of insects as food and feed. Organic Agriculture 17: 1-8. https://doi.org/10.1007/s13165-020-00290-7

Wang, G., Peng, K., Hu, J., Yi, C., Chen, X., Wu, H. and Huang, Y., 2019. Evaluation of defatted black soldier fly (*Hermetia illucens* L.) larvae meal as an alternative protein ingredient for juvenile Japanese seabass (*Lateolabrax japonicus*) diets. Aquaculture 507: 144-154. https://doi.org/10.1016/j.aquaculture.2019.04.023

Wilfart, A., Espagnol, S., Dauguet, S., Tailleur, A., Gac, A. and Garcia-Launay, F., 2016. ECOALIM: a dataset of environmental impacts of feed ingredients used in French animal production. PLoS ONE 11: e0167343. https://doi.org/10.1371/journal.pone.0167343

Use of black soldier fly and house fly in feed to promote sustainable poultry production

A. Dörper[1]*, T. Veldkamp[2] and M. Dicke[1]

[1]Laboratory of Entomology, Wageningen University & Research, P.O. Box 16, 6700 AA Wageningen, The Netherlands; [2]Wageningen Livestock Research, De Elst 1, 6700 AH Wageningen, The Netherlands; *anna.doerper@wur.nl

Abstract

The growing human population, changing dietary habits and intensifying competition between food and feed production underline the urgent need to explore novel sustainable production chains. In the past, the poultry sector has gained popularity due to its superior environmental and economic benefits compared to other livestock production systems. Therefore, it is of special interest to focus on refinement and innovation along the value chain to further improve the sector's sustainability. One major issue is the transition towards sustainable protein sources in poultry feed. In this regard, insects are the secret rising stars. Insect species such as the black soldier fly (*Hermetia illucens*) and house fly (*Musca domestica*) have been proposed for farming as multifunctional mini-livestock for feed. One major property of these flies is that larvae can convert low-quality organic waste streams into valuable body mass containing high levels of high-quality protein and fat. Furthermore, the larvae are reported to have health- and welfare-promoting effects due to bioactive compounds and poultry having a natural interest in them. The aim of the current paper is to discuss the state-of-the-art of using black soldier fly and house fly larvae as components of poultry feed and to highlight knowledge gaps, future opportunities and challenges. Some first studies have focussed on the successful partial replacement of soybean meal or fishmeal by these insects on poultry performance. However, since the sector is still in its infancy several uncertainties remain to be addressed. More research is required on identifying optimal inclusion levels, clearly differentiating between insect products based on their nutritional value and health-stimulating effects, and comparing the potential of insect products across species.

Keywords

black soldier fly – *Hermetia illucens* – health and welfare – house fly – *Musca domestica* – nutritional value – poultry feed

1   Global challenges for the poultry feed industry

Producing insects as feed has started in recent years and this industry is expected to grow exponentially in the near future, given that insects are expected to soon be authorised as pig and poultry feed in the EU (IPIFF, 2018). The new insect sector provides animal protein through a sustainable production process with low-value inputs, high-value outputs and low environmental impact and can become an important element of the circular economy (Dicke, 2018). For a novel, circular and sustainable way of producing feed, insects provide an excellent opportunity because various species can be reared on organic left-over streams. Fly larvae are important insects used as feed because they have excellent nutritional quality and can be grown sustainably on various organic waste streams (Bava *et al.*, 2019; Jucker *et al.*, 2019; Miranda *et al.*, 2020; Ocio *et al.*, 1979). Therefore, this review focuses on the application of two fly species that are currently produced as a natural ingredient in poultry feed: black soldier fly (BSF; *Hermetia illucens*) and house fly (HF; *Musca domestica*). Recent literature on BSF- and HF-based poultry feed is evaluated for effects on nutrient digestibility, growth performance, product quality, and health and welfare.

Within the last decade the global human population has grown to about 7.8 billion people (FAOSTAT, 2019) and is forecasted to increase to 9.7 billion in 2050 (FAO, 2018). This will increase the need for food and feed, while rising incomes in especially East and Southeast Asia will shift consumer patterns towards enhanced meat consumption (OECD/FAO, 2019). Therefore, one of the major challenges the food and feed sectors face is the increasing demand for protein (OECD/FAO, 2019; Van Krimpen and Hendriks, 2019). Competition between food and feed production is expected to increase even more, because considerable protein quantities in animal feed originate from sources also edible for humans (Mottet *et al.*, 2017).

The protein component of feed may originate from different sources: crops (cereals, oilseeds and pulses) and non-plant sources (e.g. former foodstuff, fishmeal and processed animal protein) (EC, 2019). Currently, most commonly used protein sources in commercial livestock and fish feed formulations are soybean meal and fishmeal (FAO, 2008; Olsen and Hasan, 2012). The biggest downside of using plant-based protein sources in animal feed, such as soybean meal, is that their cultivation requires a lot of land and water (FAO, 2009). Soybean production leads to deforestation, especially in the Amazon (Hecht, 2005) and, additionally, soybean meal requires long-distance transportation. Because global arable land is limited, increasing food and feed demands will intensify the trade-off between crop land used for food versus feed (OECD/FAO, 2019).

Fishmeal represents a highly valuable protein source in animal feed that is mainly incorporated in aquafeed, but also in feed for other livestock such as poultry and pigs (FAO, 2008; OECD/FAO, 2019). In comparison to soybean meal, fishmeal reaches about four-fold higher market prices. It is expected that fishmeal will gradually be replaced by cheaper oilseed products (OECD/FAO, 2019). Moreover, due to overexploitation of the ocean, fishmeal bears the risk of becoming even rarer in the future (Van Huis *et al.*, 2013).

The production of protein sources for feed is unevenly distributed (Van Krimpen and Hendriks, 2019). Most countries are highly dependent on international trade. The EU, for instance, imports about one quarter of its feed protein. When focussing on the EU's self-sufficiency rate of soybean meal this value even drops to 3% (EC, 2019). The combined effect of land limitation and market dominance of export countries is predicted to amplify competition and to increase market prices (Clément *et al.*, 2018; OECD/FAO, 2019).

In conclusion, feed protein sources are important, but limited resources which will become even more valuable in the next decades. The current protein sources for feed, fishmeal and soybean meal, are not sustainable. Consequently, both products are not suitable to meet with the rapidly growing demand. The development of new, sustainable, protein-rich resources is therefore of major importance.

2   Poultry production: a rising sector

Animal products commonly incorporated in human diets are pork, beef, chicken, milk, and eggs. Of these, the poultry sector recently emerged as one of the most important agricultural branches experiencing rising global attention (Augère-Granier, 2019; OECD/FAO, 2019). Several facts favour the production and consumption of these products. One reason is the increasing public attention towards higher sustainability standards in livestock husbandry (OECD/FAO, 2019). The environmental impact of poultry products is comparatively low (De Vries and De Boer, 2010; Flachowsky *et al.*, 2018). In contrast to other livestock products, chicken meat and eggs have the lowest water and carbon footprint per kg edible protein. Also, land use efficiency per kg edible protein is very high and only exceeded by dairy milk production (Flachowsky *et al.*, 2018). An effective poultry production system should pay full attention to animal health and welfare.

The attractiveness of the two sub-sectors reflects their economic growth in 2008-2018. During this period, chicken meat and egg industries recorded by far the biggest global production increase of ~41 and ~24%, respectively (FAOSTAT, 2020a). This increase is expected to continue in the future (OECD/FAO, 2019). Eggs are highly nutritious and provide essential and valuable nutrients in addition to protein (Farrell, 2013). As global population growth is expected to be accompanied by economic growth, rising incomes are expected to promote dietary shifts towards higher levels of animal protein inclusion (OECD/FAO, 2019).

This is also true for the poultry meat industry. On a global scale, poultry meat greatly impacts the daily protein supply per capita, and exceeds the protein contribution of pig and cattle meat (FAOSTAT, 2020b). The attractiveness of the sector is equally shared amongst consumers and producers (OECD/FAO, 2019). Compared to other production systems such as cattle and pig husbandry, poultry offers several advantages. First of all, it supplies lean meat, which is increasingly meaningful to consumers due to their rising awareness of healthy foods. Secondly, chickens grown for meat have low feed conversion ratios and short life cycles, which promote flexible production, limited production costs and consequently relatively low product prices (FAO, 2018; OECD/FAO, 2019). Finally, unlike most other meat sources, poultry meat consumption has no religious barriers and therefore has a unique market position (EC, 2018). However, the one-sided selection for low feed conversion and fast growth in broilers has led to severe animal welfare and health problems, such as leg and metabolic disorders (EC, 2016; Wallenbeck *et al.*, 2016).

In conclusion, the poultry sector is important for the production of protein-rich nutritious food, largely contributing to the global supply of animal protein. Its main branches are the production of chicken meat and eggs, both showing a steep upward trend in production. The poultry industry is predicted to fulfil a key position in satisfying future rising demands for animal protein. It is, therefore, an essential element to focus on refinement and innovation at every part of the value chain to improve the sector's sustainability.

3   Poultry production systems

3.1   *Broiler production*

Within the poultry sector, chickens raised for meat production are known under the term 'broilers' (EC, 2016). Broiler husbandry can be categorised into rearing systems with an intensive and less intensive production focus. Within the latter category, production is realised with slow-growing broiler strains, while fast-growing broilers dominate in intensive systems (Van Horne, 2018; Vissers *et al.*, 2019). The main difference between the two breeding strategies is the longer growth period of slow-growing broilers required to reach a marketable weight of 2-2.5 kg. In practice, the growth period varies greatly and ranges between 35 to 81 d, while fast-growing strains are slaughtered at an age of 35-42 d (EC, 2016).

The quantity of alternative production systems keeping slow growing broilers has slightly increased within the past years (Van Horne, 2018). The increase of alternative systems is primarily reasoned by welfare issues emerging in conventional systems. Commonly reported issues are leg weaknesses, reduced active behaviour, disease susceptibility, prevalence of contact dermatitis, and ascites, which can result in mortality (EC, 2016; Wallenbeck *et al.*, 2016). However, fast-growing broiler systems represent the majority. Within the EU, for instance, fast-growing broiler systems cover 90-95% of broiler production (Augère-Granier, 2019; Van Horne, 2018).

Across EU countries, market share of both systems varies greatly. For instance, within The Netherlands 30% of marketed broiler meat is based on slow-growing strains (Saatkamp *et al.*, 2019; Stadig, 2019). Alternative husbandry systems using slow-growing broilers emerged rapidly in the Netherlands from 2013 onwards. As for now, this development is unique worldwide (Saatkamp *et al.*, 2019).

Slow-growing broilers experience better welfare (Vissers *et al.*, 2019), due to lower daily weight gain and altered morphological conformation, improved walking abilities (Corr *et al.*, 2003; Wallenbeck *et al.*, 2016), and higher active behaviour (Wallenbeck *et al.*, 2016; Wilhelmsson *et al.*, 2019). Moreover, slow-growing broilers exhibit lower mortality rates (Dixon, 2020) and similar (Sakkas *et al.*, 2018) or improved immune responses (Giles *et al.*, 2019).

Sustainability in animal husbandry has many elements, including animal welfare, economics, environmental impact and social justice (Fernyhough *et al.*, 2020). Within this context one may realise that farm systems that produce slow-growing strains also negatively impact some aspects of sustainability. The improved animal welfare compared to fast-growing strains is accompanied by a higher economic and ecological burden due to higher feed conversion ratios (Rezaei *et al.*, 2018), longer life cycles, higher feed and energy requirements (Tallentire *et al.*, 2018), lower breast meat yield, and meat quality (Fanatico *et al.*, 2008; Quentin *et al.*, 2003). The market price of meat of slow-growing strains is higher and by this the economic burden is reduced. Society in general has a more critical attitude to animal meat production because of ethics (Lund *et al.*, 2016). Thus, the poultry meat sector faces the dilemma of a trade-off between different elements of sustainability.

3.2   *Laying hen production*

The European egg sector is also of great importance because it is the second producer worldwide, behind China (Augère-Granier, 2019). Based on the annual production in 2018, about 7.1 million tonnes of hen eggs were produced, of which 9.9% were produced in The Netherlands (FAOSTAT, 2020a).

Laying hens are reared within four major husbandry systems: enriched cages, barn-, free range- and organic systems (Augère-Granier, 2019). Laying hens used within these production systems can be divided into two categories which reflect strains producing eggs with either white or brown eggshell colour (Leenstra *et al.*, 2012). In practice, the eggshell colour is commonly pre-indicated by the colour of the feather dress (white or brown, Leenstra *et al.*, 2012) or earlobe (white or red, Nie *et al.*, 2016). Several regions show a specific preference for one of the two eggshell colours. For example, within the EU brown eggs dominate the market reflecting consumer preference (International Egg Commission, 2019).

The genetic trait for brown eggshell colour in laying hens negatively impacts economic and environmental aspects. White strains for instance are known to have a lighter body weight (Bean and Leeson, 2003; De Haas *et al.*, 2013; Singh *et al.*, 2009), slightly lighter eggs, equal egg productivity (Bean and Leeson, 2003; Onbaşılar *et al.*, 2015; Singh *et al.*, 2009), and more favourable feed conversion ratios (De Haas *et al.*, 2013; Fernyhough *et al.*, 2020; Onbaşılar *et al.*, 2015) in comparison to their brown companions. Hence, from an economic and environmental point of view white egg-producing strains have a distinct advantage over brown egg-producing strains. The production of brown eggs seems, therefore, mainly driven by consumer perception and preference.

In some EU countries there is a noticeable change within the market share from brown towards white layers (International Egg Commission, 2015). While the ratio in The Netherlands in 2011 was 55:45, the ratio shifted in 2019 to 35:65, respectively. Also China, the global leader in egg production, produced in 2011, 82% of their eggs with brown eggshell colour, while this proportion reduced substantially to 65% in 2019 (International Egg Commission, 2011, 2019). One reason for this development is the reduced aggressive behaviour of white layers (International Egg Commission, 2015). Hence, these strains are of special interest in countries where beak trimming is prohibited (Fernyhough *et al.*, 2020; International Egg Commission, 2015).

The literature comparing feather pecking and plumage scores between white and brown layers appears rather inconclusive. Some reports are in accordance with the previous statement and suggest less issues regarding feather loss in white layers (Damme and Urselmans, 2013; Odén *et al.*, 2002; Struthers *et al.*, 2019). Other authors even indicate worse feather conditions in white layer flocks (De Haas *et al.*, 2013; Dixon and Duncan, 2010; Leenstra *et al.*, 2012; Uitdehaag *et al.*, 2008). Variation in the literature can possibly be attributed to the multifactorial nature of feather pecking (Nicol *et al.*, 2013).

Equivalent to the debate about slow- vs fast-growing broiler strains, also eggshell colour in strains of laying hens differently affects components of sustainability. While economic and environmental characteristics favour strains laying white eggs, consumer beliefs drive production systems towards brown strains.

In summary, sustainability of the poultry sector has several dimensions (animal welfare, economics, environmental impact and social justice), that in many instances trade off. The main component appears to be a potential trade-off between animal welfare and resource friendly production. The main challenge will be to find a balanced solution or to explore new approaches combining those elements, such as environmental enrichment for resource-efficient poultry strains. Reducing pressure on the feed market by presenting an alternative and more environmentally sustainable protein source such as insects as (part of) animal feed might be an option to positively impact poultry health and welfare while production efficiency is guaranteed.

4   Insects as part of the solution

To improve sustainable production, the use of insects in animal feed has gained great interest over the past years, which is highlighted by the strong increase in academic publications on this topic (Van Huis, 2020). The increasing attention for the use of insects in animal feed is justified by their qualitatively high nutritional content, but also by their potential to act as health promoters for livestock.

4.1 Nutritional value

Currently, insect species including BSF, HF, yellow mealworm (TM), lesser mealworm, house cricket, banded cricket, and field cricket have been approved to be used in animal feeds for specific purposes (EC, 2017). Among these insects, the first three species have been proposed to be most promising for production as feed components (Sogari *et al.*, 2019; Van Huis, 2020; Veldkamp *et al.*, 2012). Although these insects have a high protein and fat content the proximate analysis demonstrates the prevalence of nutritional differences (Table 1). HF larvae for instance contain higher levels of crude protein and lower levels of crude fat than BSF (Makkar *et al.*, 2014). However, the composition can be highly variable dependent on the larval rearing substrate (Schmitt *et al.*, 2019).

When focusing on amino acid (AA) profiles, comparable values have been reported for insect sources and soybean meal (Makkar *et al.*, 2014; Veldkamp and Van Niekerk, 2019). Insects are particularly rich in lysine and methionine (Makkar *et al.*, 2014), while cystine appears to be the most limiting AA (Makkar *et al.*, 2014; Veldkamp *et al.*, 2012). Regarding animal feed production, it is essential that insects fulfil the requirements of the target animal. In poultry diets, most limiting AAs are methionine, lysine, tryptophan, and threonine (Fernandez *et al.*, 1994; Harms and Ivey, 1993). According to the essential AA index for broilers, BSF, HF, and TM larvae exceed the required quantities of essential AA (Veldkamp and Bosch, 2015). Thus, insects have a great potential to substitute soybean meal in poultry diets based on the AA profile.

TABLE 1   Proximate composition (minimum and maximum) of yellow mealworm, black soldier fly, and house fly larvae

	Dry matter (DM) g/kg fresh	Crude protein g/kg DM	Crude fat g/kg DM	References
Yellow mealworm	364-424	494-661	245-360	Azagoh *et al.*, 2016; Elahi *et al.*, 2020; Ghaly and Alkoaik, 2009; Ochoa Sanabria *et al.*, 2019
Black soldier fly	300-388	371-492	72-387	Finke, 2013; Jucker *et al.*, 2020; Schmitt *et al.*, 2019; Star *et al.*, 2020; Veldkamp and van Niekerk, 2019; Woods *et al.*, 2020
House fly	205-276	579-646	156-245	Fitches *et al.*, 2019; Qi *et al.*, 2019; Wang *et al.*, 2013

4.2 *Potential health promotors*

In addition to their excellent nutritional composition, insects recently gained further attention due to health promoting characteristics of specific components such as chitin, antimicrobial peptides and lauric acid (Figure 1).

4.2.1 Chitin

Chitin is, after cellulose, the most abundant polysaccharide in nature. Its molecular structure is nearly identical to cellulose; chitin possesses acetamide groups attached at the C-2 position, thus chitin is cellulose's 2-acetamido derivative (Rinaudo, 2006). Chitin is a main component of the exoskeleton of insects (Rinaudo, 2006), which fulfils structural functions and protects the organism against environmental hazards (Lee *et al.*, 2008). It is naturally embedded within a matrix with proteins, lipids and minerals (Józefiak *et al.*, 2016). Enzymes such as chitinase break down the molecule into chitin derivatives like chitosan (Shahidi and Abuzaytoun, 2005). Due to the polyanionic nature of chitosan, the molecule can turn into a strong chelator and bind metal ions (Rinaudo, 2006). Since most animals are unable to produce chitinase in the gastrointestinal tract (Shahidi and Abuzaytoun, 2005), chitin is considered as an indigestible fibre (Tabata *et al.*, 2017). However, mRNA expression of the gene encoding for chitinase within the glandular stomach

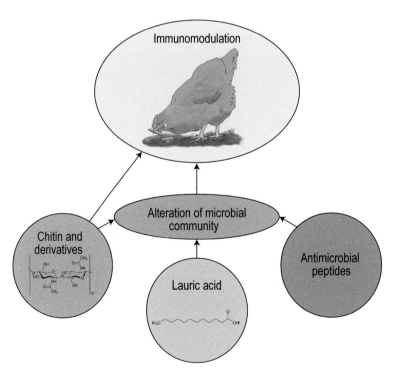

FIGURE 1   Potential direct and indirect modulation of the poultry immune system by insect chitin, lauric acid, and antimicrobial peptides

of chickens, indicates that chickens are able to digest chitin to a certain extent (Suzuki *et al.*, 2002; Tabata *et al.*, 2017).

Chitin and its derivatives have been found to positively impact the health status of animals. For instance, chitosan oligosaccharides (COS) supplemented to broiler diets significantly increased the relative weight of important immunological organs such as the spleen, bursa of Fabricius and thymus (Chi *et al.*, 2017). Similar experiments with COS in broiler diets highlighted its potential to enhance IgM levels and proinflammatory cytokine expression in blood serum (Deng *et al.*, 2008). Moreover, chitin derivatives can also serve as a substrate for caecal bacteria in chickens. The alteration of the caecal microbial community influences the health status of chickens, because microorganisms play a key role in shaping the immune system (Li *et al.*, 2007). Based on these findings, provision of chickens with insects containing chitin is expected to impact health parameters. Moreover, because the structure of chitin varies between insect species (Kaya *et al.*, 2015) and the effect of chitin on the immune system is complex and size-dependent (Lee *et al.*, 2008), their health promoting properties might differ.

To evaluate the impact of insect chitin as feed component on the immune system of animals, the analysis of ingested chitin levels is indispensable. Currently, several methods are available for quantification of chitin in whole insects (Finke, 2007, 2013; Hahn *et al.*, 2018; Woods *et al.*, 2020). However, due to the application of different methodological approaches, results are rather variable (Hahn *et al.*, 2018). For example, reported chitin contents in BSF larvae range from 54 g/kg dry matter (DM) (Finke, 2013) to 106 g/kg DM (Hahn *et al.*, 2018). Consequently, there is a need for an accurate, practically applicable and uniformly accepted method to determine chitin levels in insects. Moreover, within the cited studies neither the instar nor the age in days of the harvested larvae was mentioned. Chitin levels increase with development of the larvae (Wong *et al.*, 2019). It is of utmost importance to facilitate the comparability of chitin levels among literature.

4.2.2   Lauric acid

Lauric acid (C12:0) belongs to the group of medium-chain fatty acids (MCFA), which are known to have antibacterial properties (Skrivanova *et al.*, 2006). Insects such as BSF and HF larvae contain considerable amounts of crude fat (Veldkamp *et al.*, 2012). Especially the fat fraction of BSF larvae has been highlighted to contain great quantities of MCFAs and particularly lauric acid (Cullere *et al.*, 2019a; Gasco *et al.*, 2018). The mechanism underlying MCFA disturbance of bacterial development is not fully known to date. Yet, *in vitro* experiments indicate that MCFAs are able to enter bacterial cells through their membrane (Kim and Rhee, 2013; Skrivanova *et al.*, 2006). The entry process is assumed to lead to membrane damage, which consequently changes the cell's permeability. This damage may further promote the entry of other antibacterial compounds into the cell (Kim and Rhee, 2013).

In vitro analysis of the effects of lauric acid on *Clostridium perfringens* indeed shows that the MCFAs enter and disrupt the cytoplasmatic structure of the cell, although the bacterial membrane integrity remained intact (Skrivanova *et al.*, 2006). Additional testing with different concentrations of MCFA on *C. perfringens* development identified 0.5 mmol/l lauric acid as the minimum inhibitor concentration required to induce antibacterial activity. However, when *Escherichia coli* bacteria were treated with 0.125 to 1.5 mmol/l lauric acid, only negligible bactericidal effects were recorded (Kim and Rhee, 2013). Possibly membrane differences between Gram-negative and -positive bacteria may play a role in this. Gram-positive bacteria are more susceptible to damage caused by MCFAs, whereas it seems rather challenging to successfully inactivate Gram-negative bacteria with MCFAs (Kim and Rhee, 2013; Skrivanova *et al.*, 2006). Also, the type of MCFAs appears to play an important role. The application of the MCFA caprylic acid, for instance, had dose-dependent lethal effects on Gram-negative *E. coli* bacteria (Kim and Rhee, 2013; Skrivanova *et al.*, 2006). Furthermore, lower pH values resulting from MCFAs promoted bacterial cell death (Kim and Rhee, 2013).

Because especially Gram-positive bacteria seem to be sensitive to MCFAs, this property could be used to reduce pathogen loads in livestock. For example, in broiler chickens challenged with *C. perfringens*, supplementation of experimental diets with MCFA (C6-C12), significantly reduced the prevalence of necrotic enteritis lesions in the small intestine. The authors suggest that this was due to growth inhibition and lethal effects of MCFAs on the bacterium (Timbermont *et al.*, 2010). Thus, the provision of insects in chicken feed may modify the microbial community and reduce the prevalence of pathogens in the gastrointestinal tract.

4.2.3   Antimicrobial peptides

An important category of molecules with immunomodulatory functions contained in insects are antimicrobial peptides (AMPs; Bulet *et al.*, 1999), also known as host defence peptides (Cole and Nizet, 2016). AMPs are part of the innate immune system of various living organisms besides insects, including plants, fungi, bacteria, and animals (Józefiak and Engberg, 2017). Their specific mode of action varies on the basis of their amino acid sequence and structural organisation (Bulet *et al.*, 1999).

AMPs currently gain immense global attention, due to their antibacterial function (Gasco *et al.*, 2018) and successful action against multi-drug resistant pathogens, which are a rising threat for society (Bahar and Ren, 2013). For instance, HF larvae and pupae express AMP's such as defensin in several tissues, which act upon invading Gram-positive bacteria (Andoh *et al.*, 2018; Dang *et al.*, 2010). Experiments with HF AMPs even showed that the administration to *Salmonella pullorum* infected chickens can serve as effective treatment of the intestinal disease, without having negative effects on mucosal epithelial cells (Wang *et al.*, 2017). The novel defensin-like peptide 4 (DLP4) detected in BSF larvae exhibits activity against Gram positive bacteria including methicillin-resistant *Staphylococcus aureus* (Park *et al.*, 2015). While insects contain a wide spectrum of AMPs which have been identified to function against Gram-positive bacteria, most of them show no or little activity against Gram-negative bacteria (Dang *et al.*, 2010; Meylaers *et al.*, 2004; Park *et al.*, 2015; Vogel *et al.*, 2018).

To benefit from insect AMPs in animal feed it is important to understand their biosynthesis. For example, although defensin in HF larvae and pupae is continuously expressed at a low level, injury of HF larvae and pupae by a needle or immunised with a needle dipped in a bacterial suspension sharply elevates defensin levels (Andoh *et al.*, 2018; Dang *et al.*, 2010). Due to larval and pupal immunisation, defensin can reach values up to 300 and 600 times higher than before immunisation (Andoh *et al.*, 2018). A similar infection-dependent synthesis of DLP4 is observed in BSF larvae, though DLP4 expression is absent in unchallenged larvae (Park *et al.*, 2015). Besides, AMP expression has been found to be diet-dependent (Vogel *et al.*, 2018). AMP levels in insect products might be further influenced by larval processing. Although some AMPs have been found to be thermal stable (Hao *et al.*, 2008; Li *et al.*, 2017), heat treatment might largely impact heat labile AMPs and reduce their presence in the final product. Currently, various insect products in addition to live larvae are applied in animal feed. This includes the use of non-defatted or defatted insect meal. To the best of our knowledge, there is no information available on AMP activity in those manufactured insect products. Such information would be highly valuable.

In conclusion, current knowledge indicates that AMP production in insects is variable and is affected by injury-, infection- or diet-dependent induction. Rearing conditions and processing of insect larvae therefore likely influence their impact on immunomodulatory functions when used in livestock feed. It is, therefore, important to determine AMP levels in insect products to properly estimate their effect on chickens.

5   BSF and HF in poultry feed

Although there are yet only few studies available examining the use of BSF and HF larvae in animal feed, some studies already reported their successful inclusion in poultry diets. To prevent that conclusions are affected by nutritional imbalances between experimental diets, the following paragraphs focus only on literature explicitly indicating the comparison of isonitrogenous and isocaloric diet formulations. Table 2, 3 and 4 report the effect of dietary BSF and HF products inclusion in broiler, laying hen and other poultry feed and their effect on digestibility, performance, product quality, and animal health.

4.3   *Broilers*

4.3.1   Production performance

Few publications are available regarding the application of BSF or HF larvae in broiler nutrition. Incorporation of up to 10% of defatted BSF larval meal in broiler diets revealed an increased feed intake (FI) and average daily gain (ADG) during the first phase of rearing. On the other hand, the development of broilers receiving higher levels of partially defatted BSF larvae were rather hampered (Dabbou *et al.*, 2018) and further led to the reduction of bacterial diversity in the caeca and intestinal mucin production (Biasato *et al.*, 2020). Also inclusion of 6 and 8% BSF larval meal improved the animal performance by reducing the FI and FCR while the final BW was unaffected (Attivi *et al.*, 2020).The findings give an indication that in order to maintain favourable performance results, insect inclusion in broiler diets should be restricted to an optimal inclusion level which needs to be defined by further research (Dabbou *et al.*, 2018). Inclusion of HF larval meal lower than 2% increased the final BW, and BWG at 1.6, and 2% of inclusion. Also FI was influence which led to higher uptake of feed at 0.8 and 1.2% HF larval meal inclusion. The FCR also indicated a positive development by showing lower values at all inclusion levels (Okah and Onwujiariri, 2012). In contrast, inclusion of up to 22% HF larval meal did not alter nutrient retention (Adeniji, 2007) and performance parameters of broilers (Adeniji, 2007; Ocio *et al.*, 1979). Furthermore, results for the apparent ileal digestibility and true ileal digestibility of AA of HF larval meal and fishmeal investigated in broilers identify AA digestibility in HF larval meal comparable with AA digestibility in fishmeal (Hall *et al.*, 2018). Because diet formulation by Hall *et al.* (2018) is not in accordance with the above mentioned criterion (iso-nitrogenous and iso-caloric diet formulation), it was chosen to only report results regarding the AA digestibility. Significant positive effects on broiler development, also during later growth stages, have been reported when broilers were provided with up to 8% of HF larval meal in their diet (Khan *et al.*, 2016, 2018a,b).

Nonetheless, also between studies investigating the same insect species and using comparable inclusion levels, outcome parameters can be highly variable. For instance, broilers fed diets containing in total 4 and 8% of HF larval meal showed, in contrast to prior findings, reduced ADG during the starter phase, which resulted in higher feed conversion ratios (FCR). Darker colour of HF larval meal is assumed to be responsible for this adverse development (Elahi *et al.*, 2019). However, it is also worth noticing that the crude fat (CF) content of HF meal used within these studies varied tremendously, with values of 5.6 (Elahi *et al.*, 2019) and 27.9% (Khan *et al.*, 2016). Varying levels of CF in insect products may be responsible for differences in performance of broilers (Elwert *et al.*, 2010). This condition furthermore hampers the comparability of results between studies, because the insect products have different nutritional and likely also bioactive values.

4.3.2   Product quality

Besides affecting broiler performance, inclusion of partly defatted BSF larvae improved carcass traits, such as breast yield (Schiavone *et al.*, 2019). Also the use of
HF larval meal beneficially affected broiler dressing percentage (Khan *et al.*, 2018a),
while olfactory properties of the meat remained unaffected (Khan *et al.*, 2018a,b)
or turned out even more favourable for consumers (Khan *et al.*, 2016; Radulović *et
al.*, 2018). In some cases the analysed carcass characteristics were not significantly
affected (Elahi *et al.*, 2019). It can be concluded that low levels of inclusion of these
insect meals in broiler diets do not negatively influence carcass traits compared to
standard feed. A closer analysis of the fatty acid (FA) composition of broiler meat
revealed that partly defatted BSF larval meal increased the content of medium unsaturated FA (MUFA) and reduced the occurrence of poly unsaturated FA (PUFA).
Since the FA profile of broiler meat is dependent on the feed source (Schiavone *et
al.*, 2019) and larval FA composition is dependent on the substrate on which the
larvae were reared (Cullere *et al.*, 2019b), the unfavourable development might be
adjusted by the choice of larval substrate (Schiavone *et al.*, 2019). Also, the killing
method might have an influence on nutritional characteristics of the insect end
product (Larouche, 2019).

4.3.3   Health and welfare

Inclusion of 2, 4, and 6% of BSF larval meal in broiler diets reduced their blood
albumin levels, while 8% inclusion resulted in higher concentrations. Also triglyceride levels were affected resulting in higher values for 6 and 8% insect meal
inclusion, while uric acid was affected at any inclusion level, indicating efficient
utilisation of protein from the diet. However, protein digestibility was adversely influenced at any BSF meal inclusion level lower than 8% (Attivi *et al.*, 2020). Also the
development of intestinal organs was affected by the treatment. Higher inclusion
levels reduced the gizzard weight and intestinal length (Attivi *et al.*, 2020). On the
other hand, increased gizzard weight was reported when broilers were fed HF larval
meal. The reason for this development remains unknown (Okah and Onwujiariri,
2012). Other studies report, in contrast to the expected positive effects of bioactive
compounds in insect products on broiler health parameters, that immune organs
in broilers were not altered by the inclusion of larval meal of either of the two fly
species. Moreover, haematological parameters were within the common physiological range (Dabbou *et al.*, 2018; Elahi *et al.*, 2019). Nonetheless, albumin to globulin
ratios were not measured in the studies cited. Because in laying hens these parameters showed significant improvements (Bovera *et al.*, 2018; Marono *et al.*, 2017), it
would be an interesting addition to test the effects of BSF and HF products in broiler feed on those parameters as well. Strong positive effects on foraging and general
activity behaviour have been found when providing broilers with live BSF larvae.
Most pronounced were the effects when the larvae were included to cover 10% of
the DM intake and the provision was spread over 4 feeding moments a day. These

TABLE 2 Inclusion of black soldier fly (BSF) and house fly (HF) products in broiler diets and their effect on digestibility, performance, product quality, and animal health, as reported in studies that compare diets that are isonitrogenous and isocaloric

Age	Insect product	Target replacement	Insect inclusion %	Results (digestibility, performance, product quality and animal health)	Reference[1]
1-35 d	partially defatted BSF larval meal	soybean meal, soybean oil, corn gluten meal	0, 5, 15	Improved LW and DFI during starter period, improved final LW, no effect on FCR at <10% BSF inclusion. Lower LW during growing and finisher period and increased FCR at 15% BSF meal inclusion.	Dabbou et al., 2018
1-35 d	partially defatted BSF larval meal	soybean meal, soybean oil, corn gluten meal	0, 5, 10, 15	Mucin staining intensity in the intestinal villi reduced at 10 and 15% meal inclusion. Shift in caecal microbiota.	Biasato et al., 2020
14-56 d	BSF larval meal	fishmeal	0, 2, 4, 6, 8	Lower feed intake and FCR at 6 and 8% inclusion. Lower BWG at 2, 4 and 6% inclusion. Higher BWG at 8% inclusion. No difference in final BW. Lower Gizzard weight and intestinal length at 4, 6, and 8% inclusion. No difference in total protein, lower albumin at 2, 4, and 6% inclusion and higher values at 8% inclusion. Lower level of uric acid at any inclusion level and higher triglyceride values for 4, 6, and 8% inclusion. Lower apparent crude protein digestibility at 2, 4, and 6% inclusion level.	Attivi et al., 2020
1-35 d	defatted BSF larval meal	soybean meal, soybean oil, corn gluten meal	0, 5, 10, 15	Increased carcass weight and breast yield at <10% BSF meal inclusion. Linear increase of meat redness to a maximum at 15% BSF inclusion. Linear decrease of meat yellowness, CP, MUFA, and reduced moisture and PUFA as inclusion level was increased.	Schiavone et al., 2019
1-42 d	live BSF	–	0, 5, 5, 10, 10	Live BSF were given at inclusion levels of 5 and 10% of the dietary FI. Larvae were provided 2 (BSF52, BSF102) or 4 (BSF54, BSF104) times a day. Final BW lower in BSF102. No influence on ADMI and ADMEI. No effect on shuffling, perching, stretching and comfort behaviour. All remaining behaviours (standing idle, walking, eating, drinking, foraging, ground pecking, resting, and activity) were affected by treatment, week, and/or interaction. Higher ground pecking and total foraging at any inclusion level and provision level at any timepoint (week). Despite week 1 group BSF104 had consistently lower values for resting and higher values general activity than the control group. Food pad dermatitis was not affected. Hock burn score was higher in control group than BSF102 and BSF104. Higher gait scores in control group compared to BSF54, BSF102 and BSF104.	Ipema et al., 2020

1-35 d	HF larval meal	fishmeal	0, 0.8, 1.2, 1.6, 2	Increased BW at 0.8, 1.2, and 1.6% HF meal inclusion. Higher BWG at 1.6 and 2% inclusion. Higher feed intake at 0.8 and 1.2% inclusion. FCR lower at any inclusion level. Lower dressing % at 0.8% inclusion. Lower % of the heart at 1.2, 1.6, and 2% inclusion. Larger % of Gizzard at 0.8, 1.6, and 2% inclusion. Higher % inguinal fat at 1.2 and 1.6% of inclusion.	Okah and Onwujiari, 2012
1-42 d	HF larval meal	groundnut cake	0, 5.5, 11.0, 16.5, 22.0	No effect on nutrient retention and performance parameters.	Adeniji, 2007
1-28 d	HF larval meal	fishmeal	0, 4	No effect on any performance parameter.	Ocio et al., 1979
21-28 d	HF larval meal	fishmeal	0, 0, 20, 40, 60	Regression analysis was used to investigate the digestibility of AA. Therefore, diets contained either 20, 40, and 60% fishmeal, or 20, 40, and 60% HF larval meal. AIDC and TIDC of fishmeal and HF larval meal was comparable.	Hall et al., 2018
1-31 d	HF larval meal	soybean meal	0, 4, 5, 6	Higher DM, EE, and ash digestibility and lower crude fibre digestibility at 6% inclusion. Higher CP digestibility at 4 and 6% inclusion. Increased BW, reduced FI, and FCR, higher dressing %, and higher apparent metabolisable energy at any level of meal provision. Unchanged organoleptic meat properties.	Khan et al., 2018a
1-31 d	HF larval meal	soybean meal	0, 1, 2, 3	Apparent metabolisable energy increased at 2 and 3% meal inclusion. Digestibility of nutrients remained unchanged. Reduced FI and FCR at any inclusion level. Increased weight gain at 2 and 3% inclusion. Dressing % of meat increased at 3% meal inclusion. Improved organoleptic properties due to meal provision. Improved meat tenderness at 2% meal inclusion, higher juiciness at 1% meal inclusion. Reduced colour at 3% meal inclusion, better meat flavour at 1 and 2% meal inclusion.	Khan et al., 2016
1-35 d	HF larval meal	soybean meal	0, 8	No effect on FI, mortality, dressing %, and sensory meat characteristics. Increase in BW and reduction of FCR with administration of HF meal.	Khan et al., 2018b
1-42 d	HF larval meal	soybean meal	0, 4, 8	Linear reduction of BW, ADG, and increase in FCR during starter phase. No effect on carcass quality. Alteration of blood parameters were in physiological range.	Elahi et al., 2019
1-42 d	HF larval meal	soybean meal	0, 5 (1-7 d), 4 (8-42 d)	Increased BW, ADG, ADFI, and reduced FCR. Increased breast and leg flavour, aroma and desirability.	Radulović et al., 2018

AA = amino acid; ADG = average daily gain; ADMI = average daily dry matter intake; ADMEI = average daily metabolizable energy intake; BW = body weight; BWG = body weight gain; CP = crude protein; DFI = daily feed intake; DM = dry matter; EE = ether extract; FCR = feed conversion ratio; FI = feed intake; LW = live weight; MUFA = monounsaturated fatty acids; PUFA = poly unsaturated fatty acids; TIDC = true ileal digestibility coefficient

[1] Diet formulation by Hall et al. (2018) is not in accordance with the criterion of iso-nitrogenous and iso-caloric diet formulation. Therefore, only results regarding the AA digestibility were include in the table.

TABLE 3 Inclusion of black soldier fly (BSF) and housefly (HF) products in laying hen diets and their effect on digestibility, performance, product quality, and animal health, as reported in studies that compare diets that are isonitrogenous and isocaloric

Age	Insect product	Target replacement	Insect inclusion %	Results (digestibility, performance, product quality and animal health)	Reference
20–40 wk	partially defatted BSF larval meal	soybean meal	0, 7.3, 14.6	Lower apparent digestibility of DM and OM at 14.6% inclusion and reduced apparent CP digestibility at any inclusion level.	Bovera et al., 2018
				No effect on weight gain, FI, FCR, and egg weight. Higher lay% with 7.3% inclusion of meal and higher egg mass at all inclusion levels. Higher entire intestinal length and of the jejunum as % of LW. Higher serum magnesium, potassium, chlorine, globulin and lower albumin to globulin ratios with 14.6% meal inclusion. Lower bilirubin, cholesterol, triglyceride and blood urea nitrogen levels at any inclusion level.	
67–78 wk	live BSF larvae	soybean meal	0, 10	No effect on FI, FCR, BW, laying rate, egg weight, egg mass, mortality, and egg quality parameters. Reduction of feather damage. Higher counts of hens on the floor during the morning and lower counts in the afternoon when larvae were provided.	Star et al., 2020
24–45 wk	defatted BSF larval meal	soybean meal	0, 17	Reduced albumen weight, higher proportion of yolk, and lower albumen proportion in eggs. Redder egg yolk, higher total carotenoid and γ-tocopherol levels. Lower MUFA and higher PUFA n-6 in egg yolk.	Secci et al., 2018
20–36 wk	partially defatted BSF larval meal	soybean meal	0, 7.3, 14.6	Higher egg weight, yolk weight. Higher eggshell weight with 7.3% meal inclusion. Lower eggshell % and eggshell thickness with 14.6% inclusion. Alteration of the yolk colour. Lower egg foaming capacity of egg albumen. Modified physical and chemical characteristics of angel cake.	Secci et al., 2020
24–45 wk	defatted BSF larval meal	soybean meal	0, 17	Reduced final LW, and weight gain, FI, FRC, lay %, egg weight, and egg mass. High variation in egg size. Higher serum globulin and lower albumin to globulin ratio. Lower serum cholesterol and triglycerides. Higher serum calcium and lower chlorine levels.	Marono et al., 2017
24–45 wk	defatted BSF larval meal	soybean meal	0, 17	Increased diversity of the microbial population and altered caecal microbiota in type and relative abundance of microbes. Increased concentrations of SCFA in ceca.	Borrelli et al., 2017

Age	Insect product	Protein source replaced	Inclusion levels	Results	Reference
25-45 wk	defatted BSF larval meal	soybean meal	0, 17	Lower apparent ileal digestibility of DM, OM, and CP. Lower LW. Heavier full digestive tract as % of LW, higher specific activity of duodenum maltase, lower activity of jejunum IAP, higher activity of ileal maltase and saccharase, and lower activity of IAP and γGT. Higher villi height in duodenum, lower villi height in jejunum and ileum. Higher crypt depth and lower villi to crypt ratio in ileum Higher volatile fatty acid production in caecum. Higher acetate, butyrate and total volatile fatty acid levels in caecum.	Cutrignelli et al., 2018
18-26 wk	fresh BSF, dried BSF, BSF extract	fishmeal	0, 8, 8, 8	Inclusion of any of the BSF products increased the phagocytotic activity of peritoneal macrophages towards non-protein A Staphylococcus aureus.	Irawan et al., 2019
20-40 wk	partially defatted BSF larval meal	soybean meal	0, 7.3, 14.6	Lower villi height, villi to crypt ratio in duodenum and jejunum at any inclusion level. Higher crypt dept in jejunum at 14.6% inclusion. No effect compared to the control group on ileal level. Lower maltase activity in duodenum at 14.6% inclusion and lower IAP in jejunum at any inclusion level. Lower γGT activity in ileum at 7.3% inclusion. Higher concentration of acetate, butyrate and total VTA in caecum at 14.6% inclusion. Lower concentration of valerianic in caecum at any inclusion level. Higher level of Isobutyrate and lower level of valeriante as % of total VFA in caecum at any inclusion level. Higher level of butyrate in caecum at 14.6% inclusion.	Moniello et al., 2019
50-58 wk	undefined larval meal	fishmeal	0, 2.4, 4.7, 7.1, 9.4	No effect on FI, weight gain, FCR and mortality. Lower hen-day production and eggshell weight at highest level of meal inclusion. Lower shell thickness at 4.7% meal inclusion and shell weight at 9.4% meal inclusion. Other egg quality parameters unaffected.	Agunbiade et al. 2007
36-44 wk	HF larval meal	fishmeal	0, 0.9, 1.8, 2.7, 3.6	BW, FI, egg production, egg weight, feed efficiency, and any tested egg quality characteristics were unaffected. Egg yolk cholesterol, triglyceride, and calcium lower with inclusion of HF meal, while the concentrations of phosphorus was increased.	Akpodiete et al., 1998

BW = body weight; BWG = body weight gain; CP = crude protein; DM = dry matter; FA = fatty acid; FCR = feed conversion ratio; FI = feed intake; γGT = γ-glutamyl transpeptidase; IAP = intestine alkaline phosphatase; LW = live weight; MUFA = monounsaturated fatty acids; OM = organic matter; PUFA = poly unsaturated fatty acids; SCFA = short chain fatty acids; VFA = volatile fatty acids.

TABLE 4   Inclusion of black soldier fly (BSF) and housefly (HF) products in other poultry diets and their effect on digestibility, performance, product quality, and animal health, as reported in studies that compare diets that are isonitrogenous and isocaloric

Poultry species/type	Age	Insect product	Target replacement	Insect inclusion %	Results (digestibility, performance, product quality and animal health)	Reference
Muscovy duck	3–50 d	partially defatted BSF larval meal	corn gluten meal	0, 3, 6, 9	No effect on LW, ADG, FI, and FCR. Haematological traits were unaffected. Linear reduction of serum triglycerides, cholesterol, magnesium, alkaline phosphatase, creatinine, malondialdehyde, and nitrotyrosine. Linear increase of serum iron. No effects on histopathological scores.	Gariglio et al., 2019a
Japanese broiler quail	10–28 d	BSF larval meal	conventional protein/fat sources	0, 10, 10	Increased apparent digestibility of DM, OM and reduced apparent digestibility of ether extract of quails fed BSF2. Higher apparent digestibility of starch of quails fed meal of larvae reared on 100% layer mash (BSF1). Higher apparent digestibility of gross energy, higher metabolisable energy. Reduced slaughter weight and BWG of quails fed meal of larvae raised on 50:50 fish offal and layer mash (BSF2). Improved carcass and breast meat traits of quails fed BSF1. Lower carcass weight of quails fed BSF2. Lower FI of larval meals when given a choice.	Woods et al., 2019
Turkey	1–35 d	live BSF larvae	soybean meal	0, 12	Lower FCR, higher FI and BWG and reduced aggressive pecking towards the back and tail-base in the final week. More feed pecking during first week, less feed pecking and drinking during the third week and less object pecking during the fifth week.	Veldkamp and van Niekerk, 2019
Japanese broiler quail	10–28 d	BSF larval meal	conventional protein/fat sources	0, 10, 10	Higher meat cooking losses when fed. Higher protein, lipid, and cholesterol levels in quail meat of birds fed BSF1. Alteration of AA content and FA profile of quail meat fed BSF1. Modified meat juiciness and fibrousness and colour of quails fed BSF2.	Cullere et al., 2019b

Muscovy duck	3-50 d	partially defatted BSF larval meal	corn gluten meal	0, 3, 6, 9	Apparent digestibility of CP linearly reduced during the starter phase. Positive linear response of apparent digestibility of EE during the whole trial. Final LW, ADG, FI, and FCR was not affected. Intestinal morphology remained unaffected.	Gariglio et al., 2019b
Japanese laying quail	6-13 wk	BSF larval meal	fishmeal	0, 3.18, 6.37, 9.56	Egg production was reduced at 6.37 and 9.56% BSF meal inclusion. Egg shell thickness was reduced at 9.56% of inclusion. Egg Haugh unit, egg weight, egg yolk index, and egg yolk colour was unaffected.	Suparman et al., 2020
Japanese broiler quail	10-28 d	BSF larval meal	conventional protein/fat sources	0, 10, 10	Secondary response increased in BSF1. Lysozyme activity unaffected at day 27 but at day 37 activity was higher in BSF2. Bacterial activity unaffected at both timepoints. Wing web swelling index significantly higher in both BSF groups. At day 27 higher a_2-globulin in both BSF groups and lower γ-globulin in BSF2 group. At day 37, a_2 globulin higher in both BSF groups. Other blood parameters and microbial count in caecal content remained unaffected.	Pasotto et al., 2020

AA = amino acid; ADG = average daily gain; BW = body weight; BWG= body weight gain; CP = crude protein; DM = dry matter; EE = ether extract; FA = fatty acid; FCR = feed conversion ratio; FI = feed intake; OM = organic matter.

behavioural changes likely also reduced the incidence of hock burns and improved gait score (Ipema *et al.*, 2020).

4.4 *Laying hens*

4.4.1 Production performance

Including 10% live BSF larvae, up to 9.4% undefined larval meal and up to 14.6% of partially defatted BSF larval meal in layer nutrition showed no change in FI, weight gain and FCR (Agunbiade *et al.*, 2007; Bovera *et al.*, 2018; Star *et al.*, 2020). Moreover, egg weight remained either unaffected (Secci *et al.*, 2018) or increased by administration of BSF meal in layer diets (Secci *et al.*, 2020). However, because higher egg weight was accompanied by lower eggshell thickness, this might impact the fragility of the eggs (Secci *et al.*, 2020). Likewise administration of 9.4% undefined larval meal reduced the eggshell weight, but also hen egg production (Agunbiade *et al.*, 2007). Furthermore, performance parameters were reduced when higher inclusion levels of defatted BSF larval meal were administered (Marono *et al.*, 2017), which can be explained by either feed colour or chitin levels. Chickens are sensitive to feed colour (Heshmatollah, 2007). Hence, feed colouration may play a role at higher levels of insect meal inclusion, because BSF meal is darker than SBM (Bovera *et al.*, 2018). On the other hand, inclusion of up to 3.6% HF larval meal did not affect laying hen performance (Akpodiete *et al.*, 1998).

4.4.2 Product quality

In terms of consumer health, the binding capacity of chitin to nutrients can have favourable effects on end products for humans by reducing serum triglyceride and cholesterol (Bovera *et al.*, 2018; Marono *et al.*, 2017) and therefore egg cholesterol levels (Secci *et al.*, 2018). The same results were found when HF larval meal was administered (Akpodiete *et al.*, 1998). Further egg quality parameters influenced by the inclusion of BSF meal were the alteration of the yolk colour (Secci *et al.*, 2018, 2020), higher proportions of the yolk, lower albumen proportions in the egg, and less egg shell thickness (Secci *et al.*, 2020). The latter might be caused due to the higher egg weight, since calcium resources for shell production are limited. In contrast laying hens fed undefined larval meal showed lower egg shell weight at the highest inclusion level (Agunbiade *et al.*, 2007).

4.4.3 Health and welfare

The production of short-chain fatty acids (SCFA) in the cecum proved to have a positive effect when defatted BSF larval meal was included in the laying hen diet (Borrelli *et al.*, 2017; Cutrignelli *et al.*, 2018). The assumed mechanism is that chitin serves as a fermentation source for caecal microbiota, which leads to the production of SCFA (Borrelli *et al.*, 2017; Cutrignelli *et al.*, 2018). SCFA such as butyric acid serve as an energy source for epithelial cells in the colon and promote mucosal growth. Additionally, SCFA inhibit the growth of acid-intolerant pathogenic bacteria such as *C. perfringens*, *E. coli* and *Salmonella* (Borrelli *et al.*, 2017; Elnesr *et al.*, 2020). However, whether insect products indeed reduce the prevalence of these bacteria requires further investigation.

The better health status of laying hens has been confirmed by serum analysis. Elevated globulin levels and lower albumin to globulin ratios indicate higher resistance and improved immune responses in layers fed partially defatted or defatted BSF meal (Bovera *et al.*, 2018; Marono *et al.*, 2017). However, the inclusion of 7.6% of partially defatted BSF meal did not alter these immune parameters. Therefore, low levels of insect meal may be insufficient to induce immunological changes at the serum level (Bovera *et al.*, 2018). Besides administration of various BSF products showed that those larval products have the potential to increase the phagocytic activity of macrophages, potentially improving the innate immunity (Irawan *et al.*, 2019). Further promotion of laying hen welfare and health may be achieved by the provision of live larvae instead of larval meal. Insects are part their natural diet and may enable laying hens to express natural behaviour. Feather pecking is a well-known issue in layer flocks. Due to the provision of live larvae as a natural pecking substrate, feather conditions of laying hens were successfully improved. This achievement was likely caused by laying hens spending more time pecking on larvae rather than their pen companions. Besides pecking and ingestion of larvae might be rewarding for laying hens, thus satisfying their need to peck (Star *et al.*, 2020).

In addition to the beneficial prebiotic properties, it was also found that chitin can hinder the digestion of various nutrients (such as protein) by chitin-protein matrixes (Bovera *et al.*, 2018). This is supported by profound reduction of the apparent ileal digestibility of protein, but also organic matter, due to administration of defatted BSF larval meal (Cutrignelli *et al.*, 2018). Also morphological changes such as lower duodenal villi height and lower enzymatic activity are a result of the lower digestibility of the feed (Moniello *et al.*, 2019). Thus, to successfully include insect products in layer diets it seems crucial to determine optimal inclusion levels.

In conclusion, inclusion of BSF and HF larvae in feed improves the performance, product quality, health, and welfare of layer hens dependent on the inclusion level. Furthermore, provision levels for optimal development of hen performance and health do not necessarily coincide. Also storability of insect products may play a role in the future regarding the choice of larval product. It should be considered that processing of live insects can improve its storage properties (Veldkamp *et al.*, 2012). Therefore, insect meals are storable for longer periods than live insects and further defatting can reduce the risk of lipid oxidation (Schiavone *et al.*, 2017).

4.5 *Other poultry*

4.5.1 Production performance
BSF larvae have also been successfully incorporated in diets of other poultry species. The performance of Muscovy ducks remained unaffected by 9% inclusion of partially defatted BSF meal (Gariglio *et al.*, 2019a). Likewise, performance parameters of Japanese quails remained unchanged when fed similar levels of BSF larval meal, of larvae that had been reared on layer mash. In contrast, meal of larvae raised on 50:50 fish offal and layer mash adversely affected quail performance, likely due to the odour and palatability of the meal (Woods *et al.*, 2019). Turkeys showed higher FI and ADG, and lower FCR when they were provided with live BSF larvae (Veldkamp and van Niekerk, 2019).

4.5.2 Product quality
As reported in broilers (Schiavone *et al.*, 2019), the administration of larval meal increased MUFA and decreased PUFA in quail meat (Cullere *et al.*, 2019b). Furthermore, provision of BSF meal improved carcass and breast meat traits of quails fed meal, of larvae reared on layer mash, while meal of larvae raised on 50:50 fish offal and layer mash lowered the carcass weight of quails. In layer quails BSF meal provision affected egg production at levels higher than 3.18%, this might have occurred due to lower palatability or darker colour of the feed. Moreover, egg shell thickness was negatively influenced at the highest level of BSF meal inclusion, likely caused by the shifted calcium and phosphorus levels in BSF meal compared to fishmeal (Suparman *et al.*, 2020).

4.5.3 Health and welfare
In Japanese broiler quails, administration of BSF larval meal had a significant effect on health-related blood parameters. The wing web swelling index, an indicator for the immune system's lymphoproliferative capacity, was improved in BSF-fed groups. Also α2-globulin levels measured at day 27 and 37 were increased in the blood. Albumin-to-globulin ratio was unaffected. However, the results indicate a tendency of BSF-meal-fed groups to have lower values. Besides, the blood profile of quails was affected dependent on the prior rearing substrate of the BSF larvae. Broilers fed BSF larval meal of larvae reared on layer mash showed a higher secondary response, while quails fed larval meal of larvae reared on 50:50 layer mash and fish offal had lower γ-globulin levels at day 27 and higher lysozyme activity at day 37 compared to the control group. While blood parameters were altered, the treatments recorded no alteration of the microbial composition within the caeca. The results emphasise the immunomodulatory properties of BSF larval meal in quails (Pasotto *et al.*, 2020). Furthermore, in Muscovy ducks BSF meal had only minor effects on the digestibility of nutrients. For instance, crude protein (CP) digestibility was only reduced within the starter phase, likely due to the chitin content of the

insect meal. However, with increasing age, birds seemed to be able to adapt to the feed source (Gariglio *et al.*, 2019a). The health status of the ducks was unaffected, showing no effects on haematological traits (Gariglio *et al.*, 2019b). Also in Japanese quails digestibility parameters were affected by the diet (Woods *et al.*, 2019). When quails were given a choice between a diet containing insect meal and a control diet, birds preferred to consume the control feed. Feed colour might have played a role in this because BSF larval meal had a darker colour than SBM. Hence, these results emphasise once more the importance of appropriate feed properties when presented to poultry and highlight the power to modulate those characteristics through the choice of larval substrate (Woods *et al.*, 2019). In this regard it is important to distinguish between the attractiveness of different insect products such as live insects and insect meal. Insects are a natural feed source for poultry. Visual cues and intrinsic mechanisms trigger poultry to eagerly ingest larvae. It was reported that turkeys consumed live larvae completely within two minutes after they were provided. In future studies this eagerness of consuming live larvae might be used to improve poultry welfare by stimulating natural behaviour (Veldkamp and van Niekerk, 2019). It is, therefore, important to investigate and distinguish the effect of different insect products on poultry production and behaviour.

6   Future prospects of insects in poultry feed

Insects have a good potential to be used in animal feed. Nonetheless, some hurdles need to be overcome. Especially in terms of legislation, new regulations need to be adopted to enable insect protein to be used in commercial poultry farming. Although currently only the provision of live insects is allowed in the EU, new legislation is expected within 2020 (IPIFF, 2018).

Feed and food safety are of fundamental importance when considering insects as feed. Major concerns are addressed regarding the provision of waste materials as larval substrate as a potential entrance route of chemical and biological contaminants into the food chain (Van Huis, 2020). At present, feeding of farm animals is merely permitted using vegetable materials, apart from a few exceptions (IPIFF, 2019). To maintain sustainable production and to obviate competition with human food and animal feed production new innovative ideas are required. Latest approaches identify the valorising of plant waste streams from industrial sectors for insect rearing, for instance by recycling brewery by-products (Bava *et al.*, 2019; Jucker *et al.*, 2019; Miranda *et al.*, 2020; Ocio *et al.*, 1979). Moreover, as reported for BSF, insects may add additional value to crop products, by metabolising undesired chemical contaminants such as mycotoxins. Mycotoxins are a major feed hazard because contaminated products are no longer suitable for feeding animals (Camenzuli *et al.*, 2018). Investigations focussing on the fate of mycotoxins in insects and effects on larvae would be especially important to evaluate the suitability of the larvae for animal feed.

Recently, questions arose regarding the welfare of insects in mass production systems. In the past, insects have been considered non-sentient beings, unable to experience emotions (Van Huis, 2019). Responsible insect farming and feeding, however, includes ethical considerations. Studying insect welfare is a rather new field. While literature focussing on insect performance is available (Miranda *et al.*, 2020; Schmitt *et al.*, 2019), information considering insect welfare is scarce (but see Boppré and Vane-Wright, 2019). Also the effects of rearing condition on insect health are rarely investigated. To provide insect breeding and rearing companies with the tools to assess well-being and insect health and to develop appropriate killing methods in the future it is important to initiate studies on these aspects.

The use of plant waste products for insect rearing further promotes the establishment of a circular economy (Dicke, 2018). Various designs for business models have been discussed, including implementations on-farm to benefit primarily smallholder farmers in low and middle income countries (Chia *et al.*, 2019) and large-scale farming in Europe (Veldkamp *et al.*, 2012). Multiple options arise, benefitting different stages of the insect chain. Nonetheless, probably the most powerful factor determining the success of insects in feed is societal acceptance. Insect products are rather new to the western society. However, initial consumer studies turned out positive. While consumer attitudes for insect products as direct part of their diet are still rather low, consumers are especially willing to accept insects in animal feed (Onwezen *et al.*, 2019; Verbeke *et al.*, 2015). Consequently, the inclusion of insects in feed might pave the way for faster acceptance in human food. To support the inclusion of insects as feed, it is important to perform research into consumer acceptance and evaluate the economic robustness of using insects in feed considering the entire value chain from production to consumption.

7   Conclusions

The poultry sector is one of the most important providers of animal protein to humans. To satisfy future demands the sector is in transition towards more sustainable production. The main problems facing the industry are the need for sustainable feed protein sources and higher animal welfare, while maintaining economic viability. Both goals can potentially be met by using insects in poultry diets.

The use of insects in poultry feed is still subject to several uncertainties regarding legislation, substrate choice, ethical considerations and consumer attitude. On the other hand, insects have the potential to greatly contribute to the sustainable development goals (Barragán-Fonseca *et al.*, 2020; Dicke, 2018). For instance, the use of insects in feed contributes to food security, circular agriculture, health, economic efficiency, and animal well-being, health and nutritional satisfaction. Because the insects-as-feed sector is still in its infancy more research is required. The current state of knowledge indicates that partial replacement of fishmeal or soybean meal by fly larvae in feed is beneficial for poultry and therefore these insects can be used as a valuable replacement for conventional protein sources. In this regard, one of the rather undiscovered fields is the use of live fly larvae for welfare enhancement. To contribute to a stable sector there is a need for standardised procedures for insect production and processing, clear distinction between insect products based on their nutritional composition, determination of optimal inclusion levels for insect products per poultry species, and production aim and the direct comparison of products from different insect species.

Acknowledgements

Our research has been supported by the Netherlands Organisation for Scientific Research (NWO; NWA programme, InsectFeed project, NWA.1160.18.144). We gratefully acknowledge Wikipedia for the structural formulae for chitin and lauric acid in Figure 1. We thank Yavanna Aartsma, Andreas Baumann, Walter Jansen and Marijke de Jong for constructive comments on a previous version of the manuscript.

Conflict of interest

The authors declare no conflict of interest.

References

Adeniji, A.A., 2007. Effect of replacing groundnut cake with maggot meal in the diet of broilers. International Journal of Poultry Science 6: 822-825. https://doi.org/10.3923/ijps.2007.822.825

Agunbiade, J.A., Adeyemi, O.A., Ashiru, O.M., Awojobi, H.A., Taiwo, A.A., Oke, D.B. and Adekunmisi, A.A., 2007. Replacement of fish meal with maggot meal in cassava-based layers' diets. The Journal of Poultry Science 44: 278-282. https://doi.org/10.2141/jpsa.44.278

Akpodiete, O., Ologhobo, A. and Onifade, A., 1998. Maggot meal as a substitute for fish meal in laying chicken diet. Ghana Journal of Agricultural Science 31: 137-142.

Andoh, M., Ueno, T. and Kawasaki, K., 2018. Tissue-dependent induction of antimicrobial peptide genes after body wall injury in house fly (*Musca domestica*) larvae. Drug Discoveries & Therapeutics 12: 355-362. https://doi.org/10.5582/ddt.2018.01063

Attivi, K., Agboka, K., Mlaga, G.K., Oke, O.E., Teteh, A., Onagbesan, O. and Tona, K., 2020. Effect of black soldier fly (*Hermetia illucens*) maggots meal as a substitute for fish meal on growth performance, biochemical parameters and digestibility of broiler chickens. International Journal of Poultry Science 19: 75-80. https://doi.org/10.3923/ijps.2020.75.80

Augère-Granier, M.-L., 2019. The EU poultry meat and egg sector: Main features, challenges and prospects. European Parliamentary Research Service, Brussels, Belgium, 20 pp. https://doi.org/10.2861/33350

Azagoh, C., Ducept, F., Garcia, R., Rakotozafy, L., Cuvelier, M.-E., Keller, S., Lewandowski, R. and Mezdour, S., 2016. Extraction and physicochemical characterization of Tenebrio molitor proteins. Food Research International 88: 24-31. https://doi.org/10.2861/33350

Bahar, A. and Ren, D., 2013. Antimicrobial Peptides. Pharmaceuticals 6: 1543-1575. https://doi.org/10.3390/ph6121543

Barragán-Fonseca, K.Y., Barragán-Fonseca, K.B., Verschoor, G., Van Loon, J.J.A. and Dicke, M., 2020. Insects for peace. Current Opinion in Insect Science 40: 85-93. https://doi.org/10.1016/j.cois.2020.05.011

Bava, L., Jucker, C., Gislon, G., Lupi, D., Savoldelli, S., Zucali, M. and Colombini, S., 2019. Rearing of Hermetia illucens on different organic by-products: Influence on growth, waste reduction, and environmental impact. Animals 9: 289. https://doi.org/10.3390/ani9060289

Bean, L. and Leeson, S., 2003. Long-term effects of feeding flaxseed on performance and egg fatty acid composition of brown and white hens. Poultry Science 82: 388-394. https://doi.org/10.1093/ps/82.3.388

Biasato, I., Ferrocino, I., Dabbou, S., Evangelista, R., Gai, F., Gasco, L., Cocolin, L., Capucchio, M.T. and Schiavone, A., 2020. Black soldier fly and gut health in broiler chickens: insights into the relationship between cecal microbiota and intestinal mucin composition. Journal of Animal Science and Biotechnology 11: 11. https://doi.org/10.1186/s40104-019-0413-y

Boppré, M. and Vane-Wright, R.I., 2019. Welfare dilemmas created by keeping insects in captivity. In: Carere, C. and Mather, J. (eds) The welfare of invertebrate animals. Springer International Publishing, Cham, Switzerland, pp. 23-67. https://doi.org/10.1007/978-3-030-13947-6_3

Borrelli, L., Coretti, L., Dipineto, L., Bovera, F., Menna, F., Chiariotti, L., Nizza, A., Lembo, F. and Fioretti, A., 2017. Insect-based diet, a promising nutritional source, modulates gut microbiota composition and SCFAs production in laying hens. Scientific Reports 7: 16269. https://doi.org/10.1038/s41598-017-16560-6

Bovera, F., Loponte, R., Pero, M.E., Cutrignelli, M.I., Calabrò, S., Musco, N., Vassalotti, G., Panettieri, V., Lombardi, P., Piccolo, G., Di Meo, C., Siddi, G., Fliegerova, K. and Moniello, G., 2018. Laying performance, blood profiles, nutrient digestibility and inner organs traits of hens fed an insect meal from Hermetia illucens. Research in Veterinary Science 120: 86-93. https://doi.org/10.1016/j.rvsc.2018.09.006

Bulet, P., Hetru, C., Dimarcq, J.L. and Hoffmann, D., 1999. Antibacterial peptides in insects; structure and function. Evelopmental and Comparative Immunology 23: 329-344. https://doi.org/10.1016/s0145-305x(99)00015-4

Camenzuli, L., Van Dam, R., de Rijk, T., Andriessen, R., van Schelt, J. and Van der Fels-Klerx, H.J.I., 2018. Tolerance and excretion of the mycotoxins aflatoxin B1, zearalenone, deoxynivalenol, and ochratoxin a by Alphitobius diaperinus and Hermetia illucens from contaminated substrates. Toxins 10: 91. https://doi.org/10.3390/toxins10020091

Chi, X., Ding, X., Peng, X., Li, X. and Fang, J., 2017. Effects of chitosan oligosaccharides supplementation on the cell cycle of immune organs in broilers. Kafkas Universitesi Veteriner Fakultesi Dergisi 23: 1003-1006. https://doi.org/10.9775/kvfd.2017.17997

Chia, S.Y., Tanga, C.M., van Loon, J.J. and Dicke, M., 2019. Insects for sustainable animal feed: inclusive business models involving smallholder farmers. Current Opinion in Environmental Sustainability 41: 23-30. https://doi.org/10.1016/j.cosust.2019.09.003

Clément, T., Joya, R., Bresson, C. and Clément, C., 2018. Market developments and policy evaluation aspects of the plant protein sector in the EU. Publications Office of the European Union, Brussels, Belgium, 160 pp. https://doi.org/10.2762/022741

Cole, J.N. and Nizet, V., 2016. Bacterial evasion of host antimicrobial peptide defenses. Microbiology Spectrum 4: 0006-2015. https://doi.org/10.1128/microbiolspec.vmbf-0006-2015

Corr, S.A., Gentle, M.J., McCorquodale, C.C. and Bennett, D., 2003. The effect of morphology on walking ability in the modern broiler: a gait analysis study. Animal Welfare 12: 159-171.

Cullere, M., Schiavone, A., Dabbou, S., Gasco, L. and Dalle Zotte, A., 2019a. Meat quality and sensory traits of finisher broiler chickens fed with black soldier fly (*Hermetia illucens* L.) larvae fat as alternative fat source. Animals 9: 140. https://doi.org/10.3390/ani9040140

Cullere, M., Woods, M.J., van Emmenes, L., Pieterse, E., Hoffman, L.C. and Dalle Zotte, A., 2019b. *Hermetia illucens* larvae reared on different substrates in broiler quails: effect on physicochemical and sensory quality of the quail meat. Animals 9: 525. https://doi.org/10.3390/ani9080525

Cutrignelli, M.I., Messina, M., Tulli, F., Randazzi, B., Olivotto, I., Gasco, L., Loponte, R. and Bovera, F., 2018. Evaluation of an insect meal of the black soldier fly (*Hermetia illucens*) as soybean substitute: intestinal morphometry, enzymatic and microbial activity in laying hens. Research in Veterinary Science 117: 209-215. https://doi.org/10.1016/j.rvsc.2017.12.020

Dabbou, S., Gai, F., Biasato, I., Capucchio, M.T., Biasibetti, E., Dezzutto, D., Meneguz, M., Plachà, I., Gasco, L. and Schiavone, A., 2018. Black soldier fly defatted meal as a dietary protein source for broiler chickens: effects on growth performance, blood traits, gut morphology and histological features. Journal of Animal Science and Biotechnology 9: 49. https://doi.org/10.1186/s40104-018-0266-9

Damme, K. and Urselmans, S., 2013. Infrared beak treatment – a temporary solution? Lohmann Information 48: 61.

Dang, X.L., Wang, Y.S., Huang, Y.D., Yu, X.Q. and Zhang, W.Q., 2010. Purification and characterization of an antimicrobial peptide, insect defensin, from immunized house fly (Diptera: Muscidae). Journal of Medical Entomology 47: 1141-1145. https://doi.org/10.1603/ME10016

De Haas, E.N., Kemp, B., Bolhuis, J.E., Groothuis, T. and Rodenburg, T.B., 2013. Fear, stress, and feather pecking in commercial white and brown laying hen parent-stock flocks and their relationships with production parameters. Poultry Science 92: 2259-2269. https://doi.org/10.3382/ps.2012-02996

De Vries, M. and De Boer, I.J.M., 2010. Comparing environmental impacts for livestock products: a review of life cycle assessments. Livestock Science 128: 1-11. https://doi.org/10.1016/j.livsci.2009.11.007

Deng, X., Li, X., Liu, P., Yuan, S., Zang, J., Li, S. and Piao, X., 2008. Effect of chito-oligosaccharide supplementation on immunity in broiler chickens. Asian-Australasian Journal of Animal Sciences 21: 1651-1658. https://doi.org/10.5713/ajas.2008.80056

Dicke, M., 2018. Insects as feed and the sustainable development goals. Journal of Insects as Food and Feed 4: 147-156. https://doi.org/10.3920/JIFF2018.0003

Dixon, L.M., 2020. Slow and steady wins the race: the behaviour and welfare of commercial faster growing broiler breeds compared to a commercial slower growing breed. PLOS ONE 15: e0231006. https://doi.org/10.1371/journal.pone.0231006

Dixon, L.M. and Duncan, I.J.H., 2010. Changes in substrate access did not affect early feather-pecking behavior in two strains of laying hen chicks. Journal of Applied Animal Welfare Science 13: 1-14. https://doi.org/10.1080/10888700903369248

Elahi, U., Ma, Y., Wu, S., Wang, J., Zhang, H. and Qi, G., 2019. Growth performance, carcass characteristics, meat quality and serum profile of broiler chicks fed on housefly maggot meal as a replacement of soybean meal. Journal of Animal Physiology and Animal Nutrition 1-10. DOI: https://doi.org/10.1111/jpn.13265

Elahi, U., Wang, J., Ma, Y., Wu, S., Wu, J., Qi, G. and Zhang, H., 2020. Evaluation of yellow mealworm meal as a protein feedstuff in the diet of broiler chicks. Animals 10: 224. https://doi.org/10.3390/ani10020224

Elnesr, S.S., Alagawany, M., Elwan, H.A.M., Fathi, M.A. and Farag, M.R., 2020. Effect of sodium butyrate on intestinal health of poultry – a review. Annals of Animal Science 20: 29-41. https://doi.org/10.2478/aoas-2019-0077

Elwert, C., Knips, I. and Katz, P., 2010. A novel protein source: maggot meal of the black soldier fly (*Hermetia illucens*) in broiler feed. In: Gierus, M., Kluth, H., Bulang, M. and Kluge, H. (eds) Tagung Schweine- und Geflügelernährung. Institut für Agrar- und Ernährungswissenschaften, Universität Halle-Wittenberg, Germany, pp. 140-142.

European Commission (EC), 2016. Report from the commission to the European Parliament and the Coucil on the impact of genetic selection on the welfare of chickens kept for meat production. Brussels, Belgium. Available at: https://tinyurl.com/y3cnlsvq.

European Commission (EC), 2017. Commission Regulation (EU) 2017/893 of 24 May 2017 amending Annexes I and IV to Regulation (EC) No 999/2001 of the European Parliament and of the Council and Annexes X, XIV and XV to Commission Regulation (EU) No 142/2011 as regards the provisions on processed animal protein. Official Journal of the European Union L 138: 92-116. Available at: http://data.europa.eu/eli/reg/2017/893/oj.

European Commission (EC), 2018. EU agricultural outlook for markets and income 2018-2030. EC, Brussels, Belgium, 128 pp. Available at: https://tinyurl.com/yy8kdfup.

European Commission (EC), 2019. EU feed protein balance sheet – 2018-19. EC, Brussels, Belgium. Available at: https://tinyurl.com/y4at6cpq.

Fanatico, A.C., Pillai, P.B., Hester, P.Y., Falcone, C., Mench, J.A., Owens, C.M. and Emmert, J.L., 2008. Performance, livability, and carcass yield of slow- and fast-growing chicken genotypes fed low-nutrient or standard diets and raised indoors or with outdoor access. Poultry Science 87: 1012-1021. https://doi.org/10.3382/ps.2006-00424

FAOSTAT, 2019. Annual population. FAO, Rome, Italy. Available at: http://www.fao.org/faostat/en/#data/OA.

FAOSTAT, 2020a. Livestock primary. FAO, Rome, Italy. Available at: http://www.fao.org/faostat/en/#data/QL.

FAOSTAT, 2020b. New food balances. FAO, Rome, Italy. Available at: http://www.fao.org/faostat/en/#data/FBS.

Farrell, D., 2013. The role of poultry in human nutrition. Poultry Development Review. FAO, Rome, Italy, pp. 2-3. Available at: http://www.fao.org/3/i3531e/i3531e00.htm.

Fernandez, S.R., Aoyagi, S., Han, Y., Parsons, C.M. and Baker, D.H., 1994. Limiting order of amino acids in corn and soybean meal for growth of the chick. Poultry Science 73: 1887-1896. https://doi.org/10.3382/ps.0731887

Fernyhough, M., Nicol, C.J., Van de Braak, T., Toscano, M.J. and Tønnessen, M., 2020. The ethics of laying hen genetics. Journal of Agricultural and Environmental Ethics 33: 15-36. https://doi.org/10.1007/s10806-019-09810-2

Finke, M.D., 2007. Estimate of chitin in raw whole insects. Zoo Biology 26: 105-115. https://doi.org/10.1002/zoo.20123

Finke, M.D., 2013. Complete nutrient content of four species of feeder insects. Zoo Biology 32: 27-36. https://doi.org/10.1002/zoo.21012

Fitches, E.C., Dickinson, M., De Marzo, D., Wakefield, M.E., Charlton, A.C. and Hall, H., 2019. Alternative protein production for animal feed: Musca domestica productivity on poultry litter and nutritional quality of processed larval meals. Journal of Insects as Food and Feed 5: 77-88. https://doi.org/10.3920/JIFF2017.0061

Flachowsky, G., Meyer, U. and Südekum, K.-H., 2018. Invited review: Resource inputs and land, water and carbon footprints from the production of edible protein of animal origin. Archives Animal Breeding 61: 17-36. https://doi.org/10.5194/aab-61-17-2018

Food and Agriculture Organization of the United Nations (FAO), 2008. Poultry in the 21st century: avian influenza and beyond. Proceedings of the International Poultry Conference, held 5-7 November 2007, Bangkok, Thailand. Thieme, O. and Pilling, D. (eds) FAO Animal Production and Health Proceedings, No. 9., Italy, Rome.

Food and Agriculture Organization of the United Nations (FAO), 2009. The state of food and agriculture: Livestock in the balance. FAO, Rome, Italy. https://doi.org/10.18356/6e4ebb75-en

Food and Agriculture Organization of the United Nations (FAO), 2018. The future of food and agriculture – alternative pathways to 2050. FAO, Rome, Italy, 224 pp.

Gariglio, M., Dabbou, S., Biasato, I., Capucchio, M.T., Colombino, E., Hernández, F., Madrid, J., Martínez, S., Gai, F., Caimi, C., Oddon, S.B., Meneguz, M., Trocino, A., Vincenzi, R., Gasco, L. and Schiavone, A., 2019a. Nutritional effects of the dietary inclusion of partially defatted *Hermetia illucens* larva meal in muscovy duck. Journal of Animal Science and Biotechnology 10: 37. https://doi.org/10.1186/s40104-019-0344-7

Gariglio, M., Dabbou, S., Crispo, M., Biasato, I., Gai, F., Gasco, L., Piacente, F., Odetti, P., Bergagna, S., Plachà, I., Valle, E., Colombino, E., Capucchio, M.T. and Schiavone, A., 2019b. Effects of the dietary inclusion of partially defatted black soldier fly (*Hermetia illucens*) meal on the blood chemistry and tissue (spleen, liver, thymus, and bursa of Fabricius) histology of muscovy ducks (*Cairina moschata domestica*). Animals 9: 307. https://doi.org/10.3390/ani9060307

Gasco, L., Finke, M. and Van Huis, A., 2018. Can diets containing insects promote animal health? Journal of Insects as Food and Feed 4: 1-4. https://doi.org/10.3920/JIFF2018.x001

Ghaly, A.E. and Alkoaik, F.N., 2009. The yellow mealworm as a novel source of protein. American Journal of Agricultural and Biological Science 4: 319-331. https://doi.org/10.3844/ajabssp.2009.319.331

Giles, T., Sakkas, P., Belkhiri, A., Barrow, P., Kyriazakis, I. and Foster, N., 2019. Differential immune response to *Eimeria maxima* infection in fast- and slow-growing broiler genotypes. Parasite Immunology 41: e12660. https://doi.org/10.1111/pim.12660

Hahn, T., Roth, A., Febel, E., Fijalkowska, M., Schmitt, E., Arsiwalla, T. and Zibek, S., 2018. New methods for high-accuracy insect chitin measurement. Journal of the Science of Food and Agriculture 98: 5069-5073. https://doi.org/10.1002/jsfa.9044

Hall, H.N., Masey O'Neill, H.V., Scholey, D., Burton, E., Dickinson, M. and Fitches, E.C., 2018. Amino acid digestibility of larval meal (*Musca domestica*) for broiler chickens. Poultry Science 97: 1290-1297. https://doi.org/10.3382/ps/pex433

Hao, Y.J., Jing, Y.J., Qu, H., Li, D.S. and Du, R.Q., 2008. Purification and characterization of a thermal stable antimicrobial protein from housefly larvae, *Musca domestica*, induced by ultrasonic wave. Acta Biologica Hungarica 59: 289-304. https://doi.org/10.1556/ABiol.59.2008.3.3

Harms, R.H. and Ivey, F.J., 1993. Performance of commercial laying hens fed various supplemental amino acids in a corn-soybean meal diet. Journal of Applied Poultry Research 2: 273-282. https://doi.org/10.1093/japr/2.3.273

Hecht, S.B., 2005. Soybeans, development and conservation on the amazon frontier. Development and Change 36: 375-404. https://doi.org/10.1111/j.0012-155X.2005.00415.x

Heshmatollah, K., 2007. Preference of broiler chicks for color of lighting and feed. The Journal of Poultry Science 44: 213-219.

International Egg Commission, 2011. Annual Review 2011. International Egg Commission, London, UK. Available at: https://tinyurl.com/yxjkqmoy.

International Egg Commission, 2015. Annual Review 2015. International Egg Commission, London, UK. Available at: https://tinyurl.com/y4urqxpx.

International Egg Commission, 2019. Annual Review 2019. International Egg Commission, London, UK.

International Platform of Insects for Food and Feed (IPIFF), 2018. The European insect sector today: challanges, oppotunies and regulatory landscape. IPIFF vision paper on the future of the insect sector towards 2030. IPIFF, Brussels, Belgium. Available at: https://tinyurl.com/y305scnh.

International Platform of Insects for Food and Feed (IPIFF), 2019. Guide on good hygiene practice. IPIFF, Brussels, Belgium. Available at: https://tinyurl.com/yc978kx8.

Ipema, A.F., Gerrits, W.J.J., Bokkers, E.A.M., Kemp, B. and Bolhuis, J.E., 2020. Provisioning of live black soldier fly larvae (*Hermetia illucens*) benefits broiler activity and leg health in a frequency- and dose-dependent manner. Applied Animal Behaviour Science 230: 105082. https://doi.org/10.1016/j.applanim.2020.105082

Irawan, A.C., Rahmawati, N., Astuti, D. and Wibawan, I., 2019. Supplementation of black soldier fly (*Hermetia illucens*) on activity and capacity phagocytic macrophage of laying hens. Jurnal Ilmu Ternak dan Veteriner 24: 182-187. https://doi.org/10.14334/jitv.v24i4.2025

Józefiak, A. and Engberg, R., 2017. Insect proteins as a potential source of antimicrobial peptides in livestock production. A review. Journal of Animal and Feed Sciences 26: 87-99. https://doi.org/10.22358/jafs/69998/2017

Józefiak, D., Józefiak, A., Kierończyk, B., Rawski, M., Świątkiewicz, S., Długosz, J. and Engberg, R.M., 2016. Insects – a natural nutrient source for poultry – a review. Annals of Animal Science 16: 297-313. https://doi.org/10.1515/aoas-2016-0010

Jucker, C., Leonardi, M.G., Rigamonti, I., Lupi, D. and Savoldelli, S., 2019. Brewery's waste streams as a valuable substrate for black soldier fly *Hermetia illucens* (Diptera: Stratiomyidae). Journal of Entomological and Acarological Research 51: 8876. https://doi.org/10.4081/jear.2019.8876

Jucker, C., Lupi, D., Moore, C.D., Leonardi, M.G. and Savoldelli, S., 2020. Nutrient recapture from insect farm waste: bioconversion with *Hermetia illucens* (L.) (Diptera: Stratiomyidae). Sustainability 12: 362. https://doi.org/10.3390/su12010362

Kaya, M., Erdogan, S., Mol, A. and Baran, T., 2015. Comparison of chitin structures isolated from seven Orthoptera species. International Journal of Biological Macromolecules 72: 797-805. https://doi.org/10.1016/j.ijbiomac.2014.09.034

Khan, M., Chand, N., Khan, S., Khan, R. and Sultan, A., 2018a. Utilizing the house fly (*Musca domestica*) larva as an alternative to soybean meal in broiler ration during the starter phase. Revista Brasileira de Ciência Avícola 20: 9-14. https://doi.org/10.1590/1806-9061-2017-0529

Khan, S., Khan, R.U., Alam, W. and Sultan, A., 2018b. Evaluating the nutritive profile of three insect meals and their effects to replace soya bean in broiler diet. Journal of Animal Physiology and Animal Nutrition 102: e662-e668. https://doi.org/10.1111/jpn.12809

Khan, S., Khan, R.U., Sultan, A., Khan, M., Hayat, S.U. and Shahid, M.S., 2016. Evaluating the suitability of maggot meal as a partial substitute of soya bean on the productive traits, digestibility indices and organoleptic properties of broiler meat. Journal of Animal Physiology and Animal Nutrition 100: 649-656. https://doi.org/10.1111/jpn.12419

Kim, S.A. and Rhee, M.S., 2013. Marked synergistic bactericidal effects and mode of action of medium-chain fatty acids in combination with organic acids against *Escherichia coli* O157: H7. Applied and Environmental Microbiology 79: 6552-6560. https://doi.org/10.1128/AEM.02164-13

Larouche, J., 2019. Processing methods for the black soldier fly (*Hermetia illucens*) larvae : from feed withdrawal periods to killing methods. M.Sc. thesis, Université Laval, Québec, Canada, 100 pp.

Lee, C.G., Da Silva, C.A., Lee, J.-Y., Hartl, D. and Elias, J.A., 2008. Chitin regulation of immune responses: an old molecule with new roles. Current Opinion in Immunology 20: 684-689. https://doi.org/10.1016/j.coi.2008.10.002

Leenstra, F., Maurer, V., Bestman, M., Van Sambeek, F., Zeltner, E., Reuvekamp, B., Galea, F. and Van Niekerk, T., 2012. Performance of commercial laying hen genotypes on free range and organic farms in Switzerland, France and the Netherlands. British Poultry Science 53: 282-290. https://doi.org/10.1080/00071668.2012.703774

Li, X.J., Piao, X.S., Kim, S.W., Liu, P., Wang, L., Shen, Y.B., Jung, S.C. and Lee, H.S., 2007. Effects of chito-oligosaccharide supplementation on performance, nutrient digestibility, and serum composition in broiler chickens. Poultry Science 86: 1107-1114. https://doi.org/10.1093/ps/86.6.1107

Li, Z., Mao, R., Teng, D., Hao, Y., Chen, H., Wang, X., Wang, X., Yang, N. and Wang, J., 2017. Antibacterial and immunomodulatory activities of insect defensins-DLP2 and DLP4 against multidrug-resistant *Staphylococcus aureus*. Scientific Reports 7: 12124. https://doi.org/10.1038/s41598-017-10839-4

Lund, T.B., McKeegan, D.E.F., Cribbin, C. and Sandøe, P., 2016. Animal ethics profiling of vegetarians, vegans and meat eaters. Anthrozoös 29: 89-106. https://doi.org/10.1080/08927936.2015.1083192

Makkar, H.P.S., Tran, G., Heuzé, V. and Ankers, P., 2014. State-of-the-art on use of insects as animal feed. Animal Feed Science and Technology 197: 1-33. https://doi.org/10.1016/j.anifeedsci.2014.07.008

Marono, S., Loponte, R., Lombardi, P., Vassalotti, G., Pero, M.E., Russo, F., Gasco, L., Parisi, G., Piccolo, G., Nizza, S., Di Meo, C., Attia, Y.A. and Bovera, F., 2017. Productive performance and blood profiles of laying hens fed *Hermetia illucens* larvae meal as total replacement of soybean meal from 24 to 45 weeks of age. Poultry Science 96: 1783-1790. https://doi.org/10.3382/ps/pew461

Meylaers, K., Clynen, E., Daloze, D., DeLoof, A. and Schoofs, L., 2004. Identification of 1-lysophosphatidylethanolamine (C16:1) as an antimicrobial compound in the housefly, *Musca domestica*. Insect Biochemistry and Molecular Biology 34: 43-49. https://doi.org/10.1016/j.ibmb.2003.09.001

Miranda, C.D., Cammack, J.A. and Tomberlin, J.K., 2020. Life-history traits of house fly, *Musca domestica* L. (Diptera: Muscidae), reared on three manure types. Journal of Insects as Food and Feed 6: 81-90. https://doi.org/10.3920/JIFF2019.0001

Moniello, G., Ariano, A., Panettieri, V., Tulli, F., Olivotto, I., Messina, M., Randazzo, B., Severino, L., Piccolo, G., Musco, N., Addeo, N., Hassoun, G. and Bovera, F., 2019. Intestinal morphometry, enzymatic and mivrobial activity in laying hens fed different levels of *Hermetia illucens* larvae meal and toxic elements content in the insect meal and diets. Animals 9: 86. https://doi.org/10.3390/ani9030086

Mottet, A., de Haan, C., Falcucci, A., Tempio, G., Opio, C. and Gerber, P., 2017. Livestock: on our plates or eating at our table? A new analysis of the feed/food debate. Global Food Security 14: 1-8. https://doi.org/10.1016/j.gfs.2017.01.001

Nicol, C.J., Bestman, M., Gilani, A.-M., De Haas, E.N., De Jong, I.C., Lambton, S., Wagenaar, J.P., Weeks, C.A. and Rodenburg, T.B., 2013. The prevention and control of feather pecking: application to commercial systems. World's Poultry Science Journal 69: 775-788. https://doi.org/10.1017/S0043933913000809

Nie, C., Zhang, Z., Zheng, J., Sun, H., Ning, Z., Xu, G., Yang, N. and Qu, L., 2016. Genome-wide association study revealed genomic regions related to white/red earlobe color trait in the Rhode Island Red chickens. BMC Genetics 17: 115. https://doi.org/10.1186/s12863-016-0422-1

Ochoa Sanabria, C., Hogan, N., Madder, K., Gillott, C., Blakley, B., Reaney, M., Beattie, A. and

Buchanan, F., 2019. Yellow mealworm larvae (*Tenebrio molitor*) fed mycotoxin-contaminated wheat-a possible safe, sustainable protein source for animal feed? Toxins 11: 282. https://doi.org/10.3390/toxins11050282

Ocio, E., Viñaras, R. and Rey, J.M., 1979. House fly larvae meal grown on municipal organic waste as a source of protein in poultry diets. Animal Feed Science and Technology 4: 227-231. https://doi.org/10.1016/0377-8401(79)90016-6

Odén, K., Keeling, L.J. and Algers, B., 2002. Behaviour of laying hens in two types of aviary systems on 25 commercial farms in Sweden. British Poultry Science 43: 169-181. https://doi.org/10.1080/00071660120121364

Okah, U. and Onwujiariri, E.B., 2012. Performance of finisher broiler chickens fed maggot meal as a replacement for fish meal. Journal of Agricultural Technology 8: 471-477.

Olsen, R.L. and Hasan, M.R., 2012. A limited supply of fishmeal: Impact on future increases in global aquaculture production. Trends in Food Science and Technology 27: 120-128. https://doi.org/10.1016/j.tifs.2012.06.003

Onbaşılar, E.E., Ünal, N., Erdem, E., Kocakaya, A. and Yaranoğlu, B., 2015. Production performance, use of nest box, and external appearance of two strains of laying hens kept in conventional and enriched cages. Poultry Science 94: 559-564. https://doi.org/10.3382/ps/pev009

Onwezen, M.C., Van den Puttelaar, J., Verain, M.C.D. and Veldkamp, T., 2019. Consumer acceptance of insects as food and feed: The relevance of affective factors. Food Quality and Preference 77: 51-63. https://doi.org/10.1016/j.foodqual.2019.04.011

Organisation for Economic Co-operation and Development / Food and Agriculture Organisation (OECD/FAO), 2019. OECD-FAO agricultural outlook 2019-2028. Food and and Agriculture Organization of the United Nation, Rome, Italy. Available at: https://doi.org/10.1787/agr_outlook-2019-en.

Park, S.I., Kim, J.W. and Yoe, S.M., 2015. Purification and characterization of a novel antibacterial peptide from black soldier fly (*Hermetia illucens*) larvae. Developmental and Comparative Immunology 52: 98-106. https://doi.org/10.1016/j.dci.2015.04.018

Pasotto, D., van Emmenes, L., Cullere, M., Giaccone, V., Pieterse, E., Hoffman, L.C. and Dalle Zotte, A., 2020. Inclusion of *Hermetia illucens* larvae reared on fish offal to the diet of broiler quails: effect on immunity and caecal microbial populations. Czech Journal of Animal Science 65: 213-223. https://doi.org/10.17221/60/2020-CJAS

Qi, X., Li, Z., Akami, M., Mansour, A. and Niu, C., 2019. Fermented crop straws by *Trichoderma viride* and *Saccharomyces cerevisiae* enhanced the bioconversion rate of *Musca domestica* (Diptera: Muscidae). Environmental Science and Pollution Research 26: 29388-29396. https://doi.org/10.1007/s11356-019-06101-1

Quentin, M., Bouvarel, I., Berri, C., Le Bihan-Duval, E., Baéza, E., Jégo, Y. and Picard, M., 2003. Growth, carcass composition and meat quality response to dietary concentrations in fast-, medium- and slow-growing commercial broilers. Animal Research 52: 65-77. https://doi.org/10.1051/animres:2003005

Radulović, S., Pavlović, M., Šefer, D., Katoch, S., Hadži-Milić, M., Jovanović, D., Grdović, S. and Marković, R., 2018. Effects of housefly larvae (*Musca domestica*) dehydrated meal on production performances and sensory properties of broiler meat. Thai Journal of Veterinary Medicine 48: 63-70.

Rezaei, M., Yngvesson, J., Gunnarsson, S., Jönsson, L. and Wallenbeck, A., 2018. Feed efficiency, growth performance, and carcass characteristics of a fast- and a slower-growing broiler hybrid fed low- or high-protein organic diets. Organic Agriculture 8: 121-128. https://doi.org/10.1007/s13165-017-0178-6

Rinaudo, M., 2006. Chitin and chitosan: properties and applications. Progress in Polymer Science 31: 603-632. https://doi.org/10.1016/j.progpolymsci.2006.06.001

Saatkamp, H.W., Vissers, L.S.M., Van Horne, P.L.M. and De Jong, I.C., 2019. Transition from conventional broiler meat to meat from production concepts with higher animal welfare: experiences from the Netherlands. Animals 9: 483. https://doi.org/10.3390/ani9080483

Sakkas, P., Oikeh, I., Blake, D.P., Nolan, M.J., Bailey, R.A., Oxley, A., Rychlik, I., Lietz, G. and Kyriazakis, I., 2018. Does selection for growth rate in broilers affect their resistance and tolerance to *Eimeria maxima*? Veterinary Parasitology 258: 88-98. https://doi.org/10.1016/j.vetpar.2018.06.014

Schiavone, A., Dabbou, S., Petracci, M., Zampiga, M., Sirri, F., Biasato, I., Gai, F. and Gasco, L., 2019. Black soldier fly defatted meal as a dietary protein source for broiler chickens: effects on carcass traits, breast meat quality and safety. Animal 13: 2397-2405. https://doi.org/10.1017/S1751731119000685

Schiavone, A., De Marco, M., Martínez, S., Dabbou, S., Renna, M., Madrid, J., Hernandez, F., Rotolo, L., Costa, P., Gai, F. and Gasco, L., 2017. Nutritional value of a partially defatted and a highly defatted black soldier fly larvae (*Hermetia illucens*). Journal of Animal Science and Biotechnology 8: 1-9. https://doi.org/10.1186/s40104-017-0181-5

Schmitt, E., Belghit, I., Johansen, J., Leushuis, R., Lock, E.J., Melsen, D., Ramasamy Shanmugam, R.K., Van Loon, J. and Paul, A., 2019. Growth and safety assessment of feed streams for black soldier fly larvae: a case study with aquaculture sludge. Animals 9: 189. https://doi.org/10.3390/ani9040189

Secci, G., Bovera, F., Nizza, S., Baronti, N., Gasco, L., Conte, G., Serra, A., Bonelli, A. and Parisi, G., 2018. Quality of eggs from Lohmann Brown Classic laying hens fed black soldier fly meal as substitute for soya bean. Animal 12: 2191-2197. https://doi.org/10.1017/S1751731117003603

Secci, G., Bovera, F., Parisi, G. and Moniello, G., 2020. Quality of eggs and albumen technological properties as affected by *Hermetia illucens* larvae meal in hens' diet and hen age. Animals 10: 1-12. https://doi.org/10.3390/ani10010081

Shahidi, F. and Abuzaytoun, R., 2005. Chitin, chitosan, and co-products: chemistry, production, applications, and health effects. Advances in Food and Nutrition Research 49: 93-135. https://doi.org/10.1016/S1043-4526(05)49003-8

Singh, R., Cheng, K.M. and Silversides, F.G., 2009. Production performance and egg quality of four strains of laying hens kept in conventional cages and floor pens. Poultry Science 88: 256-264. https://doi.org/10.3382/ps.2008-00237

Skrivanova, E., Marounek, M., Benda, V. and Brezina, P., 2006. Susceptibility of *Escherichia coli*, *Salmonella* sp. and *Clostridium perfringens* to organic acids and monolaurin. Veterinarni Medicina 51: 81-88.

Sogari, G., Amato, M., Biasato, I., Chiesa, S. and Gasco, L., 2019. The potential role of insects as feed: a multi-perspective review. Animals 9: 119. https://doi.org/10.3390/ani9040119

Stadig, L., 2019. Vleeskuikenconcepten in Nederland – een vegelijking op gebied van dierenwelzijn. Nederlandse Vereniging tot Bescherming van Dieren, Den Haag, The Netherlands. Available at: https://tinyurl.com/yyh9mc9a.

Star, L., Arsiwalla, T., Molist, F., Leushuis, R., Dalim, M. and Paul, A., 2020. Gradual provision of live black soldier fly (Hermetia illucens) larvae to older laying hens: effect on production performance, egg quality, feather condition and behavior. Animals 10: 216. https://doi.org/10.3390/ani10020216

Struthers, S., Classen, H.L., Gomis, S., Crowe, T.G. and Schwean-Lardner, K., 2019. The impact of beak tissue sloughing and beak shape variation on the behavior and welfare of infrared beak-treated layer pullets and hens. Poultry Science 98: 4269-4281. https://doi.org/10.3382/ps/pez274

Suparman, Purwanti, S. and Nahariah, N., 2020. Substitution of fish meal with black soldier fly larvae (Hermetia illucens) meal to eggs production and physical quality of quail (Coturnix coturnix japonica) eggs. IOP Conference Series: Earth and Environmental Science 492: 012014. https://doi.org/10.1088/1755-1315/492/1/012014

Suzuki, M., Fujimoto, W., Goto, M., Morimatsu, M., Syuto, B. and Iwanaga, T., 2002. Cellular expression of gut chitinase mRNA in the gastrointestinal tract of mice and chickens. Journal of Histochemistry & Cytochemistry 50: 1081-1089. https://doi.org/10.1177/002215540205000810

Tabata, E., Kashimura, A., Wakita, S., Ohno, M., Sakaguchi, M., Sugahara, Y., Kino, Y., Matoska, V., Bauer, P.O. and Oyama, F., 2017. Gastric and intestinal proteases resistance of chicken acidic chitinase nominates chitin-containing organisms for alternative whole edible diets for poultry. Scientific Reports 7: 1-11. https://doi.org/10.1038/s41598-017-07146-3

Tallentire, C.W., Leinonen, I. and Kyriazakis, I., 2018. Artificial selection for improved energy efficiency is reaching its limits in broiler chickens. Scientific Reports 8: 1-10. https://doi.org/10.1038/s41598-018-19231-2

Timbermont, L., Lanckriet, A., Dewulf, J., Nollet, N., Schwarzer, K., Haesebrouck, F., Ducatelle, R. and Van Immerseel, F., 2010. Control of Clostridium perfringens -induced necrotic enteritis in broilers by target-released butyric acid, fatty acids and essential oils. Avian Pathology 39: 117-121. https://doi.org/10.1080/03079451003610586

Uitdehaag, K., Komen, H., Rodenburg, T.B., Kemp, B. and Van Arendonk, J., 2008. The novel object test as predictor of feather damage in cage-housed Rhode Island Red and White Leghorn laying hens. Applied Animal Behaviour Science 109: 292-305. https://doi.org/10.1016/j.applanim.2007.03.008

Van Horne, P.L.M., 2018. Competitiveness of the EU poultry meat sector, base year 2017: international comparison of production costs. Report 2018-116. Wageningen Economic Research, Wageningen, The Netherlands, 40 pp. Available at: https://library.wur.nl/WebQuery/wurpubs/544594.

Van Huis, A., 2019. Welfare of farmed insects. Journal of Insects as Food and Feed 5: 159-162. https://doi.org/10.3920/JIFF2019.x004

Van Huis, A., 2020. Insects as food and feed, a new emerging agricultural sector: a review. Journal of Insects as Food and Feed 6: 27-44. https://doi.org/10.3920/JIFF2019.0017

Van Huis, A., Van Itterbeeck, J., Klunder, H., Mertens, E., Halloran, A., Muir, G. and Vantomme, P., 2013. Edible insects. Future prospects for food and feed security. Food And Agriculture Organization of the United Nations Forestry Paper 171, FAO, Rome, Italy, 201 pp. https://doi.org/10.1017/CBO9781107415324.004

Van Krimpen, M.M. and Hendriks, W.H., 2019. 13: Novel protein sources in animal nutrition: considerations and examples. In: Hendriks, W.H., Verstegen, M.W.A. and Babinszky, L. (eds) Poultry and pig nutrition – Challenges of the 21st century. Wageningen Academic Publishers, Wageningen, The Netherlands, pp. 279-305. https://doi.org/10.3920/978-90-8686-884-1_13

Veldkamp, T. and Bosch, G., 2015. Insects: a protein-rich feed ingredient in pig and poultry diets. Animal Frontiers 5: 45-50. https://doi.org/10.2527/af.2015-0019

Veldkamp, T., Van Duinkerken, G., Van Huis, A., Lakemond, C.M.M., Ottevanger, E., Bosch, G. and Van Boekel, M.A.J.S., 2012. Insects as a sustainable feed ingredient in pig and poultry diets – a feasibility study. Livestock Research Report 638, Wageningen UR Livestock Research, Wageningen, The Netherlands. Available at: https://tinyurl.com/yxu35gh9.

Veldkamp, T. and Van Niekerk, T.G.C.M., 2019. Live black soldier fly larvae (*Hermetia illucens*) for turkey poults. Journal of Insects as Food and Feed 5: 301-311. https://doi.org/10.3920/JIFF2018.0031

Verbeke, W., Spranghers, T., De Clercq, P., De Smet, S., Sas, B. and Eeckhout, M., 2015. Insects in animal feed: acceptance and its determinants among farmers, agriculture sector stakeholders and citizens. Animal Feed Science and Technology 204: 72-87. https://doi.org/10.1016/j.anifeedsci.2015.04.001

Vissers, L.S.M., De Jong, I.C., Van Horne, P.L.M. and Saatkamp, H.W., 2019. Global prospects of the cost-efficiency of broiler welfare in middle-segment production systems. Animals 9: 1-17. https://doi.org/10.3390/ani9070473

Vogel, H., Müller, A., Heckel, D.G., Gutzeit, H. and Vilcinskas, A., 2018. Nutritional immunology: diversification and diet-dependent expression of antimicrobial peptides in the black soldier fly *Hermetia illucens*. Developmental and Comparative Immunology 78: 141-148. https://doi.org/10.1016/j.dci.2017.09.008

Wallenbeck, A., Wilhelmsson, S., Jönsson, L., Gunnarsson, S. and Yngvesson, J., 2016. Behaviour in one fast-growing and one slower-growing broiler (*Gallus gallus domesticus*) hybrid fed a high- or low-protein diet during a 10-week rearing period. Acta Agriculturae Scandinavica, Section A — Animal Science 66: 168-176. https://doi.org/10.1080/09064702.2017.1303081

Wang, H., Zhang, Z., Czapar, G.F., Winkler, M.K.H. and Zheng, J., 2013. A full-scale house fly (Diptera: Muscidae) larvae bioconversion system for value-added swine manure reduction. Waste Management & Research 31: 223-231. https://doi.org/10.1177/0734242X12469431

Wang, Z., Wang, J., Zhang, Y., Wang, X., ZhangG, X., Liu, Y., Xi, J., Tong, H., Wang, Q., Jia, B. and Sehn, H., 2017. Antimicrobial peptides in housefly larvae (*Musca domestica*) affect intestinal *Lactobacillus acidophilus* and mucosal epithelial cells in *Salmonella pullorum*-infected chickens. Kafkas Universitesi Veteriner Fakultesi Dergisi 23: 423-430. https://doi.org/10.9775/kvfd.2016.16901

Wilhelmsson, S., Yngvesson, J., Jönsson, L., Gunnarsson, S. and Wallenbeck, A., 2019. Welfare Quality® assessment of a fast-growing and a slower-growing broiler hybrid, reared until 10 weeks and fed a low-protein, high-protein or mussel-meal diet. Livestock Science 219: 71-79. https://doi.org/10.1016/j.livsci.2018.11.010

Wong, C., Rosli, S., Uemura, Y., Ho, Y.C., Leejeerajumnean, A., Kiatkittipong, W., Cheng, C.-K., Lam, M.-K. and Lim, J.-W., 2019. Potential protein and biodiesel sources from black soldier fly larvae: insights of larval harvesting instar and fermented feeding medium. Energies 12: 1570. https://doi.org/10.3390/en12081570

Woods, M.J., Cullere, M., Van Emmenes, L., Vincenzi, S., Pieterse, E., Hoffman, L.C. and Zotte, A.D., 2019. *Hermetia illucens* larvae reared on different substrates in broiler quail diets: effect on apparent digestibility, feed-choice and growth performance. Journal of Insects as Food and Feed 5: 89-98. https://doi.org/10.3920/JIFF2018.0027

Woods, M.J., Goosen, N.J., Hoffman, L.C. and Pieterse, E., 2020. A simple and rapid protocol for measuring the chitin content of *Hermetia illucens* (L.) (Diptera: Stratiomyidae) larvae. Journal of Insects as Food and Feed 6:285-290. https://doi.org/10.3920/JIFF2019.0030

Use of insect products in pig diets

T. Veldkamp* and A.G. Vernooij

Wageningen Livestock Research, De Elst 1, 6700 AH Wageningen,
The Netherlands; *teun.veldkamp@wur.nl

Abstract

This review is focusing on effects of inclusion of insect products in pig diets on digestibility, performance, product quality, and health parameters. In 2019 pig feed accounts for 23% of the global feed production. Soybean meal is the most common protein source in pig diets. A shift towards more sustainable feed ingredients can improve the sustainability of entire pig production. Novel protein sources currently evaluated in diets for piglets and growing pigs are insect-based ingredients. Insects are able to convert organic biomass into high-quality protein. Currently the use of insects as protein source in pig diets is not allowed due to transmissible spongiform encephalopathies regulation but it is expected that this will be allowed in the near future. Research efforts on effects of inclusion of insect products on nutrient digestibility, growth performance, product quality and pig health are therefore increasing. Nutrient digestibility of evaluated insect proteins was comparable with traditional protein sources. Nutrient digestibility of insect-based diets as well as effects on growth performance in pigs fed insect-based diets differed between studies. The differences in responses are mainly due to changes in diet ingredients and nutrient composition when insect products are included. Health related parameters were not affected by dietary inclusion of insect products. In general it can be stated that differences in results between studies may be due to different insect species and life stages being used, differences in nutritional value of the insect products, in dietary inclusion levels, in processing techniques applied, effects on palatability of the diet, (weaning) age of the animals involved and research methods applied. Overall, insect products seem to be a good alternative to partly replace traditional protein-rich ingredients in pig diets without adversely affecting growth performance, product quality and health, but more standardised research is required to reduce differences between studies.

Keywords

insect protein – feed, pigs – digestibility and growth performance – animal health

© T. VELDKAMP AND A.G. VERNOOIJ, 2025 | DOI: 10.1163/9789004707689_20

1   Introduction

This review is focusing on effects of inclusion of insect products in pig diets on nutrient digestibility of the diet, animal performance, product quality, and health parameters. The review starts with a general description of the development of pig production in the past and expected developments in the future as well as sustainability aspects in pig production mainly focussing on pig nutrition and possibilities on how to create more sustainable pig diets by use of insect-based feed ingredients. The chemical compositions of different insect species are presented and compared to the chemical composition of traditional protein sources such as soybean meal and fishmeal. The chemical composition of insect protein is important to put the effect of nutrient digestibility of insect proteins on the overall nutritional value in perspective. Only relevant papers for evaluation of nutrient digestibility, growth performance, product quality, and health parameters, published since 2000, were considered. For studies on effects of inclusion of insect products as feed ingredient in pig diets on growth performance only papers were included which met the criteria of isonitrogenously and isocalorically formulated diets. This was done to ensure that outcome parameters can be solely attributed to the inclusion of insect-based feed ingredients in the diet and not to differences in protein levels or energy levels.

2   Global meat consumption and pig production

In the past 50 years, global pig production has quadrupled and is expected to continue to grow in the next three decades. This may have a significant impact on feed use and land demand (Lassaletta *et al.*, 2019). The growing demand for pig meat results from the increase in population, as well as from the transition of the diet towards more animal protein consumption per capita (Bai *et al.*, 2018; Lassaletta *et al.*, 2014). Globally, meat consumption is generally influenced by a number of factors, such as food consumption patterns, the standard of living, meat production and animal husbandry conditions and product pricing. Also many consumers will change their eating habits, including consumption of more animal protein when urbanisation intensifies (Soare and Chiurciu, 2017). Although in Western Europe the consumption of meat is not expected to increase much (De Boer *et al.*, 2006), it has been forecasted that the worldwide demand for animal products will grow significantly in the coming decades, and that the global production of meat will double in 2050 taking 1999 as the base year (Steinfeld *et al.*, 2006). Globally, 767.5 billion pigs were produced in 2019 (USDA, 2020). Global feed production is estimated at 1,126 million metric tons in 2019 (Alltech, 2020). The production of pig feed in 2019 is estimated to be 261 million metric tons, accounting for 23.2% of the total global feed production. In 2019, pig feed production in Europe was 79.5 million metric tons, accounting for 30% of the total pig feed production (Alltech, 2020). The expected increase of global population and consumption of pig meat will require a higher demand for pig feed in the future.

3   Sustainability aspects of pig diets

Meat consumption contributes to the supply of energy, protein, essential amino acids and important micronutrients (e.g. long-chain n-3 fatty acids, copper, iron, iodine, manganese, selenium, zinc, B-vitamins) in the human food chain (De Smet and Vossen, 2016). However, despite the benefits of meat consumption, the increase in the number of livestock directly challenges the sustainability of animal production because it profoundly impacts our planet. The General Assembly of the United Nations has defined the goals of sustainable agriculture to ensure economic viability, protect natural resources, provide ecosystem services, manage rural areas, improve the quality of life in agricultural areas, ensure animal welfare and produce safe and healthy food (UN, 2015). The EU goals for animal production are in particular to increase the competitiveness and economic viability of animal production systems, to improve livestock's adaptation to diseases and increasingly extreme weather patterns related to climate change, to address issues related to diet and health, ammonia and air quality, and issues related to greenhouse gas emissions and climate change, nitrate emissions and degradation of natural resources such as water, soil and biodiversity, global food security, global trade and animal welfare (ERANET, 2020). The negative impact of livestock production on our earth can be reduced by two alternative scenario's for improving the sustainability of future pig production systems (Rauw et al., 2020). The first scenario is a high input-high output system based on sustainable intensification, maximising animal protein production efficiency on a limited land surface at the same time as minimising environmental impacts. The second scenario is a reduced input-reduced output system based on selecting animals that are more robust to climate change and are better adapted to transform low quality feed (local feeds, feedstuff co-products, food waste) into meat. Sustainability can be increased with a shift from dependence on optimally formulated feed based on imported feed ingredients and use of feed grains to use of feeds based on locally produced ingredients and co-products not fit for use as human food (Mottet et al., 2017). Soybean meal has a relatively high protein content (45-50%) with an adequate amino acid profile, low variation in nutrient composition, and contains anti-nutritional factors that are easily reduced by heat processing and is therefore a suitable ingredient in pig diets. It is an important source of the amino acid lysine, which is the first limiting amino acid in diets for pigs (De Visser et al., 2014), and other essential amino acids. Therefore, more than 50% of the total dry matter (DM) intake of pigs is based on cereals, while 9-25% is covered by oil meals, including soybean meal. In the European Union, soybean meal is the major oil seed meal consumed in animal nutrition and a large percentage of the used soybean is imported, because European production of soybean is low, compared to other oil seeds (Florou-Paneri *et al.*, 2014). The inclusion of other oilseed meals to reduce dependence on imported soybean meal, such as sunflower meal or rapeseed meal in pig diets can provide opportunities to diversify the feed matrix by using home-grown feed ingredients. These alternative feed ingredients can partially or completely replace soybean meal in pig feed, especially during growing and finishing periods. There are other protein sources that can be used for pig nutrition, such as cottonseed meal, flaxseed meal, peas and faba beans. A potential alternative protein source for animal and pig feed are also insects and insects derived products which have a lower environmental footprint related to lower land use, water use and less emissions (Van Huis and Oonincx, 2017).

4   Pig production systems and pig nutrition

Specialised pig farms generally have a large number of animals. In specialised pig farming, three types of business can be created, such as breeding farms, finishing farms, farrowing to finisher farms, and there are also farms that keep pigs as an auxiliary activity (Anonymous, 2017). Breeding farms are involved in piglet production. Sow milk is the most important source of nutrients for piglets. Milk also contains protective substances (e.g. immunoglobulins) against multiple pathogens. At weaning at about four weeks of age, piglets have to change from fluid to solid feed. This requires adjustments of the digestive system and further development of the immune system. Generally highly digestible ingredients and protein sources are used in piglets in the post-weaning phase (whey, soybean derived protein sources, cooked cereals and fish meal). Weaning also include transfer of piglets to a new environment. This transition to a change in feed is often accompanied by a decrease in feed intake (FI) and a risk for infection by pathogenic *Escherichia coli* and streptococci inducing health disorders. Piglets are transferred to a finishing pig farm when they weigh around 25 kg (age about 10 weeks). The finishing pig farm is dedicated to raise piglets to finishing pigs. The pigs stay here until they weigh about 120 kg (age about 6 months). The pigs are then slaughtered and processed into pig meat for consumption by consumers. The dietary protein and essential amino acid requirements decrease as pigs grow older. To maximise piglet growth, it is recommended that levels of crude protein (CP) be at 20-23% in pre-starter and 18-20% in starter diets. Recommended levels of crude protein in finisher diets decrease from about 19% at 10 weeks of age to 13% in the pre-slaughter phase. Most common feed ingredients used in finisher diets are cereals such as corn and wheat and the remainder mainly consists of soybean meal, sunflower seed meal, wheat middlings and palm kernel meal. Use of soybean meal, the major protein source included in pig diets, has a large environmental impact due to land and water use to grow soybeans and the large-scale and -distance transportation of this ingredient (Wiedemann *et al.*, 2016). Replacement of soybean meal with locally derived vegetal protein sources will reduce requirements for transport but land use will not reduce. By use of insects as feed ingredient associated transport and land use will reduce (Van Zanten *et al.*, 2018).

5   Chemical composition of insect species

In a feasibility study, Veldkamp *et al.* (2012) concluded that the use of insects as a sustainable protein-rich feed ingredient in pig and poultry diets is possible based on their nutritional value but their maximum inclusion levels to replace traditional protein sources should be studied further. The amino acid profile of yellow mealworm (*Tenebrio molitor*; TM), common housefly (HF), and black soldier fly (BSF) is close to the profile of soybean meal (Veldkamp and Bosch, 2015). The chemical composition of the insect species included in the present review – BSF prepupae products, HF larval meal and mealworm larvae – and the traditional protein sources soybean meal and fishmeal is presented in Table 1. The protein content of the insect species is comparable to soybean meal but lower than fishmeal and the fat content of the insect species is higher than in soybean meal and fishmeal. The amino acid profile of the insect species is close to the profile in soybean meal and fishmeal.

To compete with conventional protein sources for animal feeds and to become an interesting link in the animal feed chain to meet the growing global demand for protein, the cost price of insect production must be further reduced. However, insect production and related research are just in their infant stage. Literature on possible beneficial effects of insect products as feed ingredient on health of livestock animals is still scarce. Beneficial health effects of including insect products in pig diets may give applications in pig production an added value. Presence of chitin, lauric acid and antimicrobial peptides in insects as potential functional constituents may have positive health effects in monogastric animals (Dörper *et al.*, 2021; Gasco *et al.*, 2018; Jozefiak and Engberg, 2017).

6   Digestibility of insect products and insect-based diets

In total, five experiments were found with weaned piglets and two experiments with growing pigs. In five experiments different mealworm products were tested and in two experiments BSF products were tested. Most of the digestibility experiments are based on exchange of a traditional protein source by insect protein and so nutrient digestibility in these experiments has been evaluated at the level of the diet. Only in two papers the digestibility of the insect protein as such was evaluated. With respect to digestibility of diets as presented in Table 2 it should be noted that the nutrient digestibility of diets is affected by the nature of exchange of protein sources in the diet. Therefore, these values should be judged in this context.

TABLE 1 Chemical composition of black soldier fly prepupae products, housefly larval meal, mealworm larvae, soybean meal and fishmeal

	BSF prepupae grown on vegetable waste	Full fat BSF larval meal	Partially defatted BSF larval meal	HF larval meal	TM larvae	Soybean meal	Fishmeal
g/kg							
Dry matter	410	884	939	920	381	877	913
Crude ash	96	74	68	65	24	64	168
Crude protein	399	425	408	533	491	467	629
Crude fat	371	325	128	203	352	15	98
Ca	28.7	20.8	5.8	–	0.4	2.9	40.3
P	4.0	4.7	7.6	–	7.5	6.4	26.0
g/16 g N							
Lys	5.7	6.2	6.2	7.1	5.5	6.2	7.6
Met	1.9	2.5	2.9	2.5	1.3	1.4	2.8
Cys	0.5	0.7	1.5	2.8	0.9	1.5	0.9
Thr	3.9	4.2	4.0	5.3	5.1	3.9	4.2
Trp	1.5	–	–	6.5	4.1	1.3	1.1
Ile	4.3	5.6	4.9	3.6	5.0	4.6	4.2
Arg	5.0	5.7	5.2	4.8	5.2	7.5	5.9
Phe	4.1	5.5	4.4	6.0	3.5	5.2	3.9
His	3.1	3.7	3.0	2.9	3.2	2.7	2.6
Leu	7.0	8.6	7.6	6.1	10.6	7.7	7.3
Tyr	–	–	–	6.5	0.8	3.7	3.1
Val	6.2	7.2	6.7	4.3	7.3	4.8	4.9
Ala	6.1	7.3	8.2	5.5	8.2	4.4	6.3
Asp	9.0	11.0	10.3	9.9	8.1	11.6	9.3
Glu	10.4	11.6	15.3	13.4	11.3	17.8	13.0
Gly	5.6	7.2	6.2	4.5	5.6	4.3	6.5
Pro	5.4	7.2	7.8	3.8	7.0	5.1	4.4
Ser	3.8	4.8	5.0	2.5	5.1	5.1	4.0
Sum_AA	83.3	99.0	99.4	97.9	97.8	98.8	92.0
Reference	adapted from Spranghers et al. (2017)	adapted from Crosbie et al. (2020)	adapted from Crosbie et al. (2020)	adapted from Hall et al. (2018)	adapted from Finke (2002)	CVB (2019)	CVB (2019)

AA = amino acid; Ala = alanine; Arg = arginine; Asp = aspartate; BSF = black soldier fly (*Hermetia illucens*); Ca = calcium; Cys = cysteine; Glu = glutamate; Gly = glycine; HF = housefly (*Musca domestica*); His = histidine; Ile = isoleucine; Leu = leucine; Lys = lysine; Met = methionine; N = nitrogen; P = phosphorus; Phe = phenylalanine; Pro = proline; Ser = serine; Thr = threonine; TM = mealworm (*Tenebrio molitor*); Trp = tryptophan; Tyr = tyrosine; Val = valine.

TABLE 2 Nutrient digestibility of insect products and insect-based pig diets

Pig species/type	Age (d)	Insect product	Target replacement	Insect inclusion %	Results	Reference
Weaned piglets	21-36	Full-fat BSF and defatted BSF	Toasted soybeans	0, 4, 8	AID of CP in the 8% full-fat BSF diet was lower than that of the control diet and AID of CP in the 4% full-fat and the defatted BSF diets was higher.	Spranghers et al. (2018)
Weaned piglets ([Duroc×Yorkshire ×Landrace)	21-56	Dried mealworm (PT) powder	Fishmeal	0, 1, 2	ATTD of DM and N at 1% inclusion was lower than in the control diet.	Ao and Kim (2019)
Weaned piglets (Topigs)	21-61	Partially defatted BSF larval meal	Soybean meal	0, 5, 10	ATTD was not affected by dietary treatment.	Biasato et al. (2019)
Weaned piglets ([Yorkshire×Land-race]×Duroc)	28-63	Ground air-dried TM larvae	Soybean meal and soy oil	0, 1.5, 3.0, 4.5, 6.0	Nitrogen retention and ATTD of DM and CP linearly increased with increasing TM levels.	Jin et al. (2016)
Weaned piglets (Piétrain×[German Landrace×German Edelschwein])	35-63	TM larvae	Soybean meal	0, 5, 10	AID of all AAs, except aspartic acid, was lower at 10% inclusion than at the control diet.	Meyer et al. (2020)
Growing pigs ([Landrace×York-shire]×Duroc)	24 kg	Dried TM larvae powder	Fishmeal, meat meal, poultry meal	9.95	SID of DM, CP, total AAs, essential AAs and non-essential AAs tended to be higher than fishmeal, meat meal and poultry meal.	Yoo et al. (2019)
Growing pigs ([Landrace×York-shire]×Duroc)	29 kg	100% defatted TM larval meal and TM larvae hydrolysate	Soybean meal	0, 10	AID of DM, CP, Lys, Met and Thr in TM larvae hydrolysate was higher compared to fermented poultry by-product and hydrolysed fish soluble.	Cho et al. (2020)

AAs = amino acids; AID = apparent ileal digestibility; ATTD = apparent total tract digestibility; BSF = black soldier fly (*Hermetia illucens*); CP = crude protein; DM = dry matter; Lys = lysine; Met = methionine; N = nitrogen; PT = mealworm (*Ptecticus tenebrifer*); SID = standardised ileal digestibility; Thr = threonine; TM = mealworm (*Tenebrio molitor*).

6.1   *Weaned piglets*

Nutrient digestibility in weaned piglets was determined for 0, 1 and 2% inclusion of dried *Ptecticus tenebrifer* (PT) larvae at the expense of fishmeal (Ao and Kim, 2019). The apparent total tract digestibility (ATTD) of DM (78.8 vs 81.9%, respectively) and N (78.6 vs 81.6%, respectively) was at 1% PT inclusion lower than at the control diet at 56 d of age. Also Meyer *et al.* (2020) observed that ileal digestibility of all amino acids in weaned piglets at 63 d of age, except aspartic acid, were 6.7 to 15.6% lower at 10% TM inclusion than at 0% TM inclusion. Contrary to Ao and Kim (2019) and Meyer *et al.* (2020), nitrogen retention and apparent total tract digestibility of DM and CP linearly increased when weaned piglets were fed diets with increasing inclusion levels of ground dried TM larvae (Jin *et al.*, 2016). Digestibility of DM and CP increased from 90 to 94% and from 86 to 93%, respectively at 0% and 6% TM larvae inclusion (Jin *et al.*, 2016). Nitrogen retention increased linearly from 2.2 to 2.4 g/d at 0 and 6% TM larvae inclusion, respectively. It should be noted that this is only partly due to improved digestibility and more to the complete nutrient balance of the diets.

Biasato *et al.* (2019) concluded that ATTD was not affected in weaned piglets fed diets with 0, 5 and 10% inclusion of partially defatted BSF larval meal at the expense of soybean meal at 23 d of age as well as at 61 d of age. Spranghers *et al.* (2018) evaluated ATTD and apparent ileal digestibility (AID) of full-fat BSF and defatted BSF in weaned piglets from 21 to 36 d of age. Full-fat BSF was included in the diet at 4 and 8% and defatted BSF was included in the diet at a level supplying a similar level of protein to the diet as the diet with defatted BSF at 8% and those treatments were compared with a control diet without BSF. Full-fat BSF or defatted BSF were included in the diet at the expense of toasted soybeans. ATTD of crude protein of the control diet did not differ significantly to that of the insect-containing feed (crude protein digestibility between 77 and 78% for all treatments). Whereas the AID of crude protein in the 8% full-fat BSF diet (67.4%) was lower than that in the control diet (69.7%), the crude protein digestibility for the 4% full-fat and the defatted BSF diets was higher (73.3%) (Spranghers *et al.*, 2018).

6.2   *Growing pigs*

Yoo *et al.* (2019) determined standardised ileal digestibility (SID) of TM larvae powder, fishmeal, meat meal and poultry meal at an inclusion level of 9.95% in growing cannulated pigs of 24 kg. The SID of Arg was higher ($P < 0.05$) in pigs fed TM diet compared to that in pigs fed fish meal or meat meal diets (90.0 vs 87.6 and 88.0%, respectively). Furthermore, pigs fed poultry meal, meat meal, or TM diet showed increased ($P < 0.05$) SID of Cys compared to pigs fed fish meal diet (88.4, 86.6, 90.2 vs 83.6%, respectively). Pigs fed the TM diet tended to show increased SID of DM, total energy, CP, total amino acids (AAs), the other essential AAs (Lys, Met, Thr, Val, Ile, Leu, Phe and His), and the other non-essential AAs (Asp, Ser, Glu, Gly, Ala, Tyr and Pro) than pigs fed meat meal, poultry meal, or fish meal diet. In another study with growing pigs the AID of 100% defatted TM and TM larvae hydrolysate was determined in cannulated pigs of 29 kg (Cho et al., 2020). Pigs fed the hydrolysate of TM diet had higher SIDs of DM and CP (93.3% and 93.2%; $P < 0.05$, respectively) compared to pigs fed the other fermented poultry by-product and hydrolysed fish soluble diets. In the case of SIDs of total AAs although there was no difference between treatments ($P = 0.06$), pigs fed diets with hydrolysate of TM (83.4%) showed higher digestibility, followed by those fed with fermented poultry by-product (82.0%), defatted TM (81.9%) and hydrolysed fish soluble diets (79.2%). For SID of Lys, Met and Thr, pigs fed hydrolysate of TM and defatted TM larval meal diets showed higher SIDs ($P = 0.05$, $P < 0.05$ and $P < 0.05$, respectively) than pigs fed fermented poultry by-product and hydrolysed fish soluble diets. SIDs of non-essential amino acids (Asp, Gly, Ala) were higher ($P < 0.05$, $P < 0.05$ and $P < 0.05$, respectively) in pigs fed hydrolysate of TM, fermented poultry by-product, and defatted TM larval meal diets than those in pigs fed the hydrolysed fish soluble diet. AID and SID of Glu were higher in pigs fed hydrolysate of TM and fermented poultry by-product diets.

The two papers evaluating the SID of crude protein and amino acids in insect products are presented in Table 3 (Crosbie et al., 2020; Tan et al., 2020).

Crosbie et al. (2020) evaluated SID of amino acids and net energy contents in full fat and defatted BSF larval meals in growing pigs (Table 3). SID of CP (80.6%) and

TABLE 3   Standardised ileal digestibility of crude protein and essential amino acids in black soldier fly prepupae products and housefly prepupae meal, soybean meal and fishmeal

	Full fat BSF larval meal	Partially defatted BSF larval meal	BSF prepupae meal	HF prepupae meal	Soybean meal	Fishmeal, treated
(%)						
Crude protein	80.2	81.0	–	–	87.0	85.0
Lys	86.8	89.1	77.6	91.8	89.0	89.0
Met	90.2	79.3	91.8	98.8	90.0	89.0
Thr	87.2	86.6	79.8	91.9	85.0	88.0
Ile	87.2	89.6	77.5	87.8	88.0	90.0
Arg	92.7	95.9	86.2	97.3	93.0	92.0
Phe	95.4	97.6	76.7	90.6	89.0	87.0
His	80.7	84.3	77.8	89.6	90.0	87.0
Leu	87.2	90.7	81.0	91.8	87.0	90.0
Val	83.2	88.4	80.9	91.0	87.0	89.0
Reference	Crosbie et al. (2020)	Crosbie et al. (2020)	Tan et al. (2020)	Tan et al. (2020)	CVB (2019)	CVB (2019)

Arg = arginine; BSF = black soldier fly (*Hermetia illucens*); HF = housefly (*Musca domestica*); His = histidine; Ile = isoleucine; Leu = leucine; Lys = lysine; Met = methionine; Phe = phenylalanine; Thr = threonine; Val = valine.

Lys (88%) were not different between full-fat BSF larval meal and partially defatted BSF larval meal. SID of the essential amino acids Arg (respectively 92.7 vs 95.9%) and Val (respectively 83.2 vs 88.4%) was lower and SID of Met (respectively 90.2 vs 79.3%) was higher for full-fat BSF larval meal versus partially defatted BSF larval meal. In addition, the authors conclude that full-fat BSF larval meal was a better source of net energy for growing pigs than partially defatted BSF larval meal (3,479 vs 2,287 kcal/kg DM, respectively). Tan *et al.* (2020) determined the AID and the SID of amino acids in HF and BSF prepupae meal in growing pigs (25 kg). AID and SID of all essential amino acids in HF were higher than in BSF in growing pigs. Nutrient digestibility coefficients of defatted BSF larval meal and HF larval meal were in general comparable with digestibility coefficients of soybean meal and fishmeal except for Met which was in partially in BSF larval meal (79.3%) lower than in soybean meal (90.0%) and fishmeal (89.0%). Nutrient digestibility of HF larval meal was higher than BSF larval meal.

In summary, digestibility of dietary nutrients showed variation in weaned piglets when soybean meal and fishmeal were replaced by TM and when soybean meal and toasted soybeans were replaced by BSF larval meal. In growing pigs, TM inclusion in the diet resulted in a higher nutrient digestibility compared to fishmeal, hydrolysed fish soluble, meat meal, poultry meal and fermented poultry by-product based diets. The variability in nutrient digestibility of insect products in different studies is mainly due to changes in diet composition when insect products are included. The heterogeneity of the results in studies may also be attributed to the insect species, the insect life stage (adult, larva or pupa), the insect rearing substrate, and processing techniques and conditions (temperature at drying, extraction techniques, chitin removal) which all may influence the nutritive value of the insect products used (Barragan-Fonseca *et al.*, 2017; Cho *et al.*, 2020; Crosbie *et al.*, 2020; Gasco *et al.*, 2019; Sánchez-Muros *et al.*, 2014). Two digestion experiments were conducted in which the only protein source was the insect protein source of consideration. From these two experiments with full-fat BSF larval meal, partially defatted BSF larval meal, and HF larval meal it can be concluded that amino acid digestibility of the evaluated insect products is comparable to the amino acid digestibility of soybean meal and fishmeal. HF larval meal amino acid digestibility was higher than in BSF larval meal.

7 Growth performance

In literature, one experiment was conducted with nursing piglets, six with weaned piglets, two with growing pigs and one with finishing pigs (Table 4). In three experiments different mealworm products were tested, six experiments with BSF products and one experiment with HF.

TABLE 4   Growth performance in pigs fed insect products

Pig species/type	Age (d)	Insect product	Target replacement	Insect inclusion %	Results	Reference
Nursing piglets (Large White and Landrace)	10-28	Milled and sieved (3 mm) BSF larvae	Fishmeal	0, 3.5	Growth performance was not affected.	Driemeyer (2016)
Weaned piglets	21-36	Full-fat BSF and defatted BSF	Toasted soybeans	0, 4, 8	Growth performance was not affected.	Spranghers et al. (2018)
Weaned piglets ([Duroc×Yorkshire]×Landrace)	21-56	Dried mealworm (PT) powder	Fishmeal	0, 1, 2	Final BW at 1% inclusion was lower than at control diet. FCR at 1 and 2% inclusion was higher than at control diet. BWG at 1% inclusion was lower than at control diet during 29-42 and 21-56 d of age.	Ao and Kim (2019)
Weaned piglets (Topigs)	21-61	Partially defatted BSF larva meal	Soybean meal	0, 5, 10	FI in phase II increased linearly with increasing inclusion levels.	Biasato et al. (2019)
Weaned piglets	21-61	BSF larvae oil	Corn oil	0, 2, 4, 6	BW and BWG increased linearly and FCR decreased linearly at increasing inclusion levels.	Van Heugten et al. (2019)
Weaned piglets ([Yorkshire×Landrace]×Duroc)	28-63	Ground air-dried TM larvae	Soybean meal and soy oil	0, 1.5, 3.0, 4.5, 6.0	FI, BW and BWG increased linearly at increasing inclusion levels during phase I (28-42 d).	Jin et al. (2016)
Weaned piglets (Piétrain×[German Landrace×German Edelschwein])	35-63	TM larvae	Soybean meal	0, 5, 10	BWG at 10% inclusion was lower than at 0% and 5% inclusion.	Meyer et al. (2020)
Growing pigs (Large White)	11-29 kg	Ground sun-dried HF larvae	Fishmeal	0, 10	BWG was higher and FCR was lower for HF larvae-fed pigs compared to fishmeal-fed pigs.	Dankwa et al. (2000)
Growing pigs (Large White×Landrace)	18-53 kg	Dried BSF larval meal	Fishmeal	0, 9, 12, 14.5, 18.5	Growth performance was not affected.	Chia et al. (2019)
Finishing pigs ([Duroc×Landrace]×Large White)	76-115 kg	Dried BSF larvae powder	Soybean meal	0, 4, 8	BW and BWG at 4% inclusion was higher and FCR was lower than at 0 and 8% inclusion.	Yu et al. (2019)

BSF = black soldier fly (*Hermetia illucens*); BW = body weight; BWG = body weight gain; FCR = feed conversion ratio; FI = feed intake; HF = housefly (*Musca domestica*); PT = mealworm (*Ptecticus tenebrifer*); TM = mealworm (*Tenebrio molitor*).

7.1 *Nursing piglets*

Driemeyer (2016) studied the effect of 3.5% BSF larvae inclusion at the expense of fishmeal in pig creep diets on growth performance from 10 to 28 days of age. FI and body weight gain (BWG) were not affected by dietary treatment.

7.2 *Weaned piglets*

In an experiment of (Ao and Kim, 2019) body weight (BW) at 1% PT inclusion was lower than at control diet at 56 d of age (21.0 vs 21.8 kg, respectively). Feed conversion ratio (FCR) at 1 and 2% PT inclusion was higher from 21 to 28 d of age (1.24 and 1.23, respectively) than for the control diet (1.17). Inclusion of 5 or 10% TM in diets for weaned piglets from 35 to 63 d of age did not affect performance parameters (FI, BW, FCR) (Meyer *et al.*, 2020). Only daily BWG was lower in pigs fed diets with 10% TM compared to pigs fed control diets and 5% inclusion level (573 vs 636 and 614 g/d, respectively). The authors stated, however, that the number of pigs in the study is rather limited and the number of observations is rather low for a classical performance trial. Jin *et al.* (2016) included dried TM (0, 1.5, 3.0, 4.5 and 6.0%) at the expense of soybean meal and studied the effects on growth performance of weaned piglets. FI, BW and BWG improved linearly when the piglets were fed increasing levels of TM larvae in feeding phase I (28-42 d of age). BWG tended ($P = 0.08$) to increase linearly at increasing TM levels during phase II (42-63 d of age) and overall FCR tended ($P = 0.07$) to improve linearly at increasing TM levels up to the highest inclusion level of 6%.

Biasato *et al.* (2019) reported that growth performance was not affected in weaned piglets fed a two-phase diet except for FI in phase II. A linear response of FI to increasing BSF meal levels (0, 5 and 10% BSF larval meal) was observed (940, 950 and 970 g/d, respectively). Spranghers *et al.* (2018) studied also the effect on growth performance of weaned piglets fed diets with 4 and 8% full-fat BSF and 5.4% defatted BSF from 21 to 36 d of age. Defatted BSF was included in the diet at a level supplying a similar level of protein to the diet as the diet with defatted BSF at 8% and those treatments were compared with a control diet without BSF. BSF was included in the diet at the expense of toasted soybeans. Growth performance (FI, BWG and FCR) was not affected by dietary treatment.

7.3 *Growing pigs*

In growing pigs partial or full replacement of fishmeal in diets by BSF larval meal up to 18.5% BSF larvae inclusion at full replacement of fishmeal did not affect FI, BWG and FCR (Chia *et al.*, 2019).

When 10% ground sun-dried HF larvae was included to replace 10% fishmeal in growing pig diets, CP content of the diets was almost similar and crude fat content in the diets with HF larvae was higher than in fishmeal diets (Dankwa *et al.*, 2000). In the growth period from 11 to 29 kg daily BWG of pigs fed diets with 10% HF larvae was higher (290 vs 250 g/d, respectively) and FCR was lower (3.29 vs 3.64, respectively) compared to pigs fed diets containing 10% fishmeal (Dankwa *et al.*, 2000).

7.4   *Finishing pigs*

In finishing pigs (76-115 kg) inclusion of 4% dried BSF larvae powder at the expense of soybean meal resulted in a higher final BW (120 vs 116 and 115 kg, respectively) and BWG (980 vs 890 and 860 g/d) and a lower FCR (2.85 vs 3.21 and 3.24, respectively) compared to 0 and 8% inclusion (Yu *et al.*, 2019). FI was not affected by dietary inclusion of BSF larvae powder. The authors suggested that the underlying mechanism may be associated with the up-regulated expression of genes related to the lipogenic potential and muscle fibre composition. The lack of positive effects in the finishing pigs fed diets with 8% BSF larvae may be due to the higher level of chitin in the BSF larvae diet.

Next to insects as dietary protein source one study evaluated the inclusion of insect oil in diets for weaned piglets. The impact of increasing levels of supplemental BSF larvae oil on growth performance was studied in weaned piglets from 21 to 61 d of age (Van Heugten *et al.*, 2019). Treatments consisted of 0, 2, 4 and 6% supplemental BSF larvae oil, replacing equal amounts of corn oil. Supplementation of BSF larvae oil linearly increased BW and BWG until 46 d of age and overall. FCR was improved linearly until 46 d of age, and FI was not affected.

In summary growth performance was not affected by BSF larvae inclusion in an experiment with nursing piglets. Effects of mealworm and BSF larvae inclusion in weaned piglets diets on growth performance results were not consistent. In an experiment with growing pigs dietary inclusion of BSF larvae replacing fishmeal did not affect growth performance and in an experiment with growing pigs dietary inclusion of HF larvae replacing fishmeal improved growth performance. In finishing pigs an inclusion of 4% BSF larvae replacing soybean meal resulted in an improved growth performance compared to 0 and 8% inclusion. Including BSF larvae oil replacing corn oil up to 6% improved growth performance linearly in weaned piglets. The effect of insect products on growth performance results is highly dependent on study design, formulation of the diets and the nutritional value of insects included in the feed formulation matrix. Some of the effects observed related to the dietary inclusion of insects products are likely also related to the previous points.

8   Carcass and meat quality

Two studies are included reporting the effects of dietary insect inclusion on carcass and meat quality. In one experiment product quality was observed in growing pigs and in one experiment in finishing pigs. One study was based on HF larvae and one study on BSF larvae.

8.1   *Growing pigs*

Dankwa *et al.* (2000) replaced 10% fishmeal by ground sun-dried HF larvae in diets for growing pigs. Replacement did not affect dressing percentage and eye muscle area. Shoulder fat content in pigs fed sun-dried HF larvae was higher (3.02 vs 2.65 cm, respectively) than fat content in growing pigs fed fishmeal. This can be the result of excess energy in the diet containing HF larvae.

8.2   *Finishing pigs*

In finishing pigs (76-115 kg) dietary inclusion of dried BSF larvae powder at the expense of soybean meal affected carcass and meat quality (Yu *et al.*, 2019). Inclusion of 4% BSF larvae powder in the diet improved carcass traits and muscle chemical composition of finishing pigs and affected the meat quality via upregulating the expression of genes related to the lipogenic potential and muscle fibre composition in the longissimus thoracis of pigs. Loin-eye area (54.7 and 49.3 cm vs 45.4 cm, respectively) was increased at 4 and 8% compared with the control treatment.

9   Health

Different health related parameters were measured in eight studies reported in literature with insect inclusion in the diet (Table 5). One experiment was conducted with nursing piglets, six with weaned piglets and one experiment with growing pigs. Two studies were conducted in which TM products were included and in six studies BSF products were included.

9.1   *Nursing piglets*

Driemeyer (2016) investigated the effect of BSF larval meal supplementation on nursing piglet blood parameters of 28 litters in two treatments (a control diet with 0% larval meal inclusion and a diet with 3.5% inclusion of larval meal). Inclusion of 3.5% BSF larval meal in pig creep diets did not affect haematological and biochemical concentrations at 28 days of age. However, the author reported that the BSF larval meal diet showed an increased haemoglobin concentration and a higher haematocrit value. According to the author, these results may be considered as an indication of immunological stress, however, the animals did not show physical signs of distress when compared to the control group.

9.2   *Weaned piglets*

Meyer *et al.* (2020) concluded that TM meal can be used as a dietary source of protein in weaned pigs without causing adverse effects on intermediary metabolism. The highest TM meal inclusion level in this study was 10%. Plasma metabolomics revealed higher concentrations of Ala, Asp, Glu, Pro, Ser, Tyr and Val and a lower concentration of Asn at 10% inclusion than at control diet. Only one out of fourteen quantifiable amino acid metabolites, namely methionine sulfoxide (MetS), in plasma was elevated by 45% and 71% at 5 and 10% inclusion, respectively, compared to control diet ($P < 0.05$). Plasma concentrations of both, major carnitine/acylcarnitine species and bile acids were not different across groups. Lipidomics of liver and plasma demonstrated no differences in the concentrations of triacylglycerols, cholesterol and the main phospholipids, lysophospholipids and sphingolipids between groups. It was concluded that TM can be used as a dietary source of protein in pigs without causing adverse effects on metabolism of growing pigs. Jin *et al.* (2016) determined blood profiles in weaned piglets fed different TM inclusion levels. Blood urea nitrogen decreased linearly and insulin-like growth factor (IGF-1) increased linearly at increasing TM levels in phase II (42-63 d). High levels of blood urea nitrogen indicate that excessive amino acids are metabolised and circulate in the blood. IGF-1 as growth hormone plays an important role in controlling the structure, function of cardiovascular system and skeletal maturation (Bayes-Genis *et al.*, 2000). Blood immunoglobulin A (IgA) and G concentrations were measured as indicators of the immune response but were not affected by inclusion of TM in diets for weaned piglets.

Biasato *et al.* (2019) studied haematological, biochemical, morphometric and histopathological parameters in weaned piglets fed 0, 5 or 10% partially defatted BSF larvae at the expense of soybean meal. The parameters were not affected, except for the counts of monocytes and neutrophils, a linear and quadratic response was observed, respectively, to increasing BSF meal levels (with the maximum values corresponding to 10 and 5% BSF inclusion, respectively). This finding is difficult to explain, since none of the BSF-fed piglets showed any signs of physical distress or inflammatory diseases. Spranghers *et al.* (2018) studied intestinal health parameters in weaned piglets at 36 d of age fed diets with full-fat BSF and defatted BSF. Full-fat BSF was included at 4 and 8% and defatted BSF at a level supplying a similar level of protein to the diet as the diet with defatted BSF at 8%. pH, Lactobacilli and Streptococci in stomach, proximal small intestine and distal small intestine were not affected by dietary treatment despite higher measured lauric acid concentrations in the different segments at both inclusion levels of full-fat BSF. Alteration of intestinal specific bacterial populations and immune homeostasis in weaned piglets fed BSF larval meal as a fishmeal replacement was studied from 28 to 56 d of age (Yu *et al.*, 2020). BSF larval meal was included at 1, 2 and 4% at the expense of fishmeal. Inclusion of 2% BSF larval meal (replacement for 50% of dietary fishmeal) affected specific ileal and caecal bacterial populations and metabolic profiles, as well as the expression of mucosal immune genes. Dietary inclusion of 2% BSF larval meal selectively increased the number of certain probiotic bacteria, and the concentration of lactate and short chain fatty acids in the ileal and caecal digesta. Additionally, it selectively decreased the number of *E. coli* and the concentration of metabolites involved in nitrogen metabolism (branched chain fatty acids, biogenic amines, and phenolic and indolic compounds). The ileum mucosal mRNA expression of

TABLE 5   Health of pigs fed insect products

Pig species/type	Age (d)	Insect product	Target replacement	Insect inclusion %	Results	Reference
Nursing piglets (Large White and Landrace)	10-28	Milled and sieved (3 mm) BSF larvae	Fishmeal	0, 3.5	Evaluated haematological and biochemical parameters were not affected.	Driemeyer (2016)
Weaned piglets	21-36	Full-fat BSF and defatted BSF	Toasted soybeans	0, 4, 8	pH, Lactobacilli and D-Streptococci in stomach, proximal small intestine and distal small intestine were not affected.	Spranghers et al. (2018)
Weaned piglets (Topigs)	21-61	Partially defatted BSF larva meal	Soybean meal	0, 5, 10	Evaluated haematological and biochemical parameters were not affected, except for the monocytes and neutrophils, linear and quadratic responses were observed. Gut morphology and histological features were not affected.	Biasato et al. (2019)
Weaned piglets	21-61	BSF larvae oil	Corn oil	0, 2, 4, 6	Evaluated biochemical parameters were not affected, except cholesterol that increased linearly at higher inclusion levels. Haematological parameters were not affected, but platelet count tended to linearly increase at higher inclusion levels.	Van Heugten et al. (2019)
Weaned piglets ([Duroc × Landrace] × Large White)	28-56	Full-fat BSF larval meal	Fishmeal	0, 1, 2, 4	Supplementation with 2% BSF larval meal affected specific ileal and caecal bacterial populations and metabolic profiles, as well as the mucosal immune genes expression.	Yu et al. (2020)
Weaned piglets ([York-shire×Landrace]×Duroc)	28-63	Ground air-dried TM larvae	Soybean meal and soy oil	0, 1.5, 3.0, 4.5, 6.0	Blood urea nitrogen decreased linearly and insulin-like growth factor increased linearly at increasing inclusion levels in phase II (42-63 d). Immunoglobulin A and G concentrations were not affected.	Jin et al. (2016)
Weaned piglets (Piétrain × [German Landrace × German Edelschwein])	35-63	TM larvae	Soybean meal	0, 5, 10	Higher blood plasma concentrations of Ala, Asp, Glu, Pro, Ser, Tyr and Val and a lower concentration of Asn at 10% inclusion than at control diet. Plasma methionine sulfoxide was higher at 5% and 10% inclusion, compared to control diet.	Meyer et al. (2020)
Growing pigs (Large White × Landrace)	18-53 kg	Dried BSF larval meal	Fishmeal	0, 9, 12, 14.5, 18.5	Red or white blood cell parameters were not affected, except for neutrophil counts, which were higher at 14.5 and 18.5% inclusion compared to control diet. Platelet counts at 9, 14.5 and 18.5% inclusion were lower compared to control diet and 12% inclusion. Blood cholesterol levels were not affected.	Chia et al. (2019)

Ala = alanine; Asn = asparagine; Asp = aspartate; Glu = glutamate; Pro = proline; Ser = serine; Tyr = tyrosine; Val = valine; BSF = black soldier fly (*Hermetia illucens*); Ig = immunoglobulin; TM = mealworm (*Tenebrio molitor*).

TLR4-MyD88-NF-κB signalling pathway and proinflammatory cytokine genes, and TNF-α protein concentration were decreased, but the mRNA expression of barrier function-, development-relative, and anti-inflammatory cytokines and sIgA protein concentrations were increased.

9.3   *Growing pigs*

Chia *et al.* (2019) studied blood parameters in growing pigs fed different inclusion levels of BSF larvae as blood profiles can provide an indication of the clinical health status as well as the extent to which dietary deficiencies impact the physiological status of the animal. Red or white blood cell concentrations were not affected at any inclusion level of BSF larval meal in diets for growing pigs (Chia *et al.*, 2019) up to inclusion level of 18.5% at full replacement of fishmeal in the diet. Neutrophil counts were higher at 14.5 and 18.5% inclusion compared to control. Neutrophils are one of the first responders of inflammatory cells to migrate toward the site of inflammation. Platelet counts at 9, 14.5 and 18.5% inclusion were lower compared to 0 and 12% inclusion. Low platelet concentration implies that blood clotting might be impaired, resulting in blood loss in case of injury (Etim *et al.*, 2014).

Van Heugten *et al.* (2019) studied the impact of increasing levels of supplemental BSF larval oil on serological and haematological indices in weaned piglets from 21 to 61 d of age. Treatments consisted of 0, 2, 4 and 6% supplemental BSF larval oil, replacing equal amounts of corn oil. Supplemental BSF larval oil did not affect serological parameters, but linearly increased serum cholesterol. Haematological parameters were also not affected by BSF larval oil, but platelet count tended ($P =$ 0.082) to linearly increase at increased BSF larval oil inclusion levels.

Overall it can be summarised that health related haematological, biochemical and intestinal health parameters were not affected by dietary inclusion of insect products. Only in one study inclusion of 2% BSF larval meal (replacement 50% of dietary fishmeal) affected specific ileal and caecal bacterial populations and metabolic profiles, as well as the ileal immune status in weaned piglets.

10   Conclusions

Amino acid digestibility of full-fat BSF larval meal, partially defatted BSF larval meal and HF larval meal were comparable to the amino acid digestibility of soybean meal and fishmeal and HF larval meal amino acid digestibility was higher than in BSF larval meal. Nutrient digestibility of insect-based diets showed variation within and between studies in weaned piglets and growing pigs. The variability in nutrient digestibility of insect products is mainly due to changes in diet composition when insect products are included. The nutrient digestibility of diets depends on the source of protein that is used or exchanged. Also differences in insect sources used and processing techniques applied may have affected nutrient digestibility in diets. Effects of different insect-based diets on growth performance results were also variable. The effect of insect products on growth performance results is highly dependent on study design, formulation of the diet and the nutritional value of insects adopted during feed formulation. For further studies it is recommended to distinguish studies aimed at determining nutritional value (digestibility studies) and growth performance studies in which the effects of the insect products as dietary ingredient are observed. Health related haematological and biochemical parameters were not affected by dietary inclusion of insect products. In general it can be stated that differences in results between studies may be related to differences in the nature of the insect species, the used substrate to grow the insects and life stages used, in dietary inclusion levels, in the way of processing of the insect products, to variation in palatability of diets, age of the animal and research methods applied. Overall it can be concluded that insect products seem to be a good alternative to partly replace soybean meal or fishmeal in piglet and pig diets without adversely affecting growth performance, product quality and health status. More standardised digestibility and growth performance experiments are recommended for future research and potential positive effects of the use of insect based ingredients on animal health deserve more attention.

Acknowledgement

We thank M. Dicke, A.J.M. Jansman and A. Dörper for constructive comments on a previous version of the manuscript.

Conflict of interest

The authors declare no conflict of interest.

References

Alltech, 2020. 2020 global feed survey. Alltech, Nicholasville, KY, USA.
Anonymous, 2017. Factsheet varkenshouderij – feiten en cijfers over de Nederlandse varkenshouderij. POV, Zwolle, The Netherlands.
Ao, X. and Kim, I.H., 2019. Effects of dietary dried mealworm (*Ptecticus tenebrifer*) larvae on growth performance and nutrient digestibility in weaning pigs. Livestock Science 230: 4.
Bai, Z., Ma, W., Ma, L., Velthof, G.L., Wei, Z., Havlík, P., Oenema, O., Lee, M.R.F. and Zhang, F., 2018. China's livestock transition: driving forces, impacts, and consequences. Science Advances 4: 1-11.

Barragan-Fonseca, K.B., Dicke, M. and Van Loon, J.J.A., 2017. Nutritional value of the black soldier fly (*Hermetia illucens L.*) and its suitability as animal feed – a review. Journal of Insects as Food and Feed 3: 105-120. https://doi.org/10.3920/JIFF2016.0055

Bayes-Genis, A., Conover, C.A. and Schwartz, R.S., 2000. The insulin-like growth factor axis: a review of atherosclerosis and restenosis. Circulation Research 86: 125-130.

Biasato, I., Renna, M., Gai, F., Dabbou, S., Meneguz, M., Perona, G., Martinez, S., Lajusticia, A.C.B., Bergagna, S., Sardi, L., Capucchio, M.T., Bressan, E., Dama, A., Schiavone, A. and Gasco, L., 2019. Partially defatted black soldier fly larva meal inclusion in piglet diets: effects on the growth performance, nutrient digestibility, blood profile, gut morphology and histological features. Journal of Animal Science and Biotechnology 10: 11.

Chia, S.Y., Tanga, C.M., Osuga, I.M., Alaru, A.O., Mwangi, D.M., Githinji, M., Subramanian, S., Fiaboe, K.K.M., Ekesi, S., Van Loon, J.J.A. and Dicke, M., 2019. Effect of dietary replacement of fishmeal by insect meal on growth performance, blood profiles and economics of growing pigs in Kenya. Animals 9: 19.

Cho, K.H., Kang, S.W., Yoo, J.S., Song, D.K., Chung, Y.H., Kwon, G.T. and Kim, Y.Y., 2020. Effects of mealworm (*Tenebrio molitor*) larvae hydrolysate on nutrient ileal digestibility in growing pigs compared to those of defatted mealworm larvae meal, fermented poultry by-product, and hydrolyzed fish soluble. Asian-Australasian Journal of Animal Sciences 33: 490-500.

Crosbie, M., Zhu, C., Shoveller, A.K. and Huber, L.-A., 2020. Standardized ileal digestible amino acids and net energy contents in full fat and defatted black soldier fly larvae meals (*Hermetia illucens*) fed to growing pigs. Translational Animal Science 4: txaa104.

CVB, 2019. CVB veevoedertabel 2019 – chemische samenstellingen en nutritionele waarden van voedermiddelen. CVB, Wageningen, The Netherlands.

Dankwa, D., Oddoye, E. and Mzamo, K., 2000. Preliminary studies on the complete replacement of fishmeal by house-fly-larvae-meal in weaner pig diets: effects on growth rate, carcass characteristics, and some blood constituents. Ghana Journal of Agricultural Science 33: 223-227.

De Boer, J., Helms, M. and Aiking, H., 2006. Protein consumption and sustainability: diet diversity in EU-15. Ecological Economics 59: 267-274.

De Smet, S. and Vossen, E., 2016. Meat: the balance between nutrition and health. A review. Meat Science 120: 145-156.

De Visser, C.L.M., Schreuder, R. and Stoddard, F., 2014. The EU's dependency on soya bean import for the animal feed industry and potential for EU produced alternatives. OCL 21: D407.

Dörper, A., Veldkamp, T. and Dicke, M., 2021. Use of black soldier fly and house fly in feed to promote sustainable poultry production Journal of Insects as Food and Feed 7: 761-780, https://doi.org/10.3920/JIFF2020.0064.

Driemeyer, H., 2016. Evaluation of black soldier fly (*Hermetia illucens*) larvae as an alternative protein source in pig creep diets in relation to production, blood and manure microbiology parameters. MSc-thesis, Stellenbosch University, Stellenbosch, South Africa, 99 pp.

European Research Area on Sustainable Animal Production Systems (ERANET), 2020. SusAn. Available at: https://cordis.europa.eu/project/id/696231

Etim, N.N., Offiong, E.E.A., Williams, M.E. and Asuquo, L.E., 2014. Influence of nutrition on blood parameters of pigs. American Journal of Biology and Life Sciences 2: 7.

Finke, M.D., 2002. Complete nutrient composition of commercially raised invertebrates used as food for insectivores. Zoo Biology 21: 269-285.

Florou-Paneri, P., Christaki, E., Giannenas, I., Bonos, E., Skoufos, I., Tsinas, A., Tzora, A. and Peng, J., 2014. Alternative protein sources to soybean meal in pig diets. Journal of Food Agriculture and Environment 12: 655-660.

Gasco, L., Biasato, I., Dabbou, S., Schiavone, A. and Gai, F., 2019. Animals fed insect-based diets: state-of-the-art on digestibility, performance and product quality. Animals 9: 32.

Gasco, L., Finke, M. and Van Huis, A., 2018. Can diets containing insects promote animal health? Journal of Insects as Food and Feed 4: 1-4. https://doi.org/10.3920/JIFF2018.x001

Hall, H.N., Masey O'Neill, H.V., Scholey, D., Burton, E., Dickinson, M. and Fitches, E.C., 2018. Amino acid digestibility of larval meal (*Musca domestica*) for broiler chickens. Poultry Science 97: 8.

Jin, X.H., Heo, P.S., Hong, J.S., Kim, N.J. and Kim, Y.Y., 2016. Supplementation of dried mealworm (*Tenebrio molitor* larva) on growth performance, nutrient digestibility and blood profiles in weaning pigs. Asian-Australasian Journal of Animal Sciences 29: 979-986.

Jozefiak, A. and Engberg, R., 2017. Insect proteins as a potential source of antimicrobial peptides in livestock production. A review. Journal of Animal and Feed Sciences 26: 13.

Lassaletta, L., Billen, G., Romero, E., Garnier, J. and Aguilera, E., 2014. How changes in diet and trade patterns have shaped the N cycle at the national scale: Spain (1961-2009). Regional Environmental Change 14: 785-797.

Lassaletta, L., Estellés, F., Beusen, A.H.W., Bouwman, L., Calvet, S., Van Grinsven, H.J.M., Doelman, J.C., Stehfest, E., Uwizeye, A. and Westhoek, H., 2019. Future global pig production systems according to the shared socioeconomic pathways. Science of the Total Environment 665: 739-751.

Meyer, S., Gessner, D.K., Braune, M.S., Friedhoff, T., Most, E., Höring, M., Liebisch, G., Zorn, H., Eder, K. and Ringseis, R., 2020. Comprehensive evaluation of the metabolic effects of insect meal from *Tenebrio molitor L.* in growing pigs by transcriptomics, metabolomics and lipidomics. Journal of Animal Science and Biotechnology 11: 20.

Mottet, A., De Haan, C., Falcucci, A., Tempio, G., Opio, C. and Gerber, P., 2017. Livestock: on our plates or eating at our table? A new analysis of the feed/food debate. Global Food Security 14: 1-8.

Rauw, W.M., Rydhmer, L., Kyriazakis, I., Overland, M., Gilbert, H., Dekkers, J.C.M., Hermesch, S., Bouquet, A., Izquierdo, E.G., Louveau, I. and Gomez-Raya, L., 2020. Prospects for sustainability of pig production in relation to climate change and novel feed resources. Journal of the Science of Food and Agriculture 100: 3575-3586. https://doi.org/10.1002/jsfa.10338

Sánchez-Muros, M.-J., Barroso, F.G. and Manzano-Agugliaro, F., 2014. Insect meal as renewable source of food for animal feeding: a review. Journal of Cleaner Production 65: 16-27.

Soare, E. and Chiurciu, I.-A., 2017. Study on the pork market worldwide. Scientific Papers: Management, Economic Engineering in Agriculture & Rural Development 17: 321-326.

Spranghers, T., Michiels, J., Vrancx, J., Ovyn, A., Eeckhout, M., De Clercq, P. and De Smet, S., 2018. Gut antimicrobial effects and nutritional value of black soldier fly (*Herrnetia illucens L.*) prepupae for weaned piglets. Animal Feed Science and Technology 235: 10.

Spranghers, T., Ottoboni, M., Klootwijk, C., Ovyn, A., Deboosere, S., Meulenaer, B., Michiels, J., Eeckhout, M., De Clercq, P. and De Smet, S., 2017. Nutritional composition of black soldier fly (*Hermetia illucens*) prepupae reared on different organic waste substrates. Journal of the Science of Food and Agriculture 97: 2594-2600.

Steinfeld, H., Gerber, P., Wassenaar, T., Castel, V., Rosales, M., Rosales, M. and De Haan, C., 2006. Livestock's long shadow: environmental issues and options. FAO, Rome, Italy.

Tan, X., Yang, H.S., Wang, M., Yi, Z.F., Ji, F.J., Li, J.Z. and Yin, Y.L., 2020. Amino acid digestibility in housefly and black soldier fly prepupae by growing pigs. Animal Feed Science and Technology 263: 114446.

United Nations (UN), 2015. Resolution adopted by the General Assembly on 25 September 2015: 70/1. Transforming our world: the 2030 agenda for sustainable development. UN, New York, NY, USA.

United States Department of Agriculture (USDA), 2020. Livestock and poultry: world markets and trade. USDA, Washington, DC, USA.

Van Heugten, E., Martinez, G., McComb, A. and Koutsos, E., 2019. Black soldier fly (*Hermetia illucens*) larvae oil improves growth performance of nursery pigs. Journal of Animal Science 97: 1.

Van Huis, A. and Oonincx, D.G.A.B., 2017. The environmental sustainability of insects as food and feed. A review. Agronomy for Sustainable Development 37: 14. https://doi.org/10.1007/s13593-017-0452-8

Van Zanten, H.H.E., Bikker, P., Meerburg, B.G. and De Boer, I.J.M., 2018. Attributional versus consequential life cycle assessment and feed optimization: alternative protein sources in pig diets. International Journal of Life Cycle Assessment 23: 1-11.

Veldkamp, T. and Bosch, G., 2015. Insects: a protein-rich feed ingredient in pig and poultry diets. Animal Frontiers 5: 45-50.

Veldkamp, T., Van Duinkerken, G., Van Huis, A., Lakemond, C.M.M., Ottevanger, E., Bosch, G. and Van Boekel, T., 2012. Insects as a sustainable feed ingredient in pig and poultry diets: a feasibility study. Wageningen UR Livestock Research, Lelystad, The Netherlands.

Wiedemann, S., McGahan, E. and Murphy, C., 2016. Environmental impacts and resource use from Australian pork production assessed using life-cycle assessment. Greenhouse gas emissions. Animal Production Science 56: 14.

Yoo, J.S., Cho, K.H., Hong, J.S., Jang, H.S., Chung, Y.H., Kwon, G.T., Shin, D.G. and Kim, Y.Y., 2019. Nutrient ileal digestibility evaluation of dried mealworm (*Tenebrio molitor*) larvae compared to three animal protein by-products in growing pigs. Asian-Australasian Journal of Animal Sciences 32: 387-394.

Yu, M., Li, Z., Chen, W., Wang, G., Rong, T., Liu, Z., Wang, F. and Ma, X., 2020. *Hermetia illucens* larvae as a fishmeal replacement alters intestinal specific bacterial populations and immune homeostasis in weanling piglets. Journal of Animal Science 98: 13.

Yu, M., Li, Z.M., Chen, W.D., Rong, T., Wang, G., Li, J.H. and Ma, X.Y., 2019. Use of *Hermetia illucens* larvae as a dietary protein source: effects on growth performance, carcass traits, and meat quality in finishing pigs. Meat Science 158: 7.

Effect of using insects as feed on animals: pet dogs and cats

G. Bosch[1]* and K.S. Swanson[2,3,4]

[1]Animal Nutrition Group, Wageningen University & Research, De Elst 1, 6708 WD Wageningen, The Netherlands; [2]Department of Animal Sciences, University of Illinois at Urbana-Champaign, Urbana, IL 61801, USA; [3]Division of Nutritional Sciences, University of Illinois at Urbana-Champaign, Urbana, IL 61801, USA; [4]Department of Veterinary Clinical Medicine, University of Illinois at Urbana-Champaign, Urbana, IL 61801, USA; *guido.bosch@wur.nl

Abstract

The 'buzz' in society around insects has resulted in the appearance of insect-based pet food products on the market and more products are under development. This contribution aimed to provide background information on pet foods and the sector and to provide an overview of the current state of knowledge regarding naturalness, palatability, nutritional quality, health effects, and sustainability of insects as feed for dogs and cats. In contrast to dogs, natural diets of cats commonly contain insects but contribution to the total biomass is < 0.5% in most diets. Cats and dogs can have a different palate when it comes to insects and insect species and inclusion level influence the acceptance of the food. The apparent faecal N digestibility values for insect-based foods were in the range of foods containing conventional protein sources. Based on the indispensable amino acid (IAA) digestibility values reported for black soldier fly larvae (BSFL), housefly larvae (HFL), and yellow mealworms (YMW) in chickens and requirements of growing dogs and growing cats, the first limiting IAA were methionine (BSFL, dogs and cats; YMW, dogs and cats), threonine (BSFL, dogs), and leucine (HFL, dog and cats). More long-term studies are still required to evaluate adequacy and safety of insect-based pet foods in dogs and cats as well as studies that focus on the presence of health-promoting biofunctionalities of insects. Insect proteins have a lower environmental impact than livestock meat proteins, but this is not relevant in the context of pet foods that are largely based on animal co-products with a low environmental impact. Developments in insect rearing will make insect proteins more competitive with conventional sources. For advancing insect applications beyond hypoallergenic pet foods, it will be essential to assure insects as safe and quality ingredients as well as understanding pet owner views and values regarding insect rearing.

Keywords

health – naturalness – nutritional quality – palatability – sustainability

1   Introduction

Pet dogs (*Canis familiaris*) and cats (*Felis silvestris catus*) play an important role in many people's lives around the globe. Pet-ownership is considerable with over 170 million dogs and 190 million cats living in households in Europe and the USA (FEDIAF, 2019a; Statista, 2020a,b) and increasing numbers in particularly in Asia and Latin America. Pets provide companionship, affection and protection and owners value the unique bonds they have with their pets and their contribution to the quality of life (O'Haire, 2010; Podberscek *et al.*, 2000). Furthermore, pets are commonly viewed as family members and as equals, which has important consequences for the development and marketing of pet foods. Trends in human food, for example, are often rapidly translated into new pet food products. Insects have received considerable attention as a sustainable and sometimes even health-promoting and novel protein source for humans. The 'buzz' in society around insects has also resulted in the appearance of insect-based pet food products on the market and more products are on their way. This contribution aims to provide background information on pet foods as well as the current state of knowledge regarding some features of insects as feed for dogs and cats.

2   Pet foods

Today's pet owner's can choose from a plethora of nutritious products to support specific breeds and sizes, particular life stages, and disease predispositions, in various formats and packaging styles, and for different prices. Owners often decide to buy a specific pet food based on price, convenience, previous experience, packaging, brand reputation, and marketing claims. Furthermore, healthfulness, freshness and ingredients of a pet food are evaluated by owners (Schleicher *et al.*, 2019). Specific ingredients are associated with nutritional quality such as the presence of 'natural' ingredients (Vinassa *et al.*, 2020). Familiar phrases, popular trends and beliefs on packaging and advertisements also help to attract owners to products. These trends in pet food products often follow those in human nutrition. Examples of trends are grain-free and 'natural' pet foods (Beaton, 2014; Wall, 2018). Supposed health benefits to pets of such trendy feeding practices are often not supported by scientific studies (Schleicher *et al.*, 2019) and might even be associated with health concerns for the pet (grain-free foods were initially accused to be related to canine dilated cardiomyopathy) and social environment (raw foods potentially containing pathogenic bacteria). When consumer trends get ahead of the science and novel pet food products get associated with pet health problems, the reputation of an ingredient and/or the entire pet food industry may be negatively affected. Therefore, ingredients should undergo sufficient safety and efficacy testing before being incorporated into pet foods.

Once the food is in the bowl, its palatability is essential not only for its acceptance but also its enjoyment by the pet. For the owner, the moment of feeding contributes to the establishment and maintenance of the bond with the pet (Bradshaw and Cook, 1996; Day et al., 2009). Palatability of pet foods is therefore a key attribute. After consumption, effects on stool quality, coat condition, and more are monitored by owners and contribute to the success of a pet food. As the pet dog and cat are often fully dependent on the provision of food by their owners, it is essential that manufacturers provide appropriate and safe nutrition for health and longevity.

Of the different formats, the dry and moist foods are the most popular among pet owners in Western countries. During the manufacturing of these types of pet foods, thermal treatments like extrusion and retorting are used to improve the safety, shelf-life and nutritive properties of the foods but also to create the optimal texture and shape of the product (Hendriks et al., 1999). Although the diversity in available products is immense, the main protein sources used are similar and come from the rendering industry. For dry extruded foods, meat and by-product meals of poultry, beef, pig, lamb and fish are most commonly used. For wet retorted foods, proteins mainly originate from fresh and/or frozen meats and other animal tissues. Pet food quality attributes can be assessed with specific tests. Acceptance of a food as a measure of palatability is generally evaluated with a monadic test or a two-bowl test that evaluates the preference (relative palatability) for one food over another food (Aldrich and Koppel, 2015; Tobie et al., 2015).

The nutritional quality of a food or an ingredient depends largely on the presence of digestible and bioavailable nutrients, which should be present in amounts that assure requirements of the animal are met. Ingredients and complete foods can be chemically characterised to quantify the absolute amounts of the nutrients. Specific assays are used to estimate the degree of bioavailability of the nutrients. The standardised ileal digestibility assay, which is a routine methodology in production animal nutrition, was used in dogs decades ago but is no longer a routine procedure in the evaluation of pet foods due to ethical reasons and welfare issues. The apparent faecal digestibility assay, which is known to yield inaccurate estimates of the absorption of nutrients due to colonic fermentation (Hendriks et al., 2012), is commonly used. Standardised protocols are available for assessing apparent faecal digestibility (AAFCO, 2020; FEDIAF, 2019b). Furthermore, different in vitro assays varying in complexity (e.g. Hervera et al., 2009; Smeets-Peeters et al., 1999) and in vivo models such as the precision-fed cecectomized rooster assay (e.g. Faber et al., 2010; Johnson et al., 1998) and mink (e.g. Ahlstrøm and Skrede, 1998; Tjernsbekk et al., 2014) are also used to gain insight in nutrient digestibility or bioavailability of ingredients and complete foods. Finally, more long-term feeding protocols have been developed that aim to validate the adequacy of a pet food for the species and the life stage for which it is intended (growth, gestation/lactation, maintenance; AAFCO, 2020).

The nutritional status and veterinary care of dogs and cats has greatly improved over the past few decades, improving the quality of life and extending the lifespan of pets. These changes have increased the number of geriatric pets in the population. While this is a positive outcome, aging comes with many ailments, including osteoarthritis, oral disease, cognitive dysfunction, and chronic kidney disease. Other diseases such as obesity, diabetes mellitus, lower urinary tract diseases, gastrointestinal diseases, and food sensitivity are also quite common in the pet population. Pet food companies have responded by developing therapeutic diets intended to aid in the management of these clinical disease states. The industry has also developed over-the-counter diets and treats to target specific functional areas of healthy pets. For over-the-counter pet products, the goal is to maintain the body's organ or system functionality, thereby supporting health and wellness. The primary functional areas targeted in these products include bones and joints, oral care, immunity, skin and coat, digestive health, odour control, hairball control, and cognitive function.

3   Insects as feed for pets

3.1   *Naturalness*

Owners may prefer foods that match the ancestral history of their pet, assuming that they provide optimal nutrition. It is therefore of interest to consider the 'naturalness' of insects in the diet of dogs and cats. The dog is a direct descendent of the grey wolf (*Canis lupus*) (Leonard *et al.*, 2002; Vilà *et al.*, 1997), whereas the cat originates from the wildcat (*Felis silvestris*) in the Near East region (Driscoll *et al.*, 2007). Wolves are carnivores that hunt in packs for large ungulates (e.g. deer, wild boar), but also opportunistically feed on smaller mammals (e.g. beavers, lagomorphs) (Bosch *et al.*, 2015). Insects were reported to be present in 5 out 50 described diets of wild wolves, but with negligible contributions to the total biomass consumed (Bosch *et al.*, 2015). Like most felids, domestic cats predominantly hunt individually. They hunt a variety of prey species, with rodents (e.g. mice, voles) and lagomorphs being most common, but birds, reptiles and insects can also be part of their diet (Malo *et al.*, 2004; Pearre and Maass, 1998; Plantinga *et al.*, 2011). Though insects were not specified by Plantinga *et al.* (2011), 26 out of the 30 diets contained invertebrates. In the review of Pearre and Maass (1998), 35 diets were presented for which the consumed insect biomass was estimated. The median proportion of insect biomass was only 0.5% of the total biomass of the diet. For four of the diets, however, insects contributed 13.6, 19.1, 20 and 21.9% to the total biomass consumed. Inspection of the original data revealed that the number '19.1%' was the frequency of occurrence instead of biomass (Parmalee, 1953), the number '20%' was actually 2% as described in the text of the original article (Llewellyn and Uhler, 1952), and the number '21.9%' was not the contribution of insects to biomass eaten but the sum of dry weight of undigested fractions of grasshoppers, arachnids and beetles (Konecny, 1987). Based on the data provided by the latter author, the insects (i.e. grasshoppers + beetles) contributed 7.9% to the total dry biomass. The correctness of the remaining biomass estimates (13.6%; Bayly, 1976) could not be verified as the manuscript was not available to us. In contrast to dogs, it seems that for some cats it is natural to consume some insects, but they should not be considered as insectivores that nutritionally depend on insects and are adapted to an insect-based diet.

3.1   *Palatability*

Though many companies likely have tested insect-based pet food prototypes internally, few data have been made public. It has been reported that dogs tended to prefer dry foods containing black soldier fly larvae (*Hermetia illucens*; BSFL) meal over those containing yellow mealworm (*Tenebrio molitor*; YMW) meal (intake ratio of 60:40; n = 10), whereas cats preferred the YMW-based food (40:60; n = 10) (Beynen, 2018). Both insect meals contributed to 30% of total crude protein in these diets. Dogs have been shown to accept dry foods containing 5, 10 or 20% BSFL meal. In those 2-day studies, which might be too short to overcome a novelty effect, dogs (n = 20) readily consumed extruded foods containing 5 or 10% BSFL meal, but only 93.9% of their estimated metabolisable energy requirements for maintenance with a 20% inclusion level (Yamka *et al.*, 2019). The latter study also reported that dogs fed kibbles coated with 2.5 or 5% BSFL oil consumed 91.5 and 116.2% of their metabolisable energy requirements, respectively. Including 8, 16 or 24% banded cricket (*Gryllodes sigillatus*) meal in extruded foods did not affect food intake in dogs (n = 8) (Kilburn *et al.*, 2020).

Three out of 10 cats had refused a food containing 35% BSFL meal and three cats had an intake between 78 and 87% of the food offered (Paßlack and Zentek, 2018). For a food containing 22% BSFL meal, one cat vomited and then refused the food completely and two cats had lower food intakes (83 and 88%) (Paßlack and Zentek, 2018). In cats fed a diet containing 5 or 20% of BSFL meal for 2 days (n = 20 per diet), 38 and 54% of the 100 g of food was consumed. None of the cats rejected the 5% BSFL diet, whereas one cat rejected the 20% BSFL. Four cats fed the 5% BSFL diet ate less than 25 g on one of the 2 days, while this was observed for three cats fed the 20% BSFL diet. The latter study also evaluated the acceptance of kibbles coated with 1, 2.5 or 5% BSFL oil. Of the 20 cats fed each food, food rejection on both days was observed for two (1% BSFL oil), zero (2.5% BSFL oil) and nine (5% BSFL oil) cats, with five, seven and 16 cats consuming < 25% of their food. These studies illustrate that cats and dogs can have a different palate when it comes to insects and that insect species and inclusion level influence the acceptance of the food. Dogs seem to accept foods containing up to 10% BSFL meal and 24% banded crickets as well as kibbles coated with up to 5% BSFL oil. Cats varied more in their acceptance of foods with BSFL meal or oil and seem to accept foods containing up to 5% BSFL or coated with up to 2.5% BSFL oil.

3.3   Nutritional quality

The variation in nutrient composition of insects has been investigated in numerous studies (for reviews see Makkar *et al.*, 2014; Rumpold and Schlüter, 2013). As with conventional animal meals, nutrient composition varies among species and processing methods applied. Specific information on nutrient digestibility in dogs and cats has been published recently, though not all data originate from peer-reviewed studies (Table 1). *In vitro* N digestibility of freeze-dried ground BSFL, housefly larvae (*Musca domestica*; HFL), YMW and lesser mealworm (*Alphitobius diaperinus*) was approximately 90% and in the range of poultry meat meal (87.9%) (Bosch *et al.*, 2014, 2016).

In dogs, apparent faecal N digestibility of foods containing BSFL ranged from 73.2 to 87.2% (Beynen, 2018; Lei *et al.*, 2019; Meyer *et al.*, 2019; Yamka *et al.*, 2019) and a YMW-containing food was 83.6% (Beynen, 2018). It should be noted that in the study of Lei *et al.* (2019) the BSFL contributed minimally to the total amount of crude protein and outcomes were based on only 3 dogs, which is deemed insufficient for digestibility testing (AAFCO, 2020; FEDIAF, 2019b). Meyer *et al.* (2019) estimated that the apparent faecal N digestibility of their BSFL meal was 83.1% when included in an extruded food and 83.4% when included in a pelleted food. Feeding dogs extruded foods with 8, 16 or 24% inclusion of banded crickets resulted in apparent faecal N digestibility values of 84.8, 86.0 and 82.1%, respectively (Kilburn *et al.*, 2020). The average apparent faecal N digestibility of commercial dog foods can be considered to be close to 80% (Daumas *et al.*, 2012; Hendriks *et al.*, 2013; Hervera, 2011; Kendall *et al.*, 1982), suggesting that the insect meals are in the range of conventional protein sources. In cats, reported faecal N digestibility values of 73.4 to 79.8% have been reported for diets containing BSFL meal (Beynen, 2018; Paßlack and Zentek, 2018; Yamka *et al.*, 2019) and 80.4% for YMW (Beynen, 2018).

Dogs and cats have specific requirements for indispensable amino acids (AA). It is therefore of interest to evaluate how well the insect proteins match with the requirements of dogs and cats. As a proxy for the match, digestible indispensable AA scores (DIAAS) can be calculated in line with protein quality evaluations in human nutrition (FAO, 2013). The DIAAS-like values are calculated as (mg of digestible dietary indispensable AA in 1 g of the digestible insect protein)/(mg of the minimum requirement of the same dietary indispensable AA in 1 g of the minimum protein requirement). The digestible indispensable AA (DIAA, %) for each AA was based on the reported AA content multiplied by its ileal digestibility coefficient. The value was divided by the amount of digestible protein based on reported crude protein content and digestibility coefficient for all amino acids. The minimum protein requirements for growing (4 to 14 weeks of age) and adult dogs and growing and adult cats are taken from the National Research Council (NRC, 2006). The calculation is performed for each dietary indispensable AA and the lowest value is designated as the DIAAS and used as an indicator of dietary protein quality (FAO, 2013).

In Table 2, the digestible indispensable amino acids for BSFL, HFL and YMW are shown, which are based on ileal digestibility studies in chickens. The lowest DIAAS values for BSFL were found for methionine (0.65-1.02) when considering the requirements of growing cats and for methionine (0.81-1.01) or threonine (1.03-1.14) when considering the requirements of growing dogs. For adult cats the lowest values were for arginine (0.83-1.04) or leucine (1.03-1.14) and for adult dogs this was for methionine (0.39-0.61). For HFL the lowest values were found for leucine for both growing and adult cats and growing dogs (DIAAS of 1.25, 1.11 and 1.24, respectively) and for adult dogs this was methionine (0.93). For YMW methionine was first limiting growing and adult dogs and cats (DIAAS of 0.97, 0.55 and 0.92, respectively) and for adult cats this was leucine (0.90). For poultry meal, one of the most used protein sources in pet foods, the first limiting AA were methionine for kittens (DIAAS value of 0.53) and tryptophan for growing and adult dogs and adult cats (0.54, 0.39 and 0.66, respectively) (based on data from Deng *et al.*, 2016). Because growing dogs might receive foods that are close to their requirements (see Van Rooijen *et al.*, 2014), understanding the content and bioavailability of indispensable AA is in particular important for pet food manufacturers that formulate insect-based foods for puppies.

3.4   *Stool quality*

Owners are in close contact with the stools of their pets and may associate stool attributes (volume, consistency, odour, colour, and ease and frequency of defaecation) to intestinal health and to the overall nutritional quality of the food. Furthermore, owners seek convenience when picking up stools after their beloved dog or removing these from the litter box of their cat. Stool quality is often also assessed in studies that evaluate the acceptance or digestibility of foods. The stool consistency is commonly evaluated using a 5-point scoring system (e.g. Laflamme *et al.*, 2011; Moxham, 2001) with consistency varying from hard, dry pellets to a watery liquid that can be poured. Ideal stools are generally considered to have a firm to soft consistency and retain their shape. Stool consistency was ideal in dogs fed the extruded and pelleted foods containing 30% BSFL (Meyer *et al.*, 2019). Yamka *et al.* (2019) reported that all the foods containing either 20% BSFL meal or 5% oil resulted in ideal stool consistency in dogs and did not impact stool consistency in cats relative to the control food. Stools remained well-formed when dogs were fed extruded foods with increasing levels (8, 16 or 24%) of banded cricket meal (Kilburn *et al.*, 2020). Though the number of studies is still limited, it seems that including insect meals in dry extruded pet foods do not disturb intestinal functioning and lead to acceptable stool consistencies. Other stool quality attributes remain to be studied.

3.5   *Overall health*

Commercial pet foods that are marketed as being nutritionally complete should support the health of the dog or cat over the long-term. At this moment, only two studies have evaluated the impact of insect-based dog foods on the nutritional status and health of dogs and no studies are available that evaluated this in cats. In one study (Lei *et al.*, 2019) dogs (n = 3 per dietary treatment) were fed a dry pelleted food containing 0, 1 or 2% defatted, dried and ground BSFL meal (contributing 1.6 or 3.2% of total crude protein) for 6 weeks after which blood samples were taken for several haematological and biochemical parameters and profiling of lipids and minerals. Furthermore, these dogs were intraperitoneally challenged with *Escherichia coli* lipopolysaccharide and their immune responses were monitored. Though some changes were noted, i.e. linear increasing levels of albumin and calcium and at 6 hours after the challenge and linear decreasing levels of tumour necrosis factor-α and increasing levels of glutathione peroxidase, the number of dogs per treatment is too small to make conclusions about potential health effects of foods containing small amounts (1 or 2%) of BSFL meal. Dogs (n = 8 per dietary treatment) fed a dry extruded food with increasing levels (0, 8, 16 or 24%) of roasted and ground cricket-meal (contributing 20, 38 or 57% of total crude protein) for 29 days had on average haematology and chemistry profiles within the reference ranges for healthy dogs (Kilburn *et al.*, 2020). For some individual dogs, values were just outside the ranges, but details about the number of dogs or exact deviations were unfortunately not provided by the authors. It is clear that more long-term studies are required to evaluate adequacy and safety of insect-based pet foods in both dogs and cats.

3.6   *Biofunctionality*

Pet foods are often marketed to have specific biofunctionalities that could support pets in maintaining their health or even improve the health when the pet suffers from a specific health condition (clinical foods). Insect-based foods for dogs and cats are marketed as being 'hypoallergenic' (Beynen, 2018). Dogs and cats may develop adverse reactions to foods resulting in dermatological and/or gastro-intestinal problems like inflammation. As with food allergies, specific protein sources are commonly suspected and a change to such a hypoallergenic food with a protein source that is unfamiliar to the immune system is suggested. To the authors' knowledge, no studies are published that evaluated the effectiveness of these insect-based hypoallergenic foods in affected dogs or cats.

Apart from being a protein source for hypoallergenic foods, various health-promoting properties (e.g. hypolipidaemic, hypocholesterolaemic, immune-stimulatory, antibacterial, antiviral, antitumor) of insects or insect-derived compounds like chitin and peptides have been investigated (for reviews see Bulet *et al.*, 1999; Chernysh *et al.*, 2002; Gasco *et al.*, 2018). There are currently no data published that support the notion of a health-promoting effect of an insect-based pet food. Furthermore, translating findings in other species to dogs or cats is often not possible

TABLE 1   Evaluation of *in vitro* and *in vivo* nitrogen digestibility of various insect meals as an ingredient for dog and cat foods

Insect	Insect processing	Tested as ingredient or in a food
Black soldier fly larvae	Freeze-dried, ground	Ingredient
	Freeze-dried, ground	Ingredient
	Defatted, dried, ground	Food, extruded with unknown inclusion, contributing 30% of CP
	Defatted, dried, ground	Food, extruded with 20% inclusion, contributing 31% of CP
	Freeze-dried, ground, defatted	Food, extruded with 30% DM inclusion, contributing 52% of CP
	Freeze-dried, ground, defatted	Food, pelleted with 30% DM inclusion, contributing 52% of CP
	Defatted, dried, ground	Food, pelleted food with 1% meal, contributing 1.6% of CP
	Defatted, dried, ground	Food, pelleted food with 2% meal, contributing 3.2% of CP
	Defatted, dried, ground	Food, extruded with unknown inclusion, contributing 30% of CP
	Defatted, dried, ground	Food, extruded food with 22% meal, contributing 35% of CP
	Defatted, dried, ground	Food, extruded food with 35% meal, contributing 47% of CP
	Defatted, dried, ground	Food, extruded with 20% inclusion, contributing 31% of CP
Housefly larvae	Freeze-dried, ground	Ingredient
Banded crickets	Roasted, ground	Food, extruded with 8% inclusion, contributing 20% of CP
	Roasted, ground	Food, extruded with 16% inclusion, contributing 38% of CP
	Roasted, ground	Food, extruded with 24% inclusion, contributing 57% of CP
Yellow mealworm	Freeze-dried, ground	Ingredient
	Freeze-dried, ground	Ingredient
	Defatted, dried, ground	Food, extruded with unknown inclusion, contributing 30% of CP
	Defatted, dried, ground	Food, extruded with unknown inclusion, contributing 30% of CP
Lesser mealworm	Freeze-dried, ground	Ingredient

CP = crude protein.

Method	N digestibility	Reference
In vitro digestion for dogs	89.7%	Bosch *et al.* (2014)
In vitro digestion for dogs	87.7%	Bosch *et al.* (2016)
Faecal digestibility in dogs (n = 10)	83.9%	Beynen (2018)
Faecal digestibility in dogs (n = 6)	87.2%	Yamka *et al.* (2019)
Faecal digestibility in dogs (n = 6)	80.5%	Meyer *et al.* (2019)
Faecal digestibility in dogs (n = 6)	78.2%	Meyer *et al.* (2019)
Faecal digestibility in dogs (n = 3)	77.1%	Lei *et al.* (2019)
Faecal digestibility in dogs (n = 3)	78.5%	Lei *et al.* (2019)
Faecal digestibility in cats (n = 10)	79.8%	Beynen (2018)
Faecal digestibility in cats (n = 9)	77.0%	Paßlack and Zentek (2018)
Faecal digestibility in cats (n = 7)	73.4%	Paßlack and Zentek (2018)
Faecal digestibility in cats (n = 7)	74.9%	Yamka *et al.* (2019)
In vitro digestion for dogs	93.3%	Bosch *et al.* (2016)
Faecal digestibility in dogs (n = 8)	84.8%	Kilburn *et al.* (2020)
Faecal digestibility in dogs (n = 8)	86.0%	Kilburn *et al.* (2020)
Faecal digestibility in dogs (n = 8)	82.1%	Kilburn *et al.* (2020)
In vitro digestion for dogs	91.3%	Bosch *et al.* (2014)
In vitro digestion for dogs	92.5%	Bosch *et al.* (2016)
Faecal digestibility in dogs (n = 10)	83.6%	Beynen (2018)
Faecal digestibility in cats (n = 10)	80.4%	Beynen (2018)
In vitro digestion for dogs	91.5%	Bosch *et al.* (2014)

TABLE 2   Digestible indispensable amino acids (DIAA, %) and their scores (DIAAS) of insects calculated from data reported in the literature

	Insect				
	Black soldier fly larvae			Housefly larvae	Yellow mealworm
Reference[1] DIAA[2]	I	II	III	IV	I
Arginine	6.42	3.99	4.84-5.85	5.84	5.59
Histidine	3.65	1.85	2.44-3.17	3.40	3.17
Isoleucine	3.08	3.61	4.17-4.57	4.05	4.02
Leucine	7.27	5.64	6.50-7.08	7.10	5.73
Lysine	4.09	3.95	6.24-7.09	8.51	6.77
Methionine	1.51	1.26	1.51-1.98	3.02	1.79
Phenylalanine	3.62	3.20	3.39-4.56	7.32	3.80
Threonine	4.54	2.95	3.69-4.11	6.08	3.28
Tryptophan	–	–	0.99-1.58	7.86	5.13
Valine	5.44	5.75	5.13-7.28	4.94	5.59

	Black soldier fly larvae						Housefly larvae		Yellow mealworm	
	I		II		III		IV		I	
DIAAS[3]	Puppies	Kittens	Puppies	Kittens	Puppies	Kittens	Puppies	Kittens	Puppies	Kittens
Arginine	1.60	1.50	1.14	0.93	1.38-1.67	1.13-1.37	1.67	1.37	1.83	1.31
Histidine	1.84	2.53	1.07	1.28	1.42-1.84	1.69-2.20	1.98	2.36	2.12	2.19
Isoleucine	1.39	1.29	1.25	1.51	1.44-1.58	1.74-1.91	1.40	1.69	1.07	1.68
Leucine	1.00	1.28	0.99	1.00	1.14-1.24	1.15-1.24	1.24	1.25	1.27	1.01
Lysine	1.74	1.08	1.01	1.04	1.60-1.82	1.65-1.88	2.19	2.25	1.05	1.79
Methionine	1.15	0.78	0.81	0.65	0.97-1.28	0.78-1.02	1.94	1.55	0.97	0.92
Phenylalanine	1.31	1.63	1.11	1.44	1.17-1.58	1.53-2.05	2.53	3.29	1.25	1.71
Threonine	0.91	1.57	0.82	1.02	1.02-1.14	1.28-1.42	1.68	2.11	1.26	1.14
Tryptophan	–	–	–	–	0.99-1.58	1.37-2.19	7.86	10.88	1.81	1.81
Valine	1.71	1.92	1.92	2.03	1.71-2.43	1.81-2.57	1.65	1.74	1.83	1.31

DIAAS[4]	Cats	Dogs	Cats	Dogs	Cats	Dogs	Cats	Dogs	Cats	Dogs
Arginine	1.33	1.83	0.83	1.14	1.00-1.22	1.38-1.67	1.21	1.67	1.16	1.60
Histidine	2.24	1.95	1.14	0.99	1.50-1.95	1.30-1.69	2.10	1.82	1.95	1.69
Isoleucine	1.15	0.82	1.34	0.96	1.55-1.70	1.11-1.22	1.51	1.08	1.50	1.07
Leucine	1.14	1.08	0.89	0.84	1.02-1.11	0.96-1.05	1.11	1.05	0.90	0.85
Lysine	2.42	1.17	2.34	1.13	3.70-4.20	1.78-2.02	5.04	2.43	4.01	1.93
Methionine	1.80	0.47	1.49	0.39	1.79-2.35	0.47-0.61	3.58	0.93	2.13	0.55
Phenylalanine	1.45	0.80	1.28	0.71	1.36-1.82	0.75-1.01	2.93	1.63	1.52	0.84
Threonine	1.40	1.07	0.91	0.69	1.14-1.26	0.87-0.97	1.87	1.43	1.01	0.77
Tryptophan	–	–	–	–	1.22-1.94	0.72-1.15	9.67	5.71	–	–
Valine	1.71	1.12	1.80	1.18	1.61-2.28	1.05-1.49	1.55	1.01	1.61	1.05

[1] I = De Marco et al. (2015) reporting apparent ileal digestibility coefficients for oven-dried ground insects in 35-day old broiler chickens calculated based on the values found for the basal diet without the insect meal; II = Schiavone et al. (2017) reporting apparent ileal digestibility coefficients for oven-dried ground and partially defatted insects in 35-day old broiler chickens calculated based on the values found for the basal diet without the insect meal; III = Do et al. (2020) reporting corrected ileal digestibility coefficients for freeze-dried and ground black soldier fly larvae varying in age (0, 11, 14, 18, 23 or 29 days) in precision-fed cecectomised roosters fed the insect meal; IV = Hall et al. (2018) reporting true ileal digestibility coefficients for oven-cooked and ground insects in 28-day old broiler chickens calculated using multiple linear regression technique.

[2] Digestible indispensable amino acid content as a percentage of digestible crude protein content. The latter was based on the dietary crude protein content multiplied by the average digestibility value based on all reported amino acids.

[3] For minimal requirement data (NRC, 2006) for kittens in % of crude protein: arginine, 4.278; histidine, 1.444; isoleucine, 2.389; methionine, 1.944; leucine, 5.667; lysine, 3.778; phenylalanine, 2.222; threonine, 2.889; tryptophan, 0.722; valine, 2.833. For puppies (4 to 14 weeks of age) these values are: arginine, 3.500; histidine, 1.722; isoleucine, 2.889; methionine, 1.556; leucine, 5.722; lysine, 3.889; phenylalanine, 2.289; threonine, 4.611; tryptophan, 1.000; valine, 3.000.

[4] For minimal requirement data or adequate intake data (in italics) (NRC, 2006) for adult cats in % of crude protein: arginine, 4.813; histidine, 1.625; isoleucine, 2.688; methionine, 0.844; leucine, 6.375; lysine, 1.688; phenylalanine, 2.500; threonine, 3.250; tryptophan, 0.813; valine, 3.188. For adult dogs these values are: arginine, 3.500; histidine, 1.875; isoleucine, 3.750; methionine, 3.250; leucine, 6.750; lysine, 3.500; phenylalanine, 4.500; threonine, 4.250; tryptophan, 1.375; valine, 4.875.

because of the profound different species-specific characteristics (e.g. in physiology, metabolism and immune systems).

Lastly, the undigestible fractions of foods are important for the intestinal health as these can stimulate motility and act as a substrate that stimulates the growth of beneficial or detrimental microbiota and production metabolites that impact the host's health. The cuticle of insects contains chitin, a linear polymer of β-(1-4) N-acetyl-d-glucosamine units, which is embedded in a matrix with proteins that vary in types and degree of sclerotization (Andersen *et al.*, 1995). The cuticle properties vary within an insect and between insect species depending on the required functionalities. This part of the insect likely forms the undigestible fraction that would enter the large intestine of dogs and cats and potentially be degraded by the residing microbiota. The *in vitro* fermentability of the undigested fractions of BSFL, HFL and YMW by faecal microbiota from dogs (Bosch *et al.*, 2016) and cats (Bosch and Post, 2019) was found to be low. This could imply that the undigested fraction of these insects would act more like a bulking fibre such as cellulose. The number of faecal donors used in these studies was, however, low and the dogs and cats (and their microbiomes) were not adapted to insect-based pet foods. Also a recent study in which dogs were fed extruded foods with banded cricket meal for 29 days reported changes in only a few specific taxa of the microbiome (Jarrett *et al.*, 2019), which suggests that the prebiotic potential of the indigestible fraction of the cricket meal was low. However, more extensive studies are required to explore animal/microbiome variation and potential adaptation of microbiota to ferment compounds like chitin from different insect species as well as to explore the impact on gut health.

3.7   *Sustainability*

Of the 12 commercial hypoallergenic insect-based foods for dogs and cats evaluated by Beynen (2018), eight included a claim that insects are a sustainable protein source. Insects are energy-efficient due to their poikilothermic nature and have a relatively high proportion of edible weight. Higher feed conversion efficiencies can therefore be achieved by insects than by conventional livestock species like ruminants, pigs and poultry. Based on life cycle assessment, insect protein may have a lower environmental impact (e.g. lower land use, lower water use, less emission of CO_2-equivalents) than edible ruminant, pig or poultry protein (Van Huis and Oonincx, 2017). As with the livestock species, insect species differ in their environmental impact and, in particular, what the insects are fed has an enormous influence. For example, production of 1 kg YMW protein is estimated to generate 6 to 14 kg CO_2-equivalents (Oonincx and De Boer, 2012; Thévenot *et al.*, 2018) whereas this is approximately 3 kg when BSFL are fed a feed-grade substrate and approximately 19 kg when fed a food-grade substrate (Bosch *et al.*, 2019). Insect rearing companies are still optimising their production processes by testing, for example, the genetics and nutrition of insects. It can be expected that their production will become more efficient. Legislation might change and new low-value organic streams may be unlocked for upcycling by insects in a safe way. These developments will further reduce the environmental impact of insect protein meals.

The benchmark to which insect protein generally is compared is meat (e.g. beef, pork, chicken), which in general has indeed an overall higher environmental impact. For pet foods, however, it makes sense to not use meat, but their co-products (e.g. meat meals for dry foods and organs for wet foods) as these are the conventional protein sources used in pet foods and those that will most likely be replaced by insect meal. Peer-reviewed studies evaluating environmental impact of pet foods falsely assumed meat as main protein source (e.g. Okin, 2017; Su *et al.*, 2018) and do not provide reference data on the impact of these conventional pet food ingredients. A report of Blonk Consultants, however, estimated an impact of about 1 kg CO_2-equivalents per kg protein for a mixed meal and 2 kg per kg protein for a poultry meal (Koukouna and Broekema, 2017), which is lower than that for BSFL and YMW. Thus, the support of the claim that insects are a sustainable protein source is not unambiguous and a matter of selecting a benchmark. Furthermore, it also depends on the method of quantification and considerations of what specific aspects of sustainability (environmental, economic, societal) are of interest.

4   Concluding remarks

The 'buzz' in society around insects as sustainable and healthy new protein sources has opened the market for insect-based pet food products. The insect-based hypoallergenic foods are now an additional option to owners with dogs or cats suffering from supposed or diagnosed food sensitivity. Products like insect-based snacks are also available on the market. The latter types of products are more subject to volatile trends and are less likely to prevail when the buzz slowly silences and new trends arise. For the sake of the reputation of insects as a novel quality ingredient and the pet food industry as a whole, it is essential that safety and efficacy testing is performed and results are shared with the community. Multiple studies evaluated aspects of the nutritional quality of various insect species but the impact of long-term feeding on the nutritional status and health in dogs and cats are still largely unexplored. At the same time, there is considerable interest in the potential bioactives present in insects resulting in studies in pigs, poultry and other production animals from which pet food sector can learn from. How these may impact (promote or harm) the health of dogs and cats is presently unknown. This lack of knowledge provides an additional argument to warrant long-term studies in both dogs and cats fed foods based on insect meals and those containing specific isolated components from insects. If proven to be positive and effective, applications of insects can be expanded to clinical foods or foods supporting health and wellness.

To what extent insect meals will have even wider applications in pet foods is difficult to predict. The insect sector continues to develop as it can play a role in making the global food production system more resource-efficient and productive at the same time. In addition, Europe is striving to become more self-sufficient regarding feed proteins, further fuelling the development of alternative protein-rich feed ingredients like insects but also multiple others including algae, bacterial single cell proteins and legumes. Ongoing developments in automation and processing technologies, selective breeding and the nutrition of insects will further increase production volumes and reduce the economic cost as well as the environmental impact per unit of protein. This will make insect proteins more competitive with conventional protein sources used in the pet food industry and those used in the aquaculture and livestock sectors. At the same time the global demand for animal-derived foods will continue to grow in fast-developing countries. Meat, for example, is projected to grow from 2005/2007 to 2050 with 76% (Alexandratos and Bruinsma, 2012). The increasing volumes of animal-derived foods will also make more co-products available for pet food production. Apart from price and availability, wider applications of insects in pet foods would also likely depend on the evolving reputation that edible insects have in society. Central aspects for acceptance of edible insects in humans include trust, willingness-to-eat, overcoming disgust and neophobic reactions, and sensory attributes but it is also essential to consider ethical aspects like species-specific mass rearing conditions, transportation and killing methods (Rumpold and Langen, 2020). Furthermore, pet owners may deem specific residual organic sources to rear insects acceptable whereas they perceive other sources as unsafe or unsanitary. From a resource use efficiency point of view, however, it is of interest to grow insects on residual organic sources that are currently not used in feed for livestock or aquaculture. It is therefore of particular interest to better understand pet owner views on the use of specific organic sources to rear insects and the consequences for environmental impact, relative to animal-derived co-products conventionally used in pet foods.

Acknowledgements

This research was funded by Wageningen University & Research and University of Illinois at Urbana-Champaign. Both authors contributed fundamentally to the present manuscript.

Conflict of interest

There are no conflicts of interest.

References

Association of American Feed Control Officials (AAFCO), 2020. Official publication. AAF-CO Incorporated, Atlanta, GA, USA,

Ahlstrøm, Ø. and Skrede, A., 1998. Comparative nutrient digestibility in dogs, blue foxes, mink and rats. Journal of Nutrition 128: S2676-2677.

Aldrich, G.C. and Koppel, K., 2015. Pet food palatability evaluation: a review of standard assay techniques and interpretation of results with a primary focus on limitations. Animals 5: 43-55.

Alexandratos, N. and Bruinsma, J., 2012. World agriculture towards 2030/2050; the 2012 revision. Food and Agriculture Organization of the United Nations, Rome, Italy.

Andersen, S.O., Hojrup, P. and Roepstorff, P., 1995. Insect cuticular proteins. Insect Biochemistry and Molecular Biology 25: 153-176.

Bayly, C.P., 1976. Observations on the food of the feral cat (*Felis catus*) in an arid environment. South Australian Naturalist 51: 22-24.

Beaton, L., 2014. US petfood market update: specialty petfoods driving industry growth. Petfood Industry 56: 20-25.

Beynen, A.C., 2018. Insect-based petfood. Creature Companion: 40-41.

Bosch, G., Zhang, S., Oonincx, D.G.A.B. and Hendriks, W.H., 2014. Protein quality of insects as potential ingredients for dog and cat foods. Journal of Nutritional Science 3: e29.

Bosch, G., Hagen-Plantinga, E.A. and Hendriks, W.H., 2015. Dietary nutrient profiles of wild wolves: insights for optimal dog nutrition? British Journal of Nutrition 113: S40-S54.

Bosch, G., Vervoort, J.J.M. and Hendriks, W.H., 2016. *In vitro* digestibility and fermentability of selected insects for dog foods. Animal Feed Science and Technology 221: 174-184.

Bosch, G. and Post, M., 2019. Fermentability of undigested residues from insects in cats: an *in vitro* pilot study. In: Proceedings of the 23rd Congress of the European Society of Veterinary and Comparative Nutrition, 18-20 September 2019, Turin, Italy. pp. 224.

Bosch, G., Van Zanten, H.H.E., Zamprogna, A., Veenenbos, M., Meijer, N.P., Van der Fels-Klerx, H.J. and Loon, J.J.A., 2019. Conversion of organic resources by black soldier fly larvae: legislation, efficiency and environmental impact. Journal of Cleaner Production 222: 355-363.

Bradshaw, J.W.S. and Cook, S.E., 1996. Patterns of pet cat behaviour at feeding occasions. Applied Animal Behaviour Science 47: 61-74.

Bulet, P., Hetru, C., Dimarcq, J.L. and Hoffman, D., 1999. Antimicrobial peptides in insects; structure and function. Developmental & Comparative Immunology 23: 329-344.

Chernysh, S., Kim, S.I., Bekker, G., Pleskach, V.A., Filatova, N.A., Anikin, V.B., Platonov, V.G. and Bulet, P., 2002. Antiviral and antitumor peptides from insects. Proceedings of the National Academy of Sciences of the United States of America 99: 12628-12632.

Daumas, C., Paragon, B.-M., Thorin, C., Martin, L., Dumon, H., Ninet, S. and Nguyen, P., 2012. Evaluation of eight commercial dog diets. Journal of Nutritional Science 3: e63.

Day, J.E.L., Kergoat, S. and Kotrschal, K., 2009. Do pets influence the quantity and choice of food offered to them by their owners: lessons from other animals and the pre-verbal human infant? CAB Reviews: Perspectives in Agriculture, Veterinary Science, Nutrition and Natural Resources 4: 1-12.

De Marco, M., Martínez, S., Hernandez, F., Madrid, J., Gai, F., Rotolo, L., Belforti, M., Bergero, D., Katz, H., Dabbou, S., Kovitvadhi, A., Zoccarato, I., Gasco, L. and Schiavone, A., 2015. Nutritional value of two insect larval meals (*Tenebrio molitor* and *Hermetia illucens*) for broiler chickens: apparent nutrient digestibility, apparent ileal amino acid digestibility and apparent metabolizable energy. Animal Feed Science and Technology 209: 211-218.

Deng, P., Utterback, P.L., Parsons, C.M., Hancock, L., Swanson, K.S., 2016. Chemical composition, true nutrient digestibility, and true metabolizable energy of novel pet food protein sources using the precision-fed cecectomized rooster assay. Journal of Animal Science 94: 3335-3342.

Do, S., Koutsos, E.A., Utterback, P.L., Parsons, C.M., De Godoy, M.R.C. and Swanson, K.S., 2020. Nutrient and AA digestibility of black soldier fly larvae differing in age using the precision-fed cecectomized rooster assay. Journal of Animal Science 98: skz363.

Driscoll, C.A., Menotti-Raymond, M., Roca, A.L., Hupe, K., Johnson, W.E., Geffen, E., Harley, E.H., Delibes, M., Pontier, D., Kitchener, A.C., Yamaguchi, N., O'Brien, S.J. and Macdonald, D.W., 2007. The Near Eastern origin of cat domestication. Science 317: 519-523.

Faber, T.A., Bechtel, P.J., Hernot, D.C., Parsons, C.M., Swanson, K.S., Smiley, S. and Fahey Jr, G.C., 2010. Protein digestibility evaluations of meat and fish substrates using laboratory, avian, and ileally cannulated dog assays. Journal of Animal Science 88: 1421-1432.

FEDIAF, 2019a. Facts & figures 2019. European Pet Food Industry Federation, Brussels, Belgium.

FEDIAF, 2019b. Nutritional guidelines for complete and complementary pet food for cats and dogs. European Pet Food Industry Federation, Brussels, Belgium.

Food and Agriculture Organisation (FAO), 2013. Dietary protein quality evaluation in human nutrition. FAO, Auckland, New Zealand.

Gasco, L., Finke, M. and Van Huis, A., 2018. Can diets containing insects promote animal health? Journal of Insects as Food and Feed 4: 1-4.

Hall, H.N., Masey O'Neill, H.V., Scholey, D., Burton, E., Dickinson, M. and Fitches, E.C., 2018. Amino acid digestibility of larval meal (*Musca domestica*) for broiler chickens. Poultry Science 97: 1290-1297.

Hendriks, W.H., Emmens, M.M.A., Trass, B. and Pluske, J.R., 1999. Heat processing changes the protein quality of canned cat foods as measured with a rat bioassay. Journal of Animal Science 77: 669-676.

Hendriks, W.H., Van Baal, J. and Bosch, G., 2012. Ileal and faecal protein digestibility measurement in monogastric animals and humans: a comparative species view. British Journal of Nutrition 108: S247-S257.

Hendriks, W.H., Thomas, D.G., Bosch, G. and Fahey Jr, G.C., 2013. Comparison of ileal and total tract nutrient digestibility of dry dog foods. Journal of Animal Science 91: 3807-3814.

Hervera, M., Baucells, M.D., González, G., Pérez, E. and Castrillo, C., 2009. Prediction of digestible protein content of dry extruded dog foods: comparison of methods. Journal of Animal Physiology and Animal Nutrition 93: 366-372.

Hervera, M., 2011. Methods for predicting the energy value of dog foods. Universitat de Autònoma de Barcelona, Barcelona, Spain.

Jarrett, J.K., Carlson, A., Serao, M.R., Strickland, J., Serfilippi, L. and Ganz, H.H., 2019. Diets with and without edible cricket support a similar level of diversity in the gut microbiome of dogs. PeerJ 7: e7661.

Johnson, M.L., Parsons, C.M., Fahey Jr, G.C., Merchen, N.R. and Aldrich, C.G., 1998. Effects of species raw material source, ash content, and processing temperature on amino acid digestibility of animal by-product meals by cecectomized roosters and ileally cannulated dogs. Journal of Animal Science 76: 1112-1122.

Kendall, P.T., Blaza, S.E. and Holme, D.W., 1982. Assessment of endogenous nitrogen output in adult dogs of contrasting size using a protein-free diet. Journal of Nutrition 112: 1281-1286.

Kilburn, L.R., Carlson, A.T., Lewis, E. and Rossoni Serao, M.C., 2020. Cricket (*Gryllodes sigillatus*) meal fed to healthy adult dogs does not affect general health and minimally impacts apparent total tract digestibility. Journal of Animal Science 98: 1-8.

Konecny, M.J., 1987. Food habits and energetics of feral house cats in the Galápagos Islands. Oikos 50: 24-32.

Koukouna, E. and Broekema, R., 2017. Carbon footprint assessment of cat 3 meal for pet food applications. Blonk Consultants, Gouda, The Netherlands.

Laflamme, D.P., Xu, H. and Long, G.M., 2011. Effect of diets differing in fat content on chronic diarrhea in cats. Journal of Veterinary Internal Medicine 25: 230-235.

Lei, X.J., Kim, T.H., Park, J.H. and Kim, I.H., 2019. Evaluation of supplementation of defatted black soldier fly (*Hermetia illucens*) larvae meal in beagle dogs. Annals of Animal Science 19: 767-777.

Leonard, J.A., Wayne, R.K., Wheeler, J., Valadez, R., Guillén, S. and Vilà, C., 2002. Ancient DNA evidence for old world origin of New World dogs. Science 298: 1613-1616.

Llewellyn, L.M. and Uhler, F.M., 1952. The foods of fur animals of the Patuxent Research Refuge, Maryland. American Midland Naturalist 48: 193-203.

Makkar, H.P.S., Tran, G., Heuzé, V. and Ankers, P., 2014. State-of-the-art on use of insects as animal feed. Animal Feed Science and Technology 197: 1-33.

Malo, A.F., Lozano, J., Huertas, D.L. and Virgos, E., 2004. A change of diet from rodents to rabbits (*Oryctolagus cuniculus*). Is the wildcat (*Felis silvestris*) a specialist predator? Journal of Zoology 263: 401-407.

Meyer, L.F., Kölln, M. and Kamphues, J., 2019. Hundefutter mit Insekten? Untersuchungen zu Mischfuttermitteln mit Larven der Schwarzen Soldatenfliege als Proteinquelle. Kleintierpraxis 64: 124-135.

Moxham, G., 2001. WALTHAM feces scoring system – a tool for veterinarians and pet owners: how does your pet rate? WALTHAM Focus 11: 24-25.

National Research Council (NRC), 2006. Nutrient requirements of dogs and cats. National Academies Press, Washington, DC, USA.

O'Haire, M., 2010. Companion animals and human health: benefits, challenges, and the road ahead. Journal of Veterinary Behavior: Clinical Applications and Research 5: 226-234.

Okin, G., 2017. Environmental impacts of food consumption by dogs and cats. PLoS ONE 12: e0181301.

Oonincx, D.G.A.B. and De Boer, I.J.M., 2012. Environmental impact of the production of mealworms as a protein source for humans – a life cycle assessment. PLoS ONE 7: e51145.

Parmalee, P.W., 1953. Food habits of the feral house cat in east-central Texas. Journal of Wildlife Management 17: 375-376.

Paßlack, N. and Zentek, J., 2018. Akzeptanz, Verträglichkeit und scheinbare Nährstoffverdaulichkeit von Alleinfuttermitteln auf Basis von *Hermetia-illucens*-Larvenmehl bei Katzen. Tierärztliche Praxis Kleintiere 46: 213-221.

Pearre, S. and Maass, R., 1998. Trends in the prey size-based trophic niches of feral and house cats *Felis catus* L. Mammal Review 28: 125-139.

Plantinga, E.A., Bosch, G. and Hendriks, W.H., 2011. Estimation of the dietary nutrient profile of free-roaming feral cats: possible implications for nutrition of domestic cats. British Journal of Nutrition 106: S35-S48.

Podberscek, A.L., Paul, E.S. and Serpell, J.A., 2000. Companion animals and us: exploring the relationships between people and pets. Cambridge University Press, Cambridge, UK, 335 pp.

Rumpold, B.A. and Schlüter, O.K., 2013. Nutritional composition and safety aspects of edible insects. Molecular Nutrition and Food Research 57: 802-823.

Rumpold, B.A. and Langen, N., 2020. Consumer acceptance of edible insects in an organic waste-based bio-economy. Current Opinion in Green and Sustainable Chemistry 23: 80-84.

Schiavone, A., De Marco, M., Martínez, S., Dabbou, S., Renna, M., Madrid, J., Hernandez, F., Rotolo, L., Costa, P., Gai, F. and Gasco, L., 2017. Nutritional value of a partially defatted and a highly defatted black soldier fly larvae (*Hermetia illucens* L.) meal for broiler chickens: apparent nutrient digestibility, apparent metabolizable energy and apparent ileal amino acid digestibility. Journal of Animal Science and Biotechnology 8: 51.

Schleicher, M., Cash, S.B. and Freeman, L.M., 2019. Determinants of pet food purchasing decisions. Canadian Veterinary Journal 60: 644-650.

Smeets-Peeters, M.J.E., Minekus, M., Havenaar, R., Schaafsma, G. and Verstegen, M.W.A., 1999. Description of a dynamic *in vitro* model of the dog gastrointestinal tract and an evaluation of various transit times for protein and calcium. ATLA Alternatives to Laboratory Animals 27: 935-949.

Statista, 2020a. Number of dogs in the United States from 2000 to 2017. Available at: https://www.statista.com/statistics/198100/dogs-in-the-united-states-since-2000/

Statista, 2020b. Number of cats in the United States from 2000 to 2017. Available at: https://www.statista.com/statistics/198102/cats-in-the-united-states-since-2000/

Su, B., Martens, P. and Enders-Slegers, M.-J., 2018. A neglected predictor of environmental damage: the ecological paw print and carbon emissions of food consumption by companion dogs and cats in China. Journal of Cleaner Production 194: 1-11.

Thévenot, A., Rivera, J.L., Wilfart, A., Maillard, F., Hassouna, M., Senga-Kiesse, T., Le Féon, S. and Aubin, J., 2018. Mealworm meal for animal feed: environmental assessment and sensitivity analysis to guide future prospects. Journal of Cleaner Production 170: 1260-1267.

Tjernsbekk, M.T., Tauson, A.H. and Ahlstrøm, Ø., 2014. Ileal, colonic and total tract nutrient digestibility in dogs (*Canis familiaris*) compared with total tract digestibility in mink (*Neovison vison*). Archives of Animal Nutrition 68: 245-261.

Tobie, C., Péron, F. and Larose, C., 2015. Assessing food preferences in dogs and cats: a review of the current methods. Animals 5: 126-137.

Van Huis, A. and Oonincx, D.G.A.B., 2017. The environmental sustainability of insects as food and feed. A review. Agronomy for Sustainable Development 37: 1-14.

Van Rooijen, C., Bosch, G., Van der Poel, A.F.B., Wierenga, P.A., Alexander, L. and Hendriks, W.H., 2014. Reactive lysine content in commercially available pet foods. Journal of Nutritional Science 3: e35.

Vilà, C., Savoleinen, P., Maldonado, J.E., Amorim, I.R., Rice, J.E., Honeycutt, R.L., Crandall, K.A., Lundeberg, J. and Wayne, R.K., 1997. Multiple and ancient origins of the domestic dog. Science 276: 1687-1689.

Vinassa, M., Vergnano, D., Valle, E., Giribaldi, M., Nery, J., Prola, L., Bergero, D. and Schiavone, A., 2020. Profiling Italian cat and dog owners' perceptions of pet food quality traits. BMC Veterinary Research 16: 131.

Wall, T., 2018. Raw pet food sales growing despite health warnings. Petfood Industry 60: 24-27.

Yamka, R.M., Koutsos, E.A. and McComb, A., 2019. Evaluation of black soldier fly larvae as a protein and fat source in pet foods. Petfood Forum, Kansas City, MI, USA, pp. 8-9.

Biological contaminants in insects as food and feed

D. Vandeweyer[1,2]#, J. De Smet[1,2]#, N. Van Looveren[1,2] and
L. Van Campenhout[1,2]*

[1]KU Leuven, Department of Microbial and Molecular Systems (M²S),
Lab4Food, Geel Campus, Kleinhoefstraat 4, 2440 Geel, Belgium;
[2]KU Leuven, Leuven Food Science and Nutrition Research Centre
(LFoRCe), Kasteelpark Arenberg 20, Box 2463, 3001 Leuven, Belgium;
*leen.vancampenhout@kuleuven.be; #These authors contributed equally

Abstract

During the last decade, edible insects have successfully taken a meaningful position in the feed and food chain. To expand this position, product safety continuously needs to be warranted. This review focuses on the current knowledge and the future challenges on the prevalence of human foodborne pathogens in edible insects. The top three of the bacterial pathogens associated with insects for food are *Staphylococcus aureus*, pathogenic *Clostridium* spp. and pathogenic species of the *Bacillus cereus* group. Less is known about other types of biological contaminants, the fungi, viruses, protozoa and prions. For insects for feed, even less reports on pathogens are available so far, although the microbiota of *Hermetia illucens* is increasingly being studied in the latest years. In addition to the evaluation of endogenous microorganisms in insects, an overview is given of inoculation experiments to study the fate of specific food pathogens during rearing. Future challenges that are identified mainly relate to the fact that risk assessments directed to specific insect species are needed. Also, more research data are needed on the microbiological quality of substrates and residue, in connection with decontamination treatments. The house flora of rearing facilities has not been investigated before. The insect supply chain can generate insights in the microbiological quality of the integral chain by implementing exhaustive sampling plans and by applying predictive microbiology. Additionally, microbiological methods used in research and quality control require standardisation. Rather unexplored so far is the unculturable fraction of the insect microbial community and its importance in food safety. Last but not least, the most important microbiological challenge may well be situated in the further development of the sector: upscaling in terms of capacity and number of companies will increase the complexity of the sector. That will have implications for monitoring and control of biological contaminants.

Keywords

edible insects – microbiological pathogens – microbiological safety

1   Introduction

In addition to physical and chemical safety, feed and food also have to comply with biological safety. The legal microbiological criteria applicable for the feed and food industry rely on culture-dependent methods, in which a fresh sample is diluted or alternatively resuscitated, and then plated and incubated to determine colony counts, or incubated to observe the absence or presence of a food pathogen, respectively. Hence, depending on the biological contaminant, safety involves either its presence to be below a specific level or its complete absence in a predefined quantity of the matrix (De Loy-Hendrickx *et al.*, 2018). Biological contaminants (hazards) encompass pathogenic strains of micro-organisms (i.e. bacteria, viruses and fungi, which contain both moulds and yeasts), and of parasites (i.e. protozoa and worms), as well as the toxic substances (chemicals) they produce, i.e. bacterial toxins, such as for instance cereulide, histamine, botulin, or mycotoxins (WHO, 1995; Zwietering *et al.*, 2016).

In the last decade, and as shown in this review, increasingly more reports appear that describe the complete microbiome or subgroups of insect species reared for animal or human food, focusing on the rearing stage and/or post-harvest practices. In some of these studies, special attention is paid to – mostly bacterial – pathogens that can cause zoonoses. Conclusions with respect to microbiological safety are often difficult to draw. A first reason for this is that legislation on microbiological criteria for insects as feed or food, to be used as a reference for what can be considered as safe, is limited, as also concluded by Garofalo *et al.* (2019). Biological safety is not yet well established for insects and therefore legislation is not extensive. The extrapolation of microbiological criteria from other food types (e.g. included in Regulation (EC) No 2073/2005 on microbiological criteria for foodstuffs; EC, 2005) to insects is hardly relevant, since pathogens show a different growth pattern and physiological behaviour, such as sporulation or spore germination, in diverse foods with concomitant intrinsic and extrinsic properties (Jay *et al.*, 2005). Secondly, for some bacterial contaminants criteria exist for insects, but results reported in literature do not always involve the required amount of samples, or the results were not obtained using the methods described in the criteria.

The aim of this review is twofold. In a first part, we present an update of reviews and studies available, describing certain biological contaminants in insects for food and feed. In this review, we consider the possible occurrence of pathogenic bacteria, fungi and viruses, prions and protozoa, but we do not focus on the presence of toxic compounds produced by micro-organisms. In this first part, we make a further distinction: we first describe pathogens in insect species produced for food purposes only or for both food and feed, and secondly we describe pathogens in insects to be (generally) used in feed only. The reason for this structure is that there is a clear distinction in the amount of data available in the two domains. For insects for food or both food and feed, several studies are acquirable, and this work even has been summarised in several reviews. Hence, we will build on the most recent reviews, compare them and add reports that appeared after them, to come to the main state of the art. In contrast, for insects only used for feed, little data are available. Therefore, we provide an overview of original data, collected through searches in PubMed, Google Scholar and Web of Science, by using the search terms 'housefly', 'Musca domestica', 'black soldier fly', or 'Hermetia illucens', each combined with 'microbiota' or 'microbiome'. Only papers related to mass rearing of the insects were considered. In the second part, this review also aims at deeply discussing the future challenges in both the research on biological contaminants as well as the practical implementation of measurements to warrant biological safety in the context of the rapidly evolving insect rearing and processing sectors.

2   Biological risks associated with insects to be used in food or in both food and feed

In 2019, four literature reviews were published online that discuss studies performed so far on the microbiological quality and safety of mass reared insects for human food purposes. The reviews have in common that none of them explicitly describes biological risks for producing and processing insects into animal feed. All of them consider the industrial scale production of insects, and three of them also include insects that are wild-harvested at large scale (Table 1). While all reviews do not only cover the rearing phase but also post-harvest processing, they differ somewhat with respect to the mentioned technologies (Table 1). Finally, all reviews discuss results obtained by both culture-independent microbiological analyses (plates counts, presence absence tests), as well as by the mostly used culture-independent approach for microbial community assessment, metagenetics. Metagenetics is based on a polymerase chain reaction (PCR) to amplify and then sequence certain phylum-specific genes, typically the 16S ribosomal RNA gene for bacteria and 18S rRNA gene for fungi, from all DNA extracted from a whole microbial community (Martin et al., 2018). Below, we shortly describe the specificities of each review and then consolidate the main findings.

   Murefu et al. (2019) reviewed the food safety hazards of both reared and wild-harvested edible insects. The study revealed the lead of Africa in studying food safety of insects until 2016, yet on wild-harvested insects (Figure 1). Later, the majority of studies reporting food safety of edible insects were European, investigating other, reared species. It was concluded that the harvesting type (wild or reared) strongly

TABLE 1 Comparison of the four review publications on the microbiological safety of insects for food published online in 2019

Review characteristics	Reviews			
	Murefu *et al.* (2019)	Garofalo *et al.* (2019)	Cappelli *et al.* (2020)	Kooh *et al.* (2019)
Online publication date	6 March 2019	25 July 2019	6 September 2019	14 October 2019
Time span covered[1]	1993-2019	2000-2019	2016-2019	1994-2019
Systematic or narrative review[2]	systematic	systematic	systematic	narrative
Including wild-harvested insects?	yes	yes	no	yes
Post-harvest treatments described with respect to effect on microbiological quality	blanching (par)boiling/cooking canning chilling degutting drying[3] fermenting freezing frying milling/grinding/grounding/pulverising packing (vacuum or not) plucking rinsing/washing roasting salting smoking sterilising	blanching boiling/cooking chilling degutting drying[3] extruding fermenting (deep-)frying marinating milling/crushing plucking rinsing/washing roasting salting smoking spicing/condimenting starving sterilising	blanching boiling/(vacuum) cooking chilling cold atmospheric pressure plasma crushing/grinding drying[3] enzymatic hydrolysis fermenting fractionating freezing frying *in vitro* digestion marinating pH change pureeing rinsing/washing roasting smoking starving/fasting sterilising	boiling/cooking grinding drying[3] freezing frying high hydrostatic pressure packing rinsing roasting starving/fasting

[1] Year of oldest and newest publication included in the review.
[2] A systematic review involves a systematic search of the literature while a narrative review tends to be mainly descriptive and does not involve a systematic search of the literature (Ulman, 2011).
[3] Drying techniques can include freeze-drying, oven drying, microwave drying and/or solar drying.

affects the food safety of the insect. Regarding biological risks, the bacteria *Bacillus* (*cereus* group), *Clostridium* and *Staphylococcus* as well as the fungi *Aspergillus* and *Penicillium* were regularly mentioned for the five main edible insect species considered in this study (Table 2), as well as for several additional African species.

In the review by Garofalo *et al.* (2019), data originated from over 32 species (7

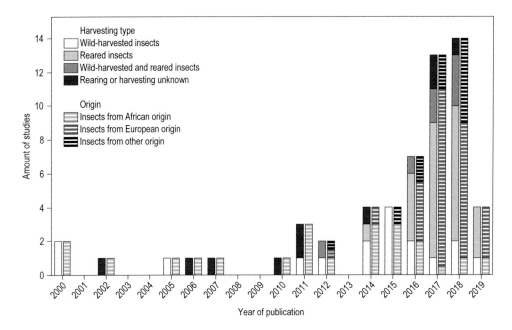

Amount of studies reporting microbiological data for insects for food between 2000 and 2019. Two graphs per year represent the reared or wild-harvested nature of the insects and the geographical origin of the insects, respectively, employed in the studies. Data were compiled from Garofalo *et al.* (2019) and supplemented with data from Kooh *et al.* (2019), Murefu *et al.* (2019) and Cappelli *et al.* (2020).

orders) of both fresh and processed insects, either harvested in the wild or (industrially) reared (Figure 1). Furthermore, also microbiological data from insect-based products were included in the review. Up till 2015, most data were obtained from wild-harvested African insect species, while as from 2016, associated with the renewed novel food regulation (Regulation (EU) No. 2015/2283; EC, 2015), a drastic increase in scientific studies on edible insects reared in Europe was observed (Figure 1). Consequently, the most studied insects for human food nowadays are the yellow mealworm (*Tenebrio molitor*), the lesser mealworm (*Alphitobius diaperinus*), the house cricket (*Acheta domesticus*), the tropical house cricket (*Gryllodes sigillatus*) and the migratory locust (*Locusta migratoria*). With an extensive metadata collection, Garofalo *et al.* (2019) were able to assess and discuss both food hygiene and food safety of edible insects. Additionally, the data provided insight into the microbial profiles associated with different insect species, to evaluate their dynamics during rearing and the effect of commonly applied treatments on those microbial profiles. Pathogenic microorganisms reported in the aforementioned most important insect species for food are listed in Table 2. The list contains both spore-forming and non-spore-forming bacteria, as well as mycotoxin-producing fungi. From the data compiled by Garofalo *et al.* (2019), it can be concluded that the bacterial

TABLE 2   Foodborne biological contaminants (bacteria and fungi) reported in edible insect species that pose a potential food safety risk. In the case only a genus name is mentioned, one or more species within that genus (other than a species that also may be listed) can be pathogenic.

Study	Insect species	Potential foodborne biological safety risks identified	
		Bacteria	Fungi
Garofalo et al. (2019)	Alphitobius diaperinus	Aeromonas, Bacillus, Pseudomonas	Aspergillus flavus, Aspergillus, Penicillium
	Tenebrio molitor	Bacillus cereus group, Bacillus, Clostridium perfringens, Clostridium, Cronobacter, Escherichia coli, Listeria[1], Pseudomonas, Salmonella[2,] Staphylococcus aureus, Staphylococcus, Vibrio, Yersinia	Penicillium
	Acheta domesticus	B. cereus group, Bacillus, C. perfringens, Clostridium, Listeria[1], Pseudomonas, Staphylococcus	Aspergillus
	Locusta migratoria	Bacillus, C. perfringens, Pseudomonas, Staphylococcus, Yersinia	Aspergillus
Kooh et al. (2019)	A. diaperinus		A. flavus
	T. molitor	B. cereus group, C. perfringens, Clostridium	
	A. domesticus	B. cereus group, C. perfringens	
	L. migratoria	C. perfringens	
	Edible insects in general	Bacillus, B. cereus group, C. perfringens, S. aureus	
Murefu et al. (2019)	A. diaperinus	Bacillus, Clostridium	A. flavus, Aspergillus, Penicillium
	T. molitor	Bacillus, B. cereus group, Clostridium, Escherichia, Listeria[1], Pseudomonas, Staphylococcus	Penicillium
	A. domesticus	Bacillus, B. cereus group, Clostridium, Listeria[1], Staphylococcus	
	L. migratoria	Staphylococcus	
Cappelli et al. (2020)	A. diaperinus		Aspergillus
	T. molitor	B. cereus group, Bacillus, C. perfringens, Clostridium, Listeria[1], Staphylococcus	
	A. domesticus	B. cereus group, Bacillus, C. perfringens, Clostridium, Listeria[1], Staphylococcus, Yersinia	

[1] Only the genus Listeria was detected so far. The pathogenic species Listeria monocytogenes has never been detected in edible insect species.
[2] Salmonella spp. were only detected by means of DNA-based analyses. Viable cells have not been detected so far.

genera *Bacillus* (including the *B. cereus* group, as discussed further below), *Clostridium* (including *Clostridium perfringens*), and *Staphylococcus* (including *Staphylococcus aureus*) are to be considered as the most relevant risks regarding food safety of edible insects. The genera *Cronobacter* (including *C. sakazakii*), *Pseudomonas* (including *Pseudomonas aeruginosa*), *Vibrio*, *Campylobacter*, *Salmonella* and *Listeria* (including *Listeria monocytogenes*) were concluded to pose a lower risk. More recently, presumptive *Cronobacter* spp. were detected sporadically in freeze-dried lesser mealworms and in house cricket meal (Greenhalgh and Amund, 2019), but also here, safety risks were concluded to be low. Regarding mycotoxin-producing fungi, *Aspergillus* spp. and *Penicillium* spp. were found to involve the highest risks.

Cappelli *et al.* (2020) provided a chemical and microbiological risk assessment applied on the insect families Tenebrionidae (darkling beetles, including *T. molitor* and *A. diaperinus*) and Gryllidae (crickets, including *A. domesticus* and *G. sigillatus*). Regarding the microbiological risks, human foodborne pathogens that were reported per insect species are again included in Table 2. The genera *Bacillus*, *Clostridium* and *Staphylococcus* were mentioned most often and also this study concluded that *Salmonella* spp. and *L. monocytogenes* are low-level risks. Further, Cappelli *et al.* (2020) reported that prions and foodborne parasites and viruses can be considered as low risks and are not described yet in reared insects for human consumption.

Finally, a review article by Kooh *et al.* (2019) described the biological health risks associated with entomophagy. Pathogens reported by this study associated with the mentioned main five insect species are also included in Table 2. Using a more general, narrative approach, the authors came to similar conclusions as Garofalo *et al.* (2019), being that edible insects and derived products pose a low risk regarding *Salmonella* spp., *Campylobacter* spp., *Yersinia* spp., *Vibrio* spp., and *L. monocytogenes*. However, *S. aureus* was concluded to be very abundant in edible insects and spore formers such as *Clostridium* spp. and *Bacillus* spp. were reported to be of major concern. Especially members of the *B. cereus* group were presented as a substantial food safety hazard for insects as human food. Information regarding other foodborne biological contaminants (parasites, prions and viruses) was concluded to be scarce.

We can compile the aforementioned reviews and the studies they cover, to make overall conclusions on the state of the art in knowledge on the microbiological safety of insects for food. First, the currently available literature shows that barely any data are available regarding prions and foodborne viruses and parasites in insects farmed for human consumption. While certain microbiological risk assessments on edible insects (EFSA Scientific Committee, 2015; Fernandez-Cassi *et al.*, 2018) claim that those biological contaminants present only a low safety risk, this has not been investigated yet. For foodborne viruses, however, a first effort was made in a very recent study (Vandeweyer *et al.*, 2020a) which investigated the presence of Hepatitis A and E virus and Norovirus genogroup II in several raw insect samples collected from industrial producers. This research could confirm the low risk regarding those three viruses for the samples investigated, since none of the viruses were detected. In view of the recent outbreak of COVID-19, Dicke *et al.* (2020) investigated the transmission potential of edible insects for the zoonotic coronavirus SARS-CoV-2 (which does not appear to be foodborne so far), and concluded that hazard to be extremely low. Nonetheless, further research investigating other foodborne viruses and other insects, as well as research regarding prions and foodborne parasites is still required. Their fate during insect processing has also not been investigated so far.

Secondly, it is clear that the three major foodborne bacterial hazards in Europe, *Salmonella* spp., *Campylobacter* spp. and *L. monocytogenes* (Van Cauteren *et al.*, 2017), are barely reported and consequently pose a low health risk for edible insects. Also *Cronobacter* spp., *Vibrio* spp. and *Yersinia* spp. are very rarely reported. However, the three main bacterial genera/species associated with food safety problems in edible insects are *S. aureus*, *Clostridium* spp., with the species *C. perfringens* and *Clostridium botulinum* being the most relevant food pathogens within this genus, and the *B. cereus* group. The *B. cereus* group, but also *Clostridium* spp., are spore-forming bacteria. This observation is noteworthy, since mitigation of the risks associated with bacterial endospores requires more drastic strategies than those for vegetative cells (Kort *et al.*, 2005). As to mycotoxin-producing fungi, *Aspergillus* spp. and *Penicillium* spp. are noted as the most hazardous species. Also in the most recent studies (Borremans *et al.*, 2019; Fernandez-Cassi *et al.*, 2020; Frigerio *et al.*, 2020; Mancini *et al.*, 2019b), (some of) the same main biological safety hazards were addressed.

From all those risks, the presence of the *B. cereus* group can probably be ranked as the highest food safety risk for insects for human consumption. As demonstrated by multiple studies (Fasolato *et al.*, 2018; Vandeweyer *et al.*, 2018, 2020a,b), the *B. cereus* group can be encountered in several insect species and derived products, sometimes in high amounts. The *B. cereus* group, or *B. cereus sensu lato* (*s.l.*), consists of a number of recognised species, including *B. cereus sensu strictu* (*s.s.*) and *Bacillus thuringiensis* (EFSA BIOHAZ Panel, 2016). All group members are genetically (and phenotypically) closely related to each other, and therefore no perfect distinction can be made between the members based on the 16S ribosomal RNA gene (Ehling-Schulz *et al.*, 2019). Moreover, their virulence for humans is very difficult to assess, since not all virulence genes (that could be used as marker gene for virulence) are known yet. Also, virulence appears to be very dependent on strain and environmental conditions (Jeßberger *et al.*, 2015) and, as the known virulence genes are situated on (possibly mobile) plasmids, they can potentially be exchanged between members (SciCom, 2018). *B. thuringiensis* possesses a similar portfolio of potential virulence genes as *B. cereus s.s.* strains and is able to express them (EFSA BIOHAZ Panel, 2016). Therefore, *B. thuringiensis*, which is typically known as entomopathogen and can be a threat for the insect yield in mass production, can also be hazardous for humans.

When we look at the insect species that have been investigated with respect to microbiological food safety so far, the most important ones are *T. molitor* and *A. domesticus*. The other mealworm and cricket species, as well as *L. migratoria* are also being considered for human food (and novel food dossiers are submitted), but they were subjected to biological risk assessments to a much lesser extent. Hence, increased efforts to fill the knowledge gap for these particular species are encouraged. Moreover, also for the black soldier fly (*H. illucens*), a novel food dossier was submitted. While rarely considered as a food source in Europe (in contrast to certain Asian and African regions (Wang and Shelomi, 2017) and the discussion of its potential as food for the Pacific Small Islands Developing States (Shelomi, 2020)), biological safety of this species has not been described regularly in the context of human food. For animal feed, however, several studies have been performed, as detailed below.

3   Biological risks associated with insects to be used in feed only

In the EU, insects are considered farm animals and hence they are subjected to the EU 'feed ban' (Regulation (EC) No 999/2001; EC, 2001), which prohibits the use of farmed animal-derived proteins in feed for ruminant and monogastric animals. This effectively limited the use of insect proteins for a long time to applications in pet food or feed for fur animals. Since the 1st of July 2017, the use of insect proteins from seven insect species in feed for aquaculture animals has been authorised (Regulation (EU) No 2017/893; EC, 2017). The exact species are black soldier fly, common housefly (*M. domestica*), yellow mealworm, lesser mealworm, house cricket, tropical house cricket and field cricket (*Gryllus assimilis*). From these seven, only the common housefly and the black soldier fly are not commonly considered as human food and thus have not yet been discussed earlier in this review. Furthermore, these two species (together with the yellow mealworm) are the most generally used insects as protein source for animal feed globally. Hence, we will focus on the microbiological risks associated with these two species to assess the biological risks for insects as feed.

To identify which bacteria or fungi are most likely to cause problems, a clear view on the insect microbiota is needed. For the black soldier fly, this was long underexplored (De Smet *et al.*, 2018), but a number of recent publications expand our understanding of the microbiota in the larvae on a variety of diets (Bruno *et al.*, 2019; Jiang *et al.*, 2019; Klammsteiner *et al.*, 2020; Shelomi *et al.*, 2020; Wu *et al.*, 2020; Wynants *et al.*, 2018; Zhan *et al.*, 2020). These studies are more and more revealing the composition of the present microbiota in the larvae and the impact of substrate on its composition. Wynants *et al.* (2018) reported distinct sets of bacteria in the larvae and residue compared to the fed substrate. The same conclusion was drawn by Klammsteiner *et al.* (2020), who even more clearly hinted at the existence of a core set of microbes that make up the gut microbiota in the larvae, which may be established early on in larval development. Other studies show that microbes from the substrate can enter the gut and, if they are able to proliferate, become part of the gut microbiota, but they also state that these microbes never fully take control of this ecological niche (Bruno *et al.*, 2019; Jiang *et al.*, 2019). Bruno *et al.* (2019) for example reported variation in the bacterial species present in distinct regions of the gut, which is most likely due to the fact that each region represents a different ecological niche with other conditions. At the same time, the consensus is growing that a set of species is present in the larva, albeit to varying extent, during rearing on most of the tested diets to date, making them the core of the black soldier fly larvae (BSFL) gut microbiota. This core appears to include *Actinomyces* sp., *Dysgonomonas* sp., *Enterococcus* sp. and *Morganella* sp. These species have been found in other insects as well and could aid in the degradation of organic substances (Klammsteiner *et al.*, 2020).

While the microbiota of the common housefly has been studied for individuals captured in the wild (Bahrndorff *et al.*, 2017; De Jonge *et al.*, 2020), to our knowledge no studies have been performed to assess its microbiota during an industrial rearing cycle, and therefore the microbiological quality cannot be compared to that of other insects for feed.

When exploring specifically the presence of food pathogens, some studies do report potential risks in BSFL. Wynants *et al.* (2018) reported no detection of *L. monocytogenes* or coagulase-positive staphylococci. In contrast, they observed one contamination with *Salmonella enterica* serovar Agona in the rearing residue, but not in the larvae. Presumptive *B. cereus* was found in the larvae with counts up to 6,000 cfu/g. Other studies report the presence of *Campylobacter* (Wu *et al.*, 2020) or *Clostridium* (Jiang *et al.*, 2019) species in their metagenetics data but they did not quantify the actual pathogen load. Hence, the *B. cereus* group seems to be a major risk for insects for feed as well. With respect to other biological risks, e.g. prions and viruses, no data could be found that address safety questions for neither of both insect species. Only one report deals with the presence of a virus (be it not a food safety issue) in the BSFL gut: Chen *et al.* (2019) isolated a novel temperate *Escherichia* phage from the BSFL gut, and different substrates caused differences in phage induction. Hence, substrates may not only shape the gut microbiota directly, but also indirectly via their impact on lytic/lysogenic switches of phages infecting gut bacteria. For foodborne viruses, no information is available to our knowledge. Also, insects are, to date, not found to be able to produce prions, but prions can enter non-processed insects from contaminated substrates, for example slaughterhouse waste (EFSA Scientific Committee, 2015). This is a low level risk, that can be avoided by substrate quality control or preventing the feeding of these larvae to the same animal species as the one present in the substrate.

4   **Risk assessments on the transfer of biological contaminants during**
         rearing of insects

Biological safety risks are not only defined by the intrinsic presence or absence of
pathogenic microorganisms. It is also important to estimate the risks for coinciden-
tal contamination during insect rearing or processing by a certain pathogen (EFSA
Scientific Committee, 2015; Wynants *et al.*, 2018). Studies that monitor the dynamics of a specific food pathogen when present in the substrate by inoculating the
pathogen in the substrate (so-called challenge tests) are emerging in recent years
(Belleggia *et al.*, 2020; Mancini *et al.*, 2019c; Wynants *et al.*, 2019).

      Almost all the available studies are about *T. molitor* and only some information
on other species is available. Mancini *et al.* (2019c) detected a transmission of 3.6-
4.6 log cfu/g of *L. monocytogenes* to mealworm larvae since the first day in the substrate at a load of 8 log cfu/g. This contamination seemed to persist for at least seven
days. According to Belleggia *et al.* (2020), *Listeria* spp. can even multiply in the gut
of the mealworm larvae. Because washing of the larvae did not cause a reduction
in microbial counts, the bacteria were assumed to be present in the interior part
of the larvae (Mancini *et al.*, 2019c). This can cause considerable health risks when
using the larvae, as a whole or after processing without degutting, as food or feed.
Also *Salmonella* sp. can be transmitted to *T. molitor* already after one day and, if
inoculated at 7 log cfu/g in the substrate, the contamination was still present after
seven days (Wynants *et al.*, 2019). BSFL have also been subjected to several challenge studies. Defilippo *et al.* (2018) determined that larvae contain the same concentration of *S. enterica* serovar Typhimurium (3-4 log cfu/g) and *L. monocytogenes*
(5-6 log cfu/g) as the inoculated substrate, but they observed a 2-log decreased concentration when reaching the pupal stage. No hypothesis for this observation was
given. Swinscoe *et al.* (2020) observed a stable concentration of *L. monocytogenes*
and *Vibrio parahaemolyticus* (6 log cfu/g) and a slightly decreasing concentration
of *Escherichia coli* and *E. coli* O157:H7 (from 7 log cfu/g to 5 log cfu/g) over a period of seven days after exposure to a contaminated substrate. The authors mention
that a selective inactivation (i.e. some strains are inhibited while others are not)
of the pathogens by the larvae occurs via exposure to antimicrobials (showing a
certain selectivity). In particular, the expression of antimicrobial peptides by BSFL
is substantial in protein-rich diets, as the seaweed-based diet used in their study.
Although pathogen transfer does not necessarily affect the viability of the larvae
in a negative way (Mancini *et al.*, 2019c), it can be a risk when using these contaminated larvae for food or feed purposes. However, it is worth noting that the results
observed seem to depend on numerous factors, such as pathogen level in the substrate and stage in the cycle of the insect, and thus general conclusions cannot be
extrapolated to other combinations of insect and pathogen species.

      Even though a number of challenge tests were performed for bacteria, transfer
of other biological risks, such as prions, viruses, fungi and parasites, is much less

investigated. One recent study on transmission of parasites via BSFL feed (Müller *et al.*, 2019) revealed a transfer of less than 1% of the parasital oocysts of *Eimeria nieschulzi*, *Eimeria tenella* or eggs of *Ascaris suum* in the larval gut. No or very low contamination was found in the prepupal stages. Nevertheless, it is still too early to neglect the potential risk of parasite transmission when using BSFL as animal feed. Next to parasites, a study from Varotto Boccazzi *et al.* (2017) investigated the transfer of fungi to BSFL reared on chicken feed and vegetable waste, with both substrates causing a different fungal community in the larvae. Larvae grown on chicken feed contained mainly *Trichosporon*, *Rhodotorula* and *Geotrichum* species, whereas in vegetable waste fed larvae, *Pichia* was the most abundant genus.

Besides challenge tests with contaminated substrates, also indirect exposure of insects to human pathogens has been studied. For the housefly (*M. domestica*), studies exist that focus on wild flies. One such study confirmed the transfer of different pathogens by exposure to facilities containing contaminated farm animals. About 20% of the flies caught on broiler farms were positive for *Salmonella* (Bailey *et al.*, 2001), and even more than 50% tested positive after 4 to 7 days exposure to environments housing *S. enterica* serovar Enteritidis challenged hens (Holt *et al.*, 2007). In another study, *Campylobacter* sp. could be isolated from about 40 to 50% of the flies sampled on chicken farms and piggeries (Rosef and Kapperud, 1983). These studies also demonstrate that rearing of the common housefly should take place in clean conditions as contaminations of the feed are very likely to enter the insect.

Despite the fact that various biological risks can be transferred to insects, some species seem to be able to reduce or even eliminate specific bacteria in their substrate. They can therefore be reasoned to decrease the biological risk. For instance, BSFL have shown to be able to reduce the amount of Enterobacteriaceae. A reduction of *S. enterica* serovar Enteritidis (3.5-5.5 log reduction after two days) and *E. coli* O157:H7 (1.5-5 log reduction after two days and even below detection limit after three days) in chicken manure has been observed (Erickson *et al.*, 2004). Also Liu *et al.* (2008) and Lalander *et al.* (2013) found a reduction of respectively *E. coli* by 5-7 log cycles in dairy manure and of *Salmonella* spp. by 7 log cycles in eight days in faecal sludge. Lalander *et al.* (2015) studied the reduction of *Salmonella* spp. and *Enterococcus* spp. in a mixture of pig manure, human faeces and dog food. For *Salmonella* spp., the concentration in the substrate was reduced from approximately 7 to less than 1 cfu/g after two weeks, while *Enterococcus* contamination remained unchanged. Those experiments could be interpreted as promising perspectives for a safe recycling of animal or human manure by BSFL. Nevertheless, according to (EC) No. 1069/2009, the use as substrate of manure, catering waste and former foodstuff containing meat or fish is forbidden for farmed animals, including insects destined for food and feed purposes. So far, the pathogen-reducing effects of BSFL have not been studied and confirmed enough to alter legislation. It is not known, for instance, whether the effects are substrate-dependent or not. More data may help to fine-tune legislation on this aspect in the future.

5   Mitigation strategies to reduce food and feed safety risks

As is clear from literature, the high microbial counts associated with untreated edible insects and the possible occurrence of certain pathogenic microorganisms require suitable mitigation strategies, i.e. strategies to reduce the microbiological risks, to be applied on the insects in order to ensure safety. The four reviews on insects for food referred to before discuss studies that report on treatments to reduce microbiological safety risks (Table 1). Altogether, thermal treatments such as boiling, roasting, (deep-)frying and blanching, even for short times, have proven to be very effective to reduce the amount of vegetative bacteria and fungi. Alternative treatments such as fermentation, cold plasma treatments, microwave drying may also reduce microbiological food safety risks. For example, a recent study investigated the efficacy of high hydrostatic pressure (HHP) to reduce certain microbiological risks in a meal derived from dried BSFL (Kashiri *et al.*, 2018). They reported that *Listeria* was not present in the extract, but *Salmonella* and *E. coli* were found at around 5 to 6 log cfu/g. The killing efficacy of a HHP treatment at 400 MPa for 7 minutes was the highest for yeasts and moulds (no survivors), but had hardly any effect on the count of naturally present total aerobic mesophilic bacteria (0.35 log reduction). No data were provided on the reduction of *Salmonella*. However, the treatment did reduce the load of inoculated *E. coli* with 6.56 log cycles (Kashiri *et al.*, 2018). The fact that this technology was less effective against naturally present bacteria than against inoculated *E. coli* is hypothesised to be due to the presence of bacterial endospores. This is an important factor, as the *B. cereus* group is considered as a high risk. Non-thermal drying methods such as freeze-drying, which is commonly used in the insect industry, only have a microbiostatic effect and are therefore not sufficient to assure microbiological food safety. Starvation of the insects at the end of the rearing cycle may be thought to have the effect of emptying the gut and hence clearing the insects from pathogens as well, but starvation should not be considered as a mitigation strategy since the effect on the microbiota appears to vary for different studies (Mancini *et al.*, 2019a,b; Wynants *et al.*, 2017).

6   Future challenges and research needs for biological safety of edible insects

6.1   *Focus needed on individual insect species*
In the same way as the gut microbiome is not equal for all traditional farm animals, it is not accurate to make general statements on the microbiota of edible insects altogether. Risk profiles and assessments should envisage individual species, such as the risk profile for *A. domesticus* by Fernandez-Cassi *et al.* in 2018 and in 2019, in order to identify and implement species-specific hygiene measures during rearing. When organising studies that focus on an individual insect species, care

should be taken to include samples obtained from industrial producers rather than from insects reared in laboratory conditions. To obtain a representative view on the microbiota and on food pathogens during mass rearing of insects, it is necessary to involve several producers, hence covering differences in substrates and production processes applied on different locations, as done for instance by Wynants *et al.* (2019). In addition, per producer several batches or rearing cycles need to be sampled using proper sampling plans, as detailed below. Batches are ideally produced over a time span of several months to capture possible variations over time in substrate quality and seasonality, environmental conditions and rearing procedures. Such studies are extensive, but allow to describe possible correlations between the microbiota including the occurrence of biological contaminants on the one side and rearing conditions (feed, environmental conditions, hygiene practices and house flora, etc.) on the other side.

To supplement and deepen species-specific risk assessments, the behaviour of certain biological contaminants during the life cycle of an insect can be studied as mentioned before by challenge tests. Deliberately infecting insects at a certain stage in their development with a known and standardised concentration of the pathogen allows to investigate if and how fast the contaminant is taken up by insect specimens (horizontal transfer). Also, it can be monitored where it is located in the gut and how passage through the gut occurs, how fast it is spread between specimens, whether it can grow in the substrate and/or insect or whether it naturally dies or maybe is actively eradicated by the insect. As discussed above, some combinations of insect and pathogen species have been investigated in this way, but a number of relevant insect-pathogen combinations have not been investigated so far. Especially for the main biological risks identified, *B. cereus*, *Clostridium* spp. and *S. aureus*, such data are still lacking for insects for food and feed. Overall, there is still a gap in the knowledge in the behaviour of biological contaminants that can be present in side or waste streams or that can be transmitted via personnel, i.e. *Bacillus*, *Campylobacter*, *Clostridium*, *Listeria*, *Salmonella*, and *Staphylococcus* during rearing of mass produced insects. *Bacillus* and *Clostridium* are two spore-forming bacterial species, and studies ideally involve both vegetative cells as well as spores. To the best of our knowledge, studies on vertical transmission of food pathogens, i.e. from one cycle to the next over the egg phase, are not yet available. It is not known so far whether the eggs can be a route by which food pathogens are introduced in a batch of larvae or nymphs. At least for *H. illucens*, it is thinkable that food pathogens may follow this transmission route, because bacterial communities including *Bacillus* sp. have been shown to be stimulating for oviposition (Yu *et al.*, 2011; Zheng *et al.*, 2013).

Also post-harvest decontamination and preservation strategies should be designed specifically for each individual species. A heat treatment comprises both a time and a temperature for the slowest heating point in the material to be treated. Inactivation kinetics cannot simply be extrapolated from one insect(-based product) to another, because: (1) heat transfer in a matrix depends on the composition (water content, fat content, etc.) and structure (whole insects, finely mixed paste or powder, or roughly grinded and hence containing air pockets) of the matrix; and (2) because individual insect species are characterised by other levels and types of target microorganisms. In the end, the time-temperature combination is to be determined to reduce the load of the most significant target organism in an insect matrix in a sufficient way, which is in the food industry often a reduction with 6, 8 or 12 log cycles. For inactivation strategies that do not use heat, also process parameters need to be defined for the matrix to be treated, i.e. a specific insect species, to obtain the same required reduction. Research to unravel the survival of spore-forming bacteria in post-harvest treatments preferably includes both vegetative cells as well as spores. While it may not be possible due to too much quality deterioration to eliminate all spores, heat treatments should not have the side effect of activating the spores, and if spores are observed to survive a treatment, then care should be taken to provide conditions during further storage and transport of the treated insect(-based product) that impede spore germination, such as (sufficient) refrigeration and/or acidification. As described before, a number of studies exist for particular species investigating the effect of heat or other treatments, but no general recommendations exist. Likely, knowledge and expertise is established in insect-processing companies and not (yet) publicly available.

6.2   *Specific attention on substrate, residue and house flora*
Not only the insects need further microbiological characterisation, also substrates should be characterised for their typical microbial profile, their core microbiota, if any, and their so-called specific spoilage organisms (SSO; Man and Jones, 2000). In the context of side stream storage, SSOs are those (subgroups of) microorganisms (bacteria, yeasts, moulds) that will dominate the microbiota when these organic streams are stored in insect production facilities and by their metabolism cause spoilage phenomena as off-odours and flavours. Side streams are often given one or more pre-treatment steps prior to be fed to the insects, such as mixing, concentration by a heat or alternative treatment, milling, acidification and so on. Those are all unit operations that can have an effect on the microbial load of the substrate, by either a killing effect or conversely improving conditions for microbial growth, and in turn on the insects reared on them. The impact on food pathogens that can occur in (mixtures of) organic side streams of the currently used preparation technologies is not yet thoroughly investigated. The effect of time-temperature conditions on survival of food pathogens inoculated in the substrate matrices at several contamination levels should be documented. Similar as to insect processing, this may have been studied to some extend by companies, but information in literature is missing.

The residue remaining after harvesting insects, containing unconsumed substrate, insect faeces or frass, and/or exuviae, offers the potential to be upcycled as soil fertiliser or plant growth supplement (Houben *et al.*, 2020). At least in the EU, legislation is not harmonised yet, and different member states impose other inactivation conditions. Several (microbiological) questions are being discussed by researchers, authorities and stakeholders. According to Regulation (EU) No 142/2011 (EC, 2011), so-called 'Method 1 to 5' (describing time, temperature and pressure combinations related to specific particle sizes of the material to be treated) or 'Method 7' may be used for the hygiene of residual fractions from insect rearing. Methods 1 to 5 apply to materials with a certain particle size, but in insect rearing the particle size of the residual fraction can vary or is not always known by producers. Method 7 allows the development of a treatment that can be shown to result in the achievement of certain microbiological criteria for *C. perfringens*, *Salmonella* and Enterobacteriaceae. While Method 7 leaves room for insect producers to optimise their own processing method, in Europe there is a general interest to equalise the processing conditions for insect residue with those valid for animal manure, i.e. heat treatment of 70 °C for one hour (IPIFF, 2019a). A next step could then be to assess whether less stringent parameters and non-thermal inactivation strategies (that better preserve the chemical quality of the residue) can still warrant compliance with the safety standards. This requires inactivation trials with inoculated residues and results will be insect-specific, since the composition of the residue is also species-specific.

In the insect sector, the impact of the house flora in a rearing and/or processing company on the microbiological safety of the end product is pretty much uncharted terrain. Numbers of microorganisms present on food contact surfaces in food companies are known to increase during production and to be reduced during proper cleaning and disinfection (Holah, 2014). While these organisms were first only thought to be responsible for spoilage problems, they are now also known to be related to safety issues. Pathogens can interact with and grow in biofilms, and in a production environment a persistent house flora that is pathogenic can develop (Holah, 2014). For example, it has been demonstrated that the composition of the resident microorganisms on a surface, or the house flora, can either promote or inhibit the growth of *L. monocytogenes* in the biofilm, and hence determine whether a surface can be a contamination route for a pathogen or not (Carpentier and Chassaing, 2004). In the insect industry, research is needed on the potential of food pathogens to reside in the production environment or not, and whether the environment can in this way be part of the transmission routes for a pathogen or not. Investigating the house flora in an insect production plant can be of support in the implementation of a hazard analysis and critical control points (HACCP) plan, which will also be discussed below. The application of a HACCP plan is based on 7 principles and consists of 12 steps (FAO/WHO, 2009). In step 6, all potential hazards should be listed that may reasonably be expected to occur in the whole process line. If the composition of the house flora is known for several surfaces in the production site, it is also known whether these communities can harbour pathogens (and which ones, at what level and at what specific locations). Next, potential transfer from the environment to the insect(-products) can be assessed. If relevant, the transfer can be identified as a hazard and implemented in the determination of critical control points, in particular related to the cleaning and disinfection practices (step 7).

6.3 *Proper sampling plans and predictive microbiology for knowledge extension*

Whether it is in the investigation of microbiological quality of insects themselves, or of substrates, residue or the production environment, a central question is always how to construct a proper sampling plan. Insect producers are responsible for the microbiological safety of their products and have to make sure their products fulfil national and international compulsory microbiological criteria, as summarised in the IPIFF guide on good hygiene practices (IPIFF, 2019b). While these criteria may still be refined in the future, the challenge behind these targets for insect producers is how to convert those requirements (legal, or maybe even more stringent Business-to-Business agreements between producer and buyer) into practices, measures and interventions, that determine the microbiological quality during production (and hence also may evolve and be fine-tuned) in order to warrant the achievement of the targets for the end product. In the context of the food industry, Jacxsens *et al.* (2009) defined such 'company specific set of control and assurance activities to realise and guarantee food safety' as a food safety management system (FSMS). An FSMS should translate good hygienic practices and the HACCP system in the specific context of the company (De Loy-Hendrickx *et al.*, 2018). Jacxsens *et al.* (2009) also described a microbial assessment scheme (MAS) to investigate the microbial performance of such FSMS, i.e. to find out whether the correct sampling locations are identified, whether the most relevant microbiological parameters are selected, and to assess the sampling and analytical methods. A MAS has been applied to assess the microbiological performance of integral production and processing chains of e.g. lamb (Osés *et al.*, 2012), pangasius (Tong Thi *et al.*, 2014) and in Kenyan fresh produce processing and export companies (Kamau Njage *et al.*, 2017), but this exercise would also be useful for insect companies. Moreover, the subdomain in food microbiology of predictive microbiology, in which mathematical models predict the growth, survival, spore germination and toxin production of pathogens in a certain food matrix, is completely unexplored so far for insects, but can certainly bring new insights.

6.4 *Microbiological methods: standardisation and dealing with the unculturable fraction*

As one of its future tasks, the Commission on Insects of the European Federation for Animal Science (formerly European Association for Animal Production or EAAP) has decided to work on the standardisation of research methods. This is also pertinent for microbiological methods. Not only the media and incubation conditions

used influence the result, but for insects it was shown that whether or not pulveris-
ing a sample prior to preparing a dilution series can make a difference in the count
of even 1.6 to 2.2 log cfu/g (Vandeweyer *et al.*, 2017). Also, for culture-dependent microbial counts, it is a prerequisite that samples are investigated immediately after
sampling or at the latest after one day storage under refrigerated conditions (rather
than after frozen storage of samples). This is a practice that is unequivocal in microbiological food analysis, but it may not yet be understood and followed by all insect
producers.

In colony counting, the incubation step entails a limitation since only a fraction of the microbiota is known to be culturable. Barcina and Irana (2009), as cited
by Fakruddin *et al.* (2013), state that in water and soil samples from nature, less
than 1% of the microorganisms is culturable. For insects used in food and feed, the
share of the culturable fraction in the total microbiota has not been estimated so
far. Moreover, within the field of food microbiology, special attention is paid to food
pathogens that can enter a viable but nonculturable (VBNC) state, implying that
their cells cannot grow on culture media but they still show metabolic activity as
described for the first time by Xu *et al.* (1982). Recently, Zhao *et al.* (2017) presented
a list of 35 foodborne pathogens that were proven to show this state, many species
of which have already been identified on edible insects. From the review by Zhao *et
al.* (2017), it appears that the VBNC state can be induced by bringing viable cells in
stress conditions, such as in several food treatments relevant for insect processing
as well (heating, cooling, drying). Based on their literature review, the authors also
conclude that some pathogens retain their pathogenicity (as shown in animal models) while others are avirulent. Microscopic counting and molecular technologies
are under development to assess VBNC food pathogens, and this is a new field in
the microbiological safety of edible insects, too.

Culture-independent methodologies do not comprise an incubation step in
which microorganisms are required to grow in order to quantify or detect them in
a next step. Culture-independent analyses can be either a metagenetic analysis, as
defined earlier, or they accomplish the amplification (and subsequent detection
or quantification) of target sequences of DNA (or RNA) of particular microbial
species, such as food pathogens (i.e. PCR technology). Metagenetics provides a
more comprehensive overview of the microbial composition of an insect sample
than plate counting does, and, as reviewed by Garofalo *et al.* (2019), the technology
is increasingly being incorporated in studies characterising the insect microbiota
during rearing and processing. Mostly metagenetics is DNA-based, although Bruno *et al.* (2019) worked on the RNA level (i.e. metatranscriptomics). Since RNA is
known to disintegrate faster after cell death in contrast to DNA, it is considered
as a good target to selectively focus on living cells. In terms of interpretation of
microbiological safety, (DNA-based) metagenetics is limited in that it only demonstrates the presence of DNA, rather than the presence of viable and virulent food
pathogens. DNA of dead cells is also detected by sequencing analyses, yet it does

not always point towards a food safety problem. Conversely, processing of insects may destroy DNA, thereby eliminating the possibility to detect pathogens, which, even though the viable cells may have been inactivated by processing, may have produced toxic metabolites that were resistant to the processing conditions applied (such as the cereulide by *B. cereus*). In addition, Filippidou *et al.* (2015) mentioned that bacterial endospores can resist to DNA extraction and remain under-detected, even when applying methods specially developed to tackle spores. Hence, meta-genetics proves its value to some extent in delivering a general overview of the microbial community composition in an insect sample. Meaningful detection (in terms of food safety assessment) of particular food pathogens can also be based on PCR techniques, preferably either Reverse Transcriptase-PCR starting from RNA, or PCR targeting DNA from living cells by first blocking DNA from dead cells using propidium monoazide (PMA) or ethidium monoazide (EMA). For example, Abd El-Aziz *et al.* (2018) examined three hundred fresh and processed samples of traditional meat purchased in Egyptian supermarkets for the presence of a variety of food pathogens. By applying PMA quantitative real-time PCR, they discovered that 90.48% of the culture-negative meat samples contained a high load of a range of VBNC food pathogens, being a strong threat to public health. To the best of our knowledge, such approach has not been applied to evaluate the microbiological safety of insect(-based feed or food)s, but as in previous PMA/EMA studies, a lot of interference of the matrix can be expected.

6.5   *Microbiological challenges related to the further development of the sector*

Edible insects are increasingly being proposed as alternative protein source, with the term 'alternative' often alluding to 'more sustainable' than traditional sources. The sector wants to maximise the sustainability of rearing and processing steps, as can be evaluated in life cycle assessment studies. One aspect in this context is the limitation or avoidance of energy consuming unit operations, such as conventional thermal treatments or freeze-drying, and their replacement by innovative thermal technologies such as microwave or ohmic heating or non-thermal treatments (high hydrostatic pressure, pulsed electric fields, low energy electron beam, intense light pulses, cold plasma, etc.). From a microbiological perspective, those technologies can involve the risk of a reduced food safety when not properly investigated and applied. Not only should process parameters be established that sufficiently reduce the target pathogens that are known so far to be relevant for insect(-product)s, but also the emergence of new microbiological food safety problems should be avoided. The design of a treatment to ensure minimal processing often comprises the combination of multiple decontamination actions to achieve the same reduction as a single lethal stress (Rosnes *et al.*, 2011). For traditional foods, during the past decades many cases are elaborated based on this combination strategy, but insects are a new matrix to be studied.

In its aim to maximally contribute to upcycling and to a circular economy, the sector hopes for green light to grow insects on low value substrates, such as unprocessed former foodstuff (preconsumer waste) containing meat and fish, postconsumer waste (i.e. catering waste), slaughterhouse waste and, for certain continents, manure. This is reflected for instance in the advice recently formulated by the Netherlands Food and Consumer Product Safety Authority (Anonymous, 2020) to rear insects on former foodstuffs containing meat, which is not allowed today. It goes without saying that feeding these lower value substrates eventually may be linked to (new) biological safety risks, when feeding these streams to insects, fresh or – implying an even larger risk – after storage. Taking this next step requires making an inventory of the food pathogens that can occur in these streams and an investigation of their behaviour in these matrices and of the efficiency of hygienisation practices.

Finally, the insect sector is rapidly growing, both in terms of number of companies as in terms of average company scale. This is demonstrated for Europe in a Factsheet of June 2020 composed by IPIFF (2020): in Europe 6,000 tonnes of insect proteins were produced by its members in 2019 and production is estimated to be around 3 million tonnes in 2030. Hence, the insect supply chain becomes more complex, involving more suppliers, intermediate B2B activities, and with an increasing intermediate and finished product portfolio, a more complex distribution network due to an increasing number and type of clients. For the traditional food industry, Martins *et al.* (2014) exhaustively elaborated on how an increasing complexity and upscaling of food production impacts the prevalence of zoonoses and complicates safety control. It is valid to extrapolate this challenge to the insect sector as well. A sector producing higher volumes and more diverse products will go hand in hand with more comprehensive sampling plans, new critical control points, a thorough traceability, longer transport and storage times, and these are all aspects to be considered from a microbiological viewpoint. This challenge may even be the most important from all challenges, since the relatively young sector definitely needs to avoid the occurrence of food crises and to show the expertise is present to accomplish the safety goals, in order to maintain and even further establish its position.

Acknowledgements

The authors wish to thank R. Smets for his assistance in the design of the figures. DV is financed by the Research Foundation – Flanders (FWO) via the SBO project ENTOBIOTA (S008519N) as well as by the European Union's Horizon 2020 Research and Innovation programme via the H2020 project SUSINCHAIN (grant agreement number 861976). JDS holds a postdoctoral fellowship grant (grant number 12V5219N) of the FWO. NVL is PhD researcher on the ERANET FACCE SURPLUS project UpWaste (ID 28) funded by FWO.

Conflict of interest

The authors declare no conflict of interest.

References

Abd El-Aziz, N.K., Tartor, Y.H., Abd El-Aziz Gharib, A. and Ammar, A.M., 2018. Propidium monoazide quantitative real-time polymerase chain reaction for enumeration of some viable but nonculturable foodborne bacteria in meat and meat products. Foodborne Pathogens and Disease 15: 226-234. https://doi.org/10.1089/fpd.2017.2356

Anonymous, 2020. Advice on animal and public health risks of insects reared on former foodstuffs as raw material for animal feed. Netherlands Food and Consumer Product Safety Authority, Utrecht, The Netherlands, 60 pp. Available at: https://english.nvwa.nl/documents/consumers/food/safety/documents/advice-on-animal-and-public-health-risks-of-insects-reared-on-former-foodstuffs-as-raw-material-for-animal-feed

Bahrndorff, S., De Jonge, N., Skovgård, H. and Nielsen, J.L., 2017. Bacterial communities associated with houseflies (*Musca domestica* L.) sampled within and between farms. PLoS ONE 12: e0169753. https://doi.org/10.1371/journal.pone.0169753

Bailey, J.S., Stern, N.J., Fedorka-Cray, P., Craven, S.E., Cox, N.A., Cosby, D.E., Ladely, S. and Musgrove, M.T., 2001. Sources and movement of *Salmonella* through integrated poultry operations: a multistate epidemiological investigation. Journal of Food Protection 64: 1690-1697. https://doi.org/10.4315/0362-028X-64.11.1690

Barcina, I. and Arana, I., 2009. The viable but nonculturable phenotype: a crossroads in the life-cycle of non-differentiating bacteria? Reviews in Environmental Science and Biotechnology 8: 245-255. https://doi.org/10.1007/s11157-009-9159-x

Belleggia, L., Milanović, V., Cardinali, F., Garofalo, C., Pasquini, M., Tavoletti, S., Riolo, P., Ruschioni, S., Isidoro, N., Clementi, F., Ntoumos, A., Aquilanti, L. and Osimani, A., 2020. *Listeria* dynamics in a laboratory-scale food chain of mealworm larvae (*Tenebrio molitor*) intended for human consumption. Food Control 114: 107246. https://doi.org/10.1016/j.foodcont.2020.107246

Borremans, A., Crauwels, S., Vandeweyer, D., Smets, R., Verreth, C., Van der Borght, M., Lievens, B. and Van Campenhout, L., 2019. Comparison of six commercial meat starter cultures for the fermentation of yellow mealworm (*Tenebrio molitor*) paste. Microorganisms 7: 540. https://doi.org/10.3390/microorganisms7110540

Bruno, D., Bonelli, M., De Filippis, F., Di Lelio, I., Tettamanti, G., Casartelli, M., Ercolini, D. and Caccia, S., 2019. The intestinal microbiota of *Hermetia illucens* larvae is affected by diet and shows a diverse composition in the different midgut regions. Applied and Environmental Microbiology 85: e01864-18. https://doi.org/10.1128/AEM.01864-18

Cappelli, A., Cini, E., Lorini, C., Oliva, N. and Bonaccorsi, G., 2020. Insects as food: a review on risks assessments of Tenebrionidae and Gryllidae in relation to a first machines and plants development. Food Control 108: 106877. https://doi.org/10.1016/j.foodcont.2019.106877

Carpentier, B. and Chassaing, D., 2004. Interactions in biofilms between *Listeria monocytogenes* and resident microorganisms from food industry premises. International Journal of Food Microbiology 97: 111-122. https://doi.org/10.1016/j.ijfoodmicro.2004.03.031

Chen, Y., Li, X., Song, J., Yang, D., Liu, W., Chen, H., Wu, B. and Qian, P., 2019. Isolation and characterization of a novel temperate bacteriophage from gut-associated *Escherichia* within black soldier fly larvae (*Hermetia illucens* L. [Diptera: Stratiomyidae]). Archives of Virology 164: 2277-2284. https://doi.org/10.1007/s00705-019-04322-w

De Jonge, N., Yssing Michaelsen, T., Ejbye-Ernst, R., Jensen, A., Elley Nielsen, M., Bahrndorff, S. and Lund Nielsen, J., 2020. Housefly (*Musca domestica* L.) associated microbiota across different life stages. Nature 10: 7842. https://doi.org/10.1038/s41598-020-64704-y

De Loy-Hendrickx, A., Debevere, J., Devlieghere, F., Jacxsens, L., Uyttendaele, M. and Vermeulen, A., 2018. Microbiological guidelines: support for interpretation of microbiological test results of foods. Die Keure, Brugge, Belgium, 478 pp.

De Smet, J., Wynants, E., Cos, P. and Van Campenhout, L., 2018. Microbial community dynamics during rearing of black soldier fly larvae (*Hermetia illucens*) and impact on exploitation potential. Applied and Environmental Microbiology 84: e02722-17. https://doi.org/10.1128/AEM.02722-17

Defilippo, F., Grisendi, A., Listorti, V., Dottori, M. and Bonilauri, P., 2018. Black soldier fly larvae reared on contaminated substrate by *Listeria monocytogenes* and *Salmonella*. Available at: https://meetings.eaap.org/wp-content/uploads/2018/Session47/47.16_aq9i52ws.pdf

Dicke, M., Eilenberg, J., Falcao Salles, J., Jensen, A.B., Lecocq, A., Pijlman, G.P., Van Loon, J.J.A. and Van Oers, M.M., 2020. Edible insects unlikely to contribute to transmission of coronavirus SARS-CoV-2. Journal of Insects as Food and Feed 6(4): 333-339. https://doi.org/10.3920/JIFF2020.0039

European Commission (EC), 2001. Regulation (EC) No 999/2001 of the European Parliament and of the Council of 22 May 2001 laying down rules for the prevention, control and eradication of certain transmissible spongiform encephalopathies. Official Journal L 147: 1-40.

European Commission (EC), 2005. Commission Regulation (EC) No 2073/2005 of 15 November 2005 on microbiological criteria for foodstuffs. Official Journal L 338: 1-26.

European Commission (EC), 2011. Commission Regulation (EU) No 142/2011 of 25 February 2011 implementing Regulation (EC) No 1069/2009 of the European Parliament and of the Council laying down health rules as regards animal by-products and derived products not intended for human consumption and implementing Council Directive 97/78/EC as regards certain samples and items exempt from veterinary checks at the border under that Directive. Official Journal L 54: 1-254.

European Commission (EC), 2015. Regulation (EU) No 2015/2283 of the European Parliament and of the Council of 25 November 2015 on novel foods. Official Journal L 327: 1-22.

European Commission (EC), 2017. Commission Regulation (EU) No 2017/893 of 24 May 2017 as regards the provisions on processed animal protein. Official Journal L 138: 92-116.

European Food Safety Authority Panel on Biological Hazards (EFSA BIOHAZ Panel), 2016. Scientific opinion on the risks for public health related to the presence of *Bacillus cereus* and other *Bacillus* spp. including *Bacillus thuringiensis* in foodstuffs. EFSA Journal 14: 4524. https://doi.org/10.2903/j.efsa.2016.4524

European Food Safety Authority Scientific Committee, 2015. Risk profile related to production and consumption of insects as food and feed. EFSA Journal 13: 4257. https://doi. org/10.2903/j.efsa.2015.4257

Ehling-Schulz, M., Lereclus, D. and Koehler, T.M., 2019. The *Bacillus cereus* group: *Bacillus Species* with pathogenic potential. Microbiology Spectrum 7(3). https://doi.org/10.1128/ microbiolspec.gpp3-0032-2018

Erickson, M.C., Islam, M., Sheppard, C., Liao, J. and Doyle, M.P., 2004. Reduction of *Escherichia coli* O157:H7 and *Salmonella enterica* serovar Enteritidis in chicken manure by larvae of the black soldier fly. Journal of Food Protection 67: 685-690. https://doi.org/10.4315/0362-028x-67.4.685

Fakruddin, M., Bin Mannan, K.S. and Andrews, S., 2013. Viable but nonculturable bacteria: food safety and public health perspective. International Scholarly Research Notices: Article ID 703813.. https://doi.org/10.1155/2013/703813

Fasolato, L., Cardazzo, B., Carraro, L., Fontana, F., Novelli, E. and Balzan, S., 2018. Edible processed insects from e-commerce: food safety with a focus on the *Bacillus cereus* group. Food Microbiology 76: 296-303. https://doi.org/10.1016/j.fm.2018.06.008

Fernandez-Cassi, X., Söderqvist, K., Bakeeva, A., Vaga, M., Dicksved, J., Vagsholm, I., Jansson, A. and Boqvist, S., 2020. Microbial communities and food safety aspects of crickets (*Acheta domesticus*) reared under controlled conditions. Journal of Insects as Food and Feed 6: 429-440. https://doi.org/10.3920/jiff2019.0048

Fernandez-Cassi, X., Supeanu, A., Jansson, A., Boqvist, S., Vagsholm, I., Boqvist, S. and Vagsholm, I., 2018. Novel foods: a risk profile for the house cricket (*Acheta domesticus*). EFSA Journal 16: e16082. https://doi.org/10.2903/j.efsa.2018.e16082

Fernandez-Cassi, X., Supeanu, A., Vaga, M., Jansson, A., Boqvist, S. and Vagsholm, I., 2019. The house cricket (*Acheta domesticus*) as a novel food: a risk profile. Journal of Insects as Food and Feed 5: 137-157. https://doi.org/10.3920/JIFF2018.0021

Filippidou, S., Junier, T., Wunderlin, T., Lo, C.C., Li, P.E., Chain, P.S. and Junier, P., 2015. Under-detection of endospore-forming firmicutes in metagenomic data. Computational and Structural Biotechnology Journal 13: 299-306. https://doi.org/10.1016/j.csbj.2015.04.002

Food and Agriculture Organisation / World Health Organisation (FAO/WHO), 2009. Hazard analysis and critical control point (HACCP) system and guidelines for its application. Food hygiene basic texts, 4th edition. Joint FAO/WHO Food Standards Programme, Codex Alimentarius Commission, Rome, Italy.

Frigerio, J., Agostinetto, G., Galimberti, A., De Mattia, F., Labra, M. and Bruno, A., 2020. Tasting the differences: microbiota analysis of different insect-based novel food. Food Research International 137: 109426.

Garofalo, C., Milanović, V., Cardinali, F., Aquilanti, L., Clementi, F. and Osimani, A., 2019. Current knowledge on the microbiota of edible insects intended for human consumption: a state-of-the-art review. Food Research International 125: 108527. https://doi.org/10.1016/j. foodres.2019.108527

Greenhalgh, J.P. and Amund, D., 2019. Examining the presence of *Cronobacter* spp. in ready-to-eat edible insects. Food Safety 7: 74-78. https://doi.org/10.14252/foodsafetyfscj. d-19-00004

Holah, J.T., 2014. Cleaning and disinfection practices in food processing. In: Lelieveld, H.L.M., Holah, J.T. and Napper, D. (eds.) Hygiene in food processing: principles and practice, 2nd edition. Woodhead Publishing, Philadelphia, PA, USA, 304 pp.

Holt, P.S., Geden, C.J., Moore, R.W. and Gast, R.K., 2007. Isolation of *Salmonella enterica* serovar enteritidis from houseflies (*Musca domestica*) found in rooms containing *Salmonella* serovar enteritidis-challenged hens. Applied and Environmental Microbiology 73: 6030-6035. https://doi.org/10.1128/AEM.00803-07

Houben, D., Daoulas, G., Faucon, M.-P. and Dulaurent, A.-M., 2020. Potential use of mealworm frass as a fertilizer: impact on crop growth and soil properties. Nature Scientific Reports 10: 4659. https://doi.org/10.1038/s41598-020-61765-x

International Platform of Insects for Food and Feed (IPIFF), 2019a. IPIFF Contribution Paper on the application of insect frass as fertilising product in agriculture. 6 pp. Available at https://ipiff.org/wp-content/uploads/2019/09/19-09-2019-IPIFF-contribution-on-insect-frass-application-as-fertilising-product-final-version.pdf

International Platform of Insects for Food and Feed (IPIFF), 2019b. IPIFF guide on good hygiene practices for European Union producers of insects as food and feed. IPIFF, Brussels, Belgium, 108 pp. Available at: https://ipiff.org/wp-content/uploads/2019/12/IPIFF-Guide-on-Good-Hygiene-Practices.pdf

International Platform of Insects for Food and Feed (IPIFF), 2020. Factsheet: Edible insects on the European market. Retrieved from https://ipiff.org/wp-content/uploads/2020/06/10-06-2020-IPIFF-edible-insects-market-factsheet.pdf

Jacxsens, L., Kussaga, J., Luning, P.A., Van der Spiegel, M., Devlieghere, F. and Uyttendaele, M., 2009. A microbial assessment scheme to measure microbial performance of food safety management systems. International Journal of Food Microbiology 134: 113-125. https://doi.org/10.1016/j.ijfoodmicro.2009.02.018

Jay, J., Loessner, M. and Golden, D., 2005. Modern food microbiology. Springer, Boston, MA, US, 790 pp. https://doi.org/10.1007/b100840

Jeßberger, N., Krey, V.M., Rademacher, C., Böhm, M.-E., Mohr, A.-K., Ehling-Schulz, M., Scherer, S. and Märtlbauer, E., 2015. From genome to toxicity: a combinatory approach highlights the complexity of enterotoxin production in *Bacillus cereus*. Frontiers in Microbiology 6: 560. https://doi.org/10.3389/fmicb.2015.00560

Jiang, C.-L., Jin, W.-Z., Tao, X.-H., Zhang, Q., Zhu, J., Feng, S.-Y., Xu, X.-H., Li, H.-Y., Wang, Z.-H. and Zhang, Z.-J., 2019. Black soldier fly larvae (*Hermetia illucens*) strengthen the metabolic function of food waste biodegradation by gut microbiome. Microbial Biotechnology 12: 528-543. https://doi.org/10.1111/1751-7915.13393

Kamau Njage, P., Sawe, C., Onyango, C., Habib, I., Njeru Njagi, E., Aerts, M. and Molenberghs, G., 2017. Microbiological performance of food safety control and assurance activities in a fresh produce processing sector measured using a microbiological scheme and statistical modeling. Journal of Food Protection 80: 177-188. https://doi.org/10.4315/0362-028X.JFP-16-233

Kashiri, M., Marin, C., Garzón, R., Rosell, C.M., Rodrigo, D. and Martínez, A., 2018. Use of high hydrostatic pressure to inactivate natural contaminating microorganisms and inoculated *E. coli* O157:H7 on *Hermetia illucens* larvae. PLoS ONE 13: e0194477. https://doi.org/10.1371/journal.pone.0194477

Klammsteiner, T., Walter, A., Bogataj, T., Heussler, C.D., Stres, B., Steiner, F.M., Schlick-Steiner, F.M., Arthofer, W. and Insam, H., 2020. The core gut microbiome of black soldier fly (*Hermetia illucens*) larvae raised on low-bioburden diets. Frontiers in Microbiology 11: 993. https://doi.org/10.3389/fmicb.2020.00993

Kooh, P., Ververis, E., Tesson, V., Boué, G. and Federighi, M., 2019. Entomophagy and public health: a review of microbiological hazards. Health 11: 1272-1290. https://doi.org/10.4236/health.2019.1110098

Kort, R., O'Brien, A.C., Van Stokkum, I.H.M., Oomes, S.J.C.M., Crielaard, W., Hellingwerf, K.J. and Brul, S., 2005. Assessment of heat resistance of bacterial spores from food product isolates by fluorescence monitoring of dipicolinic acid release. Applied and Environmental Microbiology 71: 3556-3564. https://doi.org/10.1128/AEM.71.7.3556-3564.2005

Lalander, C.H., Diener, S., Magri, M.E., Zurbrügg, C., Lindström, A. and Vinnerås, B., 2013. Faecal sludge management with the larvae of the black soldier fly (*Hermetia illucens*) – from a hygiene aspect. Science of the Total Environment 458-460: 312-318. https://doi.org/10.1016/j.scitotenv.2013.04.033

Lalander, C.H., Fidjeland, J., Diener, S., Eriksson, S. and Vinnerås, B., 2015. High waste-to-biomass conversion and efficient *Salmonella* spp. reduction using black soldier fly for waste recycling. Agronomy for Sustainable Development 35: 261-271. https://doi.org/10.1007/s13593-014-0235-4

Liu, Q., Tomberlin, J.K., Brady, J.A., Sanford, M.R. and Yu, Z., 2008. Black soldier fly (Diptera: Stratiomyidae) larvae reduce *Escherichia coli* in dairy manure. Environmental Entomology 37: 1525-1530. https://doi.org/10.1603/0046-225x-37.6.1525

Man, C.M.D and Jones, A., 2000. Shelf life evaluation of foods, 2nd edition. Aspen Publishers, Gaithersburg, MD, USA, 292 pp.

Mancini, S., Fratini, F., Tuccinardi, T., Turchi, B., Nuvoloni, R. and Paci, G., 2019a. Effects of different blanching treatments on microbiological profile and quality of the mealworm (*Tenebrio molitor*). Journal of Insects as Food and Feed 5: 225-234. https://doi.org/10.3920/JIFF2018.0034

Mancini, S., Fratini, F., Turchi, B., Mattioli, S., Dal Bosco, A., Tuccinardi, T., Nozic, S. and Paci, G., 2019b. Former foodstuff products in *Tenebrio molitor* rearing: effects on growth, chemical composition, microbiological load, and antioxidant status. Animals 9: 484. https://doi.org/10.3390/ani9080484

Mancini, S., Paci, G., Ciardelli, V., Turchi, B., Pedonese, F. and Fratini, F., 2019c. *Listeria monocytogenes* contamination of *Tenebrio molitor* larvae rearing substrate: preliminary evaluations. Food Microbiology 83: 104-108. https://doi.org/10.1016/j.fm.2019.05.006

Martin, T.C., Visconti, A., Spector, T.D. and Falchi, M., 2018. Conducting metagenomic studies in microbiology and clinical research. Applied Microbiology and Biotechnology 102: 8629-8646. https://doi.org/10.1007/s00253-018-9209-9

Martins, S.B., Häsler, B. and Rushton, J., 2014. Economic aspects of zoonoses: impact of zoonoses on the food industry. In: Sing, A. (ed.) Zoonoses – infections affecting humans and animals: focus on public health aspects. Springer, Dordrecht, The Netherlands, 1137 pp.

Müller, A., Wiedmer, S. and Kurth, M., 2019. Risk evaluation of passive transmission of animal parasites by feeding of black soldier fly (*Hermetia illucens*) larvae and prepupae. Journal of Food Protection 82: 948-954. https://doi.org/10.4315/0362-028X.JFP-18-484

Murefu, T.R., Macheka, L., Musundire, R. and Manditsera, F.A., 2019. Safety of wild harvested and reared edible insects: a review. Food Control 101: 209-224. https://doi.org/10.1016/j.foodcont.2019.03.003

Osés, S.M., Luning, P.A., Jacxsens, L., Santillana, S., Jaime, I. and Rovira, J., 2012. Microbial performance of food safety management systems implemented in the lamb production chain. Journal of Food Protection 75: 95-103. https://doi.org/10.4315/0362-028X.JFP-11-263

Rosef, O. and Kapperud, G., 1983. House flies (*Musca domestica*) as possible vectors of *Campylobacter fetus* subsp. *jejuni*. Applied and Environmental Microbiology 45: 381-383. https://doi.org/10.1128/aem.45.2.381-383.1983

Rosnes, J.T., Skåra, T. and Skipnes, D., 2011. Recent advances in minimal heat processing of fish: effects on microbiological activity and safety. Food and Bioprocess Technology 4: 833-848. https://doi.org/10.1007/s11947-011-0517-7

SciCom, 2018. Advies 23-2018 van het Wetenschappelijk Comité van 21 december 2018. Inschatting van het risico voor de consument van *Bacillus cereus* in levensmiddelen. Wetenschappelijk Comité van het Federaal Agentschap voor de Veiligheid van de Voedselketen. Available at: http://www.afsca.be/wetenschappelijkcomite/adviezen/2018/_documents/Advies23-2018_SciCom2018-04_B.cereus.pdf

Shelomi, M., 2020. Potential of black soldier fly production for pacific small island developing states. Animals 10: 1038. https://doi:10.3390/ani10061038

Shelomi, M., Wu, M.-K., Chen, S.-M., Huang, J.-J. and Burke, C.G., 2020. Microbes associated with black soldier fly (Diptera: Stratiomyidae) degradation of food waste. Environmental Entomology 49: 405-411. https://doi.org/10.1093/ee/nvz164

Swinscoe, I., Oliver, D.M., Ørnsrud, R. and Quilliam, R.S., 2020. The microbial safety of seaweed as a feed component for black soldier fly (*Hermetia illucens*) larvae. Food Microbiology 91: 103535. https://doi.org/10.1016/j.fm.2020.103535

Tong Thi, A.N., Jacxsens, L., Noseda, B., Samapundo, S., Nguyen, B.L., Heyndrickx, M. and Devlieghere, F., 2014. Evaluation of the microbiological safety and quality of Vietnamese *Pangasius hypophthalmus* during processing by a microbial assessment scheme in combination with a self-assessment questionnaire. Fisheries Science 80: 1117-1128. https://doi.org/10.1007/s12562-014-0786-y

Uman L.S., 2011. Systematic reviews and meta-analyses. Journal of the Canadian Academy of Child and Adolescent Psychiatry 20: 57-59.

Van Cauteren, D., Le Strat, Y., Sommen, C., Bruyand, M., Tourdjman, M., Jourdan-Da Silva, N., Couturier, E., Fournet, N., De Valk, H. and Desenclos, J.-C., 2017. Estimated annual numbers of foodborne pathogen-associated illnesses, hospitalizations, and deaths, France, 2008-2013. Emerging Infectious Diseases 23: 1486-1492. https://doi.org/10.3201/eid2309.170081

Vandeweyer, D., Crauwels, S., Lievens, B. and Van Campenhout, L., 2017. Microbial counts of mealworm larvae (*Tenebrio molitor*) and crickets (*Acheta domesticus* and *Gryllodes sigillatus*) from different rearing companies and different production batches. International Journal of Food Microbiology 242: 13-18. https://doi.org/10.1016/j.ijfoodmicro.2016.11.007

Vandeweyer, D., Lievens, B. and Van Campenhout, L., 2020a. Identification of bacterial endospores and targeted detection of foodborne viruses in industrially reared insects for food. Nature Food 1: 511-516. https://doi.org/10.1038/s43016-020-0120-z

Vandeweyer, D., Lievens, B. and Van Campenhout, L., 2020b. Microbiological safety of industrially reared insects for food: identification of bacterial endospores and targeted detection of foodborne viruses. BioRxiv. https://doi.org/10.1101/2020.04.22.055236

Vandeweyer, D., Wynants, E., Crauwels, S., Verreth, C., Viaene, N., Claes, J., Lievens, B. and Van Campenhout, L., 2018. Microbial dynamics during industrial rearing, processing, and storage of tropical house crickets (*Gryllodes sigillatus*) for human consumption. Applied and Environmental Microbiology 84: e00255-18. https://doi.org/10.1128/AEM.00255-18

Varotto Boccazzi, I., Ottoboni, M., Martin, E., Comandatore, F., Vallone, L., Spranghers, T., Eeckhout, M., Mereghetti, V., Pinotti, L. and Epis, S., 2017. A survey of the mycobiota associated with larvae of the black soldier fly (*Hermetia illucens*) reared for feed production. PLoS ONE 12: e0182533. https://doi.org/10.1371/journal.pone.0182533

Wang, Y.-S. and Shelomi, M., 2017. Review of black soldier fly (*Hermetia illucens*) as animal feed and human food. Foods 6: 91. https://doi.org/10.3390/foods6100091

World Health Organization, 1995. Application of risk analysis to food standards issues. Report of the Joint FAO/WHO Expert Consultation. FAO, Rome, Italy. Available at: http://www.fao.org/3/ae922e/ae922e00.htm#Contents

Wu, N., Wang, X., Xu, X., Cai, R. and Xie, S., 2020. Effects of heavy metals on the bioaccumulation, excretion and gut microbiome of black soldier fly larvae (*Hermetia illucens*). Ecotoxicology and Environmental Safety 192: 110323. https://doi.org/10.1016/j.ecoenv.2020.110323

Wynants, E., Frooninckx, L., Crauwels, S., Verreth, C., De Smet, J., Sandrock, C., Wohlfahrt, J., Van Schelt, J., Depraetere, S., Lievens, B., Van Miert, S., Claes, J. and Van Campenhout, L., 2018. Assessing the microbiota of black Soldier fly larvae (*Hermetia illucens*) reared on organic waste streams on four different locations at laboratory and large scale. Microbial Ecology 77: 913-930. https://doi.org/10.1007/s00248-018-1286-x

Wynants, E., Frooninckx, L., Van Miert, S., Geeraerd, A., Claes, J. and Van Campenhout, L., 2019. Risks related to the presence of *Salmonella* sp. during rearing of mealworms (*Tenebrio molitor*) for food or feed: survival in the substrate and transmission to the larvae. Food Control 100: 227-234. https://doi.org/10.1016/j.foodcont.2019.01.026

Wynants, E., Crauwels, S., Lievens, B., Luca, S., Claes, J., Borremans, A., Bruyninckx, L. and Van Campenhout, L., 2017. Effect of post-harvest starvation and rinsing on the microbial numbers and the bacterial community composition of mealworm larvae (*Tenebrio molitor*). Innovative Food Science and Emerging Technologies 42: 8-15. https://doi.org/10.1016/j.ifset.2017.06.004

Xu, H.S., Roberts, N., Singleton, F.L., Attwell, R.W., Grimes, D.J. and Colwell, R.R., 1982. Survival and viability of nonculturable *Escherichia coli* and *Vibrio cholerae* in the estuarine and marine environment. Microbial Ecology 8: 313-323. https://doi.org/10.1007/BF02010671

Yu, G., Cheng, P., Chen, Y., Li, Y., Yang, Z., Chen, Y. and Tomberlin, J.K., 2011. Inoculating poultry manure with companion bacteria influences growth and development of black soldier fly (Diptera: Stratiomyidae) larvae. Environmental Entomology 40: 30-35. https://doi.org/10.1603/EN10126

Zhan, S., Fang, G., Cai, M., Kou, Z., Xu, J., Cao, Y., Bai, L., Zhang, Y., Jiang, Y., Luo, X., Xu, J., Xu, X., Zheng, L., Yu, Z., Yang, H., Zhang, Z., Wang, S., Tomberlin, J.K., Zhang, J. and Huang, Y., 2020. Genomic landscape and genetic manipulation of the black soldier fly *Hermetia illucens*, a natural waste recycler. Cell Research 30: 50-60. https://doi.org/10.1038/s41422-019-0252-6

Zhao, X., Zhong, J., Wei, C., Lin, C.-W. and Ding, T., 2017. Current perspectives on viable but non-culturable state in foodborne pathogens. Frontiers in Microbiology 8: 580. https://doi.org/10.3389/fmicb.2017.00580

Zheng, L., Crippen, T., Holmes, L., Singh, B., Pimsler, M.L., Benbow, M.E., Tarone, A.M., Dowd, S., Yu, Z., Vanlaerhoven, S.L., Wood, T.K. and Tomberlin, J.K., 2013. Bacteria mediate oviposition by the black soldier fly, *Hermetia illucens* (L.), (Diptera: Stratiomyidae). Nature Scientific Reports 3: 2563. https://doi.org/10.1038/srep02563

Zwietering, M.H., Straver, J.M. and Van Asselt, E.D., 2016. The range of microbial risks in food processing. In: Lelieveld, H., Holah, J. and Gabric, D. (eds.) Handbook of hygiene control in the food industry, 2nd edition. Woodhead Publishing, Duxford, UK, pp. 43-54. https://doi.org/10.1016/B978-0-08-100155-4.00004-2

Chemical food safety hazards of insects reared for food and feed

A.M. Meyer, N. Meijer, E.F. Hoek-van den Hil and H.J. van der Fels-Klerx*

Wageningen Food Safety Research, Akkermaalsbos 2, 6708 WB Wageningen, The Netherlands; *ine.vanderfels@wur.nl

Abstract

Insects are a promising future source of sustainable proteins within a circular economy. Proving the safety of insects for food and feed is necessary prior to supplying them to the market. This literature review provides a state-of-the-art overview of the chemical food safety hazards for insects reared for food and feed, focusing mainly on transfer of contaminants from the substrate. Contaminants covered are: heavy metals, dioxins and polychlorinated biphenyls, polyaromatic hydrocarbons, pesticides, veterinary drugs, mycotoxins, and plant toxins. The twelve insect species reported as having the largest potential as feed and food in the EU are included. Transfer and bioaccumulation of contaminants depend on the chemical, insect species, life stage, and source of contaminant (spiked vs natural), as well as the particular substrate and rearing conditions. The heavy metals lead, arsenic, mercury, and cadmium can accumulate, whereas mycotoxins and polycyclic aromatic hydrocarbons (PAHs) seem not to accumulate. Mycotoxins and veterinary drugs could be degraded by insects; their metabolic routes need to be further investigated. Data are generally limited, but in particular for PAHs, plant toxins, and dioxins and dioxin-like polychlorinated biphenyls. Further research on chemical safety of different edible insects is therefore warranted.

Keywords

contaminants – edible insects – exposure – literature review

1   Introduction

Before insects are put on the European market as an ingredient for feed and food, they should be proven to be safe for livestock, pets, and humans. Insects can be efficiently reared with a minimal amount of resources on a wide range of substrates, such as organic side streams, at a high conversion rate (Van Huis *et al.*, 2013; Varclas,

2019). Insects emit lower greenhouse gasses and ammonia in comparison to conventional production animals (Oonincx *et al.*, 2011). Furthermore, insects contain high quality protein, amino acids, and vitamins for animal and human health (Rumpold and Schlüter, 2015). Insects can contain high fat fractions, including omega-3 fatty acids, that are essential for fish and human nutrition (Van Huis *et al.*, 2013). In Europe, insects are seen as a novel source of protein for feed and food production, that could help in producing enough food and feed for the growing European population (Bordiean *et al.*, 2020). The use of insect protein can partly replace the heavy import of protein sources from non-European areas.

In the European Union (EU), under Regulation (EC) No 142/2011, seven species of insects are currently legally allowed to be fed to aquaculture animals (EC, 2011a). In principle, there is no restriction on which insect species are allowed to be used in pet food (Regulation (EC) No 1069/2009) and feed for fur animals (Regulation (EC) No 999/2001) (EC, 2001, 2009). To produce and market insects for food, the producer should submit a dossier to the European Commission requesting for approval since insects were not consumed in Europe before the 15th of May 1997 (Regulation (EU) No 2015/2283) (EC, 2015a). Currently, several of the dossiers are under evaluation of EFSA (EC, 2015b). As a part of the dossier, evidence of the safety of insects for human consumption should be included. In particular, the possible presence of food safety hazards, including physical, microbiological, and chemical hazards, should not cause any short- or long-term human health problems. Information on the safety of insects for feed and food have been collected and brought together initially by EFSA (2015) and later in an extensive literature review (Van der Fels-Klerx *et al.*, 2018). Since then, additional experiments have been performed to fill the identified data gaps on the chemical safety of edible insects.

The current review aims to present an updated overview of the potential chemical food safety hazards that could be present in insects reared for feed and food. It is focused on contamination of insects by their exposure to chemicals in the substrates in the rearing phase, because of: (1) the relative importance of this contamination route for the presence of chemicals in harvested insects; (2) limited available data on effects of processing on chemicals; and (3) the relevance of food safety control at the upstream stages of the supply chain. Chemicals are generally stable and difficult to remove or reduce by processing steps, like cooking, further downstream the chain. Therefore, if contaminants do accumulate in insects, this could potentially pose a problem, either within the fat or protein fractions of the insects or the insect products themselves, through the processing, distribution, and consumption stages (Mutungi *et al.*, 2019).

2   Materials and methods

The comprehensive review by Van der Fels-Klerx *et al.* (2018) was completed in the beginning of 2017. Therefore, additional recent published studies were searched for in the current study. The review used the bibliographic databases CAB Abstracts, Web of Science, and Scopus to collect peer-reviewed papers written in the English language and published in the time period 1 January 2017 to 1 September 2020. In the case of limited data available on a particular contaminant group, earlier published papers were also included to provide a complete overview.

The literature search focused on chemical hazards in insects reared for food and/ or feed, thus excluding insects harvested from the wild. Species covered included those considered by EFSA as having the largest potential to be used as food and feed in the EU, being: black soldier fly (*Hermetia illucens*) and common housefly (*Musca domestica*); yellow mealworm (*Tenebrio molitor*), lesser mealworm (*Alphitobius diaperinus*), and giant mealworm (*Zophobas atratus*); house cricket (*Acheta domesticus*), banded cricket (*Gryllodes sigillatus*); greater wax moth (*Galleria mellonella*), lesser wax moth (*Achroia grisella*), silkworm (*Bombyx mori*), African migratory locust (*Locusta migratora migratorioides*), and American grasshopper (*Schistocerca americana*) (EFSA, 2015). Particular attention was given to bioaccumulation of chemicals in the larvae and/or cricket species from the substrate. All types of substrates were considered, also including the ones not currently legally allowed in the EU. Results are presented in the next sections for each of the main group of chemical contaminants separately, including: heavy metals, dioxins and polychlorinated biphenyls, polyaromatic hydrocarbons, pesticides, veterinary drugs, mycotoxins, and plant toxins.

3   Heavy metals

Possible bioaccumulation of several heavy metals has been investigated in various insect exposure studies held under controlled conditions. The focus of these studies was on the heavy metals currently regulated in feed and food in the EU, being; cadmium (Cd), lead (Pb), mercury (Hg), and arsenic (As) (Regulation (EC) No 1881/2006) (EC, 2006a). Table 1 summarises the results from the recent exposure studies involving a substrate contaminated by the four mentioned heavy metals with *H. illucens* and *T. molitor* larvae. Results show that the heavy metal accumulation in insects depends on the type of heavy metal, the insect species, the substrate and, possibly the packaging material of the substrate (Van der Fels-Klerx *et al.*, 2020). For example, bioaccumulation factor (BAF) of Pb in *T. molitor* fed 100% organic wheat flour was 34, whereas it was 6.1 when fed 75% organic wheat meal and 25% organic olive-pomace (Truzzi *et al.*, 2019). The authors speculated that this high accumulation may have been due to elevated levels of Pb in the carrots that had been provided as a water source to all treatments; but this had not been verified by (Truzzi *et al.*, 2019). In addition, *H. illucens* fed vegetable-based substrate in a carton had a BAF of 20.4 for Cd, but BAF for this metal was 7 when *H. illucens* were fed the same vegetable-based substrate but in a plastic container (Van der Fels-Klerx *et al.*, 2020). Furthermore, *H. illucens* fed on seaweed-enriched media accumulated Cd, Pb, Hg, and As (Biancarosa *et al.*, 2018). The two recent studies that investigated As in *H. illucens* reported BAF < 1 (Table 1). Possible As accumulation in *T. molitor* was studied in only one recent study reporting a BAF of 1.1, confirming results from (Van der Fels-Klerx *et al.*, 2016) who reported a BAF of 1.4-2.6 in *T. molitor* larvae.

Gao and co-authors (Gao *et al.*, 2019a) fed *M. domestica* with food waste including different percentages of dish waste, and reported a lower concentration of Cd in the larvae compared to other elements tested, which is congruous with previous reports (Jiang *et al.*, 2017). Currently, not all combinations of the four regulated heavy metals and the insect species considered in this review have been investigated. Therefore, species not investigated yet, such as *A. diaperinus* and *A. domesticus* among others, should be considered in further studies.

TABLE 1   Bioaccumulation of heavy metals in insect larvae

Heavy metal	Bioaccumulation factor	Species	Reference
Arsenic	0.8±0.1	*Hermetia illucens*	Proc *et al.* (2020)
	0.5-1.1	*H. illucens*	Schmitt *et al.* (2019)
	1.1±0.1	*Tenebrio molitor*	Truzzi *et al.* (2019)
	0.88-0.99	*H. illucens*	Truzzi *et al.* (2020)
Cadmium	9.1±1.4	*H. illucens*	Purschke *et al.* (2017)
	3.9±0.6	*H. illucens*	Proc *et al.* (2020)
	2.5±0.1	*H. illucens*	Schmitt *et al.* (2019)
	7-20.4	*H. illucens*	Van der Fels-Klerx *et al.* (2020)
	0.8-1.7	*T. molitor*	Truzzi *et al.* (2019)
	1.8-2.5	*T. molitor*	Mlček *et al.* (2017)
	4.2-6.9	*H. illucens*	Truzzi *et al.* (2020)
Lead	2.3±0.3	*H. illucens*	Purschke *et al.* (2017)
	0.03±0.03	*H. illucens*	Proc *et al.* (2020)
	0.8±0.1	*H. illucens*	Schmitt *et al.* (2019)
	2.1-3.1	*H. illucens*	Van der Fels-Klerx *et al.* (2020)
	5.2-34	*T. molitor*	Truzzi *et al.* (2019)
	1.6-2.3	*H. illucens*	Truzzi *et al.* (2020)
Mercury	0.5	*H. illucens*	Purschke *et al.* (2017)
	1.5±0.1	*H. illucens*	Proc *et al.* (2020)
	1.6±0.1	*H. illucens*	Schmitt *et al.* (2019)
	1.5-6.2	*T. molitor*	Truzzi *et al.* (2019)
	1.4-4.5	*H. illucens*	Truzzi *et al.* (2020)

4   Dioxins/PCBs/PAHs

Dioxins (and furans – PCDD/F) and dioxin-like polychlorinated biphenyls (dl-PCBs) are chemicals currently banned under the Stockholm Convention due to well documented toxic effects throughout the food chain. However, local contamination of these chemicals still exists in the environment, predominantly in soils, sediment, and air, which may result in contamination of crops used for feed or food (Pius et al., 2019). As insects are being reared on a variety of waste streams, these lipophilic pollutants are being investigated for safe insect rearing. The concentration of dioxins and dl-PCBs may substantially concentrate in fat extracts of insects for food/feed. Dioxins and dl-PCBs concentrations were analysed in reared insects for feed, and reported to range from 0.23 to 0.63 ng toxic equivalency factor (TEQ)/ kg dry weight (Charlton et al., 2015). The authors noted that these figures are below the EC maximum allowed content in feed materials of animal origin (1.25 ng WHO-PCDD/ F-PCB-TEQ/kg, considering 88% dry matter) (EC, 2006a). In addition, dioxins and dioxin-like PCBs were analysed in insects and insect products for food and feed. Concentrations ranged from 0.05-0.28 pg WHO-TEQ/g ww in the edible insects, which was two times higher than in the insect derived products, probably, due to dilution of other ingredients within the insect product (Poma et al., 2017). In a recent study in which H. illucens was fed on supermarket waste, larvae were analysed for dioxins and dl-PCBs (Van der Fels-Klerx et al., 2020). These treatments included substrates packaged in either plastic or carton and with or without meat. Concentrations of dioxins and dl-PCBs (WHO-2005-PCDD/F-PCB-TEQ upper bound) were 0.2-0.3 ng TEQ/kg dw for the substrates, and 0.3-0.4 ng TEQ/kg dw for both the larvae and the residual materials. Concentrations of both the substrates and reared H. illucens larvae did not exceed current EC limits for these contaminant groups in feed materials. In the same study, concentrations of polycyclic aromatic hydrocarbons (PAHs) were also analysed. Reported PAH16 (upper bound) concentrations were low, with levels in the substrates of 1.8-6.6 µg/kg dw, in the larvae of 1.9-2.1 µg/kg dw, and in the residues of 2.1-5.2 µg/kg dw (Van der Fels-Klerx et al., 2020). PAHs are chemically stable compounds in the environment which can be formed after burning. Currently, there are no legal limits for the presence of PAHs in feed materials in Europe. However, since these compounds can be carcinogenic and DNA damaging, the concentration of PAHs in insects should be as low as possible. PAHs are increasingly found in vegetable based aquafeed materials and potentially in vegetable based waste streams (Berntssen et al., 2015).

Table 2 shows that dioxins and dl-PCBs as well as PAHs did not accumulate in H. illucens larvae, or only accumulated to very little extent, up to BAF of 2 for dioxins and dl-PCBs and up to 1.2 for PAHs in the various treatments (Van der Fels-Klerx et al., 2020).

TABLE 2   Bioaccumulation of dioxins, dioxin-like polychlorinated biphenyls, and polycyclic aromatic hydrocarbons in *Hermetia illucens* larvae from supermarket returns (Van der Fels-Klerx *et al.*, 2020)

Contaminant	Bioaccumulation factor
Dioxins (WHO2005-PCDD/F-TEQ (ub))[1,2]	1.0-2.0
Sum of dioxins and dioxin-like PCBs (WHO2005-PCDD/F-PCB-TEQ (ub))[1,2]	1.0-2.0
PAH16 (ub)[2]	0.3-1.2

[1] World Health Organization (WHO) toxic equivalency factor (TEQ).
[2] ub = upper bound value (limits of detection used per congener).

Controlled experiments on possible accumulation of dioxins, dl-PCBs, and PAHs have not been identified for the other insect species under consideration of this review. Since these compounds are known to accumulate in fatty tissues of production animals, they should be investigated in future studies (Fries, 1995). Based on the results for *H. illucens* larvae, accumulation of dioxins and dl-PCBs is expected in other insect species used for food and feed.

5   Pesticides

In literature on the effects of reared insects discussed in a previous review (Van der Fels-Klerx *et al.*, 2018), it was largely concluded that tested pesticides did not accumulate in the considered species. These included azoxystrobin and propiconazole (Lalander *et al.*, 2016), chlorpyrifos and chlorpyrifos-methyl, and pirimiphos-methyl (Purschke *et al.*, 2017) in *H. illucens* larvae, and epoxiconazole (Lv *et al.*, 2014), metalaxyl (Gao *et al.*, 2014), benalaxyl (Gao *et al.*, 2013), and mycolobutanil (Lv *et al.*, 2013) in *A. diaperinus* larvae. In addition, Poma and co-authors (Poma *et al.*, 2017) confirmed the presence of pirimiphos-methyl in *Locusta migratoria* and in some insect products including a Buggie burger. Recent literature on the effects of pesticides on insect species reared specifically for food or feed purposes is scarce. Available data suggest that accumulation is most likely not a food safety concern. However, for some pesticides, the presence of low concentrations of pesticides in the substrate, below the EC legal limit, may affect insect growth and survival (unpublished data). More research is needed to determine effects of a larger variety of pesticides that are commonly found in residues of feed materials used for rearing insects.

6   Veterinary drugs

There is only limited data available on the possible presence or accumulation of veterinary drug residues in edible insects. When insects are reared on manure (though currently not allowed in Europe), they could be exposed to residues of veterinary drugs. Furthermore, veterinary drugs may also be used during the rearing of insects to prevent infections. The use of antibiotics may, however, also affect the development of edible insects and the spread of antibiotic-resistant pathogens, which could outweigh the possible benefits of using antibiotics (Grau *et al.*, 2017). Therefore, veterinary drugs could be used during insect rearing, however it is a challenge to find the optimal balance between limiting microbial growth and optimal growth and survival of the insects (Roeder *et al.*, 2010).

Some studies investigated the ability of *H. illucens* larvae to reduce pharmaceuticals in the environment. Cai and co-authors (Cai *et al.*, 2018) studied the mechanisms of tetracycline degradation by the intestinal microflora of *H. illucens*. Non-sterile substrates of moistened wheat bran were spiked with tetracycline in concentrations of 20, 40, and 80 mg/kg dw. It was shown that *H. illucens* could rapidly degrade tetracycline, which was due to the intestinal microbiota of the larvae. More than 75% of tetracycline was degraded at day 8, while at day 12 the reduction was between 95-96%. Several possible biodegradation products were identified in larval intestinal isolates (Cai *et al.*, 2018). Another study (Lalander *et al.*, 2016) showed that the half-life of the pharmaceuticals carbamazepine, roxithromycin, and trimethoprim, as spiked in the substrate, was shorter with *H. illucens* feeding off of the compost substrate and no bioaccumulation was detected in the larvae. Composting of organic waste by *H. illucens* larvae could, therefore, reduce the pharmaceuticals in the environment (Lalander *et al.*, 2016). Biodegradation of antibiotics present in swine manure and chicken manure was also observed with *M. domestica* larvae. Concentrations of nine antibiotics, including tetracyclines, sulfonamides, and fluoroquinolones were clearly decreased during a six-day larvae manure vermicomposting process. The cumulative removal of oxytetracycline, chlortetracycline, and sulfadiazine was around 70% (Zhang *et al.*, 2014). The veterinary antibiotic monensin, a feed additive according to EU regulation (EC, 2020), which is widely used in broiler feed was also reduced in a 12-day vermicomposting experiment with *M. domestica* larvae. After four days, the concentration of monensin in the chicken manure was reduced by 95%, while this reduction took twelve days for the control group. It was concluded that the reduction was due to monensin degrading bacteria in the gut of the larvae (Li *et al.*, 2019).

In other studies, some residues of veterinary drugs were detected in edible insects (Table 3). Several insects intended for human or animal consumption including *A. diaperinus*, *H. illucens*, and *A. domesticus* were screened for the possible presence of 75 (veterinary) drugs, pesticides, and mycotoxins. The veterinary drugs salicylic acid and metoprolol were detected in the three insect species (1-3 μg/kg; De

TABLE 3   Possible accumulation of veterinary drugs in insect larvae

Veterinary drugs	Results / accumulation	Species	References
Tetracycline	Degradation in the substrate by microbiota of *H. illucens*	*Hermetia illucens*	Cai *et al.* (2018)
Roxithromycin, trimethoprim, and carbamazepine	Degradation in the substrate; no accumulation in the larvae	*H. illucens*	Lalander *et al.* (2016)
Tetracycline, oxytetracycline, chlortetracycline, doxycycline, sulfadiazine, norfloxacin, ofloxacin, ciprofloxacin, and enrofloxacin	Degradation in the substrate	*Musca domestica*	Zhang *et al.* (2014)
Monensin[1]	Degradation in the substrate by the microbiota	*M. domestica*	Li *et al.* (2019)
Salicylic acid, metoprolol	Low levels found in insects intended for human or animal consumption	*H. illucens, Alphitobius diaperinus*	De Paepe *et al.* (2019)
Sulfonamide	Sulfamonomethoxine, sulfamethoxazole, and sulfamethazine were not detected, while only sulfadiazine was detected in the prepupae	*H. illucens*	Gao *et al.* (2019b)

[1] According to EU regulation monensin is a feed additive (EC, 2020).

Paepe *et al.*, 2019). Furthermore, it was shown that the antibiotic sulfonamide could affect the growth of *H. illucens*. Sulfonamide was spiked at concentrations of 0, 0.1, 1, and 10 mg/kg in the substrate. Only the highest concentration of sulfonamide affected the survival of the larvae, resulting in 30% survival. The body weight and development of the larvae were also affected by sulfonamide. Sulfamonomethoxine, sulfamethoxazole, and sulfamethazine were not detected in the prepupae, while only sulfadiazine was detected with a treatment of 1 and 10 mg/kg (0.5-0.8 mg/100 prepupae). It was concluded that *H. illucens* larvae can be used to partly remove veterinary drugs from manure to protect the environment (Gao *et al.*, 2019b).

From the limited data available, it can be concluded that residues of some veterinary drugs could be found in edible insects and could affect the larval growth, however, it has also been shown that veterinary drugs can be degraded by insects, possibly due to the gut microbiota. More research is needed on the effects of residues of specific veterinary drugs as present in potential substrates for insect rearing, in combination with the particular insect species reared on that substrate.

7   Other environmental contaminants

Poma and co-authors (Poma *et al.*, 2019) investigated the contamination levels of a large variety of insect food products derived from different insect species, purchased from five European countries (Austria, France, UK, Belgium, and the Netherlands) and three Asian countries (P.R. China, Japan, and R. Korea). A variety of species belonging to six orders, being Orthoptera, Coleoptera, Lepidoptera, Hemiptera, Odonata, and Hymenoptera, were analysed. The list of targeted compounds consisted of 31 persistent organic pollutants (POPs), including: 20 polychlorinated biphenyls (PCBs) and 11 organochlorine compounds (OCPs); 11 halogenated flame retardants (HFRs), including 9 polybrominated diphenyl ethers (PBDEs) and 2 dechlorane plus (DPs); 18 plasticisers, including 7 legacy plasticisers (LPs), and 11 alternative plasticisers (APs); 17 phosphorous flame retardants (PFRs), including 12 legacy PFR, and 5 emerging PFRs (ePFRs); 8 LP biotransformation products (LPs-BT), 11 AP biotransformation products (APs-BT), and 12 PFR biotransformation products (PFRs-BT). The authors concluded that contamination varied between insect species and products, nevertheless levels were generally low. They speculated that industrial post-harvesting handling and other ingredients may have contributed more to elevated contaminant levels in the insect products rather than the insects themselves. Bioaccumulation of contaminants was not investigated in this study.

8 Mycotoxins

Mycotoxins are secondary metabolites produced by certain fungal species, mostly from the genus *Aspergillus* spp., *Fusarium* spp., and *Penicillium* spp., that are toxic to animals and humans. Certain mycotoxins, such as aflatoxin B_1 (AFB1), deoxynivalenol (DON), zearalenone (ZEN), ochratoxin A (OTA), and fumonisins (FB1, FB2, FG1, FG2), are known to pose threats to livestock and human health (Gashaw, 2016). Therefore, the presence of these mycotoxins in feed and/or food products has been regulated or guidance levels have been established in Europe (EC, 2006a,b, 2011b).

Possible accumulation of mycotoxins in insects has been investigated in several studies with spiked and naturally contaminated substrates from different feed materials and waste streams, and with different insect species. Results from these studies lead to a better understanding of mycotoxin uptake, transformation, and excretion by insects for feed and food (Leni *et al.*, 2019). Results of recent studies showed that tolerance to mycotoxins varies, depending on insect species and mycotoxins. For example, *H. illucens* fed poultry feed spiked with AFB1 did not accumulate this toxin, and levels in the larvae were below the detection limit (0.10 µg/kg) of the analytical methods used (Table 4). However, *T. molitor* larvae fed with the same spiked substrate contained up to 1.44 µg/kg of AFB1, which is about 10% of the EU's legal limit for feed materials (Bosch *et al.*, 2017).

None of the considered mycotoxins, as presented in Table 4, accumulated in the insect species considered. *H. illucens*, *T. molitor*, and *A. diaperinus* were fed different substrates at varying concentrations of mycotoxins, with several being well above

TABLE 4   Possible accumulation of mycotoxins in insect larvae

Mycotoxins	Concentration in substrate (µg/kg)	Concentration in larvae (µg/kg)[1]	Species	References
Aflatoxin B_1	415	< LOD[2], 10% legal limit[3]	Hermetia illucens, Tenebrio molitor	Bosch et al. (2017)
	390	< LOQ[2,4]	H. illucens, Alphitobius diaperinus	Camenzuli et al. (2018)
	13	< LOD	H. illucens	Purschke et al. (2017)
Zearalenone	14.9-79.9	< LOQ	T. molitor	Niermans et al. (2019)
		< LOD[2], 60[3], < LOD[5]	H. illucens, T. molitor, Acheta domesticus	De Paepe et al. (2019)
	13,000	> LOQ[2], < LOQ[4]	H. illucens, A. diaperinus	Camenzuli et al. (2018)
	39	< LOD	H. illucens	Purschke et al. (2017)
	173	< LOD	T. molitor	Leni et al. (2019)
Deoxynivalenol	12,000	< 131	T. molitor	Ochoa Sanabria et al. (2019)
	112,000	> LOQ[2], < LOQ[4]	H. illucens, A. diaperinus	Camenzuli et al. (2018)
	698	< LOD	H. illucens	Purschke et al. (2017)
	779[2], 1,207[3]	< LOD[2], 726[3]	H. illucens, T. molitor	Leni et al. (2019)
Ochratoxin A	1,700	> LOQ[2], < LOQ[4]	H. illucens, A. diaperinus	Camenzuli et al. (2018)
	130	< LOD	H. illucens	Purschke et al. (2017)
Fumonisin B_1 and B_2	573[2], 727[3], 441[2], 294[3]	< LOD[2], < LOD[3], < LOD[2], < LOD[3]	H. illucens, T. molitor	Leni et al. (2019)

[1] LL = legal limit; LOQ = limit of quantification; LOD = limit of detection.
[2] H. illucens: initial/larval concentration.
[3] T. molitor: initial/larval concentration.
[4] A. diaperinus: initial/larval concentration.
[5] A. domesticus: initial/larval concentration.

the EC legal limit for the maximum presence of the particular toxin in feed materials (EC, 2003). For example, Camenzuli and co-authors (Camenzuli et al., 2018) fed A. diaperinus and H. illucens a wheat-based substrate contaminated with 390 µg/kg AFB1, which is 20 times the EC legal limit for AFB1 in feed materials. Not all combinations of mycotoxins and different insect species have been investigated. Based on results of studies performed so far, accumulation of mycotoxins is not expected in insects. However, this needs to be confirmed in future studies involving the species that have not been investigated so far, such as *A. domesticus, L. migratoria,* and *M. domestica.*

Even though mycotoxins were seldom found in the insect body, they were regularly found in the residues from excretion throughout these studies. Signifying, biotransformation of the mycotoxins was reported, resulting in varying levels of metabolites. For example, in a study in which *T. molitor* were fed with 79.9 µg/kg ZEN, the residual materials was found to contain 26.2 µg/kg ZEN, 6.8 µg/kg α-ZEL, and 17.3 µg/kg β-ZEL (Niermans *et al.,* 2019). In addition, when *H. illucens* were fed with 698 µg/kg DON, it resulted in 1,136 µg/kg DON found in the residual materials after excretion. This could be due to masked mycotoxins, which means that the mycotoxins could be bound to a carbohydrate or protein matrix (Purschke *et al.,* 2017). Recently, several studies on the conversion of mycotoxins with phase I and II enzymes have been performed (Meijer *et al.,* 2019). Additional studies of processes of different phase I and II enzymes regarding mycotoxin conversion are recommended to further unravel transformation of mycotoxins by insects.

9 Plant toxins

Plant toxins is referred to as the collective group of secondary metabolites in plants which have toxicological properties. The most known plant toxins include cyanogenic glycosides and alkaloids. Cyanogenic glycosides can release hydrogen cyanide when chewed or digested. Alkaloids in plants are nitrogen storage compounds that are involved to protect them against predators, function as growth regulators, and substitutes for minerals like potassium and calcium. Plants naturally synthesise alkaloid compounds based on their needs. Some of the synthesised alkaloids like pyrrolizidine, indolizidine, piperidine, and tropane alkaloids have been documented to cause toxic effects on animals and humans (Schramm *et al.,* 2019). Macel (2011) cites older studies that report that several insect species sequester Pyrrolizidine Alkaloids for their own defence, and that concentrations of these compounds in insects can exceed concentrations in the plant, from which they feed. The insect species currently being reared for food or feed are not known to exhibit this behaviour in the live stage in which they are harvested (EFSA, 2015). Bioaccumulation of plant toxins from substrates in insects for feed and food has, however, not been investigated so far.

10 Conclusions and recommendations

This review provides recent information on the chemical safety of insects reared for feed and food use, primarily focusing on accumulation of contaminants from substrates into reared insects. Generally, available data are fragmented over a wide range of contaminants and insect species, with most data collected for *H. illucens* and *T. molitor*. Factors related to the rearing phase, such as insect species, life stage, and source of the contaminant (spiked or naturally contaminated) were confirmed to affect the accumulation of contaminants in insects. Our results showed that, in addition, aspects related to the experimental setting may play a role, as well as the substrate type.

For most contaminants for which experimental data has been collected, an effect on growth and survival of insects was not observed, except for veterinary drugs and pesticides, which may lead to undesirable production effects. Bioaccumulation of some contaminants including the heavy metals, cadmium and lead can occur, and differences between species in accumulation have been observed. Dioxins and dl-PCBs and plant toxins could also potentially accumulate in insects but limited or no data, depending on the insect species, are available for these contaminants. Accumulation of mycotoxins and PAHs has not been observed so far, though, data on the later compound group are very limited. Further research is, therefore, recommended on possible accumulation of plant toxins, dioxins and dl-PCBs, and PAHs from substrates into insects. Furthermore, research on metabolic pathways of mycotoxins and veterinary drugs in insects, regarding possible detoxification/bioactivation pathways is recommended to unravel underlying mechanisms.

Acknowledgements

This project has received funding from the European Union's Horizon 2020 research and innovation programme under grant agreement No 861976. Additional financing from the Netherlands Ministry of Agriculture, Nature and Food Quality (Knowledge base program KB34, project KB-34-006-001) is acknowledged.

Conflicts of interest

The authors declare no conflict of interest.

References

Berntssen, M.H.G., Ørnsrud, R., Hamre, K. and Lie, K.K., 2015. Polyaromatic hydrocarbons in aquafeeds, source, effects and potential implications for vitamin status of farmed fish species: a review. Aquaculture Nutrition 21: 257-273. https://doi.org/10.1111/anu.12309

Biancarosa, I., Liland, N.S., Biemans, D., Araujo, P., Bruckner, C.G., Waagbø, R., Torstensen, B.E., Lock, E.-J. and Amlund, H., 2018. Uptake of heavy metals and arsenic in black soldier fly (*Hermetia illucens*) larvae grown on seaweed-enriched media. Journal of the Science of Food and Agriculture 98: 2176-2183. https://doi.org/10.1002/jsfa.8702

Bordiean, A., Krzyżaniak, M., Stolarski, M.J., Czachorowski, S. and Peni, D., 2020. Will yellow mealworm become a source of safe proteins for Europe? Agriculture 10: 1-30. https://doi.org/10.3390/agriculture10060233

Bosch, G., Van der Fels-Klerx, H.J., De Rijk, T.C. and Oonincx, D.G.A.B., 2017. Aflatoxin B_1 tolerance and accumulation in black soldier fly larvae (*Hermetia illucens*) and yellow mealworms (*Tenebrio molitor*). Toxins 9: 185. https://doi.org/10.3390/toxins9060185

Cai, M., Hu, R., Zhang, K., Ma, S., Zheng, L., Yu, Z. and Zhang, J., 2018. Resistance of black soldier fly (Diptera: Stratiomyidae) larvae to combined heavy metals and potential application in municipal sewage sludge treatment. Environmental Science and Pollution Research International 25: 1559-1567. https://doi.org/10.1007/s11356-017-0541-x

Camenzuli, L., Van Dam, R., de Rijk, T., Andriessen, R., Van Schelt, J. and Van der Fels-Klerx, H.J., 2018. Tolerance and excretion of the mycotoxins aflatoxin B_1, zearalenone, deoxynivalenol, and ochratoxin A by *Alphitobius diaperinus* and *Hermetia illucens* from contaminated substrates. Toxins 10: 91. https://doi.org/10.3390/toxins10020091

Charlton, A.J., Dickinson, M., Wakefield, M., Fitches, E., Kenis, M., Han, R., Zhu, F., Kone, N., Grant, M., Devic, E., Bruggeman, G., Prior, R. and Smith, R., 2015. Exploring the chemical safety of fly larvae as a source of protein for animal feed. Journal of Insects as Food and Feed 1: 7-16. https://doi.org/10.3920/JIFF2014.0020

De Paepe, E., Wauters, J., Van Der Borght, M., Claes, J., Huysman, S., Croubels, S. and Vanhaecke, L., 2019. Ultra-high-performance liquid chromatography coupled to quadrupole orbitrap high-resolution mass spectrometry for multi-residue screening of pesticides, (veterinary) drugs and mycotoxins in edible insects. Food Chemistry 293: 187-196. https://doi.org/10.1016/j.foodchem.2019.04.082

EFSA, 2015. Risk profile related to production and consumption of insects as food and feed. EFSA Journal 13(10): 4257. https://doi.org/10.2903/j.efsa.2015.4257

European Commission (EC), 2001. Regulation (EC) No 999/2001 laying down rules for the prevention, control and eradication of certain transmissible spongiform encephalopathies. Official Journal of the European Communities L 147: 1.

European Commission (EC), 2003. Directive 2002/32/EC of the European Parliament and of the Council of 7 May 2002 on undesirable substances in animal feed. Official Journal of the European Communities L140: 10-22.

European Commission (EC), 2006a. Commission Regulation (EC) No 1881/2006 of 19 December 2006 setting maximum levels for certain contaminants in foodstuffs. Official Journal of the European Communities L70: 1.

European Commission (EC), 2006b. Commission Recommendation (EC) No 576/2006 on the presence of deoxynivalenol, zearalenone, ochratoxin A, T-2 and HT-2 and fumonisins in products intended for animal feeding. Official Journal of the European Communities L 229: 1-3.

European Commission (EC), 2009. Regulation (EC) No 1069/2009 laying down health rules as regards animal by-products and derived products not intended for human consumption and repealing Regulation (EC) No 1774/2002 (Animal by-products Regulation). Official Journal of the European Communities L 300: 1.

European Commission (EC), 2011a. Commission Regulation (EU) No 142/2011 implementing Regulation (EC) No 1069/2009 of the European Parliament and of the Council laying down health rules as regards animal by-products and derived products not intended for human consumption and implementing Council Directive 97/78/EC as regards certain samples and items exempt from veterinary checks at the border under that Directive. Official Journal of the European Union L 54: 1.

European Commission (EC), 2011b. Commission Regulation (EU) No 574/2011 of 16 June 2011 amending Annex I to Directive 2002/32/EC of the European Parliament and of the Council as regards maximum levels for nitrite, melamine, Ambrosia spp. and carry-over of certain coccidiostats and histomonostats and consolidating Annexes I and II thereto. Official Journal of the European Union L 159: 7.

European Commission (EC), 2015a. Regulation (EU) 2015/2283 on novel foods, amending Regulation (EU) No 1169/2011 of the European Parliament and of the Council and repealing Regulation (EC) No 258/97 of the European Parliament and of the Council and Commission Regulation (EC) No 1852/2001. Official Journal of the European Union L 327: 1

European Commission (EC), 2015b. Summary of applications and notifications. European Commission, Brussels, Belgium. Available at: https://tinyurl.com/y25se4or.

European Commission (EC), 2020. Commission Implementing Regulation (EU) 2020/994 of 9 July 2020 concerning the authorisation of monensin and nicarbazin (Monimax) as a feed additive for turkeys for fattening, chickens for fattening and chickens reared for laying (holder of authorisation Huvepharma NV). Official Journal of European Union 221: 79-82.

Fries, G., 1995. A review of the significance of animal food products as potential pathways of human exposures to dioxins. Journal of animal science 73: 1639-1650. https://doi.org/10.2527/1995.7361639x

Gao, M., Lin, Y., Shi, G.-Z., Li, H.-H., Yang, Z.-B., Xu, X.-X., Xian, J.-R., Yang, Y.-X. and Cheng, Z., 2019a. Bioaccumulation and health risk assessments of trace elements in housefly (*Musca domestica* L.) larvae fed with food wastes. Science of the Total Environment 682: 485-493. https://doi.org/10.1016/j.scitotenv.2019.05.182

Gao, Q., Deng, W., Gao, Z., Li, M., Liu, W., Wang, X. and Zhu, F., 2019b. Effect of sulfonamide pollution on the growth of manure management candidate *Hermetia illucens*. PLoS ONE 14: e0216086. https://doi.org/10.1371/journal.pone.0216086

Gao, Y., Chen, J., Liu, C., Lv, X., Li, J. and Guo, B., 2013. Enantiomerization and enantioselective bioaccumulation of benalaxyl in *Tenebrio molitor* larvae from wheat bran. Journal of agricultural and food chemistry 61. https://doi.org/10.1021/jf4020125

Gao, Y., Wang, H., Qin, F., Xu, P., Lv, X., Li, J. and Guo, B., 2014. Enantiomerization and enantioselective bioaccumulation of metalaxyl in *Tenebrio molitor* larvae. Chirality 26: 88-94.

Gashaw, M., 2016. Review on mycotoxins in feeds: implications to livestock and human health. Journal of Agricultural Research and Development 5: 137-144.

Grau, T., Vilcinskas, A. and Joop, G., 2017. Sustainable farming of the mealworm *Tenebrio molitor* for the production of food and feed. Zeitschrift fur Naturforschung. Section C, Biosciences 72: 337-349.

Jiang, C., Teng, C., Li, J., Zhu, J., Fen, D., Lou, L. and Zhang, Z., 2017. The effectiveness of bioconversion of food waste by housefly larvae. Chinese Journal of Applied and Environmental Biology 23: 1159-1165. https://doi.org/10.3724/SP.J.1145.2017.03018

Lalander, C., Senecal, J., Gros Calvo, M., Ahrens, L., Josefsson, S., Wiberg, K. and Vinnerås, B., 2016. Fate of pharmaceuticals and pesticides in fly larvae composting. Science of the Total Environment 565: 279-286. 10.1016/j.scitotenv.2016.04.147

Leni, G., Cirlini, M., Jacobs, J., Depraetere, S., Gianotten, N., Sforza, S. and Dall'Asta, C., 2019. Impact of naturally contaminated substrates on *Alphitobius diaperinus* and *Hermetia illucens*: uptake and excretion of mycotoxins. Toxins 11. https://doi.org/10.3390/toxins11080476

Li, H., Wan, Q., Zhang, S., Wang, C., Su, S. and Pan, B., 2019. Housefly larvae (*Musca domestica*) significantly accelerates degradation of monensin by altering the structure and abundance of the associated bacterial community. Ecotoxicology and Environmental Safety 170: 418-426.

Lv, X., Liu, C., Li, Y., Gao, Y., Guo, B., Wang, H. and Li, J., 2013. Bioaccumulation and excretion of enantiomers of myclobutanil in *Tenebrio molitor* larvae through dietary exposure. Chirality 25: 890-896.

Lv, X., Liu, C., Li, Y., Gao, Y., Wang, H., Li, J. and Guo, B., 2014. Stereoselectivity in bioaccumulation and excretion of epoxiconazole by mealworm beetle (*Tenebrio molitor*) larvae. Ecotoxicology and environmental safety 107: 71-76.

Macel, M., 2011. Attract and deter: a dual role for pyrrolizidine alkaloids in plant-insect interactions. Phytochemistry Reviews 10: 75-82.

Meijer, N., Stoopen, G., Van der Fels-Klerx, H.J., van Loon, J., Carney, J. and Bosch, G., 2019. Aflatoxin B_1 conversion by black soldier fly (*Hermetia illucens*) larval enzyme extracts. Toxins 11: 532. https://doi.org/10.3390/toxins11090532

Mlček, J., Adámek, M., Adámková, A., Borkovcová, M., Bednářová, M. and Skácel, J., 2017. Detection of selected heavy metals and micronutrients in edible insect and their dependency on the feed using XRF spectrometry. Potravinarstvo 11(1): 725-730. https://doi.org/10.5219/850

Mutungi, C., Irungu, F.G., Nduko, J., Mutua, F., Affognon, H., Nakimbugwe, D., Ekesi, S. and Fiaboe, K.K.M., 2019. Postharvest processes of edible insects in Africa: a review of processing methods, and the implications for nutrition, safety and new products development. Critical Reviews in Food Science and Nutrition 59: 276-298. https://doi.org/10.1080/10408398.2017.1365330

Niermans, K., Woyzichovski, J., Kröncke, N., Benning, R. and Maul, R., 2019. Feeding study for the mycotoxin zearalenone in yellow mealworm (*Tenebrio molitor*) larvae – investigation of biological impact and metabolic conversion. Mycotoxin Research 35. https://doi.org/10.1007/s12550-019-00346-y

Ochoa Sanabria, C., Hogan, N., Madder, K., Gillott, C., Blakley, B., Reaney, M., Beattie, A. and Buchanan, F., 2019. Yellow mealworm larvae (*Tenebrio molitor*) fed mycotoxin-contaminated wheat-a possible safe, sustainable protein source for animal feed? Toxins 11. https://doi.org/10.3390/toxins11050282

Oonincx, D.G.A.B., Van Itterbeeck, J., Heetkamp, M.J.W., Van den Brand, H., Van Loon, J.J.A. and Van Huis, A., 2011. An exploration on greenhouse gas and ammonia production by insect species suitable for animal or human consumption. PLoS ONE 5: e14445. https://doi.org/10.1371/journal.pone.0014445

Pius, C., Sichilongo, K., Koosaletse Mswela, P. and Dikinya, O., 2019. Monitoring polychlorinated dibenzo-p-dioxins/dibenzofurans and dioxin-like polychlorinated biphenyls in Africa since the implementation of the Stockholm Convention – an overview. Environmental Science and Pollution Research 26: 101-113. https://doi.org/10.1007/s11356-018-3629-z

Poma, G., Cuykx, M., Amato, E., Calaprice, C., Focant, J.F. and Covaci, A., 2017. Evaluation of hazardous chemicals in edible insects and insect-based food intended for human consumption. Food and Chemical Toxicology 100: 70-79. https://doi.org/10.1016/j.fct.2016.12.006

Poma, G., Yin, S., Tang, B., Fujii, Y., Cuykx, M. and Covaci, A., 2019. Occurrence of selected organic contaminants in edible insects and assessment of their chemical safety. Environmental Health Perspectives 127: 127009. https://doi.org/10.1289/ehp5782

Proc, K., Bulak, P., Wiącek, D. and Bieganowski, A., 2020. *Hermetia illucens* exhibits bioaccumulative potential for 15 different elements – implications for feed and food production. Science of The Total Environment 723: 138125. https://doi.org/10.1016/j.scitotenv.2020.138125

Purschke, B., Scheibelberger, R., Axmann, S., Adler, A. and Jäger, H., 2017. Impact of substrate contamination with mycotoxins, heavy metals and pesticides on the growth performance and composition of black soldier fly larvae (*Hermetia illucens*) for use in the feed and food value chain. Food Additives and Contaminants Part A 34: 1410-1420. https://doi.org/10.1080/19440049.2017.1299946

Roeder, K.A., Kuriachan, I., Vinson, S.B. and Behmer, S.T., 2010. Evaluation of a microbial inhibitor in artificial diets of a generalist caterpillar, *Heliothis virescens*. Journal of Insect Science 10: 197. https://doi.org/10.1673/031.010.19701

Rumpold, B.A. and Schlüter, O., 2015. Insect-based protein sources and their potential for human consumption: nutritional composition and processing. Animal Frontiers 5: 20-24. https://doi.org/10.2527/af.2015-0015

Schmitt, E., Belghit, I., Johansen, J., Leushuis, R., Lock, E.-J., Melsen, D., Ramasamy Shanmugam, R.K., Van Loon, J. and Paul, A., 2019. Growth and safety assessment of feed streams for black soldier fly larvae: a case study with aquaculture sludge. Animals 9: 189. https://doi.org/10.3390/ani9040189

Schramm, S., Köhler, N. and Rozhon, W., 2019. Pyrrolizidine alkaloids: biosynthesis, biological activities and occurrence in crop plants. Molecules 24: 498. https://doi.org/10.3390/molecules24030498

Truzzi, C., Annibaldi, A., Girolametti, F., Giovannini, L., Riolo, P., Ruschioni, S., Olivotto, I. and Illuminati, S., 2020. A chemically safe way to produce insect biomass for possible application in feed and food production. International Journal of Environmental Research and Public Health 17(6): 2121. https://doi.org/10.20944/preprints202003.0056.v1

Truzzi, C., Illuminati, S., Girolametti, F., Antonucci, M., Scarponi, G., Ruschioni, S., Riolo, P. and Annibaldi, A., 2019. Influence of feeding substrates on the presence of toxic metals (Cd, Pb, Ni, As, Hg) in larvae of *Tenebrio molitor*: risk assessment for human consumption. International Journal of Environmental Research and Public Health 16(23): 4815. https://doi.org/10.3390/ijerph16234815

Van der Fels-Klerx, H.J., Camenzuli, L., Belluco, S., Meijer, N. and Ricci, A., 2018. Food safety issues related to uses of insects for feeds and foods. Comprehensive Reviews in Food Science and Food Safety 17: 1172-1183. https://doi.org/10.1111/1541-4337.12385

Van der Fels-Klerx, H.J., Camenzuli, L., Van der Lee, M.K. and Oonincx, D.G.A.B., 2016. Uptake of cadmium, lead and arsenic by *Tenebrio molitor* and *Hermetia illucens* from contaminated substrates. PLoS ONE 11: e0166186. https://doi.org/10.1371/journal.pone.0166186

Van der Fels-Klerx, H.J., Meijer, N., Nijkamp, M.M., Schmitt, E. and Van Loon, J.J.A., 2020. Chemical food safety of using supermarket returns for rearing black soldier fly larvae (*Hermetia illucens*) for feed and food. Journal of Insects as Food and Feed 6: 475-488. https://doi.org/10.3920/JIFF2020.0024

Van Huis, A., Itterbeeck, J.V., Klunder, H., Mertens, E., Halloran, A., Muir, G. and Vantomme, P., 2013. Edible insects: future prospects for food and feed security. Food and Agriculture Organization of the United Nations, Rome, Italy.

Varelas, V., 2019. Food wastes as a potential new source for edible insect mass production for food and feed: a review. Fermentation 5: 81. https://doi.org/10.3390/fermentation5030081

Zhang, Z., Shen, J., Wang, H., Liu, M., Wu, L., Ping, F., He, Q., Li, H., Zheng, C. and Xu, X., 2014. Attenuation of veterinary antibiotics in full-scale vermicomposting of swine manure via the housefly larvae (*Musca domestica*). Scientific Reports 4: 6844. https://doi.org/10.1038/srep06844

Edible insects and food safety: allergy

J.C. Ribeiro[1], B. Sousa-Pinto[2,3,4], J. Fonseca[2,4,5], S. Caldas Fonseca[1] and L.M. Cunha[1]*

[1]GreenUPorto – Sustainable Agrifood Production Research Centre, DGAOT, Faculty of Sciences, University of Porto, Campus de Vairão, Rua da Agrária 747, 4485-646 Vila do Conde, Portugal; [2]MEDCIDS – Department of Community Medicine, Information and Health Decision Sciences, Faculty of Medicine, University of Porto, Rua Dr. Plácido da Costa, 4200-450 Porto, Portugal; [3]Laboratory of Immunology, Basic and Clinical Immunology Unit, Faculty of Medicine, University of Porto, Rua Dr. Plácido da Costa, 4200-450 Porto, Portugal; [4]CINTESIS – Centre for Health Technology and Services Research, Rua Dr. Plácido da Costa, s/n 4200-450 Porto, Portugal; [5]Allergy Unit, CUF Porto Institute & Hospital, Estrada da Circunvalação 14341, 4100-180 Porto, Portugal; *lmcunha@fc.up.pt

Abstract

Edible insects are a unique food source, requiring extensive allergenic risk assessment before its safe introduction in the food market. In a recent systematic review, crustacean allergic subjects were identified as a risk group due to cross-reactivity mainly mediated by tropomyosin and arginine kinase. Immunologic co-sensitisation to house dust mites (HDM) was also demonstrated, but its clinical significance and molecular mechanisms were unclear. Furthermore, case reports of food allergy to insects were also analysed but lack of contextual information hindered the analysis. The main goal of this review is to provide an update of new information regarding food allergy caused by insects, covering relevant topics considering the guidelines for allergic risk assessment in novel foods. Newly published studies have further confirmed the role of tropomyosin as a cross-reactive allergen between edible insects and crustaceans, although there are some questions regarding the immunoglobulin E (IgE)-reactivity of this allergen in mealworm species. Furthermore, only specific treatments (enzymatic hydrolysis combined with thermal treatments) were able to eliminate IgE-reactivity of edible insects. Primary sensitisation (e.g. to *Tenebrio molitor*) has also been shown to be an important pathway for the development of food allergies, with responsible allergens being dependent on the route of sensitisation. However, more studies are necessary to better understand the potential of primary sensitisation causing cross-reactivity with other insect species, crustaceans or HDM. The clinical significance and molecular mechanisms involved in cross-reactivity between edible insects and HDM are still unclear, and a major focus should be given to better understand which allergens cause co-sensitisations between HDM and edible insects and what is the risk of HDM-only allergic subjects consuming edible insects. Contextual information about the reported cases of allergic reactions to insects have further demonstrated that insect-rearing workers and subjects with allergic diseases (in particular, food allergy to crustaceans) are the major risk groups.

Keywords

case reports – cross-reactivity – entomophagy – primary sensitisation – tropomyosin

1   Introduction

Immunoglobulin-E (IgE)-mediated food allergy can be described as an adverse reaction of the immune system to specific proteins in foods which are usually harmless (De Gier and Verhoeckx, 2018; Messina and Venter, 2020). It is estimated that 3-10% of adults and 8% of children worldwide have a food allergy, with most reactions being caused by milk, egg, peanut, tree nuts, fish, soy, wheat or crustaceans (Boyce *et al.*, 2010; Messina and Venter, 2020). These IgE-mediated reactions occur after the consumption of the food product with an onset of up to 2 hours after the consumption, with their presentations ranging from isolated cutaneous or abdominal symptoms to potential fatal reactions such as anaphylaxis (Wang and Sampson, 2011). Food allergies are developed in two phases. Firstly, in the sensitisation phase, susceptible individuals are exposed to an allergen (usually through consumption) and produce specific IgE antibodies to that allergen. Afterwards, following repeated exposure to the same allergen, IgE antibodies on the surface of mast cells recognise the specific allergen, cross-link, and activate an immunologic response (Muraro *et al.*, 2014). However, reactions can also occur due to cross-reactivity, which is defined as when IgE antibodies originally raised against one allergen bind to another structurally-related allergen. Cross-reactivity occurs frequently between allergens from taxonomically related species due to the existence of pan-allergens (proteins that are highly preserved from an evolutionary point of view, and capable of inducing allergic responses in related species) (García and Lizaso, 2011; Migueres *et al.*, 2014). In order to confirm cross-reactivity it is usually necessary to perform inhibition assays, otherwise it is recommended to use the term co-sensitisation, which consists on the simultaneous presence of different IgEs that bind to allergens that may not necessarily have common structural features (Migueres *et al.*, 2014).

Allergic reactions subsequent to insect consumption can be associated to cross-reactivity. This reaction may occur due to the phylogenetic relationship of insects with common allergen sources such as crustaceans or house dust mites

(HDM) (Pennisi, 2015). In fact, cross-reactivity with crustaceans has been demonstrated to be clinically relevant, with the main cross-reacting allergens identified being the arthropod pan-allergens tropomyosin and arginine kinase (AK). On the other hand, co-sensitisation between edible insects and HDM has been shown, but the underlying molecular mechanisms and clinical significance remain unclear. Allergic reactions to edible insects can also be associated to primary sensitisation (either through environmental (Pomés *et al.*, 2017) or occupational (Stanhope *et al.*, 2015) exposure) – several allergens have been already identified and characterised, namely AK (cockroaches, silkworm and indianmeal moth), tropomyosin (cockroaches, mosquito, termite, silverfish), aspartic protease, hemocyanin, glutathione S-transferase, troponin C, myosin light chain, serine protease and α-amylase (cockroaches). Moreover, edible insect allergens were reported to have similar behaviours to crustacean allergens in response to enzymatic and thermal treatments (De Gier and Verhoeckx, 2018; Jeong and Park, 2020; Ribeiro *et al.*, 2018). While epidemiological data and even case reports are still scarce, and often lacking in contextual information (De Gier and Verhoeckx, 2018; Ribeiro *et al.*, 2018), there have been reports indicating that insects are responsible for 4.2-19.4% of cases of food allergies in Asian countries (Ribeiro *et al.*, 2018), and that silkworm pupae is a major culprit of food allergies in China (Ji *et al.*, 2008) and Korea (Jeong and Park, 2020).

The legal status of edible insects as a novel food (Belluco *et al.*, 2017) prompts the need for an in-depth risk assessment – including the allergenic risk – so that they can be commercialised in the European Union food market. Although there is not any established protocol for allergenicity assessment of novel foods, the current guidelines are based on weight-of-evidence approach, taking into account such different issues as: (1) the history of allergic reactions to the novel food; (2) the taxonomy of the novel food (to identify possible relations with known allergic sources); and (3) the identification and characterisation of proteins of the novel food (with assessment of their allergenic potential through bioinformatics assays, comparing them to known allergens). In addition, the IgE-binding capacity of the novel food has also to be assessed, using serum form individuals allergic to other sources (for cross-reactivity) or serum from individuals sensitised to the novel food (primary sensitisation). It is also important to identify possible IgE-binding proteins and to determine the biological activity of such proteins (if they can activate an immunologic response), either through functional tests (such as basophil activation tests; BAT) or food challenges. Further tests also include the evaluation of thermal and chemical (e.g. resistance to enzymatic digestion) treatments on the allergenic properties of a novel food.(Mazzucchelli *et al.*, 2018; Verhoeckx *et al.*, 2016; Westerhout *et al.*, 2019).

The relative novelty of this theme implies that new information is being constantly published, and that the state of the art needs to be updated frequently. Therefore, in this study, we aim to update our previous review (Ribeiro *et al.*, 2018), assessing the new scientific developments related to the allergic risks of insects as food. Specifically, we aimed to cover all the relevant topics related to allergic risk assessment of edible insects including the mechanisms and allergens implied both in primary sensitisation and in cross-reactivity with crustaceans or HDM and the effects of food processing on edible insects' allergenicity. In addition, we aimed to assess epidemiological studies and case series/reports of allergic reactions following insect consumption.

2   Methods

The methodology applied in this study was based on the previous systematic review performed by the authors (Ribeiro *et al.*, 2018). In brief, a systematic search was conducted on three online databases (PubMed/Medline, Scopus and Web of Science) on May 2020 using the same query – (insect OR mite* OR carmine OR cochineal OR cockroach OR arthropod OR crustacea* OR silkworm OR locust OR grasshopper OR cricket OR mealworm OR moth OR beetle) AND (allerg* OR hypersensitiv* OR anaphyla* OR crossreactiv*) AND (food OR edible OR consumption OR entomophagy OR ingesti* OR occupati* OR consum* OR eat*). In order to avoid obtaining previously reviewed papers, only articles published since 2017 were retrieved on this database search. References of included studies and review papers concerning entomophagy were also screened.

2.1   *Inclusion and exclusion criteria*

In accordance with our aims, in this systematic review we sought to cover all the relevant topics regarding edible insects allergenicity according to current guidelines for novel food allergenic risk assessment. Therefore, we included original studies that assessed cross-reactivity or co-sensitisation between edible insects and crustaceans or HDM, as well as the molecular mechanisms in food primary sensitisation to edible insects. Moreover, articles identifying and characterising (including effects of food processing techniques) food allergens from edible insects were also included. Additionally, case reports describing allergic reactions following the intentional ingestion of insects, and studies assessing the prevalence of such reactions were also included.

We excluded articles that only assessed other types of insect allergies (e.g. respiratory allergy or reactions subsequent to stings or bites) as well as articles characterising allergens from non-edible insects (e.g. cockroaches). Additionally, articles included in our previous systematic review were also excluded.

2.2   *Study selection and data extraction*

After duplicates removal, the retrieved studies were firstly screened by title and abstract, and then by full-text reading. The full texts of studies meeting the inclusion criteria were analysed, and information was retrieved on May 2020.

The whole process for study selection and data extraction was independently performed by two authors, and any disagreement was solved by consensus.

3 Results

A total of 20 articles were included in this systematic review – 19 obtained through database research and 1 (Jiang *et al.*, 2016) obtained through screening of the references of included studies (although it was published in 2016, it was included since it was not present in our previous review) (Figure 1).

Of these 20 articles, 8 studied cross-reactivity or co-sensitisation with either crustaceans or HDM (Barre *et al.*, 2019; Beaumont *et al.*, 2019; Broekman *et al.*, 2017a; Hall *et al.*, 2018; Kamemura *et al.*, 2019; Pali-Scholl *et al.*, 2019; Palmer *et al.*, 2020; Sokol *et al.*, 2017), 5 focused on primary sensitisation (Broekman *et al.*, 2017a,b; Francis *et al.*, 2019; Jeong *et al.*, 2017; Nebbia *et al.*, 2019), 1 evaluated allergenic potential of insect tropomyosin (Klueber *et al.*, 2020), 3 studied the effects of food processing techniques on insects' allergenicity (Hall *et al.*, 2018; Hall and Liceaga, 2020; Pali-Scholl *et al.*, 2019), 4 were case reports or case series (Beaumont *et al.*, 2019; Gadisseur *et al.*, 2019; Nebbia *et al.*, 2019; Sokol *et al.*, 2017), 3 assessed the frequency of food allergies or food anaphylaxis caused by insects (Jiang *et al.*, 2016; Lee *et al.*, 2019; Rangkakulnuwat *et al.*, 2020) and 2 assessed the prevalence of allergic reaction among insect-eaters (Chomchai *et al.*, 2020; Taylor and Wang, 2018).

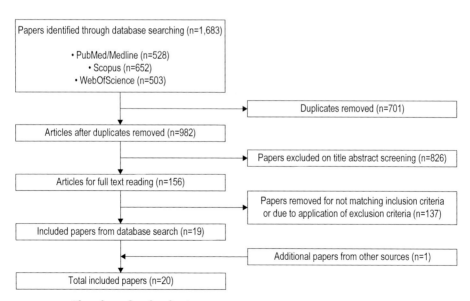

FIGURE 1   Flow chart of study selection process

3.1 *Mechanisms of immunologic co-sensitisation or cross-reactivity with crustaceans*

In our previous review (Ribeiro *et al.*, 2018), we pointed that immunologic co-sensitisation or cross-reactivity to edible insect species (such as mealworms, crickets, locusts and grasshoppers) had been shown for individuals allergic to crustaceans (or to crustaceans and HDM). The main allergens responsible for this cross-reactivity included arthropod pan-allergens tropomyosin and AK, although minor arthropod allergens (e.g. glyceraldehyde 3-phosphate dehydrogenase, myosin light chain, fructose-biphosphate aldolase, actin, α-tubulin, β-tubulin or hexamerin) were also recognised as IgE-binding proteins. Additionally, a double-blind placebo controlled food challenge further confirmed the clinical significance of the cross-reactivity between crustaceans and *Tenebrio molitor* (Broekman *et al.*, 2016).

Since then, immunologic co-sensitisation between edible insects and crustaceans was demonstrated for the first time for the following species: *Galleria mellonella*, *Hermetia illucens* (Broekman *et al.*, 2017a), *Acheta domesticus*, *Locusta migratoria* (Broekman *et al.*, 2017a; Pali-Scholl *et al.*, 2019), *Gryllodes sigillatus* (Hall *et al.*, 2018), and *Schistocerca gregaria* (Pali-Scholl *et al.*, 2019). Moreover, new reports have re-confirmed immunologic co-sensitisation for *T. molitor* (Barre *et al.*, 2019; Broekman *et al.*, 2017a; Pali-Scholl *et al.*, 2019), *Zophobas morio*, *Alphitobius diaperinus* (Broekman *et al.*, 2017a) and *Gryllus bimaculatus* (Kamemura *et al.*, 2019). Furthermore, functionality of the co-sensitisation in *T. molitor* was demonstrated through the application of BAT (Barre *et al.*, 2019; Broekman *et al.*, 2017a), while inhibition studies were used to confirm cross-reactivity with crustaceans (particularly shrimp) involving *G. bimaculatus* (Kamemura *et al.*, 2019) and *Sphenarium mexicanum* (Sokol *et al.*, 2017).

Concerning the allergens responsible for co-sensitisation or cross-reactivity, tropomyosin – whose role as an arthropod pan-allergen capable of causing cross-reactivity has been extensively reported (Wai *et al.*, 2020; Wong *et al.*, 2019) – has been identified as a cross-reacting allergen through immunoblotting in *T. molitor* (Barre *et al.*, 2019; Klueber *et al.*, 2020; Pali-Scholl *et al.*, 2019), *G. sigillatus* (Hall *et al.*, 2018), *G. bimaculatus* (Kamemura *et al.*, 2019), *S. mexicanum* (Sokol *et al.*, 2017), *S. gregaria* and *A. domesticus* (Pali-Scholl *et al.*, 2019). Moreover, Kamemura *et al.* (2019), identified the high molecular weight isoform of tropomyosin of *G. bimaculatus* as the antigen that induced shrimp-specific IgE, and it was additionally shown that this isoform had great sequence homology with both other insects species and shrimp tropomyosins. Nonetheless, there has been some conflicting information on *T. molitor* tropomyosin allergenicity – for example, Klueber *et al.* (2020) reported that this protein was capable of causing similar immunologic response (as measured by β-hexosaminidase release from rat basophilic leukaemia cells expressing the human high-affinity IgE receptor) to shrimp tropomyosin while Palmer *et al.* (2020) reported that in three mealworm species (*T. molitor*, *G. mellonella* and *Z. morio*) tropomyosin had lower IgE-reactivity than tropomyosin from *A. domesticus* or *H.*

illucens. This variation in mealworm tropomyosin IgE-reactivity may possibly be explained by individual patients characteristics, since Broekman *et al.* (2017a) reported that some shrimp-allergic patients had lower IgE-reactivity with mealworm tropomyosin than with tropomyosin from other insect species. Another possible explanation concerns small regionalised differences in protein sequence (most likely in IgE binding epitopes), since both the abundance of tropomyosin or overall sequence homology could not explain the verified diminished IgE-reactivity. Further research should be performed in order to assess if the molecular mechanisms of mealworm cross-reactivity are different from other insect species.

New studies have confirmed the role of other proteins which had been previously reported as involved in cross-reactivity with crustaceans or HDM. Such proteins include heat shock protein 70, AK (Barre *et al.*, 2019), and α-amylase (Barre *et al.*, 2019; Pali-Scholl *et al.*, 2019). In addition, larval cuticle protein, which has been identified as playing a major role in primary sensitisation to *T. molitor* (Broekman *et al.*, 2017b), was also identified as a cross-reacting protein (Barre *et al.*, 2019). Furthermore, novel IgE-binding proteins have been identified. Such proteins were apolipophorin-III and 12 kDa haemolymph protein, which have similar functions (binding and transport of hydrophobic ligands) (Barre *et al.*, 2019). Apolipophorin has already been identified as a potential allergen in mealworms (Broekman *et al.*, 2017a) due to its sequence homology with Der p 14, an allergen of HDM (Epton *et al.*, 2001). Regarding the 12 kDa haemolymph protein, it has been reported as one of the most abundant proteins in *T. molitor* supernatant (Yi *et al.*, 2016), but its allergenicity had never been previously reported.

3.2 *Mechanisms of immunologic co-sensitisation or cross-reactivity with house dust mites*

In our previous review, we reported that only one study (Van Broekhoven *et al.*, 2016) had used sera from patients solely allergic to HDM to assess cross-reactivity between edible insects and HDM. Sera from these patients was able to IgE-bind to extracts from mealworm species, and several cross-reacting proteins were identified (paramyosin, α-amylase, actin, larval cuticle protein, hexamerin and myosin heavy chain). This suggests that the molecular mechanisms of cross-reactivity between edible insects and HDM are different from the ones regulating cross-reactivity between crustaceans and HDM (with the latter being mostly related to tropomyosin) (Wong *et al.*, 2016).

Newly published studies have added additional insight into co-sensitisation between edible insects and HDM. The level of co-sensitisation to edible insects appears to be different in subjects only allergic to HDM when compared to those that are also allergic to crustaceans (or which are solely allergic to crustaceans). In fact, Barre *et al.* (2019) tested the sera of 13 HDM-allergic participants with *T. molitor* extracts, observing only 2 positive reactions. On the other hand, Pali-Scholl *et al.* (2019) reported that HDM-allergic patients had different patterns of IgE-reactivity to insects, which differed according to the tested species and body parts: there was no IgE-reactivity to *T. molitor* and to the bodies of *A. domesticus*, *L. migratoria* and *S. gregaria*; on the other hand, IgE-reactivity was found for the extremities (wings and legs) of the tested species.

Regarding allergens responsible for co-sensitisation, in a HDM-allergic patient that suffered food allergy to *T. molitor*, Beaumont *et al.* (2019) identified two allergens (tropomyosin and hexamerin 2A) by liquid chromatography coupled with tandem mass spectrometry (LC-MS/MS) while other cross-reacting proteins (larval cuticle proteins A1/A2, pupal cuticle protein G1A, α-amylase and tubulin) were also possibly present (visible in 2D-Western Blot). Of these allergens, hexamerin, α-amylase and larval cuticle protein had been previously identified as cross-reacting proteins between HDM and mealworm species (Van Broekhoven *et al.*, 2016). Interestingly, hexamerin has also been described as a cross-reacting allergen between HDM and shellfish (Giuffrida *et al.*, 2014). However, it is also important to note that, although tropomyosin was detected as a cross-reacting protein, such was based on the assessment of a subject which was not sensitised to shrimp or HDM tropomyosin, rendering unlikely that this allergen is responsible for cross-reactivity (Beaumont *et al.*, 2019).

The scarcity of studies performed with HDM-only allergic patients hinders our knowledge concerning the allergens involved in this cross-reactivity and its clinical significance. Future research should be focused on HDM/edible insects cross-reactivity, especially considering the role that HDM-sensitisation has on the development of shellfish allergy (Wai *et al.*, 2020; Wong *et al.*, 2016).

3.3   *Primary sensitisation*

Concerning primary sensitisation, in our previous review, we reported on studies which had been performed with individuals sensitised or with allergies to silkworm (Jeong *et al.*, 2016; Liu *et al.*, 2009; Wang *et al.*, 2016; Zhao *et al.*, 2015; Zuo *et al.*, 2015). These studies have identified a wide array of IgE-binding elements, such as AK (Liu *et al.*, 2009), chitinase (Zhao *et al.*, 2015; Zuo *et al.*, 2015), paramyosin (Zhao *et al.*, 2015), 27-kDa heat-stable glycoprotein (Jeong *et al.*, 2016), thiol peroxiredoxin (Wang *et al.*, 2016), vitellogenin, 30 K protein, triosephosphate isomerase, heat shock protein and chymotrypsin inhibitor (Zuo *et al.*, 2015). Of these, AK, paramyosin and chitinase were hypothesised to play a role in cross-reactivity with other arthropods (such as cockroaches or HDM), or even with shrimp due to their high sequence homology with known allergens of these species.

Newly published studies have focused on the detection and characterisation of allergens from *T. molitor* responsible for primary sensitisation and subsequent food allergy (Broekman *et al.*, 2017b; Nebbia *et al.*, 2019). Broekman *et al.* (2017b) identified larval cuticle proteins as major allergens of both respiratory and food allergy to *T. molitor*. On the other hand, Nebbia *et al.* (2019) hypothesised that cockroach allergen-like protein was the primary allergen in both respiratory and food allergies since it was present in extracts from *T. molitor* larvae and faeces (which authors proposed as the main route of sensitisation). These differences in the detected allergens could be explained by different routes of sensitisation, since the subjects reported by Nebbia *et al.* (2019) were mainly sensitised to *T. molitor* faeces while the subjects in the work by Broekman *et al.* (2017b) were not only domestic breeders of mealworm but also regular consumers of this insect. In addition, in both studies, arthropod pan-allergens such as tropomyosin, AK, myosin light and heavy chain (Broekman *et al.*, 2017b), 86 kDa early-staged encapsulation protein, and troponin C (Nebbia *et al.*, 2019) were detected as IgE-binding proteins.

However, it is still uncertain if primary sensitisation to edible insects can cause food allergies to crustaceans through cross-reactivity. In fact, Broekman *et al.* (2017b) observed that all subjects with primary sensitisation to *T. molitor* had negative oral challenges to shrimp, while the participants in the study by Nebbia *et al.* (2019) were not sensitised to shrimp or tropomyosin from other arthropods. Of note, Linares *et al.* (2008) had already previously described an individual with primary sensitisation and respiratory allergy to different species of crickets, but who had no detectable specific-IgE (sIgE) to allergic tropomyosins, and no cross-reactivity for crustaceans or mites. This lack of tropomyosin IgE-reactivity was also demonstrated in subjects allergic to silkworm pupa (Jeong *et al.*, 2017). These results seem to point out that tropomyosin might not play a major role in primary sensitisation to edible insects, which might explain the lack of cross-reactivity between individual primarily sensitised to edible insects and other arthropods. Nonetheless, it is known that there is a high degree of co-sensitisation/cross-reactivity between cockroaches and shellfish, with tropomyosin playing a major role (Wai *et al.*, 2020; Yang *et al.*, 2018), although its clinical significance is not yet established (Wong *et al.*, 2019)

Importantly, it is also possible that primary sensitisation can be somewhat species-specific, as observed in shrimp species (Jirapongsananuruk *et al.*, 2008). In a study performed with four subjects who were primarily sensitised to *T. molitor* (with respiratory and/or food allergies), it was shown that subjects had variable reactivity and sensitisation to several insect species (*Z. morio, A. diaperinus, G. mellonella, H. illucens, A. domesticus, L. migratoria*) (Broekman *et al.*, 2017a). Conversely, in the same study, most of shrimp-allergic patients were co-sensitised to all the tested insect species. Interestingly, two subjects that have had an allergic reaction to *T. molitor* consumption reported no clinical symptoms after consuming other insect species (namely greater wax moths, black soldier flies and crickets) (Nebbia *et al.*, 2019). However, some controversies remain on this question – in fact, Francis *et al.* (2019) reported that AK from *T. molitor* and from *A. domesticus* had weak conservation/homology, with apparent no cross reactivity between these species. On the contrary, Liu *et al.* (2009) not only identified AK as a major allergen of *Bombyx mori*, but also that it cross-reacts with AK from cockroaches.

3.4   *Effects of food processing technologies*

Some effects of food processing on edible insects' allergenicity had already been reported in our previous review. Overall, co-sensitisation between edible insects and crustaceans was reported not to be significantly diminished by thermal treatment (Broekman *et al.*, 2015; Van Broekhoven *et al.*, 2016), although the latter was described to have an impact on the intensity and types of allergens that are detected (Phiriyangkul *et al.*, 2015). Furthermore, *in vitro* digestion had been shown not to eliminate IgE-binding capacity of mealworm tropomyosin (Van Broekhoven *et al.*, 2016).

Food processing technologies are suggested to influence allergenicity of edible insects, and the effects of enzymatic hydrolysis (Hall *et al.*, 2018; Pali-Scholl *et al.*, 2019), microwave-assisted enzymatic hydrolysis (Hall and Liceaga, 2020), and heat treatment (Pali-Scholl *et al.*, 2019) on edible insects allergenicity have been assessed. Hall *et al.* (2018) assessed the allergenicity behaviour of tropomyosin from cricket species *G. sigillatus* and found that only a degree of hydrolysis superior to 50% with alcalase® was able to eliminate its IgE-binding capacity to shrimp-allergic sera. In a follow-up work (Hall and Liceaga, 2020) with the same species, IgG-reactivity of tropomyosin was lower with microwave-assisted enzymatic hydrolysis (also performed with alcalase and a degree of hydrolysis greater than 50%) than with just heat treatment (water bath or microwave) or water bath with enzymatic hydrolysis. These results are in line with previous studies, which had also suggested that insects' tropomyosin was able to maintain its allergenicity even after most thermal or enzymatic treatments (Broekman *et al.*, 2015; Van Broekhoven *et al.*, 2016). This behaviour is also present in shellfish tropomyosin, which has been described as resistant to most thermal treatments and even enzymatic hydrolysis (e.g. simulated gastric fluid and simulated intestinal fluid digestion systems) (Khan *et al.*, 2019), although potentially susceptible to combinations of processing techniques (Mejrhit *et al.*, 2017).

On the other hand, heat-treatment (water bath at 80 and 100 °C for 10 minutes or autoclaving at 121 and 138 °C for 20 minutes) or enzymatic hydrolysis (flavourzyme, papain, alcalase, neutrase) eliminated IgE-reactivity to *L. migratoria* in whole protein extracts of subjects with allergy to both shrimp and HDM (Pali-Scholl *et al.*, 2019). One possible explanation for these different results can be related to the protein extraction technique that was applied, which could have impacted the solubility and detection of the allergens, as shown in the work by Broekman *et al.* (2015).

3.5   *Prevalence of food allergy to insects*

The prevalence of allergic reactions caused by insect consumption has been assessed either through questionnaires directed to consumers of edible insects (Chomchai *et al.*, 2020; Taylor and Wang, 2018) (Table 1) or through retrospective analyses of series of patients assessed for food allergy or for anaphylaxis (Jiang *et al.*, 2016; Lee *et al.*, 2019; Rangkakulnuwat *et al.*, 2020) (Table 2).

TABLE 1 Description of studies assessing prevalence of food allergy amongst consumers of edible insects

Reference/Study	Country	Methodology	Total number of subjects – n	Number of self-reported allergic reactions – n (%; 95%CI)	Species (number of cases)	Other information
Taylor and Wang (2018)	Thailand	cross-sectional survey assessing, amongst others, the occurrence of side effects after eating insects	1,956	434 (22.2%; 20.4-24.0%)	water bugs – 42.9% scorpions – 30.0% grasshoppers – 22.3% crickets – 21.6% bamboo worms – 17.1% red ants – 17.0% silkworms – 16.7% red ant eggs – 11.4%	14.7% (95%CI = 13.1-16.3%; 288/1956) reported the occurrence of a single symptom and 7.4% (95%CI = 6.2-8.6%; 146/1956) reported multiple symptoms; 72.3% (95%CI = 61.4-83.2%; 47/65) of those reporting pre-existing food allergies, reported at least a single symptom following the consumption of insects
Chomchai et al. (2020)	Thailand	internet-based cross-sectional survey of people who practiced entomophagy	140	18 (12.9%; 7.3-18.5%)	silkworm larva (8 – 44.4%) grasshopper (4 – 22.2%) cricket (3 – 16.7%) bamboo caterpillar (3 – 16.7%)	allergic symptoms after insect consumption were associated with a history of respiratory allergy, skin allergy and seafood allergy

TABLE 2   Description of studies which retrospectively analysed food allergic reactions, and which included cases caused by insects

Reference	Country	Methodology	Total number of cases of food anaphylaxis/allergy	Number of cases caused by insects – n (%; 95%CI)	Species (number of cases)	Other information
Jiang et al. (2016)	China	retrospective review of outpatients diagnosed with 'anaphylaxis' or 'severe allergic reactions' in the Department of Allergy, Peking Union Medical College Hospital, from January 2000 to June 2014	1,501	5 (0.3%; 0.02-0.6%)	locusts (2) cicada (2) silkworm chrysalis (1)	
Lee et al. (2019)	Korea	retrospective review of the medical records of 812 Korean adult patients with suspected food allergy and who visited the Allergy Asthma Centre of a tertiary hospital in Korea from January 2014 to December 2018	415	13 (3.1%; 1.4-4.8%)	silkworm pupa (13	46.2% (6/13; 95%CI = 19.1-73.3%) also had food allergy to shellfish
Rangkakul- nuwat et al. (2020)	Thailand	retrospective review of electronic medical records of patients who attended the outpatient and emergency departments at Chiang Mai University Hospital from January 2007 to December 2016	209	17 (8.1%; 4.4-11.8%)	fried insects, namely grasshopper, crickets, silk worms, and bamboo worms (n not specified)	

Two studies assessed the prevalence of allergic reactions among consumers of edible insects. Such studies were performed in Thailand, and relied on self-reported symptoms following the consumption of insects. Chomchai et al. (2020) performed an Internet survey, where it was observed that 18 out of 140 assessed subjects (12.9%; 95% confidence interval [CI] = 7.3-18.5%) reported allergic reactions following the consumption of insects, of whom 4 (22.2%; 95%CI = 3.0-41.4%) reported severe symptoms. Furthermore, the occurrence of an allergic reaction to insects was found to be associated with a history of other allergies, including food allergy to seafood.

Taylor and Wang (2018) assessed the characteristics of insect-consumers in the North-Eastern region of Thailand through a cross-sectional survey delivered in public schools and hospitals, and reported that 14.7% (95%CI = 13.1-16.3%; 288 of 1,956) of insect consumers reported the occurrence of a single symptom after consuming insects, and 7.4% (95%CI = 6.2-8.6%; 146/1,956) reported multiple symptoms. Furthermore, 72.3% (95%CI = 61.4-83.2%; 47/65) of those reporting pre-existing food allergies, reported at least a single symptom following the consumption of insects. Severe reactions had allegedly been experienced by 150 participants (7.6%, 95%CI = 6.3-8.7%; 150/1,956); however, the study does not clearly specify how such reactions were defined.

A previous study performed in Laos had reported lower frequencies of allergic reactions (81/1,059; 7.6% 95%CI = 6.0-9.2%) (Barennes *et al.* (2015) (Supplementary Table S1). These differences reflect not only different consumption habits, but also different sampling methods – while Barennes *et al.* (2015) and Taylor and Wang (2018) performed their studies on populations where entomophagy is common, Chomchai *et al.* (2020) recruited participants through posters and ads in websites which could mean that people who suffered allergic reactions were more predisposed to participate in the survey. Furthermore, all these studies assessed self-reported reactions, which can also lead to an overestimation of food allergy cases. For instance, in the study performed by Taylor and Wang (2018), many of the cases classified as food allergy could be instead cases of food poisoning or allergic reaction to poison, since water bugs (*Lethocerus indicus*) (which are mostly eaten without cooking) and scorpions (*Heterometrus longimanus*) were the species that were reported to cause most allergic reactions, despite being two of the least consumed species by the participants. In fact, food poisoning – which is out of scope of this review – due to insect consumption is not rare, with several described reports of outbreaks of histamine poisoning (Chomchai and Chomchai, 2018). Histamine poisoning, also designed scombroid poisoning, is a foodborne illness that occurs due to toxic levels of histamine (caused by histidine decarboxylase formed by histamine-producing bacteria) mainly in spoiled fish and whose symptoms are very similar to IgE-mediated food allergy (Wu *et al.*, 1997).

Concerning the retrospective analyses of cases of food allergy or anaphylaxis, three different studies (Jiang *et al.*, 2016; Lee *et al.*, 2019; Rangkakulnuwat *et al.*, 2020) have been performed in Asia. Lee *et al.* (2019) retrospectively analysed medical records of 415 adult patients with suspected food allergy, reporting that 13 confirmed cases (3.1%; 95%CI = 1.4-4.8%) were caused by consumption of silkworm pupae. Additionally, six of those 13 patients (46.2%; 95%CI = 19.1-73.3%) also had food allergy to shellfish. The other studies (Jiang *et al.*, 2016; Rangkakulnuwat *et al.*, 2020) have focused on retrospective analysis of cases of food anaphylaxis and found that insect consumption caused 0.3% (5/1,501; 95%CI = 0.02-0.6%) (Jiang *et al.*, 2016) and 8.1% (17/209; 95%CI = 4.4-11.8%) (Rangkakulnuwat *et al.*, 2020) of food anaphylaxis cases. Previous studies (Supplementary Table S2) have reported widely different values of food anaphylaxis caused by the consumption of insects – 5.2% (1/24; 95%CI = 0.0-12.2%) (Jirapongsananuruk *et al.*, 2007), 17.6% (63/358; 95%CI = 13.6-21.4%) (Ji *et al.*, 2009), and 19.4% (7/36;95%CI = 6.5-32-3%) (Piromrat *et al.*, 2008). These differences in values and in the species causing most reactions can mirror the consumption habits of the regions where the studies were performed. Nonetheless, it is noteworthy to mention that there are studies assessing the prevalence of food allergy in Asia and that do not mention insects as causative agents of food allergy (Le *et al.*, 2019; Lee *et al.*, 2013). This can happen because entomophagy is a more common practice in specific regions and rural areas (Manditsera *et al.*, 2018) which can lead to several cases going unreported at national levels.

3.6   *Case reports and case series*

In our previous review, we were able to retrieve 29 cases reports of food allergies caused by insects' consumption (Supplementary Table S3). Most of the cases occurred in Asia and Africa, with the causative species mostly reflecting regional consumption habits. For example, the reported reactions that occurred in China (Ji *et al.*, 2008) were due to silkworm pupae, while the reactions occurring in Botswana were caused by mopane worms (Kung *et al.*, 2011, 2013; Okezie *et al.*, 2010). In most cases (18/29), the reactions occurred after consuming the insect for the first time, suggesting that these reactions could have occurred due to cross-reactivity with crustaceans or HDM. In fact, two of the subjects had previous history of allergic reactions to shellfish (Choi *et al.*, 2010; Piatt, 2005), while other nine had subjects were either sensitised to common aeroallergens or had an history of allergic diseases (Broekman *et al.*, 2017b; Choi *et al.*, 2010; Freye, 1996; Ji *et al.*, 2008; Kung *et al.*, 2011, 2013). Furthermore, in three cases (Broekman *et al.*, 2017b; Freye, 1996) the mechanism for food allergy was probably primary sensitisation since it occurred in subjects which were constantly exposed to the species.

In this review, we identified 16 new cases of food allergy caused by the consumption of insects (Table 3; Beaumont *et al.*, 2019; Gadisseur *et al.*, 2019; Nebbia *et al.*, 2019; Sokol *et al.*, 2017). These cases occurred in France (Beaumont *et al.*, 2019), United States of America (Sokol *et al.*, 2017), Italy (Nebbia *et al.*, 2019) and Niger (Gadisseur *et al.*, 2019). The species that caused the allergic reactions were *chapulines* (*S. mexicanum*) (Sokol *et al.*, 2017), *T. molitor* (Beaumont *et al.*, 2019; Nebbia *et al.*, 2019), and crickets (Gadisseur *et al.*, 2019).

The five cases that occurred in Western countries (Beaumont *et al.*, 2019; Nebbia *et al.*, 2019; Sokol *et al.*, 2017) represent three of the different pathways involved in food allergy to edible insects: cross-reactivity with crustaceans/HDM (Sokol *et al.*, 2017), cross-reactivity with HDM (Beaumont *et al.*, 2019), and primary sensitisation (Nebbia *et al.*, 2019). The two patients in the cases reported by Sokol *et al.* (2017) had previous history of food allergy to shellfish while also being sensitised to common aeroallergens (including HDM). The patient in the case reported by Beaumont *et al.* (2019) only had previous history of respiratory allergy to HDM and was not sensitised to shrimp. Furthermore, in these three cases, the allergic reactions occurred after consuming the insect species for the first time, as observed in most of the previously reported cases (Ribeiro *et al.*, 2018). This further suggests that these reactions occurred through cross-reactivity to HDM and/or crustaceans.

Additionally, two of the reported cases (Nebbia *et al.*, 2019) occurred due to primary sensitisation to the causative species (*T. molitor* larvae). These two cases are very similar to two other previously reported by Broekman *et al.* (2017b), since the patients were constantly exposed to *T. molitor* on their work and had no previous history of food allergy to shellfish or sensitisation to shrimp. Additionally, in both cases reported by Nebbia *et al.* (2019), the subjects also had respiratory allergies to *T. molitor*, while in the report by Broekman *et al.* (2017b) only one subject developed, respiratory allergies although he had previous history of allergies caused by HDM.

Furthermore, Gadisseur *et al.* (2019) assessed the sensitisation profile of a entomophagous population in Niger who displayed symptoms of allergy to insects and/or crustaceans/HDM. This study described 11 subjects with cases of food allergy following the consumption of crickets. In most cases (10 of 11), subjects had previous history of allergic diseases with the most common being food allergy to shellfish (7/11; 64%). Regarding sensitisation to allergens, sIgE to tropomyosin was detected in 5 individuals (5/11; 45%), while sIgE to AK was detected in 6 (6/11; 55%) subjects, respectively. It is also noteworthy to mention that, in the same study, 3 subjects who consumed crickets without developing any symptoms of food allergy were all sensitised to crickets, shrimp and cockroach. Additionally, in these subjects, sIgE was only detected for Bla g 1 (cockroach nitrile specifier), Der p 1 (mite *Dermatophagoides pteronyssinus* cysteine protease) and Der f 1 (mite *Dermatophagoides farinae* cysteine protease). These cases illustrate that sensitisation (such as positive skin prick test or sIgE to common arthropod allergens) alone is not an indication that a clinical reaction can happen.

Providing contextual information about cases of food allergy to insects' consumption is essential to better understand the mechanisms that regulate those reactions. In our previous review (Ribeiro *et al.*, 2018), lack of contextual information hindered our analysis of reported cases and only 12 of 29 individuals with food allergies to insects (41%) had previous allergic diseases or were primarily sensitised to the culprit species. On the other hand, in cases reviewed in this work, these situations occurred in 15 of 16 individuals (94%). Current literature of reported cases highlights that individual allergic to crustaceans or that are constantly exposed to edible insects appear to be the two major group risks for developing food allergies to insects. Nonetheless, several cases occurred in individuals which were sensitised to HDM or had history of allergic diseases (e.g. allergic rhinitis) which emphasises that individuals allergic to HDM may also be a risk group of food allergies to insects.

TABLE 3 Description of reported cases of allergy to insects' consumption

Reference	Age/sex/nationality	Species	Clinical symptoms	Clinical history of allergies	Other characteristics
Sokol et al. (2017)	43/M/American	Chapulines (Sphenarium mexicanum)	I, S (lips and tongue), UC, AP, D	history of allergic rhinoconjunctivitis, bronchial asthma and food allergy to shellfish	Reaction occurred after consuming chapulines for the first time Positive SPT and sIgE to grasshopper, chapulines, crickets, cockroach, mites, shellfish, cat and dog sIgE inhibition with chapulines to grasshopper, crickets, cockroach, mites, shellfish Identification of tropomyosin in immunoblot
Sokol et al. (2017)	50/F/American	Chapulines (S. mexicanum)	I (mouth, throat, generalised), S (face, perioral tissue, throat), DSw, DSp, Sy	history of allergic rhinoconjunctivitis, bronchial asthma, intermittent urticaria, moderately severe atopic dermatitis and food allergy to shellfish	Reaction occurred after consuming chapulines for the first time Positive SPT and sIgE to grasshopper, chapulines, crickets, cockroach, mites, shellfish, cat and dog sIgE inhibition with chapulines to grasshopper, crickets, cockroach and shellfish Identification of tropomyosin in Immunoblot
Beaumont et al. (2019)	31/M/French	Yellow mealworm (Tenebrio molitor)	U, A, Dys, N	rhinitis and mild asthma	Positive SPT and sIgE to dust mites and mealworm Positive sIgE to Der p 1, Der p 2 and Der p 23. Negative sIgE to Pen a 1 and Der p 10 Identification of IgE-binding proteins: tubulin α-chain, α-amylase, tropomyosin, hexamerin, pupal cuticle protein G1A and larval cuticle protein
Nebbia et al. (2019)	24/M/Italian	Yellow mealworm (T. molitor)	OAS – P (oral), T (throat)	rhinoconjunctivitis, itching and contact erythema when exposed to T. molitor	Consumed other species (greater wax moth, black soldier fly and crickets) without developing allergic reactions Reaction occurred after T. molitor hamburger for the first time Positive SPT to grass Positive SPT and BAT to T. molitor Identification of cockroach allergen-like protein, Troponin C and 86 kDa early-staged encapsulation protein as IgE-binding proteins
Nebbia et al. (2019)	27/M/Italian	Yellow mealworm (T. molitor)	OAS, P (oral), T (throat)	rhinoconjunctivitis, itching and contact erythema when exposed to T. molitor	Consumed other species (greater wax moth, black soldier fly and crickets) without developing allergic reactions Reaction occurred after T. molitor hamburger for the first time Positive SPT to Alternaria Positive SPT and BAT to T. molitor Identification of cockroach allergen-like protein, Troponin C and 86 kDa early-staged encapsulation protein as IgE-binding proteins
Gadisseur et al. (2019)	31/M/Nigerien	Crickets	OAS, U, A, GI, V	allergy to shrimp and HDM	Sensitised (positive SPT and/or sIgE) to shrimp, HDM, cockroach and cricket. Positive sIgE to HDM allergens (Der p 1, Der p 2, Der f 1, Der f 2)

Gadisseur et al. (2019)	26/F/Nigerien	Crickets	OAS, Dys, HT, V, A	allergy to shrimp	Sensitised (positive SPT and/or sIgE) to shrimp and cricket Positive sIgE to Der p 10 (HDM tropomyosin), Pen a 1, Pen m 1 (shrimp tropomyosin), Bla g 7 (cockroach tropomyosins) and HDM allergens (Der p 1, Der p 2, Der f 1, Der f 2)
Gadisseur et al. (2019)	26/F/Nigerien	Crickets	OAS, U, A, R, C, GI	allergy to shrimp and cockroach	Sensitised (positive SPT and/or sIgE) to shrimp, cockroach and cricket Positive sIgE to Der p 10 (HDM tropomyosin) Pen a 1, Pen m 1 (shrimp tropomyosin), Bla g 7 (cockroach tropomyosins), Pen m 2 (shrimp AK)
Gadisseur et al. (2019)	36/M/Nigerien	Crickets	OAS, U, A	allergy to shrimp	Sensitised (positive SPT and/or sIgE) to cockroach and cricket
Gadisseur et al. (2019)	44/M/Nigerien	Crickets	U, OAS, GI, V	allergy to shrimp	Sensitised (positive SPT and/or sIgE) to shrimp, cockroach and cricket
Gadisseur et al. (2019)	55/F/Nigerien	Crickets	OAS, U	no previous history of allergies	Sensitised (positive SPT and/or sIgE) to shrimp, HDM, cockroach and cricket Positive sIgE to Der p 10 (HDM tropomyosin), Pen a 1 (shrimp tropomyosin) and Pen m 2 (shrimp AK)
Gadisseur et al. (2019)	39/M/Nigerien	Crickets	U, GI, V	allergy to shrimp and cockroach	Sensitised (positive SPT and/or sIgE) to shrimp, cockroach and cricket Positive sIgE Pen a 1, Pen m 1 (shrimp tropomyosin), Pen m 2 (shrimp AK) and Pen m 4 (shrimp Sarcoplasmic Calcium-Binding Protein)
Gadisseur et al. (2019)	8/F/Nigerien	Crickets	OAS, U	allergy to cockroach	Sensitised (positive SPT and/or sIgE) to shrimp, cockroach and cricket Positive sIgE to Pen m 2 (shrimp AK)
Gadisseur et al. (2019)	20/M/Nigerien	Crickets	U	allergy to HDM	Sensitised (positive SPT and/or sIgE) to shrimp, HDM, cockroach and cricket Positive sIgE to HDM allergens (Der p 1, Der p 2, Der f 1, Der f 2) and Pen m 2 (shrimp AK)
Gadisseur et al. (2019)	14/F/Nigerien	Crickets	GI	allergy to cockroach	Sensitised (positive SPT and/or sIgE) to shrimp, cockroach and cricket Positive sIgE to Bla g 1
Gadisseur et al. (2019)	31/M/Nigerien	Crickets	OAS	allergy to shrimp	Positive SPT to dust mites and mopane worm Positive sIgE to Der p 10 (HDM tropomyosin), Pen a 1 (shrimp tropomyosin), Pen m 2 (shrimp AK) and Bla g 5 (cockroach Glutathione S-transferase)

Clinical symptoms: A = angioedema; AP = abdominal pain; BAT = Basophil Activation Test; C = conjunctivitis; D = diarrhoea; DSp = difficulty speaking; DSw = difficulty swallowing; Dys = dyspnoea; GI = gastrointestinal trouble; HT = hypotension; I = itchiness; N = nausea; OAS = oral allergy syndrome; P = pruritus; R = rhinitis; S = swelling; sIgE = Specific IgE; SPT = skin prick test; Sy = syncope; T = tightness; U = urticaria; UC = unconsciousness; V = vomiting.

List of allergens (WHO/IUIS, 2020); Der p 1 = cysteine protease from the European house dust mite *Dermatophagoides pteronyssinus*; Der p2 = Niemann-Pick proteins of class C2 family from *Dermatophagoides pteronyssinus*; Der p23 = peritrophin-like protein domain (PF01607) from *Dermatophagoides pteronyssinus*; Der p10 = tropomyosin from *Dermatophagoides pteronyssinus*; Der f 1 = cysteine protease from the American house dust mite *Dermatophagoides farinae*; Der f 2 = Niemann-Pick proteins of class C2 from *Dermatophagoides farinae*; Pen a 1 = tropomyosin from brown shrimp *Penaeus aztecus*; Pen m 1 = tropomyosin from the black tiger shrimp *Penaeus monodon*; Pen m 2 = arginine kinase from *Penaeus monodon*; Pen m 4 = Sarcoplasmic calcium binding protein from *Penaeus monodon*; Bla g 1 = microvilli-like protein with unknown function from the German cockroach *Blattella germanica*; Bla g 7 = tropomyosin from *Blattella germanica*.

sitised to shrimp. Furthermore, in these three cases, the allergic reactions occurred after consuming the insect species for the first time, as observed in most of the previously reported cases (Ribeiro *et al.*, 2018). This further suggests that these reactions occurred through cross-reactivity to HDM and/or crustaceans.

Additionally, two of the reported cases (Nebbia *et al.*, 2019) occurred due to primary sensitisation to the causative species (*T. molitor* larvae). These two cases are very similar to two other previously reported by Broekman *et al.* (2017b), since the patients were constantly exposed to *T. molitor* on their work and had no previous history of food allergy to shellfish or sensitisation to shrimp. Additionally, in both cases reported by Nebbia *et al.* (2019), the subjects also had respiratory allergies to *T. molitor*, while in the report by Broekman *et al.* (2017b) only one subject developed, respiratory allergies although he had previous history of allergies caused by HDM.

Furthermore, Gadisseur *et al.* (2019) assessed the sensitisation profile of a entomophagous population in Niger who displayed symptoms of allergy to insects and/or crustaceans/HDM. This study described 11 subjects with cases of food allergy following the consumption of crickets. In most cases (10 of 11), subjects had previous history of allergic diseases with the most common being food allergy to shellfish (7/11; 64%). Regarding sensitisation to allergens, sIgE to tropomyosin was detected in 5 individuals (5/11; 45%), while sIgE to AK was detected in 6 (6/11; 55%) subjects, respectively. It is also noteworthy to mention that, in the same study, 3 subjects who consumed crickets without developing any symptoms of food allergy were all sensitised to crickets, shrimp and cockroach. Additionally, in these subjects, sIgE was only detected for Bla g 1 (cockroach nitrile specifier), Der p 1 (mite *Dermatophagoides pteronyssinus* cysteine protease) and Der f 1 (mite *Dermatophagoides farinae* cysteine protease). These cases illustrate that sensitisation (such as positive skin prick test or sIgE to common arthropod allergens) alone is not an indication that a clinical reaction can happen.

Providing contextual information about cases of food allergy to insects' consumption is essential to better understand the mechanisms that regulate those reactions. In our previous review (Ribeiro *et al.*, 2018), lack of contextual information hindered our analysis of reported cases and only 12 of 29 individuals with food allergies to insects (41%) had previous allergic diseases or were primarily sensitised to the culprit species. On the other hand, in cases reviewed in this work, these situations occurred in 15 of 16 individuals (94%). Current literature of reported cases highlights that individual allergic to crustaceans or that are constantly exposed to edible insects appear to be the two major group risks for developing food allergies to insects. Nonetheless, several cases occurred in individuals which were sensitised to HDM or had history of allergic diseases (e.g. allergic rhinitis) which emphasises that individuals allergic to HDM may also be a risk group of food allergies to insects.

4   Conclusions

In conclusion, the current literature points that the two major risk groups for development of food allergy to insects' consumption are subjects allergic to crustaceans and individuals constantly exposed to edible insects. For subjects allergic to crustaceans, reactions to edible insects may occur due to cross-reactivity, which seems to be mainly mediated through tropomyosin, with tropomyosin from *T. molitor* being able to produce an allergic response in an animal model. However, other minor allergens (e.g. AK and α-amylase) may also play a role and previously unreported IgE-binding proteins (apolipophorin and 12 kDa haemolymph protein) were identified. The allergenicity of edible insects seems to be resistant to thermal treatments and digestion with enzymes (unless very specific conditions are applied), a similar behaviour to crustaceans' allergens.

On the other hand, it has also been demonstrated that individuals constantly exposed to *T. molitor* can become sensitised and subsequently develop a food allergy to this insect. Different allergens (larval cuticle protein and cockroach allergen-like protein) were identified depending on the route of sensitisation. Significantly, tropomyosin has not been identified as a significant allergen in primary sensitisation to *T. molitor or* silkworm. Additionally, it is still uncertain if this sensitisation is species-specific or if it can lead to co-sensitisation with other insect species or crustaceans.

On the other hand, co-sensitisation has been shown between HDM and edible insects, but it seems to be different from co-sensitisation between edible insects and crustaceans – a relatively small number of HDM-allergic patients are sensitised to edible insects, and IgE-binding to edible insects only occurs in specific body parts. Additionally, the clinical relevance of such co-sensitisation is not yet defined, and the underlying molecular mechanisms remain unclear (although hexamerin has been consistently identified in all studies), especially considering the apparent lack of involvement of tropomyosin.

Although substantial work has been performed within the topic of edible insects' allergenicity, there are still gaps in our current knowledge. Concerning cross-reactivity with crustaceans, future studies should assess the allergenicity of other species besides *T. molitor*, and a comparison of the molecular mechanisms between different species should be performed, namely by using extracts from different species and serum from the same patients. One of the focal points of future research should be performing studies with individuals monosensitised to HDM in order to have a better understating of HDM-edible insects cross-reactivity. The clinical significance of this co-sensitisation is still unclear and biological assays (preferably food challenges) should be performed.

One of the major flaws when studying food allergy to edible insects is the lack of reliable epidemiological data, since there still is a lack of reported cases/series of food allergy to edible insects. Prevalence studies performed in Asian countries have reported that 0.3-19.4% of food anaphylaxis/allergy cases were caused by insects' consumption. Furthermore, studies performed with populations of insect-eaters haver reported that 7.6-22.2% of individuals suffered allergic reactions after consuming insects (although these rates could be overestimated because they were bases on self-reported reactions). Despite these high prevalences, we were only able to retrieve a total of 45 cases in both reviews. This can happen because the regions where insects are consumed are mostly rural (possibly leading to sub-notification), or because several cases may have only been published in local literature (e.g. Chinese). Given the large pool of subjects allergic to the consumption of insects in these areas, it is of extreme importance for these cases to be reported so that we can better understand the characteristics of food allergy to insects, including their severity and epidemiological association with other allergy diseases. In fact, it is expected that an increasing number of cases of food allergy will be reported, due to the introduction of edible insects in the food market of Western countries. This impact is already evident, as food allergy cases have already been reported in subjects who work with *T. molitor* (Broekman *et al.*, 2017b; Nebbia *et al.*, 2019), although the number of reported cases (4) is still very small. These types of cases are essential to have a better understanding of primary sensitisation to edible insects, namely regarding the major allergens involved and whether such sensitisation is species-specific.

Acknowledgements

Author J.C. Ribeiro acknowledges Doctoral grant No. SFRH/BD/147409/2019 funded by Fundação para a Ciência e a Tecnologia (FCT). Authors J.C. Ribeiro, S.C. Fonseca and L.M. Cunha acknowledge financial support from the national funds by FCT within the scope of UIDB/05748/2020 and UIDP/05748/2020.

Conflict of interest

The authors declare no conflict of interest.

Supplementary material

Table S1. Description of studies assessing prevalence of food allergy amongst consumers of edible insects, included in both systematic reviews (current article and Ribeiro et al. 2018).

Table S2. Description of studies included in both systematic reviews (current article and Ribeiro et al. 2018), which retrospectively analysed food allergic reaction, including cases caused by insects.

Table S3. Description of reported cases included in both systematic reviews (current article and Ribeiro et al. 2018) of allergy to insects' consumption.

Supplementary material can be found online at https://doi.org/10.3920/JIFF2020.0065.

References

Barennes, H., Phimmasane, M. and Rajaonarivo, C., 2015. Insect consumption to address undernutrition, a national survey on the prevalence of insect consumption among adults and vendors in Laos. PLoS One 10: e0136458. https://doi.org/10.1371/journal.pone.0136458

Barre, A., Pichereaux, C., Velazquez, E., Maudouit, A., Simplicien, M., Garnier, L., Bienvenu, F., Bienvenu, J., Burlet-Schiltz, O., Auriol, C., Benoist, H. and Rouge, P., 2019. Insights into the allergenic potential of the edible yellow mealworm (*Tenebrio molitor*). Foods 8: 515. https://doi.org/10.3390/foods8100515

Beaumont, P., Courtois, J., Van der Brempt, X. and Tollenaere, S., 2019. Food-induced anaphylaxis to *Tenebrio molitor* and allergens implicated. Revue Francaise d'Allergologie 59: 389-393. https://doi.org/10.1016/j.reval.2019.06.001

Belluco, S., Halloran, A. and Ricci, A., 2017. New protein sources and food legislation: the case of edible insects and EU law. Food Security 9: 803-814. https://doi.org/10.1007/s12571-017-0704-0

Boyce, J.A., Assa'ad, A., Burks, A.W., Jones, S.M., Sampson, H.A., Wood, R.A., Plaut, M., Cooper, S.F., Fenton, M.J., Arshad, S.H., Bahna, S.L., Beck, L.A., Byrd-Bredbenner, C., Camargo, C.A., Jr., Eichenfield, L., Furuta, G.T., Hanifin, J.M., Jones, C., Kraft, M., Levy, B.D., Lieberman, P., Luccioli, S., McCall, K.M., Schneider, L.C., Simon, R.A., Simons, F.E.R., Teach, S.J., Yawn, B.P. and Schwaninger, J.M., 2010. Guidelines for the diagnosis and management of food allergy in the United States: report of the NIAID-sponsored expert panel. The Journal of Allergy and Clinical Immunology 126: S1-S58. https://doi.org/10.1016/j.jaci.2010.10.007

Broekman, H., Knulst, A., Den Hartog Jager, S., Monteleone, F., Gaspari, M., De Jong, G., Houben, G. and Verhoeckx, K., 2015. Effect of thermal processing on mealworm allergenicity. Molecular Nutrition and Food Research 59: 1855-1864. https://doi.org/10.1002/mnfr.201500138

Broekman, H., Knulst, A.C., De Jong, G., Gaspari, M., Den Hartog Jager, C.F., Houben, G.F. and Verhoeckx, K.C.M., 2017a. Is mealworm or shrimp allergy indicative for food allergy to insects? Molecular Nutrition and Food Research 61(9): 1601061. https://doi.org/10.1002/mnfr.201601061

Broekman, H., Knulst, A.C., Den Hartog Jager, C.F., Van Bilsen, J.H.M., Raymakers, F.M.L., Kruizinga, A.G., Gaspari, M., Gabriele, C., Bruijnzeel-Koomen, C., Houben, G.F. and Verhoeckx, K.C.M., 2017b. Primary respiratory and food allergy to mealworm. The Journal of Allergy and Clinical Immunology 140: 600-603. https://doi.org/10.1016/j.jaci.2017.01.035

Broekman, H., Verhoeckx, K.C., Jager, C.F.D., Kruizinga, A.G., Pronk-Kleinjan, M., Remington, B.C., Bruijnzeel-Koomen, C.A., Houben, G.F. and Knulst, A.C., 2016. Majority of shrimp-allergic patients are allergic to mealworm. Journal of Allergy and Clinical Immunology 137: 1261-1263. https://doi.org/10.1016/j.jaci.2016.01.005

Choi, G.S., Shin, Y.S., Kim, J.E., Ye, Y.M. and Park, H.S., 2010. Five cases of food allergy to vegetable worm (*Cordyceps sinensis*) showing cross-reactivity with silkworm pupae. Allergy: European Journal of Allergy and Clinical Immunology 65: 1196-1197. https://doi.org/10.1111/j.1398-9995.2009.02300.x

Chomchai, S. and Chomchai, C., 2018. Histamine poisoning from insect consumption: an outbreak investigation from Thailand. Clinical Toxicology (Phila) 56: 126-131. https://doi.org/10.1080/15563650.2017.1349320

Chomchai, S., Laoraksa, P., Virojvatanakul, P., Boonratana, P. and Chomchai, C., 2020. Prevalence and cluster effect of self-reported allergic reactions among insect consumers. Asian Pacific Journal of Allergy and Immunology 38: 40-46. https://doi.org/10.12932/ap-220218-0271

De Gier, S. and Verhoeckx, K., 2018. Insect (food) allergy and allergens. Molecular Immunology 100: 82-106. https://doi.org/10.1016/j.molimm.2018.03.015

Epton, M.J., Dilworth, R.J., Smith, W. and Thomas, W.R., 2001. Sensitisation to the lipid-binding apolipophorin allergen Der p 14 and the peptide Mag-1. International Archives of Allergy and Immunology 124: 57-60. https://doi.org/10.1159/000053668

Francis, F., Doyen, V., Debaugnies, F., Mazzucchelli, G., Caparros, R., Alabi, T., Blecker, C., Haubruge, E. and Corazza, F., 2019. Limited cross reactivity among arginine kinase allergens from mealworm and cricket edible insects. Food Chemistry 276: 714-718. https://doi.org/10.1016/j.foodchem.2018.10.082

Freye, H.B., 1996. Anaphylaxis to the ingestion and inhalation of *Tenebrio molitor* (mealworm) and *Zophobas morio* (superworm). Allergy and Asthma Proceedings 17: 215-219.

Gadisseur, R., Courtois, J., Laouali, S., Van der Brempt, X., Jacquier, J., Cavalier, E., Maizoumbou, D.A., Tollenaere, S. and Hamidou, T., 2019. Sensitization profile of cricket food-allergic or cricket tolerant patients in an entomophagous population in Niamey, Niger. Allergy 74: 253-253.

García, B.E. and Lizaso, M.T., 2011. Cross-reactivity syndromes in food allergy. Journal of Investigational Allergology and Clinical Immunology 21: 162-170; quiz 162 p following 170.

Giuffrida, M.G., Villalta, D., Mistrello, G., Amato, S. and Asero, R., 2014. Shrimp allergy beyond tropomyosin in Italy: clinical relevance of arginine kinase, sarcoplasmic calcium binding protein and hemocyanin. European Annals of Allergy and Clinical Immunology 46: 172-177.

Hall, F. and Liceaga, A., 2020. Effect of microwave-assisted enzymatic hydrolysis of cricket (*Gryllodes sigillatus*) protein on ACE and DPP-IV inhibition and tropomyosin-IgG binding. Journal of Functional Foods 64: 103634. https://doi.org/10.1016/j.jff.2019.103634

Hall, F., Johnson, P.E. and Liceaga, A., 2018. Effect of enzymatic hydrolysis on bioactive properties and allergenicity of cricket (*Gryllodes sigillatus*) protein. Food Chemistry 262: 39-47. https://doi.org/10.1016/j.foodchem.2018.04.058

Jeong, K.Y. and Park, J.W., 2020. Insect allergens on the dining table. Current protein and Peptide Science 21: 159-169. https://doi.org/10.2174/1389203720666190715091951

Jeong, K.Y., Han, I.S., Lee, J.Y., Park, K.H., Lee, J.H. and Park, J.W., 2017. Role of tropomyosin in silkworm allergy. Molecular Medicine Reports 15: 3264-3270. https://doi.org/10.3892/mmr.2017.6373

Jeong, K.Y., Son, M., Lee, J.Y., Park, K.H., Lee, J.H. and Park, J.W., 2016. Allergenic Characterization of 27-kDa glycoprotein, a novel heat stable allergen, from the pupa of silkworm, *Bombyx mori*. Journal of Korean Medical Science 31: 18-24. https://doi.org/10.3346/jkms.2016.31.1.18

Ji, K., Chen, J., Li, M., Liu, Z., Wang, C., Zhan, Z., Wu, X. and Xia, Q., 2009. Anaphylactic shock and lethal anaphylaxis caused by food consumption in China. Trends in Food Science and Technology 20: 227-231. https://doi.org/10.1016/j.tifs.2009.02.004

Ji, K.M., Zhan, Z.K., Chen, J.J. and Liu, Z.G., 2008. Anaphylactic shock caused by silkworm pupa consumption in China. Allergy: European Journal of Allergy and Clinical Immunology 63: 1407-1408. https://doi.org/10.1111/j.1398-9995.2008.01838.x

Jiang, N., Yin, J., Wen, L. and Li, H., 2016. Characteristics of anaphylaxis in 907 Chinese patients referred to a tertiary allergy center: a retrospective study of 1,952 episodes. Allergy, Asthma & Immunology Research 8: 353-361. https://doi.org/10.4168/aair.2016.8.4.353

Jirapongsananuruk, O., Bunsawansong, W., Piyaphanee, N., Visitsunthorn, N., Thongngarm, T. and Vichyanond, P., 2007. Features of patients with anaphylaxis admitted to a university hospital. Annals of Allergy, Asthma, & Immunology 98: 157-162. https://doi.org/10.1016/s1081-1206(10)60689-8

Jirapongsananuruk, O., Sripramong, C., Pacharn, P., Udompunturak, S., Chinratanapisit, S., Piboonpocanun, S., Visitsunthorn, N. and Vichyanond, P., 2008. Specific allergy to *Penaeus monodon* (seawater shrimp) or *Macrobrachium rosenbergii* (freshwater shrimp) in shrimp-allergic children. Clinical & Experimental Allergy 38: 1038-1047. https://doi.org/10.1111/j.1365-2222.2008.02979.x

Kamemura, N., Sugimoto, M., Tamehiro, N., Adachi, R., Tomonari, S., Watanabe, T. and Mito, T., 2019. Cross-allergenicity of crustacean and the edible insect *Gryllus bimaculatus* in patients with shrimp allergy. Molecular Immunology 106: 127-134. https://doi.org/10.1016/j.molimm.2018.12.015

Khan, M.U., Ahmed, I., Lin, H., Li, Z., Costa, J., Mafra, I., Chen, Y. and Wu, Y.N., 2019. Potential efficacy of processing technologies for mitigating crustacean allergenicity. Critical Reviews in Food Science and Nutrition 59: 2807-2830. https://doi.org/10.1080/10408398.2018.1471658

Klueber, J., Costa, J., Randow, S., Codreanu-Morel, F., Verhoeckx, K., Bindslev-Jensen, C., Ollert, M., Hoffmann-Sommergruber, K., Morisset, M., Holzhauser, T. and Kuehn, A., 2020. Homologous tropomyosins from vertebrate and invertebrate: recombinant calibrator proteins in functional biological assays for tropomyosin allergenicity assessment of novel animal foods. Clinical & Experimental Allergy 50: 105-116. https://doi.org/10.1111/cea.13503

Kung, S.J., Fenemore, B. and Potter, P.C., 2011. Anaphylaxis to Mopane worms (*Imbrasia be-*

lina). Annals of Allergy, Asthma and Immunology 106: 538-540. https://doi.org/10.1016/j.anai.2011.02.003

Kung, S.J., Mazhani, L. and Steenhoff, A.P., 2013. Allergy in Botswana. Current Allergy and Clinical Immunology 26: 202-209.

Le, T.T.K., Nguyen, D.H., Vu, A.T.L., Ruethers, T., Taki, A.C. and Lopata, A.L., 2019. A cross-sectional, population-based study on the prevalence of food allergies among children in two different socio-economic regions of Vietnam. Pediatric Allergy & Immunology 30: 348-355. https://doi.org/10.1111/pai.13022

Lee, A.J., Thalayasingam, M. and Lee, B.W., 2013. Food allergy in Asia: how does it compare? Asia Pacific Allergy 3: 3-14. https://doi.org/10.5415/apallergy.2013.3.1.3

Lee, S.C., Kim, S.R., Park, K.H., Lee, J.H. and Park, J.W., 2019. Clinical features and culprit food allergens of Korean adult food allergy patients: a cross-sectional single-institute study. Allergy, Asthma & Immunology Research 11: 723-735. https://doi.org/10.4168/aair.2019.11.5.723

Linares, T., Hernandez, D. and Bartolome, B., 2008. Occupational rhinitis and asthma due to crickets. Annals of Allergy, Asthma and Immunology 100: 566-569.

Liu, Z., Xia, L., Wu, Y., Xia, Q., Chen, J. and Roux, K.H., 2009. Identification and characterization of an arginine kinase as a major allergen from silkworm (*Bombyx mori*) larvae. International Archives of Allergy and Immunology 150: 8-14. https://doi.org/10.1159/000210375

Manditsera, F.A., Lakemond, C.M.M., Fogliano, V., Zvidzai, C.J. and Luning, P.A., 2018. Consumption patterns of edible insects in rural and urban areas of Zimbabwe: taste, nutritional value and availability are key elements for keeping the insect eating habit. Food Security 10: 561-570. https://doi.org/10.1007/s12571-018-0801-8

Mazzucchelli, G., Holzhauser, T., Cirkovic Velickovic, T., Diaz-Perales, A., Molina, E., Roncada, P., Rodrigues, P., Verhoeckx, K. and Hoffmann-Sommergruber, K., 2018. Current (food) allergenic risk assessment: is it fit for novel foods? Status quo and identification of gaps. Molecular Nutrition & Food Research 62: 1700278. https://doi.org/10.1002/mnfr.201700278

Mejrhit, N., Azdad, O., Chda, A., El Kabbaoui, M., Bousfiha, A., Bencheikh, R., Tazi, A. and Aarab, L., 2017. Evaluation of the sensitivity of Moroccans to shrimp tropomyosin and effect of heating and enzymatic treatments. Food and Agricultural Immunology 28: 969-980. https://doi.org/10.1080/09540105.2017.1323187

Messina, M. and Venter, C., 2020. Recent surveys on food allergy prevalence. Nutrition Today 55: 22-29. https://doi.org/10.1097/NT.0000000000000389

Migueres, M., Dávila, I., Frati, F., Azpeitia, A., Jeanpetit, Y., Lhéritier-Barrand, M., Incorvaia, C. and Ciprandi, G., 2014. Types of sensitization to aeroallergens: definitions, prevalences and impact on the diagnosis and treatment of allergic respiratory disease. Clinical and Translational Allergy 4: 16-16. https://doi.org/10.1186/2045-7022-4-16

Muraro, A., Werfel, T., Hoffmann-Sommergruber, K., Roberts, G., Beyer, K., Bindslev-Jensen, C., Cardona, V., Dubois, A., duToit, G., Eigenmann, P., Fernandez Rivas, M., Halken, S., Hickstein, L., Høst, A., Knol, E., Lack, G., Marchisotto, M.J., Niggemann, B., Nwaru, B.I., Papadopoulos, N.G., Poulsen, L.K., Santos, A.F., Skypala, I., Schoepfer, A., Van Ree, R., Venter, C., Worm, M., Vlieg-Boerstra, B., Panesar, S., De Silva, D., Soares-Weiser, K., Sheikh, A.,

Ballmer-Weber, B.K., Nilsson, C., De Jong, N.W. and Akdis, C.A., 2014. EAACI food allergy and anaphylaxis guidelines: diagnosis and management of food allergy. Allergy 69: 1008-1025. https://doi.org/10.1111/all.12429

Nebbia, S., Lamberti, C., Giorgis, V., Giuffrida, M.G., Manfredi, M., Marengo, E., Pessione, E., Schiavone, A., Boita, M., Brussino, L., Cavallarin, L. and Rolla, G., 2019. The cockroach allergen-like protein is involved in primary respiratory and food allergy to yellow mealworm (*Tenebrio molitor*). Clinical & Experimental Allergy 49: 1379-1382. https://doi.org/10.1111/cea.13461

Okezie, O.A., Kgomotso, K.K. and Letswiti, M.M., 2010. Mopane worm allergy in a 36-year-old woman: a case report. Journal of Medical Case Reports 4: 42. https://doi.org/10.1186/1752-1947-4-42

Pali-Scholl, I., Meinlschmidt, P., Larenas-Linnemann, D., Purschke, B., Hofstetter, G., Rodriguez-Monroy, F.A., Einhorn, L., Mothes-Luksch, N., Jensen-Jarolim, E. and Jager, H., 2019. Edible insects: cross-recognition of IgE from crustacean and house dust mite allergic patients, and reduction of allergenicity by food processing. World Allergy Organization J 12: 100006. https://doi.org/10.1016/j.waojou.2018.10.001

Palmer, L.K., Marsh, J.T., Lu, M., Goodman, R.E., Zeece, M.G. and Johnson, P.E., 2020. Shellfish tropomyosin IgE cross-reactivity differs among edible insect species. Molecular Nutrition and Food Research 64: e1900923. https://doi.org/10.1002/mnfr.201900923

Pennisi, E., 2015. All in the (bigger) family. Science 347: 220. https://doi.org/10.1126/science.347.6219.220

Phiriyangkul, P., Srinroch, C., Srisomsap, C., Chokchaichamnankit, D. and Punyarit, P., 2015. Effect of food thermal processing on allergenicity proteins in Bombay locust (*Patanga Succincta*). International Journal of Food Engineering 1: 23-28. https://doi.org/10.18178/ijfe.1.1.23-28

Piatt, J.D., 2005. Case report: urticaria following intentional ingestion of cicadas. American Family Physician 71: 2048, 2050.

Piromrat, K., Chinratanapisit, S. and Trathong, S., 2008. Anaphylaxis in an emergency department: a 2-year study in a tertiary-care hospital. Asian Pacific Journal of Allergy and Immunology 26: 121-128.

Pomés, A., Mueller, G.A., Randall, T.A., Chapman, M.D. and Arruda, L.K., 2017. New insights into cockroach allergens. Current Allergy and Asthma Reports 17: 25-25. https://doi.org/10.1007/s11882-017-0694-1

Rangkakulnuwat, P., Sutham, K. and Lao-Araya, M., 2020. Anaphylaxis: ten-year retrospective study from a tertiary-care hospital in Asia. Asian Pacific Journal of Allergy and Immunology 38: 31-39. https://doi.org/10.12932/ap-210318-0284

Ribeiro, J.C., Cunha, L.M., Sousa-Pinto, B. and Fonseca, J., 2018. Allergic risks of consuming edible insects: a systematic review. Molecular Nutrition and Food Research 62: 30. https://doi.org/10.1002/mnfr.201700030

Sokol, W.N., Wünschmann, S. and Agah, S., 2017. Grasshopper anaphylaxis in patients allergic to dust mite, cockroach, and crustaceans: is tropomyosin the cause? Annals of Allergy, Asthma and Immunology 119: 91-92. https://doi.org/10.1016/j.anai.2017.05.007

Stanhope, J., Carver, S. and Weinstein, P., 2015. The risky business of being an entomologist: a systematic review. Environmental Research 140: 619-633. https://doi.org/10.1016/j.envres.2015.05.025

Taylor, G. and Wang, N., 2018. Entomophagy and allergies: a study of the prevalence of entomophagy and related allergies in a population living in North-Eastern Thailand. Bioscience Horizons 11. https://doi.org/10.1093/biohorizons/hzy003

Van Broekhoven, S., Bastiaan-Net, S., De Jong, N.W. and Wichers, H.J., 2016. Influence of processing and *in vitro* digestion on the allergic cross-reactivity of three mealworm species. Food Chemistry 196: 1075-1083. https://doi.org/10.1016/j.foodchem.2015.10.033

Verhoeckx, K., Broekman, H., Knulst, A. and Houben, G., 2016. Allergenicity assessment strategy for novel food proteins and protein sources. Regulatory Toxicology and Pharmacology 79: 118-124. https://doi.org/10.1016/j.yrtph.2016.03.016

Wai, C.Y.Y., Leung, N.Y.H., Chu, K.H., Leung, P.S.C., Leung, A.S.Y., Wong, G.W.K. and Leung, T.F., 2020. Overcoming shellfish allergy: how far have we come? International Journal of Molecular Sciences 21: 2234. https://doi.org/10.3390/ijms21062234

Wang, J. and Sampson, H.A., 2011. Food allergy. The Journal of clinical investigation 121: 827-835. https://doi.org/10.1172/JCI45434

Wang, H., Hu, W., Liang, Z., Zeng, L., Li, J., Yan, H., Yang, P., Liu, Z. and Wang, L., 2016. Thiol peroxiredoxin, a novel allergen from *Bombyx mori*, modulates functions of macrophages and dendritic cells. American Journal of Translational Research 8: 5320-5329.

Westerhout, J., Krone, T., Snippe, A., Babe, L., McClain, S., Ladics, G.S., Houben, G.F. and Verhoeckx, K.C., 2019. Allergenicity prediction of novel and modified proteins: Not a mission impossible! Development of a random forest allergenicity prediction model. Regulatory Toxicology and Pharmacology 107: 104422. https://doi.org/10.1016/j.yrtph.2019.104422

Wong, L., Huang, C.H. and Lee, B.W., 2016. Shellfish and house dust mite allergies: is the link tropomyosin? Allergy, Asthma & Immunology Research 8: 101-106. 10.4168/aair.2016.8.2.101

Wong, L., Tham, E.H. and Lee, B.W., 2019. An update on shellfish allergy. Current Opinion in Allergy and Clinical Immunology 19: 236-242. https://doi.org/10.1097/aci.0000000000000532

World Health Organization and International Union of Immunological Societies (WHO/IUIS) Allergen Nomenclature Sub-committee, 2020. Allergen nomenclature. Available at: http://www.allergen.org.

Wu, M.L., Yang, C.C., Yang, G.Y., Ger, J. and Deng, J.F., 1997. Scombroid fish poisoning: an overlooked marine food poisoning. Veterinary and human toxicology 39: 236-241.

Yang, Z., Zhao, J., Wei, N., Feng, M., Xian, M., Shi, X., Zheng, Z., Su, Q., Wong, G.W.K. and Li, J., 2018. Cockroach is a major cross-reactive allergen source in shrimp-sensitized rural children in southern China. Allergy 73: 585-592. https://doi.org/10.1111/all.13341

Yi, L., Van Boekel, M.A.J.S., Boeren, S. and Lakemond, C.M.M., 2016. Protein identification and *in vitro* digestion of fractions from *Tenebrio molitor*. European Food Research and Technology 242: 1285-1297. https://doi.org/10.1007/s00217-015-2632-6

Zhao, X., Li, L., Kuang, Z., Luo, G. and Li, B., 2015. Proteomic and immunological identification of two new allergens from silkworm (*Bombyx mori* L.) pupae. Central European Journal of Immunology 40: 30-34. https://doi.org/10.5114/ceji.2015.50830

Zuo, J., Lei, M., Yang, R. and Liu, Z., 2015. Bom m 9 from *Bombyx mori* is a novel protein related to asthma. Microbiology and Immunology 59: 410-418. https://doi.org/10.1111/1348-0421.12271

Regulations on insects as food and feed: a global comparison

A. Lähteenmäki-Uutela[1]*, S.B Marimuthu[2] and N. Meijer[3]

[1]Finnish Environment Institute, Latokartanonkaari 11, 00790 Helsinki, Finland; [2]School of Law, University of New England, Armidale, NSW, Australia; [3]Wageningen Food Safety Research, Akkermaalsbos 2, 6708 WB, Wageningen, The Netherlands; *anu.lahteenmaki-uutela@syke.fi

Abstract

Insects, as a food and or feed source, represent an emerging protein source relevant to farmers, feed companies, food companies and food marketers globally. The growth of this industry is somewhat restricted due to outdated food and feed regulations covering insect use. The regulations also do not allow the use of all potential insects as food and feed. Governments aim to ensure food and feed safety, and each country has its own substantive and procedural rules for this purpose. However, the regulatory demands and differences between countries complicate the international marketing strategies for insect products. Food and feed regulation are separate; feed regulation may allow insect usage even when they are not allowed as food. Some countries have specific rules for novel foods, while others do not. This paper compares insect food and feed regulation of the primary production and marketing areas: the European Union, the United States, Canada, and Australia. In addition, the situation in selected countries in Central and South America, Asia and Africa is also discussed.

Keywords

food law – feed law – regulations

1   Introduction

This is a comparative review on the global regulations on insects as food and feed. Earlier reviews on legislation specific to the production and marketing of insects for food and feed have been published by Lähteenmäki-Uutela *et al.* (2017) and Lähteenmäki-Uutela *et al.* (2018). These publications are used as a starting point for this publication. The review aims to cover all main markets interesting for the global insect business. In comparison to the above-mentioned legal reviews, this paper focuses on legal developments between 2018 and 2020 as well as adds some new countries where interest in insect food and feed is on the rise and where regulations therefore need to be considered.

Legal issues can arise at various stages of the production, processing, and marketing cycle. Insects can grow on various substrates including manure and biowaste, but there are risks in bringing such insects into the food chain. Insects that are novel as food and feed may be riskier than familiar ones. The main legal question with insects is about how the risks involved may be mitigated through product safety rules. If specific government standards are lacking, companies may decide to rely on the existing industry practices or the private standards as guidance or they may decide to postpone further business investment until transparent rules appear.

This review limits itself to general food and feed safety law, because it is the most central issue for the insect business. Producers and marketers need to know which insects and insect-based ingredients are allowed for which end use. This paper does not discuss production animal welfare, genetic modification or gene editing of insects, food and feed product labelling (although allergen labelling is an important part of food safety legislation), or nutrition and health claims on insect products. Materials and methods are briefly discussed in Section 2. Sections 3 to 8 discuss regulations in the European Union, North America, Central and South America, Australia, Asia, and Africa. Section 9 presents the conclusions.

2   Materials and methods

Comparative reviews on legislation specific to the production and marketing of insects for food and feed have already been published by Lähteenmäki-Uutela *et al.* (2017) and Lähteenmäki-Uutela *et al.* (2018). These publications were used as a starting point for this publication, which is focused on legal developments between 2018 and 2020. The recent legal developments were tracked through government web pages, publications of insect industry associations, and food and feed industry news.

3   European Union

In terms of insects as food for human consumption, provisions placing insects within the scope of Regulation (EU) 2015/2283 on novel foods have been applicable since 2018. Under this new Regulation, insect food products may only be marketed when authorised after a safety assessment by the European Food Safety Authority (EFSA). At this time (October 2020), applications for food products from the following insect species have been submitted: house cricket (*Acheta domesticus*), banded/tropical house cricket (*Gryllodes sigillatus*), lesser mealworm (*Alphitobius diaperinus*), black soldier fly (*Hermetia illucens*), honey bee (*Apis mellifera*), migratory locust/grasshopper (*Locusta migratoria*), and yellow mealworm (*Tenebrio molitor*) (EC, n.d.). The first of EFSA's safety assessments are expected to be published soon.

While waiting for the EFSA assessments and thereafter Commission decisions on novel food applications, a transitional period applies for whole insects and their preparation. This means that insect foods which had been lawfully marketed on 1 January 2018, and for which an application or notification was submitted by January 2019, may continue to be marketed until the Commission hands down its decision on the respective application or notification. This means that several insect species may continue to be sold as food in Europe without novel food authorisation. On 1st October 2020, the European Court of Justice gave its ruling in the case C-526/19 stating that whole insects were *outside* the old Novel Food Regulation. This means the transitional measures for whole insects had to be expanded to all EU countries. The case was initially brought against two French ministries by company Entoma SAS, and the Supreme Administrative Court of France requested a preliminary ruling from the European Court of Justice.

Specific hygiene rules for insects intended for human consumption are currently being considered: a draft Regulation, amending Regulation (EC) No 853/2004, was published for public comment in 2018 (Ares(2019)382900). The feedback period ended in February 2019 with adoption scheduled for Q1 2019. The draft Regulation would add a new Section in Annex III of Regulation No 853/2004. At its core, the draft Regulation does not introduce any new provisions precisely for insects, but rather reiterates rules from various other legislation that already applied previously. Specifically, insects must be approved as a novel food under Regulation (EU) No 2015/2283 (Article 3 of the draft Regulation). In addition, insects may only be reared on substrates of vegetable origin or specifically allowed materials of animal origin such as fishmeal and hydrolysed proteins from non-ruminants (Article 4 of the draft Regulation), but this was already the case under Regulation (EC) No 1069/2009 and No 142/2011. Finally, according to the draft Regulation (Article 5), the 'substrate for the feeding of insects must not contain manure, catering waste or other waste' – but this was already the case under Regulation (EC) No 1069/2009. As of October 2020, this draft Regulation has yet to be implemented.

On the executive front, the European Commission's (EC) Directorate-General Health and Food Safety (DG SANTE) audited the French (DG SANTE, 2019) and Netherlands (DG SANTE, 2018) official control systems for reared insects. Although these audit reports are not legislative in nature and recommendations are primarily intended for the authorities, they do provide an insight into the legal requirements for insect rearing companies. Areas covered by the audit reports included the approval process of new establishments, microbiological testing, and required documentation. A report aimed explicit for the insect rearing industry is a 'Guide on Good Hygiene Practices', published by the International Platform of Insects for Food and Feed (IPIFF). Formal endorsement of this guidance document by the EC and Member States is pending.

In the context of insects for animal feed, one of the most critical changes in 2017 was introduced by Regulation (EU) No 2017/893. This act amended Regulations (EC) No 999/2001 and (EU) No 142/2011, allowing the feeding of seven insect species to aquaculture animals (black soldier fly (*H. illucens*), common housefly (*Musca domestica*), yellow mealworm (*T. molitor*), lesser mealworm (*A. diaperinus*), house cricket (*A. domesticus*), banded cricket (*G. sigillatus*) and field cricket (*Gryllus assimilis*)). Regulation (EU) No 2017/893 removed the requirement for reared insects that 'products of animal origin must be sourced from a registered slaughterhouse', because insect rearing facilities (where the insects are generally also 'slaughtered'), could not comply with the requirements specific to slaughterhouses. Finally, Regulation (EU) No 2019/1981 introduced a list of third countries that were authorised to export insect products complying with the mentioned Regulation (EU) No 2017/893.

Significant legislative changes regarding insects are required to be implemented in the future to overcome current barriers to the growth of the insect industry. In DG SANTE's 'strategic safety concept for insects as feed' (DG SANTE, 2017), two primary legal barriers for the insect rearing industry were identified. Firstly, it is prohibited to use former foodstuffs containing meat or fish as feed materials for insects, and secondly it is not allowed to feed processed animal proteins from insects to pigs and poultry. According to the SANTE document, before the first barrier can be lifted, it is necessary that 'operational and validated analytical techniques' for differentiating between insect material and other animals are available. Work on these methods is ongoing. For the second barrier, it is indicated that allowing any additional feed materials for insects 'would require a robust EFSA opinion assessing the risk for human and animal health and recommending measures to ensure that such risk is negligible' (DG SANTE, 2017). In February 2020, the Netherlands Food and Product Safety Authority's Bureau for Risk Assessment (NVWA-BuRo) advised the Dutch Ministry of Agriculture to propose an EU amendment to remove these two barriers. The change would allow the use of insects as feed to an animal providing that the insects have not being reared on former foodstuffs containing meat of the same species (e.g. insects reared on pig sourced meal are not subsequently fed to those same species again) (NVWA-BuRo, 2020). In a note accompanying the report, the agency's Inspector General reiterated that the prerequisites for lifting these barriers, as mentioned in the 2017 DG SANTE Safety Concept document, should be met first (IG-NVWA, 2020).

Summarising the European legal situation, the substrate options on which insects may be reared are still restricted to those materials that are also permitted for other livestock species. Developments to ease some of these restrictions are ongoing, but legislative changes are not expected to be implemented in the short-term.

Since 2017, aquaculture animals have been allowed to be fed with insect meal, the feeding insects to pets was already permitted. Legislative changes are anticipated to allow the feeding of insects to more animals such as poultry and pigs, but the timeline for these changes is unclear. Novel food applications have been submitted, but at the time of writing, no safety assessment reports have been published by the EFSA. The European Court of Justice ruling on the non-applicability of the old novel food regulation to whole insects somewhat changes the regulatory landscape as it extends the transitional period for whole insects to all EU member states.

4 North America: United States and Canada

In many parts of North America, insects have traditionally been used as a part of food culture. The production of insects for the food and feed industries started to expand after 2012. The modern insect industry had its basis in companies that had already grown crickets and mealworms for pet food. In order to avoid high labour costs, many United States and Canadian insect farms have invested in robotics, automation, sensor technology and data aggregation from the start. (Shockley *et al.*, 2018).

In the United States, governing insect food is under the stewardship of the US Food and Drug Administration (FDA). Already in 2013, the FDA gave its 'response to inquiry' that represented their thinking on insect food. It stated that insects are considered food under the Food, Drug, and Cosmetic Act (United States Code, Title 21), if that is their intended use. According to the Act, food must be clean and wholesome (i.e. free from filth, pathogens, toxins), must have been produced, packaged, stored, and transported under sanitary conditions, and must be appropriately labelled (Sec. 403). Insect-raising for human food must follow good manufacturing practices (cGMP, 21CFR110). Insects raised for animal feed cannot be diverted to human food. Insects collected from the wild cannot be sold as food. If insects are not sold as such, but altered or used as a food ingredient, they may require food additive authorisation. An insect protein is a food additive, unless it has GRAS status (generally recognised as safe). (Lähteenmäki-Uutela *et al.*, 2018). GRAS and food additive petition (FAP) are two equally legitimate and parallel processes of compliance to the US law on food ingredients. GRAS is basically a self-approval system, and FAP is government managed. A company may itself make a GRAS determination. A GRAS notice can be submitted to FDA for review, but there is no legal obligation to do so. According to Burdock Group, GRAS can offer a lower cost and a faster route to market. The burden of proof for safety is the same regardless of the process (FAP or GRAS). In both processes, the approval is for the specific use of the ingredient, not for the ingredient per se (https://burdockgroup.com/).

According to the Food, Drug and Cosmetic Act, animal foods must be safe, produced under sanitary conditions, and contain no harmful or deleterious substances. Many states base their feed regulations on recommendations on the Association of American Feed Control Officials (AAFCO). The AAFCO has established an ingredients definition for only one insect species as an animal food ingredient for livestock feed. The black soldier fly (*H. illucens*) larvae, including dried whole larvae (since 2016) and black soldier fly meal (since 2018) is permitted for use in feed for aquaculture for salmonids such as salmon, trout and char. The FDA has reviewed and approved the AAFCO's decision. Notably, the black soldier fly larvae can be reared on approved feed-grade materials, including pre-consumer food waste as a substrate, as well as other food manufacturing by-products such as spent brewery grains and other feed grade materials. (Lähteenmäki-Uutela *et al.*, 2018). Several states also allow insect-based pet foods, while other states wait for AATCO and FDA decisions. Pet treats do not have to comply with all AAFCO regulations, as they are not a source of complete nutrition.

In Canada, the safety and nutritional adequacy of novel foods must be evaluated before they enter the market (Canada Gazette Part II, Division 28: Novel Foods, October 27, 1999). Novel foods must be notified to Health Canada. If there is an international history of safe consumption, a food is not considered novel. A history of safe use means the food has been an ongoing part of the diet for a number of generations in a large, genetically diverse human population where it has been used in ways and at levels, which are similar to those expected or intended in Canada (Health Canada, 2006). Crickets are not a novel food in Canada, and there are large cricket breeders in the country.

Novel feed ingredients, i.e. feed ingredients not already listed in Schedules IV and V of the Feeds Regulations, must be authorised in Canada. Each insect species, rearing condition, and livestock species needs separate authorisation. The Canadian Food Inspection Agency is responsible for the pre-market assessment of new feed ingredients and the registrations of feed products. Black soldier fly products have been authorised to feed broiler chickens, salmonids, tilapia and poultry including chickens, ducks, geese, and turkey (Lähteenmäki-Uutela *et al.*, 2018). Pet foods are another regulatory category. Black soldier fly larvae, mealworms, and silkworm pupae are sold as pet food in Canada.

5   Central and South America: Mexico, Brazil and Argentina

In Mexico, many different insects are used as food and insects have also been used for medicinal purposes. Insects are mainly collected from the wild. (González and Contreras, 2009). Mexico's food safety laws are the responsibility of two government secretaries, SSA (Health Secretary) and SAGARPA (Agriculture, Cattle, Rural Development, Fishing and Feeding Secretary). The SSA mainly focuses on processed food and SAGARPA on primary production. The General Health Law, the Federal Vegetal Health and Animal Law, and the Products and Services Sanitary

Control Regulation are three most important regulations. (Leon and Paz, 2013.) All health-related products and services are regulated through mandatory standards called Normas Oficiales Mexicanas (NOMs). NOMs are revised at least every five years. At the time of writing, there are not yet specific NOMs for insect food or insect feed in Mexico. General NOMs apply, e.g. the Norma Oficial Mexicana NOM-251-SSA1-2009 on the hygienic standards for foods, beverages and dietary supplements.

Mexico's Organic Products Law and regulations for organic production were implemented in April 2017. There is a regulatory category for organic insect food in Mexico, including an indicative list of the species and their life stages concerned: eggs, larvae, nymphs, and adult insects of the maguey worm (*Aegiale hesperiaris*), larvae of longhorn beetles (*Cerambicidae*), larvae of 'escamoles' (*Liometopum apiculatum*), and ant eggs. These insects should be collected from areas of organic production or in ecosystems with little or no human intervention having no contact with prohibited substances. The Organic Plan or an equivalent document should demonstrate that the collection, farming, catching, limiting and processing of insects does not alter or influence the ecosystem. A record on the history of the site or an Organic Plan must be kept proving that the areas of collection, farming or catching have not been subject to treatment by prohibited substances.

Brazilian researchers, farmers and companies are increasingly interested in using insects as food and feed, particularly in replacing soybean meal with black soldier fly meal to feed the poultry. As Brazil is the leading global exporter of poultry, increasing the sustainability of the Brazilian poultry industry would be significant (Allegretti *et al.,* 2018). Brazilian feed law does not yet foresee insects as animal feed. According to Allegretti *et al.* (2018), Brazil tends to follow Codex Alimentarius standards.

In Argentina, there is interest for example in adding cricket flour to foods. The government has notified the nascent industry that insect food is currently not covered by the national Food Code. An application to add insects to the Food Code must be submitted to the National Food Commission. All packaged food must also have the approval of the National Administration of Medicines, Food and Medical Technology (Anmat) (Crespo, 2019).

6   Australia

Save for indigenous Australians, the protein consumption of the average Australian, which has been predominately meat, does not include insects. The indigenous Australian diet, also termed as 'bush tucker diet' comprises of various insects including witchery grubs, Bogong moths, termites, beetles, honeypot ants and native honey bees. (Yen, 2005).

Australia or New Zealand do not have a standalone legislation or specific government regulations on insect farming. Insect Protein Association of Australia (IPAA) has developed guidelines for its members (https://www.insectproteinassoc.com). Non-members are not bound by the rules, and they are yet to be made available for public access. Insect food is regulated under Standard 1.5.1 of the Food Standard Code as a category of novel food. These are non-traditional foods, the safety of which has not previously been established and which therefore require assessment. Three species of insects are categorised as non-novel, namely: (1) super mealworm (*Zophobas morio*); (2) house cricket (*A. domesticus*); and (3) mealworm beetle (*T. molitor*). Beetles, grasshoppers, butterflies, moths, bees, bugs and dragonflies may also be consumed, although they are not allowed to be sold.

Before a novel food is sold in Australia and New Zealand, a thorough risk-based assessment process is carried out that considers, amongst other things, 'toxicological and nutritional issues' of its chemistry and consumption patterns. Ordinarily, a novel food will have to be listed in the Standard before it can be sold as food or used as a food ingredient. Where it is not listed, an application can be forwarded to Food Standards Australia New Zealand (FSANZ), who will then include the novel food in the list after a pre-market safety assessment. There are provisions to allow 'first to market advantage': FSANZ extends, upon request being made, an exclusive permission to new owners of novel food or novel ingredients in a specific brand or class for a period of 15 months. This does not preclude other applicants seeking approval of their own different brand. Imported novel foods are governed under the 'Imported Food Control Act' 1992, 'Biosecurity Act' 2015 and 'Food Standard Code'. The Department of Agriculture enforces the 'Food Standards Code' at the border for imported food.

In Australia, the edible food industry continues to grow (Maxabella, 2019), despite there being only a limited choice of insects permitted for sale. There are more than 50 insect farmers across Australia (Jones, 2019) as well as several sellers specialising in edible insects (GrupsUP, 2020). The 'Edible Bug Shop' claims to produce and sell 200 kilograms of insects a week to the domestic market (Black, 2020). Insects are sold in a variety of forms – sweet and savoury snacks, tea leaves, candy, dukka, marshmallow, cricket powders, pasta powder, tortillas chips, white chocolates and more (https://ediblebugshop.com.au). Though in Australia insects have been traditionally consumed by the indigenous community (Yen, 2005), the culture of adapting to eating insects amongst the Australian majority is recent. Insects are increasingly consumed for their nutritional and environmental benefits (Edwards and Ranasighe, 2019).

Animal feed materials and ingredients which are fed as part of the normal diet of an animal do not require registration in Australia. Australia has strict rules on animal feed, however. Ruminant animals are not to be fed with meat, including meat and bone meal, derived from all vertebrates, including fish and birds. Such restrictions are in place to safeguard against risk of bovine spongiform encephalopathy (BSE or mad cow disease) or transmissible spongiform encephalopathies (Food Standard, 2020). This ruminant feed ban is governed under the legislation in all states and territories and is reinforced by industry-based quality assurance programs such as the FeedSafe Standard (Animal Health Australia, 2020). Feed ingredients and additive suppliers in Australia are represented by the Feed Ingredients and Additives Association Australia (FIAAA). This organisation regulates its industry via the Australian New Zealand Code of Practice for Animal Feed Ingredients and Additive Suppliers (the FIAAA Code of Practice) and the FeedSafe Standard (https://fiaaa.com.au/about/#bg). These two codes and standards are applicable to feed production in both Australia and New Zealand. Insects may be used as feed for aquaculture in all states, and as feed for poultry in NSW, ACT, Tasmania, Victoria and Western Australia. Insects used for feed are not to be fed with meat, manure and catering waste, and raw insects (live and untreated by heat) for feed are not permitted in Australia (DiGiacomo et al., 2019). According to IPAA, the insect for the feed industry in Australia comprises of five larger commercial fly farms and a variety of smaller operations that are starting research and developing processes.

Pet food in Australia is self-regulated with voluntary industry standards of the Pet Food Industry Association of Australia (PFIAA) through its Australian Standard (AS 5812-2017) for the Manufacturing and Marketing of Pet Food regulation for the pet food industry.

7 Asia: China, Japan and Thailand

The Chinese have used insects for thousands of years. Several Chinese medicines and health foods are based on insects. Honey bee larvae are used as a sedative and as an anti-inflammatory medicine; male silkmoths and male silkworms are sold for strengthening kidneys; termites are used for anti-aging purposes; and black ants for improving immunity (Lähteenmäki-Uutela et al., 2017). China has a long tradition in silk production and is the world's largest silk producer. New food raw materials, including insect protein, require authorisation from the Ministry of Health. Authorisations have general applicability: if a new food raw material is added to the Food Materials Catalogue, all food producers can use the material (Sun, 2015: 445). According to Belluco et al. (2013), silkworm pupae were authorised as a new food ingredient and according to Shen (2014), earthworm protein powder has been authorised. New feed raw materials also need to be authorised, and authorised feed materials are added to the Feed Materials Catalogue.

According to Mitsuhashi (1997), several different insects are traditional foods in Japan. These include *Oxya yezoensis* or *Oxya japonica*, the larvae and pupae of a wasp, *Vespula lewisi*, and the pupae and female adults after oviposition of the domestic silkmoth, *Bombyx mori*. These insects have been cooked with soy sauce and sugar and sold as canned foods. Larvae of the dobsonfly, *Protohermes grandis* (Neuroptera) have been used as traditional medicine (Mitsuhashi, 1997). In Japan, normal novel foods do not require pre-market authorisation, while novel additives do. Food safety is under the responsibility of the Ministry of Health, Safety and Welfare. Feed law is another issue. The Ministry of Agriculture, Forestry and Fisheries has given the Act on Safety Assurance and Quality Improvement of Feeds, which includes the maximum limits for pesticide residues, heavy metals, mycotoxins and melamine. Feed manufacturers, importers and/or dealers must submit notification prior to starting a business. (FAMIC, 2014)

Thailand is the world's biggest cricket producer. It has a Standard for cricket farming (Good Agricultural Practices for Cricket Farm, Thai Agricultural Standard 8202-2017). The Standard includes rules on farm components, feed, water, animal health, environment, and record-keeping. The aim is to produce crickets of good quality which are safe for consumers. Feed shall not be deteriorated, water must not be contaminated, equipment must be clean and hygienic, and all chemicals must be used according to the instructions. The development of the Thai cricket farming standards was connected to cricket exports, particularly to accessing the EU markets (Preteseille *et al.*, 2018: 435). Food is governed by FDA Thailand. In fish aquaculture, Thai companies are looking to replace unsustainable fishmeal with insects (Dao, 2020). Thailand also has a large broiler meat and pork industry where insect-based feed has potential. Thailand has product quality standards for each type of animal feed, but standards for insect feed are still lacking. The Ministry of Agriculture and Cooperatives regulates animal feed.

8   Africa: South Africa and Nigeria

Niassy *et al.* (2018) studied the regulatory environment for insects as food and feed in South Africa. The country has a high diversity of insect species and a high demand for insect protein. Replacing imported food and feed ingredients with local insect production can provide jobs in South Africa. Regulation is fragmented, however. Various governmental departments and various levels of government deal with food and feed safety, and there is no specific legislation on insect production or insect foods. Insects are not novel foods. The Agricultural Product Standards Act 1990 (Act 119 of 1990) governed by the Department of Agriculture, Forestry and Fishers, the Foodstuffs, Cosmetics and Disinfectants Act 1972 (Act 54 of 1972) governed by the Department of Health and the Consumer Protection Act 2008 (Act 68 of 2008) governed by the Department of Trade and Industry apply. For the large informal food markets, local rules are the most relevant. (Niassy *et al.*, 2018.)

For being able to manufacture, import, or sell farm feed or pet food in South Africa, the product must be registered according to the Fertilizers, Farm Feeds, Agricultural Remedies and Stock Remedies Act 1947 (Act 36 of 1947). Before submitting the application, one must send a sample of the farm feed or pet food to an accredited laboratory where the product is analysed. Product registrations are valid for three years, after which they must be renewed (South Africa government, n.d.).

Nigeria is another African country where interest in modern insect production is rising. Referring for example to Kelemu *et al.* (2015), Usman and Yusuf (2020) describe how several species of insects from the orders Lepidoptera, Orthoptera, Coleoptera, Isoptera and Hymenoptera are commonly consumed in Nigeria. In big cities, insects have started to be regarded as primitive foods. With the population projected to reach 400 million in 2050, Usman and Yusuf (2020) see that Nigeria must turn back to healthy and sustainable food alternatives such as insects. Indigenous knowledge on insect harvesting, preservation and use should be complemented with university-level research and development focusing e.g. on nutritional content and mass rearing (Usman and Yusuf, 2020).

According to Usman and Yusuf, insect food regulation is currently missing in Nigeria. Protecting food safety is under the competence of the National Agency for Food and Drug Administration and Control (NAFDAC). The Food and Drug Act (Cap F32 Laws of the Federal Republic of Nigeria, 2004) applies. The Standard Organization of Nigeria (SON) sets the rules for packaging materials, labelling and marketing. According to Usman and Yusuf, what is needed in Nigeria is an amendment to the NAFDAC Act to include rules for insect food safety and control.

The NAFDAC Act also applies to animal food, pet food and premixes. Both local manufacturers and importers of feed must apply for registration for each product at the NAFDAC, more specifically the Veterinary Medicine and Allied Products directorate (NAFDAC guidelines VMAP 003/13 and VMAP 004/13). Registrations are valid for five years.

9   Conclusions and discussion

Insects have traditionally been harvested and consumed in many food cultures including Mexico, China and Australia, without a specific regulatory framework. Due to concerns over climate change and food system sustainability, the industrial production of insects and their use as food and feed is now gaining momentum in developed and developing countries. Two problems emerge in parallel with this growth, the first being the lack of regulation locally, and the second is the lack of a stable and consistent set of regulations across international borders. More specifically, many local companies are interested in exporting their insect products across the world, but the regulatory demands and differences between countries complicates the initiatives to market and sell insect products.

Every major stakeholder agrees that insect production and the human food and stock feed uses of insects must be regulated in order to guarantee safety. In that respect, there is agreement to build regulatory systems around a scientific risk assessment approach. Clear and unambiguous regulation will level the playing field, encourage investments, add trust, and normalize the industry (Van der Spiegel, 2016: 213). Allegretti *et al.* (2018) suggest that public and private actors must join forces to construct a global regulatory framework for insects as a part of sustainable food systems.

For each government, the primary aim of regulating insect food and feed production is to guarantee product safety and quality. The global industry may view unharmonized regulations as simply an obstacle which generates repetitive science and repetitive administrative work. Research institutions and companies must choose where to invest their limited resources, e.g. studying familiar insects vs studying unfamiliar insects. In order to become mainstream food and feed in the Western countries, insects and their related risks need to be understood and science plays an integral part in this process. Safety concerns for insects which have traditionally been used are less than for new insects and new production technologies.

Today, much research effort goes into verifying and replicating previous experiments with insects such as crickets, mealworms, and black soldier flies. In the regulatory context in Australia, for instance, house cricket, super mealworm and mealworm beetle are not considered novel foods. In Canada, the definition of novel food excludes traditional foods from other countries. Thorough safety assessments for all insect foods are required for the European Novel Food Regulation and the American GRAS or FAP rules. In Europe, traditional foods from third countries are regulated under the novel food regulation, but as a specific category (Article 14 of Regulation (EU) No 2015/2283). A globally operating insect company will need to prepare an application or notification on the same product for each national authority, and they all have differing requirements. Many complications are obvious as demonstrated by these examples.

In theory, insects as food and feed could be regulated globally. Insects proven safe as food and feed by the scientific community could be authorized in all countries. In the absence of a global food and feed administration body, the global harmonisation of both substantive and procedural standards would be helpful for both entrepreneurs and authorities. Codex Alimentarius Commission by the FAO/WHO is the arena for developing global food and feed standards. Codex still does not have the insect standards we anticipated in Lähteenmäki-Uutela *et al.* (2018), namely a standard for insect farming practices, a list of insect species recognized as safe, a standard on fresh and processed insect product hygiene, and a standard on insect product labelling.

Whether authorisations are generic or firm-specific is an essential feature of safety regulations, impacting the market positions of companies. Larger companies may benefit, if each company needs to prove the safety of each insect product separately. The dissemination of insect innovations may be faster with generic authorisations that apply to all similar products: the followers may enter the market after the leaders have cleared the way. If the familiar insect species will eventually be authorized in all important countries, and the authorisations are generic, researchers and companies may shift their attention to new species.

In essence, much work needs to be done if we are to assist and promote this important fledgling industry. All stakeholders from government legislators through to industry players and the scientific community need to work together to achieve consistency and design a way forward locally and globally.

Acknowledgements

For Lähteenmäki-Uutela, this research was funded by the Academy of Finland under the EE-TRANS project, grant number 315898.

Conflict of interest

The authors declare no conflict of interest.

References

Allegretti, G., Talaminia, E., Schmidt, V., Bogorni, C. and Ortega, E., 2018. Insect as feed: an emergy assessment of insect meal as a sustainable protein source for the Brazilian poultry industry. Journal of Cleaner Production 171: 403-412.

Animal Health Australia, 2020. Australian ruminant feed ban. Animal Health Australia, Braddon, ACT, Australia. Available at: https://www.animalhealthaustralia.com.au/what-we-do/disease-surveillance/tse-freedom-assurance-program/australian-ruminant-feed-ban/.

Belluco, S., Losasso, C., Maggioletti, M., Alonzi, C., Paoletti, M.G. and Riggi, A., 2013. Edible insects in food safety and nutritional perspective. Comprehensive reviews in Food Science and Food Safety 12: 296-313.

Black, Z., 2020. Don't fancy eating insects? There's a good chance you already are. The New Daily, January 9, 2020. Available at: https://thenewdaily.com.au/life/eat-drink/2020/01/09/edible-insects-crickets-australia/.

Crespo, T., 2019. Eating insects, gourmet trend approaching Argentina. The trend comes from Europe to Argentina. Available at: https://www.serargentino.com/en/taste/gastronomy/eating-insects-gourmet-trend-approaching-argentina

Dao, T., 2020. Thai Union to invest in insect-based feed companies. Available at: https://www.seafoodsource.com/news/aquaculture/thai-union-considers-using-fish-feed-from-insects.

DiGiacomo, K., Akit, H. and Leury, B.J., 2019. Insects: a novel animal-feed protein source for the Australian market. Animal Production Science 59: 2037-2045. https://doi.org/10.1071/AN19301.

Directorate-General for Health and Food Safety (DG SANTE), 2017. Strategic safety concept for insects as feed, updated. DG SANTE, Brussels, Belgium. Available at: https://ec.europa.eu/food/sites/food/files/safety/docs/animal-feed_marketing_concept-paper_insects_201703.pdf.

Directorate-General for Health and Food Safety (DG SANTE), 2018. 2018-6339: final report of an audit carried out in the Netherlands from 13 to 16 November 2018 in order to evaluate the use of insects in animal feed. DG SANTE, Brussels, Belgium. Available at: https://ec.europa.eu/food/audits-analysis/audit_reports/details.cfm?rep_id=4138.

Directorate-General for Health and Food Safety (DG SANTE), 2019. 2019 2019-6647: final report of an audit carried out in France from 17 to 21 June in order to evaluate the use of insects in animal feed. DG SANTE, Brussels, Belgium. Available at: https://ec.europa.eu/food/audits-analysis/audit_reports/details.cfm?rep_id=4216.

Edwards, A. and Ranasinghe K., 2019. Does it taste like chicken? Australian insects a potential new snack. CSIROscope. Available at: https://blog.csiro.au/australian-insects-potential-new-snack/.

European Commission (EC), n.d. Novel food – authorization procedures. Summary applications and notifications. EC, Brussels, Belgium. Available at: https://ec.europa.eu/food/safety/novel_food/authorisations/summary-applications-and-notifications_en.

Food and Agricultural Materials Inspection Center (FAMIC), 2014. Act on safety assurance and quality improvement of feeds. Available at: http://www.famic.go.jp/ffis/feed/r_safety/r_feeds_safety11.html.

Food Standards Australia New Zealand, 2020. Bovine spongiform encephalopathy (BSE). Food Standards, Majura Park, ACT, Australia. Available at: https://www.foodstandards.gov.au/industry/bse/Pages/default.aspx.

González, F.C.V. and Contreras, A.T.R., 2009. La Entomofagia en México. Algunos aspectos culturales. El Periplo Sustentable 12: 57-83.

Grups UP, 2020. Super nutrition. Grups UP, Pinjarra, WA, Australia. Available at: https://grubsup.com.au.

Health Canada, 2016. Guidelines for the safety assessment of novel foods derived from plants and microorganisms. Health Canada, Ottawa, ON, Canada.

Insect Protein Association of Australia (IPAA) About – insects as feed. https://www.insect-proteinassoc.com/insects-as-feed. Accessed 5 July 2020.

Insect Protein Association of Australia (IPAA), 2020. Our mission. IPAA, Canberra, ACT, Australia. Available at: https://www.insectproteinassoc.com/.

Jones, K., 2019. Rise of the edible bugs. Sydney Herald, 7 October 2019. Available at: https://www.smh.com.au/business/small-business/rise-of-the-edible-bugs-20190927-p52vl1.html.

Kelemu, S., Niassy, S., Torto, B., Fiaboe, K., Affognon, H., Tonnang, H., Maniania, N.K. and Ekesi, S., 2015. African edible insects for food and feed: inventory, diversity, commonalities and contribution to food security. Journal of Insects as Food and Feed 1: 103-119. https://doi.org/10.3920/JIFF2014.0016

Lähteenmäki-Uutela, A., Grmelová, N., Hénault-Ethier, L., Deschamps, M.-H., Vandenberg, G. W., Zhao, A., Zhang, Y., Yang, B. and Nemane, V., 2017. Insects as food and feed: laws of the European Union, United States, Canada, Mexico, Australia, and China. European Food and Feed Law Review 12: 22-36.

Lähteenmäki-Uutela, A., Hénault-Ethier, L., Marimuthu, S.B., Talibov, S., Allen, R.N., Nemane, V., Vandenberg, G.W. and Józefiak, D., 2018. The impact of the insect regulatory system on the insect marketing system. Journal of Insects as Food and Feed 4: 187-198.

Leon, M.A. and Paz, E., 2013. A perspective of food safety laws in Mexico. Journal of the Science of Food and Agriculture 94(10): 1954-1957.

Maxabella, B., 2019. Are Australians ready yet to embrace ants, crickets and locusts for dinner? SBS, 14 February 2019. Available at: https://www.sbs.com.au/food/article/2018/06/28/edible-insects-edible-bug-shop.

Mitsuhashi, J., 1997. Insects as traditional foods in Japan. Ecology of Food and Nutrition 36: 187-199.

Nederlandse Voedsel- en Warenautoriteit (IG-NVWA), 2020. Managementreactie IG-NVWA op BuRO advies over de dier- en volksgezondheidsrisico's van op VVM gekweekte insecten als grondstof voor diervoeder. IG-NVWA, Utrecht, The Netherlands. Available at: https://www.nvwa.nl/documenten/dier/diervoeder/diervoeder/publicaties/bijlage-managementreactie-ig-nvwa-buro-advies-gekweekte-insecten-als-grondstof-voor-diervoeder.

Nederlandse Voedsel- en Warenautoriteit (NVWA-BuRo), 2020. Advies van BuRO over de risico's van op voormalige voedingsmiddelen gekweekte insecten als grondstof voor diervoeder. NVWA-BuRo, Utrecht, The Netherlands. Available at: https://www.nvwa.nl/documenten/dier/diervoeder/diervoeder/risicobeoordelingen/advies-van-buro-over-gekweekte-insecten-als-grondstof-voor-diervoeder.

Niassy, S., Ekesi, S., Hendriks, S.L. and Haller-Barker, A., 2018. Legislation for the use of insects as food and feed in the South African Context. In: Halloran, A., Flore, R., Vantomme, P. and Roos, N. (eds.) Edible insects in sustainable food systems. Springer, Berlin, Germany, pp. 457-470.

Preteseille, N., Deguerry, An., Reverberi, M. and Weigel, T., 2018. Insects in Thailand: national leadership and regional development, from standards to regulations through association. In: Halloran, A., Flore, R., Vantomme, P. and Roos, N. (eds.) Edible insects in sustainable food systems. Springer, Berlin, Germany, pp. 435-441.

Shen, R., 2014. China regulations on new food raw materials. Chemlinked. Available at: https://food.chemlinked.com/foodpedia/china-regulations-new-food-raw-materials.

Shockley, M., Lesnik, J., Nathan Allen, R. and Fonseca Munõs, A., 2018. edible insects and their uses in North America; past, present and future. In: Halloran, A., Flore, R., Vantomme, P. and Roos, N. (eds.) Edible insects in sustainable food systems. Springer, Berlin, Germany, pp. 55-78.

South Africa Government, n.d. Register farm feeds and pet food. South Africa Government, Pretoria, South Africa. Available at: https://www.gov.za/node/727474.

Sun, J., 2015. The regulation of 'Novel Food' in China: the tendency of deregulation. European Food and Feed Law Review 10: 442-448.

Usman, H.S. and Yusuf, A.A., 2020. Legislation and legal frame work for sustainable edible insects use in Nigeria. International Journal of Tropical Insect Science. https://doi.org/10.1007/s42690-020-00291-9.

Van der Spiegel, M., 2016. Safety of foods based on insects. In: Prakash, V., Martin-Belloso, O., Keener, L., Astley, S., Braun, S., McMahon, H. and Lelieveld, H. (eds.) Regulating safety of traditional and ethnic foods. Elsevier Academic Press, New York, NY, USA, pp. 205-216.

Yen, A.L., 2005. Insect and other invertebrate foods of Australian Aborigines. In: Paoletti, M.G. (ed.) Ecological implications of minilivestock. Science Publishers Inc., Enfield, NH, USA, pp. 367-387.

Industrial processing technologies for insect larvae

D. Sindermann[1]*, J. Heidhues[1], S. Kirchner[2], N. Stadermann[3] and A. Kühl[3]

[1]GEA Westfalia Separator Group GmbH, Process Technology and Innovation, Renewables, Chemicals & Pharma, Werner-Habig-Str. 1, 59302 Oelde, Germany; [2]GEA Westfalia Separator Group GmbH, Business Line Renewables, Werner-Habig-Str. 1, 59302 Oelde, Germany; [3]Maschinenfabrik Reinartz GmbH & Co. KG, Industriestraße 14, 41460 Neuss, Germany; *dirk.sindermann@gea.com

Abstract

For an economic production of safe and standardised products from commercially reared insects larvae for food and feed, industrial processing technologies for insects processing are needed. Protein meals for feed and food produced from insect larvae typically vary in fat content. Main factors influencing the fat content are the individual species, the substrate feed during rearing and the time of harvest. However, feed and food industry are looking for standardised products which can be adjusted to the customers' specifications. Separation technologies to recover insect fat and thereby reduce the residual fat content in the dry meal have been adopted from familiar applications and have been further developed for insect larvae de-fatting. Two major process technologies that are used for industrial applications are discussed in this technical report: dry and wet processing. In comparison, both technologies have their individual advantages depending on the individual application and properties aimed for. Since these processes for lipid separation are joint processes not only low-fat meal is recovered but also lipids recovered can add value as an additional product. In addition, chitin can be separated to increase the protein content in the larvae meal and add value as biomaterial for further processing, e.g. production of chitosan. Moreover, automation and cleaning of complete process lines are important considerations. Especially for future food applications.

Keywords

insect larvae – processing technology – lipids – protein – chitin – mechanical separation – drying – processing steps – processing stages – devitalisation – grinding – cleaning – automation

1   Introduction

With the expected increasing demand in food and feed, insects' leading to upscale insect production and increasing commercialisation, industrial insects larvae processing becomes more and more important for the economic production of safe and standardised products from commercially reared insects larvae for food and feed. This technical report gives an overview of the most important processing technologies and machines for recovering protein enriched meal and insects lipids from insects larvae derived from industrial insect farming. Chitin removal as a tool to increase the proportion of digestible protein in the meal is also discussed. The technologies outlined regarding processes and machinery for insects processing are mainly based on relevant industrial experience and perspective. Therefore, in this overview it is focussed on reference plants that are currently being tested or delivered in the insect industry or that have high 'technology readiness levels' (TRL) (European Commission, 2019) based on experience from very familiar applications. Processing of rearing residues is not covered in this overview.

Regarding the process technologies for insect meal and lipids recovery the important process stages for the two different technologies dry and wet processing are presented and discussed. To outline and clearly distinguish each individual process, it is focussed on differences in mechanical separation and typical drying technologies used for the two processes.

2   Motivation for insect larvae processing

Several publications about the benefits of insect larvae reared in industrial vertical farming systems to recover high value meal for feed and food have been published (De Jong, 2018; Halloran and Vantomme, 2013; Van Huis and Tomberlin, 2017; Van Huis *et al.*, 2013). For some applications the whole larvae are dried and sold as feed and sometimes food. For other applications there is a need to disintegrate the whole larvae and dry the product to recover a meal and therefore remove most of the water and retain all other ingredients of the raw material. This is a simple process and is especially interesting for start-ups entering meal markets. Disadvantageous is a relatively high fat content and therefore lower protein content. Also, transportation and storage are more difficult for meals high in fat. Higher fat contents in solids to be dried also limit the choice of suitable dryers since the fat contend in the meal can be a selection criterion for the type of dryer used.

For all these reasons, an economic removal of a significant amount of lipids from the insect larvae is considered by industrial insect meal producers. Because of a joint process, insect lipid is recovered as a separate phase which can add value to the process beside obtaining a meal containing a higher amount of protein. In some cases, the chitin fraction can also be separated and recovered as fraction for further processing (Section 7).

Common processes used in practice are derived from traditional technology for comparable applications like processing of animal by-products (Düpjohann, 1991) or fish (GEA, 1999). It is important to factor in legal requirements and regulations that are varying between regions and countries regarding the use of insect meal and derived insect lipid fractions in feed and food. This affects the use of meals and lipids for feed and especially food applications. For example, regulatory challenges are limiting the use of separation technologies when it comes to protein meal for food applications in many regions, e.g. the European Union (IPIFF, 2018). Figure 1 gives an overview over different potential insect applications and products that could be obtained by using different processing technologies.

For de-fatting insect larvae, different process technologies can be applied. In this overview the focus is on the industrial perspective for the two main processing pathways used: dry and wet processing. Therefore, not all theoretically possible approaches are discussed, but only relevant and economically feasible technologies based on practical experiences made within the insect industry and familiar industries are presented.

3   Devitalisation

Legal requirements and animal welfare aspects must be considered before larvae received from insect farming can be further processed (IPIFF, 2019). Effective and efficient killing of living larvae for further processing can be achieved with various systems. For dry processing (Section 4) devitalisation is often conducted by steam-

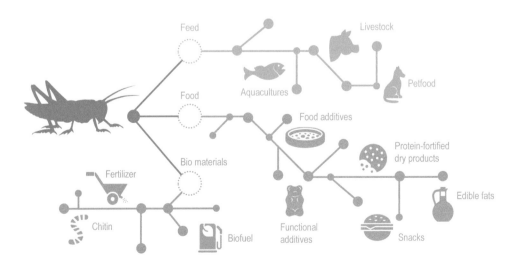

FIGURE 1   Potential applications upon processing insect larvae for food, feed and bio materials (Sindermann, 2019b)

ing. Depending on the type of dryer used for this process, steaming is performed in a separate chamber before the dryer or directly inside a chamber of the dryer. At the same time steaming increases the efficiency of the drying process significantly.

Alternatively, hot water (T > 90 °C) is applied for a rapid devitalisation. For wet processing this method has been applied using tanks for small batch processes or heat exchangers for continuous processing. When using double wall tubular heat exchangers, processing in closed systems allows efficient 'cleaning-in-place' (Section 5 'Cleaning'). This is relevant for food grade wet processing. Typically, larvae are pumped through the heat exchanger with hot water added at the feeding pump. Screening systems (e.g. vibrating screens; Figure 2) are used to remove most of the water after the devitalisation step. To safe or recycle water and recover heat, water is collected and recirculated to the feeding pump.

Insect devitalisation by hot water is not only essential for insect welfare purposes but also has advantages regarding end product qualities when applying this step. For example, the de-activation of own insect enzymes reduces the 'browning effect' of the larvae significantly. Devitalisation with hot water has proven to reduce typical browning when processing larvae (Heidhues *et al.*, 2020). Another positive effect of using hot water for devitalisation is a reduction of surface contamination due to washing the larvae before further processing (Rumpold *et al.*, 2017). This results in a reduced ash content and therefore increased protein content. Since devitalisation with hot water is resulting in some water absorption or uptake (approx. 5-10%) of the insects, it is more appropriate for wet processing (Section 5).

4   Dry processing technologies

The major feature of dry processing is a drying step to remove most of the moisture of the insect larvae prior to the mechanical separation of lipids.

This important difference to wet processing (Section 5) not only has an impact on operational costs and capital expenditure (Schmidt, 2019), it is also influences product qualities of the different fractions recovered (Böschen, 2018). Sometimes

FIGURE 2   Vibrating screen (photo by GEA Scan-Vibro A/S, Svendborg, Denmark)

this process is also referred to as 'high temperature process' when dryers with higher drying temperatures above 100 °C are used. Even at temperatures around 80 °C and retention times of approx. 30 min, temperatures and retention times are high enough to meet IPIFF guidelines for safe products (IPIFF, 2018). In some publications the process is referred to as 'screw press process' because of the press used in the lipid separation stage (Section 4 'Lipid separation').

4.1   Drying

Different types of dryers can be used for drying of insects, that cannot all be presented here. In this overview, three exemplary dryer types with a larger practical relevance are discussed in more detail. These three, totally different types can also be used to dry wet solids in the wet processing pathway described in Section 5.

4.1.1   Disc dryer

Contact dryers such as a disc dryer are the most popular dryers used for comparable purposes. In many fish oil plants or high temperature rendering plants to recover animal fats, disc dryers are common. Depending on the required throughput of insect larvae to be dried, different systems for batch or continuous operations can be utilised. Economic aspects also have to be taken into consideration.

The larvae entering this type of contact dryer are dried by rotating steam-heated discs welded on a central shaft (Figure 3). These discs increase the total contact area for larvae inside the dryer and apply indirect heat over a larger area in a relatively compact design. Larvae are pushed through the dryer from one end to the other. Paddles on the edges of the rotating discs agitate the larvae and support the movement through the dryer towards the discharge outlet. Typically, vapour is collected in an exhaust gas hood and discharged via an exhaust gas pipe towards a condensation system. The wet exhaust gases can be fully condensed or further treated in an air washing system. Heat recovery systems are recommended to make insect processing more efficient and environmentally friendly.

Product qualities are strongly influenced by process temperatures applied (Kröncke et al., 2018). If meals with high functionality and meals and lipids with light colour are required vacuum drying configurations are also available for disc dryers.

Disc dryers can also be used to dry whole larvae without further de-fatting. But the size of the whole larvae will be reduced using disc dryers. If the original structure needs to be maintained other dryers need to be selected like fluid bed dryers, microwave dryers or ring dryers.

4.1.2   Fluidised bed dryer

In contrast to contact driers such as the disc dryer, fluidised bed dryers belong to the group of air dryers. Insect larvae processed in a fluidised bed dryer float on a cushion of air. Heated (or cooled) air is flowing through the product against grav-

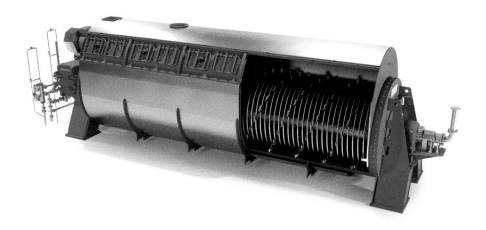

FIGURE 3   Typical disc dryer for continuous drying (Haarslev, 2020)

itation creating a fluidised bed. Process air is supplied to the bed through a perfo-
rated distributor plate and flows through the bed of solids at a sufficient velocity
to support the weight of particles in a fluidised state. Bubbles form and collapse
within the fluid bed of material promoting intense particle movement. In this state
the solids behave like a free-flowing boiling liquid. Very high heat and mass transfer
rates are obtained as a result of the intimate contact between individual particles
and the fluidising gas (GEA, 2020b).

A special feature of this type of drying are individual drying zones where insect
larvae can be dried to the required residual moisture content in a gentle way. For
insect larvae processing, because of the effect of temperature and time to the protein quality, cooling after drying is recommended. The drying and cooling section
can be two separate units or can be combined in one unit where approx. two third
of the process chamber are needed for drying and the remaining length is used for
cooling the larvae right after drying.

As illustrated in Figure 4 beside the drying and cooling chambers additional
equipment for drying and cooling air treatment and equipment for exhaust air
treatment including dust removal are required making this dryer more complex.

With fluidised bed dryers moist particles can be dried at defined temperatures
to achieve a specified residual moisture. Evaporation of large quantities of water
is possible in a relatively short period of time and under homogeneous and gentle
drying conditions.

For the insect industry another advantage is the flexible use for other end products like drying complete unprocessed insect larvae for food and feed markets. In
contrast to disc dryers the shape of larvae is retained. This also applies to microwave dryers described in the following subchapter.

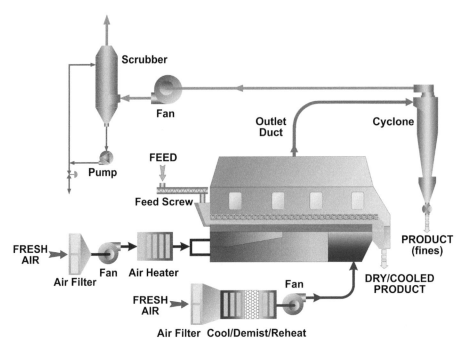

FIGURE 4   Principle of fluid bed dryer including heating and cooling section (GEA, 2020b)

4.1.3 Microwave dryer

In drying insect larvae microwave drying is a direct heating method and an alternative method (Leanarts *et al.*, 2018) with several units already installed for insects drying.

In the rapidly alternating electric field generated by microwaves, polar materials orient and reorient themselves according to the direction of the field. The rapid changes in the field cause rapid molecular reorientation of dipoles, resulting in friction and heat. For example at 2,450 MHz, the orientation of the field changes 2,450 million times per second. Different materials have different properties when exposed to microwaves, depending on the extent of energy absorption, which is characterised by the loss factor (Van den Bossche and Van Vaerenbergh, 2014).

Microwave dryers as depicted in Figure 5 are ready for operation quickly and provide a relatively short drying time with an evenly distributed energy input and well controlled drying process. Because of this energy can be saved and good organoleptic properties of the dried larvae can be achieved.

While many conventional drying systems especially for smaller capacities often work as a batch dryer microwave drying is suitable for continuous drying of the insect larvae passing a drying tunnel via a conveyor belt.

FIGURE 5   Continuous microwave dryer for insect larvae (pictures by Sairem, Décines-Charpieu, France)

4.2   *Lipid separation*

Fat and protein content of insect larvae depend strongly on species, substrate fed and developmental stage at time of harvest (Rumpold *et al.*, 2017). Typical fat contents vary e.g. for black soldier fly (BSF) larvae between 7 and 39% based on total larvae weight (Barragan-Fonseca *et al.*, 2017). Protein recovery from insect larvae is always a joint process, i.e. low-fat meal and consequently high protein products require an effective de-fatting stage. In addition, the separated lipid fraction can contribute to the added value of insect larvae processing.

An important process stage for dry processing is the mechanical de-fatting step by a screw press (Figure 6) pressing the dried insect larvae.

For BSF it is recommended to condition the product before the mechanical de-fatting. Target is to reach a product temperature, which is above the fat melting temperature, before the product gets pressed. This increases the efficiency of the press process and can be performed by using a process screw or a so-called heating cattle. The heat transfer is typically regulated via sensors and controllers.

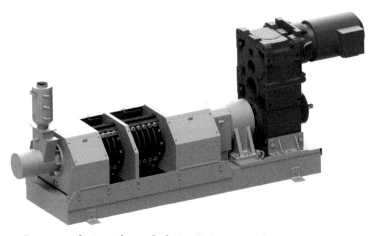

FIGURE 6   Screw press for insect larvae de-fatting (Reinartz, 2020)

Inside the screw press the insect larvae are exposed to increasing compression at the different process phases. During the first compression phase without liquid discharge, cells of the larvae are disrupted, and air can escape from hollow spaces. De-fatting takes place during the following compression stage while the pore volume is reduced, and lipids are released. After lipid separation the press cake can be formed into pellets or chips depending on specific requirements. For insect processing optimised presses are available reducing the residual fat content to approx. 6% due to a special geometry of the screw. Screw presses with temperature-controlled screw shaft for heating or cooling are available for insects processing. Additional heating improves the de-fatting process and cooling might be required if focus is retaining the functionality of the proteins in the meal. Since the lipid fraction is mainly consisting of fat, which solidifies at ambient temperature, so-called heat able pillow plates inside a fat discharge chamber keep the lipids collected in a liquid state. In combination with a fat conveying screw the discharge of the viscous fat can be controlled.

4.3   *Further processing of defatted insect meal*

4.3.1   Meal cooling
The insect press cake discharged from the screw press should be cooled after pressing to reduce an impact on protein quality and storage stability. This can be achieved with ambient air in an open evaporation screw or inside a counter current heat exchanger. Immediate subsequent cooling reduces thermal impacts on amino acid chains and therefore improves storage stability (Baltes and Matissek, 2011)

4.3.2   Meal grinding
After pressing, the press cake recovered by the screw press needs further milling to reduce its particle size based on individual requirements of the end customer. Typical particle sizes are approx. 1 mm but can be larger depending on intendent applications e.g. a more granular structure for pressing pellets for aquaculture feed.

The milling step to insect meal after the screw press pressing step typically takes place in impact crushers, flake crushers or hammer mills that are not described in detail here.

4.4   *Lipid clarification*
Lipid recovered after de-fatting of the insect larvae needs to be clarified to become a value-added product in addition to the obtained de-fatted protein-enriched insect meal. Different solutions for different plant sizes and customer requirements are available that are based on technologies used e.g. for vegetable oil or animal fat recovery.

4.4.1 Static sedimentation

For smaller quantities to be processed static sedimentation systems (Figure 7) might considered. Lipids recovered by the screw press are collected in heated sedimentation tanks for liquid clarification. After 5 to 7 days the clarified liquid is discharged by floating from the middle section of the tank. Remaining floating particles are removed by filter bags, which the fat is passing before it is collected in the clean fat tank. Sedimented solids are drained from the bottom. Despite relative low capital and operational expenditure for such systems long retention times of the lipids being in contact with residual moisture and solids might result in higher free fatty acid contents reducing the quality of the fat recovered.

4.4.2 Filters

Filter systems offer much quicker clarification overcoming the disadvantages of the static separation system using sedimentation tanks. Depending on the quantities of lipids to be clarified different filter designs should be considered. Two examples suitable for lipid clarification within dry processing systems are filter chamber presses (Figure 8) and vertical pressure leaf filters (Figure 9).

Filter Chamber Presses are mainly used for small and medium-sized production capacities. Raw fat discharged from screw presses is collected in heatable supply tanks to store the lipids in front of the discontinuous and semi-automated filter system. Filter plates can be mantled with specific filter cloths according to insect fats and specific solids content in the fat. These cloths hold back solids in different layers during filtration. The growing layer of solids collected is causing a self-filter-

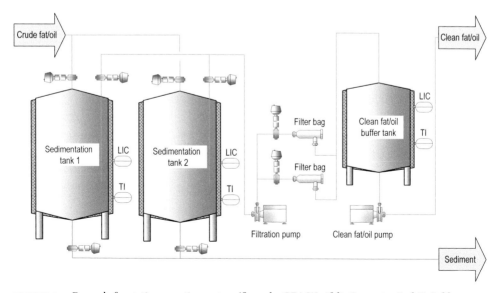

FIGURE 7   Example for static separation system (figure by GEA Westfalia Separator GmbH, Oelde, Germany)

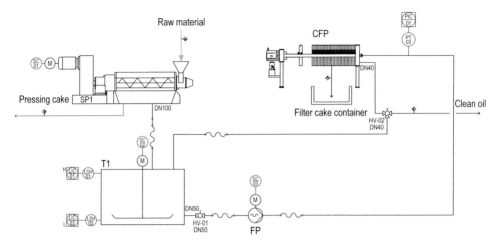

FIGURE 8   Filter chamber press (figure by Maschinenfabrik Reinartz GmbH & Co. KG, Neuss, Germany)

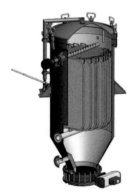

FIGURE 9   Vertical pressure leaf filter (M Miles Global, 2020)

ing layer, so that no other filtration aids are required. Depending on the design of the filter, the collected filter cake can be recirculated to the screw press.

As mentioned, another example for filters used in dry processing systems to clarify recovered lipids are vertical pressure leaf filters. These filters can be used for larger capacities of insect lipids to be clarified. Like for other clarification stages a heatable storage tank upstream of this filter system is required. The filtration process can be fully automated. Only for cleaning, filter leafs need to be removed depending on the solids load of the lipid after pressing.

4.4.3   Clarifying decanter centrifuge
Centrifuges are commonly used in familiar applications like processing of animal by-products, fish meal plants or vegetable oil plants to clarify fats and oils separated by screw presses. For medium and larger processing lines centrifuges for continu-

ous clarification of lipids from the pressing stage can be used for fully automated clarification and high flexibility if process conditions are changing.

A clarifying decanter as shown in Figure 10 is a horizontal screw centrifuge separating liquids from insoluble solids. The suspension from the liquid discharge of the press (Section 4 'Lipid separation') is fed into a relatively fast rotating cylindro-conical bowl via an inlet tube. Clarification of insect fat takes place according to differences in density. Inside the rotating bowl the heavy solids are collected at the bowl shell. The lighter liquid phase is separated as a liquid ring above the solids and is collected continuously at the liquid discharge (Hamatschek, 2016). A conveying scroll rotating at a differential speed slightly faster than bowl speed is conveying the collected solids out of the turning bowl towards solids discharge ports. In dry processing these solids can be recirculated into the feed of the screw press.

The right bowl and scroll design in terms of geometry, material and wear protection needs to be adapted for optimal clarification of insect lipids. For example, a flatter cone angle helps to convey fine and relatively light solids out of the bowl against acting centrifugal forces more easily.

Most designs allow a large flexibility regarding process variations. Relatively high solids loads can be processed with clarifying decanters (Hamatschek, 2016). This allows operators to change parameters of the screw press independently from the downstream lipid clarification. With modern systems bowl and differential speed can be adjusted according to changing throughput capacities or solids loads of suspensions to be clarified and therefore reducing energy costs, wear-and-tear if the maximum capacity is not required based on a specific production situation. In order to find an optimum clarification of the insect lipids and optimum de-fatting

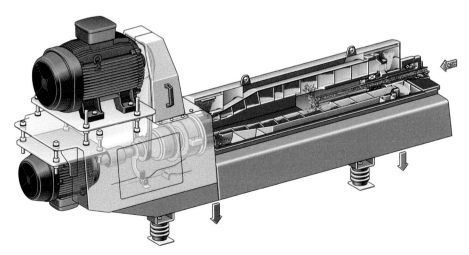

FIGURE 10   Clarifying decanter: Suspension (lipids discharged from press in dry processing) is fed into rotating bowl and separated into solids (brown arrow) and liquid (blue arrow) phase (figure by GEA Westfalia Separator GmbH, Oelde, Germany)

of the solids separated, decanters can be adjusted in terms of the liquid discharge diameter and therefore the resulting drying zone for the solids. Also, an optimum bowl and differential speed have a larger impact on the clarification results. Depending on the resulting centrifugal forces very fine solids can be separated assuring a low solid content of the insect fat.

5   Wet processing technologies

A completely different approach to recover protein-enriched meal, lipids and other fractions from insect larvae is wet or low temperature processing. In contrast to dry processing most of the water of the larvae is removed mechanically and only the separated fractions are dried afterwards. Especially for larger throughput capacities this is resulting in significant energy savings (Schmidt, 2019).

In the following an overview about the individual process steps for wet processing is provided focussing on relevant technologies to process insect larvae.

5.1   *Disintegration*

After the devitalisation stage (Section 3) a particle size reduction of the dead larvae needs to take place in wet processing. Insect lipid extraction from smaller particles is resulting in higher extraction yield and therefore lower residual fat content of the recovered protein-enriched meal (Horstmann, 2018).

5.1.1   Grinding

Grinders as depicted in Figure 11 which are commonly used in the fish and meat industry are suitable for this purpose. They have exchangeable hole plates which can be adjusted with different hole sizes for optimum fat extraction (Düpjohann, 1991).

Especially processing of wet products containing free water, lipids and proteins such as insect larvae is bearing the risk of creating emulsion when applying too much shear forces. Therefore, grinders should be operated with sharp knives and hole plates to cut the insect larvae and avoid squeezing. Moreover, high shear forces due to high speed cutters and mills for reducing the particle size of insect larvae often have the disadvantage of creating two much emulsion.

For small plants eccentric screw pumps with cutter and hole plate inside the pump housing (Figure 12) are available. These pumps enable cutting and conveying in one process step. Changing of knives and plates for re-sharpening takes more time compared with grinders designed for regular cleaning because such pumps need to be dismantled for this purpose. But relatively low capital expenditure can nonetheless be interesting for small capacity solutions.

FIGURE 11   Grinder with exchangeable hole plates (right side) for particle size reduction (figure by
            GEA Food Solutions, Bakel, The Netherlands)

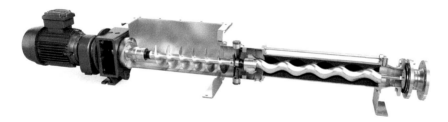

FIGURE 12   Eccentric screw pump with integrated cutting devices (Netzsch, 2020)

5.1.2   Soft separators

In the meat, poultry and fish industry soft separators are used separating soft meat
from harder shells or bones. A squeezing belt feeds the product towards a perforat-
ed rotating bowl and presses the soft components through the holes of the bowl.
Harder components remain outside the drum. The ratio between yield and quality
can be influenced by means of adjusting the pressure for squeezing the product
against the bowl shell (Baader, 2020).

For processing of insect larvae, the same technology can be applied to separate
the hard exoskeleton from the soft fraction containing water, proteins and lipids

(Laurent *et al.*, 2018). The soft fraction can be further processed like the minced larvae after the grinding stage and no further grinding is required.

The hard fraction, which is also called chitin fraction, can be further processed into chitosan. (Rumpold *et al.*, 2017). Protein, which either sticks as soft part at the exoskeleton or is bound in it, is lost for the protein recovery process. Therefore, the use of a soft separator reduces the total amount of protein which could be recovered in the downstream process.

5.2   *Pre-treatment*

In order to mechanically separate the lipids from the water and solids fractions temperatures of around 80-95 °C are recommended based on own practical experience. Arsiwalla and Aarts (2015) suggest temperatures between 70 and 100 °C. At temperatures above the melting point of insect fats separation already takes place. Lower temperatures preserve the functional properties of proteins. But due to lower fat yield and higher residual fat content in the solids fraction melting and separation take place at temperatures above 80 °C in industrial applications.

Beside the temperature the retention time can have a significant influence on the extraction of lipids. For example, in the olive oil recovery processes where the so-called malaxing is important for sufficient oil extraction (Hamatschek, 2006), longer retention times at higher temperatures are favourable to extract the lipids from its cells under continuous and gentle mixing at low impeller speed avoiding emulsions. Horstmann (2018) and Kirchner *et al.* (2019) have demonstrated the correlation between retention time and fat yield and suggest retention times of up to 120 minutes especially for *Tenebrio molitor*. To achieve this, larger processing lines can have two or more retention tanks in parallel melting in one tank and drawing product from the another for continuous processing of the downstream line. A tubular heat exchanger should be considered for pre-heating the larvae prior to the retention tanks. Especially for larger processing capacities the heating up period can be shortened. For smaller wet processing systems, a semi-continuous batch heating might be a more economical solution.

For more efficient heat transfer pre-heating of the minced raw product via tubular heat exchanger should be considered especially for medium and large capacity lines (approx. 3-10 t/h).

Addition of some hot water could be necessary depending on the amount of liquids available in the raw material and selected separation technologies (Section 5 'Mechanical separation'). This is reducing the viscosity of the suspension for further mechanical separation and helps to de-fat effectively. Since additional water is increasing drying costs the addition of water should be minimised or eliminated if possible.

Adjustment of pH has also a significant influence on lipid separation. It has been demonstrated in extensive test series with *T. molitor* (Heidhues *et al.*, 2020; Horstmann, 2018) how lower pH values (2-4) improve the creation of a larger lipid layer (by volume) compared with tests at higher pH values. Disadvantage of very low pH values are additional operating costs and the creation of salts because of acid to be added.

5.3   *Mechanical separation*

In wet processing of insect larvae, mechanical separation of the heated suspension containing lipids, water (from larvae and water added) and solids takes place mainly by using centrifuges which have different design features for the specific process stage being used for. In larger installations and especially when there is a focus on high quality of lipids recovered the combination of horizontal decanter centrifuges and vertical disc stack centrifuges is common.

5.3.1   Three-phase separating decanter

As outlined for lipid clarification in dry processing (Section 4 'Lipid separation') a clarifying decanter centrifuge separates the liquid phase from the solids phase according to differences in density continuously. For this two-phase suspension a 'liquid-solid' separation is required. In wet processing the suspension to be separated consists of at least one more phase since the water is not evaporated before separation (Figure 15). For this reason, a three-phase separating decanter for a 'liquid-liquid-solid' separation is commonly used for wet processing of insect larvae. While solid separation takes place like in a clarifying decanter described in Section 4 'Lipid clarification'.

Lipid clarification, lipids are separately discharged from the heavier water (in wet processing referred to as stickwater) according to different densities. Typical feed composition of two different samples of wet processed insect larvae are shown in Figure 13. The two liquid phases can be discharged separately as shown in the example in Figure 14.

To optimise the performance of a three-phase separating decanter the position of the separating zone can be adjusted by changing regulating rings or tubes for the heavy and light liquid discharge. This is required to minimise fat losses to the water or to reduce the amount of free water in the fat discharged to a minimum. Some centrifuge manufacturers offer systems for variable adjustment of the separating zone while the machine is in operation. This allows operators to fine tune the separation process and the insect fat yield and quality can be optimised without stopping the machine and changing parts.

For food grade solutions decanter centrifuges can be automatically cleaned with water, caustic and acid after a certain period or at the end of a production shift (Section 5 'Cleaning').

5.3.2   Disc stack separator

When the insect fat recovered needs to meet highest quality standards in terms of residual moisture and residual solids a disc stack separating centrifuge might be

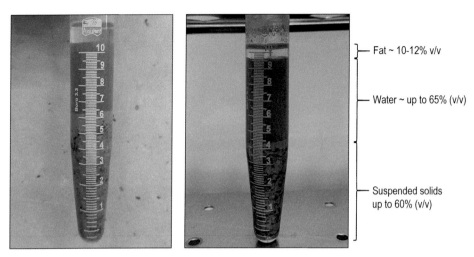

FIGURE 13   Typical spin test result of feed suspension to three-phase separating decanter centrifuge in
             wet insect larvae processing (figure by GEA Westfalia Separator GmbH, Oelde, Germany)

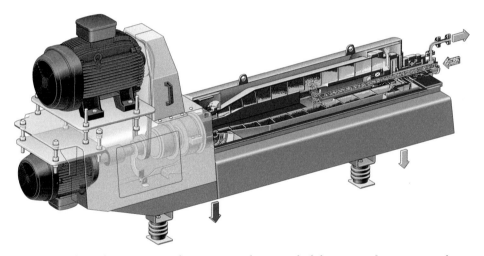

FIGURE 14   Three-phase separating decanter: Heated insect pulp fed into centrifuge is separated into
             solids (brown arrow), lipids (yellow arrow) and stickwater (blue arrow) phase (figure by
             GEA Westfalia Separator GmbH, Oelde, Germany)

required to 'polish' the fat discharge of the decanter. Typically, stack centrifuges
are utilised for suspensions which much smaller content of solids (typically up to
2% (v/v)) than possible for decanters. Due to the very high centrifugal forces and
the additional discs inside the bowl increasing the clarification area significantly
relatively small machines can easily reach equivalent clarification areas of approx.
10,000-100,000 m^2.

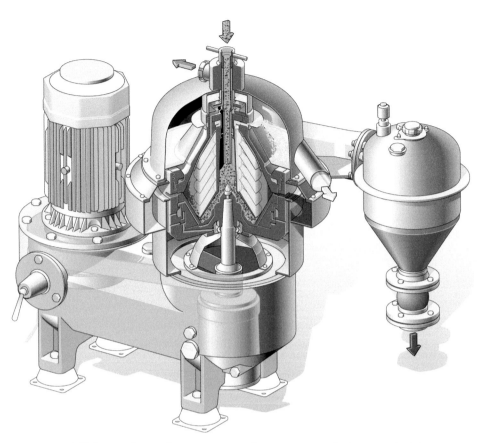

FIGURE 15 Self-cleaning disc stack separator: Insect fat containing residual water and solids is fed into the rotating bowl and is separated into clean fat (yellow arrow), stickwater (blue arrow) and solids (brown arrow) (figure by GEA Westfalia Separator GmbH, Oelde, Germany).

As depicted in Figure 15, the suspension is fed into the rotating bowl and residual solids are collected in a solids holding space inside the bowl due to their higher density. Self-cleaning machines can discharge the collected solids discontinuously out of the rotating bowl into a collection cyclone by means of a water-operated hydraulic system.

Disc stack centrifuges for insect fat polishing are designed to separate the liquid fraction into a light (clean fat) and a heavy phase (stickwater) in addition to solids separation ('liquid-liquid-solids separation'). Separation takes place according to differences in density of all three phases as discussed regarding the principle of a three-phase separating decanter before. Like for a separating decanter the position of the separation zone can be adjusted to find the optimum separation regarding fat yield and quality.

5.4   *Concentration of stickwater*

As discussed before mechanical separation of free water is the major difference of wet processing compared with dry processing of insect larvae. Due to several processing and heating steps the water separated contains soluble solids including proteins and small amounts of residual lipids. Therefore, this fraction is called stickwater like in animal fat or fish oil recovery due to its gluing properties. Especially for larger larvae processing plants the stickwater should not be discharged to the wastewater system because of its relatively high biological oxygen demand and should be further processed instead. Including a stickwater concentration stage, dissolved proteins can be recovered improving the total protein yield of a wet processing line.

5.4.1   Evaporation of the mechanically separated stickwater

One technology commonly used in familiar applications like animal by-products processing or fish oil recovery is evaporation. Basically, the dry matter of the incoming stickwater from mechanical separation (Section 5 'Concentration of stickwater') is increased by water evaporation. Especially for larger processing lines (e.g. 5 ton larvae per hour and more) this is much more efficient than direct water evaporation inside the dryer (Section 5 'Drying of solids'). Concentrated stickwater can be added more easily to the solids dryer. The addition is recommended to make use of this valuable side-stream and therefore to improve the total protein yield of the final product.

Different technical principles for evaporation are used in the food and chemical industry. Compared with many other food applications water evaporation capacities required are still relatively low for insect larvae processing. For these small capacities a suitable design for the insect processing industry is a plate evaporator as illustrated in Figure 16.

In this system stickwater and heating media (hot water or steam) are transferred in counterflow through their relevant passages. Defined plate distances in conjunction with special plate shapes generate strong turbulence, resulting in optimum heat transfer. Intensive heat transfer causes the product to boil while the vapour formed drives the residual liquid, as a rising film, into the vapour duct of the plate package. Residual liquid and vapours are separated in the downstream centrifugal separator.

Plate evaporators are more sensitive to fouling between the plates and need to be opened for manual cleaning regularly especially for food grade applications.

Currently processing lines for larvae with a capacity of up to 10 t/h are considered as being quite large plants. For such plants and the resulting larger amount of stickwater to be processed falling film evaporators could be an alternative to be considered to the plate evaporators discussed.

A falling film evaporator consists of a vertical shell-and-tube heat exchanger with a separating section for concentrate and vapour at the bottom. The stickwater

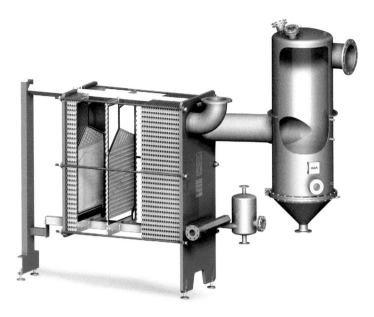

FIGURE 16   Plate evaporator (figure by GEA Wiegand GmbH, Karlsruhe, Germany)

to be concentrated is supplied at the top of the heating tubes and distributed in such a way as to flow down the inside of the tube walls as a thin film. The liquid film starts to boil due to the external heating of the tubes and is partially evaporated as a result.

If e.g. a disc dryer is used to dry the de-fatted solids recovered after mechanical separation (Section 4 'Drying') the vapours collected from the dryer can be used to heat the falling film evaporator resulting in a very energy efficient solution. Another major advantage of this type of evaporator is its possible sanitary design especially for food grade applications. But due to its significant height and more complicated internal structure the capital expenditure is significantly higher than for more simple plate evaporators. If markets for insect products will grow in future and production plants with larger capacities are considered due to economies of scale falling film evaporators will become more interesting.

5.4.2   Membrane filtration of the mechanically separated stickwater
Another option for concentration of stickwater for further processing would be membrane filtration. Here the stickwater to be concentrated flows through filter elements made from material with defined pore size. Filter elements can be made from plastics or ceramic. Some of the water passes the membranes as permeate (filtrate). The collected stickwater concentrate which could not pass the membrane layer is called retentate.

Especially because of fouling over a certain period of time this technology requires regular cleaning which can be made 'in place' (Section 5 'Cleaning'). As discussed for falling film evaporators membrane filtration will require larger capital expenditure and will only be interesting for larger quantities of stickwater to be concentrated. In addition, higher solid concentrations are possible for evaporators as concentrators for stickwater.

5.5 *Drying of solids*

Wet solids being collected during mechanical separation of the ground insects larvae (Section 5 'Mechanical separation') need to be stabilised by drying. Especially if the solids shall be used as ingredient for food applications drying to minimise microbiological problems needs to take place directly after mechanical separation of the solids. In general drying is a critical process step with strong impact on final product quality (Kröncke *et al.*, 2018) and therefore also for feed purposes an important consideration. Properties like colour, smell, taste or structure can be influenced by drying.

All three dryer types discussed for dry processing (disc, fluidised bed and microwave dryer) can be used for drying of wet solids recovered by wet processing. Especially the disc dryer has been applied quite often since it is a common technology used in familiar industries like fish meal or animal by-products processing. Since these three dryers have been discussed in detail in Section 4 'Drying', two different air dryer types especially relevant for wet processing are presented for drying wet solids.

In general drying of solids after mechanical separation in wet processing is offering a high flexibility regarding the choice of dryers since the low-fat content does not create any problems for certain pneumatic dryers or mill dryers like full-fat meals can do.

5.5.1 Ring dryer

Ring dryers employ the basic principle that solids to be dried are conveyed through the dryer in a hot air stream. Particle size reduction can be provided by a disintegrator within the dryer. In this case no subsequent milling step like after drying in a disc dryer is necessary if the structure of the insect meal shall be a fine powder. Ring dryers incorporate a centrifugal classifier allowing selective internal recirculation of semi-dried solids, effectively lengthening the retention time of larger particles in the dryer, while finer material, which dries more rapidly, exits the dryer and gets directly into the cyclone. Figure 17 shows the principle of a ring dryer.

Compared to a fluidised bed dryer as discussed in Section 4 'Drying' this pneumatic type of dryer is requiring a lower capital expenditure both for the equipment and for the space required.

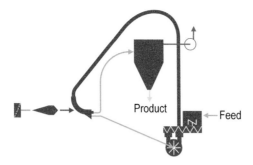

FIGURE 17   Schematic principle of ring dryer (GEA, 2020c)

5.5.2   Swirl Fluidiser®

If a more compact dryer with less height requirement than a ring dryer is needed a 'Swirl Fluidiser' (trademark of GEA) as shown in Figure 18 might be an option for insect meal producers.

The heated inlet air (e.g. directly heated by gas heater) enters a chamber where a swirl is created. The swirling air enters the drying chamber from the bottom. Wet solids are fed into the dryer by means of an agitated feed hopper and a screw conveyor. The agitator in the feed hopper is designed to de-lump the feed material and press it downwards into the screw conveyor. The feed enters the drying chamber at the upper part of a disintegrating rotor.

The fast-rotating disintegrating device, rotating vertically in the drying chamber, brakes up the feed. The dried fine product leaves the drying chamber in the top through a centrally mounted orifice. The dried solids are transported with the outlet air to a bag filter where they are discharged from the filter bottom cone by a rotary valve. The bag filter exhaust air is cleaned by a wet scrubber.

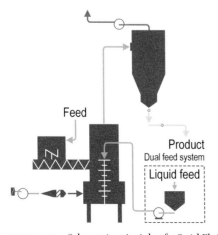

FIGURE 18   Schematic principle of a Swirl Fluidiser® (GEA, 2020c)

Both energy consumption and capital expenditure are quite similar for the swirl fluidiser and the ring dryer. The more compact swirl fluidiser is suitable for smaller buildings.

5.6 *Cleaning*

Cleaning of equipment and complete process lines needs to be considered depending on the intended use of the insect meal. While dry processing will be often used for feed, wet processing is predestined for production of ingredients for food and for feed. Therefore, cleaning is of higher importance for wet processing technologies.

Especially for food applications safe products depend on effective cleaning of processing equipment. Cleaning also needs to be efficient so that cleaning times are reduced to a minimum to have the plant available for value adding production. Constant yield and product quality can be assured with efficient cleaning.

Hamatschek (2016) lists various factors for an effective cleaning. Chemical cleaning effects are achieved by using different cleaning media like caustic and acid. A typical sequence for a cleaning procedure includes a water rinse, followed by caustic rinse, intermediate water flushing, an acid rinse and finally another water flushing step. Temperature effects are important and need to be independently adjusted by heating for each cleaning step. Mechanical cleaning effects are achieved e.g. by flushing the equipment at appropriate flow rates. Time is to be considered for sufficient swallowing and removal of protein contending dirt particles. Finally, the mechanical design of the components to be cleaned need to be hygienic in terms of geometry and surface roughness. Optimised mechanical design and cleaning procedure helps reducing the amount of water and chemical used. This contributes to reduced water footprint and therefore is an important aspect in terms of sustainability.

Most equipment for wet processing except for the grinder can be cleaned in-place (CIP). The general industry understanding on the terminology is that CIP means a totally automatic cleaning sequence with no manual involvement. CIP systems can vary regarding their complexity starting from relatively simple stand-alone units with some manual controls up to fully automated systems with full cleaning-in-place for the entire line. Once the optimum cleaning procedure has been set according to the factors mentioned above a highly effective and efficient cleaning with minimum downtime can be realised with a fully automated CIP. This can be a benefit regarding the operating costs. Because of additional tanks, pumps, valves and control technology including sensors the capital expenditure of a fully automated CIP system can be a significant factor.

6   Comparison of dry and wet processing

Both dry and wet processing technologies described within this overview have specific advantages depending on the required application and target markets and both systems have justified parallel existence. Dry processing and wet processing of larvae have been adapted from familiar applications like animal fat melting or fish oil recovery which have been applied for many decades. The equipment used for dry processing is available from familiar applications like animal by-products processing or vegetable oil recovery. In terms of the TRL (European Commission, 2019) all equipment used for the individual stages described within this work could be defined as 'TRL 9' ('proven systems in operational environment'). Reference installations from different suppliers have proven to be suitable for processing insect larvae, often larvae from BSF for the purpose of recovering protein meals for feed (GEA, 2020a).

In dry process systems most water of the processed larvae is evaporated before separation. Therefore, there is no stickwater after mechanical separation which needs to be processed or discharged. Especially for smaller capacities this can be an advantage because the higher energy input to evaporate most water in the dryer is less significant than at larger processing capacities.

By contrast during wet processing approx. 50% of the water of the processed larvae is removed by mechanical separation (three phase decanter) resulting in an increased energy efficiency compared with drying. Depending on local conditions regarding availability of heating media and energy costs, higher capital expenditure will be compensated by lower operating costs starting from throughput capacities for 3 to 5 t larvae/h (Schmidt, 2019).

As described in Section 5 'Concentration of stickwater', stickwater from wet processing should be processed to add value to the protein meal, to increase the recovered protein yield and to avoid discharging a side-stream with a relatively high amount of dissolved protein (approx. 5%) resulting in higher biological oxygen demand (BOD) if treated in waste water plants. Considering the lower amount of dissolved protein in condensate from dry processing vapours compared with stickwater from wet processing can be an advantage for dry processing.

For wet processing changes of different batches with intermediate cleaning can be done without major product losses. For the downstream dryer for drying the solids from mechanical separation this depends on the type of dryer used.

In general CIP is easier to achieve for wet processing. If meals and lipids shall be for food ingredients, cleaning is required and CIP needs to be considered (Section 5 'Cleaning').

Low process temperatures are important for high functionality of proteins and light lipid colour (Kröncke et al., 2018). This is relatively easy to achieve with wet processing. For dry processing special dryers and presses need to be selected if these factors need to be improved.

7   Chitin extraction

Chitin is included in the solids fraction separated by dry or wet processing. Removing the non-digestible chitin not only increases the protein content but also the digestibility of the finished insect meal. Chitin is a raw material for further processing, e.g. chitosan production. Therefore, removing the chitin can increase the value of the finished meal but also be an interesting by-product for additional value adding.

7.1   *Mechanical separation*

In Section 5 'Disintegration' the possibility of using a 'soft separator' has been presented.

For processing of insect larvae, the same technology applied e.g. for bone separation in the meat industry or hard-shell separation for shrimps can be applied to separate the hard exoskeleton from the soft fraction (Laurent *et al.*, 2018). The hard fraction separated by a soft separator is containing the chitin where it is bound to skleroprotein (Rumpold *et al.*, 2017). As a result, the overall protein recovery yield is slightly reduced while the finished meal is higher in protein content.

7.2   *Enzymatic hydrolysis*

Another way of removing the chitin from the protein and lipid phases is enzymatic hydrolysis. In contrast to well established chemical extraction this is an environmentally safe biological extraction using proteases (Rumpold *et al.*, 2017). After mincing of the larvae and addition of water, enzymes are added. Due to enzymatic hydrolysis most protein is dissolved in the water phase.

In the following a two-stage mechanical separation takes place. By adjusting the bowl speed, the first decanter centrifuge is 'classifying' only the heavy chitin fraction from the suspension containing dissolved but also some lighter insoluble protein. At relatively low centrifugal acceleration, the chitin is discharged as solid phase and the rest of the suspension including soft solids via the liquid phase discharge. In the second stage, a three-phase separating decanter is separating lipids, liquid protein solution and suspended solids containing protein.

The chitin phase recovered after mechanical separation is containing nearly no protein which in turn is resulting in low protein losses. On the other hand, a much more complex processing is required resulting in higher capital expenditure.

8   Scaling-up of process lines

For producers of insect meals and other insect products scaling-up is an important consideration. According to the systematic of Hamatschek (2016) to get an innovative product from an idea to an industrial production the following steps are necessary:

- lab scale trials ('test kitchen');
- test plant trials;
- pilot production;
- industrial production.

To devise a business plan very small size tests in a lab can help to determine the right process conditions and can result in an estimated mass balance which is necessary to get the economics right.

It is recommended to verify these recipes in test plants where test equipment with similar technology used for the production later on is available. When doing pilot tests, the maximum throughput for the corresponding equipment at acceptable results in terms of qualities and yields achieved should be tested to be able to scale-up to an economical feasible solution. For example, in wet processing scalability from very small test production volumes in pilot plants to larger processing capacities of up to 10,000 kg/h is relatively accurate. Since centrifuges play a major role in this wet processing, scale up factors from very small units up to larger production machines are typically higher than in other technologies like filtration.

A challenging question is the logistics for living larvae to be transported to process test centres of suppliers. In dry processing also dried larvae can be tested to scale-up e.g. presses which can make logistics much easier regarding testing.

For new products, pilot plants are reducing technical risks before larger industrial production of new products. This is to verify the economics estimated during the first development tests. Very helpful is also the production of small test batches to recover some product to be tested and verified. Many suppliers offer pre-engineered modular pilot plant solutions to limit capital expenditure.

When all steps have been passed successfully both the insect meal producer and the supplier of equipment have a relatively high security to be successful in large scale industrial production. Especially for the relatively young insect industry this should be considered for new product ideas, e.g. new ingredients for food.

9   Automation

When it comes to production lines for recovery of de-fatted insect meal as described for both dry and wet processing the level of automation becomes an important consideration. While small pilot plants can still be operated to a large extend manually a continuously operated process line needs a much higher degree of automation especially if only one to two operators have to run the complete line. Personal cost are one of the main drivers of production costs (Veldkamp *et al.*, 2012).

In general, all technologies discussed within this overview could be operated as stand-alone units. Each process stage could have its own independent electrical control unit. Especially for smaller lines which might be changed in their near future regarding set up or regarding some further adaptions of additional equipment this might be economically sensible.

Equipment in modern production lines are often liked to process control systems controlling the entire plant. Connections can be made via bus systems or other networks, e.g. ethernet. This allows flexible adoptions of the control system in case of future changes. But more important the entire process is controlled via a central system allowing the operator to have good control over qualities and yields achieved. Process data can be easily accessible for remote display or even control if required. Remote maintenance systems can be implemented reducing operating expenditure in the long term because of optimised maintenance schedules and limited number of unplanned downtimes of the line.

Automation is also an relevant consideration for cleaning of complete processing lines, especially for wet processing. Automation of the cleaning process ensures repeatability, allows validation and minimises downtime (Section 5 'Cleaning').

In future digitalisation will most probably become more important. For insect meal producers this can be interesting not only for insect farming but also for further processing, e.g. production of low-fat meals and insect fats. This might contribute to higher product qualities and efficiencies in the entire value chain.

Some research projects are looking into the possibilities for such digital technologies (IFF, 2020). An important driver is suitable sensor technology to measure the relevant values online and to optimise processes based on optimised algorithms accordingly. Especially near infrared spectroscopy (NIR) could be a vehicle to measure different relevant data like protein content, moisture, fat content, etc. This can not only help to harvest larvae at their optimum age but also to optimise the process technology for downstream processing accordingly (IFF, 2020). Research work carried out at a German vegetable oil processor has shown the significance of such technologies to optimise processes and in future to self-optimise processing systems based on online measurement solutions like NIR (Rumrich, 2020).

10   Outlook

Even though technology readiness levels for insect larvae processing are high today, the permanent market pressure and consequently the need for more economical and sustainable solutions for producers of insect meals and lipids will most likely result in further technical developments. Incremental innovations will probably play a bigger role for developing the technology for this growing industry than disruptive new and completely different technologies. Especially in the field of process automation and digitalisation existing technologies already used in familiar industries might play a larger role for insect meal producers in future than today. Technologies like NIR might contribute to optimum process control and therefore higher product qualities or increased yields for producers.

For food applications processed insect larvae added as meals or lipids to specific recipes will probably open markets for larger volumes. Unless for niche markets, higher acceptance for products with neutral ingredients in terms of e.g. smell and colour is expected than for products where insect origin can be recognised (Sindermann, 2019a). De-fatted meal with low fat content will be preferred by food and feed producers to adjust required fat contents easily. More sanitary design of processing systems including automated cleaning systems will be required when markets will be ready for de-fatted insect meals and fats as ingredients for the food industry.

11   Conclusions

Technologies for industrial production of de-fatted insect meal including insect lipids and chitin have been adapted from familiar applications and further developed for the specific needs of this industry. Both processing types, dry and wet processing, have their individual advantages for different industrial applications. For smaller and medium larvae processing capacities dry processing is often more economical. For larger capacities, wet processing has got energetic advantages because of mechanical separation of most of the water which as a consequence does not have to be evaporated directly.

All technologies discussed are used for insect larvae processing in reference plants or are used in very familiar applications and have been tested in pilot operations for insect larvae. Currently most plants for de-fatting of insect larvae are processing larvae for feed production, e.g. larvae of BSF (GEA, 2020a). For high product qualities of the recovered protein meals gentle drying technologies are available.

If in future de-fatting is requested more often for production of food ingredients depending on local legislation, food-grade design and cleaning of processing equipment becomes an important consideration for producers of such insect meals. Wet processing systems are relatively easy to clean and all equipment except the grinder for disintegration can be cleaned-in-place.

For some technologies discussed economies of scale require larger production volumes than most systems being in operation today. This has been shown e.g. for technologies like evaporation in wet processing.

Even though the insect industry will probably come up with many new product innovations within the next time, the major technologies for dry and wet processing are ready to be applied. Most equipment will be ready for use with minor adjustments and after pilot testing production of new products will be ready relatively quickly if marked market demand increases.

Conflict of interest

The authors declare no conflict of interest.

References

Arsiwalla, K. and Aarts, K., 2015. Method to convert insects and worms into nutrient streams and compositions obtained thereby. Patent: WO2014123420 A1. Available at: https://patents.google.com/patent/WO2014123420A1/en?oq=WO2014123420+A1.

Baader, 2020. What is Baader Meat? Nordischer Maschinenbau Rud. Baader GmbH+Co.KG, Lübeck, Germany. Available at: https://www.baader.com/en/products/separator_processing/.

Baltes, W. and Matiseek, R., 2011. Lebensmittelchemie, 7. Vollständig überarbeitete Auflage. Springer, Heidelberg, Germany.

Barragan-Fonseca, K.B., Dicke, M. and Van Loon, J.J.A., 2017. Nutritional value of the black soldier fly (*Hermetia illucens* L.) and its suitability as animal feed – a review. Journal of Insects as Food and Feed 3: 105-120.

Böschen, V., 2018. Comparison of wet and dry processing and fractionation for mealworms (*Tenebrio molitor*). Insecta Conference 2018. Gießen, Germany.

De Jong, B., 2018. Not to be a pest: why insects are gaining popularity as feed and food. RaboResearch, Rabobank, Utrecht, The Netherlands.

Düpjohann, J., 1991. Centrifuges, decanters and processing lines for the recovery of edible fat. Technical-scientific documentation no. 15, Westfalia Separator AG, Oelde, Germany

European Commission, 2019. HORIZON 2020 – work programme 2018-2020. European Commission Decision C(2019)4575 of 2 July 2019. European Commission, Brussels, Belgium.

GEA, 1999. Decanters and separators for industrial fish processing. GEA Westfalia Separator Industry, Oelde, Germany.

GEA, 2020a. GEA explores potential of insect protein for animal feed. GEA, Düsseldorf, Germany. Available at: https://www.gea.com/en/stories/gea-explores-potential-of-insect-protein-for-animal-feed.jsp.

GEA, 2020b. Fluid bed dryer (static). GEA, Düsseldorf, Germany. Available at: https://www.gea.com/en/products/dryers-particle-processing/fluid-beds/fluid-bed-dryer.jsp.

GEA, 2020c. GEA drying and particle formation technologies. GEA, Düsseldorf, Germany. Available at: https://www.gea.com/en/binaries/drying-spray-atomizer-fluid-bed-particle-formation-chemical-gea_tcm11-34869.pdf.

Haarslev, 2020. Disc dryer for continuous drying of de-fatted fish, animal or poultry by-products. Haarslev, Søndersø, Denmark. Available at: https://haarslev.com/products/disc-dryer-2/.

Halloran A. and Vantomme P., 2013. The contribution of insects to food security, livelihoods and the environment. Food and Agriculture Organization of the United Nations, Rome, Italy. Available at: http://www.fao.org/3/i3264e/i3264e00.pdf.

Hamatschek, J., 2016. Lebensmitteltechnologie. Eugen Ulmer, Stuttgart, Germany.

Heidhues, J., Kirchner, S. and Sindermann, D., 2020. Method for obtaining products for the food industry and/ or feed industry from insects and solid phase obtained from insects. Patent: WO 2020 / 011903 A1. Available at: https://patents.google.com/patent/WO2020011903A1/en.

Horstmann, N., 2018. Optimierung des Nassextraktionsverfahrens für die Verarbeitung von Insekten. BSc thesis, University of Applied Sciences, Osnabrück, Germany.

Internationale Forschungsgemeinschaft Futtermitteltechnik e.V. (IFF), 2020. Optimierung der Aufbereitung von Mehlkäferlarven (*Tenebrio molitor*) und daraus resultierender Produkte durch eine automatisierte Prozessführung auf Basis eines nichtinvasiven Nahinfrarot-Messsystems. IFF, Braunschweig, Germany. Available at: https://www.iff-braunschweig.de/insekten/.

International Platform of Insects for Food and Feed (IPIFF), 2018. The European insect sector today: challenges, opportunities and regulatory landscape. IPIFF, Brussels, Belgium.

International Platform of Insects for Food and Feed (IPIFF), 2019. Guide on good hygiene practices for European Union (EU) producers of insects as food and feed. IPIFF, Brussels, Belgium. Available at: https://ipiff.org/wp-content/uploads/2019/12/IPIFF-Guide-on-Good-Hygiene-Practices.pdf.

Kröncke, N., Böschen, V., Woyzichovski, J., Demtröder, S. and Benninga, R., 2018. Comparison of suitable drying processes for mealworms (*Tenebrio molitor*). Innovative Food Science and Emerging Technologies 50: 20-25. https://doi.org/10.1016/j.ifset.2018.10.009

Laurent, S., Sarton Du Jonchay, T., Levon, J.-G., Socolsky, C., Sanchez, L., Berezina, N., Armenjon, B. and Hubert, A., 2018. Method for treating insects in which the cuticles are separated from the soft part of the insects using a belt separator, Patent: WO 2018 / 122353 A1, Available at: https://patents.google.com/patent/WO2018122475A1/en.

Leanarts, S., Van der Borght, M., Callens, A. and Van Campenhaut, L., 2018. Suitability of microwave drying for mealworms (*Tenebrio molitor*) as alternative to freeze drying: impact on nutritional quality and colour, Food Chemistry 254: 129-136.

M Miles Global, 2020. Pressure filter leaf vertical & horizontal. B Miles Global Co., Ltd., Pathum Thani, Thailand. Available at: https://www.bmiles-filtration.com/product/pressure-filter-leaf-vertical-horizontal/.

Netzsch, 2020. NEMO BO/SO hopper pump with cutting device. Netzsch, Dorchester, UK. Available at: https://pumps.netzsch.com/en/products-accessories/nemo-progressing-cavity-pump/boso-with-integrated-cutting-device/.

Reinartz, 2020. Insect proteins from sustainable production. Reinartz, Neuss, Germany. Available at: https://www.reinartz.de/en/insect-proteins/.

Rumpold, B., Bussler, S., Jäger H. and Schlüter, O., 2017. Insect processing. In: Van Huis, A. and Tomberlin, J. (eds.) Insects as food and feed – from production to consumption. Wageningen Academic Publishers, Wageningen, The Netherlands, pp. 319-342.

Rumrich, F., 2020. Comparative evaluation of near and mid infrared spectroscopy in the control of process engineering processes. University of Applied Sciences Südwestfalen, Meschede, Germany.

Schmidt, J., 2019. Vergleich von Herstellkosten unterschiedlicher Verfahren zur Gewinnung von Protein und Fett aus Insektenlarven. Project Assignment, University of Applied Sciences Südwestfalen, Meschede, Germany.

Sindermann, D., 2019a. The insect revolution: too hard to swallow? Generate 19: 28-31. Available at: https://www.gea.com/en/binaries/generate-magazine-19_tcm11-52635.pdf.

Sindermann, D., 2019b. First experience with processing insects. EFPRA Congress 2019, Technical Symposium, June 13, 2019, European Fat Processors and Renderes Association (EFPRA), La Baule, France.

Van den Bossche, M. and Van Vaerenbergh, G., 2014. Microwave drying: a more efficient technology than gas-stirpping. GEA, Düsseldorf, Germany. Available at: https://www.gea.com/en/stories/microwave-drying.jsp.

Van Huis, A. and Tomberlin, J., 2017. Insects as food and feed – from production to consumption. Wageningen Academic Publishers, Wageningen, The Netherlands.

Van Huis, A., Van Itterbeeck, J., Klunder, H., Mertens, E., Halloran, A., Muir, G. and Vantomme, P., 2013. Edible insects: future prospects for food and feed security. FAO Forestry Paper 171, FAO, Rome, Italy.

Veldkamp, T., Van Duinkerken, G., Van Huis, A., Lakemond, C.M.M., Ottevanger, E., Bosch, G. and Van Boekel, M.A.J.S., 2012. Insects as a sustainable feed ingredient in pig and poultry diets – a feasibility study. Wageningen UR Livestock Research, Wageningen, The Netherlands.

Edible insect processing pathways and implementation of emerging technologies

S. Ojha[1], S. Bußler[1], M. Psarianos[1], G. Rossi[1] and O.K. Schlüter[1,2]*

[1]Quality and Safety of Food and Feed, Leibniz Institute for Agricultural Engineering and Bioeconomy (ATB), Max Eyth Allee 100, 14469 Potsdam, Germany; [2]Department of Agricultural and Food Sciences, University of Bologna, Piazza Goidanich 60, 47521 Cesena, Italy; *oschlueter@atb-potsdam.de

Abstract

The processing of insects is paramount to deliver safe and high quality raw materials, ingredients and products for large-scale food and feed applications. Depending upon the nature of the initial material and the desired end product, the processing pathways vary and may include several unit operations currently already used in food and feed processing. Insect processing pathways can involve harvesting, pre-processing, decontamination, further processing, packaging and storage. Several traditional and industrial decontamination methods have been proposed for edible insects, which include smoking, drying, blanching/boiling, marination, cooking, steaming, toasting and their combinations. Further processing steps are employed to produce insect meal, insect flour or extracted insect fractions. Each operation will have a different impact on the chemical and microbiological properties of the final product. Novel food processing technologies (e.g. high pressure processing, pulsed electric field, ultrasound and cold plasma) have shown potential to modify, complement or replace the conventional processing steps in insect processing. These technologies have been tested for microbial decontamination, enzyme inactivation, drying and extraction. Further, these are considered to be environmentally friendly and may be implemented for versatile applications to improve the processing efficiency, safety and quality of insect based products. Future research focuses in insect processing are development of efficient, environmentally friendly and low-cost processes; waste minimisation and incorporation of by-products/co-products.

Keywords

pre-processing – nonthermal technologies – extraction – novel food – microbial decontamination

1   Introduction

Edible insects have been part of the human diet throughout history (Dobermann *et al.*, 2017). Their consumption is traditionally recognised in several countries, where they are predominantly collected from their natural habitat (Dobermann *et al.*, 2017); however, with increasing demand of insect based foods in western countries, many companies have started mass rearing systems (Wade, 2020).

In order to ensure food safety, insect processing and storage should follow the same health and hygiene standards as for any other traditional food or feed (Imathiu, 2020). Hazards which are commonly found in insects comprise pathogenic microorganisms and parasites (Gałęcki and Sokół, 2019; Schlüter *et al.*, 2017), allergens (De Gier and Verhoeckx, 2018), heavy metals, toxins and other chemical contaminants (Bosch *et al.*, 2017; De Gier and Verhoeckx, 2018; Poma *et al.*, 2017, 2019; Van der Fels-Klerx *et al.*, 2016). Prions and human pathogenic virus should be considered as well (EFSA, 2015), although they have not been detected in edible insects yet.

Processing technology depends on insect species, safety hazards and type of final product. Edible insects can be marketed in three different forms: (1) whole (dried, frozen, pre-cooked); (2) processed; and (3) extracts (EFSA, 2015). Whole insects are commercially available as dried, frozen or chilled products for direct use by the consumer or food manufacturer. Processing of insects results in powder or paste, which allow the insects to be incorporated into food products or directly into dishes prepared by the consumer.

In order to improve the acceptance of insects and insect products and extend their shelf life, several traditional cooking techniques such as steaming, roasting, smoking, frying, stewing, curing (Ebenebe and Okpoko, 2015; Grabowski and Klein, 2017b; Lautenschläger *et al.*, 2017; Manditsera *et al.*, 2019; Nonaka, 2009; Obopile and Seeletso, 2013; Ramos-Rostro *et al.*, 2016; Shockley *et al.*, 2018) have been proposed. Other techniques, such as traditional sun-drying (Manditsera *et al.*, 2018), microwave processing (Lenaerts *et al.*, 2018; Vandeweyer *et al.*, 2017b), freeze-drying, oven-drying (Azzollini *et al.*, 2016; Fombong *et al.*, 2017), dry heat treatment (Bußler *et al.*, 2015), dry fractionation (Purschke *et al.*, 2018a), freezing (Melis *et al.*, 2018), marination, fermentation (Borremans *et al.*, 2018, 2020b; Patrignani *et al.*, 2020) and new processing methods as ultrasound-assisted extraction (Mishyna *et al.*, 2019; Panja, 2018; Sun *et al.*, 2018), cold atmospheric pressure plasma (CAPP) (Bußler *et al.*, 2016a), supercritical CO_2 extraction (Purschke *et al.*, 2017), enzymatic hydrolysis (Purschke *et al.*, 2018a) for protein, fat, and/or chitin extraction, three-dimensional food printing technologies (Severini *et al.*, 2018; Soares and Forkes, 2014), and several modified atmosphere packaging methods (Flekna *et al.*, 2017; Stoops *et al.*, 2017) have also been tested for insects and insect products.

Main objective of processing is to ensure food safety; however, it is also important to satisfy quality standards and consumer expectations. In this context, an important aspect to consider is the microbiological safety. In countries where insects are already recognised as food or feed, specific microbiological criteria are present. However, in the other countries, lack of specific rules may be observed. For example, insects are classified as novel foods in European Union (EU) and therefore fall within the current novel food legislation (EU Regulation 2283/2015; EFSA, 2016). Whole edible insects and their derived ingredients can only be lawfully placed on the EU market after safety assessments and authorisations. When a novel food is accepted for consumption, it will be considered as any other food, therefore it should respect the current accepted limit for microbiological contamination, stated in the EU Regulation 2073/2005 (EC, 2005). However, in the aforementioned regulation, edible insects are not considered. They could be equated to traditional meat or fish, but since their particular nature, specific microbiological criteria should be formulated. In order to overcome this problem, some European countries, such as Belgium and the Netherlands, where insects consumption is already allowed, have fixed microbiological limits for edible insect (FASFC, 2014a,b; Bureau Risicobeoordeling en Onderzoeks-programmering, 2014).

In order to guarantee high safety standard for the final consumer, suitable processing pathways should be integrated in the insect value chain. In this context, the present review aims to illustrate the current scientific knowledge in edible insect processing, focusing on the traditional pathways and providing an overview on possible future processing routes.

2   Pre-processing

Pre-processing technologies represents the first step of each edible insects processing route and mainly consist of insect harvesting/separation from the substrate residuals, insect inactivation/killing, removal of wings/legs, and washing (Rumpold and Schlüter, 2013).

Once the insects have grown to the desired size or reach a certain age, they are harvested either manually or automatically. Larvae such as mealworms and lesser mealworms are usually separated from their substrate by sieving, while crickets and grasshoppers, which typically live apart from the feeding source, can be harvested by picking them from their rearing cage or by shaking them out of the crevices they hide in. Some farms apply a starvation period (Figure 1) before harvesting by separating the insects from their food (Garofalo *et al.*, 2019). This causes the insects to empty their intestines (Finke, 2002), which, according to the breeders, results in better taste and cleaner products with reduced microbial load (Fernandez-Cassi *et al.*, 2019). However, Wynants *et al.* (2017) reported that starvation of yellow mealworms did not reduce the total viable counts (TVC) or change the bacterial community composition consistently.

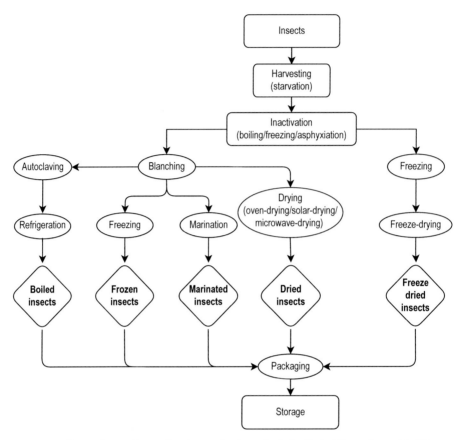

FIGURE 1   Processing pathways for the production of whole (boiled/frozen/oven-dried/freeze-dried/microwave-dried/marinated) edible insects

Following harvesting, the insects undergo several post-harvest treatments before being consumed. If not sold as living species, killing the insects is the first step. Freezing, drowning in hot or boiling water and steaming are some of the most common methods used for this purpose. Killing is a key step in insect processing since it can affect various final product quality parameters including microbial load, proximate composition, colour and taste (Adámková *et al.*, 2017; Farina, 2017; Larouche *et al.*, 2019). Larouche *et al.* (2019) compared blanching, desiccation, freezing, high hydrostatic pressure, grinding and asphyxiation (100% CO_2; 100% N_2; vacuum conditioning) as killing methods on black soldier fly larvae. They reported blanching (boiling water for 40 s) to be the most effective method in terms of time duration, final larval moisture, lipid oxidation, microbial load and colour alteration. However, due to the high TVC and presence of pathogenic microorganisms in the resulting product, a further stabilisation process was recommended (Larouche *et al.*, 2019). Variable results have been reported for effect of killing methods on nutritional composition of different insects. Adámková *et al.* (2017) observed lower fat content in mealworm killed by direct immersion in boiling water compared to freezing. However, Caligiani *et al.* (2019) and Singh *et al.* (2020) observed no difference in lipid concentration of *Hermetia illucens* and *Acheta domesticus* killed by freezing (-20 °C) or blanching (100 °C, 40 s). Additionally, higher ether extract has been recorded when crickets were killed by asphyxiation (3 hour in plastic bag at room temperature).

The killing method can influence physiochemical properties as well. For instance, higher luminosity and lower browning have been detected in house crickets killed with CO_2 (40 min) followed by blanching (40 s), while reduction in taste and lower pH have been displayed when crickets were frozen before cooking. This was explained by the *ante-mortem* stress, which is responsible for an increasing glycolytic activity, with higher conversion of glycogen into lactic acid (Farina, 2017).

In order to reduce the microbiological load, killing process can be coupled with a rinsing/washing step, as it is already applied in traditional food products such as raw fruits and vegetables (Yoon and Lee, 2018) or carcasses and fresh meat (Huffman, 2002). However, experiments conducted by Wynants *et al.* (2016, 2017) and Ng'ang'a *et al.* (2019) on mealworm larvae and wild harvested *Ruspolia differens* showed no reduction in TVC and in mesophilic aerobic count, Enterobacteriace, bacterial endospores, yeast and moulds when 1 min of washing treatment with tap water or sterile water were applied (Ng'ang'a *et al.*, 2019; Wynants *et al.*, 2016, 2017). Better results have been obtained in lesser mealworms with sterile distilled water containing antimicrobial substances such as ethanol, hydrogen peroxide or sodium hypochlorite (Crippen and Sheffield, 2006). Although these concepts can be extended to dead insects, the washing experiments aforementioned were applied on living larvae. Use of washing procedures on dead insects has been described by Fröhling *et al.* (2020). In this study, three successive washing steps with tap water were applied on frozen/thawed crickets. Although TVC of crickets was not affected by washing procedure, a reduction of approximately one logarithmic cycle was detected for total aerobic mesophilic viable count, *Enterobacteriaceae*, *Escherichia coli*, *Bacillus cereus*, *Clostridium perfringens*, *Enterococcus* spp., *Staphylococcus* spp., yeasts and moulds (Fröhling *et al.*, 2020).

3   Conventional processing routes

Although pre-processing represents the first step to connect from rearing to consumption, it cannot be considered decisive for ensuring food safety. Further processing techniques are required to obtain a product suitable for consumption. In the following section, the common conventional processing routes applied in the edible insects sector are described. Current scientific results as well as positive aspects and challenges observed for each step are discussed.

3.1   *Processing of whole insects*

Whole edible insects are usually marketed in chilled (T≤5 °C), frozen (T≤-18 °C) or dried form. Dried insects have a low moisture and water activity, therefore can be stored stable for a longer time (up to months) at room temperature (Kamau *et al.*, 2018). When cold and freezing storage are used, the cold chain should be continuously maintained during the storage (James, 1996). Even though some reduction in TVC has been detected after freezing (Grabowski and Klein, 2017a; Ssepuuya *et al.*, 2016), both cold storage and freezing cannot be considered as decontamination steps as they can only decelerate or delay microbiological growth and chemical deterioration (Kamau *et al.*, 2020). A certain microbial inactivation can be reached with drying processes, despite several studies displayed high residual microbial count on dried insects (Vandeweyer *et al.*, 2017a, 2018; Wynants *et al.*, 2018). Therefore, in order to guarantee high food safety standards and extend the shelf-life, raw edible insects should be subjected to a series of steps to ensure overall safety. An overview of possible processing pathways for production of whole edible insects is presented in Figure 1.

Every processing route should be organised in 3 steps: (1) harvesting and pre-processing; (2) decontamination; and (3) packaging and storage. Several decontamination methods have been proposed on edible insects. These comprise blanching, cooking/boiling, steaming, marination, drying, smoking, toasting and their combinations. Each operation will have a different impact on the chemical and microbiological properties of the final product. For example, Nyangena *et al.* (2020) compared the effect of toasting (5 min, 150 °C), boiling (5 min, 96 °C), solar-drying (2 days, 50-60 °C, 15-25% RH) and oven drying (2 days, 60 °C) on *A. domesticus, H. illucens, Spodoptera littoralis* and *R. differensis*. They observed that in all the species, an important reduction could be detected for TVC and *Staphilococcus aureus* (5-6 log cycles), yeasts-moulds and *Salmonella* spp. (lower than detection limits) when toasting or boiling were used. On the contrary, solar-drying and oven-drying could not guarantee the same results, if not preceded by toasting or boiling (Nyangena *et al.*, 2020). The presence of pathogenic microorganisms on edible insects after drying process represent an important safety concern, whereby dried insects should be reheated before consumption (Grabowski and Klein, 2017b). However, drying at 100 °C for 4 hours was able to reduce about 5-8 log cycles of the microbial counts (TVC, total aerobic, *E. coli, Enterobacteriaceae, Enterococcus* spp., *B. cereus, C. perfringens, Stahilococcus* spp., yeasts and moulds) in *A. domesticus* (Fröhling *et al.*, 2020). Similar reduction levels have been obtained for TVC (from 7.5 to < 1.7 log cfu/g), *Enterobacteriaceae* (from 6.8 to < 1 log cfu/g) and bacterial endospores (from 2.1 to < 1 log cfu/g) on *Tenebrio molitor, Brachytrupus* spp and *A. domesticus* after either boiling in water or roasting for 10 minutes (Klunder *et al.*, 2012).

A very common process in the food industry is blanching. Blanching allows the inactivation of vegetative microorganisms (but not bacterial spores). It refers to a process where a food is placed inside boiling water for short time (from few seconds to more than 10 minutes, depending by the product) and then cooled in cold water (Xiao *et al.*, 2014). It has been tested on edible insects by Vandeweyer *et al.* (2017b), who compared the effect of blanching for 10, 20 or 40 seconds, with or without microwave drying (16 microwave sources of 2 kW) on *T. molitor* larvae. They observed a significant reduction (from 8 log cfu/g to 1.3 log cfu/g) in total aerobic count when blanching for 40 seconds was followed by microwave treatment for at least 16 minutes (residual moisture 0.6%). Lower microbial reductions were detected when blanching was followed by shorter microwave treatments or no microwave process was applied. Blanching was able to reduce the number of *Enterobacteriaceae* and fungi (yeasts and moulds) below the detection limits (1 log cfu/g for *Enterobacteriaceae* and 2 log cfu/g for fungi), but an increase in bacterial endospores was observed (Vandeweyer *et al.*, 2017b). Similar results have been obtained by Vandeweyer *et al.* (2018) who studied the evolution in microbial counts during processing and storage of *Gryllus sigillatus.* They observed an initial reduction in TVC after smoking (80 °C, 40 min, residual moisture 5.1%) or smoking and oven-drying (80 °C, overnight, residual moisture 2.2%). However, an increase in TVC and endospores was displayed during the storage at room temperature (Vandeweyer *et al.*, 2018). Therefore, in order to obtain a final product safe for the consumer, alternative storage methods should be tested.

Still under-explored techniques in edible insects processing and storage are marination (treatment of raw meat with several ingredients to improve sensorial parameters and improve microbiological safety; Williams, 2012) and fermentation (subjecting food to the activity of microorganisms or enzymes, in order to obtain desirable sensorial and nutritional characteristics and extend the shelf life (Campbell-Platt, 1987)). While fermentation is usually conducted on insect powder or paste, marination can be realised on whole insects. In a study from Borremans *et al.* (2018), effect of marination technology was analysed on blanched (40 s) *T. molitor* larvae. The authors concluded that 6 days of marination in red wine or soy sauce could inhibit the microbial growth, allowing shelf life extension of at least 7 days. However, marination might be responsible for germination of bacterial endospores, which might represent an issue for longer storage periods (Borremans *et al.*, 2018).

Choice of the most suitable processing methods is a key and should involve the consideration of the initial raw material and the final product that would be obtained. Besides microbial safety, choice of processing technique should also consider nutritional and sensorial parameters. For example, blanching has shown to be responsible for luminosity value and non-enzymatic browning (Azzollini *et al.*, 2016) although it is effective inactivating browning enzymes (Tonneijck-Srpova *et al.*, 2019). Another common effect connected with processing is the increase in protein solubility (Womeni *et al.*, 2012). However, these effects cannot be generalised for all insect species. For example, decrease in protein digestibility was observed on *R. differens* subjected to toasting (150 °C, 5 min) and solar drying (30 °C, 40% RH), while the same process applied on *Macrotermes subhylanus* has not shown the same results (Kinyuru *et al.*, 2010). Processing could also be responsible for nutrient loss. It has been shown for *Hemijana variegate* subjected to sun drying (Egan *et al.*, 2014), while reduction in vitamin B12 has been detected in mealworm exposed to microwave drying (14 sources of 2 kW for 24 min) (Lenaerts *et al.*, 2018). Whereas, no proximate composition alteration was observed when oven drying (60 °C, 24 h) was used on *R. differens* (Fombong *et al.*, 2017). Reduction in protein and *in vitro* true dry matter digestibility have been detected in *Imbrasia belina* (Madibela *et al.*, 2007), *Eulepida mashona* and *Henicus whellani* (Manditsera *et al.*, 2019) after 1 hour of boiling process. For the last two species, boiling without roasting operation, was also responsible for reduction in zinc and iron bio-accessibility (Manditsera *et al.*, 2019). Conversely, 5-7 minutes of roasting could improve the calcium availability in mopane caterpillar (Madibela *et al.*, 2007).

3.2 *Insect meal production*

Sometimes edible insects require to be formulated into homogenous forms, such as powder or paste (Figure 2). In such cases, milling or grinding processes are required. Edible insects can be milled when still fresh or after blanching (wet milling) or after a drying process (dry milling). Wet milling is usually applied at industrial level because it results into fluid material, whereby it will be easy to handle along the production line (Dossey *et al.*, 2016). However, due to the high moisture content, the resulting material is microbially unstable (De Smet *et al.*, 2019). It should be kept at low temperature (≤5 °C) and used within a short time (De Smet *et al.*, 2019). Conversely, dry milling results in a dry powder, which shows extended shelf-life and can easily be marketed and stored at room temperature (Klunder *et al.*, 2012).

In order to extend the shelf life and have a stable product, drying operations can also be performed after the wet milling process for obtaining a powder (Dossey, 2015). When insect paste has to be dried, spray drying is usually preferred. It is a drying method where the paste is propelled into hot air. In this way, every single drop of the fluid is dried singularly; therefore, the drying process happens in a short time and results in homogeneous products and with better nutritional properties (Bhandari *et al.*, 2008; Dossey *et al.*, 2016; Son *et al.*, 2019). However, since spray-drying cannot be used for products with a high fat level, drum drying is often adopted (Dossey *et al.*, 2016). Here, the insect paste is in contact with a warm surface until it is dry (Tang *et al.*, 2003). The result is a faster drying of the layer in direct contact with the drier surface and low drying of the remaining bulk. Therefore, the final powder can be inhomogeneous and characterised by low quality (Courtois, 2013; Dossey *et al.*, 2016). Table 1 shows an overview of studies where positive and negative effects connected to traditional drying technologies have been evaluated on edible insects.

Possible processing pathways resulting in several types of insect meal are shown in Figure 2 (whole fat insect meal) and 3 (defatted insect meal). Defatted insect meal is produced by extracting fat from the insect. This operation can be performed with

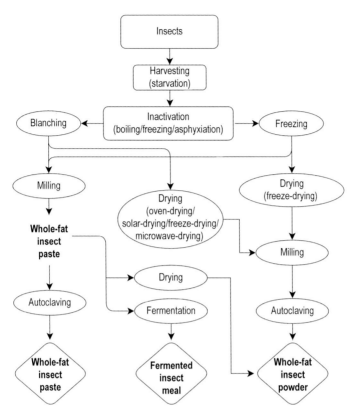

FIGURE 2   Processing pathways for the production of whole-fat and fermented insect meal

chemical (Ravi *et al.*, 2019; Son *et al.*, 2019), physical (Arsiwalla and Aarts, 2015) or mechanical (Azagoh *et al.*, 2016; Tschirner and Simon, 2015) treatments, and results in final products with different physiochemical and sensorial characteristics (Son *et al.*, 2019) and longer shelf-life (Nadarajah *et al.*, 2015; Temba *et al.*, 2017). Defatting operation can be carried out before (Figure 3, pathways B and C) or after (Figure 3, pathway A) the milling. If milling foregoes the defatting process, enzymatic hydrolysis might occur. It is due to the action of lipases and proteases present inside the insect cells and released into the meal when the insects are pulverised. Further enzymatic hydrolysis can be reached by adding commercial enzymes to the insect paste, as suggested by Arsiwalla and Aarts (2015). In order to promote the physical separation and/or the enzymatic hydrolysis, water should be present; therefore, a drying process is conducted on the defatted paste (Figure 3, pathway A). When de-fatted insect meal is produced by a mechanical defatting process (Figure 3, pathway B and C), milling is usually conducted after oil extraction (Azagoh *et al.*, 2016; Son *et al.*, 2019), while drying can be performed on the final defatted paste (Figure 3, pathway B) or on the whole insect (Figure 3, pathway C). Finally, solvent defatting

TABLE 1   Summary of studies exploring the effect of different drying processes on edible insects

Insect species	Drying methods	Notable effect	Reference
Acheta domesticus, Hermetia illucens, Spodoptera littoralis, Ruspolia differens	Oven drying Freeze drying Solar drying	Oven drying decreased bacterial populations, yeasts and mould. Solar drying had no effect. All three drying processes combined with boiling eliminated bacterial populations, yeasts, moulds.	Nyangena *et al.* (2020)
Rhynchophorus phoenicis	Solar drying Oven drying Smoke drying	Smoke drying was suggested as the preferable method as it maintains the lipid profile, while solar drying increased peroxide value.	Tiencheu *et al.* (2013)
Sternocera orissa	Oven drying Freeze drying	Oven drying method led to a higher composition of the material in compare to freeze drying	Shadung *et al.* (2012)
Imbrasia epimethea	Oven drying Solar drying	The monosaturated fatty acid content was reduced.	Lautenschläger *et al.* (2017)
Polyrhachis vicina	Solar drying	Aldehydes, increase of free fatty acid content, decrease of ketone content and hydrocarbons were observed.	Li *et al.* (2009)
R. differens	Oven drying Freeze drying	Both processes led to no significant differences in composition and product quality.	Fombong *et al.* (2017)
Tenebrio molitor	Oven drying Microwave-assisted drying	Microwave-assisted drying, combined with 40 s of blanching for less than 16 min did not reduce water activity lower than 0.6.	Vandeweyer *et al.* (2017b)
T. molitor	Oven drying Freeze drying Fluidised bed drying	Higher temperatures of drying caused darkening and shrinking due to browning reactions and tissue disruption.	Purschke *et al.* (2018a)
T. molitor	Fluidised bed drying Microwave-assisted drying Freeze drying Vacuum drying Hot air drying	Freeze drying oxidised the lipid fraction. Protein solubility was higher with vacuum drying and freeze drying and decreased with microwave-assisted drying.	Kroncke *et al.* (2018)
H. illucens	Oven drying Microwave-assisted drying	Protein digestibility was higher with oven drying, since microwave treatment polymerised protein molecules.	Huang *et al.* (2019)
Musca domestica L.	Oven drying Solar drying	Oven dried samples contained more proteins and less fat	Aniebo and Owen (2010)
Macrotermes subhylanus, R. differens	Solar drying	Solar drying of termites and grasshoppers decreased vitamin content in both species and protein digestibility of the grasshoppers.	Kinyuru *et al.* (2010)
Locusta migratoria manilensis, Bombyx mori	Freeze drying Oven drying Microwave-assisted drying	Oven and microwave drying changed the sensory characteristics of the materials in a way that they were less acceptable.	Mishyna *et al.* (2020)

operation is usually performed on dry material (Ravi *et al.*, 2019; Son *et al.*, 2019), which should be milled to ultrafine powder in order to increase the surface and to improve the solvent penetration (Son *et al.*, 2019). However, although defatting yield with solvent is higher than with other methods (Son *et al.*, 2019), solvent defatting operations are not preferred for both, environmental and sensorial aspects. Indeed, n-hexane defatted mealworm meal showed less flavour, lower colour preservation and consumer acceptability than pressure-defatted powder (Son *et al.*, 2019).

The type of blender used for meal production can also affect the final properties of the insect powder/paste. For example, Son *et al.* (2019) observed that mealworm defatted meal obtained by using a jet-mill showed lower humidity, higher luminosity, higher particle size uniformity and higher consumer acceptance than powder produced by pin, hammer or cutter mill (Son *et al.*, 2019).

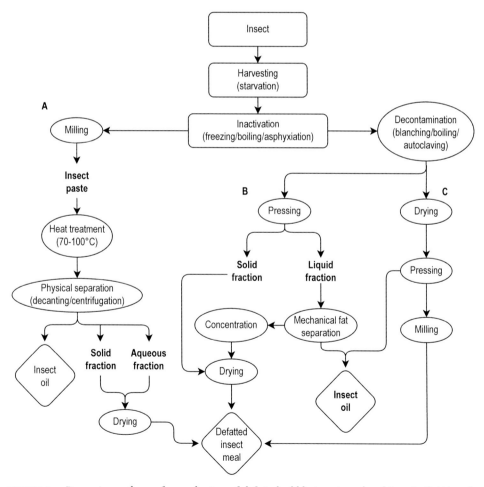

FIGURE 3   Processing pathways for production of defatted edible insect meal and insect oil: (A) as described by Arsiwalla and Aarts (2015), (B) analogous to fish meal production, (C) dry processing pathway

Apart from the three pathways already described, several other processing routes have been described for insect meal production (Borremans *et al.*, 2020a,b; Bußler *et al.*, 2016b; Fröhling *et al.*, 2020; Purschke *et al.*, 2018a,b,c; Vandeweyer *et al.*, 2018). Depending on the process goal and the insect species, the order of several operations may vary (Stoops *et al.*, 2017).

The most critical step along each processing pathway is the thermal treatment, which is important to assure food safety. Microbial decontamination can be performed on whole insects (before milling operation) (Kamau *et al.*, 2020) or directly on the powder/paste resulting from the milling process (Bußler *et al.*, 2016a,b) (Figure 2). Applying the decontamination process on meal is preferred, since milling operations break down the insect body and a release of the insect gut microbiota is obtained increasing the total microbial count (Fröhling *et al.*, 2020; Klunder *et al.*, 2012; Rumpold *et al.*, 2014). Indeed, microorganisms at the surface of food materials will be more susceptible to high temperature (Rumpold *et al.*, 2014); therefore, acceptable results can be reached with lower temperature or shorter treatment time. In this way, less impact on physio-chemical and sensorial characteristics of the final product is realised (Fröhling *et al.*, 2020). However, every time high temperatures are applied on insect meal, it may result in burnings of the powder/paste (Arsiwalla and Aarts, 2015). Concerning the physicochemical stability, the presence of browning and lipolytic enzymes should be considered. They can be responsible for browning, lipid oxidation and rancidity of insect paste. Therefore, in order to prevent undesirable reactions, a blanching process can be realised on the whole insects (Figure 3, pathways B and C) as suggested by Azagoh *et al.* (2016). Nevertheless, when blanching is performed, the adding of water can result in loss of soluble nutrients and higher energy is required for drying (Fröhling *et al.*, 2020). However, blanching of whole insect can be responsible also for microbial inactivation, resulting in low number of *Enterobacteriaceae* (Stoops *et al.*, 2017), but possible reactivation of bacterial spores should also be considered (Stoops *et al.*, 2016). For this reason, further decontamination treatment should be carried out in the next steps. Besides drying and heat based operations, an interesting stabilisation method applied on insect-based food is fermentation (Figure 2). The ability of this technology to ensure quality and safety of insect-based food has been investigated in several studies. For example, Patrignani *et al.* (2020) used *Yarrowia lipolytica* and *Debaryomyces hansenii* strains to produce cricket flour hydrolysates with high food safety, functionality, sensory and technological properties. The hydrolysates obtained by fermentation were characterised by a reduced chitin content and higher contents of antimicrobial substances (acetic acid, short chain fatty acids, chitosan, and GABA) and health-promoting molecules (arachidonic and linolenic acids, GABA, AABA, and BABA) when compared to the control. Borremans *et al.* (2018) observed that rapid acidification during fermentation, able to prevent microbial growth, could be reached within three days when blanched (40 s) paste of *T. molitor* was inoculated with a commercial starter containing *Pediococcus acidilactici, Lactobacillus curvatus* and *Staphylococcus xylosus* (Borremans *et al.*, 2018). However, the authors detected some lipid oxidation during refrigerated storage (4 °C) of the fermented paste. In order to prevent the oxidation, several additives have been tested. Borremans *et al.* (2020b) observed that sodium nitrite (150 mg/kg) and sodium lactate (60% w/w solution, 50 g/kg) are both able to reduce lipid oxidation. Different results were obtained by De Smet *et al.* (2019) who observed higher lipid oxidation of fermented insect paste stored for three weeks at 4 °C, when sodium nitrite or sodium lactate, rather than no additives, were used. Use of additive has displayed also a reduction in sulphite reducing bacteria in the fermenting paste (De Smet *et al.*, 2019). It is an important aspect because these bacteria are responsible for reduction of sulphite to hydrogen sulphite, compound associated with the unpleasant 'rotten egg' odour (Borremans *et al.*, 2018).

3.3   *Storage of insect-based food*

Although the aforementioned techniques can guarantee microbial safety, they cannot prevent the re-contamination of the product in the subsequent working steps. In order to guarantee microbiological safety for the consumer, appropriate storage methods with respect to the specific product properties should be applied. Refrigeration and freezing, as well as air containing packaging, vacuum packaging and modified atmosphere packaging (MAP) are common preservation methods used for edible insects products. Klunder *et al.* (2012) observed that crickets boiled for 1 minutes could maintain low TVC (4.1 log cfu/g) over 16 days when stored at 5-7 °C, while TVC higher than 10 log cfu/g was reached after 2 days in room temperature storage (28-30 °C) (Klunder *et al.*, 2012). Similar results have been obtained on blanched *T. molitor* larvae stored at 3 °C for 6 days (Vandeweyer *et al.*, 2017b), while stable TVC over 180 days has been found on refrigerated (about 5 °C) black soldier fly larvae (Kamau *et al.*, 2020). Refrigeration has also shown promising results for storage of fermented or marinated insects, guaranteeing microbiological and chemical stability for up to two and four weeks respectively (Borremans *et al.*, 2018, 2020b; De Smet *et al.*, 2019). Cold storage can provide good lipid stability and thus the sensorial quality of insect products (Kamau *et al.*, 2020; Lee *et al.*, 2020; Tiencheu *et al.*, 2013). However, since refrigeration temperatures are not able to inactivate lipases, enzymes responsible for lipid oxidation, freezing can be applied to inhibit enzyme activities (Tiencheu *et al.*, 2013).

Important differences in terms of lipid oxidation and microbial growth were also observed when different packaging was used: insect meal packaged in polypropylene bags and stored at room temperature showed higher lipid oxidation than insect meal packaged in polyethylene bags (Kamau *et al.*, 2020). It was attributed to the permeability of polypropylene to water, which was considered responsible for propagation of moulds able to produce lipases (Kamau *et al.*, 2020). Combined effects of packaging and storage conditions have been tested also by Ssepuuya *et al.* (2016, 2019) who observed that storage at room temperature (20 °C) could guarantee microbiological, chemical and sensorial stabilities only if opaque vacuum packaging was applied (Ssepuuya *et al.*, 2016, 2019). An alternative to vacuum packaging is MAP. It has been investigated by Stoops *et al.* (2017), who observed that MAP (60% CO_2 40% N_2) reduced microbial growth guaranteeing at least 21 days of microbiological stability for insect paste stored at 4 °C (Stoops *et al.*, 2017). However, they did not observe any reduction in bacterial spore counts.

Use of MAP may also be helpful to reduce the moulds contamination of dried whole insects or insect powder. At the best of our knowledge, no study has been conducted in this area, but this technology can be extended to insects from other dried food (Rodriguez *et al.*, 2000). MAP can represent an important step in edible insect shelf-life extension because moulds are responsible for mycotoxin production, which can be a fundamental safety issue in edible insects when the water activity is higher than 0.6 (Kamau *et al.*, 2018). In a study conducted by Kachapulula *et al.* (2018), high moulds proliferation has been observed on dried caterpillars and termites after 7 days of storage at room temperature (31 °C). It was accompanied with high aflatoxin contamination, which made the insects not safe for human or animal consumption (Kachapulula *et al.*, 2018).

3.4   *Extraction of valuable compounds from insects*

Despite a large number of studies aiming to use whole insects or insect meal as food or food ingredient, many consumers, especially from the western world, may still not accept them (Chen *et al.*, 2009). In order to overcome these limitations and promote their consumption, isolation of protein and fat-rich fractions can be performed. The resulting products may then be used as ingredients for both industrial processing and culinary home-made preparation.

Isolation of protein has been performed from insect meal obtained from several species such as migratory locust (Clarkson *et al.*, 2018; Purschke *et al.*, 2018c), desert locust (Mishyna *et al.*, 2019) house cricket (Laroche *et al.*, 2019; Ndiritu *et al.*, 2017), mealworm (Bußler *et al.*, 2016b; Kim *et al.*, 2019; Laroche *et al.*, 2019; Yi *et al.*, 2013, 2017; Zhao *et al.*, 2016) black soldier fly (Bußler *et al.*, 2016b; Caligiani *et al.*, 2018; Soetemans *et al.*, 2019). Extraction conditions, as well as solvent and insect species can strongly influence extraction yield and properties of the isolated proteins. Satisfactory protein yields have been obtained by using alkaline solutions (Laroche *et al.*, 2019; Mintah *et al.*, 2020; Mishyna *et al.*, 2019; Zhao *et al.*, 2016) since pH values between 10 and 12 have displayed to lead to higher insect protein solubility (Bußler *et al.*, 2016b; Purschke *et al.*, 2018b,c; Udomsil *et al.*, 2019; Yi *et al.*, 2017; Zhao *et al.*, 2016). However, use of chemical solvents has shown negative effects on protein functionality, impacting several parameters such as emulsion capacity, foaming capacity and foaming stability (Ndiritu *et al.*, 2017). In order to prevent it, aqueous extraction may be performed. However, lower extraction yields have been shown when aqueous solutions of ascorbic acid have been used (Amarender *et al.*, 2020; Ndiritu *et al.*, 2017; Yi *et al.*, 2013). Improved results have been reached by using

NaCl, as displayed on *T. molitor* (Yi *et al.*, 2017), or increasing pH (Purschke *et al.*, 2018b), temperature, solid/liquid ratio and/or extraction time (Bußler *et al.*, 2016b; Mintah *et al.*, 2020). Defatting can also increase the protein extract yields, as shown for several insect species (Amarender *et al.*, 2020; Kim *et al.*, 2019).

Several extraction systems have been implemented for recovery of fat from edible insects. Common fat extraction methods comprises Folch, Soxhelet, Supercritical CO_2 and aqueous extraction (Laroche *et al.*, 2019; Ramos-Bueno *et al.*, 2016; Ravi *et al.*, 2019; Tzompa-Sosa *et al.*, 2014). Choice of the method should consider the insect species because the same method applied on several species can results in different yields, as shown by Laroche *et al.* (2019), Pan *et al.* (2012), Ramos-Bueno *et al.* (2016) and Tzompa-Sosa *et al.* (2019, 2014). Moreover, different extraction systems, as well as different solvents, lead to various extraction yield when applied on the same insect species (Amarender *et al.*, 2020; Feng *et al.*, 2020; Laroche *et al.*, 2019; Mariod *et al.*, 2010). Fatty acid profiles of the extract can change with the extraction system as well (Laroche *et al.*, 2019; Purschke *et al.*, 2017; Ramos-Bueno *et al.*, 2016; Ravi *et al.*, 2019; Tzompa-Sosa *et al.*, 2014).

An interesting bioactive compound that can be extracted from insects is chitin. It is the precursor of chitosan, an amino-polysaccharide characterised by important biological activities (Mohan *et al.*, 2020). The process for chitin extraction consists in four sequential steps: delipidation, deproteinisation, demineralisation, decolorisation (Mohan *et al.*, 2020). The aim of this processing pathway is to remove all the extractible compounds from the insect powder in order to obtain the insoluble chitin fraction. This method has been implemented on a wide range of insect species, such as *Bombyx mori* (Luo *et al.*, 2019; Simionato *et al.*, 2014), *T. molitor* (Luo *et al.*, 2019; Song *et al.*, 2018), *Zophobas morio* (Shin *et al.*, 2019; Soon *et al.*, 2018), *H. illucens* (Caligiani *et al.*, 2018; Purkayastha and Sarkar, 2020), *Musca domestica* (Jing *et al.*, 2007; Kim *et al.*, 2016), *Apis mellifera* (Marei *et al.*, 2016, 2019), *Gryllus bimaculatus* (Kim *et al.*, 2017), *Schistocerca gregaria* (Marei *et al.*, 2016) and others. In order to reduce the use of chemical compounds, new eco-friendly chitin extraction systems have been developed. For example, the aforementioned method has been enhanced by using a sustainable deep eutectic solvent, which has shown attractive results on chitin extraction from black soldier fly (Zhou *et al.*, 2019). Furthermore, biological methods, based on fermentation with bacterial strains isolated from mealworm surface, have shown to be strong candidates for replacement of chemical demineralisation (Da Silva *et al.*, 2017).

4   Emerging technologies in insect processing

Recent advances in agri-food sector have shown that the emerging technologies (e.g. high hydrostatic pressures (HHP), pulsed electric fields (PEF), ultrasound (US) and CAPP) are promising alternatives for sustainable eco-friendly processing with negligible environmental impacts (Pojic *et al.*, 2018). These technologies aim to improve the safety and the quality of final products and enhance processing efficiency. However, requirement of highly trained operators and presence of remarkable initial investments may represent important hurdles for their spreading (Priyadarshini *et al.*, 2019). Legislative aspects need to be considered as well. For example, in Europe the novel food regulation (EU Regulation 2283/2015) considers products obtained by using emerging technologies as novel food if the applied technology was not used within the EU before 15 May 1997 and shows an impact on nutritional composition, structure or safety of the obtained food, which cannot be compared with the same food produced using traditional technologies. Since HHP, PEF and US are already applied in industrial food processing, the relation between emerging technologies and novel food legislation is not always clear and should be evaluated case by case every time one of these technologies is used.

In the present section, the most applied innovative technologies are described. A general overview of each technology, its fundamental effects and main applications in the food industry are given. Afterwards, possible applications of emerging technologies in the edible insects sector are discussed.

4.1   *Emerging technologies*

High hydrostatic pressure is a discontinuous or semi-continuous process that involves a fluid, submitted at a pressure in a range of 100-1000 MPa (Koutchma, 2014), surrounding the product to be treated at 0 to 120 °C (Aganovic *et al.*, 2021). However, pressure in the range of 200 up to 600 MPa at ambient or chilled temperature is usually applied in commercial food applications, with holding times rarely longer than 5 min (Aganovic *et al.*, 2021). The pressure can damage the microbial cells, leading to the inactivation of microorganisms at significantly lower temperature than conventional thermal processing and resulting in a better final product quality (Kashiri *et al.*, 2018). In food processing applications, HHP alone or in combination with heat can cause various physical, chemical, or biological changes in food, responsible for formulation of modified foods and ingredients (Bolat *et al.*, 2021; Hugas *et al.*, 2002; Rumpold *et al.*, 2014; Ugur *et al.*, 2020). Further, HHP can also be used to support the nutrient extraction from food (Preece *et al.*, 2017), to maximise the extraction yield of oils (Andreou *et al.*, 2020) and to improve product acceptability through the degradation of allergenic compounds which are naturally present in some food products (Boukil *et al.*, 2020; Li *et al.*, 2012; Penas *et al.*, 2011). However, HHP may have a negative impact on protein stability, resulting in loss of important nutritional components and changing the food texture (Boukil *et al.*, 2020; Tonneijck-Srpova *et al.*, 2019; Wang *et al.*, 2016). Moreover, when food is treated with HHP at moderate temperature, residual contamination with bacterial spores can still be detected (Campbell *et al.*, 2020; Kashiri *et al.*, 2018; Stoica *et al.*, 2013; Wang *et al.*, 2016). Therefore, the requirement of further heat operations and the high cost of HHP generating devices (Wang *et al.*, 2016), represent important challenges for using this technology for edible insect processing in food industry.

Pulsed electric field application is a relatively recent and widely employed technology in the food sector for several purposes. PEF consists of processing the food product with short high voltage electric pulses (Nowosad *et al.*, 2021). Depending on the purpose, several pulse frequencies and exposition times can be applied. The mechanism of the technology is related to triggering a change in the cell membrane integrity with consequent enhancement of its permeability (Nowosad *et al.*, 2021). This phenomenon can be exploited in several ways, including improvement of extraction efficiency of nutritional and bioactive compounds from raw materials (Alles *et al.*, 2020; Andreou *et al.*, 2020; Kaferbock *et al.*, 2020; Rahaman *et al.*, 2020) and food waste (Franco *et al.*, 2020). Further, PEF coupled with some conventional processes, like freezing, drying or mechanical extraction of juices or oils, can lead to increase in yields, efficiency optimisation (Alles *et al.*, 2020; Andreou *et al.*, 2019; Shorstkii *et al.*, in press), and milder stabilisation treatments which allow obtaining fresh-like products with texture, flavour, colour and odour closer at their natural state (Liu *et al.*, 2020; Tao *et al.*, 2019). Hurdles for PEF application in food industry include the high initial costs (Stoica *et al.*, 2013) and its lack of activity on bacterial endospores inactivation, which requires a further treatment (Reineke *et al.*, 2015; Siemer *et al.*, 2014). However, unlike HHP, no negative impact on bioactive compounds has been detected with PEF treatment (Nowosad *et al.*, 2021; Shorstkii *et al.*, in press).

Ultrasound technology consists of mechanical acoustic waves propagating in an elastic (solid or fluid) medium (Lempriere, 2003). These waves can generate compression and decompression on the matrix particles, producing a huge amount of energy (Gallo *et al.*, 2018). High intensity ultrasound operates at frequencies between 20 and 100 kHz and the intensity ranged between 10 and 1000 W/cm^2. This high intensity allows them to be disruptive by inducing cavitation inside the treated medium, with consequent alterations on physical, chemical and mechanical properties (Bhargava *et al.*, 2021). In the food industry, US is usually used to improve the efficiency of several unit operations. US can be used to enhance the extraction yield by creating cavities inside the tissues and favouring the breakdown of cell walls with a reduction of time and energy used for compounds' extraction (Choi *et al.*, 2017; Mishyna *et al.*, 2019; Ojha *et al.*, 2020; Sun *et al.*, 2018). With a similar mechanism, US can also be coupled with drying process. The micro-channels generated inside the bulk because the molecular cavitation, can favour the water elimination, reducing time and temperature of the drying processes. This can lead to a higher quality of the final product (Huang *et al.*, 2020). Besides the mass transfer, US can also be used for heat transfer. For example, US can facilitate the freezing process through improved initial nucleation (Chow *et al.*, 2005). Further, US can be also used during thawing with the goal to reduce the defrosting time in order to avoid product degradation (Li *et al.*, 2020). US coupled with heat treatments (thermosonication) or high pressure (manosonication) has found application also in sterilisation and enzyme inactivation, allowing to obtain interesting results in terms of shelf-life extension, without altering the sensorial characteristics of the food (Cameron *et al.*, 2009; Mintah *et al.*, 2019b; O'Donnell *et al.*, 2010). However, when US is not applied properly, it can be responsible for increase of temperature, with negative effects on nutritional and organoleptic characteristics (Farkas and Mohacsi-Farkas, 2011; Mintah *et al.*, 2020). Despite these drawbacks, US is the mostly used novel technology in the food sector so far (Priyadarshini *et al.*, 2019).

CAPP is a new technology with high potential, but to the authors best knowledge not applied in food industry yet. Plasma constitutes a mixture of several active chemical species, i.e. electrons, free radicals, ions, protons, excited atoms, reactive oxygen species, reactive nitrogen species, and UV radiation (Moreau *et al.*, 2008). These reactive compounds are created by applying electric or electromagnetic fields, generated at several voltages, to a gas (Ramazzina *et al.*, 2015). CAPP indicates a specific kind of plasma, generated at low temperatures (usually between 30 and 60 °C) and atmospheric pressure (Moreau *et al.*, 2008). The lower operating temperature provides interesting opportunities for shelf-life extension of several food matrices, without altering the quality parameters of temperature-sensitive products (Bußler *et al.*, 2016a; Pan *et al.*, 2019). CAPP has also shown its potential to be active not only on the viable microbial cells but also on spores, thus allowing low-temperature sterilisation of the food product (Bußler *et al.*, 2016a; Liao *et al.*, 2018; Rumpold *et al.*, 2014). The high reactivity and instability of the compounds produced during plasma treatment can also be used for inactivation of certain enzymes due to the alteration of their protein structures (Bußler *et al.*, 2016a, 2017; Han *et al.*, 2019). With the similar mechanism, CAPP can facilitate the decontamination of foods from allergens, pesticides, mycotoxins and other toxic contaminants (Gavahian and Cullen, 2020; Gavahian and Khaneghah, 2020). Another application of CAPP in the food industry is related to the modification of some food components and their physical characteristics, with the purpose to improve their quality and increase the processing efficiency (Bußler *et al.*, 2016a; Ekezie *et al.*, 2017). However, reactive oxygen species present in CAPP have shown negative activities on lipid stability, promoting lipid oxidation, which can result in off-flavour (Varilla *et al.*, 2020). CAPP can also find application in food fortification to obtain a product with specific healthy properties (Akasapu *et al.*, 2020). Furthermore, CAPP has also been applied on some by-products to extract interesting secondary nutrients as well as phenolic compounds (Bao *et al.*, 2020). Main drawbacks for CAPP application in food industry are connected with the uncertainty of find toxic residuals in food. Despite no evidences of potentially dangerous compounds in plasma treated food have been found so far (Kromm *et al.*, unpublished data; Priyadarshini *et al.*, 2019), it is recommended that each product should be assessed individually (Schlüter *et al.*, 2013).

4.2   *Innovative processing routes*

Novel technologies have shown their potential for different processes in the food industry; however, their use in edible insect processing is still scarce. Traditional methods, which are well known and easy to operate, are still preferred, due to the high variability of the insect-products and the high difficulty of process optimisation. However, novel technologies can lead to improved quality and safety of insect based products, assuring a larger consumer acceptance. Several studies have been performed in this regard. Table 2 offers an overview of these studies. Further, Figure 4 and 5 present possible innovative processing pathways, for the production of whole edible insects and insect products respectively, where novel processing technologies can modify, complement, or replace the conventional processing routes.

4.2.1   Microbial decontamination

CAPP and HHP have been studied to reach an acceptable microbial reduction grade and guarantee safety for the final consumer. Rumpold *et al.* (2014) observed that indirect plasma treatment for 10 min is very effective for surface decontamination of mealworm larvae, allowing microbial reduction of 7 log cycles. However, the same treatment has not shown any effect on gut microbial load. On the contrary, HHP is more active on gut microbial count reduction than on the surface (Kashiri *et al.*, 2018). Therefore, combination of CAPP and HHP can be a suitable alternative to traditional heat treatments in order to reach adequate microbial reduction on whole edible insects (Figure 4, pathway A). Another possibilities is to couple HHP or CAPP with traditional heat treatments realised at lower temperature and shorter time (Rumpold *et al.*, 2014). Similar reasoning can be done for insect meal. Bußler *et al.* (2016a) investigated the impact of CAPP on the microbial load of *T. molitor* flour showing reduction of the total microbial load from 7.7 to 4.3 log cfu/g. However, reduction from 6.8 log cfu/g to below detection limit has been observed when a thermal treatment for 15 min in an oven at temperatures of 120 °C was applied (Bußler *et al.*, 2016a,b). The microbial species to be inactivated should be considered as well. For example, Campbell *et al.* (2020) observed that a HHP treatment at 400 MPa for 10 min on *H. illucens* larvae is able to reduce yeasts and moulds below the detection limits, while reductions of *Enterobacteriaceae* (from 7.65 to 3.32 log cfu/g) and lactic acid bacteria (from 6.50 to 4.73 log cfu/g) were remarkable but not satisfactory. These results have confirmed the finding of Kashiri *et al.* (2018), who observed that HHP at 400 MPa for 2.5 min was responsible for a total inactivation of yeasts and moulds, but no significant effect was observed for total aerobic mesophilic microbial count, where the reduction was only 0.35 log cycles (Kashiri *et al.*, 2018).

4.2.2   Enzyme inactivation

In order to promote shelf life extension of food products, undesirable chemical reactions, such as enzymatic browning, should be prevented. Interesting results in

TABLE 2 Innovative technology in insect processing

Process goal	Technology and processing condition	Insect species and form	Main findings	Reference
Microbial decontamination	HHP treatment (35 L vessel, at room temperature) at 400 MPa and 600 MPa; 1.5 min and 10 min	*Hermetia illucens* larvae	HHP resulted in a lower inactivation for TVC and 600 MPa was needed to achieve similar reductions against Enterobacteria, LAB, and YM.	Campbell *et al.* (2020)
	HHP treatment (250 to 400 MPa, for 1.5 to 15 min)	*H. illucens* larvae	HHP was effective against natural contaminating YM, resulting more than 5 log cycle reductions at 400 MPa for all treatments; however, a low reduction of total microbial load was achieved.	Kashiri *et al.* (2018)
	CAPP generated at sinusoidal voltage of 8.8 kVPP at a frequency of 3.0 kHz using air as working gas; treatments up to 15 min	*Tenebrio molitor* flour	Total microbial load reduced from 7.72 to 4.73 $\log_{10}$ cfu/g in 15 min CAPP treatment.	Bußler *et al.* (2016a)
	HHP treatment (400, 500, and 600 MPa; up to 15 min)	*T. molitor* larvae	A HHP treatment at 600 MPa for 10 min resulted in an inactivation of 3 log cycles.	Rumpold *et al.* (2014)
	Indirect plasma treatment at frequency of 2.45 GHz and a power consumption in the range of 1.2 kW	*T. molitor* larvae	Plasma treatment (10 min) resulted in up to 7 log cycles reduction.	Rumpold *et al.* (2014)
Drying	PEF at 2 and 3 kV/cm; 5, 10 and 20 kJ/kg wet basis	Live *H. illucens* larvae	PEF pre-treatment can significantly reduce (up to 30%) drying time of insect biomass.	Alles *et al.* (2020)
Enzymatic hydrolysis	HHP treatment at 380 MPa for 1 min applied before or during the enzymatic hydrolysis by pepsin or Alcalase®	*T. molitor* larvae powder	HHP applied before the enzymatic hydrolysis improved the enzymatic hydrolysis by Alcalase® (but not by pepsin) at the very beginning of the process. Partial reduction of allergenic properties was observed. No alterations on the protein profile were observed.	Boukil *et al.* (2020)
	Ultrasound pretreatment at 600 W and 40±2 kHz for 30 min. Pulse interval of 15 s on and 5 s off. Temperature of 50 °C	Defatted *H. illucens* meal	US pretreatment enhance the enzymolysis reaction, by altering the molecular structure and arrangement and improving the enzyme-substrate interactions.	Mintah *et al.* (2019a)
	Ultrasound pretreatment at 600 W and 40±2 kHz for 10-20-30 min. Pulse interval of 15 s on and 5 s off	Dried defatted *H. illucens* meal	US assisted hydrolysis leads to hydrolysates with higher antioxidant activity.	Mintah *et al.* (2019b)
	Ultrasound pretreatment at 600 W and 40±2 kHz with a pulse interval of 15 s on and 5 s off	Dried defatted *H. illucens* larvae	Ultrasound pretreatment increase the sample lightness by reducing the protein hydrolysate molecular weight. Sonic treatments alter the secondary protein structure, reducing the hydrolysate turbidity and increasing the protein dispersibility.	Mintah *et al.* (2020)

TABLE 2   Innovative technology in insect processing (cont.)

Process goal	Technology and processing condition	Insect species and form	Main findings	Reference
Enzyme inactivation	HHP treatment (4 l vessel) at 250, 300, 400 and 500 MPa for 3 min	*T. molitor* larvae	Pressure of 400 MPa prevented the enzymatic browning.	Tonneijck-Srpo-va *et al.* (2019)
Extraction	Ultrasound assisted extraction by direct sonication at 20 kHz, amplitude of 89.9 μm in continuous pulse; 15 min	Lyophilised and grounded *Acheta domesticus* and *T. molitor*	The extraction yield was dependent on the nature insect species and the solvent of extraction. Ultrasound can be used to obtain insects extracts with improved fatty acid profile.	Otero *et al.* (2020)
	Ultrasonic assisted aqueous extraction by sonication at 20 kHz (several parameters combinations tested)	Lyophilised and grounded *Clanis bilineata* larvae	Ultrasound treatment allowed increasing oil extraction yield and oil quality. The higher yield was reached with an ultrasound treatment of 50 min at 40 °C and 400 W, with a pulsation interval of 2 s. Ultrasound treatment can increase the thermos-stability and antioxidant activity of the oil.	Sun *et al.* (2018)
	Ultrasound assisted extraction by sonication at 20 kHz, 75% of amplitude, 3 seconds of pulsation interval; 1, 15 and 20 min	Lyophilised and grounded *Bombyx mori* Microwave dried and grounded *T. molitor* and *Gryllus bimaculatus*	Ultrasound sonication significantly affects the protein extraction yield without changing the amino acid profile. Sonication time had a different impact on protein extraction yield among several insect species (maximum yield reaches after 15 min for *T. molitor* and *G. bimaculatus* and 5 min for *B. mori*) The protein extraction yield was dependent on the insect species and their life stage.	Choi *et al.* (2017)
	Ultrasound assisted protein extraction by sonication with amplitude 70% for 6 min. pulsation interval of 30 s.	Defatted powder of *Apis mellifera* larvae and pupae Defatted powder of *Schistocerca gregaria* adults	Ultrasound sonication improve the protein extraction yield Method used for protein extraction alter the molecular characteristics of protein, determining a changing on its hydrophobicity. But a specie-specific effect has to be addressed.	Mishyna *et al.* (2019)
	PEF at 2 and 3 kV/cm; 5, 10 and 20 kJ/kg wet basis	Live *H. illucens* larvae	Oil extraction yield was not affected significantly by PEF treatment. Higher amount of amino acids was recovered in oil after PEF treatment	Alles, *et al.* (2020)

CAPP = cold atmospheric pressure plasma; HHP = high hydrostatic pressure; LAB = lactic acid bacteria; PEF = pulsed eclectic field; TVC = total viable count; US = ultrasound; YM = yeasts and moulds.

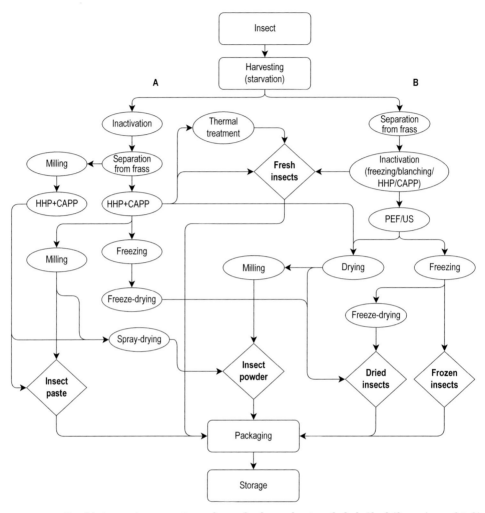

FIGURE 4   Possible innovative processing pathways for the production of whole (fresh/frozen/oven-dried/
freeze-dried/microwave-dried) edible insects and full-fat insect meal

terms of enzyme inactivation have been obtained when HHP treatments were used on *T. molitor* flour (Tonneijck-Srpova *et al.*, 2019). In this study, HHP has shown a potential to prevent enzymatic browning, which was comparable with traditional blanching (L*90 ≥ 60 after 10-11 days for HHP and L*90 ≈ 60 after 3-4 days for blanching). However, HHP treatment had a negative impact on the product texture, with a high amount of serum released from the mealworm powder. On the contrary, treatment at 500 MPa was able to prevent the enzymatic browning and at the same time to reduce the amount of released serum, obtaining a product with a weaker texture, therefore more desirable by the consumer (Tonneijck-Srpova *et al.*, 2019).

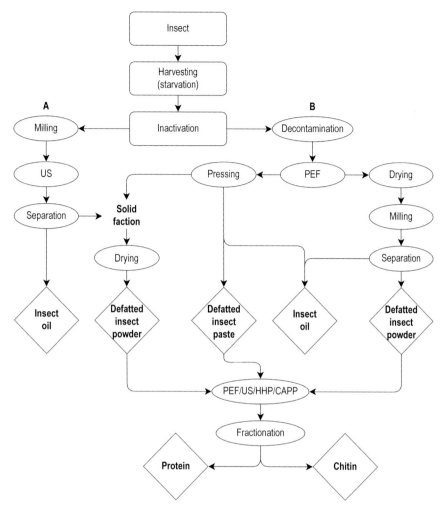

FIGURE 5   Possible innovative processing pathways for the production of defatted insect meal, insect oil, insect protein and chitin

4.2.3   Drying

Novel technologies are also known for increasing efficiency and yield of basic unit operations such as drying and freezing (Bassey, 2021; Wu, 2017) (Figure 4, pathway B). Concerning edible insects, impact of PEF pre-treatment on drying time has been evaluated by Alles *et al.* (2020). They observed that PEF at 2 kV/cm and 20 KJ/kg was able to reduce the drying time of *H. illucens* pre-pupae. However, these benefits were not observed at higher electric field strength (3 kV/cm), highlighting the importance of process optimisation (Alles *et al.*, 2020; Shorstkii *et al.*, in press).

Interesting results in terms of increase in mass transfer rate and time reduction during drying process may be obtained by applying US on insects as a pretreatment and/or in combination (Figure 4, pathway B). At the best of our knowledge, no study has investigated this potentiality on edible insects. However, basic knowledge can be transferred from other food materials (Huang *et al.*, 2020).

4.2.4 Extraction

Traditional extraction procedures are often time consuming, inefficient and non-selective (Otero *et al.*, 2020). Ultrasound and PEF have shown important solutions for these problems. For example, in a recent study, Otero *et al.* (2020) compared two different methods (i.e. ultrasound assisted extraction and pressurised liquid extraction) and two solvents (pure ethanol and a mixture of ethanol and water 1:1 v/v) to extract oil from *A. domesticus* adults and *T. molitor* larvae. They observed that ultrasound can increase the content of polyunsaturated fatty acids and decrease the content of saturated and monounsaturated fatty acids of the extract. Despite these positive results, every extraction method was found to increase the cholesterol level of the final extract. Besides, the effect of ultrasound is strictly related to the insect species and the solvent used, with better results found for *T. molitor* oil extracted with ethanol-water mixture. Important parameters that determine the extraction yield are also temperature, treatment time, pulse interval and ultrasound power. Optimisation of ultrasound treatment in order to increase the oil extracted from *Clanis bilineata* was conducted by Sun *et al.* (2018). In this study, they applied Response Surface Methodology to find out the best parameters combination in order to increase the extraction yield without altering the antioxidant properties of the oil. A possible pathway where integration of ultrasound is used for insect oil extraction is shown in Figure 5 (pathway A).

PEF has also been tested for oil extraction from insects (Figure 5, pathway B). In a recent study conducted on *H. illucens* larvae, Alles *et al.* (2020) observed that the PEF pre-treatment resulted in a size reduction and a more homogeneous distribution of fat droplets in insect cells; however, no significant increase in the extraction yield was observed. Further, the fatty acid profile of the oil extracted from PEF pre-treated sample was observed to be similar to the oil extracted without any pre-treatment.

Ultrasound, PEF, HHP and CAPP can also have an impact on protein extraction from defatted insect meal (Figure 5, pathway C). For example Choi *et al.* (2017) observed an enhanced protein extraction yield on *B. mori*, *T. molitor* and *G. bimaculatus* treated when ultrasound was used, while Mishyna *et al.* (2019) observed that ultrasound assisted extraction can enhance solubility, coagulability and foaming stability of *S. gregaria* protein extracts, while the same treatment has been responsible for reduction in solubility and coagulability in *A. mellifera* extracts.

4.2.5   Enzymatic hydrolysis

Innovative food processing methods, such as microwave, pulsed-electric field, ultrasound and HHP alone or coupled with enzymatic hydrolysis can reduce protein allergenicity of food proteins including edible insect products (Dong et al., 2021). In particular to insects, HHP and ultrasound have been studied to enhance the effect of hydrolysis of these allergenic compounds in order to increase the product quality. In a recent study, Boukil *et al.* (2020) reported that HHP at 380 MPa for 1 min applied in combination with Alcalase® enzyme could improve the protein hydrolysis degree (by up to 24%), allowing a partial reduction in allergenic activity. Ultrasonic treatment can also enhance the hydrolysis degree of insect powders (Mintah *et al.*, 2019a). For example, defatted *H. illucens* meal treated with ultrasound (600 W, frequency of 40 kHz) at 50 °C for 30 min, presented higher enzymatic activity (Mintah *et al.*, 2019a). Moreover, the damaged cells could lead to increased enzyme releases/enzyme concentrations, enhancing the hydrolysis of allergens (Mintah *et al.*, 2019a). Nevertheless, ultrasound treatment can also affect the antioxidant properties of the powder. Optimisation of the process depending on some antioxidant parameters has been conducted and validated on defatted *H. illucens* meal, concluding that sonication before the enzymatic hydrolysis allows obtaining a product with higher antioxidant power (Mintah *et al.*, 2019b). Furthermore, ultrasound can cause a change in the protein structure. Mintah *et al.* (2020) observed a reduction of β-sheet and a correspondent increase of α-helix and β-turn conformations. However, the proportion of each structure depends on the treatment parameters applied (Mintah *et al.*, 2020).

5   Conclusion and future perspectives

With the growing production of edible insects at an industrial scale, it is crucial to implement appropriate post-harvest and processing technologies for the safety, preservation, quality improvement, fractionation and storage of insects and insect products. Conventional processing pathways may include several unit operations used in traditional food and feed processing. The insect processing pathways can vary depending upon the nature of the initial material and the desired end product. In recent years, novel food processing technologies (e.g. high pressure processing, PEF, ultrasound and CAPP) have shown potential as an alternative or synergistic to the conventional technologies. The main challenges ahead of insect processing are the development of efficient, environmentally friendly and low-cost processing technologies, waste minimisation, recovery, and incorporation of by-products/co-products. Novel food processing technologies are considered to be environmentally friendly and may improve the processing efficiency, safety, quality, and sustainability of insect based products. However, the high knowledge level requirement, regulations and initial plant costs represent important hurdles to overcome.

Acknowledgements

The work of the authors was partially supported by different funding sources, i.e. by funds of the Federal Ministry of Education and Research (BMBF) based on a decision of the Parliament of the Federal Republic of Germany via the Project Management Jülich (PtJ) within the joint research project Agrarsysteme der Zukunft: F4F – Nahrung der Zukunft, Teilprojekt I (Grant no. 031B0730 I), MSCA-IF-2018 – Individual Fellowships (inTECH, Grant agreement ID: 841193) within H2020-EU.1.3.2. – Nurturing excellence by means of cross-border and cross-sector mobility and of the BlueBio ERA-NET COFUND on the Blue Bioeconomy – Unlocking the Potential of Aquatic Bioresources and The European Commission (AquaTech4Feed, Grant no. 2819ERA01M) within the Horizon 2020 programme.

Conflict of interest

The authors declare no conflict of interest.

References

Adámková, A., Adámek, M., Mlček, J., Borkovcová, M., Bednářová, M., Kouřimská, L., Skácel, J. and Vítová, E., 2017. Welfare of the mealworm (*Tenebrio molitor*) breeding with regard to nutrition value and food safety. Potravinarstvo Slovak Journal of Food Sciences 11: 460-465. https://doi.org/10.5219/779

Aganovic, K., Hertel, C., Vogel, R.F., Johne, R., Schlüter, O., Schwarzenbolz, U., Jäger, H., Holzhauser, T., Bergmair, J., Roth, A., Sevenich, R., Bandick, N., Kulling, S.E., Knorr, D., Engel, K.-H. and Heinz, V., 2021. Different aspects of high hydrostatic pressure food processing. Comprehensive Reviews in Food Science and Food Safety 20: 3225-3266. https://doi.org/10.1111/1541-4337.12763

Akasapu, K., Ojah, N., Gupta, A.K., Choudhury, A.J.M. and Mishra, P., 2020. An innovative approach for iron fortification of rice using cold plasma. Food Research International 136: 109599. https://doi.org/10.1016/j.foodres.2020.109599

Alles, M.C., Smetana, S., Parniakov, O., Shorstkii, I., Toepfl, S., Aganovic, K. and Heinz, V., 2020. Bio-refinery of insects with Pulsed electric field pre-treatment. Innovative Food Science & Emerging Technologies 64: 102403.

Amarender, R.V., Bhargava, K., Dossey, A.T. and Gamagedara, S., 2020. Lipid and protein extraction from edible insects – crickets (Gryllidae). LWT – Food Science and Technology 125: 109222. https://doi.org/10.1016/j.lwt.2020.109222

Andreou, V., Dimopoulos, G., Dermesonlouoglou, E. and Taoukis, P., 2019. Application of pulsed electric fields to improve product yield and waste valorization in industrial tomato processing. Journal of Food Engineering 270: 109778. https://doi.org/10.1016/j.jfoodeng.2019.109778

Andreou, V., Psarianos, M., Dimopoulos, G., Tsimogiannis, D. and Taoukis, P., 2020. Effect of pulsed electric fields and high pressure on improved recovery of high-added-value compounds from olive pomace. Journal of Food Science 85: 1500-1512. https://doi.org/10.1111/1750-3841.15122

Aniebo, A.O. and Owen, O.J., 2010. Effects of age and method of drying on the proximate composition of housefly larvae (*Musca domestica* Linnaeus) meal (HFLM). Pakistan Journal of Nutrition 9: 485-487.

Arsiwalla, T. and Aarts, K., 2015. Method to convert insects or worms into nutrient streams and compositions obtained thereby. U.S. Patent Application No. 14/408,371.

Azagoh, C., Ducept, F., Garcia, R., Rakotozafy, L., Cuvelier, M.E., Keller, S., Lewandowski, R. and Mezdour, S., 2016. Extraction and physicochemical characterization of *Tenebrio molitor* proteins. Food Research International 88: 24-31. https://doi.org/10.1016/j.foodres.2016.06.010

Azzollini, D., Derossi, A. and Severini, C., 2016. Understanding the drying kinetic and hygroscopic behaviour of larvae of yellow mealworm (*Tenebrio molitor*) and the effects on their quality. Journal of Insects as Food and Feed 2: 233-243.

Bao, Y., Reddivari, L. and Huang, J.Y., 2020. Enhancement of phenolic compounds extraction from grape pomace by high voltage atmospheric cold plasma. LWT – Food Science and Technology 133: 109970. https://doi.org/10.1016/j.lwt.2020.109970

Bassey, E.J., Cheng, J.H. and Sun, D.W., 2021. Novel nonthermal and thermal pretreatments for enhancing drying performance and improving quality of fruits and vegetables. Trends in Food Science & Technology 112: 137-148. https://doi.org/10.1016/j.tifs.2021.03.045

Bhandari, B.R., Patel, K.C. and Chen, X.D., 2008. Spray drying of food materials-process and product characteristics. Drying Technologies in Food Processing 4: 113-157.

Bhargava, N., Mor, R.S., Kumar, K. and Sharanagat, V.S., 2021. Advances in application of ultrasound in food processing: a review. Ultrasonics Sonochemistry 70: 105293. https://doi.org/10.1016/j.ultsonch.2020.105293

Bolat, B., Ugur, A.E., Oztop, M.H. and Alpas, H., 2021. Effects of high hydrostatic pressure assisted degreasing on the technological properties of insect powders obtained from *Acheta domesticus* & *Tenebrio molitor*. Journal of Food Engineering 292: 110359.

Borremans, A., Bussler, S., Sagu, S.T., Rawel, H., Schlüter, O.K. and Van Campenhout, L., 2020a. Effect of blanching plus fermentation on selected functional properties of mealworm (*Tenebrio molitor*) powders. Foods 9: 917. https://doi.org/10.3390/foods9070917

Borremans, A., Lenaerts, S., Crauwels, S., Lievens, B. and Van Campenhout, L., 2018. Marination and fermentation of yellow mealworm larvae (*Tenebrio molitor*). Food Control 92: 47-52.

Borremans, A., Smets, R. and Van Campenhout, L., 2020b. Fermentation versus meat preservatives to extend the shelf life of mealworm (*Tenebrio molitor*) paste for feed and food applications. Frontiers in Microbiology 11: 1510. https://doi.org/10.3389/fmicb.2020.01510

Bosch, G., Van der Fels-Klerx, H.J., De Rijk, T.C. and Oonincx, D.G.A.B., 2017. Aflatoxin B1 tolerance and accumulation in black soldier fly larvae (*Hermetia illucens*) and yellow mealworms (*Tenebrio molitor*). Toxins 9: 185. https://doi.org/10.3390/toxins9060185

Boukil, A., Perreault, V., Chamberland, J., Mezdour, S., Pouliot, Y. and Doyen, A., 2020. High hydrostatic pressure-assisted enzymatic hydrolysis affect mealworm allergenic proteins. Molecules 25: 2685. https://doi.org/10.3390/molecules25112685

Bußler, S., Ehlbeck, J. and Schlüter, O.K., 2017. Pre-drying treatment of plant related tissues using plasma processed air: Impact on enzyme activity and quality attributes of cut apple and potato. Innovative Food Science & Emerging Technologies 40: 78-86. https://doi.org/10.1016/j.ifset.2016.05.007

Bußler, S., Rumpold, B.A., Fröhling, A., Jander, E., Rawel, H.M. and Schlüter, O.K., 2016a. Cold atmospheric pressure plasma processing of insect flour from *Tenebrio molitor*: impact on microbial load and quality attributes in comparison to dry heat treatment. Innovative Food Science & Emerging Technologies 36: 277-286.

Bußler, S., Rumpold, B.A., Jander, E., Rawel, H.M. and Schlüter, O.K., 2016b. Recovery and techno-functionality of flours and proteins from two edible insect species: meal worm (*Tenebrio molitor*) and black soldier fly (*Hermetia illucens*) larvae. Heliyon 2: e00218. https://doi.org/10.1016/j.heliyon.2016.e00218

Bußler, S., Steins, V., Ehlbeck, J. and Schlüter, O., 2015. Impact of thermal treatment versus cold atmospheric plasma processing on the techno-functional protein properties from *Pisum sativum* 'Salamanca'. Journal of Food Engineering 167: 166-174. https://doi.org/10.1016/j.jfoodeng.2015.05.036

Caligiani, A., Marseglia, A., Leni, G., Baldassarre, S., Maistrello, L., Dossena, A. and Sforza, S., 2018. Composition of black soldier fly prepupae and systematic approaches for extraction and fractionation of proteins, lipids and chitin. Food Research International 105: 812-820. https://doi.org/10.1016/j.foodres.2017.12.012

Cameron, M., McMaster, L.D. and Britz, T.J., 2009. Impact of ultrasound on dairy spoilage microbes and milk components. Dairy Science & Technology 89: 83-98. https://doi.org/10.1051/dst/2008037

Campbell-Platt, G., 1987. Fermented foods of the world. A dictionary and guide. Butterworth-Heinemann, London, UK, 291 pp.

Campbell, M., Ortuño, J., Stratakos, A.C., Linton, M., Corcionivoschi, N., Elliott, T., Koidis, A. and Theodoridou, K., 2020. Impact of thermal and high-pressure treatments on the microbiological quality and *in vitro* digestibility of black soldier fly (*Hermetia illucens*) larvae. Animals 10: 682.

Chen, X., Feng, Y. and Chen, Z., 2009. Common edible insects and their utilization in China. Entomological Research 39: 299-303. https://doi.org/10.1111/j.1748-5967.2009.00237.x

Choi, B.D., Wong, N.A.K. and Auh, J.H., 2017. Defatting and sonication enhances protein extraction from edible insects. Korean Journal for Food Science of Animal Resources 37: 955-961. https://doi.org/10.5851/kosfa.2017.37.6.955

Chow, R., Blindt, R., Chivers, R. and Povey, M., 2005. A study on the primary and secondary nucleation of ice by power ultrasound. Ultrasonics 43: 227-230. https://doi.org/10.1016/j.ultras.2004.06.006

Clarkson, C., Mirosa, M. and Birch, J., 2018. Potential of extracted locusta migratoria protein fractions as value-added ingredients. Insects 9: 20. https://doi.org/10.3390/insects9010020

Courtois, F., 2013. Roller and drum drying for food powder production. In: Bhandari, B., Bansal, N., Zhang, M and Schuck, P. (eds.) Handbook of food powders. Woodhead Publishing, Sawston, UK, pp. 85-104. https://doi.org/10.1533/9780857098672.1.85

Crippen, T.L. and Sheffield, C., 2006. External surface disinfection of the lesser mealworm (Coleoptera: Tenebrionidae). Journal of Medical Entomology 43(5): 916-923. https://doi.org/10.1093/jmedent/43.5.916

Da Silva, F.K.P., Brück, D.W. and Brück, W.M., 2017. Isolation of proteolytic bacteria from mealworm (*Tenebrio molitor*) exoskeletons to produce chitinous material. FEMS Microbiology Letters 364. https://doi.org/10.1093/femsle/fnx177

De Gier, S. and Verhoeckx, K., 2018. Insect (food) allergy and allergens. Molecular Immunology 100: 82-106. https://doi.org/10.1016/j.molimm.2018.03.015

De Smet, J., Lenaerts, S., Borremans, A., Scholliers, J., Van der Borght, M. and Van Campenhout, L., 2019. Stability assessment and laboratory scale fermentation of pastes produced on a pilot scale from mealworms (*Tenebrio molitor*). LWT – Food Science and Technology 102: 113-121. https://doi.org/10.1016/j.lwt.2018.12.017

Dobermann, D., Swift, J.A. and Field, L.M., 2017. Opportunities and hurdles of edible insects for food and feed. Nutrition Bulletin 42: 293-308. https://doi.org/10.1111/nbu.12291

Dong, X., Wang, J., & Raghavan, V. (2021). Critical reviews and recent advances of novel non-thermal processing techniques on the modification of food allergens. Critical reviews in food science and nutrition, *61*(2), 196-210.

Dossey, A.T., 2015. Insect products and methods of manufacture and use thereof. U.S. Patent Application No. 14/537,960.

Dossey, A.T., Tatum, J.T. and McGill, W.L., 2016. Modern insect-based food industry: current status, insect processing technology, and recommendations moving forward. Insects as Sustainable Food Ingredients: Production, Processing and Food Applications: 113-152. https://doi.org/10.1016/B978-0-12-802856-8.00005-3

Ebenebe, C.I. and Okpoko, V., 2015. Edible insect consumption in the South Eastern Nigeria. International Journal of Science and Engineering 6: 171-177.

European Commission (EC), 2005. Commission Regulation (EC) No 2073/2005 on microbiological criteria for foodstuffs. Official Journal of the EU L 338: 1-26. Available at: http://data.europa.eu/eli/reg/2005/2073/oj.

European Food Safety Authority (EFSA), 2015. Risk profile related to production and consumption of insects as food and feed. EFSA Journal 13: 4257.

European Food Safety Authority (EFSA), 2016. Guidance on the preparation and presentation of the notification and application for authorisation of traditional foods from third countries in the context of regulation (EU) 2015/2283. EFSA Journal 14(11): 4590.

Egan, B.A., Toms, R., Minter, L.R., Addo-Bediako, A., Masoko, P., Mphosi, M. and Olivier, P.A.S., 2014. Nutritional significance of the edible insect, Hemijana variegata Rothschild (Lepidoptera: Eupterotidae), of the Blouberg Region, Limpopo, South Africa. African Entomology 22: 15-23. https://doi.org/10.4001/003.022.0108

Ekezie, F.G.C., Sun, D.W. and Cheng, J.H., 2017. A review on recent advances in cold plasma technology for the food industry: current applications and future trends. Trends in Food Science & Technology 69: 46-58. https://doi.org/10.1016/j.tifs.2017.08.007

Farina, M.F., 2017. How method of killing crickets impact the sensory qualities and physiochemical properties when prepared in a broth. International Journal of Gastronomy and Food Science 8: 19-23. https://doi.org/10.1016/j.ijgfs.2017.02.002

Farkas, J. and Mohacsi-Farkas, C., 2011. History and future of food irradiation. Trends in Food Science & Technology 22: 121-126. https://doi.org/10.1016/j.tifs.2010.04.002

Federal Agency for the Safety of the Food Chain (FASFC), 2014a. Circular concerning the breeding and marketing of insects and insect-based food for human consumption. FASFC, Brussels, Belgium.

Federal Agency for the Safety of the Food Chain (FASFC), 2014b. Common advice SciCom 14-2014 and SHC Nr. 9160 – subject: food safety aspects of insects intended for human consumption. FASFC, Brussels, Belgium.

Feng, W., Xiong, H., Wang, W., Duan, X., Yang, T., Wu, C., Yang, F., Wang, T. and Wang, C., 2020. A facile and mild one-pot process for direct extraction of lipids from wet energy insects of black soldier fly larvae. Renewable Energy 147: 584-593.

Fernandez-Cassi, X., Supeanu, A., Vaga, M., Jansson, A., Bogyist, S. and Vagsholm, I., 2019. The house cricket (*Acheta domesticus*) as a novel food: a risk profile. Journal of Insects as Food and Feed 5: 137-157. https://doi.org/10.3920/Jiff2018.0021

Finke, M.D., 2002. Complete nutrient composition of commercially raised invertebrates used as food for insectivores. Zoo Biology 21: 269-285. https://doi.org/10.1002/zoo.10031

Flekna, G., Mäke, D., Bauer, S. and Bauer, F., 2017. Microbiological and chemical changes during storage of edible insects. Book of Abstracts INSECTA 2019. ATB, Potsdam, Germany, pp. 99. Available at: https://tinyurl.com/4vjbv4f3.

Fombong, F.T., Van der Borght, M. and Vanden Broeck, J., 2017. Influence of freeze-drying and oven-drying post blanching on the nutrient composition of the edible insect *Ruspolia differens*. Insects 8: 102.

Franco, D., Munekata, P.E.S., Agregan, R., Bermudez, R., Lopez-Pedrouso, M., Pateiro, M. and Lorenzo, J.M., 2020. Application of pulsed electric fields for obtaining antioxidant extracts from fish residues. Antioxidants 9: 90. https://doi.org/10.3390/antiox9020090

Fröhling, A., Bußler, S., Durek, J. and Schlüter, O.K., 2020. Thermal impact on the culturable microbial diversity along the processing chain of flour from crickets (*Acheta domesticus*). Frontiers in Microbiology 11: 884. https://doi.org/10.3389/fmicb.2020.00884

Gałęcki, R. and Sokół, R., 2019. A parasitological evaluation of edible insects and their role in the transmission of parasitic diseases to humans and animals. PLoS ONE 14: e0219303. https://doi.org/10.1371/journal.pone.0219303

Gallo, M., Ferrara, L. and Naviglio, D., 2018. Application of ultrasound in food science and technology: a perspective. Foods 7: 164. https://doi.org/10.3390/foods7100164

Garofalo, C., Milanovic, V., Cardinali, F., Aquilanti, L., Clementi, F. and Osimani, A., 2019. Current knowledge on the microbiota of edible insects intended for human consumption: a state-of-the-art review. Food Research International 125: 108527. https://doi.org/10.1016/j.foodres.2019.108527

Gavahian, M. and Cullen, P.J., 2020. Cold plasma as an emerging technique for mycotoxin-free food: efficacy, mechanisms, and trends. Food Reviews International 36: 193-214. https://doi.org/10.1080/87559129.2019.1630638

Gavahian, M. and Khaneghah, A.M., 2020. Cold plasma as a tool for the elimination of food contaminants: recent advances and future trends. Critical Reviews in Food Science and Nutrition 60: 1581-1592. https://doi.org/10.1080/10408398.2019.1584600

Grabowski, N.T. and Klein, G., 2017a. Microbiological analysis of raw edible insects. Journal of Insects as Food and Feed 3: 7-14. https://doi.org/10.3920/Jiff2016.0004

Grabowski, N.T. and Klein, G., 2017b. Microbiology of processed edible insect products – results of a preliminary survey. International Journal of Food Microbiology 243: 103-107. https://doi.org/10.1016/j.ijfoodmicro.2016.11.005

Han, Y.X., Cheng, J.H. and Sun, D.W., 2019. Activities and conformation changes of food enzymes induced by cold plasma: a review. Critical Reviews in Food Science and Nutrition 59: 794-811. https://doi.org/10.1080/10408398.2018.1555131

Huang, C., Feng, W., Xiong, J., Wang, T., Wang, W., Wang, C. and Yang, F., 2019. Impact of drying method on the nutritional value of the edible insect protein from black soldier fly (*Hermetia illucens* L.) larvae: amino acid composition, nutritional value evaluation, *in vitro* digestibility, and thermal properties. European Food Research and Technology 245: 11-21. https://doi.org/10.1007/s00217-018-3136-y

Huang, D., Men, K.Y., Li, D.P., Wen, T., Gong, Z.L., Sunden, B. and Wu, Z., 2020. Application of ultrasound technology in the drying of food products. Ultrasonics Sonochemistry 63: 104950. https://doi.org/10.1016/j.ultsonch.2019.104950

Huffman, R.D., 2002. Current and future technologies for the decontamination of carcasses and fresh meat. Meat Science 62: 285-294. https://doi.org/10.1016/S0309-1740(02)00120-1

Hugas, M., Garriga, M. and Monfort, J.M., 2002. New mild technologies in meat processing: high pressure as a model technology. Meat Science 62: 359-371. https://doi.org/10.1016/S0309-1740(02)00122-5

Imathiu, S., 2020. Benefits and food safety concerns associated with consumption of edible insects. NFS Journal 18: 1-11. https://doi.org/10.1016/j.nfs.2019.11.002

James, S., 1996. The chill chain 'from carcass to consumer'. Meat Science 43: S203-S216. https://doi.org/10.1016/0309-1740(96)00066-6

Jing, Y.J., Hao, Y.J., Qu, H., Shan, Y., Li, D.S. and Du, R.Q., 2007. Studies on the antibacterial activities and mechanisms of chitosan obtained from cuticles of housefly larvae. Acta Biologica Hungarica 58: 75-86. https://doi.org/10.1556/ABiol.57.2007.1.7

Kachapulula, P.W., Akello, J., Bandyopadhyay, R. and Cotty, P.J., 2018. Aflatoxin contamination of dried insects and fish in Zambia. Journal of Food Protection 81: 1508-1518. https://doi.org/10.4315/0362-028x.Jfp-17-527

Kaferbock, A., Smetana, S., De Vos, R., Schwarz, C., Toepfl, S. and Parniakov, O., 2020. Sustainable extraction of valuable components from Spirulina assisted by pulsed electric fields technology. Algal Research-Biomass Biofuels and Bioproducts 48: 101914. https://doi.org/10.1016/j.algal.2020.101914

Kamau, E., Mutungi, C., Kinyuru, J., Imathiu, S., Affognon, H., Ekesi, S., Nakimbugwe, D. and Fiaboe, K.K.M., 2020. Changes in chemical and microbiological quality of semi-processed black soldier fly (*Hermetia illucens* L.) larval meal during storage. Journal of Insects as Food and Feed 6: 417-428. https://doi.org/10.3920/Jiff2019.0043

Kamau, E., Mutungi, C., Kinyuru, J., Imathiu, S., Tanga, C., Affognon, H., Ekesi, S., Nakim-bugwe, D. and Fiaboe, K.K.M., 2018. Moisture adsorption properties and shelf-life estimation of dried and pulverised edible house cricket *Acheta domesticus* (L.) and black soldier fly larvae *Hermetia illucens* (L.). Food Research International 106: 420-427. https://doi.org/10.1016/j.foodres.2018.01.012

Kashiri, M., Marin, C., Garzón, R., Rosell, C.M., Rodrigo, D. and Martínez, A., 2018. Use of high hydrostatic pressure to inactivate natural contaminating microorganisms and inoculated *E. coli* O157: H7 on *Hermetia illucens* larvae. PLoS ONE 13: e0194477.

Kim, M.W., Han, Y.S., Jo, Y.H., Choi, M.H., Kang, S.H., Kim, S.A. and Jung, W.J., 2016. Extraction of chitin and chitosan from housefly, *Musca domestica*, pupa shells. Entomological Research 46: 324-328. https://doi.org/10.1111/1748-5967.12175

Kim, M.W., Song, Y.S., Han, Y.S., Jo, Y.H., Choi, M.H., Park, Y.K., Kang, S.H., Kim, S.A., Choi, C. and Jung, W.J., 2017. Production of chitin and chitosan from the exoskeleton of adult two-spotted field crickets (*Gryllus bimaculatus*). Entomological Research 47: 279-285. https://doi.org/10.1111/1748-5967.12239

Kim, T.K., Yong, H.I., Jeong, C.H., Han, S.G., Kim, Y.B., Paik, H.D. and Choi, Y.S., 2019. Technical functional properties of water- and salt-soluble proteins extracted from edible insects. Food Science of Animal Resources 39: 643-654. https://doi.org/10.5851/kosfa.2019.e56

Kinyuru, J.N., Kenji, G.M., Njoroge, S.M. and Ayieko, M., 2010. Effect of processing methods on the *in vitro* protein digestibility and vitamin content of edible winged termite (*Macrotermes subhylanus*) and grasshopper (*Ruspolia differens*). Food and Bioprocess Technology 3: 778-782. https://doi.org/10.1007/s11947-009-0264-1

Klunder, H.C., Wolkers-Rooijackers, J., Korpela, J.M. and Nout, M.J.R., 2012. Microbiological aspects of processing and storage of edible insects. Food Control 26: 628-631. https://doi.org/10.1016/j.foodcont.2012.02.013

Koutchma, T., 2014. Adapting high hydrostatic pressure (HPP) for food processing operations. Academic Press, Cambridge, MA, USA.

Kroncke, N., Boschen, V., Woyzichovski, J., Demtroder, S. and Benning, R., 2018. Comparison of suitable drying processes for mealworms (*Tenebrio molitor*). Innovative Food Science & Emerging Technologies 50: 20-25. https://doi.org/10.1016/j.ifset.2018.10.009

Laroche, M., Perreault, V., Marciniak, A., Gravel, A., Chamberland, J. and Doyen, A., 2019. Comparison of conventional and sustainable lipid extraction methods for the production of oil and protein isolate from edible insect meal. Foods 8: 572. https://doi.org/10.3390/foods8110572

Larouche, J., Deschamps, M.-H., Saucier, L., Lebeuf, Y., Doyen, A. and Vandenberg, G.W., 2019. Effects of killing methods on lipid oxidation, colour and microbial load of black soldier fly (*Hermetia illucens*) larvae. Animals 9: 182.

Lautenschläger, T., Neinhuis, C., Kikongo, E., Henle, T. and Förster, A., 2017. Impact of different preparations on the nutritional value of the edible caterpillar *Imbrasia epimethea* from northern Angola. European Food Research and Technology 243: 769-778. https://doi.org/10.1007/s00217-016-2791-0

Lee, H., Bang, W.Y., Kim, Y., Bae, S., Hahn, D. and Jung, Y.H., 2020. Storage characteristics of two-spotted cricket (*Gryllus bimaculatus* De Geer) powder according to drying method and storage temperature. Entomological Research 50: 517-524. https://doi.org/10.1111/1748-5967.12472

Lempriere, B.M., 2003. Ultrasound and elastic waves: frequently asked questions. Elsevier, Amsterdam, The Netherlands.

Lenaerts, S., Van der Borght, M., Callens, A. and Van Campenhout, L., 2018. Suitability of microwave drying for mealworms (*Tenebrio molitor*) as alternative to freeze drying: Impact on nutritional quality and colour. Food Chemistry 254: 129-136.

Li, B., Qiao, M.Y. and Lu, F., 2012. Composition, nutrition, and utilization of Okara (soybean residue). Food Reviews International 28: 231-252. https://doi.org/10.1080/87559129.2011.595023

Li, D., Sihamala, O., Bhulaidok, S. and Shen, L.R., 2009. Changes in the organic compounds following sun drying of edible black ant (*Polyrhachis Vicina* Roger). Acta Alimentaria 38: 493-501. https://doi.org/10.1556/AAlim.38.2009.4.9

Li, D.N., Zhao, H.H., Muhammad, A.I., Song, L.Y., Guo, M.M. and Liu, D.H., 2020. The comparison of ultrasound-assisted thawing, air thawing and water immersion thawing on the quality of slow/fast freezing bighead carp (*Aristichthys nobilis*) fillets. Food chemistry 320: 126614. https://doi.org/10.1016/j.foodchem.2020.126614

Liao, X.Y., Li, J., Muhammad, A.I., Suo, Y.J., Chen, S.G., Ye, X.Q., Liu, D.H. and Ding, T., 2018. Application of a dielectric barrier discharge atmospheric cold plasma (Dbd-Acp) for *Escherichia coli* inactivation in apple juice. Journal of Food Science 83: 401-408. https://doi.org/10.1111/1750-3841.14045

Liu, C.Y., Pirozzi, A., Ferrari, G., Vorobiev, E. and Grimi, N., 2020. Effects of pulsed electric fields on vacuum drying and quality characteristics of dried carrot. Food and Bioprocess Technology 13: 45-52. https://doi.org/10.1007/s11947-019-02364-1

Luo, Q., Wang, Y., Han, Q.Q., Ji, L.S., Zhang, H.M., Fei, Z.H. and Wang, Y.Q., 2019. Comparison of the physicochemical, rheological, and morphologic properties of chitosan from four insects. Carbohydrate Polymers 209: 266-275. https://doi.org/10.1016/j.carbpol.2019.01.030

Madibela, O.R., Seitiso, T.K., Thema, T.F. and Letso, M., 2007. Effect of traditional processing methods on chemical composition and *in vitro* true dry matter digestibility of the Mophane worm (*Imbrasia belina*). Journal of Arid Environments 68: 492-500. https://doi.org/10.1016/j.jaridenv.2006.06.002

Manditsera, F.A., Lakemond, C.M., Fogliano, V., Zvidzai, C.J. and Luning, P.A., 2018. Consumption patterns of edible insects in rural and urban areas of Zimbabwe: taste, nutritional value and availability are key elements for keeping the insect eating habit. Food Security 10: 561-570.

Manditsera, F.A., Luning, P.A., Fogliano, V. and Lakemond, C.M.M., 2019. Effect of domestic cooking methods on protein digestibility and mineral bioaccessibility of wild harvested adult edible insects. Food Research International 121: 404-411. https://doi.org/10.1016/j.foodres.2019.03.052

Marei, N., Elwahy, A.H.M., Salah, T.A., El Sherif, Y. and Abd El-Samie, E., 2019. Enhanced antibacterial activity of Egyptian local insects' chitosan-based nanoparticles loaded with ciprofloxacin-HCl. International Journal of Biological Macromolecules 126: 262-272. https://doi.org/10.1016/j.ijbiomac.2018.12.204

Marei, N.H., Abd El-Samie, E., Salah, T., Saad, G.R. and Elwahy, A.H.M., 2016. Isolation and characterization of chitosan from different local insects in Egypt. International Journal of Biological Macromolecules 82: 871-877. https://doi.org/10.1016/j.ijbiomac.2015.10.024

Mariod, A.A., Abdelwahab, S.I., Gedi, M.A. and Solati, Z., 2010. Supercritical carbon dioxide extraction of sorghum bug (*Agonoscelis pubescens*) oil using response surface methodology. Journal of the American Oil Chemists Society 87: 849-856. https://doi.org/10.1007/s11746-010-1565-2

Melis, R., Braca, A., Mulas, G., Sanna, R., Spada, S., Serra, G., Fadda, M.L., Roggio, T., Uzzau, S. and Anedda, R., 2018. Effect of freezing and drying processes on the molecular traits of edible yellow mealworm. Innovative Food Science & Emerging Technologies 48: 138-149.

Mintah, B.K., He, R.H., Dabbour, M., Agyekum, A.A., Xing, Z., Golly, M.K. and Ma, H.L., 2019a. Sonochemical action and reaction of edible insect protein: influence on enzymolysis reaction-kinetics, free-Gibbs, structure, and antioxidant capacity. Journal of Food Biochemistry 43. https://doi.org/10.1111/jfbc.12982

Mintah, B.K., He, R.H., Dabbour, M., Golly, M.K., Agyekum, A.A. and Ma, H.L., 2019b. Effect of sonication pretreatment parameters and their optimization on the antioxidant activity of *Hermetia illucens* larvae meal protein hydrolysates. Journal of Food Processing and Preservation 43: e12982. https://doi.org/10.1111/jfpp.14093

Mintah, B.K., He, R.H., Dabbour, M., Xiang, J.H., Jiang, H., Agyekum, A.A. and Ma, H.L., 2020. Characterization of edible soldier fly protein and hydrolysate altered by multiple-frequency ultrasound: structural, physical, and functional attributes. Process Biochemistry 95: 157-165. https://doi.org/10.1016/j.procbio.2020.05.021

Mishyna, M., Haber, M., Benjamin, O., Martinez, J.I. and Chen, J., 2020. Drying methods differentially alter volatile profiles of edible locusts and silkworms. Journal of Insects as Food and Feed 6(4): 405-415. https://doi.org/10.3920/JIFF2019.0046

Mishyna, M., Martinez, J.-J.I., Chen, J. and Benjamin, O., 2019. Extraction, characterization and functional properties of soluble proteins from edible grasshopper (*Schistocerca gregaria*) and honey bee (*Apis mellifera*). Food Research International 116: 697-706.

Mohan, K., Ganesan, A.R., Muralisankar, T., Jayakumar, R., Sathishkumar, P., Uthayakumar, V., Chandirasekar, R. and Revathi, N., 2020. Recent insights into the extraction, characterization, and bioactivities of chitin and chitosan from insects. Trends in Food Science & Technology 105: 17-42. https://doi.org/10.1016/j.tifs.2020.08.016

Moreau, M., Orange, N. and Feuilloley, M.G.J., 2008. Non-thermal plasma technologies: new tools for bio-decontamination. Biotechnology Advances 26: 610-617. https://doi.org/10.1016/j.biotechadv.2008.08.001

Nadarajah, S., Margot, F. and Secomandi, N., 2015. Relaxations of approximate linear programs for the real option management of commodity storage. Management Science 61: 3054-3076. https://doi.org/10.1287/mnsc.2014.2136

Ndiritu, A.K., Kinyuru, J.N., Kenji, G.M. and Gichuhi, P.N., 2017. Extraction technique influences the physico-chemical characteristics and functional properties of edible crickets

(*Acheta domesticus*) protein concentrate. Journal of Food Measurement and Characterization 11: 2013-2021. https://doi.org/10.1007/s11694-017-9584-4

Ng'ang'a, J., Imathiu, S., Fombong, F., Ayieko, M., Broeck, J.V. and Kinyuru, J., 2019. Microbial quality of edible grasshoppers *Ruspolia differens* (Orthoptera: Tettigoniidae): From wild harvesting to fork in the Kagera Region, Tanzania. Journal of Food Safety 39: e12549. http://doi.org/10.1111/jfs.12549

Nonaka, K., 2009. Feasting on insects. Entomological Research 39: 304-312. https://doi.org/10.1111/j.1748-5967.2009.00240.x

Nowosad, K., Sujka, M., Pankiewicz, U. and Kowalski, R., 2021. The application of PEF technology in food processing and human nutrition. Journal of Food Science and Technology 58: 397-411. https://doi.org/10.1007/s13197-020-04512-4

Nyangena, D.N., Mutungi, C., Imathiu, S., Kinyuru, J., Affognon, H., Ekesi, S., Nakimbugwe, D. and Fiaboe, K.K.M., 2020. Effects of traditional processing techniques on the nutritional and microbiological quality of four edible insect species used for food and feed in East Africa. Foods 9: 574. https://doi.org/10.3390/foods9050574

O'Donnell, C.P., Tiwari, B.K., Bourke, P. and Cullen, P.J., 2010. Effect of ultrasonic processing on food enzymes of industrial importance. Trends in Food Science & Technology 21: 358-367. https://doi.org/10.1016/j.tifs.2010.04.007

Obopile, M. and Seeletso, T.G., 2013. Eat or not eat: an analysis of the status of entomophagy in Botswana. Food Security 5: 817-824. https://doi.org/10.1007/s12571-013-0310-8

Ojha, K.S., Aznar, R., O'Donnell, C. and Tiwari, B.K., 2020. Ultrasound technology for the extraction of biologically active molecules from plant, animal and marine sources. Trends in Analytical Chemistry 122: 115663. https://doi.org/10.1016/j.trac.2019.115663

Bureau Risicobeoordeling en Onderzoeks-programmering, 2014. Advisory report on the risks associated with the consumption of mass-reared insects. Bureau Risicobeoordeling en Onderzoeks-programmering, Utrecht, The Netherlands.

Otero, P., Gutierrez-Docio, A., Del Hierro, J.N., Reglero, G. and Martin, D., 2020. Extracts from the edible insects *Acheta domesticus* and *Tenebrio molitor* with improved fatty acid profile due to ultrasound assisted or pressurized liquid extraction. Food chemistry 314: 126200.

Pan, W.J., Liao, A.M., Zhang, J.G., Dong, Z. and Wei, Z.J., 2012. Supercritical carbon dioxide extraction of the oak silkworm (*Antheraea pernyi*) pupal oil: process optimization and composition determination. International Journal of Molecular Sciences 13: 2354-2367. https://doi.org/10.3390/ijms13022354

Pan, Y.Y., Cheng, J.H. and Sun, D.W., 2019. Cold plasma-mediated treatments for shelf life extension of fresh produce: a review of recent research developments. Comprehensive Reviews in Food Science and Food Safety 18: 1312-1326. https://doi.org/10.1111/1541-4337.12474

Panja, P., 2018. Green extraction methods of food polyphenols from vegetable materials. Current Opinion in Food Science 23: 173-182.

Patrignani, F., Del Duca, S., Vannini, L., Rosa, M., Schlüter, O. and Lanciotti, R., 2020. Potential of *Yarrowia lipolytica* and *Debaryomyces hansenii* strains to produce high quality food ingredients based on cricket powder. LWT – Food Science and Technology 119: 108866. https://doi.org/10.1016/j.lwt.2019.108866

Penas, E., Gomez, R., Frias, J., Baeza, M.L. and Vidal-Valverde, C., 2011. High hydrostatic pressure effects on immunoreactivity and nutritional quality of soybean products. Food Chemistry 125: 423-429. https://doi.org/10.1016/j.foodchem.2010.09.023

Pojic, M., Misan, A. and Tiwari, B., 2018. Eco-innovative technologies for extraction of proteins for human consumption from renewable protein sources of plant origin. Trends in Food Science & Technology 75: 93-104. https://doi.org/10.1016/j.tifs.2018.03.010

Poma, G., Cuykx, M., Amato, E., Calaprice, C., Focant, J.F. and Covaci, A., 2017. Evaluation of hazardous chemicals in edible insects and insect-based food intended for human consumption. Food and Chemical Toxicology 100: 70-79. https://doi.org/10.1016/j.fct.2016.12.006

Poma, G., Yin, S.S., Tang, B., Fujii, Y., Cuykx, M. and Covaci, A., 2019. Occurrence of selected organic contaminants in edible insects and assessment of their chemical safety. Environmental Health Perspectives 127. https://doi.org/10.1289/Ehp5782

Preece, K.E., Hooshyar, N., Krijgsman, A.J., Fryer, P.J. and Zuidam, N.J., 2017. Intensification of protein extraction from soybean processing materials using hydrodynamic cavitation. Innovative Food Science & Emerging Technologies 41: 47-55. https://doi.org/10.1016/j.ifset.2017.01.002

Priyadarshini, A., Rajauria, G., O'Donnell, C.P. and Tiwari, B.K., 2019. Emerging food processing technologies and factors impacting their industrial adoption. Critical Reviews in Food Science and Nutrition 59: 3082-3101. https://doi.org/10.1080/10408398.2018.1483890

Purkayastha, D. and Sarkar, S., 2020. Physicochemical structure analysis of chitin extracted from pupa exuviae and dead imago of wild black soldier fly (*Hermetia illucens*). Journal of Polymers and the Environment 28: 445-457. https://doi.org/10.1007/s10924-019-01620-x

Purschke, B., Brüggen, H., Scheibelberger, R. and Jäger, H., 2018a. Effect of pre-treatment and drying method on physico-chemical properties and dry fractionation behaviour of mealworm larvae (*Tenebrio molitor* L.). European Food Research and Technology 244: 269-280.

Purschke, B., Sanchez, Y.D.M. and Jager, H., 2018b. Centrifugal fractionation of mealworm larvae (*Tenebrio molitor*, L.) for protein recovery and concentration. LWT – Food Science and Technology 89: 224-228. https://doi.org/10.1016/j.lwt.2017.10.057

Purschke, B., Stegmann, T., Schreiner, M. and Jäger, H., 2017. Pilot-scale supercritical CO_2 extraction of edible insect oil from *Tenebrio molitor* L. larvae – influence of extraction conditions on kinetics, defatting performance and compositional properties. European Journal of Lipid Science and Technology 119: 1600134. https://doi.org/10.1002/ejlt.201600134

Purschke, B., Tanzmeister, H., Meinlschmidt, P., Baumgartner, S., Lauter, K. and Jager, H., 2018c. Recovery of soluble proteins from migratory locust (Locusta migratoria) and characterisation of their compositional and techno-functional properties. Food Research International 106: 271-279. https://doi.org/10.1016/j.foodres.2017.12.067

Rahaman, A., Zeng, X.A., Farooq, M.A., Kumari, A., Murtaza, M.A., Ahmad, N., Manzoor, M.F., Hassan, S., Ahmad, Z., Chen, B.R., Zhan, J.J. and Siddeeg, A., 2020. Effect of pulsed electric fields processing on physiochemical properties and bioactive compounds of apricot juice. Journal of Food Process Engineering 43: e13449. https://doi.org/10.1111/jfpe.13449

Ramazzina, I., Berardinelli, A., Rizzi, F., Tappi, S., Ragni, L., Sacchetti, G. and Rocculi, P., 2015. Effect of cold plasma treatment on physico-chemical parameters and antioxidant activity of minimally processed kiwifruit. Postharvest Biology and Technology 107: 55-65. https://doi.org/10.1016/j.postharvbio.2015.04.008

Ramos-Bueno, R.P., González-Fernández, M.J., Sánchez-Muros-Lozano, M.J., García-Barroso, F. and Guil-Guerrero, J.L., 2016. Fatty acid profiles and cholesterol content of seven insect species assessed by several extraction systems. European Food Research and Technology 242: 1471-1477.

Ramos-Rostro, B., Ramos-Elorduy Blásquez, J., Pino-Moreno, J.M., Viesca-González, F.C., Martínez-Maya, J.J., Sierra-Gómez Pedroso, L.d.C. and Quintero-Salazar, B., 2016. Calidad sanitaria de alimentos elaborados con gusano rojo de agave (Comadia redtembacheri H.) en San Juan Teotihuacán, Estado de México, México. Agrociencia 50: 391-402.

Ravi, H.K., Vian, M.A., Tao, Y., Degrou, A., Costil, J., Trespeuch, C. and Chemat, F., 2019. Alternative solvents for lipid extraction and their effect on protein quality in black soldier fly (*Hermetia illucens*) larvae. Journal of Cleaner Production 238: 117861. https://doi.org/10.1016/j.jclepro.2019.117861

Reineke, K., Schottroff, F., Meneses, N. and Knorr, D., 2015. Sterilization of liquid foods by pulsed electric fields-an innovative ultra-high temperature process. Frontiers in Microbiology 6: 400. https://doi.org/10.3389/fmicb.2015.00400

Rodriguez, M., Medina, L.M. and Jordano, R., 2000. Effect of modified atmosphere packaging on the shelf life of sliced wheat flour bread. Nahrung-Food 44: 247-252. https://doi.org/10.1002/1521-3803(20000701)44:4<247::Aid-Food247>3.0.Co;2-I

Rumpold, B.A., Fröhling, A., Reineke, K., Knorr, D., Boguslawski, S., Ehlbeck, J. and Schlüter, O., 2014. Comparison of volumetric and surface decontamination techniques for innovative processing of mealworm larvae (*Tenebrio molitor*). Innovative Food Science & Emerging Technologies 26: 232-241.

Rumpold, B.A. and Schlüter, O.K., 2013. Potential and challenges of insects as an innovative source for food and feed production. Innovative Food Science & Emerging Technologies 17: 1-11. https://doi.org/10.1016/j.ifset.2012.11.005

Schlüter, O., Ehlbeck, J., Hertel, C., Habermeyer, M., Roth, A., Engel, K.H., Holzhauser, T., Knorr, D. and Eisenbrand, G., 2013. Opinion on the use of plasma processes for treatment of foods. Molecular Nutrition & Food Research 57: 920-927. https://doi.org/10.1002/mnfr.201300039

Schlüter, O., Rumpold, B., Holzhauser, T., Roth, A., Vogel, R.F., Quasigroch, W., Vogel, S., Heinz, V., Jager, H., Bandick, N., Kulling, S., Knorr, D., Steinberg, P. and Engel, K.H., 2017. Safety aspects of the production of foods and food ingredients from insects. Molecular Nutrition & Food Research 61: 1600520. https://doi.org/10.1002/mnfr.201600520

Severini, C., Azzollini, D., Albenzio, M. and Derossi, A., 2018. On printability, quality and nutritional properties of 3D printed cereal based snacks enriched with edible insects. Food Research International 106: 666-676.

Shadung, K.G., Mphosi, M.S. and Mashela, P.W., 2012. Influence of drying method and location on proximate chemical composition of African metallic wood boring beetle, Sternocera orissa (Coleoptera: Buprestidae) in Republic of South Africa. African Journal of Food Science 6: 155-158.

Shin, C.S., Kim, D.Y. and Shin, W.S., 2019. Characterization of chitosan extracted from mealworm beetle (*Tenebrio molitor, Zophobas morio*) and rhinoceros beetle (*Allomyrina dichotoma*) and their antibacterial activities. International Journal of Biological Macromolecules 125: 72-77. https://doi.org/10.1016/j.ijbiomac.2018.11.242

Shockley, M., Lesnik, J., Allen, R.N. and Muñoz, A.F., 2018. Edible insects and their uses in North America; past, present and future. In: Halloran, A., Flore, R., Vantomme, P. and Roos, N. (eds.) Edible insects in sustainable food systems. Springer International Publishing, Cham, Switzerland, pp. 55-79. https://doi.org/10.1007/978-3-319-74011-9_4

Shorstkii, I., Alles, M.C., Parniakov, O., Smetana, S., Aganovic, K., Sosnin, M., Toepfl, S. and Heinz, V., in press. Optimization of pulsed electric field assisted drying process of black soldier fly (*Hermetia illucens*) larvae. Drying Technology. https://doi.org/10.1080/073739 37.2020.1819825

Siemer, C., Toepfl, S. and Heinz, V., 2014. Inactivation of *Bacillus subtilis* spores by pulsed electric fields (PEF) in combination with thermal energy – I. Influence of process- and product parameters. Food Control 39: 163-171. https://doi.org/10.1016/j.foodcont.2013.10.025

Simionato, J.I., Villalobos, L.D.G., Bulla, M.K., Coro, F.A.G. and Garcia, J.C., 2014. Application of chitin and chitosan extracted from silkworm chrysalides in the treatment of textile effluents contaminated with remazol dyes. Acta Scientiarum – Technology 36: 693-698. https://doi.org/10.4025/actascitechnol.v36i4.24428

Singh, Y., Cullere, M., Kovitvadhi, A., Chundang, P. and Dalle Zotte, A., 2020. Effect of different killing methods on physicochemical traits, nutritional characteristics, in vitro human digestibility and oxidative stability during storage of the house cricket (*Acheta domesticus* L.). Innovative Food Science & Emerging Technologies 65: 102444.

Soares, S. and Forkes, A., 2014. Insects Au gratin-an investigation into the experiences of developing a 3D printer that uses insect protein based flour as a building medium for the production of sustainable food, DS 78. In: Proceedings of the 16th International Conference on Engineering and Product Design Education (E&PDE14), Design Education and Human Technology Relations. September 4-5, 2014. University of Twente, Enschede, The Netherlands, pp. 426-431.

Soetemans, L., Uyttebroek, M., D'Hondt, E. and Bastiaens, L., 2019. Use of organic acids to improve fractionation of the black soldier fly larvae juice into lipid- and protein-enriched fractions. European Food Research and Technology 245: 2257-2267. https://doi. org/10.1007/s00217-019-03328-7

Son, Y.J., Lee, J.C., Hwang, I.K., Nho, C.W. and Kim, S.H., 2019. Physicochemical properties of mealworm (*Tenebrio molitor*) powders manufactured by different industrial processes. LWT – Food Science and Technology 116: 108514. https://doi.org/10.1016/j.lwt.2019.108514

Song, Y.S., Kim, M.W., Moon, C., Seo, D.J., Han, Y.S., Jo, Y.H., Noh, M.Y., Park, Y.K., Kim, S.A., Kim, Y.W. and Jung, W.J., 2018. Extraction of chitin and chitosan from larval exuvium and whole body of edible mealworm, *Tenebrio molitor*. Entomological Research 48: 227-233. https://doi.org/10.1111/1748-5967.12304

Soon, C.Y., Tee, Y.B., Tan, C.H., Rosnita, A.T. and Khanna, A., 2018. Extraction and physico-chemical characterization of chitin and chitosan from Zophobas mono larvae in varying sodium hydroxide concentration. International Journal of Biological Macromolecules 108: 135-142. https://doi.org/10.1016/j.ijbiomac.2017.11.138

Ssepuuya, G., Aringo, R.O., Mukisa, I.M. and Nakimbugwe, D., 2016. Effect of processing, packaging and storage-temperature based hurdles on the shelf stability of sauteed ready-to-eat Ruspolia nitidula. Journal of Insects as Food and Feed 2: 245-253. https://doi.org/10.3920/jiff2016.0006

Ssepuuya, G., Smets, R., Nakimbugwe, D., Van der Borght, M. and Claes, J., 2019. Nutrient composition of the long-horned grasshopper Ruspolia differens Serville: effect of swarming season and sourcing geographical area. Food Chemistry 301: 125305. https://doi.org/10.1016/j.foodchem.2019.125305

Stoica, M., Mihalcea, L., Borda, D. and Alexe, P., 2013. Non-thermal novel food processing technologies. An overview. Journal of Agroalimentary Processes and Technologies 19: 212-217.

Stoops, J., Crauwels, S., Waud, M., Claes, J., Lievens, B. and Van Campenhout, L., 2016. Microbial community assessment of mealworm larvae (Tenebrio molitor) and grasshoppers (Locusta migratoria migratorioides) sold for human consumption. Food Microbiology 53: 122-127. https://doi.org/10.1016/j.fm.2015.09.010

Stoops, J., Vandeweyer, D., Crauwels, S., Verreth, C., Boeckx, H., Van der Borght, M., Claes, J., Lievens, B. and Van Campenhout, L., 2017. Minced meat-like products from mealworm larvae (Tenebrio molitor and Alphitobius diaperinus): microbial dynamics during production and storage. Innovative Food Science & Emerging Technologies 41: 1-9.

Sun, M., Xu, X., Zhang, Q., Rui, X., Wu, J. and Dong, M., 2018. Ultrasonic-assisted aqueous extraction and physicochemical characterization of oil from Clanis bilineata. Journal of Oleo Science 67: 151-165.

Tang, J., Feng, H. and Shen, G.Q., 2003. Drum drying. In: Encyclopaedia of Agricultural, Food, and Biological Engineering. Marcel Dekker, Inc., New York, NY, USA, pp. 211-214.

Tao, Y., Han, M.F., Gao, X.G., Han, Y.B., Show, P.L., Liu, C.Q., Ye, X.S. and Xie, G.J., 2019. Applications of water blanching, surface contacting ultrasound-assisted air drying, and their combination for dehydration of white cabbage: drying mechanism, bioactive profile, color and rehydration property. Ultrasonics Sonochemistry 53: 192-201. https://doi.org/10.1016/j.ultsonch.2019.01.003

Temba, M.C., Njobeh, P.B. and Kayitesi, E., 2017. Storage stability of maize-groundnut composite flours and an assessment of aflatoxin B-1 and ochratoxin A contamination in flours and porridges. Food Control 71: 178-186. https://doi.org/10.1016/j.foodcont.2016.06.033

Tiencheu, B., Womeni, H.M., Linder, M., Mbiapo, F.T., Villeneuve, P., Fanni, J. and Parmentier, M., 2013. Changes of lipids in insect (Rhynchophorus phoenicis) during cooking and storage. European Journal of Lipid Science and Technology 115: 186-195. https://doi.org/10.1002/ejlt.201200284

Tonneijck-Srpova, L., Venturini, E., Humblet-Hua, K.N.P. and Bruins, M.E., 2019. Impact of processing on enzymatic browning and texturization of yellow mealworms. Journal of Insects as Food and Feed 5: 267-277. https://doi.org/10.3920/Jiff2018.0025

Tschirner, M. and Simon, A., 2015. Influence of different growing substrates and processing on the nutrient composition of black soldier fly larvae destined for animal feed. Journal of Insects as Food and Feed 1: 249-259. https://doi.org/10.3920/jiff2014.0008

Tzompa-Sosa, D.A., Yi, L., Van Valenberg, H.J.F. and Lakemond, C.M.M., 2019. Four insect oils as food ingredient: physical and chemical characterisation of insect oils obtained by an aqueous oil extraction. Journal of Insects as Food and Feed 5: 279-292. https://doi.org/10.3920/Jiff2018.0020

Tzompa-Sosa, D.A., Yi, L.Y., Van Valenberg, H.J.F., Van Boekel, M.A.J.S. and Lakemond, C.M.M., 2014. Insect lipid profile: aqueous versus organic solvent-based extraction methods. Food Research International 62: 1087-1094. https://doi.org/10.1016/j.foodres.2014.05.052

Udomsil, N., Imsoonthornruksa, S., Gosalawit, C. and Ketudat-Cairns, M., 2019. Nutritional values and functional properties of house cricket (*Acheta domesticus*) and field cricket (*Gryllus bimaculatus*). Food Science and Technology Research 25: 597-605. https://doi.org/10.3136/fstr.25.597

Ugur, A.E., Bolat, B., Oztop, M.H. and Alpas, H., 2020. Effects of high hydrostatic pressure (HHP) processing and temperature on physicochemical characterization of insect oils extracted from *Acheta domesticus* (house cricket) and *Tenebrio molitor* (yellow mealworm). Waste and Biomass Valorization. https://doi.org/10.1007/s12649-020-01302-z

Van der Fels-Klerx, H.J., Camenzuli, L., Van der Lee, M.K. and Oonincx, D.G.A.B., 2016. Uptake of cadmium, lead and arsenic by *Tenebrio molitor* and *Hermetia illucens* from contaminated substrates. PLoS ONE 11: e0166186. https://doi.org/10.1371/journal.pone.0166186

Vandeweyer, D., Crauwels, S., Lievens, B. and Van Campenhout, L., 2017a. Microbial counts of mealworm larvae (*Tenebrio molitor*) and crickets (*Acheta domesticus* and *Gryllodes sigillatus*) from different rearing companies and different production batches. International Journal of Food Microbiology 242: 13-18. https://doi.org/10.1016/j.ijfoodmicro.2016.11.007

Vandeweyer, D., Lenaerts, S., Callens, A. and Van Campenhout, L., 2017b. Effect of blanching followed by refrigerated storage or industrial microwave drying on the microbial load of yellow mealworm larvae (*Tenebrio molitor*). Food Control 71: 311-314.

Vandeweyer, D., Wynants, E., Crauwels, S., Verreth, C., Viaene, N., Claes, J., Lievens, B. and Van Campenhout, L., 2018. Microbial dynamics during industrial rearing, processing, and storage of tropical house crickets (*Gryllodes sigillatus*) for human consumption. Applied and Environmental Microbiology 84: e00255-18. https://doi.org/10.1128/AEM.00255-18

Varilla, C., Marcone, M. and Annor, G.A., 2020. Potential of cold plasma technology in ensuring the safety of foods and agricultural produce: a review. Foods 9: 1435. https://doi.org/10.3390/foods9101435

Wade, M. and Hoelle, J., 2020. A review of edible insect industrialization: scales of production and implications for sustainability. Environmental Research Letters 15: 123013. https://doi.org/10.1088/1748-9326/aba1c1

Wang, C.Y., Huang, H.W., Hsu, C.P. and Yang, B.B., 2016. Recent advances in food processing using high hydrostatic pressure technology. Critical Reviews in Food Science and Nutrition 56: 527-540. https://doi.org/10.1080/10408398.2012.745479

Williams, J.B., 2012. Marination: processing technology. Handbook of meat and meat processing. CRC Press, Boca Raton, FL, USA, pp. 495-504.

Womeni, H.M., Tiencheu, B., Linder, M., Nabayo, E.M.C., Tenyang, N., Mbiapo, F.T., Villeneuve, P., Fanni, J. and Parmentier, M., 2012. Nutritional value and effect of cooking,
drying and storage process on some functional properties of Rhynchophorus phoenicis.
Advance Research in Pharmaceuticals and Biologicals 2: L203-L219.

Wynants, E., Bruyninckx, L. and Van Campenhout, L., 2016. Effect of rinsing on the overall
microbial load of mealworm larvae (*Tenebrio molitor*). In: Proceedings of the 21st Conference on Food Microbiology. September 15-16, 2016. Brussels, Belgium, pp. 15-16.

Wynants, E., Crauwels, S., Lievens, B., Luca, S., Claes, J., Borremans, A., Bruyninckx, L. and
Van Campenhout, L., 2017. Effect of post-harvest starvation and rinsing on the microbial
numbers and the bacterial community composition of mealworm larvae (*Tenebrio molitor*). Innovative Food Science & Emerging Technologies 42: 8-15. https://doi.org/10.1016/j.
ifset.2017.06.004

Wynants, E., Crauwels, S., Verreth, C., Gianotten, N., Lievens, B., Claes, J. and Van Campenhout, L., 2018. Microbial dynamics during production of lesser mealworms (*Alphitobius
diaperinus*) for human consumption at industrial scale. Food Microbiology 70: 181-191.
https://doi.org/10.1016/j.fm.2017.09.012

Wu, X.F., Zhang, M., Adhikari, B. and Sun, J., 2017. Recent developments in novel freezing and
thawing technologies applied to foods. Critical Reviews in Food Science and Nutrition 57:
3620-3631. https://doi.org/10.1080/10408398.2015.1132670

Xiao, H.W., Bai, J.W., Sun, D.W. and Gao, Z.J., 2014. The application of superheated steam
impingement blanching (SSIB) in agricultural products processing – a review. Journal of
Food Engineering 132: 39-47. https://doi.org/10.1016/j.jfoodeng.2014.01.032

Yi, L., Van Boekel, M.A.J.S. and Lakemond, C.M.M., 2017. Extracting *Tenebrio molitor* protein
while preventing browning: effect of pH and NaCl on protein yield. Journal of Insects as
Food and Feed 3: 21-31. https://doi.org/10.3920/Jiff2016.0015

Yi, L.Y., Lakemond, C.M.M., Sagis, L.M.C., Eisner-Schadler, V., Van Huis, A. and Van Boekel,
M.A.J.S., 2013. Extraction and characterisation of protein fractions from five insect species. Food Chemistry 141: 3341-3348. https://doi.org/10.1016/j.foodchem.2013.05.115

Yoon, J.H. and Lee, S.Y., 2018. Review: comparison of the effectiveness of decontaminating
strategies for fresh fruits and vegetables and related limitations. Critical Reviews in Food
Science and Nutrition 58: 3189-3208. https://doi.org/10.1080/10408398.2017.1354813

Zhao, X., Vazquez-Gutierrez, J.L., Johansson, D.P., Landberg, R. and Langton, M., 2016. Yellow
mealworm protein for food purposes – extraction and functional properties. PLoS ONE
11: e0147791. https://doi.org/10.1371/journal.pone.0147791

Zhou, P.F., Li, J.B., Yan, T., Wang, X.P., Huang, J., Kuang, Z.S., Ye, M.Q. and Pan, M.S., 2019. Selectivity of deproteinization and demineralization using natural deep eutectic solvents
for production of insect chitin (*Hermetia illucens*). Carbohydrate Polymers 225: 115255.
https://doi.org/10.1016/j.carbpol.2019.115255

The new packaged food products containing insects as an ingredient

M. Reverberi

Bugsolutely (www.bugsolutely.com), Plento (www.plentofoods.com),
20-19 Sukhumvit soi 39, Bangkok 10110, Thailand; massimo.reve@gmail.com

Abstract

This article focuses on a newly defined category of insect as food: packaged processed insects (PPIs). PPIs integrate dry insects (in pieces or in powder) in packaged food products such as snacks, pasta, or baked goods. PPIs have been on the market for a few years, and they are still far from being mainstream. The commercial challenges they face, all of which are addressed in this article, include production costs, certifications and regulations, marketing communication, and retail distribution and consumer targeting. The western preconception of insect eating, which has been extensively covered in the existing literature, is also taken into account, but from the food makers' angle: how to deal with it when selling PPIs. Eleven PPI makers have contributed to this article by agreeing to a standardised interview protocol about their PPI products. Their names and a summary of the information collected is found in Supplementary material S1.

Keywords

packaged processed insect products (PPIs) – edible insect products – cricket flour – insect food development

1   Introduction

This article proposes a fundamental conceptual distinction: that the traditional, culturally-sanctioned practice of eating whole insects is a categorically different notion from the more recent use of mainly farmed insects as processed ingredients in packaged foods.

Traditional entomophagy mostly relates to the local consumption of whole insects collected from the wild and is found mostly in non-Western countries. In contrast, the practice which for brevity we will call packaged processed insects (PPI), is taking place on a light-industrial scale, largely in Western countries. PPIs use ingredients made from farmed insects (usually mealworm flour or cricket flour) to create foods such as crackers, pasta, energy bars, and snacks, as well as burgers and meatballs.

PPIs differ from traditional entomophagy for other reasons, too. One is consumer motivation: Traditional insect consumption is often driven by taste preferences within an informal economy. But PPI's key selling points also include the consumer's perception of the virtues of sustainability and nutrition.

Distribution and production channels are different, too. PPIs require at least two steps: both the farming and processing of the insect-based ingredients. In some cases one more intermediate step, the cricket-flour making, is performed by a company which buys the crickets from farmers, processes them into flour, and sells it as an ingredient it to food manufacturers.

The need to distinguish between whole insects and processed insects as an ingredient has been raised in very recent years, after considering the differences between the two categories (Orsi *et al.*, 2019). A study of European start-up websites has noted how narrow the processed packaged food category is (Pippinato *et al.*, 2020).

PPIs are a very new food category, with almost all processed-insect products appearing on the market in the past ten years. Among the pioneers are cricket-flour items such as cookies by Bitty Food (USA), chips by Six Foods (USA), meatballs and schnitzel by Damhert (Belgium), and cricket pasta by Bugsolutely (Thailand). Aldento (Belgium) offers a mealworm pasta, and in 2016 the Canadian start-up One Hop Kitchen launched pasta sauces containing mealworms and crickets.

Cricket flour has been used extensively in energy bars, with more than 20 USA and European brands (Reverberi, 2020a). Most of these bars contain a small percentage of cricket flour (5 to 10%), presumably to maintain a low retail price in view of the high cost of cricket flour in the USA and Europe. Recent trends have raised the percentage to 10 to 15% in new products, with peaks at 20%, with the result of more evident nutritional advantages. Depending on the food category, there might be limits to increasing the insect powder percentage due to taste, texture and binding issues. North American insect-food makers seem to believe that crickets are the best choice in terms of customer acceptance, while in Europe a growing number of start-ups are choosing mealworm powder for their new products (possibly as a result of the lower cost of mealworm powder compared to cricket powder).

In this article 'insect powder' and 'insect flour' are considered synonyms. Most PPI producers working with processed farmed insects call the resulting ingredient 'insect flour', while a minority refers to it as 'insect powder'. Some experts dispute the use of the term 'flour' (Dossey, 2016) arguing that flours are made from vegetable, not animal, sources. Still, for a marketing usage, the word flour may resonate better with consumers.

2   Methodology

PPI makers were selected from the public list on the blog BugBurger (www.bug-burger.se), known for containing a large number of insect start-ups (more than 300, including specialised retailers and farms), and which is frequently updated. The author of the list provided his suggestions for the selection.

Eleven start-ups have been selected among 70 PPI makers, using criteria that focused on those with an innovative product, and which reflected different food categories (e.g. snacks, energy bars, burgers, crackers, etc.).

The selected PPI entities were then sent a questionnaire consisting of 10 standardised questions. Three of the 11 interviews were conducted by email, with the 8 others interviews conducted as teleconferences, which allowed for a more open conversation and exchange of ideas. Although the interviewees were generally reluctant to disclose turnover or sales data, they freely answered questions about their distribution channels and retail strategy, which greatly assisted the understanding the commercial aspects of PPIs.

One product (Bug Recipes), a home-cooking subscription box which includes whole insects, does not completely fit the methodological definition of PPI but was considered relevant, since the promotion of insects in food through home-cooking is new and interesting.

3   Targeting consumers and selling points

Since the early years of their emergence (2014-2018), PPIs faced the dilemma of trying to go mainstream or aim at a market niche. According to the PPI makers interviewed, PPIs have the potential to be a mass market product. However, bias against eating insects, high prices due to small-scale production, and other obstacles novelty foods usually come across may work against the ambition of targeting the full spectrum of consumers. The most commonly selected consumer groups to start with are people with a high degree of appeal for diet, fitness, health and sustainability. One of the interviewees mentioned 'flexitarians' (those who rarely eat meat, and believe in a semi-vegetarian diet with a touch of 'flexibility') as a very promising target group. Vegetarians were also seen as a target group by most of the start-ups, too. The three key reasons influencing consumers towards a vegetarian diet are: promoting sustainability, reducing animal suffering and eating healthier food (Ruby, 2012). PPIs makers often make the same arguments for eating their products.

A similar demographic is described as 'lifestyles of health and sustainability', a particular market segment related to sustainable living, and generally composed of a relatively upscale and well-educated population segment. Another large target group is comprised of 'foodies', i.e. people looking for new food experiences. In terms of consumer age, the range 20-40 is often cited by PPI makers. While there is little literature on the socio-demographic profile of potential Western edible insect consumers, higher education seems to be an important factor (Cicatiello *et al.*, 2016). This may be connected with elevated awareness of the nutritional and environmental benefits of insect eating. Gender seems to have some relevance, too, with males slightly more inclined to try insect food (Verbeke, 2014; Wendin, 2017). Age is also related to edible insect consumption, and there are indicators suggesting younger people should be targeted more than people above their 40s (Videbaek, 2020).

In terms of what to communicate to the target, the PPI makers surveyed for this article classified taste as the key factor. In food, taste ranks as the top buyer's driver (IFIC Foundation, 2019), and without being delicious, a product has little chance of generating repeated sales.

Nutrition is usually considered the second key point helping generate sales. While taste is subjective, and most consumers are unlikely to be swayed to buy a product by a bald claim that it's 'yummy!', a food's nutritional value is an objective fact. PPIs high in proteins, minerals or vitamins can leverage these nutritional benefits as selling points.

Four of the PPI makers interviewed emphasised that environmental concerns are becoming more important to their customers. This is confirmed in a market research study (Nielsen, 2019) which reveals a clear consumer trend toward sustainability in food products. Thanks to the vast media coverage on edible insects in the past five years, most consumers perceive of PPIs as an environmentally friendly product. The sustainability of farmed crickets as food has been investigated (Halloran *et al.*, 2017) and confirmed, although little similar research is available about the sustainability for farmed insects like mealworms and silkworms.

4   Packaging and consumer trust

The image of eating whole insects can trigger negative emotions in Western consumers, according to a number of surveys and authors (for example, Collins, 2019). This is likely the reason why most PPI packaging does not display photos of insects (although logos and package labels sometimes contain stylised drawings of insects).

Obtaining consumer trust is a critical point with PPIs (Lensvelt, 2014), especially as Western consumers often associate insects with pests and diseases (Orsi *et al.*, 2019; Van Huis *et al.*, 2013), paying little mind to whether the insects were farmed and processed in a safe plant. From this perspective, modern, well designed industrial packaging would generate more confidence than the paper bags that have been used (and are still in use) in some PPI packaging. Paper bags are obviously a cheaper option for small-scale productions, and are frequently used with natural, organic food products, but given the negative perceptions about eating insects, the adoption of packaging aligned with other mass-market consumer products might foster the consumer perception of PPI as a 'normal' product made in a certified, standardised factory. PPI pioneers like Exo or Chirps followed this approach, and changed packaging a number of times to make the product look more mainstream. They also made very clear it is an insect-based product. Conversely, any attempt to hide the insect ingredient – for example, by naming it 'alternative protein' – could backfire in terms of consumer trust, if it is perceived as an attempt to disguise an unhealthy or unsavoury ingredient. In brief, the best approach to market an insect ingredient may be to be clear about the insect content, but with the text, not visually.

5   Protein source or superfood?

In the period 2015-2020, cricket energy bars have been the most common choice among makers of PPIs. Targeting the fitness niche with an alternative protein makes sense as a market strategy, but also results in competing in an established, very competitive market where customers are often aware of nutritional facts and the comparative value for money when looking for protein. When compared with commoditised sources of protein like whey, insects might be too expensive to appeal to consumers. The expression 'insect protein' is quite common, but it may undermine the potential of PPIs as a 'superfood' containing other nutritional benefits.

There is no legal barrier in using the word 'superfood' to describe PPIs (Schiemer *et al.,* 2018). Claims on the product packaging are subject to regulatory restrictions in many countries, but the word 'superfood' may still be used when communicating to the market and the media.

6   PPIs are not 'survival food'

While PPIs seem to have reached higher consumer acceptance in Western countries, the consumption of whole insects is still more likely to trigger a negative emotional reaction within the consumer audience (Videbaek *et al.,* 2020; Wendin *et al.,* 2017). In the popular imagination, insects are considered as a 'last-resort' food source to be consumed only during periods of food insecurity. Indeed, popular entertainment has perpetuated this idea, with reality TV shows like 'Man vs. Wild' depicting insect consumption as a near desperate alternative to starvation. Similarly, the post-apocalyptic blockbuster movie and TV series 'Snowpiercer' depicts an underclass being force-fed distasteful insect-based foods.

Despite these popular distortions and biases, it is true that entomophagy is a valid solution to food insecurity. Collecting or farming insects in poor countries represents a common practice in many countries and a very interesting opportunity as a protein source. Many initiatives are in place to develop this approach to eating insects, for example the Dutch-Kenyan-Ugandan initiative called the Flying Food project, or the LAO project 'Improving Livelihoods and Food Security in Laos and Cambodia' (Weigel *et al.,* 2018).

But the perception of insects as a last-resort survival food represents a challenge to convincing Western consumers to eat PPIs. This aversion appears to be deeply rooted: most Westerners seem to show this bias before they ever encounter or taste insect-based foods. And it is interesting to note that the bias may be increasing even in countries where eating insects has been a tradition. Modernisation, globalisation and westernisation may have a key role in changes regarding entomophagy (Muller, 2019).

Marketers of PPIs should therefore try to counter this perception by focusing on concepts like taste, nutrition and sustainability. In brief, PPIs should be presented as a positive choice, not as a necessity, and reject the image of insect eating as 'primitive people's practice' (Verbeke, 2014).

7   Allergy and COVID-19 from a marketing perspective

Despite the widespread usage of crickets as human food, cases of food allergy to crickets are relatively rare (Pener, 2016). Most of the PPI products have an allergen alert on the label, which is also a requirement by the US Food and Drug Administration (FDA). From a business perspective, the risk of consumers suffering allergic reactions to PPIs, and the likely huge media backlash that would follow, cannot be dismissed. Although edible insects have thus far proven to be an appealing narrative for the media, given the Western public's underlying bias against eating insects, a deadly outcome of an allergic reaction could trigger wide media coverage, and possibly negative consequences on the market growth of PPIs.

According to research by Dicke *et al.* (2020), the hazard of edible insects being a transmission vector of COVID-19 is extremely low. This finding is very important from a business perspective: edible insects are already the target of some consumer fears, and any perceived connection to the global pandemic would exponentially magnify those emotional negative reactions. It will be crucial for insect start-ups to communicate to the market that certified insect products are as safe, if not even safer, than other animal-based food.

8   The difficulty of knowing the size of the market

At least five market surveys with economical data have been published in the past few years, from Meticulous Research, Global Market Inside, Persistence Market Research, Arcluster and Credence Research. All the projections predict strong growth in edible insects in the next ten to twenty years. However, there are a number of reasons to question the validity of these data. First, traditional entomophagy is mostly a local countryside tradition, with insects collected from the wild or in very few cases raised in small scale-farms and consumed locally, for example as a street food or shared within a village. Even in countries like Thailand, where farming crickets has turned into a profitable side business for thousands of farmers (Hanboonsong *et al.*, 2013), the market still relies on an informal distribution in food markets where no receipts or invoices are ever released. The possibility to track traditional insects eating is close to impossible via official sales records.

When it comes to international commerce of insect flours, there are few ways to accurately calculate the size of the market, because the World Customs Organization (WCO) has not yet created any code for edible insects. These codes, known as harmonised system, are used internationally to categorise goods and apply duties, and it takes years to the WCO to release an update when new product categories appear. Further complicating the valuation of the PPI market, PPI products are often sold directly from manufacturers through their online shops and shipped individually, which makes it difficult to gather reliable data on overall market sales.

Given these barriers to accurate measurement, one should be wary of predictions such as the recent forecast that 'the global edible insects market is expected to reach USD 7.96 billion by 2030' (Meticulous Research, 2019). It is almost impossible at present to collect accurate data on the insect-based foods market at the level of consumer sales.

When considering Europe only, where insects are farmed and processed in a certified and controlled environment, it is possible to have more reliable market data, at least on the quantity of food-grade cricket and mealworm powder produced. According to a survey of 33 companies conducted by the International Platform of Insects for Food and Feed (IPIFF), the EU non-profit organisation focusing on insects for food and feed (IPIFF, 2020), consumption of PPI and whole-insect foods in 2019 totalled an estimated 500 tons. In the food industry, that is a negligible amount, corresponding to a dozen shipping containers.

Only a few food-grade insect farms (Entomo Farm, Aspire Food Group, Protifarm) have reached a post-seed-start-up status. According to data from Linkedin, Aspire's staffing level is almost 60 and the last funding they received in a Series A round was 7.3 million USD. Entomo Farm affirms they have 35 million crickets growing at any time in their 2,000 m² farm (Canadian Food Business, 2020). Protifarm, the largest European insect for food farm, has already reached the Series B with investors pouring in the company more than 10 million USD (Pothering, 2019). But dozens of other farms are much smaller, which is also the case for PPI makers.

In terms of staff, the companies interviewed for this article have an average of 5 full time employees, including the founders. The interview with the PPI representative for Micarna revealed that even through the parent company has a workforce of 3,400, only three employees are focused on edible insects. The survey by IPIFF

(IPIFF, 2020) of 33 European edible insect companies reported that 81% of them are micro-companies with less than 10 employees.

Between 2014 and 2018 there had been a few attempts to create reliable databases of new edible-insect companies (those not belonging to traditional entomophagic cultures and mostly located in Western countries). Among these efforts two were notables: the Bugs Feed website, developed in conjunction with the launch of the Danish documentary 'Bugs', and the Maggot Master database created by Ilkka Tapponen, now converted into a Wiki page with around 400 entities. Both lists appear to be inactive now, although still useful for research purposes. The list handled by Anders Engstrom at www.bugburger.se seems to be the best option to get a real-time picture of the edible insect business at the moment, and it contains 320 entries. About half of these may fall into the PPI or the 'whole packaged insects' categories, while the others are farmers/processors and retailers specialising in edible insects, plus a few bug restaurants. According to Engstrom (who provided suggestions for this article), the mortality rate of the insect-based food start-ups is high, and while it is easy to find new entities when they launch a product, it is much harder to know when they cease activities. There are many recent examples of small scale businesses whose scarce sales resulted in failure after a couple of years, including One Hop Kitchen (Canada), Mophagy (UK), and Lithic Nutrition (USA).

9   The challenge of entering supermarkets

Start-up-made novelty foods often hit the market through online shops, because of the lower cost and relative simplicity of e-commerce. But PPIs still need to win consumer trust, with one important avenue being exposure on the shelves of supermarkets or corner shops. These traditional food retailers expect products to sell quickly, which might be a challenge as the PPI market ramps up. But obtaining such shelf exposure will undoubtably improve consumers' emotional confidence in PPIs. Consumers need to touch the physical product and read its label claims and facts, and they tend to trust a supermarket chain more than an online shop. Until PPIs are widely distributed in normal retail points of purchase, consumers will not fully embrace it as credible and safe.

The companies interviewed here confirmed that PPIs perform better through physical retailers than through e-commerce. In the early stages (between 2015 and 2018), start-ups focused on e-commerce to reduce distribution and retail costs, but now PPIs makers, particularly those in the European Union, are often targeting supermarket chains. A few European supermarkets are testing the products in some stores. It is the case with Sainsbury's (UK), Irma (Denmark), Metro (Germany), Coop (Switzerland), Rewe (Germany). Some start-ups also reported exploring alternative and easily approachable channels, such as specialised organic and novelty-food shops, food trucks and food services. Organic shops seemed a particularly good fit for the PPI customer profile, but it is hard to get an organic certification for PPI products. In the case of crickets, farms are struggling to find the organic cricket feed needed to have a complete organic rearing process. Furthermore, their business is often too small to make the cost of an organic certification feasible.

Access to supermarket chains is also harder if PPIs do not have industrial certifications like ISO or BRC. Using a reliable co-packer (third party) for the production of the PPI often ensures a reasonable level of food safety standards.

10   Product development, production cost and retail price

Start-ups have two options for product design: make the prototype in a professional kitchen or engage an R&D centre. Shared kitchens are becoming a popular way to develop recipes, an option that is viable for simple products like bread or an energy bar, but less feasible if the product needs machinery like an extruder or a vacuum fryer. Expensive large industrial machines can be found in co-packer facilities, and the co-packer might be willing to test the product formulation in the factory.

Universities often work for private companies in product development. For example, Plento (www.plentofoods.com) outsourced the development of an insect snack to the Singaporean Polytechnic's Food Innovation Research Center. R&D can be performed also by private specialised centres. Although this solution can be a major financial commitment for a start-up, it secures a methodological approach and expertise which is not always available within the start-up team.

When it comes to setting the product price, consumers who are early adopters usually accept to pay a premium price for innovation, nutrition and sustainability, but only to a certain point. And PPI products are often quite expensive. Where does the premium price of PPI products come from?

First, small companies are confronted with higher costs of running small production batches. Small volumes means co-packers will charge the start-ups more for the production.

If the PPI is meant for export, either informally through basic postal services, or officially with a proper export operation, the product price can easily double because of shipment and/or custom fees. The custom and freight fees of an international cargo, even when as small as half a ton (just a couple of pallets or cubic meters) exceed 1000 USD, and the final product price will be highly affected.

The insect ingredient cost is also critical, of course. Despite insect powder farmers and processors are scaling up production and lowering the insect flour price every year, insects are still an expensive ingredient. At the dawn of PPIs, around 2015, the North American cricket flour wholesale price was close to 100 USD/kg and in Thailand was averaging 30 USD/kg. In 2020 it is below 40 USD/kg in North America and below 20 USD/kg in Thailand. Western farms like Entomo (Canada), Aspire Food (USA) and Protifarm (The Netherlands) are testing automation and vertical farming to reduce the cost of labour, while Thailand's estimated 7.5 tonnes per year production (Hanboonsong, 2013) is largely based on small, family farms, with very low operating costs, but also low productivity. It is yet to be seen what insect-farming approach will prevail between the Thai model of Small Scale Livestock Farm (as defined by Halloran *et al.,* 2018) and Western-model factory farming. European cricket flour is even more expensive (between 60 and 90 USD/kg) than the North American one but they offer a more reasonable price for mealworm flour and dry whole mealworms (around 25 USD/kg), arguably one of the reasons for the growth of European PPIs made with mealworms. The insect price largely determines the choice of insect percentage in PPIs, which averaged between 6 and 12% five years ago, and is growing every year with a number of PPIs featuring 15 to 20% insect content in 2020. The PPI makers interviewed reported a typical percentage of 10 to 20%, with two exceptions (one product with 5%, and another with 30%).

11   Production, certification and regulations

A contract packager, commonly called a co-packer or an OEM, is a manufacturer that will produce, package, and/or label a product for clients. A co-packer is often a factory that not only manufactures its own products, but also makes private-label products for other companies. In addition to manufacturing, packaging, and labelling, co-packers also offer start-ups the benefit of certifications as well as other services like storage and shipping (O'Donoughue, 2020). The advantages for a start-up with limited resources and a concept still to prove are evident, and this is the most common option chosen by consumer-product start-ups in the early stages of their companies. The PPI makers interviewed confirmed this approach: 8 out of 10 are using a co-packer.

In terms of food safety international standards, HACCP is the most common certification adopted by insect flour processors and normally accepted by food agencies, like the US FDA. But supermarket chains usually require higher certifications level (ISO, BRC or IFS) therefore cricket flour manufacturer are expected to raise their certification level soon, otherwise their customers – the PPIs makers – won't be able to access mainstream food retailers. 9 of the companies interviewed derive their product certification from the co-packer, and only one has the lowest level of certification (GMP).

Even when having Internationally recognised safety certifications, PPIs may face great difficulties with food regulations in the world, in terms of clearing customs or being sold on the market. Regulations about insects as food differ widely (or are simply absent) in most countries. The topic is too complex to be discussed in detail in this article, but has been treated extensively elsewhere (Shockley *et al.,* 2017, and for the European Union situation, Reverberi, 2020b). It should be noted that regulations are one of the critical barriers to PPI. For example, in the EU and (with the exception of silkworms) in China insects are considered novel foods subject to a long, expensive approval process. Crickets and mealworms are also absent from the United Nations FAO Codex Alimentarius where food standards are set internationally. Both the FAO and the WCO would require a member state to initiate the lengthy process of adding such a new category. As of 2018 no such request has been registered and therefore it may be years before any insect-related customs and safety standards are set globally.

12   The bias about eating insects

Several explanations for the Western bias against eating insects have been offered, some connected to the perception of insects as pests and the cause of agricultural damage. Also, insects in tropical areas are larger and easier to harvest (Van Huis, 2013), while livestock like cattle prefers a colder climate, which may be an additional economical difference between the two areas. But most of the reasoning revolves around psychological and emotional factors. In particular, food neophobia – the fear for unfamiliar foods – which is an innate human reaction against potential physical hazards. According to some authors (Baker *et al.,* 2018; Verbeke, 2014) this might be the most critical barrier to edible insect acceptance.

The distaste towards insect eating sounds like a contradiction, as Western people don't know what an insect *taste* like. It is clearly challenging for PPI makers fighting against a preconception with little rational basis. They must convince consumers that the taste of the products is good, while consumers often perceive it as bad, not because it is, but because the food taboo translates into a taste preconception.

Luckily, a few recent analysis of the problem include the distinction between whole insects and processed products, and they focus on Western countries, including Germany (Orsi *et al.,* 2019), the United States (Baker *et al.,* 2018), and Italy (Cicatiello *et al.,* 2016). These recent studies can help PPI makers' to make good product development and marketing choices. Simply pointing out that many non-Western populations have always eaten insects won't help much in convincing Europeans and North American consumers to try edible insect products (Deroy *et al.,* 2015).

13   Conclusions

In order to correctly frame the new food products made with insects as an ingredient, a new category has been used in this research: packaged processed insects or PPI. About ten years after the first products appeared on the market, PPIs have generated a wide media coverage, but they are still not mainstream in terms of sales and presence in supermarkets and the makers are still quite small companies.

On the other hand, packaging has improved, some European retail chain are featuring PPIs on the shelves and the insect content percentage has increased, thanks to more affordable insect powder prices. The number of start-ups and the product variety has also grown. In the past few years Europe has also seen the raise of mealworm as ingredient and now many start-ups manufacture both mealworm – and cricket-based PPIs. Start-ups are targeting specific consumer groups (such as health – and fitness-conscious people and foodies) but in the interviews they affirmed they aim at a mass market in the future, thanks to PPIs main selling points: high nutrition and sustainability. Overcoming the regulatory obstacles and entering larger retail chains might be the next key step for the success of PPIs.

Conflict of interest

The author has a potential conflict of interest by being the founder of two edible insect start-ups (Plento and Bugsolutely). Products made by these companies may be in competition with other PPI. Plento and Bugsolutely are not among the companies interviewed for this article.

Acknowledgements

Anders Engström, insect bloggers and entrepreneur, offered guidance and valuable insights. His list of PPI products (https://www.bugburger.se/guide/the-big-list-of-edible-insect-products) is constantly updated and a precious reference. Andrea Marianelli, F&B consultant and nutritionist, conducted some of the interviews and provided precious feedback. Without the help of the PPI company founders who accepted to be interviewed writing this article would not have been possible.

Supplementary material

Supplementary material S1. Short summary of the packaged processed insect (PPI) product makers who have answered the questionnaire.

Supplementary material can be found online at https://doi.org/10.3920/JIFF2020. 0111.

References

Baker, M.A., Shin, J.T. and Kim, Y.W., 2018. Customer acceptance, barriers, and preferences in the U.S. Edible insects in sustainable food systems. Springer International Publishing, Berlin, Germany, pp. 387-399. https://doi.org/10.1007/978-3-319-74011-9

Canadian Food Business, 2020. Crickets: the gateway bug. Available at: https://canadian-foodbusiness.com/2020/02/11/crickets-the-gateway-bug/

Cicatiello, C., De Rosa, B., Franco, S. and Lacetera, N., 2016. Consumer approach to insects as food: barriers and potential for consumption in Italy. British Food Journal 118(9): 2271-2286. https://doi.org/10.1108/BFJ-01-2016-0015

Collins, C., Vaskou, P. and Kountouris, Y., 2019. Insect food products in the western world: assessing the potential of a new 'green' market. Annals of the Entomological Society of America 112: 518-528. https://doi.org/10.1093/aesa/saz015

Deroy, O., Reade, B. and Spence, C., 2015. The insectivore's dilemma, and how to take the West out of it. Food Quality and Preference 44: 44-55. https://doi.org/10.1016/j.food-qual.2015.02.007

Dicke, M., Eilenberg, J., Falcao Salles, J., Jensen, A.B., Lecocq, A., Pijlman, G.P., Van Loon, J.J.A. and Van Oers, M.M., 2020. Edible insects unlikely to contribute to transmission of coronavirus SARS-CoV-2. Journal of Insects as Food and Feed 6: 333-339. https://doi.org/10.3920/JIFF2020.0039

Dossey, A., Tatum, J. and McGill, W., 2016. Modern insect-based food industry: current status, insect processing technology, and recommendations moving forward. In: Dossey, A.T., Morales-Ramos, J.A. and Guadalupe Rojas, M. (eds.) Insects as sustainable food ingredients: production, processing and food applications. Elsevier, New York, NY, USA, pp. 113-152.

Halloran, A., Flore, R., Vantomme, P. and Roos, N., 2018. Edible insects in sustainable food systems. Springer, Berlin, Germany. https://doi.org/10.1007/978-3-319-74011-9

Halloran, A., Hanboonsong, Y., Roos, N. and Bruun, S., 2017. Life cycle assessment of cricket farming in north-eastern Thailand. Journal of Cleaner Production 156: 83-94. https://doi.org/10.1016/j.jclepro.2017.04.017

Hanboonsong, Y., Jamjanya, T. and Durst, P., 2013. Six-legged livestock: edible insect farming, collection and marketing in Thailand. FAO Regional Office for Asia and the Pacific, Bang-kok, Thailand. Available at: http://www.fao.org/3/i3246e/i3246e.pdf

IFIC Foundation, 2019. 2019 food & health survey. Available at: https://foodinsight.org/2019-food-and-health-survey

International Platform of Insects for Food and Feed (IPIFF), 2020. Questionnaire on the EU market. IPIFF, Brussels, Belgium. Available at: https://ipiff.org/wp-content/up-loads/2020/06/10-06-2020-IPIFF-edible-insects-market-factsheet.pdf

Lensvelt, E. and Steenbekkers, L., 2014. Exploring consumer acceptance of entomophagy: a survey and experiment in Australia and the Netherlands. Ecology of Food and Nutrition 53: 543-561. https://doi.org/10.1080/03670244.2013.879865

Meticulous Research, 2019. Edible insects market – global opportunity analysis and industry forecast. Available at: https://www.meticulousresearch.com/product/edible-insects-market-forecast/

Muller, A., 2019. Insects as food in Laos and Thailand – a case of 'westernisation'? Journal of Social Science 47: 204-223. https://doi.org/10.1163/15685314-04702003

Nielsen, 2019. Sustainable food products on the rise. Available at: https://www.meatpoultry.com/articles/20722-nielsen-sustainable-food-products-on-the-rise

O'Donoughue, A., Jennings, W. and Ahn, S., 2020. Finding and using a co-packer. University of Florida, Gainesville, FL, USA. https://doi.org/10.32473/edis-fs380-2020

Orsi, L., Voege, L.L., Stranieri, S., 2019. Eating edible insects as sustainable food? Exploring the determinants of consumer acceptance in Germany. Food Research International 125: 108573. https://doi.org/10.1016/j.foodres.2019.108573

Pener, M., 2016. Allergy to crickets: a review. Journal of Orthoptera Research 25: 91-95. https://doi.org/10.1665/034.025.0208

Pippinato, L., Gasco, L., Di Vita, G. and Mancuso, T., 2020. Current scenario in the European edible-insect industry: a preliminary study. Journal of Insects as Food and Feed 6: 371-381. https://doi.org/10.3920/JIFF2020.0008

Pothering, J., 2019. Protifarm raises Series B to scale robot-grown beetle 'tofu'. AgFunderNews. Available at: https://agfundernews.com/protifarm-raises-series-b-to-scale-robotized-beetle-tofu-farm.html

Reverberi, M., 2020a. Edible insects: cricket farming and processing as an emerging market. Journal of Insects as Food and Feed 6: 211-220. https://doi.org/10.3920/JIFF2019.0052

Reverberi, M., 2020b. Current EU novel food regulation. Bugsolutely Blog, Bangkok, Thailand. Available at: https://www.bugsolutely.com/novelfood/

Ruby, M.B., 2012. Vegetarianism. A blossoming field of study. Appetite 58: 141-150. https://doi.org/10.1016/j.appet.2011.09.01

Schiemer, C., Halloran, A., Jespersen, K. and Kaukua, P., 2018. Marketing insects: superfood or solution-food? Edible insects in sustainable food systems. Springer, Berlin, Germany. https://doi.org/10.1007/978-3-319-74011-9_14

Shockley, M., Allan, R.N. and Gracer, D., 2017. Product development and promotion. In: Van Huis, A. and Tomberlin, J.K. (eds.) Insects as food and feed: from production to consumption. Wageningen Academic Publishers, Wageningen, The Netherlands.

Van Huis, A., Van Itterbeeck, J., Klunder, H., Mertens, E., Halloran, A., Muir, G. and Vantomme, P., 2013. Edible insects: future prospects for food and feed security. FAO Forestry Paper 171. FAO, Rome, Italy, pp. 35-43. Available at: http://www.fao.org/docrep/018/i3253e/i3253e.pdf

Verbeke, W., 2014. Profiling consumers who are ready to adopt insects as a meat substitute in a Western society. Food Quality and Preference 39: 147-155. https://doi.org/10.1016/j.foodqual.2014.07.008

Videbaek, P. and Grunert, K., 2020. Disgusting or delicious? Examining attitudinal ambivalence towards entomophagy among Danish consumers. Food Quality and Preference 83: 103913. https://doi.org/10.1016/j.foodqual.2020.103913

Weigel, T., Fèvre, S., Berti, P.R., Sychareun, V., Thammavongsa, V., Dobson, E. and Kongmanila, D., 2018. The impact of small-scale cricket farming on household nutrition in Laos. Journal of Insects as Food and Feed 4: 89-99. https://doi.org/10.3920/JIFF2017.0005

Wendin, K., Norman, C., Forsberg, S., Langton, M., Davidsson, F., Josell, Å., Prim, M. and Berg, J., 2017. Eat 'em or not? Insects as a culinary delicacy. In: Proceedings of the 10th International Conference on Culinary Arts and Sciences, July 5-7, 2017, Copenhagen, Denmark.

Correlates of the willingness to consume insects: a meta-analysis

B. Wassmann*, M. Siegrist and C. Hartmann

Department of Health Sciences and Technology, ETH Zürich, Universitaetstrasse 22, 8092 Zurich, Switzerland;
*bwassmann@hest.ethz.ch

Abstract

Although insects are a sustainable meat alternative, the willingness to consume (WTC) them remains generally low. We synthesised the effects of WTC correlates reported in 37 studies and also investigated the moderating effects of certain study characteristics. Across a large number of studies, affect-based factors, such as 'food neophobia', 'disgust' and 'the expected unpleasant taste of insects', were consistently strongly correlated with WTC ($\bar{r}$ = -0.33-0.55). Information-based factors, such as 'the perceived sustainability of insects as food' and 'the perceived nutritiousness of insects as food', also impacted WTC ($\bar{r}$ = 0.32-0.55). However, the number of contributing studies in this regard was low. Curiosity appears to be relevant to WTC because 'food sensation and innovation seeking' ($\bar{r}$ = 0.29) positively impacted WTC. 'Age, education and gender' were relatively unrelated to WTC ($\bar{r}$ = -0.14-0.00) across a large number of studies. Combatting affective barriers through gradual and/or early exposure – i.e. increasing the 'familiarity with the concept of eating insects' ($\bar{r}$ = 0.10) and allowing consumers' experiences with 'insect consumption' ($\bar{r}$ = 0.35) to develop over time – will help foster entomophagy acceptance in the long run. In comparison, information-based interventions may have limited effectivity, but they can be implemented in the short term. As meta-regressions have shown, future researchers must consider whether the presentation of the edible insects has moderating effects, e.g. presenting specific as compared to non-specific products (β = -0.47) or actual products (β = -0.56) and pictures of such products (β = -0.55) as compared to mere verbal descriptions. Classical psychological entomophagy factors have been explored comprehensively, and research should also adopt a more market-oriented focus.

Keywords

entomophagy – consumer acceptance – food neophobia – sustainable protein – meat alternative

1   Introduction

The world demand for animal protein continues to increase along with a rapidly growing population. However, conventional meat production systems cannot sufficiently meet these needs, and they also carry high environmental costs. Alternative, more sustainable protein sources are therefore needed (Boland *et al.*, 2013). According to a UN report, one answer to these global concerns could be the more widespread adoption of insects into the human diet (Van Huis *et al.*, 2013). Firstly, insects are high in protein and key macronutrients and lower in cholesterol than many other meat products (Belluco *et al.*, 2013; Nowak *et al.*, 2016; Payne *et al.*, 2016). Secondly, the production of insects may have lower costs as compared to livestock production in terms of feed conversion efficiency, greenhouse gas and ammonia emissions, water and land use and animal welfare (Halloran *et al.*, 2016; Oonincx *et al.*, 2010; Smetana *et al.*, 2015). In fact, the production of insects can have a lower environmental impact than many other meat alternatives, such as cultured meat or milk-, gluten and myco-based proteins (Smetana *et al.*, 2015).

Although there is a growing awareness of these nutritional and environmental benefits, the willingness to consume (WTC) insects remains low for the majority of the population, especially in Western countries (Hartmann *et al.*, 2015; Verbeke, 2015). To identify the psychological barriers responsible for this, research on the acceptance of entomophagy (e.g. the practice of eating insects) has expanded in recent years. While there have been several reviews compiling these studies qualitatively (Hartmann and Siegrist, 2017b; Kim *et al.*, 2019; Mancini *et al.*, 2019a; Sogari *et al.*, 2019a), the current meta-analysis is one of the first attempts at a quantitative synthesis. We therefore present not only an overview of factors related to entomophagy acceptance but also estimates of their effect sizes. This allows for an approximate quantification of the importance of various WTC correlates (both as individual factors and in comparison with one another) and certain study characteristics of past entomophagy acceptance research (in terms of moderating effects). Furthermore, we provide an overview of methodological aspects of previous entomophagy studies and identify which variables have been investigated frequently and – in turn – which variables should be given more attention. Our findings aim to consolidate the existing body of knowledge on the barriers to and potential avenues for the acceptance of edible insects and may also illuminate the next steps appropriate for entomophagy acceptance research.

2   Methods

2.1   *Selection of relevant studies*
A literature search of Web of Science (Core Collection) was conducted in May 2020. The Advanced Search tool was used with the following search string: TI=(insect*

OR bug* OR entomophagy) AND TS=(substitute OR alternative OR sustainable OR replac* OR entomophagy OR 'eating insects' OR 'insects as food') AND TS=(-consum* OR behav* OR accept* OR perception* OR attitude* OR eat*). The search was restricted to articles published in the English language.

The study selection process is depicted in Figure 1; the inclusion and exclusion criteria are summarised in Table 1. The literature search yielded 1,023 records. In a first step, the titles and abstracts of these records were screened and 968 records that did not meet the inclusion criteria were eliminated. In a second step, the remaining 55 records were read in full, and eight more records were eliminated because they did not meet the inclusion criteria. Most of these eliminated records were excluded because they did not provide effect size(s) specifically related to consumers' WTC insects as food. For example, Barsics *et al.* (2017) investigated the overall liking of insects as food, not WTC. As another example, Ebenebe *et al.* (2017) was not included, because their findings were given in frequencies, not as effect sizes. The identification and screening of the records were performed independently by two parties. Interrater agreement was high (r = 0.90), and discrepancies between the parties were resolved by a more thorough review of the eligibility of the articles in question. Overall, 47 records met the inclusion criteria.

Because the current review aims to analyse the correlational effect between the WTC and various variables (WTC correlates), these specific correlation coefficients had to be extracted from the 47 records meeting the inclusion criteria. This occurred in two ways: in 14 records, Pearson's *r* correlation coefficients relevant to the current meta-analytic objective were reported. These were taken directly into the analysis. For the remaining 33 records, statistical values were not given as Pearson's r correlation coefficients, and the corresponding authors were therefore contacted to obtain the correlation coefficients. Through this method, the relevant correlation coefficients of 23 records were obtained and included in our analysis. Eight records were eliminated because the authors did not respond, and three records were eliminated because the provided data were unusable. One not-yet-published record was identified through personal communication with the author. Ultimately, 37 records were included in the review.

2.2   *Study information*

Table 2 shows the general study characteristics (sample size, country of origin, and data collection method) and more detailed study information. Specifically, it shows the WTC type assessed in the studies (WTT: willingness to try insects as food; WTE: willingness to eat insects as food; WTP: willing to pay for insects as food; WTA: willingness to adopt insects as food), the presentation of the insect food (descriptions, pictures or the real products) and the type of insect food investigated in the studies (specific insect products were mentioned such as 'insect patties'; or there was no specification given concerning the food type and insects were referred to simply as 'edible insects'). Additionally, brief descriptions of the insect food presented in the studies are given.

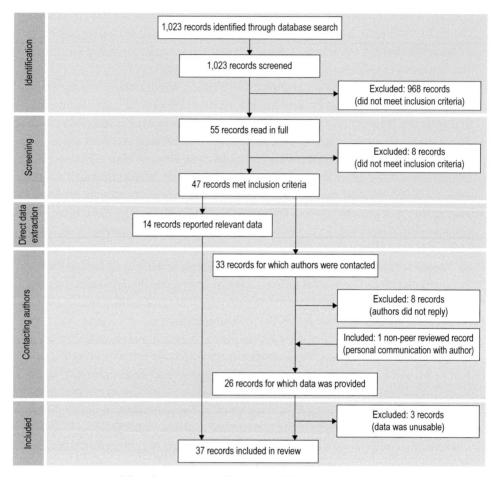

FIGURE 1   Summary of the selection process of the included literature

TABLE 1   Inclusion and exclusion criteria used for the article selection

Inclusion criteria	Exclusion criteria
Quantitative study.	Qualitative studies, review articles, opinion papers
Full-text paper in the English language.	and outlooks, conference papers and abstracts,
Investigates consumers' willingness to consume	concept articles.
insects as food or related measures (such as	Not related to consumer behaviour (e.g. insect-
willingness to try, buy, eat, pay, purchase or adopt	based food technology development, human
insects as food).	digestion of insect proteins, environmental impact
Provides (an) effect size(s) correlating at least one	of insect production systems).
variable to consumers' willingness to consume	Focus on sensory perception of insects as food.
insects.	Focus on consumer perception of insects primarily
	as feed.
	Does not provide (an) effect size(s) correlating
	at least one variable to consumers' willingness to
	consume insects as food.

2.3   *Selection of relevant effect sizes*

From the 37 records included in this review, 195 correlations between WTC and 49 variables (WTC correlates) were extracted. However, we decided to exclude the WTC correlates (n = 15) measured by only one study because doing so would not have allowed the calculation of a mean effect size. Ultimately, this analysis included 180 correlations between WTC insects and 34 WTC correlates.

These 34 WTC correlates (Table 3) can be divided into the following three subgroups: 'sociodemographic variables and general attitudes' ('gender', 'age', 'education', 'environmental concern', 'health concern') (Figure 2), 'variables related to eating' ('food neophobia', 'meat consumption') (Figure 3) and 'variables related to eating insects' ('disgust at eating insects', 'perceived sustainability of insects as food') (Figure 4).

2.4   *Data analysis*

The current analysis reports all effect sizes as Pearson's r correlation coefficients. Some articles (n = 5) reported more than one correlation for the same relationship between WTC and a certain WTC correlate. For example, some studies reported the correlation between WTC and a certain WTC correlate for a vegetarian and a non-vegetarian group (e.g. Elorinne *et al.*, 2019). For these cases, a composite correlation coefficient was calculated based on the data provided by the authors (Hunter and Schmidt, 2004). Some articles (n = 2) contained multiple studies (e.g. Chan, 2019). If only one study within a multi-study article was relevant, only the study information and results of the relevant study were included. If multiple studies within one multi-study article were relevant, composite correlation coefficients (Hunter and Schmidt, 2004) and the average sample size across the studies were calculated.

To estimate mean effect sizes and their variability in the metanalysis, the random effects method (Hedges and Olkin, 2014) was applied because random differences across studies were likely (significant Q statistic; Table 3). This methodology incorporates the influence of sample size and weighs effect sizes accordingly. Baujat plots and corresponding diagnostics were inspected to check for potential outliers and influential cases. Funnel plots, Egger's regression tests and the Rank correlation tests indicated that publication bias was not influential. The resulting mean effect sizes (Table 3) were interpreted according to Funder and Ozer (2019), whereby $\bar{r}$ = 0.10 represents a small effect, $\bar{r}$ = 0.20 a medium effect and $\bar{r}$ = 0.30 a large effect. Corresponding confidence and credibility intervals were given to communicate the precision (or uncertainty) of the summary estimate. Table 3 also depicts heterogeneity analyses. A significant Q statistic was interpreted as evidence that heterogeneity is present across the reported results of the studies. I^2 was reported as a quantification of these inconsistencies across studies and interpreted according to Deeks *et al.* (2011), whereby an I^2 of 0-40% indicates unimportant heterogeneity, an I^2 of 30-60% indicates moderate heterogeneity, an I^2 of 50-90% indicates substan-

TABLE 2   Overview of included studies[1]

Study	N	Sample country	Data collection[2]	WTC-measure[3]	Insect food presentation[4]	Insect food type[5]	Insect food description	Measured WTC-correlate
Baker et al. (2016)	207	USA	OS	WTP	P	S	insect spice-mix, insect fried-rice	23, 25
Brunner and Nuttavuthisit (2019)	1,042	Switzerland, Thailand	PPS	WTA	D	S	insect patties, insect chips, etc.	1, 2, 3, 5, 6, 7, 17, 20, 21, 23, 24, 27, 28, 29
Chan (2019)	202	Mechanical Turk	OS	WTT	D	S	'insects as meat substitute', deep-fried insects, insect cookies	22, 30
De Boer et al. (2013)	1,083	Netherlands	OS	WTE	D	S	insect snack	1, 2, 3, 7, 9
Dupont and Fiebelkorn (2020)	718	Germany	PPS	WTE	P	S	insect patties	1, 2, 6, 7, 13, 14, 20
Dupont, Hagedorn and Fiebelkorn, pers. comm.	497	Germany	PPS	WTE	P	S	insect patties	1, 2, 3, 6, 7, 8, 13, 14, 19, 20, 21
Elorinne et al. (2019)	567	Finland	OS	WTE	D	NS	'foods of insect origin'	6, 8, 16, 22, 23, 24, 25, 27, 28, 29, 31, 32, 33, 34
Fischer and Steenbekkers (2018)	140	Netherlands	E	WTT	D	NS	whole insects	1, 2, 6, 22
Gere et al. (2017)	400	Hungary	OS	WTE	D	NS	'food containing insect ingredients'	6, 20, 21
Gmuer et al. (2016)	428	Switzerland	OS	WTE	P	S	insect chips, deep-fried insects	1, 2, 6, 7, 21, 22, 23, 26
Grasso et al. (2019)	1,825	EU countries	OS	WTE	D	NS	'foods containing insect-based protein'	1, 2, 3, 9, 10, 11, 12, 15, 16
Hartmann and Siegrist (2016)	104	Switzerland	E	WTE	R	S	insect chips	1, 3, 6, 7, 8, 20, 22, 23, 26
Hartmann et al. (2015)	995	Germany, China	OS	WTE	P	S	'insects as meat substitute', deep-fried insects, insect cookies	1, 2, 3, 6, 9, 11, 12, 15, 17, 20, 21, 23, 24, 30
Jensen and Lieberoth (2019)	189	Denmark	OS	WTE	R	S	roasted insects, insect spring rolls, insect soup	6, 21, 22, 24
Komher et al. (2019)	311	Germany	E	WTP	P	S	insect patties	1, 2, 3, 6, 7, 9, 10, 11, 12, 22
La Barbera et al. (2018)	160	Western countries	CAS	WTA	D	NS	'insect-based food'	6, 22, 26
Lammers et al. (2019)	516	Germany	OS	WTE	P	S	insect patties, whole insects	1, 2, 3, 4, 6, 7, 8, 13, 14, 19, 20, 21
Legendre et al. (2019)	337	USA	OS	WTP	D	NS	'edible insects'	20, 23
Mancini et al. (2019b)	165	Italy	PPS	WTE	R	S	'insects', insect bread	6, 24

Study	N	Country	Method	Measure	Type	Sig.	Product	WTC correlates
Megido et al. (2014)	189	Belgium	PPS	WTE	R	S	flavoured whole insects	1, 2, 20
Megido et al. (2016)	159	Belgium	E	WTE	R	S	insect patties	1, 20, 21, 23, 33
Menozzi et al. (2017)	213	Italy	OS	WTE	D	NS	'products containing insects'	1, 24, 27, 29, 34
Orsi et al. (2019)	293	Germany	OS	WTE	P	S	insect patties, insect protein bars, insect pasta, insect granola	1, 2, 3, 4, 5, 6, 7, 20, 21, 22, 25
Piha et al. (2018)	887	EU countries	OS	WTP	P	S	roasted insects, insect nuggets, insect snack, insect wok, insect seasoning	6, 21
Powell et al. (2019)	510	UK	E	WTP	P	S	various insect foods (e.g. insect patties)	1, 2, 3, 4, 7, 18, 23, 25, 28, 29, 30, 32, 33
Rozin and Ruby (2019)	675	USA, India	OS	WTE	D	S	various whole and roasted insects (e.g. crickets)	28
Ruby and Rozin (2019)	692	India, USA	OS	WTE	P	S	insect tacos, insect dosas, insect lollipops, insect cookies, insect parathas	22, 25, 28
Ruby et al. (2015)	399	USA, India	OS	WTT	P	S	insect tacos, insect dosas, insect lollipops, insect cookies, insect parathas	6, 8, 22, 25, 27, 30, 32
Schäufele et al. (2019)	342	Germany	CAS	WTT	P	S	insect risotto rice	1, 2, 3, 6, 7, 20, 21, 24, 31
Schösler et al. (2012)	1,083	Netherlands	OS	WTT	P	S	insect pizza, chocolate-coated insects, insect salad	1, 2, 3, 7
Sogari et al. (2019b)	88	Italy	PPS	WTE	D	NS	'insect products and insect-based products'	1, 2, 6, 21, 23
Tan et al. (2016)	976	Netherlands	OS	WTP	P	S	insect stew, insect curry, insect brownies, insect cakes	6, 23, 26, 31
Tan et al. (2015)	103	Netherlands	E	WTE	R	S	insect patties	26
Tan et al. (2017)	135	Netherlands	OS	WTP	R	S	whole insects, insect meal balls, insect shakes	26, 34
Verbeke (2015)	368	Belgium	OS	WTA	D	NS	'insects as substitute for meat'	1, 2, 3, 10, 20
Verneau et al. (2016)	282	Denmark, Italy	E	WTE	R	S	insect chocolate bars	20, 27
Videbæk and Grunert (2020)	975	Denmark	E	WTA	D	S	whole insects, pureed insects, insects with fish, insect bread, insect products on the Danish market	6, 8, 18, 22

[1] The measured WTC correlate's enumeration can be found in Table 3 and Figures 2-4.

[2] E = experiment, PPS = paper pencil survey, CAS = computer-administered survey, OS = online survey.

[3] WTT = willingness to try, WTE = willingness to eat, WTP = willingness to pay, WTA = willingness to adopt.

[4] D = description, P = picture, R = real product.

[5] S = specific insect food product (e.g. insect patties), NS = no specification given regarding the insect as food (e.g. 'insects as food', 'edible insects').

tial heterogeneity and an I^2 of 75-100% indicates considerable heterogeneity. Similarly, τ was reported to indicate the extent of variation, or heterogeneity, among the reported results of the studies. Forest plots are presented to display the (composite) correlation coefficients of the WTC correlates for each of the three subgroups (Figures 2-4). Lastly, meta-regressions were conducted to test whether certain study characteristics had explanatory value concerning heterogeneity.

All data analyses were conducted in R (R Core Team, 2010; SPSS Version 24.0, IBM Corp, Armonk, NY, USA) with the 'metafor' (Viechtbauer, 2010) and 'robumeta' (Fisher and Tipton, 2015) packages, while Figures 2-4 were created using Tableau (Tableau Software Inc., 2003).

3   Results

3.1   Descriptive study characteristics

Table 2 shows the most important study characteristics. Most studies were surveys (online, paper-pencil or computer-administered) conducted with participants from European countries. Only eight studies involved an experiment, and only six studies involved participants not from Europe (e.g. the USA, India, China or Thailand). As the WTC measure, the majority (n = 21) of the studies assessed WTE, while seven studies assessed WTP, five studies assessed WTT and four studies assessed WTA. To assess these measures, mostly descriptions (n = 14) or pictures (n = 15) of the insect foods in question were presented, while in eight studies, the real insect food product was presented to participants. For most studies (n = 29) participants were asked to indicate their willingness to consume specific insect-based food products (e.g. mealworm patties, insect-flour protein bars or deep-fried crickets). In the other eight studies, a specific type of insect food was not given. Participants were asked to indicate their willingness to consume 'foods of insect origin' or 'insect-based foods'.

3.2   Results of the meta-analysis

3.2.1   Heterogeneity analyses

For 25 of the 34 WTC correlates, the Q statistic provided evidence for significant heterogeneity across studies. Correspondingly, the I^2 and τ results for these WTC correlates were high (with I^2 values ranging from 53 to 96%), indicating that the effect sizes extracted from the studies varied greatly. For nine of the 34 WTC correlates, there was no evidence for significant heterogeneity across studies. However, because most of these low-heterogeneity WTC correlates had a low number of contributing studies (k < 6), there is a high level of uncertainty regarding these results (Deeks *et al.*, 2011).

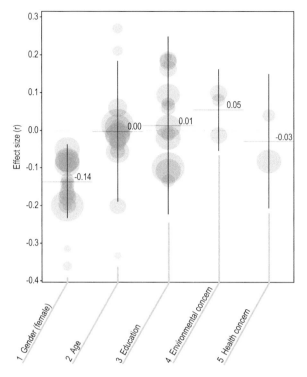

FIGURE 2   Forest plot showing the mean effect sizes between the willingness to eat insects as food insects and the willingness to consume correlates 1". Error bars represent the 80% CR credibility intervals of the effect size. Grey circles represent the effect size for included studies. The size of the circle indicates the study sample size. Sorted by k (number of studies contributing to mean effect size).

3.2.2   Mean estimated effect sizes

3.2.2.1   *Sociodemographic variables and general attitudes*

This subgroup consists of WTC correlates 1-5 (Figure 2). There was relatively large heterogeneity concerning the reported effect sizes for the sociodemographic variables and 'education', as well as the attitudes 'environmental concern' and 'health'. This heterogeneity was especially pronounced for 'education'. Considering the fact that the composite effect sizes for these four WTC correlates were also close to 0, these results suggest no correlation between these variables and WTC insects. The number of contributing studies should be taken into account when considering these results however: While the inconsistent effects of 'age' and 'education' were found across a large number of studies, the composite effects of the attitudes (i.e. 'environmental concern' and 'health concern') were calculated from only two to three studies. It is therefore uncertain whether the inconsistent effect of the attitudes would persist if there were more data from more studies to draw from.

TABLE 3　Summary of the meta-analyses for the 34 WTC correlates (respectively enumerated)[1]

WTC-correlate	k	N	$\bar{r}$	95% CI	80% CR	Q	τ	I²
Sociodemographic variables and general attitudes								
1 Gender (female)	20	10,323	-0.14	[-0.16, -0.27]	[-0.23, -0.03]	44.42***	0.05	53.68
2 Age	17	9,847	0.00	[-0.05, 0.04]	[-0.19, 0.18]	61.23***	0.09	82.73
3 Education	13	8,919	0.04	[-0.02, 0.10]	[-0.17, 0.25]	99.01***	0.11	87.96
4 Environmental concern	3	1,319	0.05	[-0.01, 0.12]	[-0.05, 0.16]	3.38	0.04	42.37
5 Health concern	2	1,335	-0.03	[-0.14, 0.08]	[-0.20, 0.14]	3.28	0.07	69.47
Variables related to eating in general								
6 Food neophobia	21	8,919	-0.33	[-0.37, -0.26]	[-0.53, -0.07]	179.15***	0.13	88.56
7 Meat consumption	12	6,396	0.08	[0.04, 0.11]	[-0.01, 0.17]	21.61*	0.12	50.16
8 Food sensation and innovation seeking	6	3,058	0.29	[0.25, 0.32]	[0.25, 0.32]	2.59	< 0.01	< 0.01
9 Importance of taste for food choice	4	4,164	0.01	[-0.09, 0.12]	[-0.22, 0.24]	45.57***	0.11	91.73
10 Importance of sustainability for food choice	3	3,961	0.10	[-0.01, 0.22]	[-0.12, 0.32]	19.17***	0.10	91.99
11 Importance of healthiness for food choice	3	3,081	0.05	[-0.14, 0.23]	[-0.30, 0.39]	50.26***	0.16	95.57
12 Importance of convenience for food choice	3	3,081	0.06	[-0.09, 0.21]	[-0.23, 0.34]	39.61***	0.13	93.52
13 Intention to reduce meat consumption	3	1,200	0.04	[-0.10, 0.18]	[-0.22, 0.30]	12.86*	0.12	83.28
14 Food disgust	3	1,200	-0.44	[-0.50, -0.36]	[-0.54, -0.32]	4.16	0.05	52.00
15 Importance of affordability for food choice	2	2,770	0.07	[-0.14, 0.28]	[-0.29, 0.41]	30.60***	0.15	96.73
16 Food fussiness	2	2,392	-0.21	[-0.26, -0.15]	[-0.28, -0.13]	1.71	0.03	41.64
17 Importance of social acceptability for food choice	2	1,987	0.36	[0.33, 0.40]	[0.33, 0.40]	0.59	< 0.01	< 0.01
18 Disgust sensitivity	2	1,485	-0.23	[-0.37, -0.07]	[-0.45, 0.02]	8.35*	0.11	88.04
19 Food technology neophobia	2	1,013	-0.27	[-0.32, -0.20]	[-0.32, -0.20]	0.72	< 0.01	< 0.01

Variables related to eating insects

	k	N	$\bar{r}$	CI	CR	Q	I²
20 Familiarity with the concept of eating insects	14	5,899	**0.10**	[-0.11, 0.31]	[-0.63, 0.74]	883.85***	98.61
21 Experience with eating insects	12	5,786	0.35	[0.22, 0.46]	[-0.12, 0.69]	276.23***	96.45
22 Disgust at eating insects	12	4,460	-0.53	[-0.66, -0.36]	[-0.86, 0.13]	543.89***	97.79
23 Perceived tastiness of insects as food	11	5,363	0.55	[0.33, 0.71]	[-0.31, 0.91]	600.68***	97.99
24 Perceived social acceptability of insects as food	7	3,463	0.35	[0.24, 0.45]	[0.05, 0.59]	52.69***	90.68
25 Perceived risk of insects as food	6	2,669	-0.33	[-0.44, -0.21]	[-0.58, -0.02]	44.04***	90.68
26 Preference for carrier food²	6	1,906	0.10	[0.05, 0.14]	[0.05, 0.14]	3.17	< 0.01
27 Perceived sustainability of insects as food	5	2,788	0.57	[0.29, 0.75]	[-0.17, 0.89]	335.37***	98.69
28 Perceived ethicalness of insects as food	4	2,976	0.27	[-0.01, 0.51]	[-0.35, 0.72]	162.04***	98.44
29 Perceived healthiness of insects as food	4	2,332	0.42	[0.37, 0.46]	[0.33, 0.49]	5.26	42.48
30 Perceived nutritiousness of insects as food	4	2,056	0.32	[0.20, 0.42]	[0.08, 0.52]	16.72*	85.03
31 Preference for visibility of insects in food	3	1,885	0.41	[0.19, 0.58]	[0.02, 0.71]	39.21***	95.83
32 Perceived naturalness of insects as food	3	1,476	0.41	[0.20, 0.57]	[-0.01, 0.70]	40.34***	94.79
33 Perceived visual appeal of insects as food	3	1,236	0.27	[0.00, 0.50]	[0.26, 0.68]	40.16***	95.54
34 Familiarity with carrier food²	3	915	0.35	[-0.14, 0.69]	[-0.56, 0.87]	127.17***	97.93

[1] Sorted by k and N (within each subgroup). Significant $\bar{r}$ in bold. k = number of studies contributing to meta-analysis; N = total sample size; $\bar{r}$ = mean observed correlation; CI = confidence interval around $\bar{r}$; CR = credibility interval around $\bar{r}$; Q = test for homogeneity; τ = estimated standard deviation of the distribution of the true effects across studies; I² = proportion of heterogeneity due to between-study differences. *P < 0.05, ***P<0.001.

[2] Carrier food refers to the food product that contains the edible insect (e.g. a burger patty or a chocolate bar).

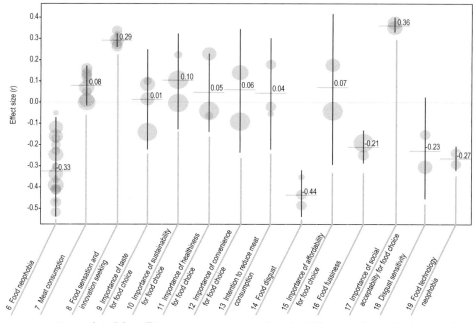

FIGURE 3   Forest plot of the willingness to consume correlates 'variables related to eating in general'.
              For a description of the figure elements, view the description of Figure 2.

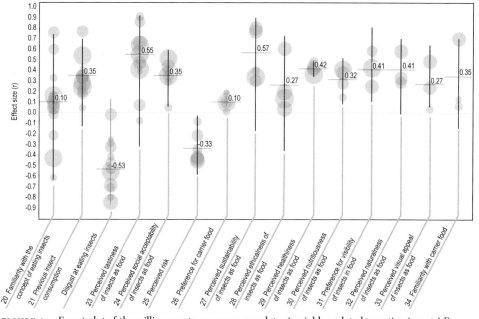

FIGURE 4   Forest plot of the willingness to consume correlates 'variables related to eating insects'. For
              a description of the figure elements, view the description of Figure 2.

Within this subgroup, 'gender' was the only variable with any correlation to WTC. For this variable, the heterogeneity across a large number of studies was relatively low, and a small composite effect size was calculated. The female participants in the identified literature, therefore, were consistently less willing to consume insects than the male participants; however, this effect was small.

3.2.2.2   *Variables related to eating in general*

This subgroup consists of WTC correlates 6-19 (Figure 3). Most of the food choice motives ('importance of taste', 'healthiness', 'sustainability', 'convenience', and 'affordability for food choice') had composite effect sizes close to o, suggesting no link between these variables and consumers' WTC insects. Interestingly, the 'importance of social acceptability for food choice' was the only food choice motive strongly correlated with WTC. Again, however, it should be taken into account that all of these food choice motive composite effect sizes only included two or three studies. Thus, there is little certainty that the calculated mean effect sizes represent the 'true' mean effect sizes for these variables.

In contrast to the food choice motives, we have 'food neophobia' and 'food sensation and innovation seeking'. The composite effect of these WTC correlates can be considered large, and it included a relatively large number of homogeneous correlation coefficients, indicating a high level of certainty that these factors are closely linked to WTC insects. Such large composite effect sizes were also found for the related constructs 'food technology neophobia' and 'food fussiness'. However, the number of contributing studies for these factors was small, limiting the credibility of these results. Also, WTC correlates related to disgust ('food disgust' and 'disgust sensitivity') had large negative composite effect sizes. Again, however, the number of contributing studies for these variables was small.

Overall, therefore, it appears that more general food choice motives ('importance of taste', 'sustainability', 'healthiness', 'nutritiousness', and 'convenience for food choice') are weakly linked to WTC insects. Variables, however, which are related to the newness and acceptability of food ('food neophobia', 'food sensation and innovation seeking', 'food technology neophobia', 'food fussiness' and the 'social acceptability of food'), as well as affect ('food disgust' and 'disgust sensitivity'), are more strongly linked to WTC insects.

3.2.2.3   *Variables related to eating insects*

In this subgroup (Figure 4), the WTC correlates with the largest composite effect sizes were the 'perceived sustainability of insects as food' ($\bar{r}$ = 0.57), the 'perceived tastiness of insects as food' ($\bar{r}$ = 0.55) and 'disgust at eating insects as food' ($\bar{r}$ = -0.53). While the disgust factor is unsurprisingly the greatest barrier to the acceptance of entomophagy, an awareness of its environmental benefits – more so than the awareness of health and nutrition related benefit – appears to be the most important driver of the acceptance of insects as food. Furthermore, the results suggest that making insect food appear tasty to consumers could be highly important for consumers' WTC insects.

Interestingly, the reported effect sizes of 'familiarity with the concept of eating insects' were very heterogeneous ($\bar{r}$ = 0.10) across the identified literature, which may be a reason for the small mean effect size that was obtained for this WTC correlate.

Other WTC correlates that were strongly correlated to WTC and also had a relatively large number of contributing studies were 'experience with eating insects' ($\bar{r}$ = 0.35, k = 12) and the 'perceived social acceptability of insects as food' ($\bar{r}$ = 0.35, k = 7). The other variables from this subgroup (i.e. WTC correlates 31-34) only include two or three studies; the credibility of the calculated composite effect for these variables is therefore highly limited.

3.2.3   Meta-regressions

Meta-regressions were conducted to determine whether the study characteristics described in Table 2 can explain the high levels of heterogeneity shown for many WTC correlates (Table 3). Specifically, the moderating effects of the type of data collection (experiment, paper pencil survey, computer-administered survey or online survey), the type of WTC measure (WTT, WTE, WTP or WTA), the insect food presentation (description, picture or real product) and the insect food type (specific insect food product or non-specific type of insect food) were tested. These meta-regressions were conducted for WTC correlates fulfilling two specifications, as recommended by Deeks *et al.* (2011): a meta-regression was only performed for WTC correlates that had considerably high heterogeneity (i.e. I^2 > 75%) and also included ten or more studies (i.e. k > 10). Thus, the WTC correlates examined through this procedure were 'gender', 'age', 'education', 'food neophobia', 'familiarity with the concept of eating insects', 'experience with eating insects', 'disgust at eating insects', and the 'perceived tastiness of eating insects'.

As can be seen in Table 4, moderating effects were found for two WTC correlates. Firstly, the correlation between WTC and 'familiarity with the concept of eating insects' was moderated by insect food presentation, explaining 27.5% of variance. Specifically, presenting participants with pictures of the insect food and the real insect food product negatively affected the link between the correlate and WTC (as opposed to presenting only a description). Secondly, the correlation between WTC and the 'experience with eating insects' was moderated by insect food type, explaining 40.6% of variance. Specifically, asking participants about specific insect foods negatively affected the link between the correlate and the WTC (as opposed to not mentioning any specific type of insect food).

TABLE 4   Results of meta-regressions

WTC correlate	Moderator	k	β	SE	z	95% CI	R[1]
Familiarity with the concept of eating insects	Insect food presentation	14					
	Picture (vs description)		-0.55*	0.22	-2.44	[-1.10, -0.02]	27.5%
	Real product (vs description)		-0.56*	0.28	-2.44	[-0.99, -0.11]	27.5%
Previous insect consumption	Insect food type	12					
	Specific (vs no specification given)[2]		-0.47*	0.14	3.27	[-0.76, -0.18]	40.6%

k = number of studies contributing to meta-regression; β = regression estimate; SE = standard error; z = z-value; 95% CI = confidence interval around β; R² = percentage of explained variance. *P < 0.05.
[1] Specific insect food products: e.g. insect patties or insect chocolate bars; No specification given concerning the type of food product: e.g. 'insects as food' or 'edible insects'.

4   Discussion

4.1   *Main findings and implications*

To better understand the generally low WTC insects, we synthesised the effect sizes of various WTC correlates reported in previous studies. We present not only an overview of factors related to entomophagy acceptance – as previous qualitative reviews have already done – but also estimates of their effect sizes. Although there was a high level of heterogeneity for some WTC correlates, our results provide an approximate quantification of the importance of various WTC correlates (both as individual factors and in comparison with one another) and certain study characteristics of past entomophagy acceptance research (in terms of moderating effects).

Our main conclusion is that affect-based factors are most relevant to WTC insects. For instance, the 'classical' food choice motives (Steptoe *et al.*, 1995) (i.e. *the importance of sustainability, healthiness, convenience* and *affordability for food choice*), which are not necessarily related to affect but rather to more general food quality measures, were hardly correlated to WTC ($r^- = 0.01$-0.10). Instead, across all 37 included studies, the factors with consistently large effect sizes ($r^- = -0.33$-0.55) were factors strongly related to affect. Specifically, factors related to the fear of the unfamiliar, disgust, pleasure and social acceptability (i.e. *food neophobia, food technology neophobia, food disgust, food disgust sensitivity, food fussiness, disgust at eating insects*, the *perceived risk of insects as food*, the *perceived social acceptability of insects as food* and the *perceived tastiness of insects as food*). Evolutionarily, these factors stem from human protection mechanisms to prevent the consumption of potentially harmful substances (Chapman and Anderson, 2012; Martins and Pliner, 2006; Tuorila *et al.*, 1994). Naturally, these can also be barriers to other novel foods, such as genetically-modified foods (Costa-Font *et al.*, 2008), cultured meat (Siegrist and Hartmann, 2020) or other novel meat substitutes, e.g. mycoprotein (Hartmann and Siegrist, 2017a). Compared to most of these foods, however, consumers appear to be more averse toward insects (Dupont and Fiebelkorn, 2020; Grasso *et al.*, 2019).

4.1.1   Approaches to increasing entomophagy acceptance

Pliner and Salvy (2006) propose various approaches to increase the acceptance of novel foods. In combination with our findings, these could be applicable to entomophagy, potentially to varying degrees of effectivity.

The first and perhaps most effective approach is the reduction of the main barriers to entomophagy acceptance – 'food neophobia' ($r^- = -0.33$), 'disgust' ($r^- = -0.53$, $r^- = -0.44$) and the 'expected (unpleasant) taste of insects' ($r^- = -0.55$). In general, 'food neophobia' and 'disgust' are decreased through gradual and/or early exposure to unfamiliar food (Birch *et al.*, 1987; Loewen and Pliner, 1999; Pliner, 1982; Sullivan and Birch, 1990; Wardle *et al.*, 2003). The emergence of insect food products in supermarkets, 'bug banquets' (Looy and Wood, 2006) and the media (Legendre *et al.*, 2019) have all contributed to the public's steady familiarisation with entomophagy. Indeed, we found that 'familiarity with the concept of entomophagy' ($r^- = 0.10$) and 'previous insect consumption' ($r^- = 0.35$) both positively impact WTC (Gere *et al.*, 2017; Hartmann *et al.*, 2015; Legendre *et al.*, 2019). As a next step, entomophagy should be promoted specifically to children. Dupont and Fiebelkorn (2020), who conducted one of the first entomophagy acceptance studies with a sample of children and adolescents, recommend tasting sessions, teaching units and methods for including edible insects in class, e.g. in biology or geography class (Fiebelkorn and Kuckuck, 2019; Fiebelkorn and Puchert, 2018). As an emotion connected to the oral sense, disgust is linked to taste (Rozin *et al.*, 2009). Pelchat and Pliner (1995) found that providing individuals with the verbal information that a novel food tasted good increased their willingness to try it. Thus, to combat the 'perceived (unpleasant) taste of insects as food', insect food advertisements and packaging could explicitly create positive expectations. Future studies should examine how familiarisation with and positive expectations regarding edible insects can be fostered more concretely.

As a second approach, the benefits of insects as food should be emphasised. We found that the 'perceived sustainability of insects as food' ($r^- = 0.55$) appears to be their most compelling benefit. In second and third places, consumers are also compelled by the 'perceived healthiness' ($r^- = 0.42$) and 'nutritiousness ($r^- = 0.32$) of insects as food'. Insects may therefore have the most success in the 'green consumer' market (e.g. as insect-based meat substitutes for environmentally motivated consumers) and, alternatively, among fitness and health-oriented consumers (e.g. insect protein bars and shakes; insects as a 'healthier' meat alternative). Indeed, informing consumers about either the individual or societal benefits of edible insects differentially impacts WTC (La Barbera *et al.*, 2018; Verneau *et al.*, 2016). More research on this topic is needed to better understand the various insect consumer segments (Brunner and Nuttavuthisit, 2019). However, strong emotional aversions may block information effects (Martins *et al.*, 1997; Pliner and Salvy, 2006). Thus, as Pliner and Salvy (2006) have concluded, cognitive interventions tend to have limited impact on affect-based food aversions. In the case of entomophagy, therefore, emphasising the positives of edible insects ('Approach 2') may be less effective than directly combatting the negative affective factors ('Approach 1').

Lastly regarding Approach 3, humans not only exhibit aversion but also curiosity regarding novel foods (Rozin and Rozin, 1981). Indeed, we found that consumers who are 'food sensation and innovation seeking' were likely to have a higher WTC ($\bar{r}$ = 0.29), a result in line with the finding that 'food neophobia', 'food technology neophobia' and the 'importance of social acceptability for food choice' are all negatively correlated with WTC. Brunner and Nuttavuthisit (2019) propose that insect food products could be marketed as something unique and exciting, e.g. at special 'insect bars' similar to sushi bars.

4.2   *Other findings and implications*

In accordance with past research (e.g. Lammers *et al.*, 2019), we found certain sociodemographic variables to be unrelated to entomophagy aversions: the effects of 'age' ($\bar{r}$ = 0.00) and 'education' ($\bar{r}$ = 0.04) were highly inconsistent across studies, with mean effects sizes close to 0. There was a negative mean effect for 'gender', i.e. being female ($\bar{r}$ = -0.14); however, this effect was small and could have been influenced by the tendency of women to generally have higher disgust sensitivity than men (Egolf et al., 2018). Entomophagy acceptance, however, varies cross-culturally (Gómez-Luciano *et al.*, 2019; Verneau *et al.*, 2016), especially when comparing Western with Asian countries (Brunner and Nuttavuthisit, 2019; Hartmann *et al.*, 2015; Ruby and Rozin, 2019; Ruby *et al.*, 2015). We were not able to investigate national differences, because many studies with non-Western participants had to be eliminated during the literature search and, as a result, most of the included studies consisted of primarily Western samples. A sociocultural perspective should be incorporated in future syntheses of entomophagy acceptance research.

In addition to sociodemographic variables, an individual's dietary behaviour may also predict the WTC insects. Indeed, 'meat consumption' was positively associated with WTC. This estimated effect, however, included studies with vegans and vegetarians. Naturally, most vegans and vegetarians are less willing to eat insects (Elorinne *et al.*, 2019), and their inclusion influenced our estimated effect size. Therefore, our findings do not represent how omnivore's WTC insects correlates with their meat consumption and – more importantly – 'their intention to reduce' meat consumption. Dupont *et al.* (pers. comm.) and Verbeke (2015), for example, found that individuals who intended to or had already reduced their meat consumption were more WTC insects, perhaps because they saw them as a sustainable meat replacement. Despite this, Lammers *et al.* (2019) showed that omnivores would prefer to merely try insects as opposed to adopting them as meat-substitutes. More studies focused on insects specifically as meat-substitutes are needed because it may be easier to encourage the 'switch' from meat to insects than persuade consumers to adopt insects into their diet without any further framing.

Multiple studies show that consumers prefer edible insects to be invisible and processed rather than visible and whole (Gmuer *et al.*, 2016; Hartmann and Siegrist, 2016; Orsi *et al.*, 2019). Unfortunately, the experimental design of most of these studies did not allow us to calculate an effect size estimating how 'the degree of processing' impacts WTC insects. Still, we found that the 'preference for the visibility of insects in food' ($\bar{r}$ = 0.41) was positively correlated with WTC.

Both the 'preference for' ($\bar{r}$ = 0.10) and 'familiarity of the carrier food' ($\bar{r}$ = 0.35) – the 'main' food product carrying the insect ingredient – were positively correlated with WTC. However, only a few studies (Elorinne *et al.*, 2019; Hartmann and Siegrist, 2016; La Barbera *et al.*, 2018; Menozzi *et al.*, 2017; Tan *et al.*, 2015, 2016, 2017) contributed to these findings. Thus, carrier food variables should be investigated more in depth, for example, the product type, e.g. pasta vs chocolate bar (Lombardi *et al.*, 2019; Orsi *et al.*, 2019); the flavour, e.g. savoury vs sweet (Schäufele *et al.*, 2019; Tan *et al.*, 2016), or the serving context, e.g. snack vs meal vs dessert (Brunner and Nuttavuthisit, 2019; Elorinne *et al.*, 2019).

In the same vein, other variables related to marketing and product development must be explored in more depth, such as price (Lombardi *et al.*, 2019), convenience (Brunner and Nuttavuthisit, 2019; Elorinne *et al.*, 2019), insect species (Fischer and Steenbekkers, 2018; Rozin and Ruby, 2019) and the packaging (Baker *et al.*, 2016). The 'classical' variables of entomophagy, as Lammers *et al.* (2019) call them, such as 'food neophobia' or 'disgust', have been explored sufficiently, and future research should be more market oriented.

4.2.1   Meta-regression results

The meta-regressions showed that study characteristics such as insect food presentation may have moderating effects. Specifically, the positive effects of 'familiarity with entomophagy' and 'previous insect consumption' on WTC insects were stifled when participants were shown specific products as compared to non-specific products (β = -0.47) or actual products (β = -0.56) and pictures of the product (β = -0.55) as compared to mere verbal descriptions. Baker *et al.* (2016) found similar study characteristics to be influential because they showed that providing images of processed (vs whole) insects positively influenced perceptions in retail settings, while providing vague (vs explicit) descriptions did so in restaurant settings. In our analysis, moderating effects were not found for the other study characteristics. However, because meta-regressions were conducted for only a small number of WTC correlates due to methodological restrictions, other study characteristics – such as offering tasting samples (Mancini *et al.*, 2019b; Megido *et al.*, 2016; Tan *et al.*, 2015) or asking participants to try vs buy insects as food (Tan *et al.*, 2016) – may actually have moderating effects as well.

4.2.2   Limitations

Regarding the main limitation, many studies and effect sizes concerning entomophagy acceptance could not be included in this analysis due to the inclusion criteria. Our findings therefore do not include the entirety of the existing literature on this topic. Specifically, we only included studies that measured WTC edible insects (and not, for example, the sensory liking of edible insects) and studies for which we were able to obtain these effects as Pearson's r. Furthermore, we only included WTC correlates measured in at least two papers. For example, variables such as consumers' attitudes toward organic production methods (Kornher *et al.*, 2019), purchase activism (Legendre *et al.*, 2019) and nutritional knowledge (Brunner and Nuttavuthisit, 2019), in relation to WTC, had only been measured by single studies and were therefore not included in this analysis. These may be interesting variables for future research to consider.

Furthermore, many other study characteristics which we have not tested for could have potential moderating effects. For example, the sample age group (i.e. if the sample consisted of students/general population, or young adults/adults/senior citizens), or the date of data collection (as opposed to the date of publication) could be of relevance.

5   Conclusions

Aversions to edible insects appear to be most strongly linked to affect-based factors, which are best reduced through gradual and/or early exposure to entomophagy. Unsurprisingly, therefore, steady familiarisation over a longer period of time is the safest bet to foster the acceptance of entomophagy in the long term. Concerning interventions that may already show effects in the short term, information-based approaches (e.g. emphasising the benefits of insects as food) may convince consumers to at least try edible insects. However, the influence of these cognitive factors may be blocked by the strong emotional insect aversions, limiting the success of insects with consumers compelled 'merely' by their sustainability and health related benefits. As an alternative approach, human curiosity appears to be decisive for WTC insects. It is likely therefore that food sensation and innovation seekers could be among the first adopters of entomophagy.

Many general psychological factors have been investigated comprehensively in the context of entomophagy acceptance. Future research should focus on the nuances of how to concretely bring insects to consumers. With this, it becomes even more important that researchers consider the influence of certain study characteristics (e.g. the utilised stimuli) to ensure unbiased results. Furthermore, future research should include product development and the marketing of more specific insect food products.

Acknowledgements

We thank all the authors who upon request provided us with the necessary data to conduct the meta-analysis. We also thank Luana Giacone for her valuable help during the study identification and screening process as well as Victoria Cologna for her help with the construction of the figures using the Tableau Software.

Conflict of interest

The authors declare no conflict of interest.

References

Baker, M.A., Shin, J.T. and Kim, Y.W., 2016. An exploration and investigation of edible insect consumption: the impacts of image and description on risk perceptions and purchase intent. Psychology & Marketing 33: 94-112.

Barsics, F., Megido, R.C., Brostaux, Y., Barsics, C., Blecker, C., Haubruge, E. and Francis, F., 2017. Could new information influence attitudes to foods supplemented with edible insects? British Food Journal 119: 2027-2039.

Belluco, S., Losasso, C., Maggioletti, M., Alonzi, C.C., Paoletti, M.G. and Ricci, A., 2013. Edible insects in a food safety and nutritional perspective: a critical review. Comprehensive Reviews in Food Science and Food Safety 12: 296-313.

Birch, L.L., McPheee, L., Shoba, B., Steinberg, L. and Krehbiel, R., 1987. 'Clean up your plate': effects of child feeding practices on the conditioning of meal size. Learning and Motivation 18: 301-317.

Boland, M.J., Rae, A.N., Vereijken, J.M., Meuwissen, M.P., Fischer, A.R., Van Boekel, M.A., Rutherfurd, S.M., Gruppen, H., Moughan, P.J. and Hendriks, W.H., 2013. The future supply of animal-derived protein for human consumption. Trends in Food Science & Technology 29: 62-73.

Brunner, T.A. and Nuttavuthisit, K., 2019. A consumer-oriented segmentation study on edible insects in Switzerland and Thailand. British Food Journal 122: 482-488.

Chan, E.Y., 2019. Mindfulness and willingness to try insects as food: the role of disgust. Food Quality and Preference 71: 375-383.

Chapman, H.A. and Anderson, A.K., 2012. Understanding disgust. Annals of the New York Academy of Sciences 1251: 62-76.

Costa-Font, M., Gil, J.M. and Traill, W.B., 2008. Consumer acceptance, valuation of and attitudes towards genetically modified food: review and implications for food policy. Food Policy 33: 99-111.

De Boer, J., Schösler, H. and Boersema, J.J., 2013. Motivational differences in food orientation and the choice of snacks made from lentils, locusts, seaweed or 'hybrid' meat. Food Quality and Preference 28: 32-35.

Deeks, J.J., Higgins, J., Altman, D.G. and Green, S., 2011. Cochrane handbook for systematic reviews of interventions version 5.1. Cochrane, London, UK.

Dupont, J. and Fiebelkorn, F., 2020. Attitudes and acceptance of young people toward the consumption of insects and cultured meat in Germany. Food Quality and Preference 85: 103983.

Ebenebe, C.I., Amobi, M.I., Udegbala, C., Ufele, A.N. and Nweze, B.O., 2017. Survey of edible insect consumption in south-eastern Nigeria. Journal of Insects as Food and Feed 3: 241-252. https://doi.org/10.3920/JIFF2017.0002

Egolf, A., Siegrist, M. and Hartmann, C., 2018. How people's food disgust sensitivity shapes their eating and food behaviour. Appetite 127: 28-36.

Elorinne, A.-L., Niva, M., Vartiainen, O. and Väisänen, P., 2019. Insect consumption attitudes among vegans, non-vegan vegetarians, and omnivores. Nutrients 11: 292.

Fiebelkorn, F. and Kuckuck, M., 2019. Insekten oder *in-vitro*-Fleisch – Was ist nachhaltiger? Eine Beurteilung mithilfe der Methode des „Expliziten Bewertens". Praxis Geographie 6: 14-21.

Fiebelkorn, F. and Puchert, N., 2018. Aufgetischt: Mehlwurm statt Rindfleisch. Unterricht Biologie 42: 12-16.

Fischer, A.R. and Steenbekkers, L.B., 2018. All insects are equal, but some insects are more equal than others. British Food Journal 120: 852-863.

Fisher, Z. and Tipton, E., 2015. Robumeta: an R-package for robust variance estimation in meta-analysis. Available at: https://arxiv.org/abs/1503.02220

Funder, D.C. and Ozer, D.J., 2019. Evaluating effect size in psychological research: sense and nonsense. Advances in Methods and Practices in Psychological Science 2: 156-168.

Gere, A., Székely, G., Kovács, S., Kókai, Z. and Sipos, L., 2017. Readiness to adopt insects in Hungary: a case study. Food Quality and Preference 59: 81-86.

Gmuer, A., Guth, J.N., Hartmann, C. and Siegrist, M., 2016. Effects of the degree of processing of insect ingredients in snacks on expected emotional experiences and willingness to eat. Food Quality and Preference 54: 117-127.

Gómez-Luciano, C.A., De Aguiar, L.K., Vriesekoop, F. and Urbano, B., 2019. Consumers' willingness to purchase three alternatives to meat proteins in the United Kingdom, Spain, Brazil and the Dominican Republic. Food Quality and Preference 78: 103732.

Grasso, A.C., Hung, Y., Olthof, M.R., Verbeke, W. and Brouwer, I.A., 2019. Older consumers' readiness to accept alternative, more sustainable protein sources in the European Union. Nutrients 11: 1904.

Halloran, A., Roos, N., Eilenberg, J., Cerutti, A. and Bruun, S., 2016. Life cycle assessment of edible insects for food protein: a review. Agronomy for Sustainable Development 36: 57.

Hartmann, C., Shi, J., Giusto, A. and Siegrist, M., 2015. The psychology of eating insects: a cross-cultural comparison between Germany and China. Food Quality and Preference 44: 148-156.

Hartmann, C. and Siegrist, M., 2016. Becoming an insectivore: results of an experiment. Food Quality and Preference 51: 118-122.

Hartmann, C. and Siegrist, M., 2017a. Consumer perception and behaviour regarding sustainable protein consumption: a systematic review. Trends in Food Science & Technology 61: 11-25.

Hartmann, C. and Siegrist, M., 2017b. Insects as food: perception and acceptance. Findings from current research. Ernahrungs Umschau 64: 44-50.

Hedges, L. and Olkin, I., 2014. Statistical methods for meta-analysis. Academic press, Cambridge, MA, USA.

Hunter, J.E. and Schmidt, F.L., 2004. Methods of meta-analysis: correcting error and bias in research findings. Sage, Thousand Oaks, CA, USA.

Jensen, N.H. and Lieberoth, A., 2019. We will eat disgusting foods together – evidence of the normative basis of Western entomophagy-disgust from an insect tasting. Food Quality and Preference 72: 109-115.

Kim, T.-K., Yong, H.I., Kim, Y.-B., Kim, H.-W. and Choi, Y.-S., 2019. Edible insects as a protein source: a review of public perception, processing technology, and research trends. Food Science of Animal Resources 39: 521.

Kornher, L., Schellhorn, M. and Vetter, S., 2019. Disgusting or innovative-consumer willingness to pay for insect based burger patties in Germany. Sustainability 11: 1878.

La Barbera, F., Verneau, F., Amato, M. and Grunert, K., 2018. Understanding Westerners' disgust for the eating of insects: the role of food neophobia and implicit associations. Food Quality and Preference 64: 120-125.

Lammers, P., Ullmann, L.M. and Fiebelkorn, F., 2019. Acceptance of insects as food in Germany: is it about sensation seeking, sustainability consciousness, or food disgust? Food Quality and Preference 77: 78-88.

Legendre, T.S., Jo, Y.H., Han, Y.S., Kim, Y.W., Ryu, J.P., Jang, S.J. and Kim, J., 2019. The impact of consumer familiarity on edible insect food product purchase and expected liking: the role of media trust and purchase activism. Entomological Research 49: 158-164.

Loewen, R. and Pliner, P., 1999. Effects of prior exposure to palatable and unpalatable novel foods on children's willingness to taste other novel foods. Appetite 32: 351-366.

Lombardi, A., Vecchio, R., Borrello, M., Caracciolo, F. and Cembalo, L., 2019. Willingness to pay for insect-based food: the role of information and carrier. Food Quality and Preference 72: 177-187.

Looy, H. and Wood, J.R., 2006. Attitudes toward invertebrates: are educational' bug banquets' effective? The Journal of Environmental Education 37: 37-48.

Mancini, S., Moruzzo, R., Riccioli, F. and Paci, G., 2019a. European consumers' readiness to adopt insects as food. A review. Food Research International 122: 661-678.

Mancini, S., Sogari, G., Menozzi, D., Nuvoloni, R., Torracca, B., Moruzzo, R. and Paci, G., 2019b. Factors predicting the intention of eating an insect-based product. Foods 8: 270.

Martins, Y., Pelchat, M.L. and Pliner, P., 1997. 'Try it; it's good and it's good for you': effects of taste and nutrition information on willingness to try novel foods. Appetite 28: 89-102.

Martins, Y. and Pliner, P., 2006. 'Ugh! That's disgusting!': identification of the characteristics of foods underlying rejections based on disgust. Appetite 46: 75-85.

Megido, R.C., Gierts, C., Blecker, C., Brostaux, Y., Haubruge, É., Alabi, T. and Francis, F., 2016. Consumer acceptance of insect-based alternative meat products in Western countries. Food Quality and Preference 52: 237-243.

Megido, R.C., Sablon, L., Geuens, M., Brostaux, Y., Alabi, T., Blecker, C., Drugmand, D., Haubruge, É. and Francis, F., 2014. Edible insects acceptance by Belgian consumers: promising attitude for entomophagy development. Journal of Sensory Studies 29: 14-20.

Menozzi, D., Sogari, G., Veneziani, M., Simoni, E. and Mora, C., 2017. Eating novel foods: an application of the theory of planned behaviour to predict the consumption of an insect-based product. Food Quality and Preference 59: 27-34.

Nowak, V., Persijn, D., Rittenschober, D. and Charrondiere, U.R., 2016. Review of food composition data for edible insects. Food Chemistry 193: 39-46.

Oonincx, D.G.A.B., Van Itterbeeck, J., Heetkamp, M.J., Van den Brand, H., Van Loon, J.J.A. and Van Huis, A., 2010. An exploration on greenhouse gas and ammonia production by insect species suitable for animal or human consumption. PLoS ONE 5: e14445.

Orsi, L., Voege, L.L. and Stranieri, S., 2019. Eating edible insects as sustainable food? Exploring the determinants of consumer acceptance in Germany. Food Research International 125: 108573.

Payne, C.L., Scarborough, P., Rayner, M. and Nonaka, K., 2016. A systematic review of nutrient composition data available for twelve commercially available edible insects, and comparison with reference values. Trends in Food Science & Technology 47: 69-77.

Pelchat, M.L. and Pliner, P., 1995. 'Try it. You'll like it'. Effects of information on willingness to try novel foods. Appetite 24: 153-165.

Piha, S., Pohjanheimo, T., Lähteenmäki-Uutela, A., Křečková, Z. and Otterbring, T., 2018. The effects of consumer knowledge on the willingness to buy insect food: an exploratory cross-regional study in Northern and Central Europe. Food Quality and Preference 70: 1-10.

Pliner, P., 1982. The effects of mere exposure on liking for edible substances. Appetite 3: 283-290.

Pliner, P. and Salvy, S., 2006. Food neophobia in humans. Frontiers in Nutritional Science 3: 75.

Powell, P.A., Jones, C.R. and Consedine, N.S., 2019. It's not queasy being green: the role of disgust in willingness-to-pay for more sustainable product alternatives. Food Quality and Preference 78: 103737.

R Core Team, 2010. R: a language and environment for statistical computing. R Foundation for Statistical Computing, Vienna, Austria.

Rozin, E. and Rozin, P., 1981. Culinary themes and variations. Natural History 90: 6-14.

Rozin, P., Haidt, J. and Fincher, K., 2009. From oral to moral. Science 323(5918): 1179-1180.

Rozin, P. and Ruby, M.B., 2019. Bugs are blech, butterflies are beautiful, but both are bad to bite: admired animals are disgusting to eat but are themselves neither disgusting nor contaminating. Emotion 20: 854-865.

Ruby, M.B. and Rozin, P., 2019. Disgust, sushi consumption, and other predictors of acceptance of insects as food by Americans and Indians. Food Quality and Preference 74: 155-162.

Ruby, M.B., Rozin, P. and Chan, C., 2015. Determinants of willingness to eat insects in the USA and India. Journal of Insects as Food and Feed 1: 215-225.

Schäufele, I., Albores, E.B. and Hamm, U., 2019. The role of species for the acceptance of edible insects: evidence from a consumer survey. British Food Journal 121: 2190-2204.

Schösler, H., De Boer, J. and Boersema, J.J., 2012. Can we cut out the meat of the dish? Constructing consumer-oriented pathways towards meat substitution. Appetite 58: 39-47.

Siegrist, M. and Hartmann, C., 2020. Perceived naturalness, disgust, trust and food neophobia as predictors of cultured meat acceptance in ten countries. Appetite 155: 104814.

Smetana, S., Mathys, A., Knoch, A. and Heinz, V., 2015. Meat alternatives: life cycle assessment of most known meat substitutes. The International Journal of Life Cycle Assessment 20: 1254-1267.

Sogari, G., Menozzi, D., Hartmann, C. and Mora, C., 2019a. How to measure consumers acceptance towards edible insects? A scoping review about methodological approaches. In: Sogari, G., Mora, C. and Menozzi, D. (eds.) Edible insects in the food sector. Springer, Berlin, Germany, pp. 27-44.

Sogari, G., Menozzi, D. and Mora, C., 2019b. The food neophobia scale and young adults' intention to eat insect products. International Journal of Consumer Studies 43: 68-76.

Steptoe, A., Pollard, T.M. and Wardle, J., 1995. Development of a measure of the motives underlying the selection of food: the food choice questionnaire. Appetite 25: 267-284.

Sullivan, S.A. and Birch, L.L., 1990. Pass the sugar, pass the salt: experience dictates preference. Developmental Psychology 26: 546.

Tableau Software Inc, 2003. Tableau desktop (version 2018.1.10). Mountain view. Tableau Software Inc., Palo Alto, CA, USA. Available at: https://www.tableau.com/ products/ desktop.

Tan, H.S.G., Fischer, A.R., Van Trijp, H.C. and Stieger, M., 2015. Tasty but nasty? Exploring the role of sensory-liking and food appropriateness in the willingness to eat unusual novel foods like insects. Food Quality and Preference 48: 293-302.

Tan, H.S.G., Van den Berg, E. and Stieger, M., 2016. The influence of product preparation, familiarity and individual traits on the consumer acceptance of insects as food. Food Quality and Preference 52: 222-231.

Tan, H.S.G., Verbaan, Y.T. and Stieger, M., 2017. How will better products improve the sensory-liking and willingness to buy insect-based foods? Food Research International 92: 95-105.

Tuorila, H., Meiselman, H.L., Bell, R., Cardello, A.V. and Johnson, W., 1994. Role of sensory and cognitive information in the enhancement of certainty and linking for novel and familiar foods. Appetite 23: 231-246.

Van Huis, A., Van Itterbeeck, J., Klunder, H., Mertens, E., Halloran, A., Muir, G. and Vantomme, P., 2013. Edible insects: future prospects for food and feed security. Food and Agriculture Organization of the United Nations, Rome, Italy.

Verbeke, W., 2015. Profiling consumers who are ready to adopt insects as a meat substitute in a Western society. Food Quality and Preference 39: 147-155.

Verneau, F., La Barbera, F., Kolle, S., Amato, M., Del Giudice, T. and Grunert, K., 2016. The effect of communication and implicit associations on consuming insects: an experiment in Denmark and Italy. Appetite 106: 30-36.

Videbæk, P.N. and Grunert, K.G., 2020. Disgusting or delicious? Examining attitudinal ambivalence towards entomophagy among Danish consumers. Food Quality and Preference
83: 103913.

Viechtbauer, W., 2010. Conducting meta-analyses in R with the metafor package. Journal of
Statistical Software 36: 1-48.

Wardle, J., Cooke, L.J., Gibson, E.L., Sapochnik, M., Sheiham, A. and Lawson, M., 2003. Increasing children's acceptance of vegetables; a randomized trial of parent-led exposure.
Appetite 40: 155-162.

Profitability of insect farms

H.H. Niyonsaba*, J. Höhler, J. Kooistra, H.J. Van der Fels-Klerx and M.P.M. Meuwissen

Business Economics Group, Wageningen University, P.O. Box 8130, 6700 EW Wageningen, The Netherlands; *hilde.niyonsaba@wur.nl

Abstract

Despite growing interest from entrepreneurs, knowledge on the profitability of commercial-scale insect production is scarce. Insight into the economic figures of insect production is needed by farmers aiming to start insect farms, by banks seeking to provide financing, and by governments planning policy interventions. This review provides an overview of the profitability and underlying economic figures relating to the production of *Hermetia illucens, Alphitobius diaperinus, Tenebrio molitor* and *Acheta domesticus*. To enhance data interpretation, we also provide a brief overview of the global insect sector, with specific attention to farm-level operational practices. Sales prices refer to fresh larvae, dried larvae or larvae meal, whereas operational costs include costs for feed, labour, electricity, water and gas. Operational cost components differ per insect species, and therefore the relevant margins are specified for three insect species. The energy, feed, and labour margin for production of *H. illucens* ranges from € -798 to 15,576 per tonne of dried larvae. The feed and labour margin for production of *T. molitor* ranges from € 7,620 to 13,770 per tonne of fresh larvae. For production of *A. domesticus* the feed margin ranges from € 12,268 to 78,676 per tonne of larvae meal. The margin range for *A. diaperinus* cannot be estimated, due to a lack of data in the literature. The ranges mainly reflect the differences in sales prices, which are found to heavily depend on the geographical market location, type of market (feed or food) and quantity sold. Major operational costs include feed and labour, with feed costs varying substantially within and between insect species. The economic figures and margins presented in this article provide a foundation for the further development of the insect production sector.

Keywords

insect production – farming – edible insects – economics – *H. illucens* – *A. diaperinus* – *T. molitor* – *A. domesticus* – sales prices – costs

1 Introduction

The production of insects is promising from the perspective of both food security and sustainability (Aiking and de Boer, 2019), and it is believed that insects can fill at least part of the projected protein gap (Henchion *et al.*, 2017). For farmers to start producing insects, however, insight is needed into the profitability and underlying economic figures of insect production. If it is not profitable, farmers are not likely to switch to insect production or to start insect farms; banks will not be eager to provide financing, and actors within the value chain will also hesitate to invest (Hillier *et al.*, 2016). Government officials and scientists also need insight into farm costs and margins for policy interventions and further empirical research.

The insect sector distinguishes itself from existing agricultural sectors by its young and dynamic nature and the proliferation of small-scale farms[11] (Derrien and Boccuni, 2018; Marberg *et al.*, 2017). Given that each insect species is unique, with specific farming requirements, a species-specific approach is needed (Heckmann *et al.*, 2019). The European insect sector is in a developmental stage. Although there are some large farms, the majority is still operating on a small scale (Mancuso *et al.*, 2019). To date, most research has focused on the technical aspects of insect production, such as food safety and quality (Van der Fels-Klerx *et al.*, 2018), nutritional values (Roos and Van Huis, 2017), production and processing practices (Rumpold and Schlüter, 2013) and environmental benefits (Oonincx, 2017). Knowledge concerning the profitability of insect production remains scarce.

This study intends to provide a comprehensive overview of the profitability of insect production and underlying economic figures. It includes an analysis of peer-reviewed articles and other literature on costs and prices involved in the production of the following insect species: *Hermetia illucens* (black soldier fly), *Tenebrio molitor* (yellow mealworm), *Alphitobius diaperinus* (lesser mealworm), and *Acheta domesticus* (house cricket). These four species are amongst the species that have received the most interest within the context of insect production in the Western world in recent decades (Van Huis, 2020). To enhance data interpretation, Section 2 provides an overview of the current status of insect production globally, with specific attention to the European insect sector and its operational practices. We focus on the European context, due to its high potential for sustainable protein provision and its growing commercial insect industry (Aiking and De Boer, 2019). The literature search strategy and data processing approach are presented in Section 3, followed by the literature results (Section 4), including an overview of sales prices[12] and costs of insect production, as well as estimated margin ranges. We present sales prices only for products produced in compliance with EU legislative requirements or

---

11 Due to scale differences in the insect sector the terms 'farms', 'companies' and 'businesses' are used interchangeably in this article, with the latter relating primarily to large-scale production facilities.
12 Sales prices refer to both output prices and retail prices.

comparable, including results from China, Egypt, Kenya and Thailand. For example, data from farms using human excreta as feed are excluded, as this practice is not allowed in the EU. Because the literature review is based on data that have been sourced in different ways, Section 4 also provides details on the research approach and context. The article ends with a general discussion and conclusions (Section 5).

2 The global insect sector and European operational production practices

In some parts of the world, especially in the tropics, insects are caught in the wild and have been part of human diets for a long time (Marcucci, 2020). Since it became clear that insects could also be reared on larger scale to serve as a protein source for feed and food, commercial insect production has been receiving additional interest. Insects are already being produced commercially in many parts of the world, with regional variations in insect species, scale of production and farm types – as well as in insect-eating habits. For example, insects are commonly reared and consumed on both larger and smaller scales in Asia, Africa and Latin America (Kelemu *et al.*, 2015; Raheem *et al.*, 2019). In the past decade, the number of farms producing insects on a larger scale has been increasing in America and Europe as well, with the former focusing on *A. domesticus* and the latter producing mainly *H. illucens* and *T. molitor*. These Western insect industries were initiated largely by entrepreneurial companies, many of which have obtained financial support through crowdfunding campaigns and large investors (Dossey *et al.*, 2016). In addition, many small enterprises are selling their insect products in niche markets as specialty items (Macombe *et al.*, 2019). The International Platform for Food and Feed (IPIFF), a non-profit organisation based in the EU, reports that its members have raised € 600 million for investments in insect production through September 2019, and that its members are already producing 6,000 tonnes of insects for food and feed each year. Production could reach up to five million tonnes by 2030 (IPIFF, 2019). It has even been predicted that the global insect feed market will reach a value of USD 1.4 billion by 2024 (Research and Markets, 2019). In Europe, insects are produced mostly as a protein ingredient for fish feed and pet food, as these are the applications that are currently allowed under European law (Sogari *et al.*, 2019). Expected relaxations in European legislation are likely to include allowing the incorporation of insect proteins in pig and chicken feed, thereby increasing both the demand for and the number of insect-based feeds. The current growth in the insect-feed market is accompanied by increased interest in the use of insects for food in Europe (IPIFF, 2020a). IPIFF (IPIFF, 2019) indicates that aquaculture absorbs more than half of the insect production. For the other market segments no market shares could be found.

To date, the majority of Western insect farms are quite small, with much less technology and mechanisation than other agricultural farms (Dossey *et al.*, 2016).

In terms of operations, European insect farms generally have two separate production units: one for the maintenance of breeding colonies and another one for the production of larvae from the eggs (Halloran *et al.*, 2018). Many insect farms also perform first-step processing after harvesting the insects by drying them. An additional unit is needed for farms that also capture maturing of insects (as is the case for *A. domesticus*). In general, the indoor nature of insect production presents an ideal opportunity for farms to specialise rather than doing all aspects of the insect production steps themselves (Dossey *et al.*, 2016). There are roughly three different business set-ups for insect farms: (1) farms that purchase eggs or small larvae from a supplier and focus on the fattening and, if relevant, the further maturation of the larvae; (2) farms that cover the entire production process, from laying eggs to harvesting and the first-processing (i.e. drying) of larvae; and (3) large-scale production facilities, which cover all steps of the production of insect larvae, as well as further processing steps (e.g. milling, de-fattening and the fractioning of proteins or fats). This literature review focuses on insect production and the processing steps drying and milling, and not on further processing (i.e. de-fattening meal or fractioning of proteins or fats) for incorporation of insect produce into feed or food.

The material and machine inputs required for an insect farm depend on the level of mechanisation. Basic inputs include the plastic trays in which insects are reared and fed. Depending on the level of mechanisation, these inputs can also include drying machines (Pleissner and Smetana, 2020), conveyor belts for internal transport, and continuous feeding systems (Ites *et al.*, 2020). Operational inputs are divided into material input and labour. Within the category of material inputs, the feed for insects is generally regarded as a main component, given that the quality and composition of feed has a substantial influence on the growth of insects and the quality of the end product (Jensen *et al.*, 2017). Furthermore, the production environment is a crucial factor for the survival and growth of insects (Rumpold and Schlüter, 2013). For this reason, the material inputs of insect farms also include the energy and water used for insect production, as well as for first-step and further processing (Ortiz *et al.*, 2016). Another important operational input is labour, given the highly labour-intensive character of insect farms, and especially the current small-scale, largely non-mechanised insect-production process (Meuwissen, 2011). More specifically, the majority of the farms operate largely with manual production systems for the feeding, housing, harvesting and cleaning of insects (Dobermann *et al.*, 2017; Rumpold and Schlüter, 2013).

With regard to the output of insect farms, the main product types include small larvae, grown larvae and mature insects. The processing formats for these products include fresh larvae and dried larvae. Larger production facilities also produce insect meal, which requires further processing. Insect frass is also regarded an important output in terms of volume, but is not yet commercialised on the European market. Additionally, the regulatory framework is 'fragmented' and often restricts insect farmers from upcycling insect frass to high quality fertiliser (IPIFF, 2020b).

3   Literature search strategy and approach

3.1   *Search strategy*
The literature search initially focused on peer-reviewed articles published in the
period 2010 -May 2020 in three databases: Scopus, CAB Abstracts and Web of Science. Search strings were pre-defined and included the following: 'black soldier fly'
OR '*Hermetia illucens*' OR 'lesser mealworm' OR '*Alphitobius diaperinus*' OR 'yellow mealworm' OR '*Tenebrio molitor*' OR 'cricket' OR '*A. domesticus*' AND 'feed OR
food' AND 'financ*' OR 'econom*'. An additional constraint was added for *A. domesticus*: AND NOT 'bean'.
    The articles obtained through the search were screened for relevance according
to selection criteria. First, to be included, a paper had to contain relevant economic
quantitative data (e.g. economic figures on operational costs or sales prices for insect production). Second, the production system investigated had to focus on the
commercial production of insects for food or feed, and not for cosmetics or medicines. Third, the production processes addressed were required to comply with EU
regulations, or at least fulfil similar conditions (e.g. insects should not be fed on
human or animal waste streams). Fourth, the article had to be written in either
English or Dutch. No limitation was set on the geographical area in which the study
had been performed, as long as production circumstances were compatible with
the European context.
    Due to the relatively low number of hits from refereed databases, snowballing
was used to identify additional resources. Snowballing implies using the references
of retrieved papers to identify additional papers relevant to our study aims. The
initial hits obtained from refereed databases and snowballing are listed in Table 1
and their data are discussed below. For *H. illucens*, six peer-reviewed articles and
one report were included in the analysis. The search for *A. diaperinus* led to the
inclusion of one report and two non-peer-reviewed articles. For *T. Molitor*, three
peer-reviewed articles and one report were analysed. The search for *A. domesticus*
led to the inclusion of five peer-reviewed articles and one market-analysis report.

3.2   *Processing retrieved data*
The economic structure of a business (e.g. an insect farm) can be divided into three
main components: operations (with gross margin as a key figure), investments (for
which capital expenditures are an important parameter) and financing in the form
of debt and equity (Brealey, 2001). Given the scarcity of data on the capital costs
of and sources of financing for actual insect farms, this review further focused on
operational costs, with operational inputs divided into production-related material
costs and labour. Sales prices were specified for three processing formats, differing
in dry matter contents of insect products: fresh larvae, dried larvae and larvae meal.
Also, the end-market (i.e. feed, pet food, and human consumption) for each sales
price was identified. Insect frass was not included in this review, since the use of in-

TABLE 1   Initial hits and final number of articles and other sources included

Number of refereed articles after initial search	Insect species			
	Hermetia illucens	*Alphitobius diaperinus*	*Tenebrio molitor*	*Acheta domesticus*
SCOPUS	57	6	24	32
CAB Abstracts	84	17	64	43
(overlap with SCOPUS)	34	4	15	10
Web of science	58	8	31	18
(overlap with SCOPUS)	42	5	18	15
Total number of unique articles	123	22	86	68
Number of refereed articles included	6	0	3	5
Number of other sources included	1	2	1	1

sect frass as fertiliser on a commercial level is impeded by the lack of a harmonised EU regulatory framework (IPIFF, 2020b). All costs and sales prices were recalculated into euros, applying an exchange rate of 1/1.10 (EUR/USD), as retrieved on 28 May 2020 (Investing.com, 2020), and presented in EUR (€) per tonne of product. If the costs and prices in an article were not expressed as €/tonne, we calculated these from the data given in the article. Margins for each insect species were calculated based on the sales prices and averages of operational costs per tonne of product across the references available.

For example, for each tonne of larvae, we used the following equation:

Margin per tonne of larvae = sales price per tonne of larvae – average operational costs per tonne of larvae

Since sales prices differed considerably between end-markets, while cost components were comparable within the different insect species, margins were calculated based on sales prices and average operational costs. Based on the availability of information on operational costs, three different margins were specified: energy, feed, and labour margin for production of dried *H. illucens* larvae, feed and labour margin for production of fresh *T. molitor* larvae, and feed margin for production of *A. domesticus* meal. These margins were combined into a margin range for the different insect species, with the range mainly expressing the variety in sales prices related to different end-markets.

4   Profitability of insect production

Economic figures for the production of *H. illucens*, *A. diaperinus*, *T. molitor* and
A. domesticus are presented in Table 2-5. The sources are listed in these tables, along
with their research contexts, including country, study method, underlying source of
the economic figures and targeted end-market (i.e. feed, pet food, or human consumption). All included studies were conducted in an experimental or normative
modelling context. In other words, none of the studies presents factual data on operational costs from operating farms. With regard to sales prices, the process presented in most of the studies are based on retail prices or personal communications.
In addition, most of the studies focus on topics other than economics. Sales prices
are investigated primarily within the context of feeding trials or consumer-acceptance studies. The economic figures presented in Table 2-5 include sales prices for
the various processing formats. Operational costs include feed, water, electricity, labour and gas. In the following sections, we provide details on all four insect species,
starting with sales prices, and followed by operational costs. Estimated margins and
their ranges are presented in Table 6 in the last section.

4.1   *Profitability of* Hermetia illucens *production*
Economic figures for the production of *H. illucens* from the selected studies are presented in Table 2. Reported sales prices for dried *H. illucens* larvae vary considerably,
ranging from € 1,816 to € 18,900 per tonne of product. All of these sales prices are
based on retail prices from around 2018, although they differ according to the type
of market in which they were sold. For example, the sales price for *H. illucens* larvae
sold in small quantities for pet food (€ 18,190 per tonne of product) is very high,
compared to the others in the same category. The authors describe this price as not
so realistic, compared to another sales price mentioned in the article and ascribe
it to the small quantity sold (Ites *et al.*, 2020). Sales prices for larvae meal range
from € 427 to € 5,091 per tonne of product. Sales prices in Egypt and Kenya are low,
which could be explained by inter alia the low operational costs in these countries.
In one study (Llagostera *et al.*, 2019), the authors explain a reduction in sales prices
between 2016 and 2018 in terms of increased competition over time. Sales prices for
fresh larvae (i.e. not dried or processed) – ranging from € 2,000 to € 3,000 per tonne
of product – are obtained from only one report (Hilkens *et al.*, 2016). Contrary to
reported observations, fresh larvae could be expected to have the lowest sales price,
as they require no additional processing steps.
      Only two articles report on operational costs for dried *H. illucens* larvae. One
study estimates costs for *H. illucens* production in Germany based on a hypothetical
design for an insect production facility. The other study also suggests a hypothetical
(modular) design system within the German context. The large differences between
the two articles with regard to operational costs per tonne of product could be ex
plained as follows. First, the costs per unit for water, electricity and labour differ

TABLE 2   Research context and economic figures for *Hermetia illucens* production

Source Refereed literature	Research context Country	Study method	Underlying source of economic figures	Targeted end-market	Price (€/tonne of product) Dried larvae[1]	Larvae meal	Fresh larvae	Operational costs[2] (€/tonne of dried larvae)
Abdel-Tawwab et al., 2020	Egypt	Feeding trial to implement larvae meal	Local market prices 2020	feed	n/a	427	n/a	n/a
Chia et al., 2019	Kenya	Feeding trial to implement larvae meal	Local market prices 2019	feed	n/a	464	n/a	n/a
Ites et al., 2020	Germany	Modular system design and economic analysis of three different types of food wastes for insect production, applied to German conditions	Retail price Illucens GmbH (2018)	pet food	18,190[a]	n/a	n/a	n/a
			Estimated price based on hypothetical farm design (24-51 tonnes of dry larvae/year)	n/a	n/a	n/a	n/a	1,760[c]; 317[d,h]; 1,274[e,i]; 426fj
			Retail price Illucens GmbH (2018)	pet food	6,500[b]	n/a	n/a	n/a
Llagostera et al., 2019	Spain	Discrete choice experiment to assess preferences of 215 Spanish consumers on the incorporation of insect meal into fish feed	Personal communication (2016)	n/a	n/a	5,091	n/a	n/a
			Personal communication Entomb AgroIndustrial Platform (2018)	n/a	n/a	2,273	n/a	n/a
Mancuso et al., 2019	Italy	SWOT analysis of 28 European insect companies	Personal communication with an EU company Nextalim (2018)	feed	n/a	2,000	n/a	n/a
			Based on personal communication (L. Gasco, 2018)	n/a	2,500	n/a	n/a	n/a
Pleissner and Smetana, 2020	Germany	Economic feasibility analysis of hypothetical farm design for insect production	Retail market price (place and year n/a)	n/a	1,816	n/a	n/a	n/a
			Estimated price based on hypothetical farm design (1,092 tonnes of dried larvae per year)	n/a	n/a	n/a	n/a	21[c]; 160[d,l]; 126[e,l]; 135[f,m]; 1,009[g,n]
Other								
Hilkens et al., 2016	Netherlands	Market-analysis report	Company data (Protix, 2015)	pet food	n/a	n/a	2,000-3,000	n/a

[a]Sold in small quantities; [b]sold in bigger quantities. [2] Operational costs: [c]feed; [d]water; [e]electricity, [f]labour, [g]gas; calculation basis: [h]4.34/m3; [i]€ 0.5/kWh; [j]€ 21.41/hour; [k]1.54/m3; [l]€ 0.034/kWh; [m]€ 30.80/hour; [n]€ 0.0528/kWh.

considerably. This could be related to the production volume, which is higher for the design with the lower operational costs per tonne of product. Second, the designs differ in terms of feed costs. Pleissner and Smetana (2020) report that their low feed costs include only a fee that might possibly be charged for the treatment of food waste, in addition to transport and collection costs. In contrast, Ites *et al.* (2020) also account for the costs of the feed itself. Although the authors provide no information on the quality of feed, it is known that the optimal conditions and type of food waste used to produce *H. illucens* have a major impact on factors that increase both profitability (e.g. the growing time of the larvae) and the quality of the product (Law and Wein, 2018).

4.2   *Profitability of* Alphitobius diaperinus *production*

Economic figures for the production of *A. diaperinus* are presented in Table 3. The number of articles is low, and the sales prices reported represent different processing formats. The three-fold price difference between fresh larvae and larvae meal could be explained by the higher dry-matter content of meal as compared to fresh product. Moreover, additional processing steps are needed to produce larvae meal. The high sales price for freeze-dried *A. diaperinus* reported by Meuwissen (2011) was originally sourced from a website advertising such products as delicacies for human consumption, where they were likely sold in small quantities with a high standard of quality.

TABLE 3   Research context and economic figures for *Alphitobius diaperinus* production

Source	Research context and set-up			
Other	Country	Study method	Underlying source of economic figures	Targeted end-market
Hilkens *et al.*, 2016	Netherlands	Market-analysis report	Company data (Protifarm, 2016)	pet food
Meuwissen, 2011	Netherlands	Market-analysis report	Retail price (Netherlands, 2011)	human consumption
		Report on protein transition	Price analysis Worldbank 2011	n/a

	Price (€/tonne of product)			Operational costs (€/tonne of dried larvae)
	Dried larvae[1]	Larvae meal	Fresh larvae	
Hilkens *et al.*, 2016	n/a	15,000	n/a	n/a
Meuwissen, 2011	118,000[a]	n/a	n/a	n/a
	n/a	n/a	4,750	n/a

[1] [a]Processing format: freeze-dried.

4.3 Profitability of Tenebrio molitor production

Economic figures for the production of *T. molitor* are presented in Table 4. Reported sales prices for *T. molitor larvae* range from € 5,727 to € 97,000 per tonne of product. The low sales prices originating from China could be explained by the well-developed market in China, where *T. molitor* is produced industrially for food and feed, for both domestic use and export (Llagostera *et al.*, 2019). The low price could also be related to economies of scale, as a larger production scale often implies a higher level of mechanisation and lower labour costs (Ortiz *et al.*, 2016). The European retail price obtained from Ortiz *et al.* (2016) includes only dried *T. molitor* larvae dedicated to human consumption, thus possibly explaining the higher sales prices for Europe. A high sales price of € 97,000 per tonne of dried larvae could be explained by the high quality of the product sold in retail. Sales prices for fresh *T. molitor* larvae range from € 10,850 to € 17,000 per tonne of product. The price difference observed between 2011 (Meuwissen, 2011) and 2019 (Mancuso *et al.*, 2019) might be related to the increasing competition and/or technical progress within the insect sector. The difference in sales prices from one source reported in Mancuso *et al.* (2019) is related to the processing format, as the larvae with higher sales prices are sold in frozen form.

With regard to the operational costs for fresh *T. molitor*, only feed and labour costs could be obtained, and from only one report. The labour costs of € 2,140 per tonne of product mentioned in this report are comparatively high, and the author attributes them to the low level of mechanisation in insect production, which further implies low productivity (Meuwissen, 2011).

4.4 Profitability of Acheta domesticus production

Economic figures for the production of *A. domesticus* are presented in Table 5, consisting largely of sales prices for *A. domesticus* meal and fresh product. The only sales price available for dried product is high relative to the other processing formats. Sales prices for *A. domesticus* range from € 18,182 to € 84,590 per tonne of product. The lower prices are largely from Thailand, where operational costs are generally low. The Thai market for *A. domesticus* is also well-known and further developed. Remarkably, the sales prices obtained from Morales-Ramos *et al.* (2020), which originated from the United States of America, are quite high compared to those reported by Reverberi (2020) who also includes sales prices from this region. Morales-Ramos *et al.* (2020) note that the costs presented in their study do not include the costs of labour or diet mixing. Three different articles present sales prices for fresh *A. domesticus* sold in Thailand, ranging from € 1,867 to € 3,952 per tonne of product. The authors attribute this price range to price decreases over time due to increasing competition. In addition, prices are higher in rural areas than they are in urban areas, with differences between small and large-scale farms.

With regard to operational costs, Morales-Ramos *et al.* (2020) investigate different feeding substrates for crickets and their associated feed costs, with an average of

TABLE 4 Research context and economic figures for *Tenebrio molitor* production

Source		Research context			Price (€/tonne of product)			Operational costs (€/tonne of fresh larvae)
Refereed literature	Country	Study method	Underlying source of economic figures	Targeted end-market	Dried larvae	Larvae meal	Fresh larvae	
Mancuso et al., 2019	Italy	SWOT analysis of 28 European insect companies	Company data – Krecafeed (Proti-farm, 2019)	pet food	n/a	n/a	10,850	n/a
			Company data – Krecafeed (Proti-farm, 2019)	n/a	n/a	n/a	17,000[a]	n/a
Ortiz et al., 2016	European Union	Literature review	Retail prices of insects for food (2015)	human consumption	45,454	n/a	n/a	n/a
	China			human consumption	5,727	n/a	n/a	n/a
Rumpold and Schlüter, 2013	Germany	Literature review	Commercial website	human consumption	32,330[b]	n/a	n/a	n/a
Other								
Meuwissen, 2011	Netherlands	Market-analysis report	Retail price (The Netherlands, 2011)	human consumption	97,000[a]	n/a	15,800	n/a
			(De Bakker and Dagevos, 2010)	n/a	n/a	n/a	n/a	1,090[c]; 2,140[d]

[a]Frozen; [b]freeze-dried; [c]feed costs; [d]labour.

TABLE 5   Research context and economic figures for *Acheta domesticus* production

Article Refereed literature	Country	Study method	Research context Underlying source of economic figures	Targeted end-market	Price (€/tonne of product) Dried *A. domesticus*	*A. domesticus* meal	Fresh *A. domesticus*	Operational costs (€/tonne of larvae meal)
Halloran *et al.*, 2016	Thailand	Assessment of actors in the cricket industry	Sales prices Thailand	human consumption	n/a	26,363	1,867–3,952	n/a
Halloran *et al.*, 2017	Thailand	Analysis impact of cricket farming on livelihood	Sales prices from five Northern provinces Thailand	human consumption	n/a	n/a	2,018–2,950	n/a
Hanboonsong *et al.*, 2013	Thailand	Review of primary and secondary interview data	Sales prices from Thailand	human consumption	n/a	n/a	2,363–3,272	n/a
Morales-Ramos *et al.*, 2020	USA	Feeding trial with different feed formulations	Average sales price of nine companies; feed costs based on internet prices	n/a	n/a	84,590	n/a	5,914[a,b]
Reverberi, 2020	Canada	Review on the development of cricket farming	Cricket Farm Inc.	n/a	n/a	18,182	n/a	n/a
	USA		Founder of Aspire Food Group	n/a	n/a	36,364	n/a	n/a
	USA		USA cricket farming consultant	human consumption	n/a	30,000	n/a	n/a
	Belgium		Founder of the Belgian edible insect association	n/a	45,455	n/a	n/a	n/a
Other								
Meuwissen, 2011	The Netherlands	Market-analysis report	Retail price (The Netherlands, 2011)	human consumption	200,000[c]	n/a	n/a	n/a

[a]Operational costs are based on an average calculation of five different feed formulations; [b]feed costs; [c]freeze-dried.

€ 5,914 per tonne. Higher feed costs are not always reflected in higher prices. For example, the feeding substrate with the highest feed costs yields the lowest revenue. It should be noted, however, that the prices for four similar feed formulations are relatively low compared to a special type of formulation. No data could be found on operational costs other than feed. In a general comment, however, Morales-Ramos *et al.* (2020) note that the costs of mass-production are high, due to the primitive production techniques that are used, which involve a high amount of labour. In addition, the production of *A. domesticus* requires expensive commercial feed substrates.

4.5   *Margins*

Based on the prices and average operational costs for *H. illucens, T. molitor* and *A. domesticus* production (Table 2-5), the margins are calculated and shown in Table 6. Due to the lack of economic figures on the operational costs of *A. diaperinus* production, it was not possible to calculate the margin for this insect species. The energy, feed, and labour margin for production of *H. illucens* is € -798 to 15,576 per tonne of dried larvae. The feed and labour margin for production of *T. molitor* was € 7,620 to 13,770 per tonne of fresh larvae. The feed margin for production of *A. domesticus* was € 12,268 to 78,676 per tonne of larvae meal.

Since prices differ between the specified end-markets, individual margins are calculated. The operational costs are composed of different components, as for *T. molitor* only feed and labour are included, and for *A. domesticus* production only feed. Regarding the prices, some prices (i.e. retail prices) refer to prices further along the supply chain, but are used as proxies for sales prices. Sales prices are highest for the food market, followed by pet food, and were lowest for (aqua) feed. Also, the size of outlet influences the price, as small sold quantities in niche markets result in higher prices. In addition, a difference is seen in prices between Western and non-Western countries.

5   Discussion and conclusions

This review aims to present current data on the profitability of insect production and an overview of the economic figures underlying its profitability. Before stating our conclusions, we discuss the interpretation and content of the economic figures.

The number of available sources from which unique economic data could be obtained is low. In many cases, data were cross-referenced, leading back to the same original sources. It is difficult to assess the reliability of some of these original sources, as they have been retrieved from commercial websites, reports, and other grey literature. To ensure the reliability of our overview, we have only included data from sources which presented original references for economic data.

Second, the research methods and set-ups used in the included studies are heter-

TABLE 6   Margins (Energy, feed and labour – Labour and feed – Feed) specified per sales price and estimated margin ranges

	Targeted end market	Country or Region (prices)	Operational costs include	Sales price	Average operational costs	Margin[1]	Margin range[2]
Hermetia illucens (dried larvae)	pet food (small quantities)	Germany	feed, water, electricity, labour, gas.	€18,190	€2,614	€15,576[a]	€-798-€15,576[a]
	pet food (bigger quantities)	Germany	feed, water, electricity, labour, gas	€6,500		€3,886[a]	
	n/a	European Union	feed, water, electricity, labour, gas	€2,500		€-114[a]	
	n/a	European Union	feed, water, electricity, labour, gas	€1,816		€-798[a]	
Tenebrio molitor (fresh larvae)	n/a	European Union	feed, labour	€17,000	€3,230	€13,770[b]	€7,620-€13,770[b]
	human consumption	The Netherlands	feed, labour	€15,800		€12,570[b]	
	n/a	European Union	feed, labour	€10,850		€7,620[b]	
Acheta domesticus (larvae meal)	n/a	USA	feed	€84,590	€5,914	€78,676[c]	€12,268-€78,676[c]
	n/a	USA	feed	€45,455		€30,450[c]	
	n/a	USA	feed	€30,000		€24,086[c]	
	n/a	Canada	feed	€18,182		€12,268[c]	

1 [a]Energy, feed, and labour margin; [b]feed and labour margin; [c]feed margin.
2 [a]Energy, feed, and labour margin range; [b]feed and labour margin range; [c]feed and labour margin range.

ogenous, and mostly focus on topics other than economics. In particular, prices are investigated primarily within the context of feeding trials or consumer-acceptance studies, which present retail prices (and not sales prices). None of the studies includes factual farm data on operational costs. Such data are based predominantly on hypothetical designs (for *H. illucens*) and placed within the context of experimental feed trials (for *A. domesticus*). For this reason, they do not necessarily reflect actual practices. These factors made it also difficult to translate the economic figures into margins. It is also important to note that, for actual farm set-ups, profitability entails more than sales prices and operational costs alone. In addition, it is necessary to consider capital expenditures and financing costs, neither of which is reported in any of the studies. Costs and prices may not necessarily reflect individual farms, instead they provide an overview of the available data ranges. Third, the majority of the studies included in the review present only sales or retail prices. The use of retail prices is likely to result in an overestimation of margins. These prices differed for end-markets too, with in general highest prices for human consumption followed by pet food, and feed. Also, the calculation of average operational costs across references was impeded by the scarcity of data on operational costs. For example, the average operational costs for *H. illucens* production could be calculated only based on data obtained from two sources, and only one source was available for *T. molitor* and *A. domesticus*. In addition, economic figures obtained for operational costs differed according to the species of insects. For *T. molitor* and *A. domesticus*, the literature provides information only on the costs of feed and/or labour, and not on any other material input costs (e.g. electricity, water, gas). These differences in the operational cost items for different studies prevented us from comparing margin ranges across insects species. Therefore, we have calculated individual margins and presented the margins in ranges. The aforementioned considerations should be taken into account when interpreting these margins and their ranges.

Fourth, the prices and feed costs for *A. domesticus* meal are considerably higher than those for *H. illucens*, *A. diaperinus* and *T. molitor* meal. This difference is in line with Brynning *et al.* (2020), who note that the prices of *A. diaperinus* and *T. molitor* larvae are lower than those of *A. domesticus*, which might be one of the reasons that they are more attractive for food and feed companies to include in their products. Remarkably, most of the economic figures obtained for *H. illucens*, *A. diaperinus*, and *T. molitor* are from Europe and China, while data from *A. domesticus* production originate from the United States of America and Thailand. In general, the literature included in the overview reflects a wide range of sales prices for comparable outputs (e.g. the same processing format for same insects), and they include several outliers.

Our findings on variations in sales prices are in line with the findings of Mancuso *et al.* (2019), who conclude that sales prices depend on many factors, including size of outlet, insect species, product type (larvae, adult, pupae), chain stage and processing format. It is interesting to note that, with the exception of the article by Morales-Ramos *et al.* (2020), most of the studies included in this overview provide no information on the quality of feed and dry matter content, which is known to influence the composition – and thus the value – of the product (Law and Wein, 2018).

In conclusion, this study provides an up-to-date overview of the profitability of insect production and its underlying economic figures. The margins are based on sales prices and average operational costs, expressed as margin ranges specified for three insect species. Energy, feed and labour margin for *H. illucens* production, feed and labour margin for *T. molitor* production, and feed margin for *A. domesticus* production are respectively: € -798 to 15,576 per tonne of dried larvae, € 7,620 to 13,770 per tonne of fresh larvae and € 12,268 to 78,676 per tonne of meal. Reliable margins for the production of *A. diaperinus* could not be calculated, due to a lack of economic figures on operational costs in the literature. Margin ranges cannot be compared between the different insect species, as they are calculated for different processing formats and include different cost-components. The ranges of margins for all insect species are wide, which can be led back to the wide variety and differences in sales prices.

The reported sales prices depend on geographical location and type of market (i.e. feed, pet food or human consumption), as well as on the quantity sold in niche markets. Sales prices are relatively low in non-Western countries, relative to those in Western countries, which might be related to inter alia lower operational costs in the former. Operational cost is, however, not the only component determining the sales price, as it is also dependent on, for example, the type of market sold and the size of outlet. In general, sales prices for products intended for food are higher than those for products intended for feed. This difference could be due to the higher quality required for food production and the small quantities sold. The price difference over time that has been observed for the production of *H. illucens* meal could be explained by an increase in competition or technical progress.

The amount of information available with regard to operational costs varies with species. For example, labour costs are subject to differences in the hourly fees applied. Considering the labour-intensive process of insect production, which is related to the low level of mechanisation on most insect farms, these labour costs are seen as an important cost component. Although we could not draw any conclusions on cost-reduction or profit increase of insect production, it has been suggested that increasing levels of mechanisation will reduce labour costs and that the use of low-value feed substrates will reduce operational costs. Regarding the sales of farm output, commercialisation of insect frass as fertiliser could provide an additional source of income for insect farmers. Additional research is nevertheless needed with regard to the additional costs and consequences of larger-scale production. Further research on the profitability of insect farms is also highly recommended, in order to make a constructive contribution to the development of the insect sector and to increase the availability of economic data on insect production. The focus should be on collecting factual farm data, in order to establish an overview of the profitability of practicing farms. First, farm outlet prices are necessary to calculate a reliable gross margin for insect rearing companies. Second, the inclusion of capital expenditures and financing costs in the calculation of profitability would provide a more differentiated calculation at the company level. This would provide farmers with a more reliable source of data to consult when starting new farms, while helping and financial institutions and governments to facilitate access to financing and other services. Another valuable future contribution would be a clear overview of the different prices and costs associated with different geographical locations, which could be linked to different production systems. This could help to identify which production systems would be most suitable for which countries (e.g. in terms of the amount of labour needed).

Acknowledgements

This project has received funding from the European Union's Horizon 2020 research and innovation programme under grant agreement No 861976.

Conflicts of interest

The authors declare no conflict of interest.

References

Abdel-Tawwab, M., Khalil, R.H., Metwally, A.A., Shakweer, M.S., Khallaf, M.A. and Abdel-Latif, H.M., 2020. Effects of black soldier fly (*Hermetia illucens* L.) larvae meal on growth performance, organs-somatic indices, body composition, and hemato-biochemical variables of European sea bass, *Dicentrarchus labrax*. Aquaculture 522: 735136. https://doi.org/10.1016/j.aquaculture.2020.735136

Aiking, H. and De Boer, J., 2019. Protein and sustainability – the potential of insects. Journal of Insects as Food and Feed 5: 3-7. https://doi.org/10.3920/JIFF2018.0011

Brealey, R.A., 2001. Fundamentals of corporate finance. McGraw Hill, New York, NY, USA.

Brynning, G., Bækgaard, J.U. and Heckmann, L.-H.L., 2020. Investigation of consumer acceptance of foods containing insects and development of non-snack insect-based foods. Industrial Biotechnology 16: 26-32. https://doi.org/10.1089/ind.2019.0028

Chia, S.Y., Tanga, C.M., Osuga, I.M., Alaru, A.O., Mwangi, D.M., Githinji, M., Subramanian, S., Fiaboe, K.K., Ekesi, S. and Van Loon, J.J., 2019. Effect of dietary replacement of fishmeal by insect meal on growth performance, blood profiles and economics of growing pigs in Kenya. Animals 9: 705. https://doi.org/10.3390/ani9100705

De Bakker, H. and Dagevos, H., 2010. Meat lovers, meat miners and meat avoiders: sustainable protein consumption in a carnivorous food culture. Vleesminnaars, vleesminderaars en vleesmijders: duurzame eiwitconsumptie in een carnivore eetcultuur. LEI Wageningen UR, Wageningen, The Netherlands.

Derrien, C. and Boccuni, A., 2018. Current status of the insect producing industry in Europe, Edible insects in sustainable food systems. Springer, Berlin, Germany, pp. 471-479.

Dobermann, D., Swift, J. and Field, L., 2017. Opportunities and hurdles of edible insects for food and feed. Nutrition Bulletin 42: 293-308. https://doi.org/10.1111/nbu.12291

Dossey, A., Tatum, J. and McGill, W., 2016. Modern insect-based food industry: current status, insect processing technology, and recommendations moving forward. Insects as sustainable food ingredients. Elsevier, Amsterdam, The Netherlands, pp. 113-152. https://doi.org/10.1016/B978-0-12-802856-8.00005-3

Halloran, A., Flore, R., Vantomme, P. and Roos, N., 2018. Edible insects in sustainable food systems. Springer, Berlin, Germany.

Halloran, A., Roos, N., Flore, R. and Hanboonsong, Y., 2016. The development of the edible cricket industry in Thailand. Journal of Insects as Food and Feed 2: 91-100. https://doi.org/10.3920/JIFF2015.0091

Halloran, A., Roos, N. and Hanboonsong, Y., 2017. Cricket farming as a livelihood strategy in Thailand. The Geographical Journal 183: 112-124. https://doi.org/10.1111/geoj.12184

Hanboonsong, Y., Jamjanya, T. and Durst, P.B., 2013. Six-legged livestock: edible insect farming, collection and marketing in Thailand. Food and Agriculture Organisation of the United Nations, Regional Office for Asia and the Pacific, Bangkok, Thailand, 57 pp.

Heckmann, L.-H., Andersen, J., Eilenberg, J., Fynbo, J., Miklos, R., Jensen, A.N., Nørgaard, J.V. and Roos, N., 2019. A case report on inVALUABLE: insect value chain in a circular bioeconomy. Journal of Insects as Food and Feed 5: 9-13. https://doi.org/10.3920/JIFF2018.0009

Henchion, M., Hayes, M., Mullen, A.M., Fenelon, M. and Tiwari, B., 2017. Future protein supply and demand: strategies and factors influencing a sustainable equilibrium. Foods 6: 53. https://doi.org/10.3390/foods6070053

Hilkens, W., De Klerk, B. and Van Gestel, D., 2016. Insect farming: small sector with big opportunities. Insectenkweek: kleine sector met grote kansen. ABN AMRO/Brabantse Ontwikkelings Maatschappij, Amsterdam/Tilburg, The Netherlands, 37 pp.

Hillier, D., Ross, S., Westerfield, R., Jaffe, J. and Jordan, B., 2016. Corporate finance. McGraw Hill, New York, NY, USA.

Investing.com, 2020. Exchange rate EUR/USD. Available at: https://nl.investing.com/currencies/eur-usd

International Platform of Insects for Food and Feed (IPIFF), 2019. The European insect sector today: challenges, opportunities and regulatory landscape. IPIFF vision paper on the future of the insect sector towards 2030. IPIFF, Brussels, Belgium.

International Platform of Insects for Food and Feed (IPIFF), 2020a. Edible insects on the European Market. IPIFF, Brussels, Belgium.

International Platform of Insects for Food and Feed (IPIFF), 2020b. The insect sector milestones towards sustainable food supply chains. IPIFF, Brussels, Belgium.

Ites, S., Smetana, S., Toepfl, S. and Heinz, V., 2020. Modularity of insect production and processing as a path to efficient and sustainable food waste treatment. Journal of Cleaner Production 248: 119248. https://doi.org/10.1016/j.jclepro.2019.119248

Jensen, K., Kristensen, T.N., Heckmann, L.-H. and Sørensen, J.G., 2017. Breeding and maintaining high-quality insects. In: Van Huis, A. and Tomberlin, J.K. (eds.) Insects as food and feed: from production to consumption. Wageningen Academic Publishers, Wageningen, The Netherlands, pp. 174-198.

Kelemu, S., Niassy, S., Torto, B., Fiaboe, K., Affognon, H., Tonnang, H., Maniania, N. and Ekesi, S., 2015. African edible insects for food and feed: inventory, diversity, commonalities and contribution to food security. Journal of Insects as Food and Feed 1: 103-119. https://doi.org/10.3920/JIFF2014.0016

Law, Y. and Wein, L., 2018. Reversing the nutrient drain through urban insect farming – opportunities and challenges. Bioengineering 5: 226-237. https://doi.org/10.3934/bioeng.2018.4.226

Llagostera, P.F., Kallas, Z., Reig, L. and De Gea, D.A., 2019. The use of insect meal as a sustainable feeding alternative in aquaculture: current situation, Spanish consumers' perceptions and willingness to pay. Journal of Cleaner Production 229: 10-21. https://doi.org/10.1016/j.jclepro.2019.05.012

Macombe, C., Le Feon, S., Aubin, J. and Maillard, F., 2019. Marketing and social effects of industrial scale insect value chains in Europe: case of mealworm for feed in France. Journal of Insects as Food and Feed 5: 215-224. https://doi.org/10.3920/JIFF2018.0047

Mancuso, T., Pippinato, L. and Gasco, L., 2019. The European insects sector and its role in the provision of green proteins in feed supply. Calitatea 20: 374-381.

Marberg, A., Van Kranenburg, H. and Korzilius, H., 2017. The big bug: the legitimation of the edible insect sector in the Netherlands. Food Policy 71: 111-123. https://doi.org/10.1016/j.foodpol.2017.07.008

Marcucci, C., 2020. Food frontiers: Insects as food, is the future already here? Mediterranean Journal of Nutrition and Metabolism 13(1): 43-52. https://doi.org/10.3233/MNM-190348

Meuwissen, P., 2011. Insects as new protein source. A scenario exploration of market opportunities. Insecten als nieuwe eiwitbron. Een scenarioverkenning van de marktkansen. ZLTO, 's Hertogenbosch, The Netherlands.

Morales-Ramos, J.A., Rojas, M.G., Dossey, A.T. and Berhow, M., 2020. Self-selection of food ingredients and agricultural by-products by the house cricket, *Acheta domesticus* (Orthoptera: Gryllidae): a holistic approach to develop optimized diets. PLoS ONE 15: e0227400. https://doi.org/10.1371/journal.pone.0227400

Oonincx, D., 2017. Environmental impact of insect production. In: Van Huis, A. and Tomberlin, J.K. (eds.) Insects as food and feed: from production to consumption. Wageningen Academic Publishers, Wageningen, The Netherlands, pp. 79-93.

Ortiz, J.C., Ruiz, A.T., Morales-Ramos, J., Thomas, M., Rojas, M., Tomberlin, J., Yi, L., Han, R., Giroud, L. and Jullien, R., 2016. Insect mass production technologies. In: Rojas, M.G., Morales-Ramos, J.A. and Dossey, A.T. (eds.) Insects as sustainable food ingredients. Elsevier, Amsterdam, The Netherlands, pp. 153-201.

Pleissner, D. and Smetana, S., 2020. Estimation of the economy of heterotrophic microalgae-and insect-based food waste utilization processes. Waste Management 102: 198-203. https://doi.org/10.1016/j.wasman.2019.10.031

Raheem, D., Carrascosa, C., Oluwole, O.B., Nieuwland, M., Saraiva, A., Millán, R. and Raposo, A., 2019. Traditional consumption of and rearing edible insects in Africa, Asia and Europe. Critical Reviews in Food Science and Nutrition 59: 2169-2188. https://doi.org/10.1080/10408398.2018.1440191

Research and Markets, 2019. Insect feed market – growth, trends and forecasts (2020-2025). Available at: https://www.researchandmarkets.com/reports/4904389/insect-feed-market-growth-trends-and-forecasts

Reverberi, M., 2020. Edible insects: cricket farming and processing as an emerging market. Journal of Insects as Food and Feed 6: 211-220. https://doi.org/10.3920/JIFF2019.0052

Roos, N. and Van Huis, A., 2017. Consuming insects: are there health benefits? Journal of Insects as Food and Feed 3: 225-229. https://doi.org/10.3920/JIFF2017.x007

Rumpold, B.A. and Schlüter, O.K., 2013. Potential and challenges of insects as an innovative source for food and feed production. Innovative Food Science & Emerging Technologies 17: 1-11. https://doi.org/10.1016/j.ifset.2012.11.005

Sogari, G., Amato, M., Biasato, I., Chiesa, S. and Gasco, L., 2019. The potential role of insects as feed: a multi-perspective review. Animals 9: 119. https://doi.org/10.3390/ani9040119

Van der Fels-Klerx, H., Camenzuli, L., Belluco, S., Meijer, N. and Ricci, A., 2018. Food safety issues related to uses of insects for feeds and foods. Comprehensive Reviews in Food Science and Food Safety 17: 1172-1183. https://doi.org/10.1111/1541-4337.12385

Van Huis, A., 2020. Insects as food and feed, a new emerging agricultural sector: a review. Journal of Insects as Food and Feed 6: 27-44. https://doi.org/10.3920/JIFF2019.0017

Advancing edible insects as food and feed in a circular economy

A. van Huis[1]*, B.A. Rumpold[2], H.J. van der Fels-Klerx[3] and J.K. Tomberlin[4]

[1]*Laboratory of Entomology, Wageningen University & Research, P.O. Box 16, 6700 AA, Wageningen, The Netherlands;* [2]*Department of Education for Sustainable Nutrition and Food Science, Technische Universität Berlin, Marchstr. 23, 10587 Berlin, Germany;* [3]*Wageningen Food Safety Research, Akkermaalsbos 2, 6708 WB, Wageningen, The Netherlands;* [4]*Department of Entomology, Texas A&M University, College Station, TX 77843-2475, USA;* **editor-in-chief@insectsasfoodandfeed.com*

Abstract

An overview is given of the special issue on edible insects covering a number of aspects along the value change. The articles presented cover topics about producing insects both as food for humans and feed for animals, ranging from environmental impact, facility design, (left-over) substrates, the role of microbes, genetics, diseases, nutrition, to insect welfare. Possible health benefits of insects for humans and animals are discussed as well as the potential dangers in terms of allergies and chemical/biological contaminants. Regulatory frameworks are examined and assessed for remaining obstacles. The technologies dealing with the processing and extraction of proteins, lipids, and chitin were also reviewed. Consumers' perception of insect-derived food products is discussed as well. A unique aspect of this special issue within the 'Journal of Insects as Food and Feed' is a first attempt to discuss the economics of the industry. The special issue concludes with a discussion of policy and challenges facing the sector.

Keywords

edible insects – environment – nutrition – food safety – consumer attitudes – processing

1   Introduction

According to the International Platform of Insects as Food and Feed (IPIFF), 1 billion US\$ has been invested in the European insect industry with exciting results. For example, in 2019, its members produced more than 6,000 tonnes of insect protein (https://ipiff.org/; accessed 11 June 2021). In December 2020, Meticulous Research (2021) expect the value of the edible insects market worldwide to increase at a compound annual growth rate of 26.5% from 2020 to 2027 to reach US\$4.63 billion by 2027. In 2019, Barclays (2021) estimated the insect protein market would be worth up to US\$8 billion by 2030. In conjunction with the increased value, the number of estimated insect producers in the world now stands at 325 (BugBurger, 2021).

The academic interest is also very much recognised. A search of the phrase 'edible insects' in Web of Science (accessed 11 June 2021) resulted in 48 hits from 1945 to 2010, and 805 hits from 2011 to 2020 with more than half in 2019 and 2020 alone. The sector is diversifying, which is evidenced by the number of pet food companies that are now developing insect-based products. In fact, the first product (dried powder of yellow mealworm, *Tenebrio molitor*) was allowed to be marketed by the European Union (EU) in early 2021, with more products of edible insect species anticipated in the near future. This expansion and diversification may snowball, resulting in other companies becoming involved in the industry.

The articles in this special issue cover a large range of topics related to the production chain of insects as food and feed. Currently there are three types of insect farms: (1) farms purchasing eggs or small larvae from a supplier and rearing them to the harvestable stage; (2) farms covering the entire production process, from eggs to dried insects; and (3) farms that cover both production and processing (fractionating the larvae into protein, fat, and chitin) (Niyonsaba *et al.*, 2021). As the sector develops, parts of the value chain are expected to be dealt with by separate companies. Along the chain, research and innovation will be key drivers in accelerating the transition to sustainable, healthy, and inclusive food systems from primary production to consumption (Riccaboni *et al.*, 2021).

2   Insect production and facility design

Robert Kok provides four articles reviewing preliminary project design, explained as:

> the initial stage in project development that makes it possible for an entepreneur to gain insight into the feasibility and potential profitability of setting up an insect production facility.

Preliminary project design is presented as a central activity linking early concept development, data acquisition and concept crystallisation to formal engineering design, construction, and ultimate plant operation. The first article in this issue (Kok, 2021a) deals with overall mass and energy/heat balances of a rudimentary larval rearing setup. A model is proposed to make it easy for the entrepreneur to use a trial-and-error approach to investigate the effect of different parameter values on system operation. Thus, they can enter values for parameters such as feed composition, temperature of the cooling air, etc. and instantly see the effect on system productivity, conversion efficiency, energy requirements, etc. (in the supplementary material there is a spreadsheet where the reader can change the parameters). This approach facilitates the overall procedure of reaching final decisions about the organism, the feed, the processing approach, the scale of operation, etc. The entopreneur can consider various alternatives prior to making major commitment decisions. His second article in this issue (Kok, 2021b) examines the system dynamics as related to overall heat and mass balances. As demonstrated, a modelling and simulation approach based on research data allows for the integration of knowledge about organism kinetics, reactor configuration, process performance and control system activity. This may lead to a greater accuracy and precision with regards to production. Such data provides for greater understanding of all quantitative aspects of the project and the selected process type, including heat and mass flows, kinetics, inputs and outputs, dynamics, equipment requirements, etc. It should be noted in a prior publication by Kok (2017) in the book by Van Huis and Tomberlin (2017) that there are a number of basic process types already in use for industrial insect production. These topics are updated in the third article in this issue (Kok, 2021c). Sub-processes suitable for rearing insect larvae on dry and semi-dry feeds are reviewed, as well as the type of larval rearing to be employed, the reactor configuration and the operational approach to be used. Furthermore, several sub-process types, reactors, and much more are discussed. By accounting for these factors, the entopreneur should have an idea of the scope and scale of the project being proposed. In the fourth article in this issue by Kok (2021d), the project as a whole should be taken into account such as: project location issues, to what degree the facility is to be integrated or segregated into different units, what functionality is to be housed within the various building envelopes, and how safety and hygiene concerns should be addressed.

3   Environmental impact

Insects as food and feed have many environmental advantages. This recognition is in part due to lower greenhouse gas and ammonia emissions, reduced water and land requirements, and the ability to convert organic side streams into high-value products. To evaluate the environmental impact of insect production, both direct and indirect impacts need to be considered. This accountability can be done using a life cycle assessment (LCA). In this issue, Smetana et al. (2021) compared 24 LCA studies to determine the sustainability of insect production chains; most dealt with the black soldier fly Hermetia illucens, yellow mealworm T. molitor and house fly Musca domestica. The majority of the 24 studies assessed used the attributional LCA approach, which only takes a specific product life cycle into account; a few

conducted a consequential LCA, which also considers alternative uses of a product (e.g. using manure as biogas). Most focused on cradle-to-gate approaches, not
taking transportation and distribution of insect biomass/products into account.
Most impacts are associated with use of energy (e.g. electricity, fuel, natural gas)
and consequently global warming potential, non-renewable energy use, as well as
water and land use. Type of feed and modelling of its assessment was in many cases
decisive for the determination of the environmental impact of insects. Selection of
by-product allocation rules, substitution criteria and waste scenarios determined
the wide range of environmental impacts presented for food processing by-products, food waste and manure.

4   Insect welfare

Animal welfare is a critical issue for the insects as food and feed sector (Van Huis,
2021). The central question in this debate is whether insects are 'sentient beings'.
This may be the case when considering: (1) the very efficient functional way in
which their brains are organised; and (2) their capacity for social and associative
learning, and diverse ways of communication. When insects are farmed and killed,
the precautionary principle is used, which assumes that they can experience pain.
Van Huis (2021) used the following five freedoms of Brambell (1965) as a framework
to discuss the consequences of welfare issues for the edible insect sector: freedom
from hunger and thirst, from discomfort, from pain, injury or disease, from fear and
distress, and freedom to express normal behaviour.

5   Organic side streams as feed

The kind of material (i.e. feed) that is received by an insect mass-rearing facility
dictates what is produced (i.e. insect biomass, frass). This statement particularly
applies to the insects as feed sector. Organic materials utilised to mass produce insects as feed or food can be highly diverse. Consequently, output levels, quality, and
in the end, value, are highly dependent on these materials. This special issue covers
the rearing of insects on manure, the role of microbes, the impact of substrate on
the nutritional quality of the insect and finally how the use of leftovers (insect faeces and left-over substrate) can be used as promotor or protector of plants.

5.1   *Rearing insects on manure*
Cammack *et al.* (2021) examine the potential for animal waste (i.e. faeces) to be used
as substrate for a number of insects that can serve as feed, e.g. the lesser mealworm
Alphitobius diaperinus, the house fly, and the black soldier fly. While this material
is not currently approved for mass production in the EU, United States, or other

Western nations, it holds great potential. An overview of each species is presented as well as a case study with the black soldier fly and what is known about its abilities for recycling animal waste.

5.2   Role of microbes

Zhang *et al.* (2021) present a complimentary aspect of this process by discussing the use of microbes to enhance organic waste recycling in a general sense, while expanding to include other insect species, such as waxworm, *Plodia interpunctella*, and yellow mealworm. Furthermore, they demonstrate the possibility of using insects to recycle other waste streams than those commonly used and not yet heavily considered as substrates, such as plastics.

5.3   Impact of substrate

Pinotti and Ottoboni (2021) provide a suitable conclusion for the previous two articles discussed, as they present an overview of the impact of substrate variability on the resulting quality and value generated by the insects. They also use the black soldier fly as a case study by diving into detail on many studies that have independently explored select waste streams on the production of this insect. Now, through this study, an appreciation for the relationship between nutritional variation and optimised production can be considered.

5.4   Waste as fertiliser

Of course, the insects mass produced are just one of the primary products manufactured. The other is the digested waste mixed with insect frass that remains. Chavez and Uchanski (2021) discuss these wastes resulting from such systems (e.g. black soldier fly, house fly and mealworm) and their similarities to inorganic fertilisers. However, unlike traditional inorganic fertilisers, the waste resulting from mass producing insects contains valuable additives – mainly chitin. This material, which is a key component of insect exoskeleton, could promote plant health by reducing the likelihood of disease. They also discuss the topic that not all insect waste is the same. Differences between species, but also diets provided to these insects, influence the 'character' of the waste. Therefore, the expectation that generalisations can be drawn across systems, or even within a population that experiences variations in the diet provided to them, should be tempered.

6   Optimising rearing processes

One the major concerns of insect-rearing companies is the possibility that colonies will be affected by pathogens, so the first article deals with pathogen (i.e. disease) management. Another issue that companies must deal with is whether they can optimise the genetic make-up of their population to enhance production, waste conversion, and nutritional values. The other question for companies is the nutritional composition of the insects (whether this can be improved; see the article of Oonincx and Finke (2021).

6.1   *Disease management*

Until a few years ago the knowledge on pests and diseases in insect-rearing systems was very limited. However, insect diseases in rearing facilities, from small to industrial set-ups, are feared, as they may compromise the whole enterprise. With the growing value of this market, the increased attention to this topic for preventing and curing diseases is fully justified. Although black soldier fly larvae seem resistant to infection and disease (Joosten *et al.*, 2020), future expansions of the industry open the door to possible disease-related issues. Safe, hygienic operating procedures and continuous monitoring will be required. Also, the development of strains resistant to pathogens should be considered, such as with viruses in crickets (De Miranda *et al.*, 2021). In this special issue Maciel-Vergara *et al.* (2021) review insect pathogens that are causing disease in the most common insect species reared or collected to be used as food or feed. For the prevention of diseases, they discuss biotic and abiotic factors that may potentially trigger insect diseases. They also look at how to prevent stress factors, which are critical for the development of disease outbreaks, and often more than one pathogen is involved. The complexity of multifactorial relations requires a holistic approach to understand the various aspects related to insect diseases. For natural enemies other than pathogens (e.g. predators and parasitoids) in insect rearing systems we refer to the publication of Eilenberg *et al.* (2017).

6.2   *Insect breeding*

While diet can certainly be used to manipulate production of insects that are mass reared, Eriksson and Picard (2021) cover this topic from a different perspective by focusing on genetics. The value of deciphering the genetic makeup of mass-produced insect species has benefitted other insect-based systems (e.g. sericulture, apiculture, biological control). By understanding these other systems, guidelines, or a framework, could be developed for these 'newer' insect systems (e.g. black soldier fly, mealworm). After discussing these more traditional systems, they provide an overview of the vast toolset (i.e. old versus new methods) available for assessing a population for strengths and weaknesses. They discuss the value of such information when breeding a population in captivity while attempting to optimise its genetic makeup as a means to enhance production, waste conversion, and nutritional value.

6.3   *Manipulating nutritional values*

Oonincx and Finke (2021) in this issue provide details of how insects, much like other livestock, can be manipulated to a certain extent in terms of their nutritional composition. The same authors previously gave an overview of the (a)biotic factors that may influence nutrition (Finke and Oonincx, 2014). The protein content of insects can be high and very well digested with suitable amino acid profiles (methionine being the first limiting one) for both production animals and humans. Fat content and fatty acid composition vary depending on species, life stage, diet, and sex. Most minerals are present in adequate concentrations to meet nutrient requirements of animals and humans, although calcium levels may be low depending on the model used. In insects, B vitamins are adequately present, vitamin A levels low, vitamin E concentrations depend on the diet, while vitamin D content depends on UV-B radiation during insect development. Insect processing may destroy vitamins and denature proteins due to heat. Oonincx and Finke (2021) point out that the slogan 'you are what you eat' also relates to negative aspects of the process – whereby contaminates, such as heavy metals, can bioaccumulate in insects. In conclusion, diet can strongly affect the concentration of most nutrients, but has much less of an impact on amino acids and minerals.

7 Health aspects of using edible insects on humans and animals

Information on the possible positive effects of insects on the health of humans and of animals is becoming more readily available; however, this advancement is only in its infant stages. This may cover all aspects from immune and microbiota responses to effects on disease resistance.

7.1 Human health

Stull (2021) discusses the impact of insects on human health. While the use of insects as food might seem novel, the practice has been in existence for as long as humanity has recorded history. Stull (2021) did an extensive literature review using SCOPUS, Web of Science, and PubMed. What she uncovered was quite surprising, as not much is known about the impact of consuming insects on human health. Results generated from her search indicate that select insects can serve as a substitute for more traditional plant-based proteins and can have a positive impact on human health by: (1) providing amino acids similar to soya; (2) improving gut health; and (3) meeting micronutrient deficiencies. However, as Stull (2021) points out, research published to date has barely scratched the surface.

7.2 Animal health

As previously mentioned, the quality of substrates used to rear insects impacts the quality of the insects produced. Consequently, the value of the insect mass produced for livestock, poultry, aquaculture, and petfood sectors can vary. However, consistency, predictability, and safety are critical if the insects are to truly be established as a dependable feed stream for farmed vertebrates for the feed industry.

Gasco *et al.* (2021) provide an extensive overview of insects as feed across a broad spectrum of animals that are mass produced, including fish and crustaceans, poultry, pigs, and even rabbits. Initial aspects of the publication provide gross level assessment of the nutritional makeup (e.g. protein, fat, amino acid composition) of select insects that are mass produced as feed. They cover aspects ranging from immune and microbiota responses to influence on disease resistance. Other studies published in this compilation provide greater details for mass produced animals.

8   Using insects as feed

Insects can be used to feed fish, poultry, pigs and pets. In the articles mentioned below the effects of using insect species on these animals is discussed.

8.1   *Fish*

Liland *et al.* (2021) provide an extensive evaluation of published literature examining the use of insects as feed in the aquaculture industry. This assessment evaluated 91 publications, 415 experimental diets across 35 fish species, while considering 14 insect species. Assessment of the studies indicates a high degree of breadth across these criteria demonstrating the diversity of the aquaculture industry. However, while of tremendous value simply for amassing such an extensive, and diverse literature, Liland *et al.* (2021) demonstrate a critical limitation. In general, the industry still appears to be lacking depth of understanding for a targeted fish being reared on insect-based diets as well as the value of a given insect species as a diet for the diverse global aquaculture industry.

8.2   *Poultry*

Dörper *et al.* (2021) produce a similar evaluation of insects as feed for the poultry industry. They also explored various insect models that can be used presently (e.g. black soldier fly), or possibly if approved (e.g. house fly, yellow mealworm) as a feed. Unlike the aquaculture industry, fewer data are available on the use of these insects as feed. Dörper *et al.* (2021) provide an overview of the benefits of insects as feed for the poultry industry ranging from improved animal health to conversion rates across the different sectors (e.g. layers, broilers). Their extensive summary table highlights studies where insects have been used as poultry feed: an overview of the associated results as well as the references.

8.3   *Pigs*

Similar to other domestic stock previously discussed, Veldkamp and Vernooij (2021) provide an overview of the use of insects as feed for pigs. They discuss the pig industry and its value as well as its reliance on traditional row crops for feed. This initial discussion sets the stage for their presentation of data generated from studies that examined the inclusion of insects as part of the feed used to rear pigs. They demonstrate aspects of the pig diet that can be replaced. However, they recognise variability in the results generated (i.e. pig response to being fed such diets) from such studies. They conclude by offering their support for including insects in the pig diet as they do not adversely impact associated growth; however, additional research is needed to determine if differences between studies are due to the insect diet or some other covariate.

8.4   *Pets*

A recent advancement in terms of regulations in various parts of the world has been the approval of insects for use in pet food. Bosch and Swanson (2021) discuss the use of such products as part of dog and cat food. The stage for their discussion is set by providing an overview of the pet food industry. They recognise each has its own parameters with regard to formulations (e.g. palate, nutrition, and – indirectly – owner preference for pet food). Unlike other studies published in this special issue, they focus on the amino acids that are key drivers in the dog and cat food industries. Methionine, threonine and leucine levels are key when making decisions about which components to include in dog and cat food. They then transition into discussing what is known about insects as dog and cat food by providing extensive tables and reviewing previous studies published.

9   Food safety and legislation

When talking about food safety of insects as food and feed, there are a number of issues that need to be considered, such as, but not limited to: allergy, as insects are closely related to crustaceans (Pennisi, 2015), biological and chemical contaminants, and of course the legal framework.

9.1   *Allergy*

Ribeiro *et al.* (2021) reviewed 20 articles and provide an update on scientific developments related to the allergic risks of insects as food. They specifically looked at the molecular mechanisms and major allergens implied both in primary sensitisation and in cross-reactivity with crustaceans or house dust mites. They also discuss the effect of food processing on allergenicity of edible insects. Food allergies are defined as adverse immune responses to food proteins that result in typical clinical symptoms involving the dermatologic, respiratory, gastrointestinal, cardiovascular, and/or neurologic systems (Anvari *et al.*, 2019). The most common type, which occurs shortly after eating, is immunoglobulin E (IgE)-mediated. When immune cells encounter the allergenic protein, IgE antibodies are produced, which initiate the allergic reaction. Ribeiro *et al.* (2021) concluded that: (1) tropomyosin is confirmed to be a cross-reactive allergen between edible insects and crustaceans; (2) allergenicity of edible insects seems to be resistant to thermal treatments and digestion with enzymes, and only the use of very specific conditions can eliminate the IgE-reactivity of edible insects; (3) clinical significance in cross-reactivity between edible insects and house dust mites is still unclear, in particular for subjects who are only allergic to house dust mites; (4) major risk groups are those who are involved in rearing the insects and those that are allergic to crustaceans. An effective allergen management procedure is the correct labelling of insect-containing foods (Garino *et al.*, 2020).

9.2   *Biological contaminants*

Biological contaminants encompass pathogenic strains of microorganisms (i.e. bacteria, viruses and fungi), parasites (i.e. protozoa and worms), as well as toxic substances (chemicals) they produce, i.e. mycotoxins. Vandeweyer *et al.* (2021) review first the possible occurrence of biological contaminants (excluding toxic compounds produced) in mass-rearing systems and then discuss future challenges. Concerning wild-harvested insects, they refer to other publications. From the review they conclude that: (1) data regarding prions and foodborne viruses and parasites in insects farmed for human consumption are scarce; (2) the most important bacterial pathogens associated with insects for food are *Staphylococcus aureus*, pathogenic *Clostridium* spp. and pathogenic species of the *Bacillus cereus* group. The last group are spore-forming bacteria and drastic strategies are required to mitigate the risks associated with bacterial endospores. Concerning insects for feed, few reports on pathogens seem to be available. Prions cannot be produced by insects, but could be acquired from slaughterhouse waste (EFSA Scientific Committee, 2015). Vandeweyer *et al.* (2021) also made an assessment on the transfer of biological contaminants during the rearing (e.g. contamination from the diet). They also mention further challenges which we discuss in Section 12.

9.3   *Chemical contaminants*

Meyer *et al.* (2021) review the hazards of chemical contaminants and focus on substrates for twelve edible insect species. Most data are available for the black soldier fly and the yellow mealworm. Chemical contaminants investigated so far include heavy metals, dioxins and polychlorinated biphenyls, polyaromatic hydrocarbons, pesticides, veterinary drugs, mycotoxins, and plant toxins. Those chemicals are generally stable, and it is difficult to remove or reduce them by processing. Transfer and bioaccumulation depend on the insect species, life stage, substrate type, chemical involved, source (spiked vs natural), and rearing conditions. Accumulation occurs with lead, arsenic, mercury, and cadmium, but apparently not for mycotoxins and polycyclic aromatic hydrocarbons. Although mycotoxins and veterinary drugs can be degraded by insects, the metabolites formed and their possible toxicity is not clear. Few data are available for possible accumulation (or degradation) of polyaromatic hydrocarbons, plant toxins, and dioxins and dioxin-like polychlorinated biphenyls.

9.4   *Regulatory issues*

Lähteenmäki-Uutela *et al.* (2021) compare insect food and feed regulation in the EU, the United States, Canada, and Australia and to some extent in countries of Central and South America, Asia and Africa, in particular for the period from 2018 to 2020. Food and feed regulations are often separate; feed regulation may allow insect usage even when they are not allowed as food. In the EU applications for food products from seven insect species have been submitted, and the first assessment by the European Food Safety Authority was adopted on 20 November 2020 for dried powder of the yellow mealworm (EFSA, 2021). The other insects may continue to be marketed until the EU makes a decision. In 2017, the EU decided to allow the use of seven insect species as aquafeed. However, insect proteins are not allowed to be fed to pigs and poultry nor are foodstuffs containing meat or fish allowed to be used as feed materials for insects. In Canada black soldier fly products have been authorised as feed for poultry and certain fish species. In China silkworm pupae are authorised as a food ingredient. However, insects as food and feed should be regulated globally as the regulatory demands and differences between countries complicate the international marketing strategies for insect products. Insects proven safe as food and feed by the scientific community should be authorised in all countries (Lähteenmäki-Uutela *et al.*, 2021).

10   Processing

Several industrial technologies are used for processing, such as grinding, separation and drying (Sindermann *et al.*, 2021). Then in the chapter of Ojha *et al.* (2021).in this issue the extraction and utilisation of insect protein, lipid and chitin as well as novel processing technologies are discussed

10.1   *Technologies*

The adoption of insects as food and feed on an industrial level is, besides the issue of acceptance, hindered by regulatory aspects and by the challenge of up-scaling production at a competitive price. Up-scaling also concerns processing of insects into food and feed products. This aspect requires species-specific, cost-effective, and targeted technologies and machines. It is not necessary to reinvent the wheel; existing processing technologies can be adapted and optimised for the separation and/or processing of insects into intermediates and products for use in food and feed. An overview of existing industrial processing technologies for insect larvae is given by Sindermann *et al.* (2021). The authors distinguish between wet and dry processing, which refers to the use or lack of use of water and a drying step before or after the potential lipid removal. Several machines for grinding, separation and drying were presented. It was concluded that automation and cleaning in situ are additional aspects to be considered when designing a processing line. It was stated that the adaption and adjustment of existing technologies will meet the processing demand of an up-scaled insect production set-up in due time.

10.2   *Processing and extraction pathways and emerging technologies*
Insect processing is crucial to ensure safe and high-quality raw materials and food products based on edible insects. In addition to unit operations already traditionally used in food processing, emerging technologies such as high hydrostatic pressure processing, ultrasound, cold plasma or pulsed electric fields have shown potential for insect processing. In their review, Ojha *et al.* (2021) give an overview of traditional and future processing pathways and technologies for insect processing. This ranges from pre-processing (harvesting, cleaning, killing) through decontamination to packaging and storage. Conventional processing routes and their processing steps to produce whole insects, whole-fat meal, fermented insect meal, defatted insect meal, insect oil and other insect fractions such as protein and chitin are presented. In addition, potential processing pathways using emerging technologies are shown. The authors conclude that the main challenges of insect processing include the development of low-cost and sustainable processing technologies with a holistic approach concerning by-products and production waste. While emerging technologies offer a number of benefits, there are still hurdles such as a necessary high level of knowledge and the plant costs.

11   Consumer and marketing

Despite the nutritional and environmental benefits of insects, consumer acceptance is still low, especially in Western countries (Motoki *et al.*, 2021). For a successful promotion of insects as food, an in-depth understanding of the consumers' perception is fundamental. Consumer attitudes regarding the acceptance of insects as food have been reviewed in a number of articles among which Hartmann and Siegrist (2017), Mancini *et al.* (2019), Motoki *et al.* (2021) and Sogari *et al.* (2019). The studies vastly differ in methodology and approach. A systematic analysis of these studies and pooling of their results could elucidate the consumers' perception of insects as food and highlight research gaps, something that has been done by Wassmann *et al.* (2021) in this issue. Consumer attitudes may also be influenced by how the products appear on the market (Reverberi, 2021). Finally, the economics of insect production has until now scarcely been addressed. A first attempt has been made by Niyonsaba *et al.* (2021) to shed more light on the profitability of insect farms.

11.1   *Consumer attitudes*
In a meta-analysis, (Wassmann *et al.*, 2021) assessed 37 research publications with the objective of investigating the correlation between willingness to consume and several (34) variables. A small correlation was found with gender, female partic-

ipants being less likely to consume insects. Strong links were found for variables related to the novelty and acceptability of insects as food as well as to affective factors such as disgust, the latter being the greatest barrier. Wassmann *et al.* (2021) conclude that one key to overcoming disgust is tastiness, as disgust is closely related to taste (Rozin *et al.*, 2009). The most positive impact on the consumers' willingness to taste insects as food is awareness of the environmental benefits of such products (Wassmann *et al.*, 2021). Moderating effects on willingness to consume were the presentation of the insect product (description, picture or real product).

11.2   *Economics*

Data on the profitability of commercial-scale insect production are scarce. Niyonsaba *et al.* (2021) made a first attempt to obtain insight into the economics of insect production. They looked at four insect species: the black soldier fly, the yellow mealworm, the house cricket *Acheta domesticus*, and the lesser mealworm, though few data are available for the latter. There is a huge variation in sales prices (Table 1), due to: (1) the market type (feed, pet food or human consumption; pet food often concerns small quantities); (2) large differences in operational costs (per unit of water, electricity, labour and feed) in particular among regions (costs in developing countries often being low); (3) fresh or processed larvae (processing requires extra costs); (4) differences in level of mechanisation (often related to economies of scale); (5) differences between rural and urban areas. The prices and feed costs for house cricket meal are considerably higher than those for the black soldier fly and mealworms. This may be the reason that food and feed companies consider mealworms and black soldier fly more attractive to include in their products than crickets (Niyonsaba *et al.*, 2021). This is because price is considered an obstacle in using insect products (Brynning *et al.*, 2020). Major operational costs are those of feed and labour, with feed costs varying substantially within and between insect species.

11.3   *Packaged processed food*

Even with a willingness to consume insects, most Western consumers still lack the skills to prepare insect-containing meals at home. Ready-to-eat or ready-to-cook

TABLE 1   Sales prices and average operational costs according to edible insect species and regions; data from Niyonsaba *et al.* (2021)

Larvae	Range of sales prices (€/tonne)		Average operational costs (€/tonne)[1]
	World	Canada, EU and Canada	
Hermetia illucens (dried)	1,816-18,900		2,614
Tenebrio molitor (fresh)	5,727-97,000	10,850-17,000 (EU)	3,230
Acheta domesticus (meal)	18,182-84,590	18,182-84,590 (USA, Canada)	5,914

Operational costs include feed, labour, electricity, water and gas.

insect products can facilitate their use and consumption. Emerging insect products are cookies, bars, crisps, pasta, meat balls, schnitzel, pasta sauces, or even whole insects. Most of them contain cricket flour or mealworm larvae flour. Despite the vast media coverage, insect products have not yet succeeded in becoming mainstream; they still remain a niche product (Reverberi, 2021). For start-ups, a major challenge is the identification of target groups and selling points. Producers target LOHAS (Lifestyle of Health and Sustainability) (Pittner, 2017) consumers, foodies, flexitarians, and even vegetarians (Reverberi, 2021). Potential selling points, apart from taste, include nutritional and environmental benefits and using the claim 'superfood'. Packaging should be modern and 'mainstream' with a clear indication that the product contains insects. In addition, an allergen warning must be included on the label. Further challenges include market size estimates, supermarket outlets, and competitive prices. Co-packers are used by 80% of start-ups to facilitate manufacturing, packaging and labelling, certification, storage and shipping (Reverberi, 2021). Based on the interviews, they conclude that bias against consumption of insects, high production costs, regulatory barriers, and access to retail markets are the main challenges when marketing insect products.

12   Challenges

What are the challenges that lie ahead in the sector of insects as food and feed? The editors of the special issue discuss how these challenges can be tackled, using among other things the recommendations of the authors of the special issue.

12.1   *Exploring other insect species*

The most commonly reared insect groups – cricket, mealworm, and black soldier fly – derive protein from the substrate. This means that the limitation is the amount of protein in the feed. In the drive towards a circular economy, the substrates that contain some protein will increasingly compete with other uses. For example, Pinotti *et al.* (2021) showed that food leftovers (former food products and bakery by-products – also targeted by the insect industry) can be used as an alternative to traditional feed ingredients in pig and ruminant nutrition. Hence, the livestock sector is also focussing on food waste as feed, avoiding competition for natural resources, and reducing the environmental impact of the animal production systems. Organic side streams are competed for and this will make them increasingly expensive. Low-cost feeds are often those composed of cellulose and lignin and low in protein. To be able to use those as feed, can we use organisms able to digest it or able to synthesise protein? Several beetles, some cockroaches and most termites can do much of this, aided by bacteria and protozoa living symbiotically in their guts. The combination of microorganisms and insects needs full attention, e.g. some microbial symbionts of arthropod guts are able to synthesise protein de novo from atmospheric nitrogen

(Douglas, 2009; Nardi *et al.*, 2002). One possibility is also to feed the insects with fungi-fermented high cellulose and lignin substrates (Qi *et al.*, 2019).

12.2   *Environmental impact and insect welfare*

Smetana *et al.* (2021), who compared several environmental impact studies of insects as food and feed, state that in order to achieve consistency and comparability of results, future studies should carefully determine system boundaries, scale of production, the reference (functional) unit, and the methodology used for impact assessment, e.g. what type of life cycle analysis is used (attributional or consequential).

Concerning insect welfare, an international effort should be made to establish guidelines or standards of care for the industry about well-being and slaughter (Van Huis, 2021), while also considering limitations that developing nations face. Any guidelines developed should be sensitive to diverse socio-economic standards that exist globally, and care should be taken to avoid excluding other nations that are financially disadvantaged. Also, more information is needed on the cognitive and emotional capacity of invertebrate species. The choice of species should reflect the enormous diversity in nervous structures and the quantity of the species reared.

12.3   *Organic side streams to rear insects*

In general terms, between 55 and 95% of the nitrogen (N) and about 70% of the phosphorus (P) ingested by livestock are excreted through urine or faeces (Leip *et al.*, 2019). When not disposed of, wasted or applied in excess of crop nutrient needs, can manure be considered a co-product and used sustainably? One of the possibilities is biodegradation/biotransformation of manure by insects, and Cammack *et al.* (2021) explains it as follows:

> first from waste to fly larvae as feed for chickens; then the manure from the chickens can be upcycled by black soldier fly into purified proteins and amino acids for aquaculture feeds, oil for biodiesel, and the frass for soil amendment.

They pose the question whether the lesser mealworm, house fly, and other insects will also have such abilities.

The role of microbes in this process is crucial as explained by (Zhang *et al.*, 2021). Even plastic biodegradation can be accelerated by mealworm and their gut microbiome (Brandon *et al.*, 2021). Also, Peng *et al.* (2019) demonstrated that *Tenebrio* species and their gut microbes were able to biodegrade polystyrene.

The left-over substrates from insect mass-rearing farms can be used to promote plant growth and health, the latter probably through chitin (Chavez and Uchanski, 2021). They also indicated that levels of bioavailable phosphorous (P) in insect frass are high, which is important considering the limited availability of P for agriculture in the future.

12.4   *Optimising rearing processes*

Maciel-Vergara *et al.* (2021) made several recommendations about disease management, such as having a clear understanding of the biology of insect pathogens and the interactions with their hosts and knowing the correlation between production variables and host-pathogen dynamics. They also stress that protocols need to be developed for the management, prevention, and control of diseases.

Eriksson and Picard (2021), while discussing genetic and genomic selection, predict that next-generation molecular technologies, such as the CRISPR/Cas genome editing toolkit, will fuel future innovations. They mention approaches to isolate and characterise genetic variants linked to phenotypes. What has not yet been researched is the role of endosymbionts, such as Wolbachia, which for example could have a role in the degradation of waste. Kaya *et al.* (2021) are concerned that the genetically highly uniform domesticated strains of the black soldier fly threaten the genetic integrity of unique local resources through introgression.

Oonincx and Finke (2021) discuss the manipulation of the nutritional composition of insects, which can be done by: (1) harvesting a specific life stage; (2) diet composition; (3) environmental manipulation (e.g. vitamin D3 synthesis in yellow mealworm by UV-B emission; Oonincx *et al.* (2018)); (4) processing (mostly in a negative way, e.g. destruction of vitamins and denaturation of proteins due to heat).

12.5   *Health aspects of insects as food and feed*

The health benefits of insects as food and feed for people, livestock, poultry, fish, and pets has received increasing attention. If there are health benefits, it will certainly boost the interest of potential users. However, concerning those benefits for human health, Stull (2021) recommend rigorous and well-controlled human intervention trials not only to confirm health benefits but also to assess the associated risks. What also needs to be addressed are issues like nutrient bioavailability, the fate of dietary chitin, and *in vivo* activity of bioactive peptides.

Gasco *et al.* (2021) consider that insect use as animal feed is impaired by high prices and low availability. These issues are due to the lack of large-scale production units and legislation uncertainties. Insect consumption by animals seems to stimulate the immune system, improving disease resistance. This action of insects as a prebiotic in animals is of interest because antibiotic resistance is a major public health concern: between 2010 and 2030, the global consumption of antimicrobials is estimated to increase by 67% (Van Boeckel *et al.*, 2015); and resistant bacteria in animals could be pathogenic to humans (Manyi-Loh *et al.*, 2018). According to Gasco *et al.* (2018) what still needs to be determined for each mass-produced insect species is the dose and duration of administering the insect product and which kind of diseases the product protects against in aquatic and terrestrial farmed animals.

12.6   Insects as feed

The feed market represents about 1,040 million tons, of which poultry uses 45%, pigs and ruminants each another 24-25%, and aquaculture and pets each another 3-4% (Table 2). The global commercial feed manufacturing sector generates an estimated annual turnover of over US $400 billion (IFIF, 2021).

In most countries the use of insects as aquaculture feed or for pets is allowed. If it were allowed for poultry and pigs in other parts of the world outside the United States, where it is already approved for black soldier fly, this would create an enormous market. In four articles dealing with aquafeeds, poultry, pigs and pets, the prospects of using insects as feed are examined.

Aquaculture is growing about 5% a year and in 2018 produced 82 million tonnes, valued at USD 250 billion (FAO, 2020). Half of all global seafood production for human consumption is farmed, and this demand is expected to rise to over 60% by 2030 (CAIA/FCC, 2021). Aquafeed is about 4% of all feed produced worldwide (Table 2). In the development of sustainable aquafeeds, a major challenge is to lower its dependence on fish meal and fish oil. According to Liland *et al.* (2021) about a quarter of the feed can contain insects without reducing the performance of the aquaculture species. The limitation to using insects as aquafeed is reduced protein digestibility, imbalanced amino acid profile and increasing levels of saturated fatty acid. Another problem is the high variation in quality and compositions of insect meals for the non-defatted meals and insect oils.

Close to a half of all the feed produced goes to poultry (Table 2). There are legislative hurdles in a number of countries regarding its use. Partial replacement of fishmeal or soybean meal by fly larvae is possible. As the animals need to forage to obtain the insect, the welfare of the animals could be enhanced. Dörper *et al.* (2021) recommended: (1) standardised procedures for insect production and processing;

TABLE 2   Feed production estimates in million metric tons of feed in 2019 (Alltech, 2020)

	Million metric tons		%
Poultry	465.0		44.7
Layer		157.7	
Broiler		307.3	
Pig	260.9		25.1
Ruminant	245.3		23.6
Beef		115.4	
Dairy		129.9	
Aquaculture	41.0		3.9
Pet	27.7		2.7
Total	1,039.9		100

(2) distinction between insect products based on their nutritional composition; (3) determination of optimal inclusion levels for insect products per poultry species; (4) comparison of products from different insect species.

One quarter of all feed produced goes to pigs (Table 2). Veldkamp and Vernooij (2021) suggest that in order to understand the effects of the insect products in diets a distinction should be made between studies to determine nutritional value (digestibility studies) and growth performance. They also recommend more standardised digestibility and growth performance experiments. The potential effects of the use of insect-based ingredients on animal health are interesting and deserve more attention.

Although only 3% of the feed produced goes to pets (Table 2), in 2020 and 2021 an increasing number of pet feed companies marketed insect-based feed products. The insect-containing hypoallergenic foods are now an option for owners with dogs or cats suffering from supposed or diagnosed food sensitivity. Bosch and Swanson (2021) recommend for insect-based ingredients in pet food: (1) safety and efficacy testing; (2) a study of the impact of long-term feeding on the nutritional status and health in dogs and cats; (3) an exploration of the health benefits of insect ingredients in feed on pets; (4) a survey of consumer attitudes specifically when the insects are grown on organic side streams.

12.7   *Food safety and legislation*

Vandeweyer *et al.* (2021) consider the following as major challenges for mitigating the risk of biological contaminants: (1) risk assessments directed at specific insect species; (2) more data required on the microbiological quality of substrates and residue and in relation to this the development of decontamination treatments; (3) investigating the house flora of rearing facilities; (4) implementing exhaustive sampling plans by applying predictive microbiology; (5) standardisation of microbiological research methods and quality control; (6) exploring the unculturable fraction of the insect microbial community and its importance in food safety; (7) developing monitoring and control protocols of biological contaminants in the upscaling of the sector. Concerning the chemical contaminants Meyer *et al.* (2021) recommend further research on the metabolic pathways of mycotoxins and veterinary drugs in insects, as detoxification/bioactivation pathways are not clear. The growth of the insect sector is constrained by a lack of (1) local regulations, and (2) a stable and consistent set of regulations across international borders (Lähteenmäki-Uutela *et al.*, 2021). A future perspective given by Lähteenmäki-Uutela *et al.* (2021) is that if the familiar insect species are eventually authorised in all major countries, and the authorisations are generic, researchers and companies may shift their attention to new species.

12.8   *Processing*

For some of the technologies discussed, economies of scale require larger production volumes than most systems in operation today (Sindermann *et al.*, 2021). Even though the insect industry will probably come up with many new product innovations within the near future, the major technologies for dry and wet processing are ready to be applied. In upscaling, some equipment may require minor adjustments.

The main challenges of insect processing include developing efficient, sustainable, and low-cost processing technologies, waste minimisation, recovery, and incorporation of by-products/co-products (Ojha *et al.*, 2021). Novel food-processing technologies should improve the processing efficiency, safety, quality, and sustainability of insect-based products, although the know-how required and the high costs are still hurdles.

12.9   *Consumer attitudes and marketing*

Wassmann *et al.* (2021) feel that classical psychological factors have been sufficiently researched and suggest that future studies on willingness to consume should be more market-oriented, focussing on marketing and product development, and on variables like price, convenience, species, and packaging. However, there is a psychological factor that may need to be further explored, i.e. Youssef and Spence (2021) proposed a multisensory environment for a gastronomic dining experience, e.g. when having an insect dish, incorporate natural sounds (singing crickets), project nature videos, and provide ambient aromas.

Insect products are not yet mainstream, although they have been on the market for about 10 years. According to Reverberi (2021) the commercial challenges include lowering production costs, facilitating certifications and regulations, improving marketing, increasing retail distribution and expanding consumer targeting.

In the study by Niyonsaba *et al.* (2021) on the profitability of insect farms, there was a wide variety of sales prices and operational costs. They recommend: (1) research on the effect of scaling-up on profitability; (2) more knowledge about farm outlet prices to calculate a reliable gross margin; (3) inclusion of capital expenditures and financing costs in the calculation of profitability; (4) obtaining an overview of prices and costs per geographical location to identify which production systems are most appropriate for which country (e.g. in terms of the amount of labour needed).

13   Conclusions and policy

The IPIFF with 76 members in 2021 (https://ipiff.org/; accessed 11 June 2021) has been productive and impactful. With its secretariat in Brussels, the organisation can represent the interests of the insect production sector to EU policy makers, European stakeholders, and citizens. Their publications are high standard and cover a range of topics such as hygiene, sustainability, and the use of insect frass, etc. On other continents there are also stakeholder organisations such as the North American Coalition of Insect Agriculture (NACIA). Its role is to encourage the positive use of insects in North America, and educate and help build awareness and acceptance for insects as a solution to up-cycle nutrients back into the food chain. In Australia in 2021 a roadmap was published as a strategy to advance the edible insect industry (Ponce-Reyes and Lessard, 2021). In this country there is the Insect Protein Association Australia (IPAA) promoting the use and role of insects within the Australian food and feed ecosystem to government, academia, and industry. In Asia, there is the Asian Food and Feed Insect Association (AFFIA) representing the interests of its members from the industry and from academia working with insects as food and feed in Asia; China, however, is not included. In Africa there is the African Association of Insect Scientists (AAIS), however their mandate is much broader than edible insects. It may be time to coordinate efforts at a global level.

The reason for the creation of IPIFF was the lack of a regulatory framework for edible insects, which was considered a major hurdle in advancing the industry. With the expected easing of the legislation for insects as food and feed, it is anticipated that the sector will develop quickly. With a conducive regulatory framework, companies will no longer face uncertainties and academics will no longer be restricted by legislation. In fact, they can be a major contributor to the easing of legislation. For example, Belghit *et al.* (2021) proposed a molecular method for improved detection, differentiation and tracing of prohibited material in processed animal proteins (PAPs) derived from insects in feed chains. This may contribute to lifting the feed ban rules as the transmissible spongiform encephalopathy regulation in the EU prohibits the use of PAPs derived from insect in feed for pigs and poultry.

Research in insects as food and feed can initiate new developments such as: genetically improving breeding stocks; promoting health for humans, animals, and plants; safe use of substrates such as manure or catering waste; reducing environmental impact by lowering emissions of greenhouse gases, ammonia, and nitrogen; biotransforming waste; creating new insect products, for example, for cosmetics; and many more developments can be added in the production, processing and consumption of edible insects.

Maybe it is also time for an academic society through which researchers from all over the globe could communicate on edible insect issues. There are exciting times ahead for all of us, rich with opportunities to use insects for our benefit.

Conflict of interest

The authors declare no conflict of interest.

References

Alltech, 2020. 2020 global feed survey. Alltech, Nicholasville, KY, USA. Available at: https://www.alltech.com/sites/default/files/GFS_Brochure_2020.pdf.

Anvari, S., Miller, J., Yeh, C. and Davis, C., 2019. IgE-mediated food allergy. Clinical Reviews in Allergy & Immunology 57: 244-260. https://doi.org/10.1007/s12016-018-8710-3

Barclays, 2021. Insect protein: bitten by the bug. Barclays, London, UK. Available at: https://www.investmentbank.barclays.com/our-insights/insect-protein-bitten-by-the-bug.html

Belghit, I., Varunjikar, M., Lecrenier, M.C., Steinhilber, A.E., Niedzwiecka, A., Wang, Y.V., Dieu, M., Azzollini, D., Lie, K., Lock, E.J., Berntssen, M.H.G., Renard, P., Zagon, J., Fumière, O., Van Loon, J.J.A., Larsen, T., Poetz, O., Braeuning, A., Palmblad, M. and Rasinger, J.D., 2021. Future feed control – tracing banned bovine material in insect meal. Food Control 128: 108183. https://doi.org/10.1016/j.foodcont.2021.108183

Bosch, G. and Swanson, K.S., 2021. Effect of using insects as feed on animals: pet dogs and cats. Journal of Insects as Food and Feed 7: 795-805. https://doi.org/10.3920/JIFF2020.0084

Brambell, F.W.R., 1965. Report of the Technical Committee to enquire into the welfare of animals kept under intensive livestock husbandry systems. Cmnd. 2836, December 3, 1965. Her Majesty's Stationery Office, London, UK.

Brandon, A.M., Garcia, A.M., Khlystov, N.A., Wu, W.-M. and Criddle, C.S., 2021. Enhanced bioavailability and microbial biodegradation of polystyrene in an enrichment derived from the gut microbiome of *Tenebrio molitor* (mealworm Larvae). Environmental Science & Technology 55: 2027-2036. https://doi.org/10.1021/acs.est.0c04952

Brynning, G., Bækgaard, J.U. and Heckmann, L.-H.L., 2020. Investigation of consumer acceptance of foods containing insects and development of non-snack insect-based foods. Industrial Biotechnology 16: 26-32. https://doi.org/10.1089/ind.2019.0028

BugBurger, 2021. The eating insects startups: here is the list of entopreneurs around the world! Available at: http://www.bugburger.se/foretag/the-eating-insects-startups-here-is-the-list-of-entopreneurs-around-the-world

Canadian Aquaculture Industry Alliance & Fisheries Council of Canada (CAIA/FCC), 2021. Canada's blue economy strategy 2040. CAIA/FCC, Ottawa, ON, USA. Available at: http://fisheriescouncil.com/wp-content/uploads/2020/10/Canadas-Blue-Economy-Strategy_FINAL-9-15-2020.pdf

Cammack, J.A., Miranda, C.D., Jordan, H.R. and Tomberlin, J.K., 2021. Upcycling of manure with insects: current and future prospects. Journal of Insects as Food and Feed 7: 605-619 https://doi.org/10.3920/JIFF2020.0093

Chavez, M. and Uchanski, M., 2021. Insect left-over substrate as plant fertiliser. Journal of Insects as Food and Feed 7: 683-694. https://doi.org/10.3920/JIFF2020.0063

De Miranda, J.R., Granberg, F., Onorati, P., Jansson, A. and Berggren, Å., 2021. Virus prospecting in crickets – discovery and strain divergence of a novel iflavirus in wild and cultivated *Acheta domesticus*. Viruses 13: 364. https://doi.org/10.3390/v13030364

Dörper, A., Veldkamp, T. and Dicke, M., 2021. Use of black soldier fly and house fly in feed to promote sustainable poultry production. Journal of Insects as Food and Feed 7: 761-780. https://doi.org/10.3920/JIFF2020.0064

Douglas, A.E., 2009. The microbial dimension in insect nutritional ecology. Functional Ecology 23: 38-47. https://doi.org/10.1111/j.1365-2435.2008.01442.x

European Food Safety Authority (EFSA), 2021. Safety of dried yellow mealworm (*Tenebrio molitor* larva) as a novel food pursuant to Regulation (EU) 2015/2283. EFSA Journal 19(1): 6343. https://doi.org/10.2903/j.efsa.2021.6343

European Food Safety Authority (EFSA) Scientific Committee, 2015. Risk profile related to production and consumption of insects as food and feed. EFSA Journal 13: 4257. https://doi.org/10.2903/j.efsa.2015.4257

Eilenberg, J., Gasque, S.N. and Ros, V.I.D., 2017. Natural enemies in insect production systems. In: Van Huis, A. and Tomberlin, J.K. (eds.) Insects as food and feed: from production to consumption. Wageningen Academic Publishers, Wageningen, The Netherlands, pp. 201-223. https://doi.org/10.3920/978-90-8686-849-0

Eriksson, T. and Picard, C.J., 2021. Genetic and genomic selection in insects as food and feed. Journal of Insects as Food and Feed 7: 661-682. https://doi.org/10.3920/JIFF2020.0097

Food and Agriculture Organisation (FAO), 2020. The state of world fisheries and aquaculture 2020 – sustainability in action. FAO, Rome, Italy. https://doi.org/10.4060/ca9229en

Finke, M.D. and Oonincx, D., 2014. Insects as food for insectivores. In: Morales-Ramos, J.A., Guadalupe Rojas, M. and Shapiro-Ilan, D. (ed.) Mass production of beneficial organisms. Academic Press, San Diego, CA, USA, pp. 583-616. https://doi.org/10.1016/B978-0-12-391453-8.00017-0

Garino, C., Mielke, H., Knüppel, S., Selhorst, T., Broll, H. and Braeuning, A., 2020. Quantitative allergenicity risk assessment of food products containing yellow mealworm (*Tenebrio molitor*). Food and Chemical Toxicology 142: 111460. https://doi.org/10.1016/j.fct.2020.111460

Gasco, L., Finke, M. and Van Huis, A., 2018. Can diets containing insects promote animal health? Journal of Insects as Food and Feed 4: 1-4. https://doi.org/10.3920/JIFF2018.x001

Gasco, L., Józefiak, A. and Henry, M., 2021. Beyond the protein concept: health aspects of using edible insects on animals. Journal of Insects as Food and Feed 7: 715-741. https://doi.org/10.3920/JIFF2020.0077

Hartmann, C. and Siegrist, M., 2017. Consumer perception and behaviour regarding sustainable protein consumption: a systematic review. Trends in Food Science & Technology 61: 11-25. https://doi.org/10.1016/j.tifs.2016.12.006

The International Feed Industry Federation (IFIF), 2021. Global feed statistics. IFIF, Wiehl, Germany. Available at: https://ifif.org/global-feed/statistics/

Joosten, L., Lecocq, A., Jensen, A.B., Haenen, O., Schmitt, E. and Eilenberg, J., 2020. Review of insect pathogen risks for the black soldier fly (*Hermetia illucens*) and guidelines for reliable production. Entomologia Experimentalis et Applicata 168: 432-447. https://doi.org/10.1111/eea.12916

Kaya, C., Generalovic, T.N., Ståhls, G., Hauser, M., Samayoa, A.C., Nunes-Silva, C.G., Roxburgh, H., Wohlfahrt, J., Ewusie, E.A., Kenis, M., Hanboonsong, Y., Orozco, J., Carrejo, N., Nakamura, S., Gasco, L., Rojo, S., Tanga, C.M., Meier, R., Rhode, C., Picard, C.J., Jiggins, C.D., Leiber, F., Tomberlin, J.K., Hasselmann, M., Blanckenhorn, W.U., Kapun, M. and Sandrock, C., 2021. Global population genetic structure and demographic trajectories of the black soldier fly, *Hermetia illucens*. BMC Biology 19: 94. https://doi.org/10.1186/s12915-021-01029-w

Kok, R., 2017. Insect production and facility design. In: Van Huis, A. and Tomberlin, J.K. (eds.) Insects as food and feed: from production to consumption. Wageningen Academic Publishers, Wageningen, The Netherlands, pp. 143-172. https://doi.org/10.3920/978-90-8686-849-0

Kok, R., 2021a. Preliminary project design for insect production: part 1 – overall mass and energy/heat balances. Journal of Insects as Food and Feed 7: 499-509. https://doi.org/10.3920/JIFF2020.0055

Kok, R., 2021b. Preliminary project design for insect production: part 2 – organism kinetics, system dynamics and the role of modelling & simulation. Journal of Insects as Food and Feed 7: 511-523. https://doi.org/10.3920/JIFF2020.0146

Kok, R., 2021c. Preliminary project design for insect production: part 3 – sub-process types and reactors. Journal of Insects as Food and Feed 7: 525-539. https://doi.org/10.3920/JIFF2020.0145

Kok, R., 2021d. Preliminary project design for insect production: part 4 – facility considerations. Journal of Insects as Food and Feed 7:541-551. https://doi.org/10.3920/JIFF2020.0164

Lähteenmäki-Uutela, A., Marimuthu, S.B. and Meijer, N., 2021. Regulations on insects as food and feed: a global comparison. Journal of Insects as Food and Feed 7: 849-856. https://doi.org/10.3920/JIFF2020.0066

Leip, A., Ledgard, S., Uwizeye, A., Palhares, J.C.P., Aller, M.F., Amon, B., Binder, M., Cordovil, C.M.d.S., De Camillis, C., Dong, H., Fusi, A., Helin, J., Hörtenhuber, S., Hristov, A.N., Koelsch, R., Liu, C., Masso, C., Nkongolo, N.V., Patra, A.K., Redding, M.R., Rufino, M.C., Sakrabani, R., Thoma, G., Vertès, F. and Wang, Y., 2019. The value of manure – manure as co-product in life cycle assessment. Journal of Environmental Management 241: 293-304. https://doi.org/10.1016/j.jenvman.2019.03.059

Liland, N.S., Araujo, P., Xu, X.X., Lock, E.-J., Radhakrishnan, G., Prabhu, A.J.P. and Belghit, I., 2021. A meta-analysis on the nutritional value of insects in aquafeeds. Journal of Insects as Food and Feed 7: 743-759. https://doi.org/10.3920/JIFF2020.0147

Maciel-Vergara, G., Jensen, A.B., Lecocq, A. and Eilenberg, J., 2021. Diseases in edible insect rearing systems. Journal of Insects as Food and Feed 7: 621-638. https://doi.org/10.3920/JIFF2021.0024

Mancini, S., Moruzzo, R., Riccioli, F. and Paci, G., 2019. European consumers' readiness to adopt insects as food. A review. Food Research International 122: 661-678. https://doi.org/10.1016/j.foodres.2019.01.041

Manyi-Loh, C., Mamphweli, S., Meyer, E. and Okoh, A., 2018. Antibiotic use in agriculture and its consequential resistance in environmental sources: potential public health implications. Molecules 23: 795. https://doi.org/10.3390/molecules23040795

Meticulous Research, 2021. Edible insects market to reach $4.63 billion by 2027. Meticulous Research, Redding, CA, USA. Available at: https://www.meticulousresearch.com/press-release/184/edible-insects-market-2027

Meyer, A.M., Meijer, N., Hoek-Van den Hil, E.F. and Van der Fels-Klerx, H.J., 2021. Chemical food safety hazards of insects reared for food and feed. Journal of Insects as Food and Feed 7: 823-831. https://doi.org/10.3920/JIFF2020.0085

Motoki, K., Ishikawa, S.-I. and Park, J., 2021. Review and future directions of consumer acceptance of insect-based foods. The Japanese Journal of Psychology 92: 52-67. https://doi.org/10.4992/jjpsy.92.20402

Nardi, J.B., Mackie, R.I. and Dawson, J.O., 2002. Could microbial symbionts of arthropod guts contribute significantly to nitrogen fixation in terrestrial ecosystems? Journal of Insect Physiology 48: 751-763. https://doi.org/10.1016/S0022-1910(02)00105-1

Niyonsaba, H.H., Höhler, J., Kooistra, J., Van der Fels-Klerx, H.J. and Meuwissen, M.P.M., 2021. Profitability of insect farms. Journal of Insects as Food and Feed 7: 923-934. https://doi.org/10.3920/JIFF2020.0087

Ojha, S., Bußler, S., Psarianos, M., Rossi, G. and Schlüter, O.K., 2021. Edible insect processing pathways and implementation of emerging technologies. Journal of Insects as Food and Feed 7: 877-900. https://doi.org/10.3920/JIFF2020.0121

Oonincx, D.G.A.B. and Finke, M.D., 2021. Nutritional value of insects and ways to manipulate their composition. Journal of Insects as Food and Feed 7: 639-659. https://doi.org/10.3920/JIFF2020.0050

Oonincx, D.G.A.B., Van Keulen, P., Finke, M.D., Baines, F.M., Vermeulen, M. and Bosch, G., 2018. Evidence of vitamin D synthesis in insects exposed to UVb light. Scientific Reports 8: 10807. https://doi.org/10.1038/s41598-018-29232-w

Peng, B.-Y., Su, Y., Chen, Z., Chen, J., Zhou, X., Benbow, M.E., Criddle, C., Wu, W.-M. and Zhang, Y., 2019. Biodegradation of polystyrene by dark (*Tenebrio obscurus*) and yellow (*Tenebrio molitor*) mealworms (Coleoptera: Tenebrionidae). Environmental Science & Technology 53: 5256-5265. https://doi.org/10.1021/acs.est.8b06963

Pennisi, E., 2015. All in the (bigger) family. Revised arthropod tree marries crustacean and insect fields. Science of the Total Environment 347: 220-221. https://doi.org/10.1126/science.347.6219.220

Pinotti, L., Luciano, A., Ottoboni, M., Manoni, M., Ferrari, L., Marchis, D. and Tretola, M., 2021. Recycling food leftovers in feed as opportunity to increase the sustainability of livestock production. Journal of Cleaner Production 294: 126290. https://doi.org/10.1016/j.jclepro.2021.126290

Pinotti, L. and Ottoboni, M., 2021. Substrate as insect feed for bio-mass production. Journal of Insects as Food and Feed 7: 585-596. https://doi.org/10.3920/JIFF2020.0110

Pittner, M., 2017. Consumer segment LOHAS. Nachhaltigkeitsorientierte Dialoggruppen im Lebensmitteleinzelhandel. Springer Fachmedien, Wiesbaden, Germany.

Ponce-Reyes, R. and Lessard, B., 2021. Edible insects – a roadmap for the strategic growth of an emerging Australian industry. Commonwealth Scientific and Industrial Research Organisation, Canberra, Australia. Available at: https://research.csiro.au/edibleinsects/wp-content/uploads/sites/347/2021/04/CSIRO-Edible-Insect-Roadmap.pdf

Qi, X., Li, Z., Akami, M., Mansour, A. and Niu, C., 2019. Fermented crop straws by *Trichoderma viride* and *Saccharomyces cerevisiae* enhanced the bioconversion rate of *Musca do-*

mestica (Diptera: Muscidae). Environmental Science and Pollution Research 26: 29388-29396. https://doi.org/10.1007/s11356-019-06101-1

Reverberi, M., 2021. The new packaged food products containing insects as an ingredient. Journal of Insects as Food and Feed 7: 901-908. https://doi.org/10.3920/JIFF2020.0111

Ribeiro, J.C., Sousa-Pinto, B., Fonseca, J., Fonseca, S.C. and Cunha, L.M., 2021. Edible insects and food safety: allergy. Journal of Insects as Food and Feed 7: 833-847. https://doi.org/10.3920/JIFF2020.0065

Riccaboni, A., Neri, E., Trovarelli, F. and Pulselli, R.M., 2021. Sustainability-oriented research and innovation in 'farm to fork' value chains. Current Opinion in Food Science 42: 102-112. https://doi.org/10.1016/j.cofs.2021.04.006

Rozin, P., Haidt, J. and Fincher, K., 2009. Psychology. From oral to moral. Science 323: 1179-1180. https://doi.org/10.1126/science.1170492

Sindermann, D., Heidhues, J., Kirchner, S., Stadermann, N. and Kühl, A., 2021. Industrial processing technologies for insect larvae. Journal of Insects as Food and Feed 7: 857-875 https://doi.org/10.3920/JIFF2020.0103

Smetana, S., Spykman, R. and Heinz, V., 2021. Environmental aspects of insect mass production. Journal of Insects as Food and Feed 7: 553-571. https://doi.org/10.3920/JIFF2020.0116

Sogari, G., Menozzi, D., Hartmann, C. and Mora, C., 2019. How to measure consumers acceptance towards edible insects? – a scoping review about methodological approaches. In: G. Sogari, C. Mora and D. Menozzi (eds.) Edible insects in the food sector: methods, current applications and perspectives. Springer International Publishing, Cambridge, UK, pp. 27-44. https://doi.org/10.1007/978-3-030-22522-3_3

Stull, V.J., 2021. Impacts of insect consumption on human health. Journal of Insects as Food and Feed 7: 695-713. https://doi.org/10.3920/JIFF2020.0115

Van Boeckel, T.P., Brower, C., Gilbert, M., Grenfell, B.T., Levin, S.A., Robinson, T.P., Teillant, A. and Laxminarayan, R., 2015. Global trends in antimicrobial use in food animals. Proceedings of the National Academy of Sciences 112: 5649-5654. https://doi.org/10.1073/pnas.1503141112

Van Huis, A., 2021. Welfare of farmed insects. Journal of Insects as Food and Feed 7: 573-584. https://doi.org/10.3920/JIFF2020.0061

Van Huis, A. and Tomberlin, J., 2017. Insects as food and feed: from production to consumption. Wageningen Academic Publishers, Wageningen, The Netherlands, 448 pp. https://doi.org/10.3920/978-90-8686-849-0

Vandeweyer, D., De Smet, J., Van Looveren, N. and Van Campenhout, L., 2021. Biological contaminants in insects as food and feed. Journal of Insects as Food and Feed 7: 807-822. https://doi.org/10.3920/JIFF2020.0060

Veldkamp, T. and Vernooij, A.G., 2021. Use of insect products in pig diets. Journal of Insects as Food and Feed 7: 781-793. https://doi.org/10.3920/JIFF2020.0091

Wassmann, B., Siegrist, M. and Hartmann, C., 2021. Correlates of the willingness to consume insects: a meta-analysis. Journal of Insects as Food and Feed 7: 909-922. https://doi.org/10.3920/JIFF2020.0130

Youssef, J. and Spence, C., 2021. Náttúra by kitchen theory: an immersive multisensory dining concept. International Journal of Gastronomy and Food Science 24: 100354. https://doi.org/10.1016/j.ijgfs.2021.100354

Zhang, J.B., Yu, Y.Q., Tomberlin, J.K., Cai, M.M., Zheng, L.Y. and Yu, Z.N., 2021. Organic side streams: using microbes to make substrates more fit for mass producing insects for use as feed. Journal of Insects as Food and Feed 7: 597-604. https://doi.org/10.3920/JIFF2020.0078

Printed in the United States
by Baker & Taylor Publisher Services